MW01659606

Student Solutions Manual

for

Tussy and Gustafson's

Intermediate Algebra

Third Edition

Donna Flint
South Dakota State University

Betty Larson
South Dakota State University

THOMSON
BROOKS/COLE

Australia • Canada • Mexico • Singapore • Spain • United Kingdom • United States

Printed in the United States of America
1 2 3 4 5 6 7 09 08 07 06 05

Printer: Thomson/West

ISBN: 0-534-49407-2

For more information about our products, contact us at:
Thomson Learning Academic Resource Center
1-800-423-0563

For permission to use material from this text or product, submit a request online at **http://www.thomsonrights.com.**
Any additional questions about permissions can be submitted by email to **thomsonrights@thomson.com.**

Thomson Higher Education
10 Davis Drive
Belmont, CA 94002-3098
USA

Asia (including India)
Thomson Learning
5 Shenton Way
#01-01 UIC Building
Singapore 068808

Australia/New Zealand
Thomson Learning Australia
102 Dodds Street
Southbank, Victoria 3006
Australia

Canada
Thomson Nelson
1120 Birchmount Road
Toronto, Ontario M1K 5G4
Canada

UK/Europe/Middle East/Africa
Thomson Learning
High Holborn House
50–51 Bedford Road
London WC1R 4LR
United Kingdom

Latin America
Thomson Learning
Seneca, 53
Colonia Polanco
11560 Mexico
D.F. Mexico

Spain (including Portugal)
Thomson Paraninfo
Calle Magallanes, 25
28015 Madrid, Spain

Preface

This *Student Solutions Manual* has been prepared to accompany the textbook, *Intermediate Algebra*, Third Edition by Alan S. Tussy and R. David Gustafson. Included in the manual are solutions to all the odd exercises for each section of the book, as well as all solutions for Chapter Reviews, Chapter Tests, and the Appendices.

Any feedback regarding errors, appropriateness of method and style would be appreciated. You may send these directly to us at the address below.

We would like to thank the authors of the textbook for the opportunity to be involved in this project as well as Rebecca Subity, Assistant Editor at Thomson Higher Education, for her enthusiastic support and advice. In addition, we would like to thank our families (Neil, Robert, Jenni, Toby and Quincy) for their patience as we worked on this project.

To the students, we wish you joy and excitement as you accept the challenge of a new math course. We hope you will find that our solutions and study tips help you to enjoy a successful and valuable math course.

Best wishes!

Donna Flint
Betty Larson

South Dakota State University
Department of Mathematics and Statistics
Box 2220
Brookings, SD 57007

Table of Contents

Contents

Chapter 1 A Review of Basic Algebra

Study Tip

How NOT to use this book:
After you have done a problem, check the answer in this book.
If you got it wrong, look at the correct solution.
Say to yourself, "Oh, now I know how to do it," and skip to the next type of problem

How TO use this book:
After you have done a problem, check the answer in this book.
If you got it wrong, look at the correct solution.
Close this book.
Redo the problem you missed.
Do a similar problem and repeat the process.
Do this until you are able to do that type of problem right on the first try.

You learn mathematics by DOING mathematics.

Section 1.1 The Language of Algebra

VOCABULARY

1. equation 3. expressions 5. formula 7. multiplication

CONCEPTS

9. expression 11. equation 13. equation 15. expression

17. a. a line graph b. 1 hour; 2 in. c. 7 in.; 0 in.

19. $c = 13u + 24$ where c represents the cost each semester and u represents the number of units taken.

21. $w = \frac{c}{75}$ where w represents the number of social workers and c represents the number of clients.

23. $A = t + 15$ where A represents the adjusted test score and t represents the original test score.

25. $c = 12b$ where c represents the number of crayons in a case and b represents the number of boxes of crayons in a case.

27. $h = t - 10$. In each situation, the height of the base, h, is 10 feet less than the tower height, t.

NOTATION

29. a. D b. 2, 4 c. multiplication, addition

PRACTICE

31.

Number of packages, p	$c = \frac{p}{12}$	Number of cartons, c
24	$c = \frac{24}{12}$	2
72	$c = \frac{72}{12}$	6
180	$c = \frac{180}{12}$	15

33.

K	$n = 22.44 - K$	n
0	$n = 22.44 - 0$	22.44
1.01	$n = 22.44 - 1.01$	21.43
22.44	$n = 22.44 - 22.44$	0

35. 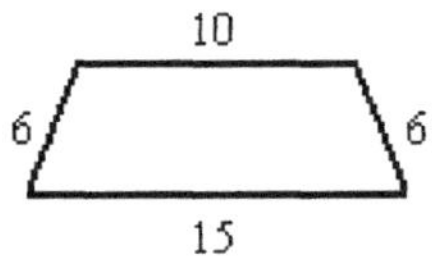

Trapezoid: $a = 6$, $b = 10$, $c = 6$, $d = 15$
$P = a + b + c + d = 6 + 10 + 6 + 15 = 37$ in.

37. 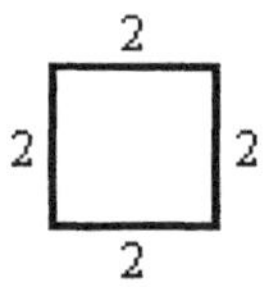

Square: $s = 2$ yards
$P = 4s = 4(2) = 8$ yards

APPLICATIONS

39. a. $C = 10h + 20$ Where C represents the Cost of renting the carpet-cleaning system and h represents the number of hours it is rented.

b.

h	$C = 10h + 20$	C
1	$C = 10(1) + 20$	30
2	$C = 10(2) + 20$	40
3	$C = 10(3) + 20$	50
4	$C = 10(4) + 20$	60
8	$C = 10(8) + 20$	100

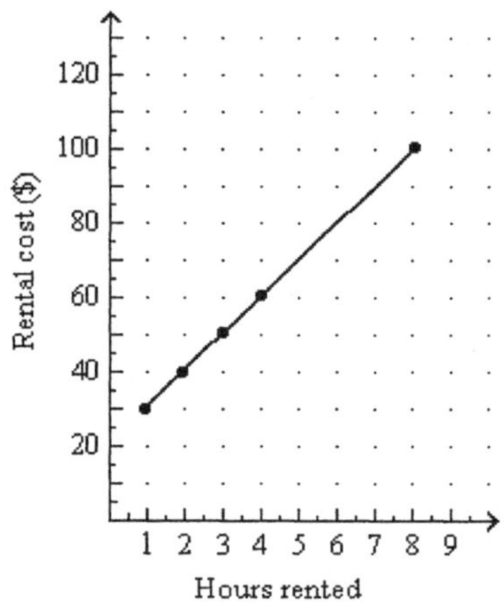

41.

f	$s = 180 - f$	s
30	$s = 180 - 30$	150
45	$s = 180 - 45$	135
90	$s = 180 - 90$	90
135	$s = 180 - 135$	45
150	$s = 180 - 150$	30

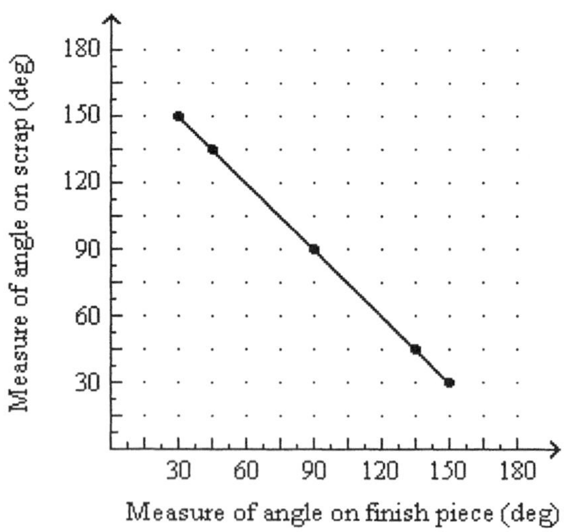

WRITING

43. An equation is a statement of equality between two or more expressions.
Expressions: $3x + 5$, $4x - 7$
Equation: $3x + 5 = 4x - 7$

45. Answers will vary.

Section 1.2 The Real Number System

VOCABULARY

1. rational
3. absolute value
5. Irrational
7. composite

CONCEPTS

9. 1, 2, 9
11. −3, 0, 1, 2, 9
13. $\sqrt{3}$, π
15. 2
17. 2
19. 9
21. nonrepeating; irrational
23. repeating; rational

25. $7=\frac{7}{1}$; $-7\frac{3}{5}=-\frac{38}{5}$; $0.007=\frac{7}{1000}$; $700.1=700\frac{1}{10}=\frac{7001}{10}$

27. 3.5 or −3.5, because |−3.5| also equals 3.5

29.

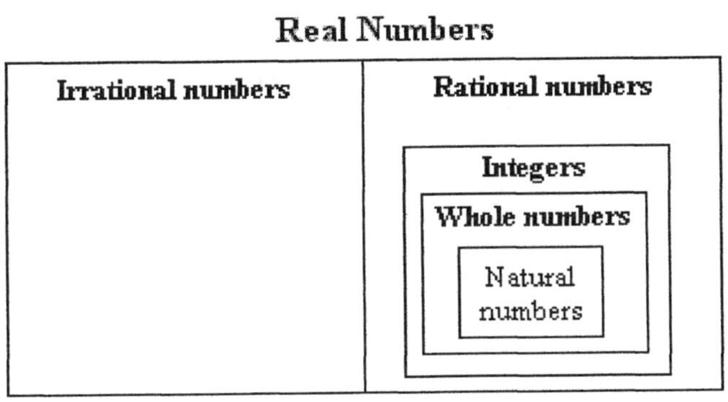

NOTATION

31. is less than
33. braces

PRACTICE

35. $\frac{7}{8}=0.875$; terminating

37. $-\frac{11}{15}=-7.333\ldots=-0.7\overline{3}$; repeating

39.

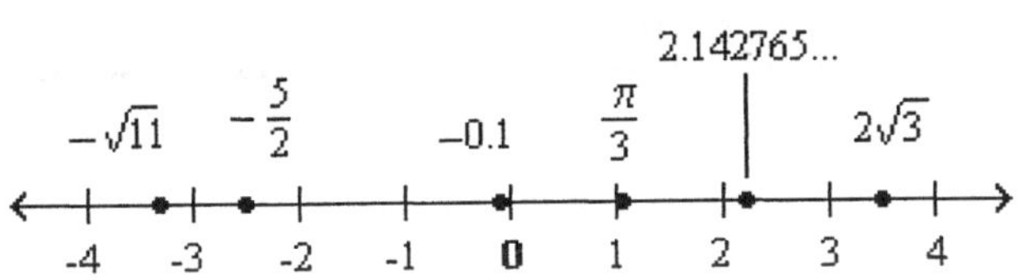

41.

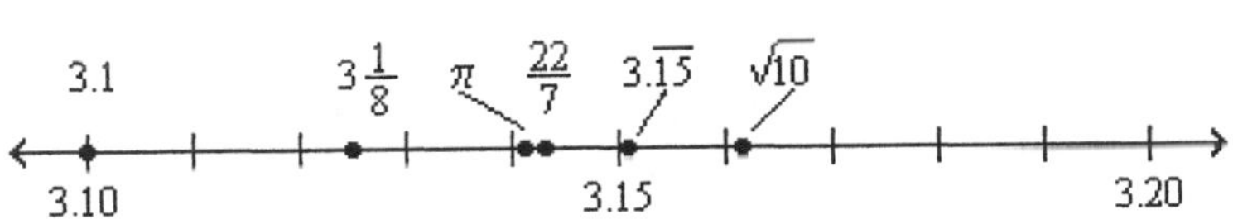

43.

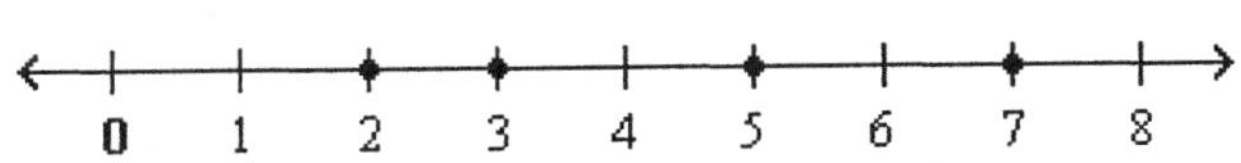

45.

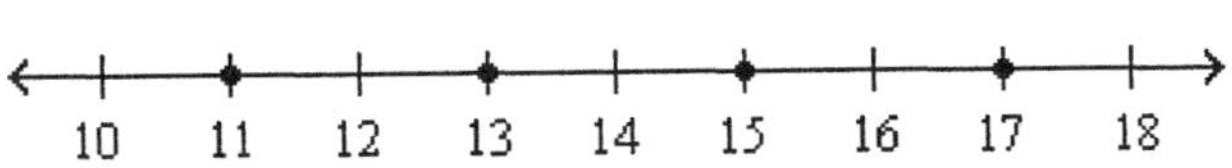

47.

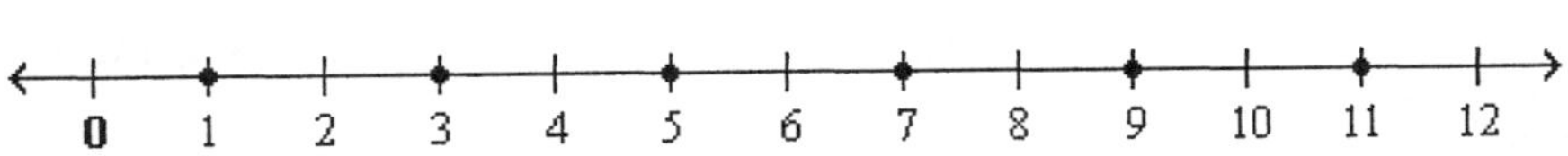

49. <, since 8 is to the left of 9

51. >, since −(−5) = 5 and 5 is to the right of −10 on the number line

53. <, since −7.1 = −7.100 and −7.999 is to the left of −7.100 on the number line

55. >, since 6.1111… is to the right of 6 on the number line

In exercises 57 through 60, exchange the two numbers first and then enter the reversed inequality symbol.

57. $12 < 19$

59. $-5 \geq -6$

61. $|20| = 20$

63. $-|-6| = -(6) = -6$

65. $|-5.9| = 5.9$

67. $\left|\frac{5}{4}\right| = \frac{5}{4}$

APPLICATIONS

69. Using a calculator: $3\frac{2}{25} = 3.0800$, $\frac{77}{50} = 1.5400$, $\frac{15}{16} = 0.9375$, $2\frac{5}{8} = 2.6250$, $\frac{\pi}{4} \approx 0.7854$, $\sqrt{8} \approx 2.8284$

71. The tip of the rotor blade will travel in a circle when it makes one complete revolution. Calculate the circumference of that circle, where $D = 2r = 2(18) = 36$ ft.

$C = \pi D = \pi(36) = 113.10$ ft

WRITING

73. The whole numbers are the non-negative integers.

75. There are no even prime numbers greater than 2 because 2 is a factor of all even numbers.

REVIEW

77. expression

79.

x	$T = x - 1.5$	T
3.7	$T = 3.7 - 1.5$	2.2
10	$T = 10 - 1.5$	8.5
30.6	$T = 30.6 - 1.5$	29.1

Section 1.3 Operations with Real Numbers

VOCABULARY

1. sum; difference
3. evaluate
5. squared; cubed
7. base; exponent
9. negative

CONCEPTS

11. a. add first, then multiply: $6 + 3 \cdot 2 = 9 \cdot 2 = 18$
    b. 12; Multiplication is to be done before addition. multiply first, then add: $6 + 3 \cdot 2 = 6 + 6 = 12$

13. a. area b. volume c. perimeter

NOTATION

15.
$$\begin{aligned} 60 - 20 \cdot 2^3 &= 60 - 20 \cdot \mathbf{8} \\ &= 60 - \mathbf{160} \\ &= -100 \end{aligned}$$

17. radical symbol

PRACTICE

19 a. -6 b. 6

21. $-3 + (-5) = -8$, like signs

23. $-7.1 + 2.8 = -4.3$, unlike signs

In exercises 25 through 28, change the sign of the number subtracted and then add.

25. $-3 - 4 = -3 + (-4) = -7$

27. $-3.3 - (-3.3) = -3.3 + 3.3 = 0$

29. $-2(6) = -12$, unlike signs

31. $-0.3\,(5) = -1.5$

33. $-5\,(6)\,(-2) = 60$

35. $\frac{-8}{4} = -2$

In exercises 33 through 36 , express in terms of the lowest common denominator, then add.

37. $\frac{1}{2}+\left(-\frac{1}{3}\right)=\frac{3}{6}+\left(-\frac{2}{6}\right)=\frac{1}{6}$

39. $\frac{1}{2}-\left(-\frac{3}{5}\right)=\frac{5}{10}-\left(-\frac{6}{10}\right)=\frac{5}{10}+\frac{6}{10}=\frac{11}{10}$

In exercise 41, multiply the numerators and denominators, apply the correct sign, and then divide out common factors if possible.

41. $\left(-\frac{3}{5}\right)\left(\frac{10}{7}\right)=-\frac{3\cdot 10}{5\cdot 7}=-\frac{3\cdot 2\cdot \cancel{5}}{\cancel{5}\cdot 7}=-\frac{6}{7}$

In exercise 43, multiply by the reciprocal, apply the correct sign, and then divide out common factors if possible.

43. $-\frac{16}{5}\div\left(-\frac{10}{3}\right)=-\frac{16}{5}\cdot\left(-\frac{3}{10}\right)=+\frac{16\cdot 3}{5\cdot 10}=\frac{\overset{1}{\cancel{2}}\cdot 8\cdot 3}{5\cdot \underset{1}{\cancel{2}}\cdot 5}=\frac{24}{25}$

45. $12^2 = 12 \cdot 12 = 144$

47. $-5^2 = -(5 \cdot 5) = -25$

49. $4 \cdot 2^3 = 4 \cdot (2 \cdot 2 \cdot 2) = 4 \cdot 8 = 32$

51. $(1.3)^2 = (1.3)(1.3) = 1.69$

53. $\sqrt{64} = 8$, because $8^2 = 64$.

55. Since $\sqrt{\frac{9}{16}}=\frac{3}{4}$, $-\sqrt{\frac{9}{16}}=-\frac{3}{4}$.

57. $3 - 5 \cdot 4 = 3 - 20 = -17$

59. $\left(-3-\sqrt{25}\right)^2=(-3-5)^2=(-8)^2=(-8)(-8)=64$

61.	$2+3\left(\frac{25}{5}\right)+(-4)=2+3(5)+(-4)$	Do the division within the parentheses.
	$=2+15-4$	Do the multiplication.
	$=17-4$	Do the addition.
	$=13$	Do the subtraction.
63.	$\frac{-\sqrt{49}-3^2}{2\cdot 4}=\frac{-7-9}{8}$	In the numerator, evaluate the square root and the power. In the denominator, do the multiplication.
	$=\frac{-16}{8}$	Do the subtraction.
	$=-2$	Do the division.
65.	$-2\lvert 4-8\rvert=-2\lvert -4\rvert$	Do the subtraction within the absolute value symbol.
	$=-2\cdot 4$	Find the absolute value.
	$=-8$	Do the multiplication.
67.	$(4+2\cdot 3)^4=(4+6)^4$	Do the multiplication within the parentheses.
	$=(10)^4$	Do the addition within the parentheses.
	$=10{,}000$	Evaluate the power.
69.	$3+2[-1-4(5)]=3+2(-1-20)$	Do the multiplication within the brackets.
	$=3+2(-21)$	Do the addition within the parentheses.
	$=3-42$	Do the multiplication.
	$=-39$	Do the subtraction.
71.	$3-[3^3+(3-1)^3]=3-[3^3+(2)^3]$	Do the addition within the parentheses.
	$=3-(27+8)$	Evaluate the powers within the brackets.
	$=3-35$	Do the addition within the parentheses.
	$=-32$	Do the subtraction.

73. $\dfrac{|-25|-2(-5)}{2^4-9}=\dfrac{25+10}{16-9}$ — In the numerator, find the absolute value and do the multiplication. In the denominator, evaluate the power.

$=\dfrac{35}{7}$ — Do the addition and subtraction.

$=5$ — Do the division.

75. $\dfrac{3[-9+2(7-3)]}{(8-5)(9-7)}=\dfrac{3[-9+2(4)]}{(3)(2)}$ — Do the additions within the parentheses.

$=\dfrac{3(-9+8)}{6}$ — Do the multiplications.

$=\dfrac{3(-1)}{6}$ — Do the addition within the parentheses.

$=-\dfrac{3}{6}=-\dfrac{\overset{1}{\cancel{3}}}{\underset{1}{\cancel{3}}\cdot 2}$ — Do the multiplication.

Divide out the common factors.

$=-\dfrac{1}{2}$

77. $\dfrac{(6-5)^4+21}{\left|\left(\sqrt{16}\right)^2-27\right|}=\dfrac{1^4+21}{\left|(4)^2-27\right|}$ — In the numerator, do the addition within the parentheses. In the denominator, evaluate the square root.

$=\dfrac{1+21}{|16-27|}$ — Evaluate the powers.

$=\dfrac{22}{|-11|}$ — Do the addition and subtraction.

$=\dfrac{22}{11}$ — Find the absolute value.

$=2$ — Do the division.

79. $54^3-16^4+19(3)=157{,}464-65{,}536+19(3)$ — Evaluate the powers using a calculator.

$=157{,}464-65{,}536+57$ — Do the multiplication.

$=91{,}985$ — Do the addition and subtraction from left to right.

81. $-\dfrac{2}{3}a^2=-\dfrac{2}{3}(-6)^2$ — Substitute -6 for a.

$=-\dfrac{2}{3}(36)$ — Evaluate the power.

$=-\dfrac{72}{3}$ — Do the multiplication.

$=-24$ — Do the division.

83. $\frac{y_2 - y_1}{x_2 - x_1} = \frac{(-4) - (12)}{(5) - (-3)}$ Substitute –3 for x_1, 5 for x_2, 12 for y_1, and –4 for y_2.

$= \frac{-16}{5+3}$ In the numerator, do the subtraction. In the denominator, simplify $-(-3) = +3$.

$= \frac{-16}{8}$ Do the addition in the denominator.

$= -2$ Do the division.

85. $(x + y)(x^2 - xy + y^2) = (-4 + 5)[(-4)^2 - (-4)(5) + (5)^2]$ Substitute –4 for x and 5 for y.

$= 1[16 - (-20) + 25]$ Evaluate the powers within the brackets, and do the addition within the parentheses.

$= 1(16 + 20 + 25)$ Simplify $-(-20) = +20$.
$= 1(61)$ Do the addition within the parentheses.
$= 61$ Do the multiplication.

87. $\frac{x^2}{a^2} + \frac{y^2}{b^2} = \frac{(-3)^2}{5^2} + \frac{(-4)^2}{(-5)^2}$ Substitute –3 for x, –4 for y, 5 for a, and –5 for b.

$= \frac{9}{25} + \frac{16}{25}$ Evaluate all the powers.

$= \frac{25}{25} = 1$ Do the addition and then divide out the common factor leaving 1.

89. $\sqrt{(x_2 - x_1)^2 + (y_2 - y_1)^2} = \sqrt{[4 - (-2)]^2 + [(-4) - 4)]^2}$ Substitute –2 for x_1, 4 for x_2, 4 for y_1, and –4 for y_2.

$= \sqrt{(4 + 2)^2 + (-4 - 4)^2}$ Simplify within the parentheses.

$= \sqrt{6^2 + (-8)^2}$ Do the addition and subtraction within the parentheses.

$= \sqrt{36 + 64}$ Evaluate the powers

$= \sqrt{100}$ Do the addition.

$= 10$ Find the square root.

91. $A = \frac{1}{2}bh$ The formula for the area of a triangle.

$= \frac{1}{2}(2.75)(8.25)$ Substitute 2.75 for b and 8.25 for h.

$= 11.34375 \approx 11.3 \text{ cm}^2$ Do the multiplication and round to the nearest tenth.

93. $V = lwh$ The formula for the volume of a rectangular solid.

$= (2.5)(3.7)(10.2)$ Substitute 2.5 for l, 3.7 for w, and 10.2 for h.

$= 94.35 \text{ cm}^3$ Do the multiplication and round to the nearest hundredth.

95.

$V = \frac{4}{3}\pi r^3$	The formula for the volume of a sphere.
$= \frac{4}{3}\pi(5.7)^3$	Substitute 5.7 for r.
$= \frac{4}{3}\pi(185.193)$	Evaluate the power.
$= \frac{740.772\pi}{3}$	Do the multiplication.
$= 775.73462 \approx 775.73 \text{ m}^3$	Use a calculator and round to the nearest hundredth.

APPLICATIONS

97. Find the area of a rectangular piece of aluminum foil in square feet. Since the dimensions are in yards and inches, convert the length and width to feet and then substitute into the formula for area.

$l = 8\frac{1}{3} \text{ yards} = \left(\frac{25}{3} \text{ yards}\right)\left(\frac{3 \text{ feet}}{1 \text{ yard}}\right) = 25 \text{ feet}$	Convert yards to feet for the length.
$w = 12 \text{ inches} = 1 \text{ foot}$	Convert inches to feet for the width.
$A = lw = (25)(1) = 25 \text{ ft}^2$	Substitute 25 for l and 1 for w into the formula for area of a rectangle.

99.

$\frac{1}{2}h_1(b_1 + b_2)$	The area of the bottom flap.
b_3h_3	The combined area of the two side flaps.
$\frac{1}{2}b_1h_2$	The area of the top flap.
b_1b_3	The area of the face.
$A = \frac{1}{2}h_1(b_1 + b_2) + b_3h_3 + \frac{1}{2}b_1h_2 + b_1b_3$	The given formula.
$= \frac{1}{2}(2)(6+2) + (3)(3) + \frac{1}{2}(6)(2.5) + (6)(3)$	Substitute $b_1 = 6$, $b_2 = 2$, $b_3 = 3$, $h_1 = 2$, $h_2 = 2.5$, and $h_3 = 3$ into the given formula.
$= 1(8) + 9 + 3(2.5) + 18$	
$= 17 + 7.5 + 18 = 42.5 \text{ in.}^2$	

101. Sum the column of numbers.
Balance = 5889 + 927 + (−2928) + 1645 + (−894) + 715 + (−6321) = \$(967)

103. Calculate the volume of the two scoops of ice cream and the volume of the cone. Then compare the two volumes.

$2V = 2\left(\frac{4}{3}\pi r^3\right) = 2\left(\frac{4}{3}\pi(1)^3\right) = \frac{8}{3}\pi = 8.4 \text{ in.}^3$	Substitute $r = (2 \text{ in.}) \div 2 = 1$ into the formula for the volume of a sphere.
$V = \frac{1}{3}\pi r^2 h$	Substitute $r = (2 \text{ in.}) \div 2$ and $h = 6$ in. into the formula for the volume of a cone
$= \frac{1}{3}\pi(1)^2(6) = \frac{1}{3}\pi(1)(6) = 2\pi = 6.3 \text{ in.}^3$	

Yes, the cone will overflow because there is a larger volume of ice cream than the cone can contain.

105. Convert all the dimensions into inches.
h = 11 inches, b_1 = 1 ft 8 in. = 12 in. + 8 in. = 20 in., b_2 = 1 ft 2 in. = 12 in. + 2 in. = 14 in.

$$A = \frac{1}{2}h(b_1 + b_2)$$
$$= \frac{1}{2}(11)(20+14) = \frac{1}{2}(11)(34) = 187 \text{ in.}^2$$

Substitute $h = 11$, $b_1 = 20$, and $b_2 = 14$ into the formula for the area of a trapezoid.

107.
$$\text{child's dose} = \frac{\text{Age of child}}{\text{Age of child} + 12}\left(\begin{matrix}\text{average}\\ \text{adult dose}\end{matrix}\right)$$
$$= \frac{6}{6+12}(27) = \frac{6}{18}(27) = 9 \text{ units}$$

Substitute 6 for the age of the child and 27 for the average adult dosage.

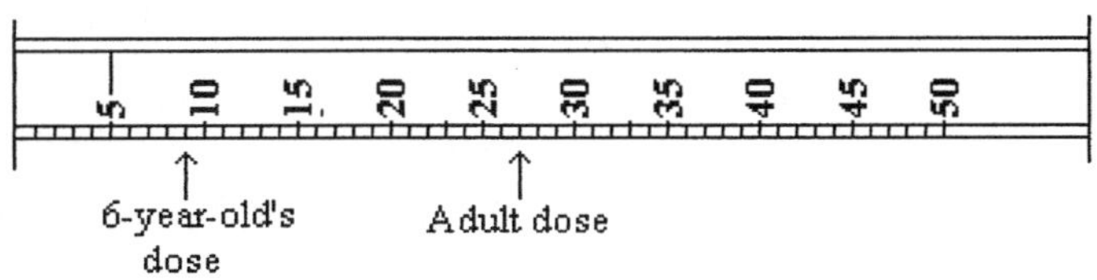

WRITING

109. The difference between x and y is the same as the sum of x and negative y.

111. It would be incorrect because the placement of the exponent would indicate that the 25 should be squared, not that the feet are squared. The area of the figure would be (5 ft)(5 ft) = (5 · 5)(1ft · 1ft) = 25 ft^2.

REVIEW

113. −7 and 3

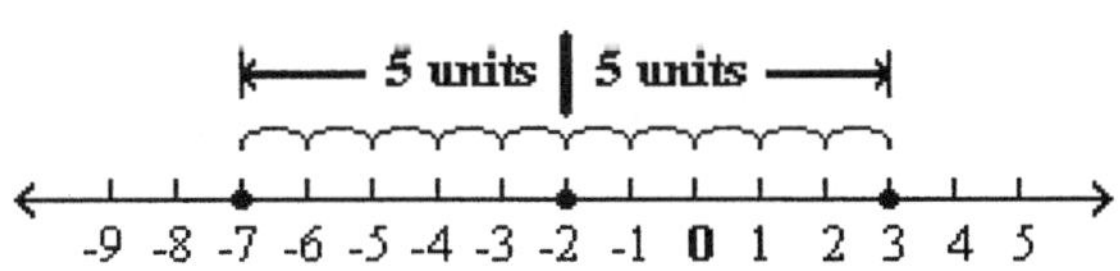

115. {…, −2, −1, 0, 1, 2, …}

117. true, because all decimals would include terminating and nonterminating decimals, as well as, repeating and nonrepeating decimals.

Section 1.4 — Simplifying Algebraic Expressions

VOCABULARY

1. Like 3. term 5. simplify 7. undefined

CONCEPTS

9. a. $(x + y) + z = x + (y + z)$ b. $xy = yx$ c. $r(s + t) = rs + rt$

11 a. 5 b. −5

13. a. $\frac{16}{15}$ b. $-\frac{1}{20}$ c. 2 d. $\frac{1}{x}$

15. a. $0 + 2x = 2x$ b. $-3a + 0 = -3a$ c. $0 - 9t = -9t$
 d. $0 - (-22x) = -1(-22x) = 22x$

17. a. $2x^2, -x, 6$ b. 2, −1, 6 19. yes, $8x$ 21. no, different variables

23. yes, $-2x^2$ 25. no, different variables 27. multiplication by −1

PRACTICE

29. $3+7 = \underline{7+3}$

31. $3(2+d) = \underline{3\cdot 2 + 3d}$

33. $c+0 = \underline{c}$

35. $25\cdot\frac{1}{25} = \underline{1}$

37. $8+(7+a) = \underline{(8+7)+a}$

39. $(x+y)2 = \underline{2(x+y)}$

41.
$$\begin{aligned} (37.9+25.2)+14.3 &= 37.9+(25.2+14.3) \\ (63.1)+14.3 &= 37.9+(39.5) \quad \text{Do the addition within the parentheses} \\ 77.4 &= 77.4 \end{aligned}$$
➔ Associative property of addition

43.
$$\begin{aligned} 2.73(4.534+57.12) &= 2.73\cdot 4.534 + 2.73\cdot 57.12 \\ 2.73(61.654) &= 12.37782+155.9376 \\ 168.31542 &= 168.31542 \end{aligned}$$
➔ Distributive property

45.
$$\begin{aligned} -4(t-3) &= -4(t)-(-4)(3) && \text{Distribute the multiplication by } -4. \\ &= -4t-(-12) && \text{Do the multiplications.} \\ &= -4t+12 \end{aligned}$$

47.
$$\begin{aligned} -(t-3) &= -1(t-3) && \text{Write the } - \text{ sign in front of the parentheses as } -1. \\ &= -1(t)-(-1)(3) && \text{Distribute the multiplication by } -1. \\ &= -t-(-3) && \text{Do the multiplications.} \\ &= -t+3 \end{aligned}$$

49.
$$\begin{aligned} \frac{2}{3}(3s-9) &= \frac{2}{3}(3s)-\frac{2}{3}(9) && \text{Distribute the multiplication by 2/3.} \\ &= \frac{2\cdot 3}{3}s-\frac{2\cdot 9}{3} && \text{Do the multiplications.} \\ &= \frac{2\cdot \overset{1}{\cancel{3}}}{\underset{1}{\cancel{3}}}s-\frac{2\cdot 3\cdot \overset{1}{\cancel{3}}}{\underset{1}{\cancel{3}}} && \text{Divide out the common factors.} \\ &= 2s-6 \end{aligned}$$

51.
$$\begin{aligned} 0.7(s+2) &= 0.7(s)+0.7(2) && \text{Distribute the multiplication by 0.7.} \\ &= 0.7s+1.4 && \text{Do the multiplications.} \end{aligned}$$

53.
$$\begin{aligned} 3\left(\frac{4}{3}x-\frac{5}{3}y+\frac{1}{3}\right) &= 3\left(\frac{4}{3}x\right)-3\left(\frac{5}{3}y\right)+3\left(\frac{1}{3}\right) && \text{Distribute the multiplication by 3.} \\ &= \frac{3\cdot 4}{3}x-\frac{3\cdot 5}{3}y+\frac{3\cdot 1}{3} && \text{Do the multiplications.} \\ &= \frac{\overset{1}{\cancel{3}}\cdot 4}{\underset{1}{\cancel{3}}}x-\frac{\overset{1}{\cancel{3}}\cdot 5}{\underset{1}{\cancel{3}}}y+\frac{\overset{1}{\cancel{3}}\cdot 1}{\underset{1}{\cancel{3}}} && \text{Divide out the common factors.} \\ &= 4x-5y+1 \end{aligned}$$

55. $(9m+n) = 72\,m + 8\,n$

57. $(-4r-3s)(-6) = -30s+5t$

59.
$$\begin{aligned} 9(8m) &= (9\cdot 8)m && \text{Use the associative property of multiplication} \\ &= 72m && \text{Do the multiplication.} \end{aligned}$$

61.
$$\begin{aligned} 5(-9q) &= [(5)\cdot(-9)]q && \text{Use the associative property of multiplication} \\ &= -45q && \text{Do the multiplication.} \end{aligned}$$

63.
$$\begin{aligned} (-5p)(-6b) &= [(-5)(-6)](p\cdot b) && \text{Use the commutative and associative properties to group the numbers and group the variables.} \\ &= 30pb && \text{Do the multiplications.} \end{aligned}$$

65. $-5(8r)(-2y) = [-5(8)(-2)](r \cdot y)$ Use the commutative and associative properties to group the numbers and group the variables.
$= 80ry$ Do the multiplications.

67. $3x + 15x = (3 + 15)x$ Add the coefficients.
$= 18x$ Keep the variable x.

69. $18x^2 - 5x^2 = 13x^2$ Subtract the coefficients and keep the variable x^2.

71. $-9x + 9x = 0x$ Add the coefficients and keep the variable x.
$= 0$ Do the multiplication.

73. $-b^2 + b^2 = 0b^2$ Add the coefficients and keep the variable b^2.
$= 0$ Do the multiplication.

75. $8x + 5x - 7x = (8 + 5 - 7)x$ Add all the coefficients.
$= 6x$ Keep the variable x.

77. $3x^2 + 2x^2 - 5x^2 = (3 + 2 - 5)x^2$ Add all the coefficients.
$= 0x^2$ Keep the variable x^2.
$= 0$ Do the multiplication.

79. $3.8h - 0.7h = 3.1h$ Add the coefficients and keep the variable h.

81.
$$\frac{2}{5}ab - \left(-\frac{1}{2}ab\right) = \frac{2 \cdot 2}{5 \cdot 2}ab - \left(-\frac{1 \cdot 5}{2 \cdot 5}ab\right)$$ Express each fraction in terms of the LCD.
$$= \frac{4}{10}ab - \left(-\frac{5}{10}ab\right)$$ Do the multiplications.
$$= \left(\frac{4}{10} + \frac{5}{10}\right)ab$$ Add the coefficients.

Keep the variables ab.
$$= \frac{9}{10}ab$$

83.
$$\frac{3}{5}t + \frac{1}{3}t = \frac{3 \cdot 3}{5 \cdot 3}t + \frac{1 \cdot 5}{3 \cdot 5}t$$ Express each fraction in terms of the LCD.
$$= \frac{9}{15}t + \frac{5}{15}t$$ Do the multiplications.
$$= \frac{14}{15}t$$ Add the coefficients and keep the variable t.

85. $4(y + 9) - 8y = 4y + 36 - 8y$ Distributive the multiplication by 4.
$= -4y + 36$ Combine like terms: $4y + (-8y) = -4y$.

87. $2z + 5(z - 4) = 2z + 5z - 20$ Distribute the multiplication by 5.
$= 7z - 20$ Combine like terms: $2z + 5z = 7z$.

89. $8(2c + 7) - 2(c - 3) = 16c + 56 - 2c + 6$ Use the distributive property twice.
$= 14c + 62$ Combine like terms: $16c + (-2c) = 14c$, $56 + 6 = 62$.

91. $2x^2 + 4(3x - x^2) + 3x = 2x^2 + 12x - 4x^2 + 3x$ Distribute the multiplication by 4.
$= -2x^2 + 15x$ Combine like terms

93. $-(a + 2) - (a - b) = -a - 2 - a + b$ Distribute the multiplication by -1 twice.
$= -2a + b - 2$ Combine like terms.

95. $-3(p-2)+2(p+3)-5(p-1) = -3p+6+2p+6-5p+5$ — Distributive property three times.
$= -6p+17$ — Combine like terms.

APPLICATIONS

97. a. $l = 20$ meters, w = self parking width + valert parking width $= x + 6$ meters

Total area $= l \cdot w = 20$ meters $\cdot\ (x+6)$ meters
$= 20(x+6)\ \text{m}^2$

b. Area by sum = area of self–parking + area of valet parking
$= (20\text{ m})(x\text{ meters}) + (6\text{ meters})(20\text{ meters})$
$= (20x+120)\ \text{m}^2$

c. Total area = Area by sum
$20(x+6) = 20x+120$; Distributive property.

99. Calculate the length of the ribbon where $l = (x+1)$ in., $w = (x-6)$ in., $h = (x-11)$ in., and bow $= x$ in.

Ribbon	=	2(top/bottom length)	+	2(top/bottom width)	+	4(sides)	+	bow
	=	$2(l)$	+	$2(w)$	+	$4(h)$	+	bow
	=	$2(x+1)$	+	$2(x-6)$	+	$4(x-11)$	+	x
	=	$2x+2$	+	$2x-12$	+	$4x-44$	+	x
	=	$(9x-54)$ in.						

WRITING

101. The distributive property is for multiplication over addition. This example is only multiplication.
103. Like terms are terms in which the variable part of the terms are identical.

REVIEW

105. $-5.6-(-5.6) = -5.6+5.6 = 0$

107. $\left(-\frac{3}{2}\right)\left(\frac{7}{12}\right) = -\frac{\overset{1}{\cancel{3}}\cdot 7}{2\cdot\underset{1}{\cancel{3}}\cdot 4} = -\frac{7}{8}$

109. $(4+2\cdot 3)^3 = (4+6)^3 = 10^3 = 1000$

111. $\frac{-\sqrt{64}-5^2}{2\cdot 4+3} = \frac{-8-25}{8+3} = \frac{-33}{11} = -3$

Section 1.5 Solving Linear Equations and Formulas

VOCABULARY

1. equation
3. satisfies
5. contradiction

CONCEPTS

7. c; c

9. a. $5y+2-3y = 2y+2$ — Combine like terms.

b. $5y+2-3y = 8$
$2y+2 = 8$ — Combine like terms.
$2y+2-2 = 8-2$ — Subtract 2 from both sides.
$2y = 6$ — Simplify each side of the equation.
$y = 3$ — To isolate x, divide both sides by 2.

c. $5y+2-3y = 5(8)+2-3(8)$ — Substitute 8 for y in the expression.
$= 40+2-24$ — Do the multiplications.
$= 18$ — Do the addition.

11. All real numbers (⊕).

13. To clear the equation of fractions.

NOTATION

15.
$$\begin{aligned} -2(x+7) &= 20 \\ \mathbf{-2x} - 14 &= 20 \\ -2x - 14 + \mathbf{14} &= 20 + \mathbf{14} \\ -2x &= \mathbf{34} \\ \frac{-2x}{-2} &= \frac{34}{-2} \\ x &= -17 \end{aligned}$$

17. a. -1 b. $\frac{2}{3}$

PRACTICE

19.

$3x+2 = 17$	The original equation.
$3(5)+2 \stackrel{?}{=} 17$	Substitute 5 for x.
$15+2 \stackrel{?}{=} 17$	Do the multiplication.
$17 = 17$	Do the addition. Yes, 5 is a solution.

21.

$3(2m-3) = 15$	The original equation.
$3[2(5)-3] \stackrel{?}{=} 15$	Substitute 5 for m.
$3(10-3) \stackrel{?}{=} 15$	Do the multiplication.
$3(7) \stackrel{?}{=} 15$	Do the addition.
$21 \neq 15$	Do the multiplication. No, 5 is not a solution.

23.

$\frac{x}{4} = 7$	The original equation.
$4\left(\frac{x}{4}\right) = 4(7)$	Multiply both sides by 4.
$x = 28$	Do the multiplication.

25.

$-\frac{4}{5}s = 16$	The original equation.
$5\left(-\frac{4}{5}s\right) = 5(16)$	Multiply both sides by 5.
$-4s = 80$	Do the multiplication.
$\frac{-4x}{-4} = \frac{80}{-4}$	Divide both sides by –4.
$x = -20$	Do the division.

27.

$2x-1=0$	
$2x-1+1=0+1$	Add 1 to both sides.
$2x=1$	Do the addition.
$\frac{2x}{2}=\frac{1}{2}$	Divide both sides by 2.
$x=\frac{1}{2}$	Simplify.

29.

$5y+6=0$	
$5y+6-6=0-6$	Subtract 6 from both sides.
$5y=-6$	Do the subtraction.
$\frac{5y}{5}=\frac{-6}{5}$	Divide both sides by 5.
$y=-\frac{6}{5}$	Simplify.

31. $8x = x$

$8x - x = x - x$ Subtract x from both sides.

$7x = 0$ Do the subtraction.

$\frac{7x}{7} = \frac{0}{7}$ Divide both sides by 7.

$x = 0$ Simplify.

33. $0 = -x$

$\frac{0}{-1} = \frac{-x}{-1}$ Divide both sides by -1.

$0 = x$ or $x = 0$

35. $2x + 2(1) = 6$

$2x + 2 = 6$ Do the multiplication.

$2x + 2 - 2 = 6 - 2$ Subtract 2 from both sides.

$2x = 4$ Do the subtraction.

$x = 2$ Divide both sides by 2.

37. $4(3) + 2y = -6$

$12 + 2y = -6$ Do the multiplication.

$12 + 2y - 12 = -6 - 12$ Subtract 12 from both sides.

$2y = -18$ Do the subtraction.

$y = -9$ Divide both sides by 2.

39. $200 = 34 - t$

$200 - 34 = 34 - t - 34$ Subtract 34 from both sides.

$166 = -t$ Do the subtraction.

$-166 = t$ Multiply both sides by -1.

$t = -166$ Rewrite with t on the left side.

41. $1.6a = 4.032$

$a = 2.52$ Divide both sides by 1.6.

43. $3x + 1 = 3$

$3x + 1 - 1 = 3 - 1$ Subtract 1 from both sides.

$3x = 2$ Do the subtraction.

$\frac{3x}{3} = \frac{2}{3}$ Divide both sides by 3.

$x = \frac{2}{3}$ Simplify.

45. $3(k - 4) = -36$

$3k - 12 = -36$ Distribute the multiplication by 3.

$3k - 12 + 12 = -36 + 12$ Add 12 to both sides.

$3k = -24$ Do the addition.

$k = -8$ Divide both sides by 3.

47. $4j + 12.54 = 18.12$

$4j + 12.54 - 12.54 = 18.12 - 12.54$ Subtract 12.54 from both sides.

$4j = 5.58$ Do the subtraction (Hint: use a calculator.).

$j = 1.395$ Divide both sides by 4.

49.

$$\begin{aligned} 4a - 22 - a &= -2a - 7 \\ 3a - 22 &= -2a - 7 && \text{Combine like terms on the left side.} \\ 3a - 22 + 2a + 22 &= -2a - 7 + 2a + 22 && \text{Add } 2a \text{ and 22 to both sides.} \\ 5a &= 15 && \text{Combine like terms on each side.} \\ a &= 3 && \text{Divide both sides by 5.} \end{aligned}$$

51.

$$\begin{aligned} 2(2x + 1) &= x + 15 + 2x \\ 4x + 2 &= 3x + 15 && \text{Distribute on the left side, combine like terms on the right.} \\ 4x + 2 - 3x - 2 &= 3x + 15 - 3x - 2 && \text{Subtract } 3x \text{ and 2 from both sides.} \\ x &= 13 && \text{Combine like terms on each side.} \end{aligned}$$

53.

$$\begin{aligned} 2(a - 5) - (3a + 1) &= 0 \\ 2a - 10 - 3a - 1 &= 0 && \text{Distribute the 2 and the } -1 \text{ on the left side.} \\ -a - 11 &= 0 && \text{Combine like terms on the left side.} \\ -a - 11 + 11 &= 0 + 11 && \text{Add 11 to both sides.} \\ -a &= 11 && \text{Do the addition.} \\ a &= -11 && \text{Multiply both sides by } -1. \end{aligned}$$

55.

$$\begin{aligned} 9(x + 2) &= -6(4 - x) + 18 \\ 9x + 18 &= -24 + 6x + 18 && \text{Distribute the multiplication by the 9 and the } -6. \\ 9x + 18 &= -6 + 6x && \text{Combine like terms on the right side.} \\ 9x + 18 - 6x - 18 &= -6 + 6x - 6x - 18 && \text{Subtract } 6x \text{ and 18 from both sides.} \\ 3x &= -24 && \text{Combine like terms on each side.} \\ x &= -8 && \text{Divide both sides by 3.} \end{aligned}$$

57.

$$\begin{aligned} \frac{1}{2}x - 4 &= -1 + 2x \\ 2\left(\frac{1}{2}x - 4\right) &= 2(-1 + 2x) && \text{Multiply both sides by 2.} \\ x - 8 &= -2 + 4x && \text{Do the multiplication.} \\ x - 8 - 4x + 8 &= -2 + 4x - 4x + 8 && \text{Subtract } 4x \text{ and add 8 to both sides.} \\ -3x &= 6 && \text{Combine like terms on each side.} \\ x &= -2 && \text{Divide both sides by } -3. \end{aligned}$$

59.

$$\begin{aligned} \frac{b}{2} - \frac{b}{3} &= 4 \\ 6\left(\frac{b}{2} - \frac{b}{3}\right) &= 6(4) && \text{Multiply both sides by the LCD 6.} \\ 6\left(\frac{b}{2}\right) - 6\left(\frac{b}{3}\right) &= 6(4) && \text{Distribute the multiplication by 6.} \\ 3b - 2b &= 24 && \text{Do the multiplication.} \\ b &= 24 && \text{Combine like terms on the left side.} \end{aligned}$$

61.

$$\frac{a+1}{3}+\frac{a-1}{5}=\frac{2}{15}$$

$$15\left(\frac{a+1}{3}+\frac{a-1}{5}\right)=15\left(\frac{2}{15}\right)$$ Multiply both sides by the LCD 15.

$$15\left(\frac{a+1}{3}\right)+15\left(\frac{a-1}{5}\right)=15$$ Distribute the multiplication by 15.

$$5(a+1)+3(a-1)=2$$ Do the multiplication.

$$5a+5+3a-3=2$$ Distribute on the left side.

$$8a+2=2$$ Combine like terms on the left side.

$$8a+2-2=2-2$$ Subtract 2 from both sides.

$$8a=0$$ Do the addition.

$$a=0$$ Divide both sides by 8.

63.

$$\frac{5a}{2}-12=\frac{a}{3}+1$$

$$6\left(\frac{5a}{2}-12\right)=6\left(\frac{a}{3}+1\right)$$ Multiply both sides by the LCD 6.

$$6\left(\frac{5a}{2}\right)-6(12)=6\left(\frac{a}{3}\right)+6(1)$$ Distribute the multiplication by 6.

$$3(5a)-72=2a+6$$ Do the multiplication.

$$15a-72=2a+6$$ Distribute on the left side.

$$15a-72-2a+72=2a+6-2a+72$$ Subtract $2a$ and add 72 to both sides.

$$13a=78$$ Combine like terms on each side.

$$a=6$$ Divide both sides by 13.

65.

$$\frac{3+p}{3}-4p=1-\frac{p+7}{2}$$

$$6\left(\frac{3+p}{3}-4p\right)=6\left(1-\frac{p+7}{2}\right)$$ Multiply both sides by the LCD 6.

$$6\left(\frac{3+p}{3}\right)-6(4p)=6(1)-6\left(\frac{p+7}{2}\right)$$ Distribute the multiplication by 6.

$$2(3+p)-24p=6-3(p+7)$$ Do the multiplication by 6.

$$6+2p-24p=6-3p-21$$ Do the multiplications by 2 and 3.

$$6-22p=-3p-15$$ Combine like terms on both sides.

$$6-22p+3p-6=-3p-15+3p-6$$ Add $3p$ and subtract 6 from both sides.

$$-19p=-21$$ Combine like terms on each side.

$$\frac{-19p}{-19}=\frac{-21}{-19}$$ Divide both sides by -19.

$$p=\frac{21}{19}$$ Simplify.

67.
$$\begin{aligned}
0.45 &= 16.95 - 0.25(75 - 3x) \\
100(0.45) &= 100[16.95 - 0.25(75 - 3x)] && \text{Multiply both sides by 100.} \\
45 &= 100(16.95) - 100[0.25(75 - 3x)] && \text{Distribute the multiplication by 100.} \\
45 &= 1695 - 25(75 - 3x) && \text{Do the multiplications.} \\
45 &= 1695 - 1875 + 75x && \text{Do the multiplication by 25.} \\
45 &= -180 + 75x && \text{Combine like terms on the right side.} \\
45 + 180 &= -180 + 75x + 180 && \text{Add 180 to both sides.} \\
225 &= 75x && \text{Combine like terms on both side.} \\
3 &= x && \text{Divide both sides by 75.}
\end{aligned}$$

69.
$$\begin{aligned}
0.04(12) + 0.01t - 0.02(12 + t) &= 0 \\
0.48 + 0.01t - 0.24 - 0.02t &= 0 && \text{Distribute the multiplications on the left side.} \\
-0.01t + 0.24 &= 0 && \text{Combine like terms on the left side.} \\
-0.01t + 0.24 - 0.24 &= 0 - 0.24 && \text{Subtract 0.24 from both sides.} \\
-0.01t &= -0.24 && \text{Combine like terms on both sides.} \\
t &= 24 && \text{Divide by } -0.01.
\end{aligned}$$

71.
$$\begin{aligned}
4(2 - 3t) + 6t &= -6t + 8 \\
8 - 12t + 6t &= -6t + 8 && \text{Distribute the multiplication on the left side.} \\
8 - 6t &= -6t + 8 && \text{Combine like terms on the left side.} \\
8 - 6t + 6t &= -6t + 8 + 6t && \text{Add } 6t \text{ to both sides.} \\
8 &= 8 && \text{Combine like terms on each side.}
\end{aligned}$$

This equation is an **identity**; all real values for y will make the original equation true.

73.
$$\begin{aligned}
3(x - 4) + 6 &= -2(x + 4) + 5x \\
3x - 12 + 6 &= -2x - 8 + 5x && \text{Distribute the multiplications on each side.} \\
3x - 6 &= 3x - 8 && \text{Combine like terms on each side.} \\
3x - 6 - 3x &= 3x - 8 - 3x && \text{Subtract } 3x \text{ from both sides.} \\
-6 &\neq -8 && \text{Combine like terms on each side.}
\end{aligned}$$

This is a **contradiction**; there is no value for x that will make the original equation true.

75.
$$\begin{aligned}
2y + 1 &= 5(0.2y + 1) - (4 - y) \\
2y + 1 &= 1y + 5 - 4 + y \\
2y + 1 &= 2y + 1
\end{aligned}$$

This equation is an **identity;** all real values for y will make the original equation true.

77.
$$V = \frac{1}{3}Bh \qquad \text{Solving for } B.$$

$$3(V) = 3\left(\frac{1}{3}Bh\right) \qquad \text{Multiply both sides by the LCD 3.}$$

$$3V = Bh \qquad \text{Do the multiplication.}$$

$$\frac{3V}{h} = \frac{Bh}{h} \qquad \text{Divide both sides by } h.$$

$$\frac{3V}{h} = B \text{ or } B = \frac{3V}{h} \qquad \text{Simplify and rewrite.}$$

79.
$$I = Prt \qquad \text{Solving for } t.$$

$$\frac{I}{Pr} = \frac{Prt}{Pr} \qquad \text{Divide both sides by } Pr.$$

$$\frac{I}{Pr} = t \text{ or } t = \frac{I}{Pr} \qquad \text{Simplify and rewrite.}$$

81.

$$P = 2l + 2w$$ Solving for w.

$$P - 2l = 2l + 2w - 2l$$ Subtract $2l$ from both sides.

$$P - 2l = 2w$$ Combine like terms on the right side.

$$\frac{P-2l}{2} = w \text{ or } w = \frac{P-2l}{2}$$ Divide each side by 2 and rewrite.

83.

$$A = \frac{1}{2}h(B+b)$$ Solving for B.

$$2(A) = 2\left[\frac{1}{2}h(B+b)\right]$$ Multiply both sides by the LCD 2.

$$2A = h(B+b)$$ Do the multiplication.

$$2A = hB + hb$$ Distribute the multiplication by h.

$$2A - hb = hB + hb - hb$$ Subtract hb from both sides.

$$2A - hb = hB$$ Combine like terms.

$$\frac{2A-hb}{h} = \frac{hB}{h}$$ Divide each side by h.

$$\frac{2A-hb}{h} = B \text{ or } B = \frac{2A-hb}{h}$$ Simplify and rewrite.

85.

$$y = mx + b$$ Solving for x.

$$y - b = mx + b - b$$ Subtract b from both sides.

$$y - b = mx$$ Combine like terms.

$$\frac{y-b}{m} = \frac{mx}{m}$$ Divide both sides by m.

$$\frac{y-b}{m} = x \text{ or } x = \frac{y-b}{m}$$ Simplify and rewrite.

87.

$$\overline{v} = \frac{1}{2}(v+v_0)$$ Solve for v_0.

$$2\overline{v} = 2\left[\frac{1}{2}(v+v_0)\right]$$ Multiply both sides by the LCD 2.

$$2\overline{v} = v + v_0$$ Do the multiplication.

$$2\overline{v} - v = v + v_0 - v$$ Subtract v from both sides.

$$2\overline{v} - v = v_0 \text{ or } v_0 = 2\overline{v} - v$$ Simplify and rewrite.

89.

$$S = \frac{a - lr}{1 - r}$$ Solving for l.

$$(1 - r)S = (1 - r)\left(\frac{a - lr}{1 - r}\right)$$ Multiply both sides by LCD $(1 - r)$.

$$S - rS = a - lr$$ Do the multiplication.

$$S - rS - a = a - lr - a$$ Subtract a from both sides.

$$S - rS - a = -lr$$ Combine like terms.

$$\frac{S - rS - a}{r} = \frac{-lr}{r}$$ Divide both sides by r.

$$\frac{S - rS - a}{r} = -l$$ Simplify.

$$\frac{-S + rS + a}{r} = l \text{ or } l = \frac{a + rS - S}{r}$$ Multiply both sides by -1 and rewrite.

91.

$$S = \frac{n(a + l)}{2}$$ Solving for l.

$$2S = 2\left(\frac{n(a + l)}{2}\right)$$ Multiply both sides by the LCD 2.

$$2S = n(a + l)$$ Do the multiplication.

$$2S = na + nl$$ Distribute the multiplication by n.

$$2S - na = na + nl - na$$ Subtract na from both sides.

$$2S - na = nl$$ Combine like terms.

$$\frac{2S - na}{n} = \frac{nl}{n}$$ Divide both sides by n.

$$\frac{2S}{n} - \frac{na}{n} = l \text{ or } l = \frac{2S - na}{n}$$ Simplify and rewrite.

APPLICATIONS

93.

$$F = \frac{9}{5}C + 32$$ Solving for C.

$$F - 32 = \frac{9}{5}C + 32 - 32$$ Subtract 32 from both sides.

$$F - 32 = \frac{9}{5}C$$ Combine like terms.

$$\frac{5}{9}(F - 32) = \frac{5}{9}\left(\frac{9}{5}C\right)$$ Multiply by the reciprocal of $\frac{9}{5}$.

$$\frac{5}{9}(F - 32) = C \text{ or } C = \frac{5}{9}(F - 32)$$ Do the multiplication and rewrite.

Mercury:	High	$C = \frac{5}{9}(810 - 32) = \frac{5}{9}(778) = 432.22 \approx 432°\text{C}$
	Low	$C = \frac{5}{9}(-290 - 32) = \frac{5}{9}(-322) = -178.88 \approx -179°\text{C}$
Earth:	High	$C = \frac{5}{9}(136 - 32) = \frac{5}{9}(104) = 57.77 \approx 58°\text{C}$
	Low	$C = \frac{5}{9}(-129 - 32) = \frac{5}{9}(-161) = -89.44 \approx -89°\text{C}$
Mars:	High	$C = \frac{5}{9}(63 - 32) = \frac{5}{9}(31) = 17.22 \approx 17°\text{C}$
	Low	$C = \frac{5}{9}(-87 - 32) = \frac{5}{9}(-119) = -66.11- \approx -66°\text{C}$

95.

$$A=\frac{d\pi(r_1^2-r_2^2)}{360}$$ Solving for d.

$$\frac{360}{\pi(r_1^2-r_2^2)}A=\frac{360}{\pi(r_1^2-r_2^2)}\left(\frac{d\pi(r_1^2-r_2^2)}{360}\right)$$ Multiply by the reciprocal of $\frac{\pi(r_1^2-r_2^2)}{360}$.

$$\frac{360A}{\pi(r_1^2-r_2^2)}=d \text{ or } d=\frac{360A}{\pi(r_1^2-r_2^2)}$$ Do the multiplication and rewrite.

Luxury car: Substitute 513 for A, 22 for r_1, and 8 for r_2.

$$d=\frac{360A}{\pi\left(r_1^2-r_2^2\right)}=\frac{360(513)}{\pi(22^2-8^2)}=\frac{184{,}680}{\pi(484-64)}=\frac{184{,}680}{\pi(420)}=139.96\approx 140°$$

Sport utility vehicle: Substitute 586 for A, 22 for r_1, and 8 for r_2.

$$d=\frac{360A}{\pi\left(r_1^2-r_2^2\right)}=\frac{360(586)}{\pi(22^2-8^2)}=\frac{210{,}960}{\pi(484-64)}=\frac{210{,}960}{\pi(420)}=159.88\approx 160°$$

97.

$$PV=nR(T+273)$$ Solving for n.

$$\frac{PV}{R(T+273)}=\frac{nR(T+273)}{R(T+273)}$$ Divide both sides by $R(T+273)$.

$$\frac{PV}{R(T+273)}=n \text{ or } n=\frac{PV}{R(T+273)}$$ Simplify and rewrite.

Trial 1: Substitute 0.082 for R, 0.900 for P, 0.250 for V, and 90 for T.

$$n=\frac{PV}{R(T+273)}=\frac{0.900(0.250)}{0.082(90+273)}=\frac{0.225}{0.082(363)}=0.00755\approx 0.008$$

Trial 2: Substitute 0.082 for R, 1.250 for P, 1.560 for V, and −10 for T.

$$n=\frac{PV}{R(T+273)}=\frac{1.250(1.560)}{0.082(-10+273)}=\frac{1.950}{0.082(263)}=0.09042\approx 0.090$$

99.

$$C=0.07n+6.50$$ Solving for n.

$$C-6.50=0.07n+6.50-6.50$$ Subtract 6.50 from both sides.

$$C-6.50=0.07n$$ Combine like terms.

$$\frac{C-6.50}{0.07}=\frac{0.07n}{0.07}$$ Divide both sides by 0.07.

$$\frac{C-6.50}{0.07}=n \text{ or } n=\frac{C-6.50}{0.07}$$ Simplify and rewrite.

April: substitute 49.97 for C.
$$n=\frac{C-6.50}{0.07}=\frac{49.97-6.50}{0.07}=\frac{43.47}{0.07}=621 \text{ kwh}$$

May: substitute 76.50 for C.
$$n=\frac{C-6.50}{0.07}=\frac{76.50-6.50}{0.07}=\frac{70}{0.07}=1000 \text{ kwh}$$

June: substitute 125.00 for C.
$$n=\frac{C-6.50}{0.07}=\frac{125.00-6.50}{0.07}=\frac{118.5}{0.07}=1692.857\approx 1692.9 \text{ kwh}$$

101.

$$A = 2\pi r^2 + 2\pi rh$$ Solving for h.

$$A - 2\pi r^2 = 2\pi r^2 + 2\pi rh - 2\pi r^2$$ Subtract $2\pi r^2$ from both sides.

$$A - 2\pi r^2 = 2\pi rh$$ Combine like terms.

$$\frac{A - 2\pi r^2}{2\pi r} = \frac{2\pi rh}{2\pi r}$$ Divide both sides by $2\pi r$.

$$\frac{A - 2\pi r^2}{2\pi r} = h \text{ or } h = \frac{A - 2\pi r^2}{2\pi r}$$ Simplify and rewrite.

WRITING

103. To *solve an equation* means to find a value for the variable that, when substituted into the original equation, will make the equation true.

105. An identity is an equation for which any real number will make the equation true. One way you can recognize an identity is to reduce the equation to 0=0. Example: $3 + x = 3 + x$

REVIEW

107. $-(4 + t) + 2t = -4 - t + 2t$ Distribute the multiplication by –1.
$= -4 + t$ Combine like terms.
$= t - 4$ Rewrite with the variable first.

109. $4(b + 8) - 8b = 4b + 32 - 8b$ Distribute the multiplication by 4.
$= -4b + 32$ Combine like terms.

111. $3.8b - 0.9b = 2.9b$ Combine like terms.

113. $\frac{3}{5}t + \frac{2}{5}t = \frac{5}{5}t$ Combine like terms.
$= t$ Simplify.

Section 1.6 Using Equations to Solve Problems

VOCABULARY

1. acute 3. complementary 5. right 7. angles

CONCEPTS

9.

	Decibels	Compared to conversation
Conversation	d	—
Vacuum cleaner	$d + 15$	15 decibels more
Circular saw	$2d - 10$	10 decibels less than twice
Jet takeoff	$2d + 20$	20 decibels more than twice
Whispering	$\frac{d}{2} - 10$	10 decibels less than half
Rock band	$2d$	twice the decibel level

11. a. $\frac{2}{3}x$ b. $2x$ c. $x + 2x + \frac{2}{3}x$ 13. $26.5 - x$

APPLICATIONS

15. **Analysis**: The combined sales of General Mills' Cheerios and Kellogg's Frosted Flakes is \$1,003 million. Frosted Flakes sales were \$324 million less than sales of Cheerios. Find sales for each cereal.

Equation: Let x represent the sales of Cheerios, then x -324 represents the sales of Frosted Flakes.

Cheerios Sales	plus	Frosted Flakes Sales	is	total sales.
x	$+$	$(x-324)$	$=$	1003

Solve:

$x+x-324$	$=$	1003	
$2x-324$	$=$	1003	Combine like terms.
$2x-324+324$	$=$	$1003+324$	Add 324 to both sides.
$2x$	$=$	1327	Combine like terms.
x	$=$	663.5	Divide both sides by 2.

Frosted Flakes sales= $x-324 = 663.5-324 = 339.5$

Conclusion: Sales of Frosted Flakes is \$339.5 million and for Cheerios is \$663.5 million.

17. **Analysis**: The chaperons will have the \$1810 cost of the tour reduced by \$15.50 for each student. Find how many students the chaperons need to supervise so that the cost is \$1500.

Equation: Let x represent the number of students a chaperon would supervise.

Original cost	minus	number of students	times	15.50	is	desired cost.
1810	$-$	x	$\bullet$	15.50	$=$	1500

Solve:

$1810-15.50x$	$=$	1500	
$1810-15.50x-1810$	$=$	$1500-1810$	Subtract 1810 from both sides.
$-15.50x$	$=$	-310	Combine like terms.
x	$=$	20	Divide both sides by 15.50.

Conclusion: The chaperon would have to supervise 20 students.

19. **Analysis**: Find how many miles a man may drive in 1 day and stay within his budget of \$75.

Equation: Let x represent the number of miles.

Daily charge	plus	charge per mile	times	number of miles	is	budget amount.
29.25	$+$	0.15	$\bullet$	x	$=$	75.00

Solve:

$29.25+0.15x$	$=$	75.00	
$29.25+0.15x-29.25$	$=$	$75.00-29.25$	Subtract 29.25 from both sides.
$0.15x$	$=$	45.75	Combine like terms.
x	$=$	305	Divide both sides by 0.15.

Conclusion: He could travel 305 miles and stay within his budget.

21. **Analysis**: Find how many shares of each stock a couple owns if the total value is \$53,900.

Equation: Let x represent the number of shares of Big Bank. Since the couple has a total of 500 shares in the account, then $500 - x$ would represent the number of Safe Savings shares.

Type of stock	Value per share	Number of shares	Total value of stock
Big Bank Corporation	\$115	x	\$115x
Safe Savings & Loan	\$97	$500 - x$	\$97$(500 - x)$

Value of Big Bank	plus	value of Safe Savings	is	total value.
$115x$	$+$	$97(500 - x)$	$=$	53,900

Solve:

$115x + 97(500 - x)$	$=$	53,900	
$115x + 48{,}500 - 97x$	$=$	53,900	Distribute the multiplication by 97.
$18x + 48{,}500$	$=$	53,900	Combine like terms.
$18x + 48{,}500 - 48{,}500$	$=$	$53{,}900 - 48{,}500$	Subtract 48,500 from both sides.
$18x$	$=$	5400	Combine like terms.
x	$=$	300	Divide both sides by 18.

Conclusion: The couple would have 300 shares of Big Bank Corporation and $500 - 300 = 200$ shares of Safe Savings and Loan.

23. **Analysis**: Find how many of each type of calculator was sold when the total sales were \$4980.

Equation: Let x represent the number of scientific calculators sold and then $x + 15$ would represent the number of graphing calculators sold.

Calculator	Value per calculator	Number of calculators	Total value of calculators
Scientific	\$18	x	\$18x
Graphing	\$87	$x + 15$	\$87$(x + 15)$

Value of scientific	plus	value of graphing	is	total value.
$18x$	$+$	$87(x + 15)$	$=$	4980

Solve:

$18x + 87(x + 15)$	$=$	4980	
$18x + 87x + 1305$	$=$	4980	Distribute the multiplication by 87.
$105x + 1305$	$=$	4980	Combine like terms.
$105x + 1305 - 1305$	$=$	$4980 - 1305$	Subtract 1305 from both sides.
$105x$	$=$	3675	Combine like terms.
x	$=$	35	Divide both sides by 105.

Conclusion: The bookstore sold 35 scientific calculators and $x + 15 = 50$ graphing calculators.

25. **Analysis**: Find the length of each piece of board when a board that is 22 feet long is cut into two pieces.

Equation: Let x represent the length of the shorter piece, then $2x + 1$ will represent the length of the other piece.

Length of shorter piece	plus	length of other piece	is	length of board.
x	$+$	$2x + 1$	$=$	22

Solve:

$x + 2x + 1$	$=$	22	
$3x + 1$	$=$	22	Combine like terms.
$3x + 1 - 1$	$=$	$22 - 1$	Subtract 1 from both sides.
$3x$	$=$	21	Combine like terms.
x	$=$	7	Divide both sides by 3.

Conclusion: The shorter piece is 7 ft long and the other piece is $2x + 1 = 15$ ft long.

27. **Analysis**: Find the measures of the angles in the illustration.

Equation: Let x represent the measure of the acute angle and $5x$ represents the measure of the obtuse angle. Together the two angles form a straight angle whose measure is 180°.

Measure of acute angle	plus	measure of obtuse angle	is	straight angle.
x	$+$	$5x$	$=$	180°

Solve:

$$x + 5x = 180$$
$$6x = 180 \quad \text{Combine like terms.}$$
$$x = 30 \quad \text{Divide both sides by 6.}$$

Conclusion: The measure of the acute angle is 30° and the measure of the obtuse angle is $5x = 150°$.

29. **Analysis**: Find how many degrees the tower is leaning from the vertical.

Equation: Let x represent the measure of the angle that the tower is from the vertical and $17x$ represents the measure of the other angle. Together the two angles form a right angle whose measure is 90°.

Measure of angle from vertical	plus	measure of other angle	is	right angle.
x	$+$	$17x$	$=$	90°

Solve:

$$x + 17x = 90$$
$$18x = 90 \quad \text{Combine like terms.}$$
$$x = 5 \quad \text{Divide both sides by 18.}$$

Conclusion: The tower is 5 degrees from vertical.

31. **Analysis**: Find the value of x when the angles are supplementary.

Equation: The measure of the angle 1 is represented by $x + 50$ and the measure of the angle 2 is represented by $2x - 20$. The two angles are supplementary which means their sum is 180°.

Measure of angle 1	plus	measure of angle 2	is	180°.
$x + 50$	$+$	$2x - 20$	$=$	180

Solve:

$$x + 50 + 2x - 20 = 180$$
$$3x + 30 = 180 \quad \text{Combine like terms.}$$
$$3x + 30 - 30 = 180 - 30 \quad \text{Subtract 30 from both sides.}$$
$$3x = 150 \quad \text{Combine like terms.}$$
$$x = 50 \quad \text{Divide both sides by 3.}$$

Conclusion: The value of x is 50°.

33. **Analysis**: Find the value of x when two lines intersect with the given measures for their angles.

Equation: The measure of angle 1 is represented by $3x + 10$ and the measure of angle 3 is represented by $5x - 10$. The measures of the two angles are equal to each other.

Measure of angle 1	equals	measure of angle 3.
$3x + 10$	$=$	$5x - 10$

Solve:

$$3x + 10 = 5x - 10$$
$$3x + 10 - 5x - 10 = 5x - 10 - 5x - 10 \quad \text{Subtract } 5x \text{ and 10 from both sides.}$$
$$-2x = -20 \quad \text{Combine like terms.}$$
$$x = 10 \quad \text{Divide both sides by } -2.$$

Conclusion: The value of x is 10°.

35. **Analysis**: Find the measure of two equal base angles in a quadrilateral when the other two angles are given.

Equation: The sum of the measures of the four angles is 360°.

m(∠1)	plus	m(∠2)	plus	m(∠3)	plus	m(∠4)	is	360°.
x	+	x	+	140°	+	100°	=	360

Solve:

$x + x + 140 + 100$	=	360	
$2x + 240$	=	360	Combine like terms.
$2x + 240 - 240$	=	$360 - 240$	Subtract 240 from both sides.
$2x$	=	120	Combine like terms.
x	=	60	Divide both sides by 2.

Conclusion: The value of x is 60°.

37. **Analysis:** Measure the length and width of both rectangles. Rectangle i is approximately 1.7 inches long by 0.8 inches wide. Rectangle ii is approximately 1.7 inches long by 1 inch wide. A golden rectangle would have a length of 1.618 inches and a width of 1 inch.

Conclusion: Rectangle ii is the closest in measurement to the golden rectangle.

39. **Analysis:** The length of the fence will enclose one length and two widths of the rectangular pasture.

Equation: Let w represent the width of the pasture, then $2w$ would represent the length l. The length of the fence will be 624 feet. Use $P = l + 2w$ as the formula for the length of the fence.

P	=	l	+	$2w$
624	=	$2w$	+	$2w$

Solve:

624	=	$2w + 2w$	
624	=	$4w$	Combine like terms.
156	=	w	Divide both sides by 4

Conclusion: The dimensions of the pasture would be 156 feet wide and $2w = 312$ feet long.

41. **Analysis:** The fence will enclose two lengths and two widths of the rectangular swimming pool plus the walkway of uniform width. The length of the pool will be 30 feet plus two widths of the walkway. Similarly, the width of the enclosure will be 20 feet plus two widths of the walkway.

Equation: Let x represent the uniform width of the walkway, then $30 + 2x$ would represent the length l of the enclosure. The width w of the enclosure would be $20 + 2x$. The length of the fence will be 180 feet. Use $P = 2l + 2w$ as the formula for the length of the fence.

P	=	$2l$	+	$2w$
180	=	$2(30 + 2x)$	+	$2(20 + 2x)$

Solve:

180	=	$2(30 + 2x) + 2(20 + 2x)$	
180	=	$60 + 4x + 40 + 4x$	Distribute the multiplications by 2.
180	=	$8x + 100$	Combine like terms.
$180 - 100$	=	$8x + 100 - 100$	Subtract 100 from both sides.
80	=	$8x$	Combine like terms.
10	=	x	Divide both sides by 8.

Conclusion: The walkway would be 10 feet wide.

43. **Analysis:** The drawer will be divided into three sections by a partition that is ½ inch thick. The second section will be 3 inches longer than the first section and the third section will be 3 inches longer than the second.

Equation: Let x represent the length of the space for the first section. The length of the space for the second section will be represented by $x + 3$ and the length of the space for the third section will be represented by $(x + 3) + 3 = x + 6$. There will be two partitions of ½ inch thickness, $2\left(\frac{1}{2}\right) = 1$ in. The total depth of the drawer is 28 inches.

First space	plus	second space	plus	third space	plus	2 partitions	=	drawer depth.
x	+	$x + 3$	+	$x + 6$	+	1	=	28

Solve:

$$x + x + 3 + x + 6 + 1 = 28$$
$$3x + 10 = 28 \quad \text{Combine like terms.}$$
$$3x = 18 \quad \text{Combine like terms.}$$
$$x = 6 \quad \text{Divide both sides by 3.}$$

Conclusion: The length of the first space will be 6 inches.

WRITING

45. Answers will vary.

REVIEW

47. repeating.

49. $\{\ldots, -4, -3, -2, -1, 0, 1, 2, 3, 4, \ldots\}$

51.
$$2x^2 + 5x - 3 = 2(-3)^2 + 5(-3) - 3 \quad \text{Substitute } -3 \text{ for } x.$$
$$= 2(9) - 15 - 3 \quad \text{Evaluate the power.}$$
$$= 18 - 15 - 3 \quad \text{Do the multiplication.}$$
$$= 0 \quad \text{Do the addition.}$$

Section 1.7 More Problem Solving

VOCABULARY

1. principal
3. median
5. amount; base

CONCEPTS

7. a. Circulation increased 272,616

   b. 272,616 is what percent of 2,172,923 ?
   $$272{,}616 = x \cdot 2{,}172{,}923$$
   $$x < 0.\ 1255 = 12.55\%$$

9. a. \$30,000 – \$5,000 = \$25,000

   b. $\$(30{,}000 - x)$

11.

	Price	Pounds	Value (price • pounds)
M & M plain	\$2.49	30	\$74.70
M & M peanut	\$2.79	p	\$2.79p
Mixture	\$2.59	$p + 30$	\$2.59$(p + 30)$

NOTATION

13. What number is 5% of 10.56?
$$x = 5\% \cdot 10.56$$
$$x = 0.05 \cdot 10.56$$

15. 32.5is 74% of what number?
$$32.5 = 74\% \cdot x$$
$$32.5 = 0.74 \cdot x$$
$$32.5 = 0.74x$$

17.
$$\begin{aligned} 0.09x + 0.08(2{,}000 - x) &= 400 \\ \mathbf{100}\ [0.09x + 0.08(2000 - x) &= \mathbf{100}\ (400) \\ \mathbf{9x}\ + 8(2000 - x) &= \mathbf{40{,}000} \\ 9x + \mathbf{16{,}000} - 8x &= 40{,}000 \\ x &= 24{,}000 \end{aligned}$$

19. $I = Prt$

21. $v = pn$

APPLICATIONS

23. **Analysis:** The United States used 97.6 quadrillion Btu which was 23.5% of the energy used in the world. Find the total energy used in the world.

Equation: Let x represent the energy used in the world.

Energy used by U.S	is	23.5%	of	energy used by world
97.6	=	23.5%	•	x

Solve:
$$\begin{aligned} 97.6 &= 23.5\% \cdot x \\ 97.6 &= 0.235x && \text{Write 25\% as a decimal.} \\ 415.32 &< x && \text{Use a calculator to divide both sides by 0.235.} \end{aligned}$$

Conclusion: The world consumption of energy in 1998 was about 415 quadrillion Btu.

25. **Analysis:** The regular price is \$826, the sale price is \$660.80, and the markdown is the product of \$826 and the percent of markdown.

Equation: Let r represent the percent of markdown, expressed as a decimal.

Sale price	is	regular price	minus	markdown.
660.80	=	826	–	$r \cdot 826$

Solve:
$$\begin{aligned} 660.80 &= 826 - r \cdot 826 \\ 660.80 &= 826 - 826r && \text{Rewrite } r \bullet 826 \text{ as } 826r. \\ -165.20 &= -826r && \text{Subtract 826 from both sides.} \\ 0.2 &= r && \text{Use a calculator to divide both sides by } -826. \\ 20\% &= r && \text{Write 0.2 as a percent.} \end{aligned}$$

Conclusion: The percent markdown is 20%.

27. **Analysis:** The retail price is \$65 and the markup is the product of the wholesale price and the 30% profit.

Equation: Let x represent the wholesale price.

Retail price	is	wholesale price	plus	markup.
65	=	x	+	$30\% \cdot x$

Solve:
$$\begin{aligned} 65 &= x + 30\% \cdot x \\ 65 &= x + 0.30x && \text{Write 30\% as a decimal.} \\ 65 &= 1.3x && \text{Combine like terms.} \\ 50 &= x && \text{Use a calculator to divide both sides by 1.3.} \end{aligned}$$

Conclusion: The wholesale price is \$50.

29. **Analysis:** The stock engine has 118 hp and the engine with the chip has 129 hp. The horsepower increase is $129 - 118 = 11$ hp.

Equation: Let x represent the percent increase in horsepower.

What percent	of	stock engine	is	increased horsepower?
x	•	118	=	11

Solve:
$$\begin{aligned} x \bullet 118 &= 11 \\ 118x &= 11 && \text{Rewrite } x \cdot 118 \text{ as } 118x. \\ x &= 0.0932\ldots && \text{Use a calculator to divide both sides by 118.} \\ x &= 9\% && \text{Write the decimal as a percent.} \end{aligned}$$

Conclusion: The horsepower increased by 9%.

31.

Season	Broadway attendance	Attendance increase	% increase (increase ÷ previous year)
1996–97	10.6 million	—	—
1997–98	11.5 million	0.9 million	8.5%
1998–99	11.7 million	0.2 million	1.7%

32. **Analysis**: The number in of drive-in theaters 1958 was 4,063 and in 2003 was 401. To find the percent decrease, first find the amount of decrease. 4,063-401=3662.

Equation: Let x represent the percent decrease in number of theaters.

What percent	of	1958 number	is	difference in number of theaters?
x	•	4,063	=	3662

Solve:

$$x \bullet 4{,}063 = 3662$$

$$4{,}063x = 3662 \quad \text{Rewrite } x \cdot 4{,}063 \text{ as } 4{,}063x.$$

$$x = 0.9013\ldots \quad \text{Use a calculator to divide both sides by 4,063.}$$

$$x = 90\% \quad \text{Write the decimal as a percent.}$$

Conclusion: The number of drive-in theaters decreased by about 90%.

33. **City**: To find the mean, we add the values and divide by the number of values, which is 10.

$$\text{mean} = \frac{61+52+36+34+32+32+32+32+30+29}{10} = \frac{370}{10} = 37$$

To find the median, arrange the values in increasing order: 29, 30, 32, 32, **32, 32**, 34, 36, 52, 61 There are an even number of values, so sum the middle two values and divide by 2.

$$\text{median} = \frac{32+32}{2} = \frac{64}{2} = 32$$

Mode = 32 since it occurs most often (4 times).

Hwy: To find the mean, we add the values and divide by the number of values, which is 10.

$$\text{mean} = \frac{68+45+42+41+41+41+39+39+37+40}{10} = \frac{433}{10} = 43.3$$

To find the median, arrange the values in increasing order: 37, 39, 39, 40, **41, 41**, 41, 42, 45, 68
There are an even number of values, so sum the middle two values and divide by 2.

$$\text{median} = \frac{41+41}{2} = \frac{82}{2} = 41$$

Mode = 41 since it occurs most often (3 times).

35. **Analysis:** To find the average mean score add all the test scores and then divide by the number of tests. The mean of the 4 tests needs to be 85 or higher.

Equation: Let x represent the lowest score the candidate needs on the written test.

The sum of the 4 tests divided by 4	is	the average score.
$\frac{78+91+87+x}{4}$	=	85

35 cont.

Solve:

$$\frac{78+91+87+x}{4}=85$$

$$\frac{256+x}{4}=85 \quad \text{Combine like terms in the numerator.}$$

$$4\left(\frac{256+x}{4}\right)=4(85) \quad \text{Multiply both sides by 4.}$$

$$256+x=340$$

$$x=84 \quad \text{Subtract 256 from both sides.}$$

Conclusion: A minimum score of 84 on the written test would be needed.

37. **Analysis:** An investment of \$12,000 split between two accounts. The interest received will be \$1060 per year.

Equation: Let x represent the number of dollars invested in the Money Market at 8.0%. Then \$(12,000 – x) is the number of dollars invested in the 5–year CD at 9.0%.

	P •	r •	t =	I
Money Market	x	0.08	1	$0.08x$
5–year CD	$12{,}000 - x$	0.09	1	$0.09(12{,}000 - x)$

Interest earned at 8%	plus	interest earned at 9%	is	total interest earned.
$0.08x$	+	$0.09(12{,}000 - x)$	=	1060

Solve:

$$\begin{aligned} 0.08x + 0.09(12{,}000 - x) &= 1060 \\ 0.08x + 1080 - 0.09x &= 1060 && \text{Distribute the multiplication by 0.09.} \\ -0.01x + 1080 &= 1060 && \text{Combine like terms.} \\ -0.01x &= -20 && \text{Subtract 1080 from both sides.} \\ x &= 2000 && \text{Use a calculator to divide both sides by } -0.01. \end{aligned}$$

Conclusion: \$2000 should be invested in the Money Market and \$(12,000 – 2,000) = \$10,000 should be invested in the 5–year CD.

39. **Analysis:** A inheritance of an unknown amount is split between two investments. The interest received will be \$4050 per year.

Equation: Let x represent the number of dollars invested at 7.0%. Then $2x$ is the number of dollars invested at 10.0%.

	P •	r •	t =	I
Certificate of Deposit	x	0.07	1	$0.07x$
Biotech company	$2x$	0.10	1	$0.10(2x)$

Interest earned at 7%	plus	interest earned at 10%	is	total interest earned.
$0.07x$	+	$0.10(2x)$	=	4050

Solve:

$$\begin{aligned} 0.07x + 0.10(2x) &= 4050 \\ 0.07x + 0.20x &= 4050 && \text{Do the multiplication by 0.10.} \\ 0.27x &= 4050 && \text{Combine like terms.} \\ x &= 15{,}000 && \text{Use a calculator to divide both sides by 0.27.} \end{aligned}$$

Conclusion: \$15,000 was invested at 7% and \$$2x$ = \$30,000 was invested at 10%. The total inheritance was \$15,000 + \$30,000 = \$45,000.

41. **Analysis:** Two investments were made. $300,000 was deposited in a Swiss bank at 8% and an unknown amount was deposited in the Cayman Islands at 5%. The interest earned was 7.25% of the total amount.

Equation: Let x represent the number of dollars invested in the Cayman Islands at 5.0%. Then $300,000 + x$ represents the number of dollars invested in total.

	P •	r •	t =	I
Swiss bank	300,000	0.08	1	24,000
Cayman Islands	x	0.05	1	$0.05x$

Interest earned at 8%	plus	interest earned at 5%	is	total interest earned.
24,000	+	$0.05x$	=	$0.0725(300,000 + x)$

Solve:

$$24,000 + 0.05x = 0.0725(300,000 + x)$$

$$24,000 + 0.05x = 21,750 + 0.0725x \quad \text{Distribute the multiplication by 0.0725.}$$

$$-0.0225x = -2250 \quad \text{Subtract 24,000 and 0.0725x from each side.}$$

$$x = 100,000 \quad \text{Use a calculator to divide both sides by } -0.0225.$$

Conclusion: $100,000 was invested in the Cayman Islands bank.

43. **Analysis:** The vehicles are traveling toward each other for the same amount of time.

Equation: Let t represent the time that each vehicle travels. The distance between the office and the home is 20 miles.

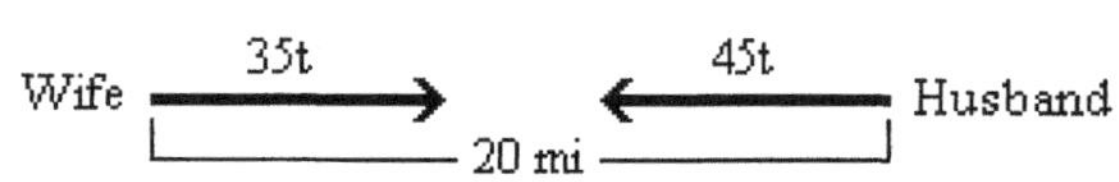

	r •	t =	d
Wife	35	t	$35t$
Husband	45	t	45t

Wife's distance	plus	Husband's distance	is	distance between work and home.
$35t$	+	$45t$	=	20

Solve:

$$35t + 45t = 20$$

$$80t = 20 \quad \text{Combine like terms.}$$

$$t = 0.25 \quad \text{Use a calculator to divide both sides by 80.}$$

Conclusion: The wife and husband will meet in 0.25 hour or 15 minutes.

45. **Analysis:** The vehicles are traveling in the same direction. The cyclist will travel for one more hour than the support staff.

Equation: Let t represent the time that the support staff travels. Then $t + 1$ represents the time the cyclist travels.

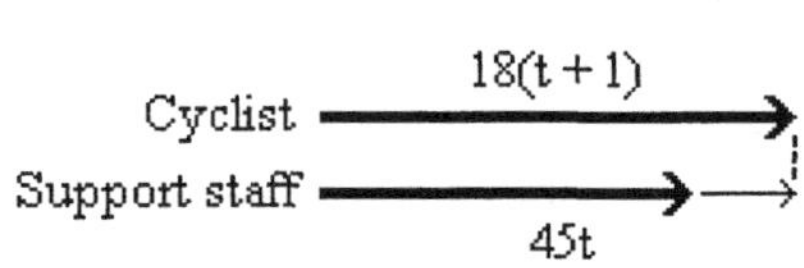

	r •	t =	d
Cyclist	18	$t + 1$	$18(t + 1)$
Support staff	45	t	45t

Cyclist distance	equals	support staff distance.
$18(t + 1)$	=	$45t$

Solve:

$$18(t + 1) = 45t$$

$$18t + 18 = 45t \quad \text{Distribute the multiplication by 18.}$$

$$18 = 27t \quad \text{Subtract } 18t \text{ from both sides.}$$

$$t = \frac{2}{3} \quad \text{Divide both sides by 27.}$$

Conclusion: The support staff will catch up with the cyclist in $\frac{2}{3}$ hour.

47. **Analysis:** The vehicles are traveling away from each other for the same amount of time starting from the same point.

Equation: Let t represent the time that each convoy travels. The distance between the two convoys will be 135 miles.

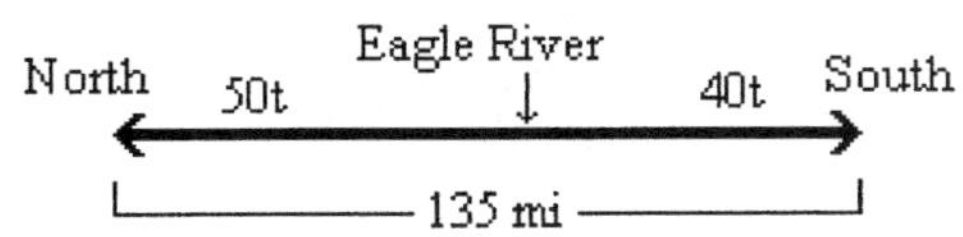

	r	• t	$= d$
North	50	t	$50t$
South	40	t	$40t$

North distance plus South distance is distance between two convoys.

$50t + 40t = 135$

Solve:

$$50t + 40t = 135$$
$$90t = 135 \quad \text{Combine like terms.}$$
$$t = 1.5 \quad \text{Use a calculator to divide both sides by 90.}$$

Conclusion: The convoys will be 135 miles apart in 1.5 hours which would be 2 P.M. + 1.5 hours = 3:30 P.M.

49. **Analysis:** The jet ski goes upstream and then returns the same distance downstream. When the jet ski goes upstream the current will decrease the actual speed and when the jet ski goes downstream the current will increase the actual speed.

Equation: Let t represent the time that the jet ski travels downstream.

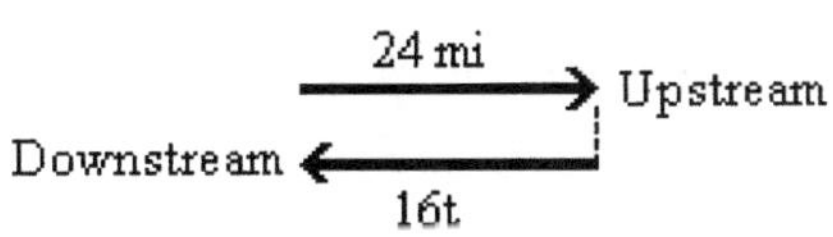

	r	• t	$= d$
Upstream	$12 - 4$	3	$8(3) = 24$
Downstream	$12 + 4$	t	$(12 + 4)t = 16t$

Upstream distance equals downstream distance.

$24 = 16t$

Solve:

$$24 = 16t$$
$$t = 1.5 \quad \text{Divide both sides by 16.}$$

Conclusion: The rider returns in $1\frac{1}{2}$ hours.

51. **Analysis:** A 30–pound mixture that contains red licorice and lemon gumdrops will sell for \$2 per pound.

Equation: Let x represent the number of pounds of red licorice. Then $30 - x$ will represent the number of pounds of lemon gumdrops.

	p	• n	$= v$
Red licorice	1.90	x	$1.90x$
Lemon gumdrops	2.20	$30 - x$	$2.20(30 - x)$
Mixture	2.00	30	60

Red licorice value plus lemon gumdrop value is value of mixture.

$1.90x + 2.20(30 - x) = 60$

Solve:

$$1.90x + 2.20(30 - x) = 60$$
$$1.90x + 66 - 2.20x = 60 \quad \text{Distribute the multiplication by 2.20.}$$
$$-0.30x + 66 = 60 \quad \text{Combine like terms.}$$
$$-0.30x = -6 \quad \text{Subtract 66 from both sides.}$$
$$x = 20 \quad \text{Use a calculator to divide both sides by } -0.30.$$

Conclusion: The mixture would need 20 pounds of red licorice at \$1.90 per pound and $30 - x = 10$ pounds of lemon gumdrops at \$2.20 per pound.

53. **Analysis:** A planting mix that contains premium organic mix and sawdust will sell for $1.08 per cubic foot.

Equation: Let x represent the number of cubic feet of premium organic planting mix. Then $6000 - x$ will represent the number of cubic feet of sawdust.

	p	• n	= v
Premium organic	1.57	x	$1.57x$
Sawdust	0.10	$6000 - x$	$0.10(6000 - x)$
Mixture	1.08	6000	6480

Premium organic value	plus	sawdust value	is	value of mixture.
$1.57x$	+	$0.10(6000 - x)$	=	6480

Solve:

$$\begin{aligned} 1.57x + 0.10(6000 - x) &= 6480 \\ 1.57x + 600 - 0.10x &= 6480 && \text{Distribute the multiplication by 0.10.} \\ 1.47x + 600 &= 6480 && \text{Combine like terms.} \\ 1.47x &= 5880 && \text{Subtract 600 from both sides.} \\ x &= 4000 && \text{Use a calculator to divide both sides by 1.47.} \end{aligned}$$

Conclusion: The mixture would need 4000 cubic feet of premium organic mix at $1.57 per cubic foot and $6000 - x = 2000$ cubic feet of sawdust at $0.10 per pound.

55. **Analysis:** Water will be added to a 15% solution of alcohol to dilute it to a 10% solution.

Equation: Let x represent the number of ounces of water added. Then $20 + x$ will represent the number of ounces of the 10% solution.

	percent alcohol	• total quantity	= quantity of alcohol
15% solution	0.15	20	3
Water	0.00	x	0
10% solution	0.10	$20 + x$	$0.10(20 + x)$

Quantity of alcohol in 15% solution	plus	quantity of alcohol in water	is	quantity of alcohol in 10% solution.
3	+	0	=	$0.10(20 + x)$

Solve:

$$\begin{aligned} 3 + 0 &= 0.10(20 + x) \\ 3 &= 2 + 0.10x && \text{Distribute the multiplication by 0.10.} \\ 1 &= 0.10x && \text{Subtract 2 from both sides.} \\ 10 &= x && \text{Use a calculator to divide both sides by 0.10.} \end{aligned}$$

Conclusion: 10 ounces of water should be added to the 15% solution of alcohol.

57. **Analysis:** Cream will be mixed with milk to get 20 gallons of milk containing 4% butterfat.

Equation: Let x represent the number of gallons of cream added. Then $20 - x$ will represent the number of gallons of the 2% milk solution.

	percent butterfat	• total quantity	= quantity of butterfat
Cream	0.22	x	$0.22x$
2% Milk	0.02	$20 - x$	$0.02(20 - x)$
4% Milk	0.04	20	0.80

Quantity of butterfat in 22% cream	plus	quantity of butterfat in 2% milk	is	quantity of butterfat in 4% milk
$0.22x$	+	$0.02(20 - x)$	=	0.80

Solve:

$$\begin{aligned} 0.22x + 0.02(20 - x) &= 0.80 \\ 0.22x + 0.40 - 0.02x &= 0.80 && \text{Distribute the multiplication by 0.02.} \\ 0.20x + 0.40 &= 0.80 && \text{Combine like terms.} \\ 0.20x &= 0.40 && \text{Subtract 0.40 from both sides.} \\ x &= 2 && \text{Use a calculator to divide both sides by 0.20.} \end{aligned}$$

Conclusion: 2 gallons of cream should be added to the 2% milk.

WRITING

59. The car speed is measured using miles per **hour** while the time is measured using **minutes**.

61. Answers will vary.

REVIEW

63.
$$\begin{aligned} 9x &= 6x \\ 3x &= 0 \\ x &= 0 \end{aligned}$$

65.
$$\begin{aligned} \frac{8(y-5)}{3} &= 2(y-4) \\ 3\left(\frac{8(y-5)}{3}\right) &= 3[2(y-4)] \\ 8(y-5) &= 6(y-4) \\ 8y-40 &= 6y-24 \\ 2y &= 16 \\ y &= 8 \end{aligned}$$

Chapter 1 — Key Concept: "Let $x =$"

Analyzing the Problem

1. Given: 48 states, 4 more lie east of the Mississippi River than west. Find: how many states lie west.
2. Given: one angle is 5° more than twice another, one angle is 90°, the sum of the measures of the angles is 180°. Find: the measure of the smallest angle.

Letting a Variable Represent an Unknown Quantity

3. Let x = the length of the shortest piece.
4. Let x = the number of miles he can drive.

Forming an Equation

5. a. subtraction b. multiplication c. addition d. division
6. a. $d = rt$ b. $I = Prt$ c. $v = pn$ d. $P = 2l + 2w$

7.

	% alcohol	Amount	Amount alcohol
Spray 1	0.15	15 oz	**0.15(15)**
Spray 2	0.50	x oz	**0.50x**
Mixture	0.40	$(15 + x)$ oz	**0.40(15 + x)**

The equation would be: $0.15(15) + 0.50x = 0.40(15 + x)$.

8. $450x + 500x = 2850$

Chapter 1 Chapter Review

Section 1.1

1. $C = 2t + 15$

2. $l = \dfrac{25}{w}$

3. $P = u - 3$

4.

p	$T = 30p$	T
6.0	$T = 30(6.0)$	180
6.5	$T = 30(6.5)$	195
7.0	$T = 30(7.0)$	210
7.5	$T = 30(7.5)$	225
8.0	$T = 30(8.0)$	240

5. a.

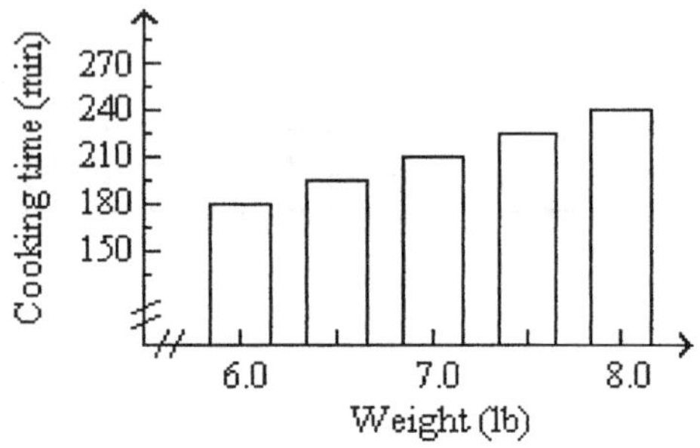

b.

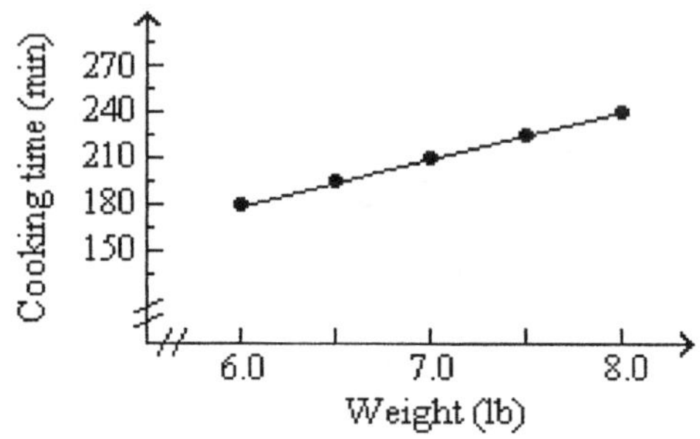

6.

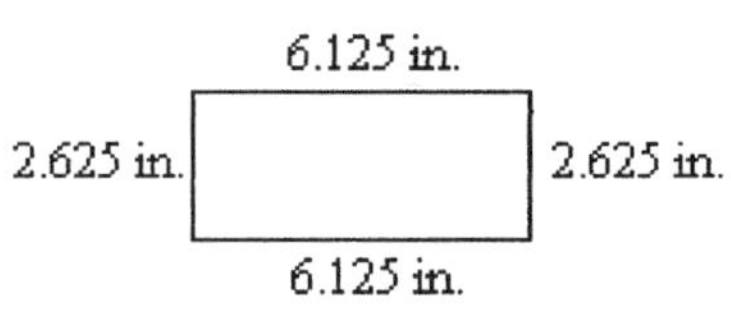

Rectangle: $l = 6.125$ in., $w = 2.625$ in.

$P = 2l + 2w = 2(6.125) + 2(2.625) = 12.25 + 5.25 = 17.5$ in.

Section 1.2

7. 7

8. 0, 7

9. −5, 0, 7

10. $-5, 0, 2.4, 7, -\dfrac{2}{3}, -3.\overline{6}, \dfrac{15}{4}$

11. $-\sqrt{3}$, π, 0.13242368…

12. all

13. $-5, -\sqrt{3}, -\dfrac{2}{3}, -3.\overline{6}$

14. $2.4, 7, \pi, \dfrac{15}{4}$, 0.13242368…

15. 7

16. none

17. 0

18. −5, 7

19. >, since −16 is to the right of −17.

20. <, since 1.8 is to the left of 2.5.

21. false, since 23.000001 is to the left of 23.100000

22. true

23.

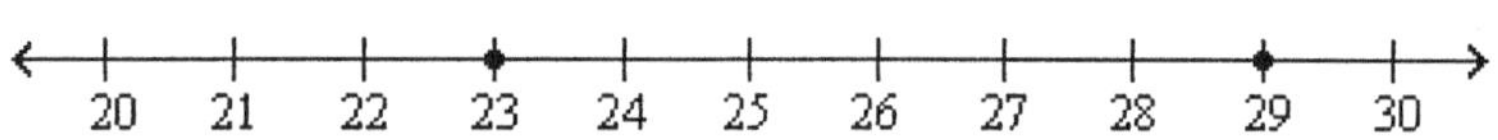

24.

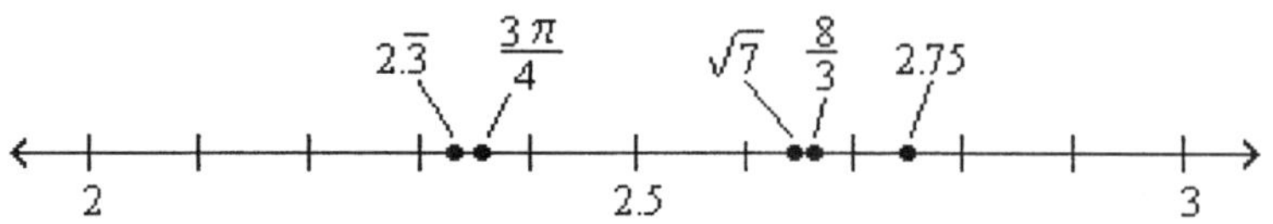

25. $|-18| = 18$

26. $-|6.26| = -(6.26) = -6.26$

Section 1.3

27. $-3 + (-4) = -7$

28. $-70.5 + 80.6 = 10.1$

39. $-\frac{1}{2} - \frac{1}{4} = -\frac{2}{4} - \frac{1}{4} = -\frac{3}{4}$

30. $-6 - (-8) = -6 + 8 = 2$

31. $(-4.2)(-3.0) = 12.6$

32. $-\frac{1}{10} \cdot \frac{5}{16} = -\frac{1 \cdot \cancel{5}^{1}}{2 \cdot \cancel{5}_{1} \cdot 16} = -\frac{1}{32}$

33. $\frac{-2.2}{-11} = 0.2$

34. $-\frac{9}{8} \div 21 = -\frac{9}{8} \cdot \frac{1}{21} = -\frac{3 \cdot \cancel{3}^{1}}{8 \cdot \cancel{3}_{1} \cdot 7} = -\frac{3}{56}$

35. $15 - 25 - 23 = -10 - 23 = -33$

36. $-3.5 + (-7.1) + 4.9 = -10.6 + 4.9 = -5.7$

37. $-3(-5)(-8) = 15(-8) = -120$

38. $-1(-1)(-1)(-1) = 1(-1)(-1) = -1(-1) = 1$

39. $(-3)^5 = -243$

40. $\left(-\frac{2}{3}\right)^2 = \left(-\frac{2}{3}\right)\left(-\frac{2}{3}\right) = \frac{4}{9}$

41. 0.3 cubed $= (0.3)^3 = 0.027$

42. $-5^2 = -(5)(5) = -25$

43. $\sqrt{4} = \sqrt{2 \cdot 2} = 2$

44. $-\sqrt{100} = -\sqrt{10 \cdot 10} = -10$

45. $\sqrt{\frac{9}{25}} = \sqrt{\frac{3}{5} \cdot \frac{3}{5}} = \frac{3}{5}$

46. $\sqrt{0.64} = \sqrt{0.8 \cdot 0.8} = 0.8$

47. $-6 + 2(-5)^2 = -6 + 2(25) = -6 + 50 = 44$

48. $\frac{-20}{4} - (-3)(-2)(-\sqrt{1}) = -5 - (6)(-1) = -5 - (-6) = -5 + 6 = 1$

49. $4 - (5 - 9)^2 = 4 - (-4)^2 = 4 - 16 = -12$

50.
$$\begin{aligned} 4 + 6[-1 - 5(25 - 3^3)] &= 4 + 6[-1 - 5(25 - 27)] \\ &= 4 + 6[-1 - 5(-2)] \\ &= 4 + 6(-1 + 10) \\ &= 4 + 6(9) \\ &= 4 + 54 \\ &= 58 \end{aligned}$$

51. $2|-1.3 + (-2.7)| = 2|-1.3 - 2.7| = 2|-4.0| = 2(4) = 8$

52. $\frac{(7-6)^4 + 32}{36 - (\sqrt{16} + 1)^2} = \frac{1^4 + 32}{36 - (4+1)^2} = \frac{1 + 32}{36 - 5^2} = \frac{33}{36 - 25} = \frac{33}{11} = 3$

53. $(-10)^3\left(\frac{-6}{-2}\right)(-1) = (-1000)(3)(-1) = 3000$

54. $-(-2 \cdot 4)^2 = -(-8)^2 = -64$

55. $(x + y)(x^2 - xy + y^2)$ for $x = -2$ and $y = 4$: $(-2 + 4)((-2)^2 - (-2)(4) + 4^2) = 2(4 + 8 + 16) = 2(28) = 56$

56.

$$\frac{-b-\sqrt{b^2-4ac}}{2a}=\frac{-(-3)-\sqrt{(-3)^2-4(2)(-2)}}{2(2)}$$ Substitute 2 for a, –3 for b, and –2 for c.

$$=\frac{3-\sqrt{9+16}}{4}$$ Evaluate the power and do the multiplications.

$$=\frac{3-\sqrt{25}}{4}$$ Do the addition under the square root symbol.

$$=\frac{3-5}{4}$$ Evaluate the square root.

$$=\frac{-2}{4}$$ Do the addition.

$$=-\frac{1}{2}$$ Simplify.

57. $A = s^2$ The formula for the area of a square.
$= (10)^2 = 100 \text{ in.}^2$ Substitute the 10 for the length of a side.

58. To find the diameter of the cone subtract the 2 edges of 1 in. from the total base length. $d = (10 - 2) = 8$ in.

$$V=\frac{1}{3}\pi r^2 h$$ The formula for the volume of a cone.

$$=\frac{1}{3}\pi(4)^2(15)$$ Substitute4 for r and 15 for h.

$$=\frac{1}{3}\pi(16)(15)$$ Evaluate the power.

$$\approx 251.327\ldots \approx 251.3 \text{ in.}^3$$ Do the multiplication and round to the nearest tenth.

Section 1.4

59. $3x + 21$

60. $5t$

61. 0

62. $27 + (1 + 99)$

63. 1

64. m

65. 1

66. 0

67. $-3(5 \cdot 2)$

68. $(z + t) \cdot t$

69. $\frac{102}{102} = 1$

70. $\frac{0}{6} = 0$

71. $\frac{-25}{1} = -25$

72. $\frac{5.88}{0} =$ is undefined

73. $8(x + 6) = 8x + 48$

74. $-6(x - 2) = -6x + 12$

75. $-(-4 + 3y) = -1(-4 + 3y) = 4 - 3y$

76. $(3x - 2y)1.2 = 1.2(3x - 2y) = 3.6x - 2.4y$

77. $10(2c^2 - c + 1) = 20c^2 - 10c + 10$

78. $\frac{2}{3}(3t + 9) = \frac{2}{3}(3t) + \frac{2}{3}(9) = 2t + 2(3) = 2t + 6$

79. $8(6k) = 8 \cdot 6 \cdot k = 48k$

80. $(-7x)(-10y) = (-7)(-10)(x)(y) = 70xy$

81. $-9(-3p)(-7) = -9(-3)(-7)(p) = 27(-7)(p) = -189p$

82. $15a + 30a + 7 = 45a + 7$

83. $3g^2 - 3g^2 = 0$

84. $-m + 4(m - 12) = -m + 4m - 48 = 3m - 48$

85. $\frac{7}{8}x + \frac{1}{8}x = \frac{8}{8}x = x$

86. $21.45l - 45.99l = -24.54l$

87. $5t^3 + 6t^2 - 2t^2 + t^3 = 6t^3 + 4t^2$

88. $8(2h + 9) - 5(h - 1) = 16h + 72 - 5h + 5 = 11h + 77$

Section 1.5

Substitute −6 for x.

89.
$$\begin{aligned} 6 - x &= 2x + 24 \\ 6 - (-6) &\stackrel{?}{=} 2(-6) + 24 \\ 6 + 6 &\stackrel{?}{=} -12 + 24 \\ 12 &= 12; \text{ Yes} \end{aligned}$$

90.
$$\begin{aligned} \frac{5}{3}(x - 3) &= -12 \\ \frac{5}{3}(-6 - 3) &\stackrel{?}{=} -12 \\ \frac{5}{3}(-9) &\stackrel{?}{=} -12 \\ 5(-3) &\stackrel{?}{=} -12 \\ -15 &\neq -12; \text{ No} \end{aligned}$$

91.

$\frac{x}{5} = -45$	The original equation.
$5\left(\frac{x}{5}\right) = 5(-45)$	Multiply both sides by 5.
$x = -225$	Do the multiplication. Solution set: {−225}

92.

$t - 3.67 = 4.23$	The original equation.
$t = 4.23 + 3.67$	Add 3.67 to both sides.
$t = 7.9$	Do the addition. Solution set: {7.9}

93.

$0.0035 = 0.25g$	The original equation.
$0.014 = g$	Divide both sides by 0.25. Solution set: {0.014}

94.

$0 = x + 4$	The original equation.
$-4 = x$	Subtract 4 from both sides. Solution set: {−4}

95.

$5x + 12 = 0$	
$5x = -12$	Subtract 12 from both sides.
$x = -\frac{12}{5}$	Divide both sides by 5.

96.

$-3x - 7 + x = 6x + 20 - 5x$	
$-2x - 7 = x + 20$	Combine like terms.
$-2x - 7 - x + 7 = x + 20 - x + 7$	Subtract x and add 7 to both sides.
$-3x = 27$	Combine like terms.
$x = -9$	Divide both sides by −3.

97.

$4(y - 1) = 28$	
$4y - 4 = 28$	Distribute the multiplication by 4.
$4y = 32$	Add 4 both sides.
$y = 8$	Divide both sides by 4.

98.

$2 - 13(x - 1) = 4 - 6x$	
$2 - 13x + 13 = 4 - 6x$	Distribute the multiplication by −13.
$15 - 13x = 4 - 6x$	Combine like terms.
$15 - 13x + 6x - 15 = 4 - 6x + 6x - 15$	Add $6x$ and subtract 15 from both sides.
$-7x = -11$	Combine like terms.
$x = \frac{11}{7}$	Divide both sides by −7.

99.

$$\begin{aligned} \frac{8(x-5)}{3} &= 2(x-4) \\ 3\left(\frac{8(x-5)}{3}\right) &= 3[2(x-4)] && \text{Multiply both sides by the LCD 3.} \\ 8(x-5) &= 6(x-4) && \text{Do the multiplication.} \\ 8x-40 &= 6x-24 && \text{Distribute the multiplications.} \\ 2x &= 16 && \text{Subtract } 6x \text{ and add 40 to both sides.} \\ x &= 8 && \text{Divide both sides by 2.} \end{aligned}$$

100.

$$\begin{aligned} \frac{3y}{4}-14=-\frac{y}{3}-1 & && \text{The original equation.} \\ 12\left(\frac{3y}{4}-14\right)=12\left(-\frac{y}{3}-1\right) & && \text{Multiply both sides by the LCD 12.} \\ 3(3y)-12(14) &= 4(-y)-12(1) && \text{Do the multiplication.} \\ 9y-168 &= -4y-12 && \text{Do the multiplication.} \\ 13y &= 156 && \text{Combine like terms.} \\ y &= 12 && \text{Divide both sides by 13.} \end{aligned}$$

101.

$$\begin{aligned} -k &= -0.06 \\ k &= 0.06 && \text{Divide both sides by } -1. \end{aligned}$$

102.

$$\begin{aligned} \frac{5}{4}p &= -10 \\ 4\left(\frac{5}{4}\right)p &= 4(-10) && \text{Multiply both sides by the LCD 4.} \\ 5p &= -40 && \text{Do the multiplication.} \\ p &= -8 && \text{Divide both sides by 5.} \end{aligned}$$

103.

$$\begin{aligned} \frac{4t+1}{3}-\frac{t+5}{6}=\frac{t-3}{6} & && \text{The original equation.} \\ 6\left(\frac{4t+1}{3}-\frac{t+5}{6}\right)=6\left(\frac{t-3}{6}\right) & && \text{Multiply both sides by the LCD 6.} \\ 2(4t+1)-(t+5) &= (t-3) && \text{Do the multiplication.} \\ 8t+2-t-5 &= -t-3 && \text{Distribute the multiplications.} \\ 7t-3 &= -t-3 && \text{Combine like terms.} \\ 6t &= 0 && \text{Subtract } t \text{ and add 3 to both sides.} \\ t &= 0 && \text{Divide both sides by 6.} \end{aligned}$$

104.

$$\begin{aligned} 33.9-0.5(75-3x) &= 0.9 \\ 33.9-37.5+1.5x &= 0.9 && \text{Distribute the multiplication by 0.5.} \\ -3.6+1.5x &= 0.9 && \text{Combine like terms.} \\ 1.5x &= 4.5 && \text{Add 3.6 to both sides.} \\ x &= 3 && \text{Divide both sides by 1.5.} \end{aligned}$$

105.

$$\begin{aligned} 2(x-6) &= 10+2x \\ 2x-12 &= 10+2x && \text{Distribute the multiplication by 2.} \\ -12 &\neq 10 && \text{Subtract } 2x \text{ from both sides.} \end{aligned}$$

This is a contradiction; there is no value for x that will make the original equation true.

106.

$$\begin{aligned} -5x+2x-1 &= -(3x+1) \\ -3x-1 &= -3x-1 && \text{Combine like terms and distribute the } -1. \\ -1 &= -1 && \text{Add } 3x \text{ both sides.} \end{aligned}$$

This equation is an **identity** and indicates that all real values for y will make the original equation true.

107.
$$V = \pi r^2 h$$
$$\frac{V}{\pi r^2} = h \qquad \text{Divide both sides by } \pi r^2 h.$$

108.
$$Y + 2g = m$$
$$2g = m - Y \qquad \text{Subtract } Y \text{ from both sides.}$$
$$g = \frac{m - Y}{2} \qquad \text{Divide both sides by 2.}$$

109.
$$\frac{T}{6} = \frac{1}{6}ab(x+y)$$
$$6\left(\frac{T}{6}\right) = 6\left(\frac{1}{6}ab(x+y)\right) \qquad \text{Multiply both sides by the LCD 6.}$$
$$T = ab(x+y) \qquad \text{Do the multiplication.}$$
$$T = abx + aby \qquad \text{Distribute the multiplication by } ab.$$
$$T - aby = abx \qquad \text{Subtract } aby \text{ from both sides.}$$
$$\frac{T - aby}{ab} = x \qquad \text{Divide both sides by } ab.$$

110.
$$V = \frac{4}{3}\pi r^3 \qquad \text{The original equation.}$$
$$\frac{3}{4\pi}V = \frac{3}{4\pi}\left(\frac{4}{3}\pi r^3\right) \qquad \text{Multiply both sides by the reciprocal of } \frac{4}{3}\pi.$$
$$\frac{3V}{4\pi} = r^3 \qquad \text{Do the multiplication.}$$

Section 1.6

111. **Analysis**: Find the number of passengers served by each of two airports.

Equation: Let x represent the number of passengers that Chicago served. Then $x + 5.4$ represents the number of passengers that Atlanta served.

Atlanta passengers	plus	Chicago passengers	is	total passengers.
$x + 10$	$+$	x	$=$	143

Solve:

$$x + 10 + x = 143$$
$$2x + 10 = 143 \qquad \text{Combine like terms.}$$
$$2x + 10 - 10 = 143 - 10 \qquad \text{Subtract 10 from both sides.}$$
$$2x = 133 \qquad \text{Combine like terms.}$$
$$x = 66.5 \qquad \text{Divide both sides by 2.}$$

Conclusion: The Chicago airport served 66.5 million passengers and the Atlanta airport served $x + 10 = 76.5$ million passengers.

112. Tuition cost less 5 times number of children is $245 - 5 \cdot c$.

113. 5(\$1000) = \$5000; x(\$1000) = \$1000x.

114. **Analysis:** The cable will be divided into four pieces. Each successive piece will be 3 feet longer than the previous piece. The cable is 186 feet long.

Equation: Let x represent the length of the first piece. The next piece will be represented by $x + 3$ and the next piece will be represented by $x + 6$, etc.

First piece	plus	second piece	plus	third piece	plus	fourth piece	is	cable length.
x	$+$	$x + 3$	$+$	$x + 6$	$+$	$x + 9$	$=$	186

Solve:

$4x + 18$	$=$	186	Combine like terms.
$4x$	$=$	168	Subtract 18 from both sides.
x	$=$	42	Divide both sides by 4.

Conclusion: The length of the first piece will be 42 feet. The length of the other pieces are $x + 3 = 45$ feet, $x + 6 = 48$ feet, and $x + 9 = 51$ feet.

115. **Analysis**: The two angles in the problem are supplementary.

Equation: Let x represent the measure of the obtuse angle. Then $x \div 2 - 15$ represents the measure of the acute angle. Together the two angles form a straight angle whose measure is 180°.

Measure of obtuse angle	plus	measure of acute angle	is	straight angle.
x	$+$	$\frac{x}{2} - 15$	$=$	180°

Solve:

$x + \frac{x}{2} - 15$	$=$	180	
$2x + x - 30$	$=$	360	Multiply both sides by the LCD 2.
$3x - 30$	$=$	360	Combine like terms.
$3x$	$=$	390	Add 30 to both sides.
x	$=$	130	Divide both sides by 3.

Conclusion: The obtuse angle has a measure of 130°. The acute angle has a measure of $\frac{x}{2} - 15 = 50°$.

Section 1.7

116. **Analysis:** An investment of $25,000 will be split between two accounts. The interest received will be $2430 per year.

Equation: Let x represent the number of dollars invested in the first investment at 10.0%. Then $(25,000 – x) is the number of dollars invested in the second investment at 9.0%.

	P •	r •	t =	I
First investment	x	0.10	1	$0.10x$
Second investment	$25,000 - x$	0.09	1	$0.09(25,000 - x)$

Interest earned at 10%	plus	interest earned at 9%	is	total interest earned.
$0.10x$	$+$	$0.09(25,000 - x)$	$=$	2430

Solve:

$0.10x + 0.09(25,000 - x)$	$=$	2430	
$0.10x + 2250 - 0.09x$	$=$	2430	Distribute the multiplication by 0.09.
$0.01x + 2250$	$=$	2430	Combine like terms.
$0.01x$	$=$	180	Subtract 2250 from both sides.
x	$=$	18,000	Use a calculator to divide both sides by 0.01.

Conclusion: The first investment will be $18,000 at 10% and the second investment will be $25,000 - x = \$7,000$ at 9%.

117. **Analysis:** Find the percent of increase (decrease) by first finding the difference, and calculating that as a percent of the number of cars sold in 2001.

Toyota Camrys: increase: 434,145 – 390,449 = 43,696

What percent	of	2001 sales	is	increase in sales?
x	•	390,449	=	43,696

Solve:

$390{,}449\,x = 43{,}696$

$x = 0.1119\ldots$ Use a calculator to divide both sides by 390,449.

$x = 11.2\%$ Write the decimal as a percent.

Conclusion: Sales of Toyota Camrys increased by 11.2%.

Honda Accord: increase: 414,718 – 398,980 = 15,738

What percent	of	2001 sales	is	decrease in sales?
x	•	414,718	=	15,738

Solve:

$414{,}718\ x = 15{,}738$

$x = .0.3795\ldots$ Use a calculator to divide both sides by 390,449.

$x = 3.8\%$ Write the decimal as a percent.

Conclusion: Sales of Honda Accords decreased by 3.8%.

118. The find the mean, we add the values and divide by the number of values, which is 11.

$$\text{mean} = \frac{500+499+504+505+505+505+505+505+506+504+507}{11} = \frac{5545}{11} = 504$$

To find the median, arrange the values in increasing order:

$500, 499, 504, 505, 505, 505, 505, 505, 506, 504, 507$ The middle (6th) term is 505

Mode = 55 since it occurs most often (5 times).

119. **Analysis:** The vehicles are traveling in the same direction. The celebrity will travel for one more minute than the photographer.

Equation: Let t represent the time that the photographer travels. Then $t + 1$ represents the time the celebrity travels.

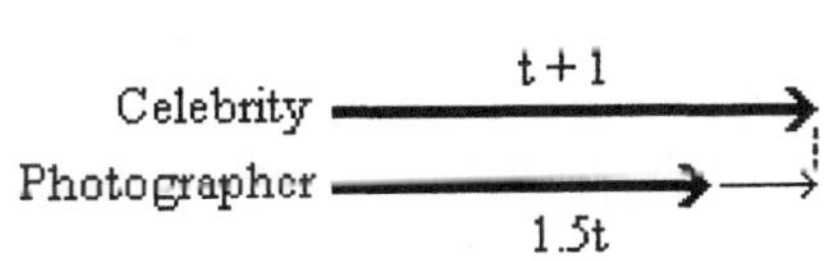

	r	• t	= d
Celebrity	1	$t + 1$	$1(t + 1)$
Photographer	1.5	t	1.5t

Celebrity distance	equals	photographer distance.
$1(t + 1)$	=	$1.5t$

Solve:

$t + 1 = 1.5t$

$1 = 0.5t$ Subtract t from both sides.

$t = 2$ Divide both sides by 0.5.

Conclusion: The photographer will catch up with the celebrity in 2 minutes.

120. **Analysis:** Water will be added to a 12% solution of alcohol to dilute it to an 8% solution.

Equation: Let x represent the number of gallons of water added. Then $20 + x$ will represent the number of gallons of the 8% solution.

	percent alcohol •	total quantity	= quantity of alcohol
12% solution	0.12	20	2.4
Water	0.00	x	0
8% solution	0.08	$20 + x$	$0.08(20 + x)$

Quantity of pesticide in 12% solution	plus	quantity of pesticide in water	is	quantity of pesticide in 8% solution.
2.4	+	0	=	$0.08(20 + x)$

Solve:

$$\begin{aligned} 2.4 + 0 &= 0.08(20 + x) \\ 2.4 &= 1.6 + 0.08x && \text{Distribute the multiplication by 0.08.} \\ 0.8 &= 0.08x && \text{Subtract 1.6 from both sides.} \\ 10 &= x && \text{Use a calculator to divide both sides by 0.08.} \end{aligned}$$

Conclusion: 10 gallons of water should be added to the 12% solution of alcohol.

121. The number of pounds of the mixture will be $x + 3$. The value will be the *price* times the *number of pounds*. The expression is $\$3.95(x + 3)$.

Chapter 1 Chapter Test

1. $s = T + 10$
2. $A = \frac{1}{2}bh$
3. $-2, 0, 5$
4. $-2, 0, -3\frac{3}{4}, 9.2, \frac{14}{5}, 5$
5. $\pi, -\sqrt{7}$
6. all
7. 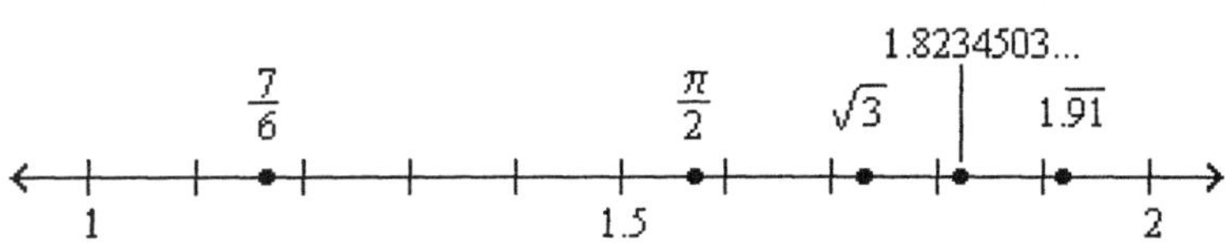

8. 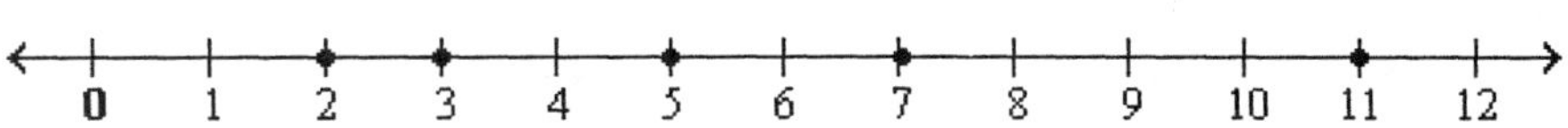

9. $-|8| = -(8) = -8$
10. $|-5.5| = 5.5$
11. $7 - (-5.3) = 7 + 5.3 = 12.3$
12. $-\frac{5}{3}\left(-\frac{4}{25}\right) = \frac{(-5)(-4)}{3(25)} = \frac{\overset{1}{\cancel{5}} \cdot 4}{3 \cdot \underset{1}{\cancel{5}} \cdot 5} = \frac{4}{15}$
13. $\frac{1}{2} - \left(-\frac{3}{5}\right) = \frac{5}{10} + \frac{6}{10} = \frac{11}{10}$
14. $(-4)^3 = (-4)(-4)(-4) = -64$
15. $\frac{2[-4-2(3-1)]}{3(\sqrt{9})(2)} = \frac{2[-4-2(2)]}{3(3)(2)} = \frac{\overset{1}{\cancel{2}}(-4-4)}{9(\underset{1}{\cancel{2}})} = \frac{-8}{9}$
16. $7 + 2[-1 - 4(5)[= 7 + 2(-1 - 20) = 7 + 2(-21) = 7 - 42 = -35$

17. $\dfrac{-3b+a}{ac-b}=\dfrac{-3(-3)+(2)}{(2)(4)-(-3)}=\dfrac{9+2}{8+3}=\dfrac{11}{11}=1$

18. $$\text{child's dose}=\frac{\text{Age of child}}{\text{Age of child}+12}\left(\begin{array}{c}\text{average}\\ \text{adult dose}\end{array}\right)$$
Substitute 8 for the age of the child and 250 for the average adult dosage.
$$=\frac{8}{8+12}(250)=\frac{8}{20}(250)=100$$

The dose for the 8–year old child would be 100 mg.

19. Commutative property of addition

20. Associative property of multiplication

21. Distributive property

22. Multiplicative inverse property

31. $-y+3y+9y=2y+9y=11y$

42. $5(-9q)=-45q$

25. $-(4+t)+t=-4-t+t=-4$

26. $12\left(-\frac{4}{3}+\frac{1}{6}a+\frac{3}{2}b\right)=12\left(-\frac{4}{3}\right)+12\left(\frac{1}{6}a\right)+12\left(\frac{3}{2}b\right)=-16+2a+6(3b)=-16+2a+18b$

27.

$9(x+4)+4$	$=$	$4(x-5)$	
$9x+36+4$	$=$	$4x-20$	Distribute the multiplications.
$9x+40$	$=$	$4x-20$	Combine like terms.
$5x$	$=$	-60	Subtract 4x and 40 from both sides.
x	$=$	-12	Divide both sides by 5.

28.

$\dfrac{y-1}{5}+2$	$=$	$\dfrac{2y-3}{3}$	
$15\left(\dfrac{y-1}{5}+2\right)$	$=$	$15\left(\dfrac{2y-3}{3}\right)$	Multiply both sides by the LCD 15.
$3(y-1)+15(2)$	$=$	$5(2y-3)$	Distribute.
$3y-3+30$	$=$	$10y-15$	Distribute.
$3y+27$	$=$	$10y-15$	Combine like terms.
$-7y$	$=$	-42	Subtract 10y and 27 from both sides.
y	$=$	6	Divide both sides by –7.

29.

$6-(x-3)-5x$	$=$	$3[1-2(x+2)]$	
$6-x+3-5x$	$=$	$3(1-2x-4)$	Distribute
$9-6x$	$=$	$3-6x-12$	Combine like terms and distribute
$9-6x$	$=$	$-9-6x$	Combine like terms
$9-6x+6x$	$=$	$-9-6x+6x$	Add 6x to both sides
9	$=$	-9	Since equation is not true, contradiction (no solution)

30. The find the mean, we add the values and divide by the number of values, which is 13.
$$\frac{5.6+6.8+7.5+6.9+6.1+5.6+5.4+4.9+4.5+4.2+4.0+4.7+5.8}{13}=\frac{72}{13}=5.538\ldots\approx 5.5$$

31. To find the median, arrange the values in increasing order:

4.0, 4.2, 4.5, 4.7, 4.9, 5.4, 5.6, 5.6, 5.8, 6.1, 6.8, 6.9, 7.5

There are an odd number of values, so use the middle (7th) value. The median is 5.6.

32. The mode = 5.6 since it occurs most often (2 times).

33.

$$P = L + \frac{s}{f}i$$

$P - L = \frac{s}{f}i$	Subtract L from both sides.
$f(P - L) = si$	Multiply both sides by f.
$\frac{f(P-L)}{s} = i$	Divide both sides by s.

34. **Analysis:** Find the area of the rectangular viewing window of a calculator that has a perimeter of 26 cm and is 5 cm longer than it is wide. Substitute these values into the formula for perimeter.

Equation: Let w represent the width of the window. Then $w + 5$ would represent the length l. The perimeter of the window would be 26 cm. Use the formula for perimeter $P = 2l + 2w$.

$$P = 2l + 2w$$
$$26 = 2(w+5) + 2w$$

Solve:

$26 = 2(w+5) + 2w$	
$26 = 2w + 10 + 2w$	Do the multiplication
$26 = 4w + 10$	Combine like terms.
$16 = 4w$	Subtract 10 from both sides.
$4 = w$	Divide both sides by 4.

Conclusion: The width of the window is 4 cm and the length is $w + 5 = 9$ cm.

35. **Analysis:** The national debt in 1960 was \$291 billion. The debt increased to \$5776 billion by 1999. This is an increase of \$5776 – \$291 = \$5485 billion.

Equation: Let x represent the percent increase in debt from 1960 to 1999.

What percent	of	1960 debt	is	increased debt?
x	•	291	=	5485

Solve:

$x \cdot 291 = 5485$	
$291x = 5485$	Rewrite $x \cdot 291$ as $291x$.
$x = 18.8487\ldots$	Use a calculator to divide both sides by 291.
$x = 1885\%$	Write the decimal as a percent.

Conclusion: The national debt increased by 1885%.

36. **Analysis:** An investment of \$10,000 is split between two accounts. The interest received will be \$860 per year.

Equation: Let x represent the number of dollars invested at 8.0%. Then \$(10,000 – x) is the number of dollars invested at 9.0%.

	P •	r •	t =	I
8% investment	x	0.08	1	$0.08x$
9% investment	$10{,}000 - x$	0.09	1	$0.09(10{,}000 - x)$

Interest earned at 08%	plus	interest earned at 9%	is	total interest earned.
$0.08x$	+	$0.09(10{,}000 - x)$	=	860

Solve:

$0.08x + 0.09(10{,}000 - x) = 860$	
$0.08x + 900 - 0.09x = 860$	Distribute the multiplication by 0.09.
$-0.01x + 900 = 860$	Combine like terms.
$-0.01x = -40$	Subtract 900 from both sides.
$x = 4000$	Use a calculator to divide both sides by –0.01.

Conclusion: The investment at 8% was \$4000 .

37. **Analysis:** Two different gold alloys are mixed to obtain an alloy that is 25% gold.

Equation: Let x represent the number of ounces of 40% gold alloy used. Then $10 + x$ will represent the number of ounces of the 25% alloy.

	percent gold •	total quantity =	quantity of gold
40% Alloy	0.40	x	$0.40x$
10% Alloy	0.10	10	1
25% Alloy	0.25	$10 + x$	$0.25(10 + x)$

Quantity of gold in 40% alloy	plus	quantity of gold in 10% alloy	is	quantity of gold in 25% alloy.
$0.40x$	+	1	=	$0.25(10 + x)$

Solve:

$$0.40x + 1 = 0.25(10 + x)$$

$$0.40x + 1 = 2.5 + 0.25x \quad \text{Distribute the multiplication by 0.25.}$$

$$0.15x = 1.5 \quad \text{Subtract } 0.25x \text{ and 1 from both sides.}$$

$$x = 10 \quad \text{Use a calculator to divide both sides by 0.15.}$$

Conclusion: 10 ounces of the 40% gold alloy should be added to the 10% gold alloy.

38. When 3 is substituted for x into the equation, it forms a true statement.

Chapter 2 Graphs, Equations of Lines, and Functions

Study Tip

If Math is your mountain, your instructor is your mountain guide.

Politely stop your instructor and ask questions as soon as they arise.
Visit your instructor in their office as soon as you think you might be confused.
If you have a different method to do a problem, ask your instructor about it.
If your instructor offers review sessions- ATTEND!
Do the assignments recommended by your instructor- EVEN IF THEY AREN'T GRADED!

Section 2.1 The Rectangular Coordinate System

VOCABULARY

1. ordered
3. origin
5. rectangular
7. midpoint

CONCEPTS

9. origin; right; down
11. Quadrant II

13.

x	y
0	**−3**
1	**−1**
2	**1**
3	3

14.

x	y
−2	**2**
−1	**−1**
0	−2
1	**−1**
2	**2**

15. one

NOTATION

17. t: horizontal; d: vertical
19. x sub 1

PRACTICE

21-28.

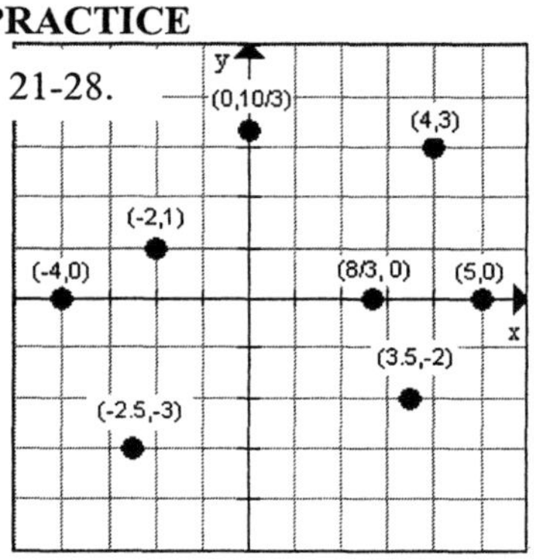

29. $A(2,4)$
31. $C(-2.5,-1.5)$
33. $E(3, 0)$
35. $G(0, 0)$

37. a. on the surface
 b. diving
 c. 1000 ft.
 d. 500 ft.

39. a. 1993
 b. Imports exceeded production by about 3.5 million barrels per day.

41. $P(x_1, y_1) = (0, 0)$, $Q(x_2, y_2) = (6, 8)$

$$x_M = \frac{x_1 + x_2}{2} = \frac{0+6}{2} = \frac{6}{2} = 3 \qquad y_M = \frac{y_1 + y_2}{2} = \frac{0+8}{2} = \frac{8}{2} = 4$$

The midpoint is (3, 4)

43. $P(x_1, y_1) = (6, 8)$, $Q(x_2, y_2) = (12, 16)$

$$x_M = \frac{x_1 + x_2}{2} = \frac{6+12}{2} = \frac{18}{2} = 9$$

$$y_M = \frac{y_1 + y_2}{2} = \frac{8+16}{2} = \frac{24}{2} = 12$$

The midpoint is (9, 12)

45. $P = (2, 4)$, $Q = (5, 8)$

$$x_M = \frac{2+5}{2} = \frac{7}{2}$$

$$y_M = \frac{4+8}{2} = \frac{12}{2} = 6$$

The midpoint is $\left(\frac{7}{2}, 6\right)$

47. $P = (-2, -8)$, $Q = (3, 4)$

$$x_M = \frac{-2+3}{2} = \frac{1}{2}$$

$$y_M = \frac{-8+4}{2} = \frac{-4}{2} = -2$$

The midpoint is $\left(\frac{1}{2}, -2\right)$

49. $P = (-3, 5)$, $Q = (-5, -5)$

$$x_M = \frac{-3+(-5)}{2} = \frac{-8}{2} = -4$$

$$y_M = \frac{5+(-5)}{2} = \frac{0}{2} = 0$$

The midpoint is $\left(1, \frac{1}{2}\right)$

51. $M(x_M, y_M) = M(-2, 3)$, $P(x_1, y_1) = (-8, 5)$
Find $Q(x_2, y_2)$

$$x_m = \frac{x_1 + x_2}{2}$$
$$-2 = \frac{-8 + x_2}{2}$$
$$-4 = -8 + x_2$$
$$4 = x_2$$

$$y_M = \frac{y_1 + y_2}{2}$$
$$3 = \frac{5 + y_2}{2}$$
$$6 = 5 + y_2$$
$$1 = y_2$$
$$Q = (4,1)$$

53. $M = (-7, -3)$, $Q = (6, -3)$. Find P.

$$x_m = \frac{x_1 + x_2}{2}$$
$$-7 = \frac{x_1 + 6}{2}$$
$$-14 = x_1 + 6$$
$$-20 = x_1$$

$$y_M = \frac{y_1 + y_2}{2}$$
$$-3 = \frac{y_1 + (-3)}{2}$$
$$-6 = y_1 - 3$$
$$-3 = y_1$$
$$P = (-20, -3)$$

APPLICATIONS

55. Jonesville (5, B), Easley (1, B), Hodges (2, E), Union (6, C)

57. a. (2, −1) b. no c. yes

59. a. 6 b. 7 strokes from Holes 3 through 10 c. the 16th hole d. the 18th hole

61. a. \$2 b. \$4 c. \$7 d. \$9

63. a. $(x, y, z) = (0, 181, 56)$: $x = 0$ would be in the center left/right of the plane, $y = 181$ would be the farthest back, and $z = 56$ would be up. This would describe the tip of the tail.

b. $(x, y, z) = (-46, 48, 19)$: $x = -46$ would be on the negative left/right of the plane, $y = 8$ would be partially back, and $z = 19$ would be up slightly. This would describe the front of the engine.

c. $(x, y, z) = (84, 94, 24)$: $x = 84$ would be on the positive left/right of the plane, $y = 94$ would be farther back than the engine, and $z = 24$ would be up. This would describe the tip of the wing.

65. Begin at the origin. Move to the left two units, then move up 5 units. Make a dot.

REVIEW

67. $-5 - 5(-5) = -5 + 25 = 20$

69. $$\frac{-3 + 5(2)}{9 + 5} = \frac{-3+10}{14} = \frac{7}{14} = \frac{1}{2}$$

71.
$$-4x + 0.7 = -2.1$$
$$-4x = 2.8$$
$$x = 0.7$$

Section 2.2 Graphing Linear Equations

VOCABULARY

1. satisfy
3. x–intercept ; y-intercept
5. vertical

CONCEPTS

7. a. Plug in $x = -1,\ y = 3$ $\quad 3 = -5(-1) - 2 = 5 - 2$ since this is true, $(-1,3)$ is a solution.
   b. Plug in $x = 3,\ y = -13$ $\quad -13 \leftarrow -5\,(3) - 2 = -15 - 2 = -17$ since this is not true, $(3, -13)$ is not a solution.
9. x–intercept: $(-6, 0)$; y–intercept: $(0, 3)$
10. a. y–axis b. x–axis
11. a. x–intercept: $(-3, 0)$; y–intercept: $(0, 4)$ b. False, point M is not on the line.
13. Substitute 10 for x and compare the y value that is calculated: $y = -4x - 1$.

NOTATION

15.
$$\begin{aligned} 2x + 2y &= -8 \\ 2(\ \mathbf{-3}\ ) + 2(-1) &\overset{?}{=} -8 \\ -6 + (\ \mathbf{-2}\ ) &\overset{?}{=} -8 \\ -8 &= -8 \end{aligned}$$
The point $(-3,-1)$ is a solution

17. the y-axis

PRACTICE

19.

x	$y = -x + 4$	y
−1	$y = -(-1) + 4$	5
0	$y = -(0) + 4$	4
2	$y = -(2) + 4$	2

21.

x	$y = -\frac{1}{3}x - 1$	y
−3	$y = -\frac{1}{3}(-3) - 1$	0
0	$y = -\frac{1}{3}(0) - 1$	−1
3	$y = -\frac{1}{3}(3) - 1$	−2

23.

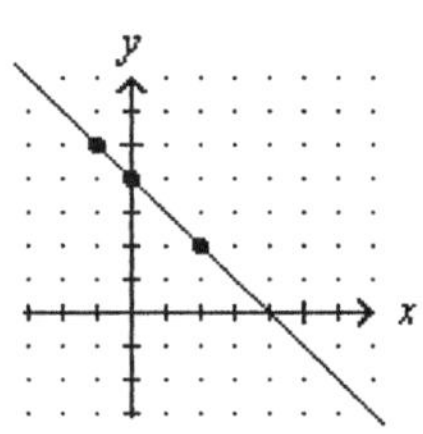

25.

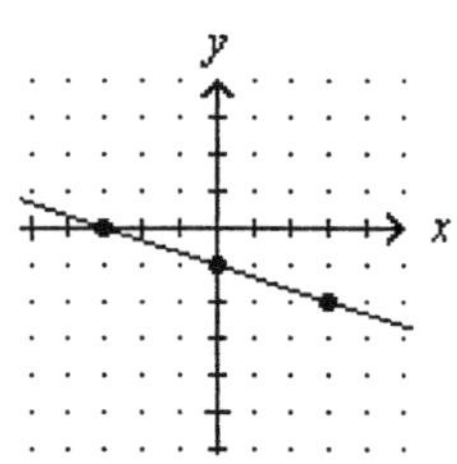

27.

x	$y = x$	y
−3	$y = -3$	−3
0	$y = 0$	0
2	$y = 2$	2

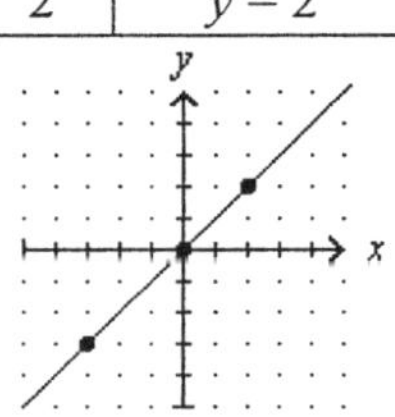

29.

x	$y = -3x + 2$	y
−1	$y = -3(-1) + 2$	5
0	$y = -3(0) + 2$	2
2	$y = -3(2) + 2$	−4

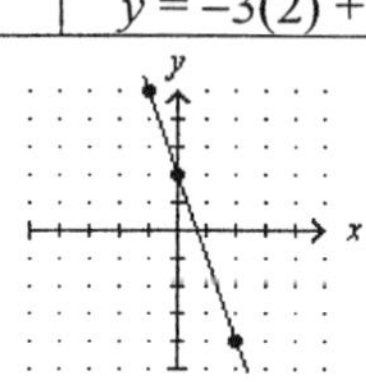

31.

x	$y = 3 - x$	y
−1	$y = 3 - (-1)$	4
0	$y = 3 - (0)$	3
2	$y = 3 - (2)$	1

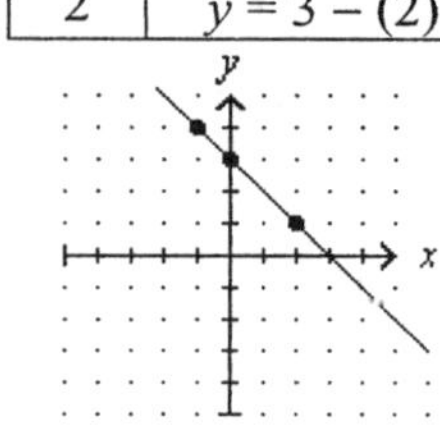

33.

x	$y = \frac{x}{4} - 1$	y
-4	$y = \frac{-4}{4} - 1$	-2
0	$y = \frac{0}{4} - 1$	-1
4	$y = \frac{4}{4} - 1$	0

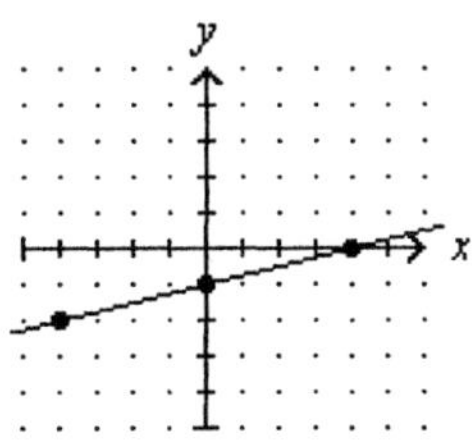

35.

x	$x = 3$	y
3	The value of y can be any number	-1
3		2
3		5

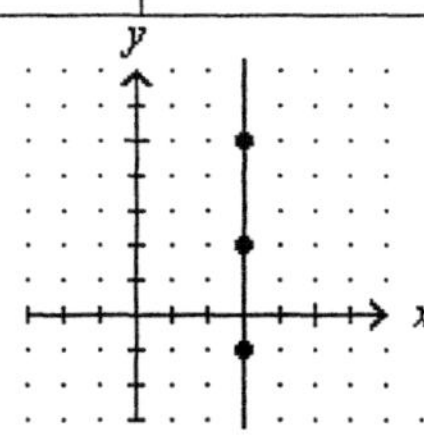

37.

x	$y - 2 = 0$	y
-3	$y = 2$ x can be any number	2
0		2
3		2

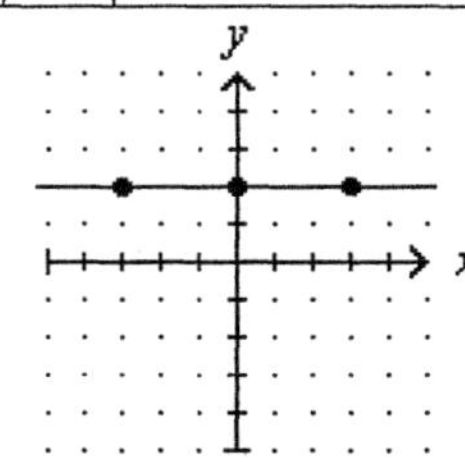

39.

x	$-3y + 2 = 5$	y
-2	$-3y = 3$ or $y = -1$	-1
1		-1
4		-1

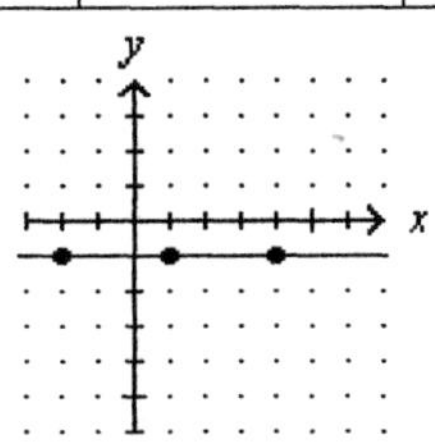

41.

x	$3x + 4y = 12$	y
0	$3(0) + 4y = 12$	3
4	$3x + 4(0) = 12$	0
-4	$3(-4) + 4y = 12$ $4y = 24$	6

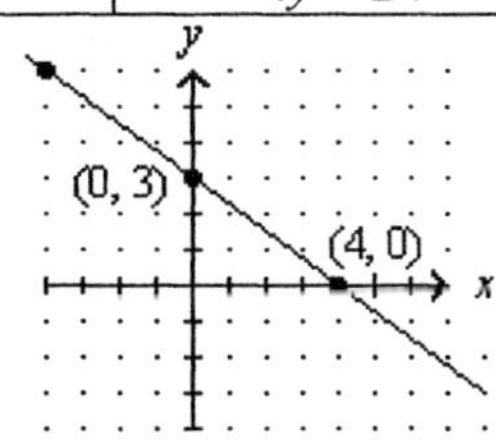

43.

x	$3y = 6x - 9$	y
0	$3y = 6(0) - 9$	-3
1.5	$3(0) = 6x - 9$ $9 = 6x$	0
3	$3y = 6(3) - 9$ $3y = 9$	3

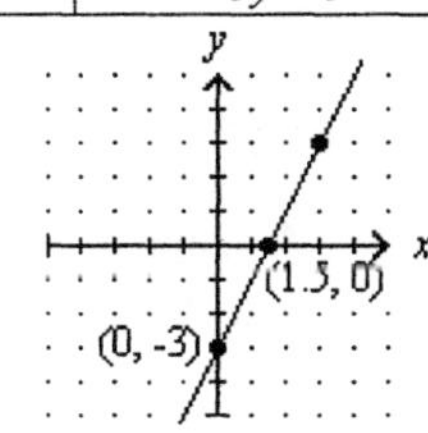

45.

x	$2y + x = -2$	y
0	$2y + (0) = -2$	-1
-2	$2(0) + x = -2$	0
-4	$2y + (-4) = -2$ $2y = 2$	1

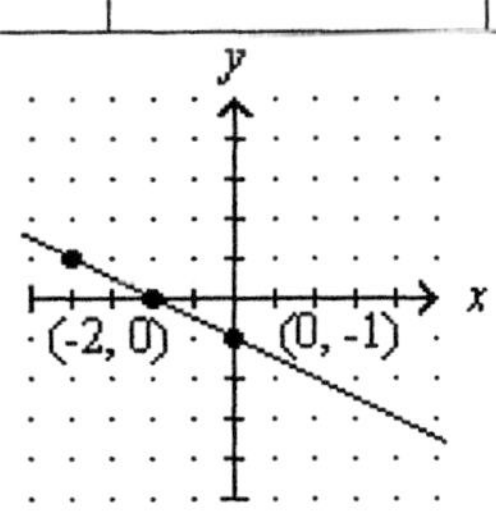

47.

x	$3x + 4y - 8 = 0$	y
0	$3(0) + 4y - 8 = 0$ $4y = 8$	2
$2\frac{2}{3}$	$3x + 4(0) - 8 = 0$ $3x = 8$	0
-4	$3(-4) + 4y - 8 = 0$ $4y = 20$	5

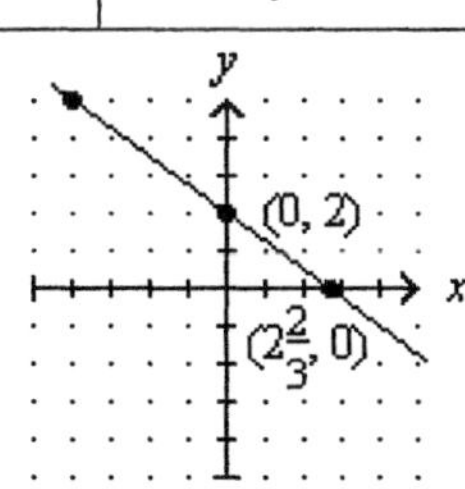

49.

x	$3x = 4y - 11$	y
0	$3(0) = 4y - 11$ $11 = 4y$	$2\frac{3}{4}$
$3\frac{2}{3}$	$3x = 4(0) - 11$ $3x = -11$	0
3	$3(3) = 4y - 11$ $9 = 4y - 11$ $20 = 4y$	5

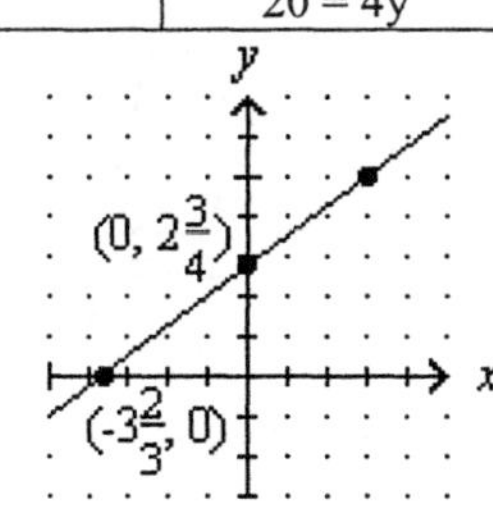

51.

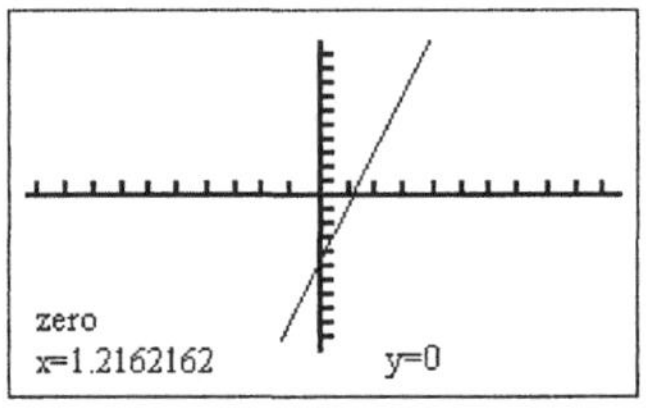

53.

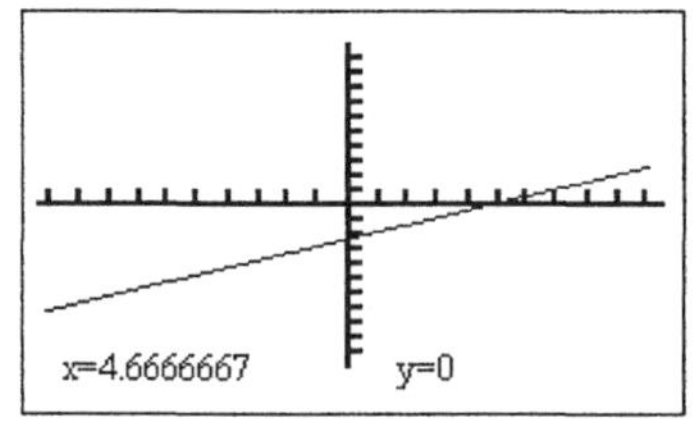

55.

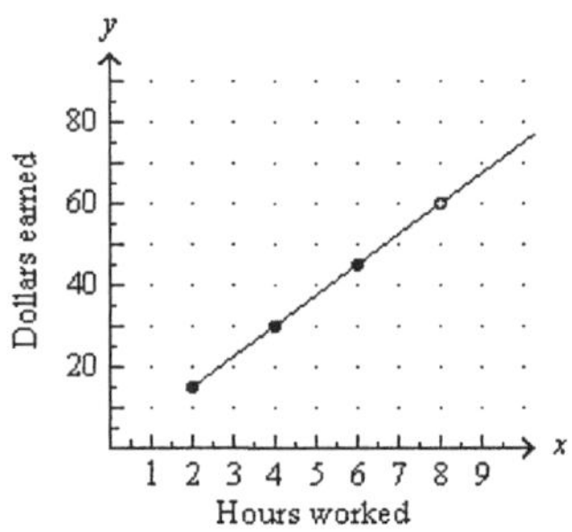

The student would earn \$60 for working 8 hours.

57.

t	$s = -0.9t + 65.5$	s
0	$s = -0.9(0) + 65.5$	65.5
5	$s = -0.9(5) + 65.5$	61.0
10	$s = -0.9(10) + 65.5$	56.5

The number of swimmers in 1998 would be about 58.2 million.

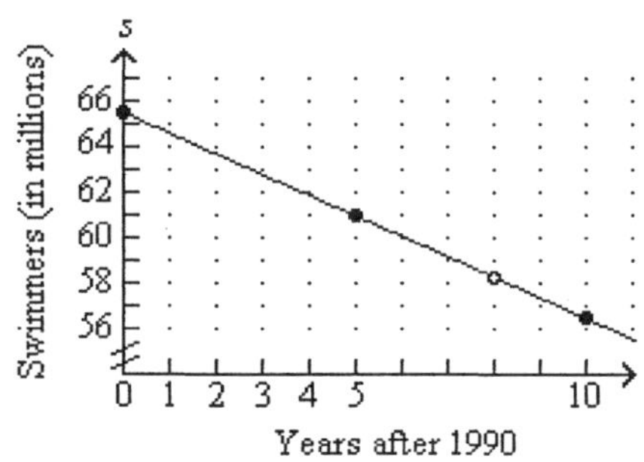

59.
$$\begin{aligned} y &= 7500x + 125{,}000 \\ &= 7500(5) + 125{,}000 \quad \text{Substitute 5 for } x. \\ &= 37500 + 125{,}000 \\ &= \$162{,}500 \end{aligned}$$

61.
$$\begin{aligned} y &= -1360x + 17{,}000 \\ 0 &= -1360x + 17{,}000 \quad \text{Substitute 0 for } y \\ 1360x &= 17{,}000 \\ x &= 12.5 \text{ years} \end{aligned}$$

63. a. $c = 10t + 2$

b.

t	$c = 10t + 2$	c
1	$c = 10(1) + 2$	12
2	$c = 10(2) + 2$	22
3	$c = 10(3) + 2$	32
4	$c = 10(4) + 2$	42

c. The cost of 6 tickets would be \$62.

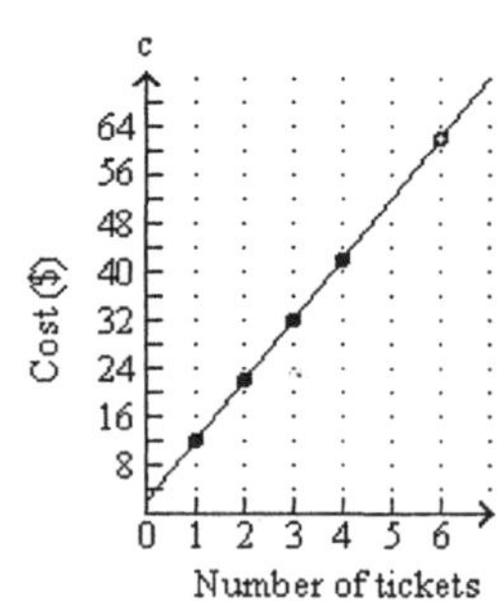

WRITING

65. Let $y = 0$ and find the x–intercept. Let $x = 0$ and find the y–intercept. Plot these two points. Connect them with a straight line.

REVIEW

67. 11, 13, 17, 19, 23, 29

69. Quadrant III

71. $-4(-20s) = 80s$

73. $-(-3x - 8) = 3x + 8$

Section 2.3 Rate of Change and the Slope of a Line

VOCABULARY

1. Slope 3. change 5. reciprocals

CONCEPTS

7. a. l_3; 0 b. l_2; undefined

c. l_1; Calculate the slope by looking at two points on the line and substituting in slope formula. Sample points: (–1, 2) and (–2, 0). Substitute 0 for y_2, 2 for y_1, –2 for x_2, and –1 for x_1.

$$\text{slope} = \frac{\text{change in y}}{\text{change in x}} = \frac{y_2 - y_1}{x_2 - x_1} = \frac{0-2}{-2-(-1)} = \frac{-2}{-2+1} = \frac{-2}{-1} = 2$$

d. l_4; Calculate the slope by looking at two points on the line and substituting in slope formula. Sample points: (1, 4) and (2, 1). Substitute 1 for y_2, 4 for y_1, 2 for x_2, and 1 for x_1.

$$\text{slope} = \frac{\text{change in y}}{\text{change in x}} = \frac{y_2 - y_1}{x_2 - x_1} = \frac{1-4}{2-1} = \frac{-3}{1} = -3$$

9. a. Rate of increase $= \dfrac{\text{change in number of CDs}}{\text{change in time}} = \dfrac{943-286}{9} = \dfrac{657}{9} = 73$ million units/yr.

b. Rate of decrease $= \dfrac{\text{change in number of cassettes}}{\text{change in time}} = \dfrac{126-441}{9} = \dfrac{-315}{9} = -35$ million units/yr.

11. a. Calculate the slope by looking at two points on the line and substituting in slope formula. Sample points: (0, 4) and (3, 0). Substitute 0 for y_2, 4 for y_1, 3 for x_2, and 0 for x_1.

$$\text{slope} = \frac{\text{change in y}}{\text{change in x}} = \frac{y_2 - y_1}{x_2 - x_1} = \frac{0-4}{3-0} = \frac{-4}{3} = -\frac{4}{3}$$

b. Calculate the slope by looking at two points on the line and substituting in slope formula. Sample points: (–3, 0) and (0, –2). Substitute –2 for y_2, 0 for y_1, 0 for x_2, and –3 for x_1.

$$\text{slope} = \frac{\text{change in y}}{\text{change in x}} = \frac{y_2 - y_1}{x_2 - x_1} = \frac{-2-0}{0-(-3)} = \frac{-2}{3} = -\frac{2}{3}$$

NOTATION

13. $\text{slope} = m = \dfrac{y_2 - y_1}{x_2 - x_1}$

15. a. Δy represents the change in y, or the rise of the line between two points. In this case, $\Delta y = 6$.

b. Δx represents the change in x, or the run of the line between two points. In this case, $\Delta x = 8$.

c. $\dfrac{\Delta y}{\Delta x} = \dfrac{\text{rise}}{\text{run}} = \dfrac{6}{8} = \dfrac{3}{4}$

PRACTICE

17. Points: $(x_1, y_1) = (-2, 1)$ and $(x_2, y_2) = (1, -7)$

$$m = \frac{y_2 - y_1}{x_2 - x_1} = \frac{-7-1}{1-(-2)} = \frac{-8}{3} = -\frac{8}{3}$$

19. Points: $(x_1, y_1) = (6, 1)$ and $(x_2, y_2) = (-2, -6)$

$$m = \frac{y_2 - y_1}{x_2 - x_1} = \frac{-6-1}{-2-6} = \frac{-7}{-8} = \frac{7}{8}$$

21. Points: $(x_1, y_1) = (0, 0)$ and $(x_2, y_2) = (3, 9)$

$$m = \frac{y_2 - y_1}{x_2 - x_1} = \frac{9-0}{3-0} = \frac{9}{3} = 3$$

23. Points: $(x_1, y_1) = (-1, 8)$ and $(x_2, y_2) = (6, 1)$

$$m = \frac{y_2 - y_1}{x_2 - x_1} = \frac{1-8}{6-(-1)} = \frac{-7}{7} = -1$$

25. Points: $(x_1, y_1) = (3, -1)$ and $(x_2, y_2) = (-6, 2)$

$$m = \frac{y_2 - y_1}{x_2 - x_1} = \frac{2-(-1)}{-6-3} = \frac{3}{-9} = -\frac{1}{3}$$

27. Points: $(x_1, y_1) = (7, 5)$ and $(x_2, y_2) = (-9, 5)$

$$m = \frac{y_2 - y_1}{x_2 - x_1} = \frac{5-5}{-9-7} = \frac{0}{-14} = 0$$

29. Points: $(x_1, y_1) = (-7, -5)$ and $(x_2, y_2) = (-7, -2)$

$$m = \frac{y_2 - y_1}{x_2 - x_1} = \frac{-2-(-5)}{-7-(-7)} = \frac{3}{0} \text{ is undefined}$$

31. Points: $(x_1, y_1) = (a, b)$ and $(x_2, y_2) = (b, a)$

$$m = \frac{y_2 - y_1}{x_2 - x_1} = \frac{a-b}{b-a} = \frac{a-b}{-1(-b+a)} = -\frac{a-b}{a-b} = -1$$

33.

x	$3x + 2y = 12$	y
0	$3(0) + 2y = 12$	6
4	$3x + 2(0) = 12$	0

$(x_1, y_1) = (0, 6)$ and $(x_2, y_2) = (4, 0)$

$$m = \frac{y_2 - y_1}{x_2 - x_1} = \frac{0-6}{4-0} = \frac{-6}{4} = -\frac{3}{2}$$

35.

x	$3x = 4y - 2$ $3x - 4y = -1$	y
0	$3(0) - 4y = -2$	$\frac{1}{2}$
$-\frac{2}{3}$	$3x - 4(0) = -2$	0

$(x_1, y_1) = \left(0, \frac{1}{2}\right)$ and $(x_2, y_2) = \left(-\frac{2}{3}, 0\right)$

$$m = \frac{y_2 - y_1}{x_2 - x_1} = \frac{0 - \frac{1}{2}}{-\frac{2}{3} - 0} = \left(-\frac{1}{2}\right) \div \left(-\frac{2}{3}\right) = \frac{3}{4}$$

37.

x	$y = \frac{x-4}{2}$ $2y = x - 4$ $-x + 2y = -4$	y
0	$(0) + 2y = 4$	2
-4	$-x + 2(0) = 4$	0

$(x_1, y_1) = (0, 2)$ and $(x_2, y_2) = (-4, 0)$

$$m = \frac{y_2 - y_1}{x_2 - x_1} = \frac{0-2}{-4-0} = \frac{-2}{-4} = \frac{1}{2}$$

39.

x	$4y = 3(y + 2)$ $4y = 3y + 6$ $y = 6$	y
0	$y = 6$	6
3	$y = 6$	6

$(x_1, y_1) = (0, 6)$ and $(x_2, y_2) = (3, 6)$

$$m = \frac{y_2 - y_1}{x_2 - x_1} = \frac{6-6}{3-0} = \frac{0}{3} = 0$$

41. The two slopes are negative reciprocals, therefore the lines are perpendicular

43. The two slopes are reciprocals, but not negative reciprocals, thus the lines are neither parallel nor perpendicular.

45. The two slopes are reciprocals, but not negative reciprocals, thus the lines are neither parallel nor perpendicular.

47. Points: $(x_1, y_1) = (3, 4)$ and $(x_2, y_2) = (4, 2)$

$$m = \frac{y_2 - y_1}{x_2 - x_1} = \frac{2-4}{4-3} = \frac{-2}{1} = -2$$

Parallel. Slopes are the same.

49. Points: $(x_1, y_1) = (-2, 1)$ and $(x_2, y_2) = (6, 5)$

$$m = \frac{y_2 - y_1}{x_2 - x_1} = \frac{5-1}{6-(-2)} = \frac{4}{8} = \frac{1}{2}$$

Perpendicular. Slopes are negative reciprocals

51. Points: $(x_1, y_1) = (5, 4)$ and $(x_2, y_2) = (6, 6)$

$$m = \frac{y_2 - y_1}{x_2 - x_1} = \frac{6-4}{6-5} = \frac{2}{1} = 2$$

Neither.

APPLICATIONS

53. Part 1: $m = \frac{\Delta y}{\Delta x} = \frac{\text{change in elevation}}{\text{distance}} = \frac{2600-2000}{28{,}000} = \frac{600}{28{,}000} = \frac{3}{140}$ or 0.0214...

Part 2: $m = \frac{\Delta y}{\Delta x} = \frac{\text{change in elevation}}{\text{distance}} = \frac{2000-1440}{8400} = \frac{560}{8400} = \frac{1}{15}$ or 0.0666...

Part 3: $m = \frac{\Delta y}{\Delta x} = \frac{\text{change in elevation}}{\text{distance}} = \frac{1440-560}{17{,}600} = \frac{880}{17{,}600} = \frac{1}{20}$ or 0.05

Convert the three slopes to decimals to compare the steepness. Part 2 is the steepest.

55. The vertical distance for each contour line is 50 feet. The west face has four contour lines, or 4(50) = 200 feet of elevation. The east face has five contour lines, or 5(50) = 250 feet of elevation.

West face slope: $m = \frac{\Delta y}{\Delta x} = \frac{\text{change in elevation}}{\text{change in horizontal distance}} = \frac{200\text{ ft}}{2000\text{ ft}} = \frac{1}{10}$

East face slope: $m = \frac{\Delta y}{\Delta x} = \frac{\text{change in elevation}}{\text{change in horizontal distance}} = \frac{250\text{ ft}}{1000\text{ ft}} = \frac{1}{4}$

57. Convert the horizontal distance to feet. 2.5(5280) = 13,200 feet

Find the slope: $m = \frac{\Delta y}{\Delta x} = \frac{\text{change in elevation}}{\text{change in horizontal distance}} = \frac{528\text{ ft}}{13{,}200\text{ ft}} = \frac{1}{25} = 0.04$ *or* 4%.

59. Cross–brace: (–7, –4); (–3, –2) $\quad m = \frac{y_2 - y_1}{x_2 - x_1} = \frac{-2-(-4)}{-3-(-7)} = \frac{2}{4} = \frac{1}{2}$

Support 1: (–9, 0); –7, –4) $\quad m = \frac{y_2 - y_1}{x_2 - x_1} = \frac{-4-0}{-7-(-9)} = \frac{-4}{2} = -2$

Support 2: (–5, 0); (–3, –2) $\quad m = \frac{y_2 - y_1}{x_2 - x_1} = \frac{-2-0}{-3-(-5)} = \frac{-2}{2} = -1$

The cross–brace is perpendicular to Support 1 because their slopes are negative reciprocals.

WRITING

61. The Democrats could say that projections show Medicare spending should be \$288 billion in 2002 but only \$247 billion will be spent. Thus spending has been cut. The Republicans could say that they changed the rate of growth but the program spending still has a positive slope so spending is increasing.

63. In a vertical line there is no horizontal change. Since the denominator measures horizontal change and this is zero, the slope is a fraction with a 0 denominator. Such a fraction is undefined.

REVIEW

65. **Analysis:** A 60–pound mixture that contains black licorice bits and orange gumdrops will sell for $2 per pound.

Equation: Let x represent the number of pounds of black licorice bits. Then $60 - x$ will represent the number of pounds of orange gumdrops.

	p	• n	= v
Black licorice bits	1.90	x	$1.90x$
Orange gumdrops	2.20	$60 - x$	$2.20(60 - x)$
Mixture	2.00	60	120

Black licorice value plus orange gumdrop value is value of mixture.

$$1.90x + 2.20(60 - x) = 120$$

Solve:

$$1.90x + 2.20(60 - x) = 120$$

$$1.90x + 132 - 2.20x = 120 \quad \text{Distribute the multiplication by 2.20.}$$

$$-0.30x + 132 = 120 \quad \text{Combine like terms.}$$

$$-0.30x = -12 \quad \text{Subtract 132 from both sides.}$$

$$x = 40 \quad \text{Use a calculator to divide both sides by } -0.30.$$

Conclusion: The mixture would need 40 pounds of black licorice bits at $1.90 per pound and $60 - x = 20$ pounds of orange gumdrops at $2.20 per pound.

67. Reading from the graph, when $d = 10$, $t = 4$ hours.

Section 2.4 Writing Equations of Lines

VOCABULARY/ CONCEPTS

1. $y - y_1 = m(x - x_1)$

3. perpendicular

5. No, a point is also needed.

7. The line passes through $(-2, -3)$ and $(1, -1)$.

$$m = \frac{y_2 - y_1}{x_2 - x_1} = \frac{-1-(-3)}{1-(-2)} = \frac{2}{3}$$

$(x_1, y_1) = (-2, -3)$ and $(x_2, y_2) = (1, -1)$

$$y - y_1 = m(x - x_1)$$

$$y - (-3) = \frac{2}{3}[x - (-2)]$$

$$y + 3 = \frac{2}{3}(x + 2)$$

9. Compare $y = -\frac{2}{3}x + 1$ to $y = mx + b$. Slope $= m = -\frac{2}{3}$. The y–intercept $= (0, b) = (0, 1)$.

11. Rewrite each equation in slope-intercept form.

$$y - 2 = 3(x - 2)$$
$$y - 2 = 3x - 6$$
$$y = 3x - 4$$

$$y = 3x - 4$$

$$3x - y = 4$$
$$-y = -3x + 4$$
$$y = 3x - 4$$

Yes, all three lines are the same.

13. a. Compare $y = 2x$ to $y = mx + b$. The y–intercept $= (0, b) = (0, 0)$.
b. $x = -3$ is a vertical line and will not cross the y–axis, therefore there is no y–intercept.

15. No, the slopes are not negative reciprocals. Their product is not -1: $1(-0.9) = -0.9$.

NOTATION

17.
$$\begin{aligned} y+2 &= \frac{1}{3}(x+3) \\ y+2 &= \frac{1}{3}\mathbf{x}+1 \\ y+2-\mathbf{2} &= \frac{1}{3}x+1-\mathbf{2} \\ y &= \frac{1}{3}x-\mathbf{1} \end{aligned}$$

$m=\frac{1}{3}$, $b=\mathbf{-1}$

PRACTICE

19. $m=5$, $(x_1, y_1)=(0, 7)$
$$\begin{aligned} y-y_1 &= m(x-x_1) \\ y-(7) &= 5[x-(0)] \\ y-7 &= 5x \\ y &= 5x+7 \end{aligned}$$

21. $m=-3$, $(x_1, y_1)=(2, 0)$
$$\begin{aligned} y-y_1 &= m(x-x_1) \\ y-(0) &= -3[x-(2)] \\ y &= -3(x-2) \\ y &= -3x+6 \end{aligned}$$

23. $(x_1, y_1)=(0, 0)$ and $(x_2, y_2)=(4, 4)$
$$m=\frac{y_2-y_1}{x_2-x_1}=\frac{4-0}{4-0}=\frac{4}{4}=1$$
$$\begin{aligned} y-y_1 &= m(x-x_1) \\ y-(0) &= 1[x-(0)] \\ y &= x \end{aligned}$$

25. $(x_1, y_1)=(3, 4)$ and $(x_2, y_2)=(0, -3)$
$$m=\frac{y_2-y_1}{x_2-x_1}=\frac{-3-4}{0-3}=\frac{-7}{-3}=\frac{7}{3}$$
$$\begin{aligned} y-y_1 &= m(x-x_1) \\ y-(4) &= \frac{7}{3}[x-(3)] \\ y-4 &= \frac{7}{3}x-7 \\ y &= \frac{7}{3}x-3 \end{aligned}$$

27. $(x_1, y_1)=(-1, 3)$ and $(x_2, y_2)=(2, 5)$
$$m=\frac{y_2-y_1}{x_2-x_1}=\frac{5-3}{2-(-1)}=\frac{2}{3}$$
$$\begin{aligned} y-y_1 &= m(x-x_1) \\ y-(3) &= \frac{2}{3}[x-(-1)] \\ y &= \frac{2}{3}x+\frac{11}{3} \end{aligned}$$

29. $m=3$, $b=17$
$$\begin{aligned} y &= mx+b \\ y &= 3x+17 \end{aligned}$$

31. $m=-7$, passing through (7, 5)

Find b:
$$\begin{aligned} y &= mx+b \\ 5 &= -7(7)+b \\ 5 &= -49+b \\ 54 &= b \end{aligned}$$

The equation is: $y=-7x+54$

33. $m=0$, passing through (2, –4)

Find b:
$$\begin{aligned} y &= mx+b \\ -4 &= 0(2)+b \\ -4 &= b \end{aligned}$$

The equation is: $y=0x-4$ or $y=-4$

35. passing through (6, 8) and (2, 10)

$$m = \frac{y_2 - y_1}{x_2 - x_1} = \frac{10-8}{2-6} = \frac{2}{-4} = -\frac{1}{2}$$

Find b:

$$\begin{aligned} y &= mx + b \\ 8 &= -\frac{1}{2}(6) + b \\ 8 &= -3 + b \\ 11 &= b \end{aligned}$$

The equation is: $y = -\frac{1}{2}x + 11$

37.

$$\begin{aligned} 3x - 2y &= 8 \\ -2y &= -3x + 8 \\ \frac{-2y}{-2} &= \frac{-3x}{-2} + \frac{8}{-2} \\ y &= \frac{3}{2}x - 4 \end{aligned}$$

slope $= \frac{3}{2}$, y–intercept = (0, –4)

39.

$$\begin{aligned} -2(x + 3y) &= 5 \\ -2x - 6y &= 5 \\ -6y &= 2x + 5 \\ \frac{-6y}{-6} &= \frac{2x}{-6} + \frac{5}{-6} \\ y &= -\frac{1}{3}x - \frac{5}{6} \end{aligned}$$

slope $= -\frac{1}{3}$, y–intercept $= \left(0, -\frac{5}{6}\right)$

41. $y = x - 1$

slope = 1

y–intercept = (0, –1)

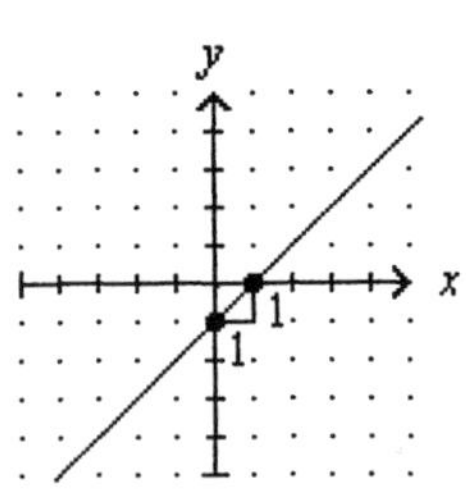

43. $y = \frac{2}{3}x + 2$

slope $= \frac{2}{3}$

y–int. = (0, 2)

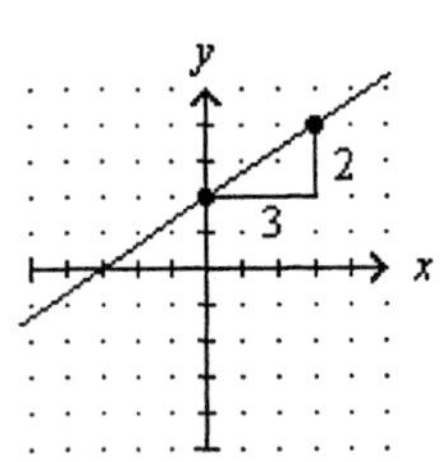

45.

$$\begin{aligned} 4y - 3 &= -3x - 11 \\ 4y &= -3x - 8 \\ \frac{4y}{4} &= \frac{-3x}{4} - \frac{8}{4} \\ y &= -\frac{3}{4}x - 2 \end{aligned}$$

slope $= -\frac{3}{4}$, y–intercept = (0, –2)

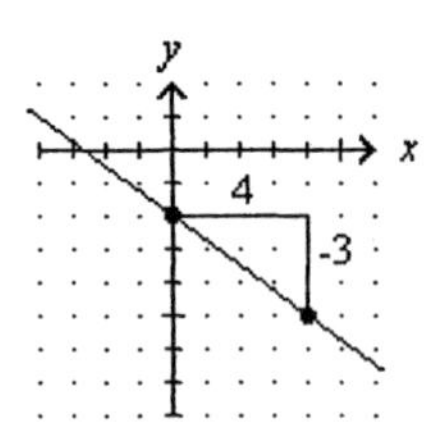

47. The slope of both lines is 3. The lines are parallel.

49. Solve $x + y = 2$ for y: $y = -x + 2$. The slopes are negative reciprocals, therefore the lines are perpendicular.

51. Solve $3x + 6y = 1$ for y. The slopes are opposites, but not negative reciprocals, therefore the lines are neither.

$$\begin{aligned} 3x + 6y &= 1 \\ 6y &= -3x + 1 \\ \frac{6y}{6} &= \frac{-3x}{6} + \frac{1}{6} \\ y &= -\frac{1}{2}x + \frac{1}{6} \end{aligned}$$

53. $y = 3$ is a horizontal line and $x = 4$ is a vertical line, therefore the lines are perpendicular.

55. Find the slope of $y = 4x - 7$.
New equation: $m = 4$, $(x_1, y_1) = (0, 0)$.

$$\begin{aligned} y - y_1 &= m(x - x_1) \\ y - (0) &= 4[x - (0)] \\ y &= 4x \end{aligned}$$

57. Find the slope of $4x - y = 7$.

$$\begin{aligned} 4x - y &= 7 \\ -y &= -4x + 7 \\ \frac{-y}{-1} &= \frac{-4x}{-1} + \frac{7}{-1} \\ y &= 4x - 7 \end{aligned}$$

New equation: $m = 4$, $(x_1, y_1) = (2, 5)$.

$$\begin{aligned} y - y_1 &= m(x - x_1) \\ y - (5) &= 4[x - (2)] \\ y - 5 &= 4(x - 2) \\ y - 5 &= 4x - 8 \\ y &= 4x - 3 \end{aligned}$$

59. Find the slope of $x = \frac{5}{4}y - 2$.

$$\begin{aligned} x &= \frac{5}{4}y - 2 \\ \frac{5}{4}y &= x + 2 \\ \frac{4}{5}\left(\frac{5y}{4}\right) &= \frac{4}{5}x + \frac{4}{5} \cdot 2 \\ y &= \frac{4}{5}x + \frac{8}{5} \end{aligned}$$

New equation: $m = \frac{4}{5}$, $(x_1, y_1) = (4, -2)$

$$\begin{aligned} y - y_1 &= m(x - x_1) \\ y - (-2) &= \frac{4}{5}[x - (4)] \\ y + \frac{10}{5} &= \frac{4}{5}x - \frac{16}{5} \\ y &= \frac{4}{5}x - \frac{26}{5} \end{aligned}$$

61. Find the slope of $y = 4x - 7$.
See exercise 55. Use the neg. reciprocal of $m = 4$.

New equation: $m = -\frac{1}{4}$, $(x_1, y_1) = (0, 0)$

$$\begin{aligned} y - y_1 &= m(x - x_1) \\ y - (0) &= -\frac{1}{4}[x - (0)] \\ y &= -\frac{1}{4}x \end{aligned}$$

63. Find the slope of $4x - y = 7$.
See exercise 57. Use the neg. reciprocal of $m = 4$.

New equation: $m = -\frac{1}{4}$, $(x_1, y_1) = (2, 5)$

$$\begin{aligned} y - y_1 &= m(x - x_1) \\ y - (5) &= -\frac{1}{4}[x - (2)] \\ y - 5 &= -\frac{1}{4}(x - 2) \\ y - \frac{10}{2} &= -\frac{1}{4}x + \frac{1}{2} \\ y &= -\frac{1}{4}x + \frac{11}{2} \end{aligned}$$

65. Find the slope of $x = \frac{5}{4}y - 2$.

See exercise 59. Use the neg. reciprocal of $m = \frac{4}{5}$.

New equation: $m = -\frac{5}{4}$, $(x_1, y_1) = (4, -2)$

$$\begin{aligned} y - y_1 &= m(x - x_1) \\ y - (-2) &= -\frac{5}{4}[x - (4)] \\ y + 2 &= -\frac{5}{4}x + 5 \\ y &= -\frac{5}{4}x + 3 \end{aligned}$$

APPLICATIONS

67. When the TV was new, its age x was 0, and its value y was \$1750. After 3 years, $x = 3$ and its value y is \$800. The points would be (0, 1750) and (3, 800).

$$m = \frac{y_2 - y_1}{x_2 - x_1} = \frac{800 - 1750}{3 - 0} = -\frac{950}{3},$$

y–intercept ($x = 0$) would be 1750.

Depreciation equation (using $y = mx + b$) is $y = -\frac{950}{3}x + 1750$

69. When the painting was originally sold, its age x was 0, and its value y was \$36,225,000. After 20 years, $x = 20$ and its value y is estimated to be \$72,450,000. The points would be (0, 36,225,000) and (20, 72,340,000).

$$m = \frac{y_2 - y_1}{x_2 - x_1} = \frac{72{,}450{,}000 - 36{,}225{,}000}{20 - 0} = \frac{36{,}225{,}000}{20} = 1{,}811{,}250,$$

y–intercept ($x = 0$) would be 36,225,000.

Appreciation equation (using $y = mx + b$) is $y = 1{,}811{,}250x + 36{,}225{,}000$

71. a. When the population p was 77,000, the number of burglaries B was 575. The rate of increase, or slope, was 1 for every 100 new residents. The point $(p, B) = (77{,}000, 575)$ and $m = \frac{1}{100}$.

$$\begin{aligned} y - y_1 &= m(x - x_1) \\ B - 575 &= \frac{1}{100}(p - 77{,}000) && \text{Substituting the known point.} \\ B - 575 &= \frac{1}{100}p - 770 \\ B &= \frac{1}{100}p - 195 && \text{The equation.} \end{aligned}$$

b. Substitute 110,000 for p in the equation. $B = \frac{1}{100}p - 195 = \frac{1}{100}(110{,}000) - 195 = 1100 - 195 = 905$

73. a. The points (t, E) are (1, 10) and (19, 1)

Find b: $y = mx + b$

$$\begin{aligned} 10 &= -\frac{1}{2}(1) + b \\ 10 &= -\frac{1}{2} + b \\ \frac{21}{2} &= b \end{aligned}$$

The equation is: $E = -\frac{1}{2}t + \frac{21}{2}$

b. The slope indicates that the number of errors is reduced by 1 for every 2 trials.

c. The x–intercept indicates that on the 21st trial, the rat should make no errors.

75. a. The points are (35, 16) and (15, –11)

$$m = \frac{y_2 - y_1}{x_2 - x_1} = \frac{-11 - 16}{15 - 35} = \frac{-27}{-20} = 1.35$$

Find b:

$$\begin{aligned} y &= mx + b \\ 16 &= 1.35(35) + b \\ 16 &= 47.25 + b \\ -31.25 &= b \end{aligned}$$

The equation is: $y = 1.35x - 31.25$

b. When the actual temperature is 0°F, the wind–chill temperature is about –31°F.

WRITING

77. First, determine the slope of the line connecting the two points. Then, use the point–slope form of the equation of a line with either point to determine the equation of the line.

79. m represents the slope of the line. b represents the y–coordinate of the y–intercept.

REVIEW

81. **Analysis:** An unknown amount was split into three investments. The interest received will be \$2037 per year.

Equation: Let x represent the equal amount of dollars invested in each investment. The investments earn 6%, 7%, and 8% interest.

	P •	r •	t =	I
First	x	0.06	1	$0.06x$
Second	x	0.07	1	$0.07x$
Third	x	0.08	1	$0.08x$

First interest	plus	second interest	plus	third interest	is	total interest.
$0.06x$	+	$0.07x$	+	$0.08x$	=	2037

Solve:

$$0.06x + 0.07x + 0.08x = 2037$$

$$0.21x = 2037 \quad \text{Combine like terms.}$$

$$x = 9700 \quad \text{Use a calculator to divide by 0.21.}$$

Conclusion: Each investment was \$9700 so the total investment was 3(9700) = \$29,100.

83.

$$3x = 2x$$

$$3x - 2x = 2x - 2x \quad \text{Subtract } 2x \text{ from both sides}$$

$$x = 0$$

Section 2.5 Introduction to Functions

VOCABULARY

1. function; one 3. range 5. x 7. y

CONCEPTS

9. $f(-1)$

11. a. Label the arrow going upwards with "Range" and the arrow going to the right with "Domain."

b. Domain is all real numbers greater than or equal to 0. Range is all real numbers greater than or equal to 2.

NOTATION

13.

$$f(x) = x^2 - 3x$$

$$f(-5) = (-5)^2 - 3(\ \mathbf{-5}\ )$$

$$= \mathbf{25} + 15$$

$$= 40$$

15. of

PRACTICE

17. yes 19. yes 21. yes 23. no

25.

$$f(x) = 3x$$

$$f(3) = 3(3) \quad \text{Replace } x \text{ with } 3$$

$$= 9$$

$$f(x) = 3x$$

$$f(-1) = 3(-1) \quad \text{Replace } x \text{ with } -1.$$

$$= -3$$

27.

$$f(x) = 2x - 3$$

$$f(3) = 2(3) - 3 \quad \text{Replace } x \text{ with } 3$$

$$= 3$$

$$f(x) = 2x - 3$$

$$f(-1) = 2(-1) - 3 \quad \text{Replace } x \text{ with } -1.$$

$$= -5$$

29.

$$f(x) = 7 + 5x$$

$$f(3) = 7 + 5(3) \quad \text{Replace } x \text{ with } 3$$

$$= 22$$

$$f(x) = 7 + 5x$$

$$f(-1) = 7 + 5(-1) \quad \text{Replace } x \text{ with } -1.$$

$$= 2$$

31.
$$\begin{aligned} f(x) &= 9 - 2x \\ f(3) &= 9 - 2(3) && \text{Replace } x \text{ with 3} \\ &= 3 \end{aligned} \qquad \begin{aligned} f(x) &= 9 - 2x \\ f(-1) &= 9 - 2(-1) && \text{Replace } x \text{ with } -1. \\ &= 11 \end{aligned}$$

33.
$$\begin{aligned} g(x) &= x^2 \\ g(2) &= (2)^2 && \text{Replace } x \text{ with 2.} \\ &= 4 \end{aligned} \qquad \begin{aligned} g(x) &= x^2 \\ g(3) &= (3)^2 && \text{Replace } x \text{ with 3.} \\ &= 9 \end{aligned}$$

35.
$$\begin{aligned} g(x) &= x^3 - 1 \\ g(2) &= (2)^3 - 1 && \text{Replace } x \text{ with 2.} \\ &= 7 \end{aligned} \qquad \begin{aligned} g(x) &= x^3 - 1 \\ g(3) &= (3)^3 - 1 && \text{Replace } x \text{ with 3.} \\ &= 26 \end{aligned}$$

37.
$$\begin{aligned} g(x) &= (x + 1)^2 \\ g(2) &= [(2) + 1]^2 && \text{Replace } x \text{ with 2.} \\ &= (3)^2 = 9 \end{aligned} \qquad \begin{aligned} g(x) &= (x + 1)^2 \\ g(3) &= [(3)+ 1]^2 && \text{Replace } x \text{ with 3.} \\ &= (4)^2 = 16 \end{aligned}$$

39.
$$\begin{aligned} g(x) &= 2x^2 - x \\ g(2) &= 2(2)^2 - (2) && \text{Replace } x \text{ with 2.} \\ &= 2(4) - 2 = 6 \end{aligned} \qquad \begin{aligned} g(x) &= 2x^2 - x \\ g(3) &= 2(3)^2 - (3) && \text{Replace } x \text{ with 3.} \\ &= 2(9) - 3 = 15 \end{aligned}$$

41.
$$\begin{aligned} h(x) &= |x| + 2 \\ h(2) &= |2| + 2 && \text{Replace } x \text{ with 2.} \\ &= 4 \end{aligned} \qquad \begin{aligned} h(x) &= |x| + 2 \\ h(-2) &= |-2| + 2 && \text{Replace } x \text{ with } -2. \\ &= 4 \end{aligned}$$

43.
$$\begin{aligned} h(x) &= x^2 - 2 \\ h(2) &= (2)^2 - 2 && \text{Replace } x \text{ with 2.} \\ &= 2 \end{aligned} \qquad \begin{aligned} h(x) &= x^2 - 2 \\ h(-2) &= (-2)^2 - 2 && \text{Replace } x \text{ with } -2 \\ &= 2 \end{aligned}$$

45.
$$\begin{aligned} h(x) &= \frac{1}{x+3} \\ h(2) &= \frac{1}{(2)+3} = \frac{1}{5} && \text{Replace } x \text{ with 2.} \end{aligned} \qquad \begin{aligned} h(x) &= \frac{1}{x+3} \\ h(-2) &= \frac{1}{(-2)+3} = 1 && \text{Replace } x \text{ with } -2 \end{aligned}$$

47.
$$\begin{aligned} h(x) &= \frac{x}{x-3} \\ h(2) &= \frac{(2)}{(2)-3} = -2 && \text{Replace } x \text{ with 2.} \end{aligned} \qquad \begin{aligned} h(x) &= \frac{x}{x-3} \\ h(-2) &= \frac{(-2)}{(-2)-3} = \frac{2}{5} && \text{Replace } x \text{ with } -2 \end{aligned}$$

49.

t	$f(t) = \|t - 2\|$	$f(t)$
-1.7	$f(-1.7) = \|(-1.7) - 2\|$ $= \|-3.7\|$	3.7
0.9	$f(0.9) = \|(0.9) - 2\|$ $= \|-1.1\|$	1.1
5.4	$f(5.4) = \|(5.4) - 2\|$ $= \|3.4\|$	3.4

51.

Input	$g(x) = x^3$	Output
$\frac{-3}{4}$	$g(x) = \left(\frac{-3}{4}\right)^3$	$-\frac{27}{64}$
$\frac{1}{6}$	$g(x) = \left(\frac{1}{6}\right)^3$	$\frac{1}{216}$
$\frac{5}{2}$	$g(x) = \left(\frac{5}{2}\right)^3$	$\frac{125}{8}$

53.
$$\begin{aligned} g(x) &= 2x \\ g(w) &= 2(w) && \text{Replace } x \text{ with } w \\ &= 2w \end{aligned} \qquad \begin{aligned} g(x) &= 2x \\ g(w + 1) &= 2(w + 1) && \text{Replace } x \text{ with } w + 1. \\ &= 2w + 2 \end{aligned}$$

55.
$$\begin{aligned} g(x) &= 3x - 5 \\ g(w) &= 3(w) - 5 && \text{Replace } x \text{ with } w \\ &= 3w - 5 \end{aligned} \qquad \begin{aligned} g(x) &= 3x - 5 \\ g(w + 1) &= 3(w + 1) - 5 && \text{Replace } x \text{ with } w + 1. \\ &= 3w + 3 - 5 \\ &= 3w - 2 \end{aligned}$$

57. The denominator $(6 - x)$ cannot be 0. So, D = the set of all real numbers except 6.

59. D = the set of all real numbers, because there are no real numbers that need to be excluded.

61. D = the set of all real numbers, because there are no real numbers that need to be excluded.

63. D = the set of all real numbers, because there are no real numbers that need to be excluded.

65. D = {−2, 4, 6}, the set of the given values of x. R = {3, 5, 7}, the set of the given values of y.

67. D = the set of all real numbers except 4, because 4 would make the denominator 0.
R = the set of all real numbers except 0, because a fraction with a numerator other than 0 cannot equal 0.

69. Not a function, because some vertical lines would intersect the graph more than once.

71. A function, because every vertical line would intersect the graph only one time.

73. A function, because every vertical line would intersect the graph only one time.

75. A function, because every vertical line would intersect the graph only one time.

77.

x	$f(x) = 2x - 1$	$f(x)$
−1	$f(-1) = 2(-1) - 1$	−3
0	$f(0) = 2(0) - 1$	−1
2	$f(2) = 2(2) - 1$	3

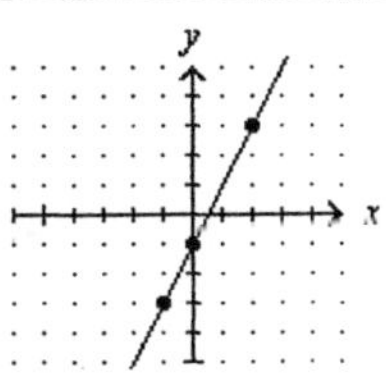

D = R = the set of all real numbers

79.

x	$f(x) = \frac{2}{3}x - 2$	$f(x)$
−3	$f(-3) = \frac{2}{3}(-3) - 2$	−4
0	$f(0) = \frac{2}{3}(0) - 2$	−2
3	$f(3) = \frac{2}{3}(3) - 2$	0

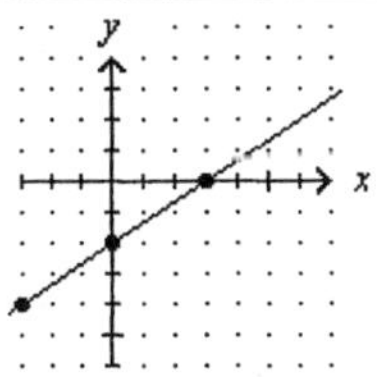

D = R = the set of all real numbers

81. No, linear functions do not have exponents.

83. Yes, this equation is in the form $y = mx + b$.

APPLICATIONS

85. $C(68) = \frac{5}{9}(68 - 32) = \frac{5}{9}(36) = 20°\text{C}$

$C(77) = \frac{5}{9}(77 - 32) = \frac{5}{9}(45) = 25°\text{C}$ Between 20°C and 25°C.

87. a. $I(b) = 1.75b - 50$ b. $I(110) = 1.75(110) - 50 = 192.50 - 50 = \$142.50$

89. a. (200, 25), (200, 90), (200, 105)   b. It does not pass the vertical line test.

91. a. $T(25{,}000) = 700 + 0.15(25{,}000 - 7{,}000) = 700 + 0.15(18{,}000) = 700 + 2700 = \$3{,}400$
This is the tax on an income of $25,000.

b. $T(a) = 3{,}910 + 0.25(a - 28400)$

93. a. $f(3) = -16(3)^2 + 256(3) = -16(9) + 768 = -144 + 768 = 624$ ft.
b. $f(16) = -16(16)^2 + 256(16) = -4096 + 4096 = 0$. The toy rocket strikes the ground 16 seconds after being shot.

WRITING

95. Answers will vary. Example: graph, equation, table, described in words

REVIEW

97. $-3\frac{3}{4} = -\left[3\left(\frac{4}{4}\right)+\frac{3}{4}\right] = -\left(\frac{12}{4}+\frac{3}{4}\right) = -\frac{15}{4}$

99. $0.333\ldots = \frac{1}{3}$ by recognition

Section 2.6 Graphs of Functions

VOCABULARY

1. squaring
3. absolute value
5. vertical
7. reflection

CONCEPTS

9. 4; left
11. 5: up

13. a. The line $y = 3x - 2$ crosses $y = 4$ at the point (2, 4), therefore $x = 2$.

 b. The line $y = 3x - 2$ crosses $y = -2$ at the point (0, –2), therefore $x = 0$.

15. The line crosses the x-axis at (–6, 0), therefore $x = -6$.

17. Bottom: Translated 3 units down, so $k = -3$.

 Middle: Translated 1 unit up, so $k = 1$.

 Top: Translated 2 units up, so $k = 2$.

19. a. $g(-2) = 3$ b. $g(1) = 0$ c. $g(2.5) = 1.5$

PRACTICE

21.

x	$f(x) = x^2 - 3$	$f(x)$
–2	$f(x) = (-2)^2 - 3$	1
–1	$f(x) = (-1)^2 - 3$	–2
0	$f(x) = (0)^2 - 3$	–3
1	$f(x) = (1)^2 - 3$	–2
2	$f(x) = (2)^2 - 3$	1

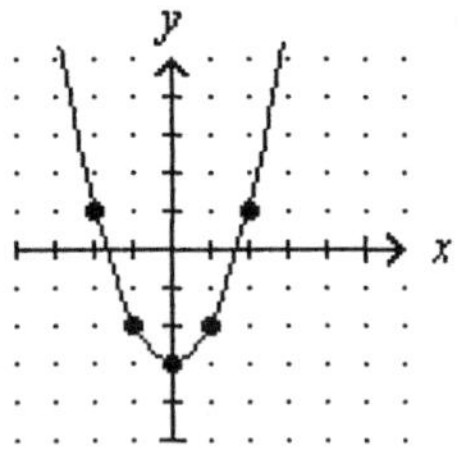

D = the set of real numbers. R = the set of all real numbers greater than or equal to –3.

23.

x	$f(x) = (x - 1)^3$	$f(x)$
–2	$f(x) = [(-2) - 1]^3$	–27
–1	$f(x) = [(-1) - 1]^3$	–8
0	$f(x) = [(0) - 1]^3$	–1
1	$f(x) = [(1) - 1]^3$	0
2	$f(x) = [(2) - 1]^3$	1

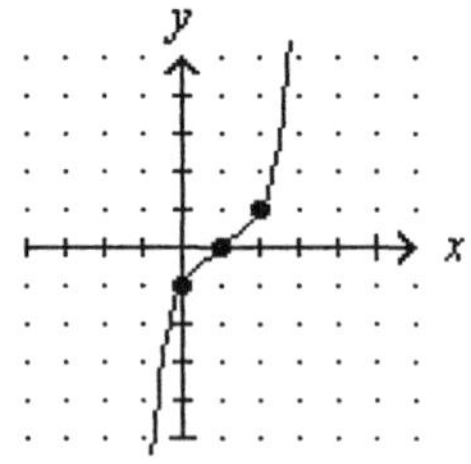

D = R = the set of all real numbers.

25.

x	$f(x) = \|x\| - 2$	$f(x)$
-2	$f(x) = \|-2\| - 2$	0
-1	$f(x) = \|-1\| - 2$	-1
0	$f(x) = \|0\| - 2$	-2
1	$f(x) = \|1\| - 2$	-1
2	$f(x) = \|2\| - 2$	0

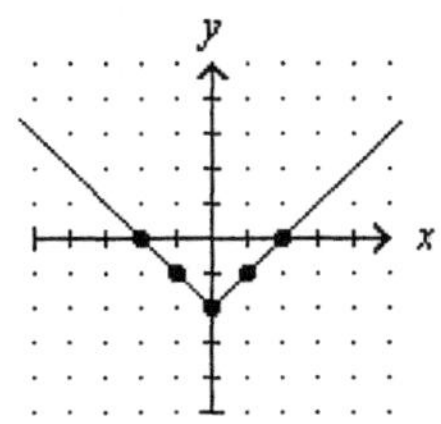

D = the set of real numbers. R = the set of all real numbers greater than or equal to -2.

27.

x	$f(x) = \|x - 1\|$	$f(x)$
-2	$f(x) = \|(-2) - 1\|$	3
-1	$f(x) = \|(-1) - 1\|$	2
0	$f(x) = \|(0) - 1\|$	1
1	$f(x) = \|(1) - 1\|$	0
2	$f(x) = \|(2) - 1\|$	1

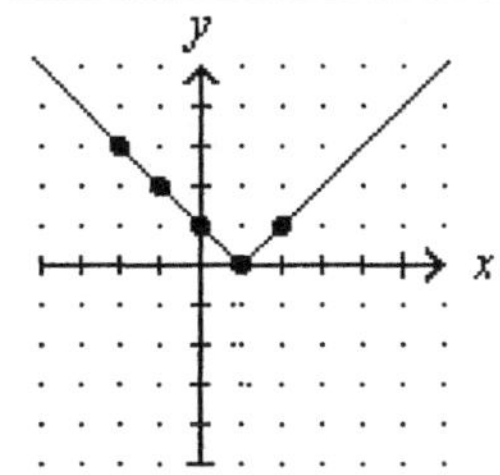

D = the set of real numbers. R = the set of all real numbers greater than or equal to 0.

29. Original graph of $f(x) = x^2 + 8$ with window settings of $[-4, 4]$ for x and $[-4, 4]$ for y.

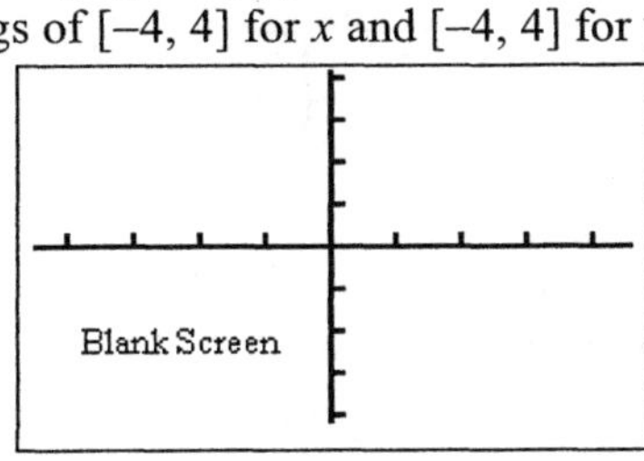

Viewing window of $[-4, 4]$ for x and $[-2, 12]$ for y.

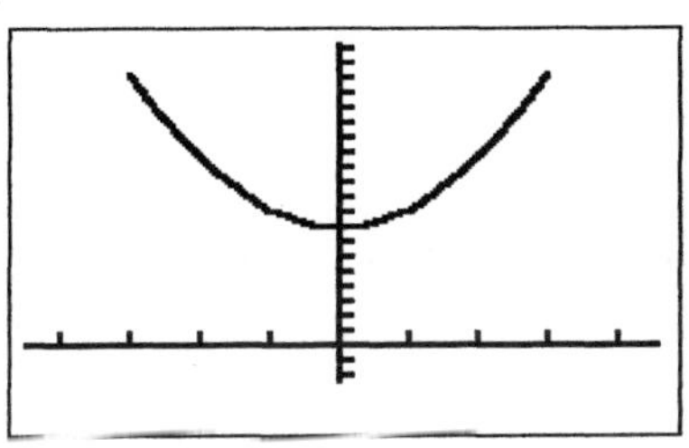

31. Original graph of $f(x) = |x + 5|$ with window settings of $[-4, 4]$ for x and $[-4, 4]$ for y.

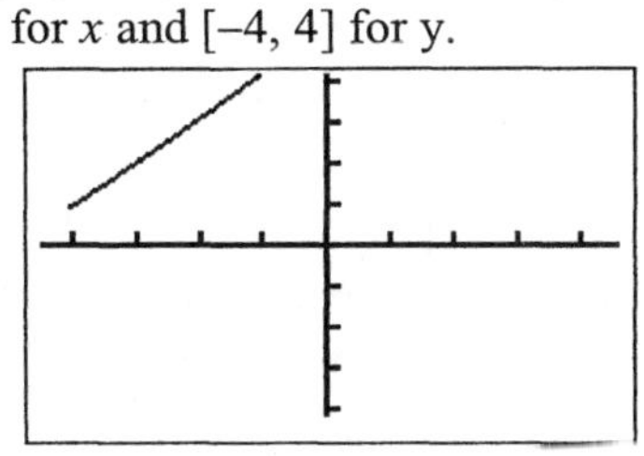

Viewing window of $[-10, 2]$ for x and $[-4, 4]$ for y.

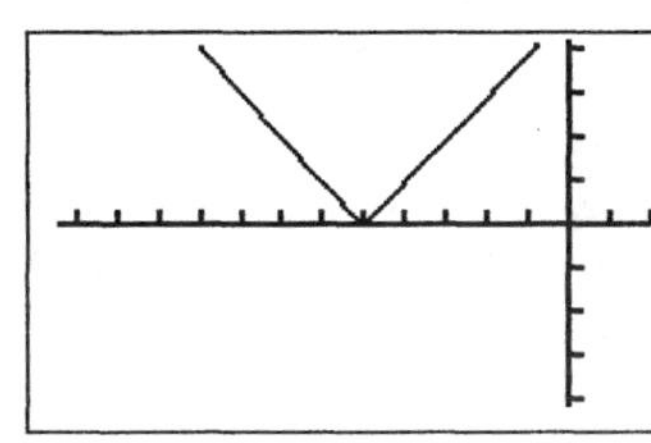

33. Original graph of $f(x) = (x - 6)^2$ with window settings of $[-4, 4]$ for x and $[-4, 4]$ for y.

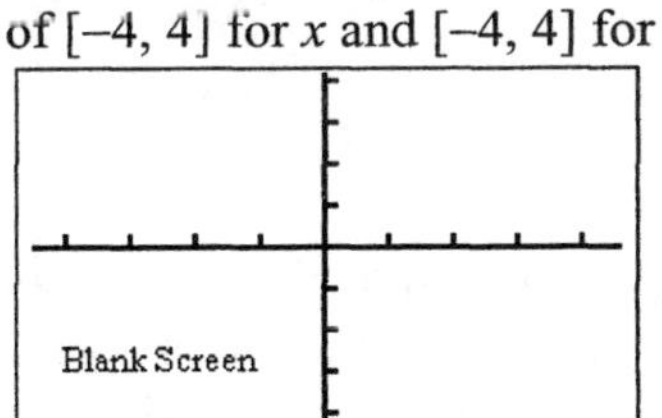

Viewing window $[2, 12]$ for x; $[-2, 10]$ for y.

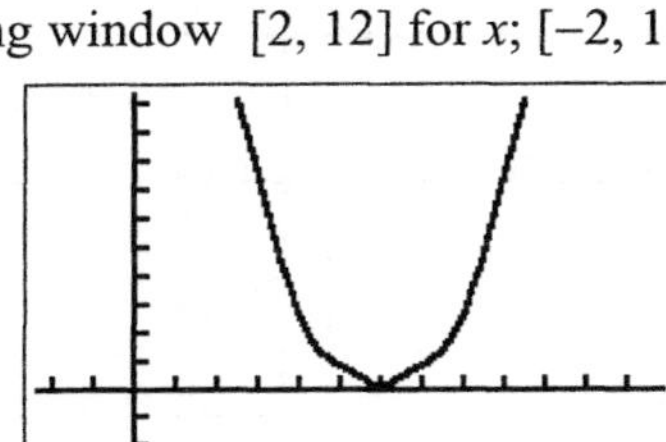

35. Original graph of $f(x) = x^3 + 8$ with window settings of $[-4, 4]$ for x and $[-4, 4]$ for y.

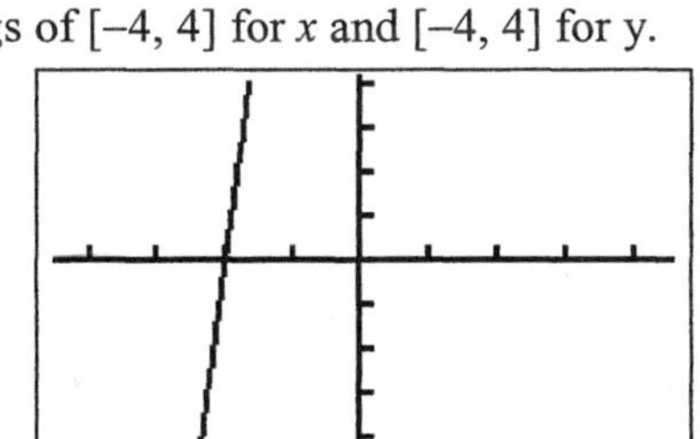

Viewing window $[-4, 4]$ for x; $[-2, 20]$ for y.

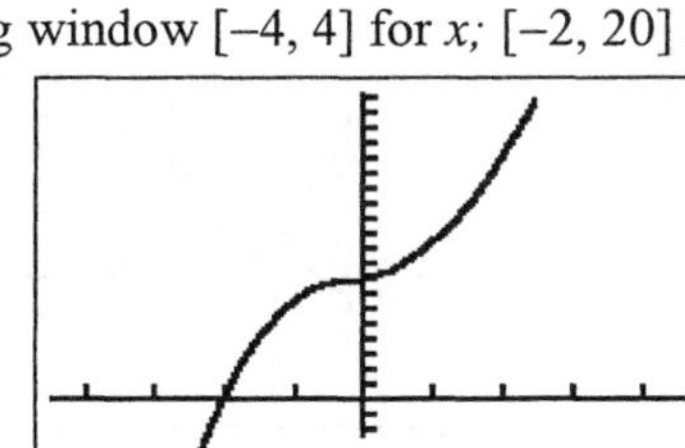

In exercises 37 through 48, translate or reflect the following associated functions.

$f(x) = x^2$	$f(x) = x^3$	$f(x) = \lvert x \rvert$
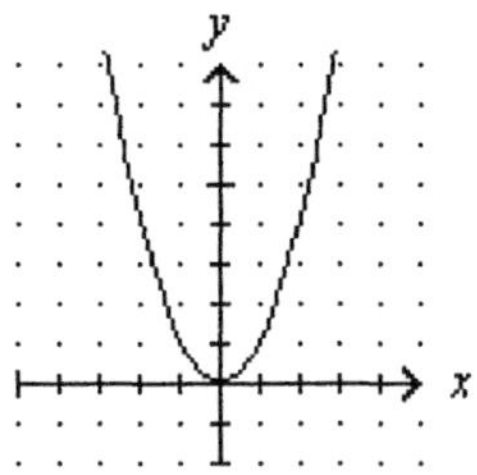	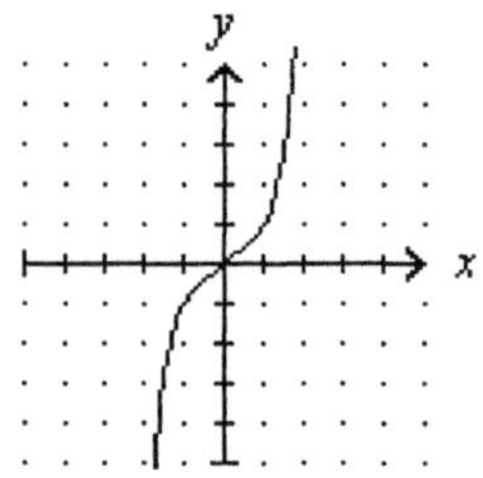	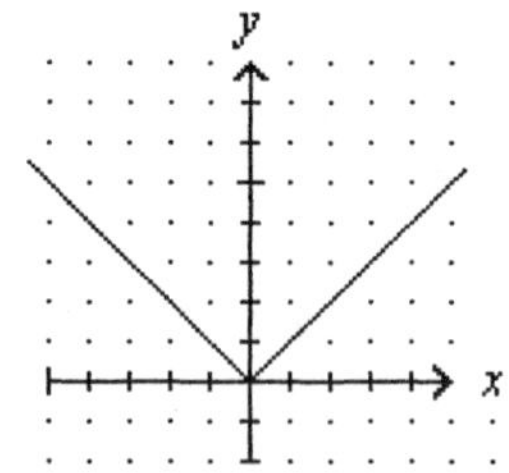

37. $f(x) = x^2 - 5$
Translate the associated function, $f(x) = x^2$, down 5 units.

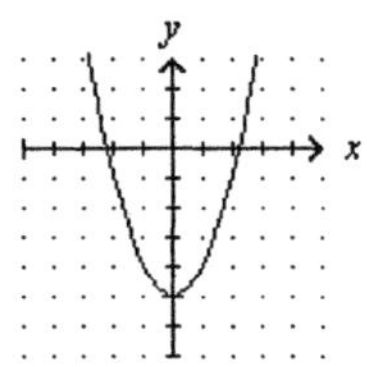

39. $f(x) = (x - 1)^3$
Translate the associated function, $f(x) = x^3$, right 1 unit.

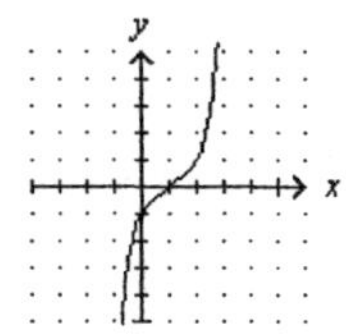

41. $f(x) = \lvert x - 2 \rvert - 1$
Translate the associated function, $f(x) = \lvert x \rvert$, right 2 units and down 1 unit.

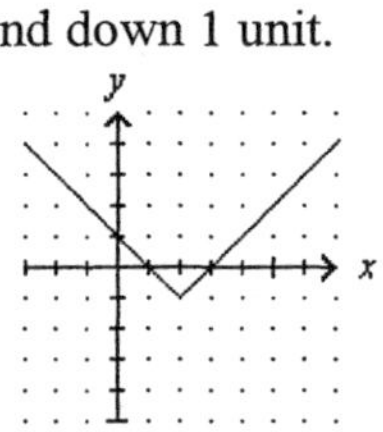

43. $f(x) = (x + 1)^3 - 2$
Translate the associated function, $f(x) = x^3$, left 1 unit and down 2 units.

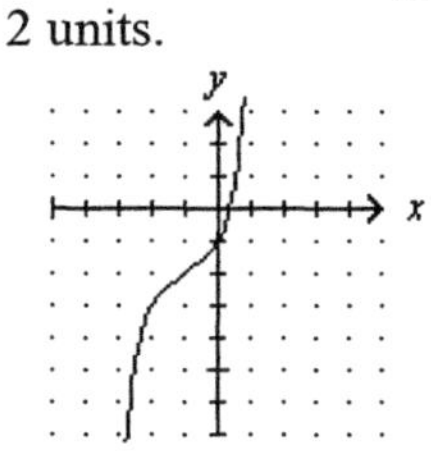

45. $f(x) = -x^3$
Reflect the associated function, $f(x) = x^3$, about the x–axis.

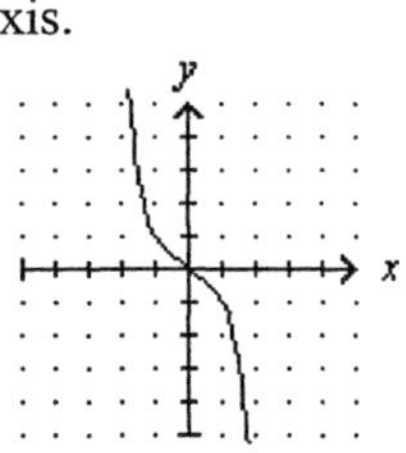

47. $f(x) = -x^2$
Reflect the associated function, $f(x) = x^2$, about the x–axis.

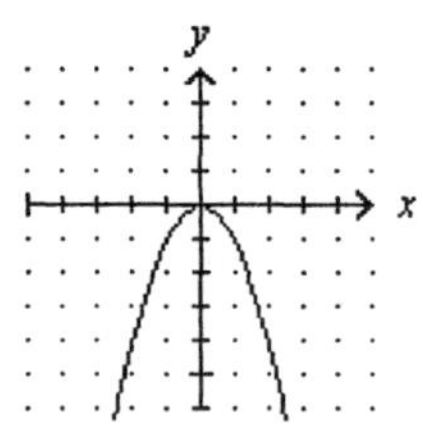

49. Graph $y_1 = 4(x - 1)$ and $y_2 = 3x$ on the same screen. The intersection of the two lines is at the point (4, 12). Therefore $x = 4$.

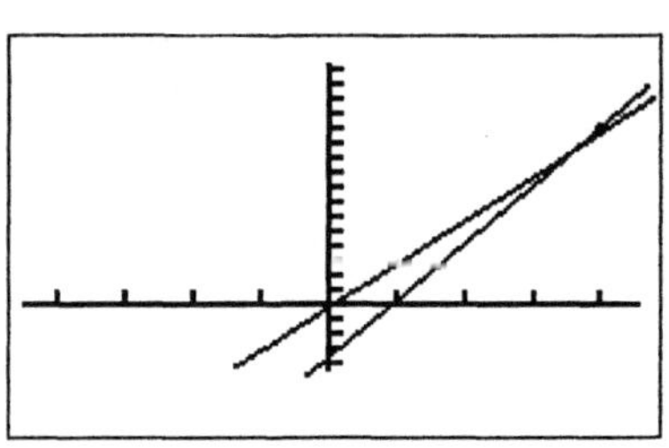

51. Graph $y_1 = 11x + 6(3 - x)$ and $y_2 = 3$ on the same screen. The intersection of the two lines is at the point (−3, 3). Therefore $x = -3$.

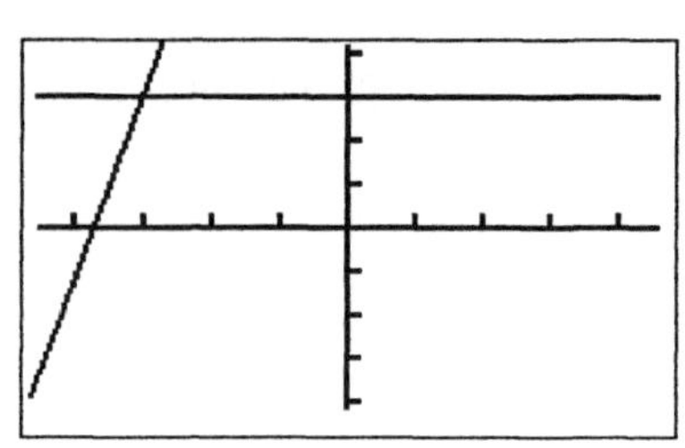

APPLICATIONS

53. The graph resembles the absolute value function, $f(x) = |x|$.

55. The shape is a parabola.

WRITING

57. Assign values to the independent variable, usually x, and use the equation to find values of the dependent variable, usually y. Plot the ordered pairs on a coordinate system and connect them with a smooth curve.

59. Graph $y = 2x + 6$ and see where the graph crosses the x–axis ($y = 0$).

REVIEW

61.
$$\begin{aligned} T - W &= ma \\ -W &= ma - T \\ W &= -ma + T \\ W &= T - ma \end{aligned}$$

63.
$$\begin{aligned} s &= \frac{1}{2}gt^2 + vt \\ 2s &= gt^2 + 2vt \\ 2s - 2vt &= gt^2 \\ g &= \frac{2s - 2vt}{t^2} \end{aligned}$$

65. **Analysis:** The budget was increased by 20% over the last year. Find the new budget for this year.

Equation: Let x represent the budget for this year.

Last year's budget	plus	20% increase	is	this year's budget.
4.5	+	20% • 4.5	=	x

Solve:
$$\begin{aligned} 4.5 + 0.20(4.5) &= x \\ 4.5 + 9 &= x \quad \text{Do the multiplication.} \\ 5.4 &= x \quad \text{Combine like terms.} \end{aligned}$$

Conclusion: This year's budget is \$5.4 million.

Chapter 2 — Key Concept: Functions

Functions

1. a. correspondence; input; range; one; domain
   b. dependent; independent

2. The number 6, which is a member of the *domain*, is *input* into the function machine. When the function or rule is applied, the *output* is 19. The number 19 is a member of the *range* of the function.

Four Ways to Represent a Function

3. $y = 2x + 3 = 2(-10) + 3 = -17$

4. $A = \pi r^2$

5. Find $t = 1.5$ in the table and read the associated value for h. The value of h would be 60 ft.

6. Find $x = 1$ and read the associated value for y. $y = 2$.

Function Notation

7. $f(x) = 2x + 3$; $f(0) = 2(0) + 3 = 3$

8. $A(r) = \pi r^2$

9. $h(4) = -16(4)^2 + 64(4) = -16(16) + 256 = -256 + 256 = 0$
   The projectile will strike the ground 4 seconds after being shot into the air.

10. Find $x = -3$ and read the associated value for $f(-3)$. $f(-3) = -2$.
    Find $x = 3$ and read the associated value for $f(3)$. $f(3) = 4$.

Chapter 2 Chapter Review

Section 2.1

1-5.

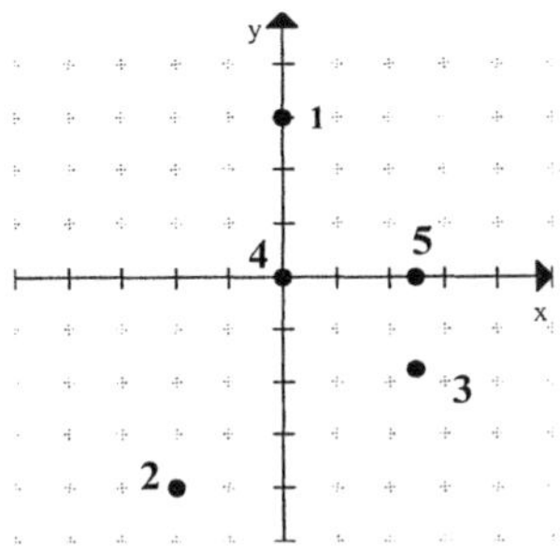

6.

x	y	(x, y)
−3	−5	(−3, −5)
−2	0	(−2, 0)
−1	3	(−1, 3)
0	4	(0, 4)
1	3	(1, 3)
2	0	(2, 0)
3	−5	(3, −5)

7. 1 ft below its normal level

8. decreased by 3 ft

9. from day 3 to beginning of day 4

10. $10 increments

11. $800

12. $P(x_1, y_1) = (8, -2)$, $Q(x_2, y_2) = (-4, 6)$

$$x_M = \frac{x_1 + x_2}{2} = \frac{8+(-4)}{2} = \frac{4}{2} = 2 \qquad y_M = \frac{y_1 + y_2}{2} = \frac{-2+6}{2} = \frac{4}{2} = 2$$

The midpoint is (2, 2)

13. $M(x_M, y_M) = M\left(-\frac{1}{2}, \frac{1}{2}\right)$, $P(x_1, y_1) = P(-3, 7)$. Find $Q(x_2, y_2)$.

$$x_m = \frac{x_1 + x_2}{2} \quad \text{and} \quad y_M = \frac{y_1 + y_2}{2}$$

$$-\frac{1}{2} = \frac{-3 + x_2}{2} \qquad \frac{1}{2} = \frac{7 + y_2}{2}$$

$$-1 = -3 + x_2 \qquad 1 = 7 + y_2$$

$$2 = x_2 \qquad -6 = y_2$$

$Q(x_2, y_2) = (2, -6)$

Section 2.2

14.

x	$y = 3x + 4$	y
−2	$y = 3(-2) + 4$	−2
0	$y = 3(0) + 4$	4
1	$y = 3(1) + 4$	7

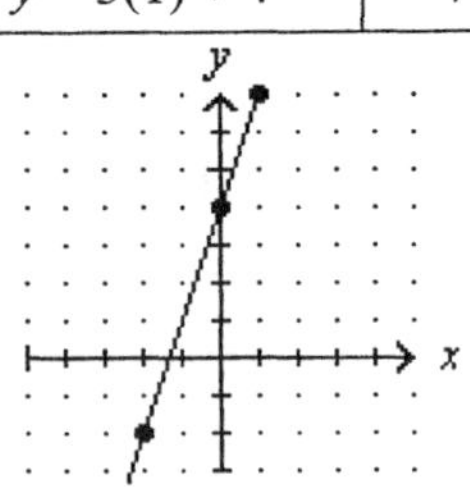

15.

x	$y = -\frac{1}{3}x - 1$	y
−3	$y = -\frac{1}{3}(-3) - 1$	0
0	$y = -\frac{1}{3}(0) - 1$	−1
3	$y = -\frac{1}{3}(3) - 1$	−2

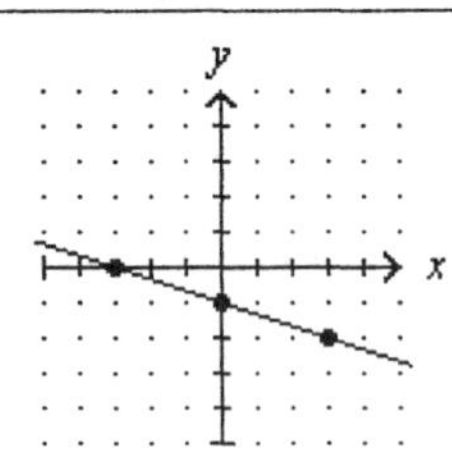

16.

x	$2x + y = 4$	y
0	$2(0) + y = 4$	4
2	$2x + (0) = 4$	0
3	$2(3) + y = 4$ $6 + y = 4$	−2

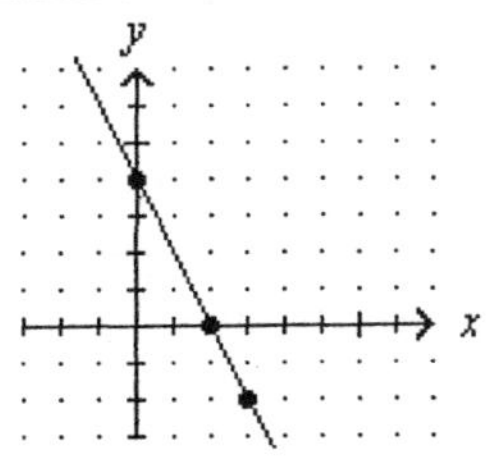

17.

x	$3x - 4y - 8 = 0$	y
0	$3(0) - 4y - 8 = 0$ $-4y = 8$	−2
$2\frac{2}{3}$	$3x - 4(0) - 8 = 0$ $3x = 8$	0
−4	$3(-4) - 4y - 8 = 0$ $-12 - 4y = 8$ $-4y = 20$	−5

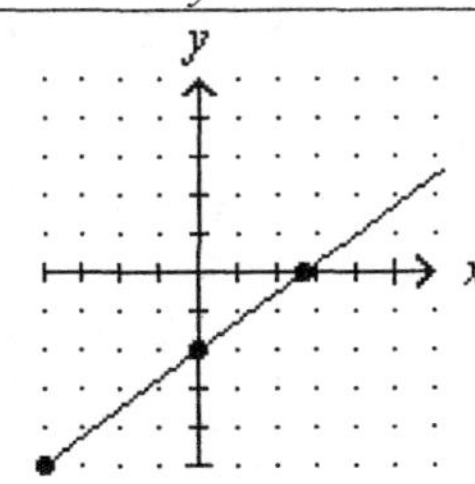

18.

x	$y = 4$	y
−1	The value of x	4
2	can be any	4
4	number	4

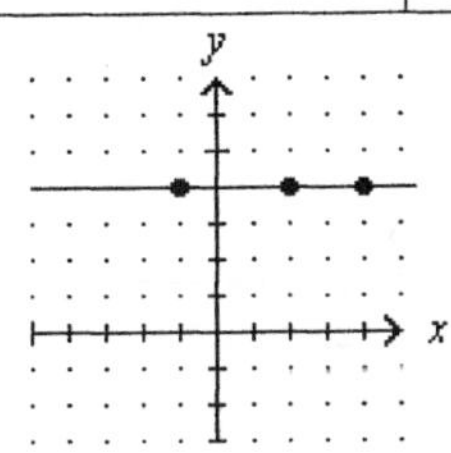

19.

x	$x = -2$	y
−2	The value of y	−2
−2	can be any	1
−2	number	4

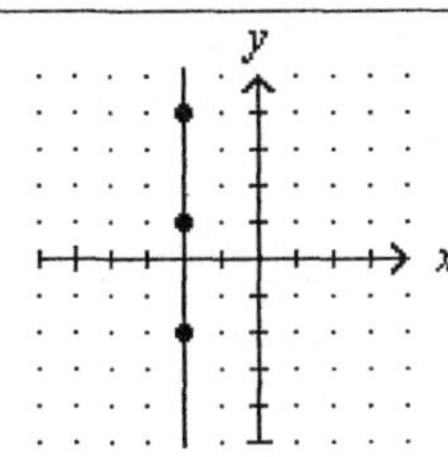

20.

x	$y = -3x$	y
−3	$y = -3(-3)$	**9**
0	$y = -3(0)$	**0**
3	$y = -3(3)$	**−9**

21.

x	$y = \frac{1}{2}x - \frac{5}{2}$	y
−3	$y = \frac{1}{2}(-3) - \frac{5}{2} = -\frac{8}{2}$	**−4**
0	$y = \frac{1}{2}(0) - \frac{5}{2} = -\frac{5}{2}$	$-\frac{5}{2}$
3	$y = \frac{1}{2}(3) - \frac{5}{2} = -\frac{2}{2}$	**−1**

Section 2.3

22. l_1; Calculate the slope by looking at two points on the line and substituting in slope formula.
Sample points: (−3, −2) and (2, 2). Substitute 2 for y_2, −2 for y_1, 2 for x_2, and −3 for x_1.

$$\text{slope} = \frac{\text{change in y}}{\text{change in x}} = \frac{y_2 - y_1}{x_2 - x_1} = \frac{2-(-2)}{2-(-3)} = \frac{4}{5}$$

l_2; Calculate the slope by looking at two points on the line and substituting in slope formula.
Sample points: (0, 4) and (5, −4). Substitute −4 for y_2, 4 for y_1, 5 for x_2, and 0 for x_1.

$$\text{slope} = \frac{\text{change in y}}{\text{change in x}} = \frac{y_2 - y_1}{x_2 - x_1} = \frac{-4-(4)}{5-(0)} = \frac{-8}{5} = -\frac{8}{5}$$

23. Rate of decrease $= \dfrac{\text{change in amount of market share}}{\text{change in time}} = \dfrac{48.6 - 21.2}{20} = \dfrac{27.4}{20} = 1.37\text{ \%/yr.}$

24. Points: $(x_1, y_1) = (2, 5)$ and $(x_2, y_2) = (5, 8)$

$$m = \frac{y_2 - y_1}{x_2 - x_1} = \frac{8-5}{5-2} = \frac{3}{3} = 1$$

25. Points: $(x_1, y_1) = (3, -2)$ and $(x_2, y_2) = (-6, 12)$

$$m = \frac{y_2 - y_1}{x_2 - x_1} = \frac{12-(-2)}{-6-3} = \frac{14}{-9} = -\frac{14}{9}$$

26. Points: $(x_1, y_1) = (-2, 4)$ and $(x_2, y_2) = (8, 4)$

$$m = \frac{y_2 - y_1}{x_2 - x_1} = \frac{4-4}{8-(-2)} = \frac{0}{10} = 0$$

27. Points: $(x_1, y_1) = (-5, -4)$ and $(x_2, y_2) = (-5, 8)$

$$m = \frac{y_2 - y_1}{x_2 - x_1} = \frac{8-(-4)}{-5-(-5)} = \frac{12}{0} \text{ is undefined}$$

28.

x	$y = \frac{2}{3}x + 18$	y
0	$y = \frac{2}{3}(0) + 18$	18
−27	$0 = \frac{2}{3}x + 18$ $0 = 2x + 54$ $-2x = 54$	0

$(x_1, y_1) = (0, 18)$ and $(x_2, y_2) = (-27, 0)$

$$m = \frac{y_2 - y_1}{x_2 - x_1} = \frac{0-18}{-27-0} = \frac{-18}{-27} = \frac{2}{3}$$

29.

x	$4x + 2y = 8$	y
0	$4(0) + 2y = 8$	4
2	$4x + 2(0) = 8$	0

$(x_1, y_1) = (0, 4)$ and $(x_2, y_2) = (2, 0)$

$$m = \frac{y_2 - y_1}{x_2 - x_1} = \frac{0-4}{2-0} = \frac{-4}{2} = -2$$

30.

x	$x = 10$	y
10	$x = 10$	6
10	$x = 10$	0

$(x_1, y_1) = (10, 6)$ and $(x_2, y_2) = (10, 0)$

$$m = \frac{y_2 - y_1}{x_2 - x_1} = \frac{0-6}{10-10} = \frac{-6}{0} \text{ is undefined}$$

31.

x	$y = 7$	y
0	$y = 7$	7
3	$y = 7$	7

$(x_1, y_1) = (0, 7)$ and $(x_2, y_2) = (3, 7)$

$$m = \frac{y_2 - y_1}{x_2 - x_1} = \frac{7-7}{3-0} = \frac{0}{3} = 0$$

32. The two slopes are negative reciprocals, therefore the lines are perpendicular

33. Since $\frac{1}{2} = 0.5$, the two slopes are equal. The lines are parallel.

Section 2.4

34. $m = 3$, passing through $(-8, 5)$

Find b:
$$y = mx + b$$
$$5 = 3(-8) + b$$
$$5 = -24 + b$$
$$29 = b$$

The equation is: $y = 3x + 29$

35. passing through $(-2, 4)$ and $(6, -9)$

$$m = \frac{y_2 - y_1}{x_2 - x_1} = \frac{-9-4}{6-(-2)} = \frac{-13}{8} = -\frac{13}{8}$$

Find b:
$$y = mx + b$$
$$4 = -\frac{13}{8}(-2) + b$$
$$\frac{16}{4} = \frac{13}{4} + b$$
$$\frac{3}{4} = b$$

The equation is: $y = -\frac{13}{8}x + \frac{3}{4}$

36. Find the slope of $3x - 2y = 7$

$$\begin{aligned} -2y &= -3x + 7 \\ \frac{-2y}{-2} &= \frac{-3x}{-2} + \frac{7}{-2} \\ y &= \frac{3}{2}x - \frac{7}{2} \end{aligned}$$

New equation: $m = \frac{3}{2}$, $(x_1, y_1) = (-3, -5)$

$$\begin{aligned} y - y_1 &= m(x - x_1) \\ y - (-5) &= \frac{3}{2}[x - (-3)] \\ 2(y + 5) &= 3(x + 3) \\ 2y + 10 &= 3x + 9 \\ 2y &= 3x - 1 \end{aligned}$$

The equation is: $3x - 2y = 1$

37. Find the slope of $3x - 2y = 7$
See part a. Use the negative reciprocal of $m = \frac{3}{2}$.

New equation: $m = -\frac{2}{3}$, $(x_1, y_1) = (-3, -5)$

$$\begin{aligned} y - y_1 &= m(x - x_1) \\ y - (-5) &= -\frac{2}{3}[x - (-3)] \\ 3(y + 5) &= -2(x + 3) \\ 3y + 15 &= -2x - 6 \\ 3y &= -2x - 21 \end{aligned}$$

The equation is: $2x + 3y = -21$

38.

$$\begin{aligned} 3x + 4y &= -12 \\ 4y &= -3x - 12 \\ \frac{4y}{4} &= -\frac{3x}{4} - \frac{12}{4} \\ y &= -\frac{3}{4}x - 3 \end{aligned}$$

slope $= -\frac{3}{4}$, y-intercept $= (0, -3)$

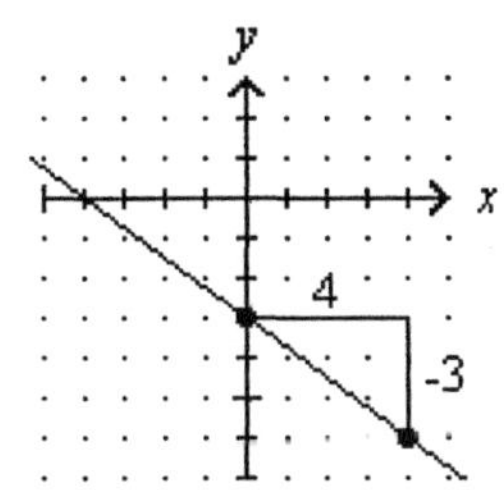

39. When the copy machine was new, its age x was 0, and its value y was \$8700. After 5 years, $x = 5$ and its value y is \$100. The points would be (0, 8700) and (5, 100).

$$m = \frac{y_2 - y_1}{x_2 - x_1} = \frac{100 - 8700}{5 - 0} = -\frac{8600}{5} = -1720,$$

y-intercept ($x = 0$) would be 8700.

Depreciation equation (using $y = mx + b$) is $y = -1720x + 8700$

Section 2.5

40. yes 41. yes 42. no 43. no

44. $f(x) = 3x + 2$; $f(-3) = 3(-3) + 2 = -9 + 2 = -7$ (Replace x with -3)

45. $g(x) = \dfrac{x^2 - 4x + 4}{2}$; $g(8) = \dfrac{(8)^2 - 4(8) + 4}{2} = \dfrac{64 - 32 + 4}{2} = \dfrac{36}{2} = 18$ (Replace x with 8)

46. $g(x) = \dfrac{x^2 - 4x + 4}{2}$; $g(-2) = \dfrac{(-2)^2 - 4(-2) + 4}{2} = \dfrac{4 + 8 + 4}{2} = \dfrac{16}{2} = 8$ (Replace $x = -2$)

47. $f(x) = 3x + 2$; $f(t) = 3(t) + 2 = 3t + 2$ (Replace x with t)

48. D = the set of all real numbers. R = the set of all real numbers.

49. D = the set of all real numbers. R = the set of all real numbers greater than or equal to 1.

50. D = the set of all real numbers except 2, because 2 would make the denominator 0.
R = the set of all real numbers except 0, because a fraction with a numerator other than 0 cannot equal 0.

51. D = the set of all real numbers. R = the set of all nonpositive real numbers.

52. A function, because every vertical line would intersect the graph only one time.

53. Not a function, because some vertical lines would intersect the graph more than once.

54. $C(x) = 105x + 175$, where x represents the number of hours taped.

55. $C(6) = 105(6) + 175 = 630 + 175 = \$805$

56. Yes, this equation is in the form $y = mx + b$.

57. No, linear functions do not have exponents.

Section 2.6

58. $f(x) = x^2 - 3$
Translate the associated function, $f(x) = x^2$, down 3 units.

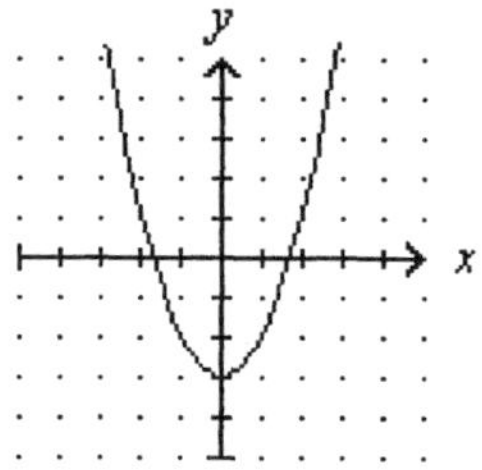

59. $f(x) = (x - 2)^3 + 1$
Translate the associated function $f(x) = x^3$, right 2 units and up 1 units.

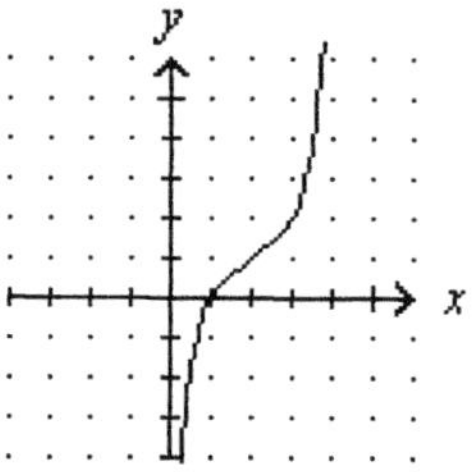

60 $f(x) = |x + 2|$
Translate the associated function, $f(x) = |x|$, left 2 units.
$g(x) = |x - 2|$
Translate the associated function, $f(x) = |x|$ right 2 units.
$h(x) = -|x|$
Reflect the associated function, $f(x) = |x|$ about the x–axis.

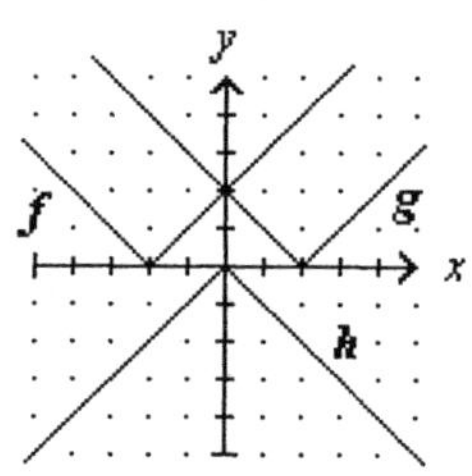

61. $f(-2) = -4$ b. $f(3) = 3$

62. The line $y = 2(2 - x) + x$ crosses $y = 0$ at the point (4, 0), therefore $x = 4$.

63. The line $y = 2(2 - x) + x$ crosses $y = 5$ at the point (–1, 5), therefore $x = -1$.

64. The line $y = 2(2 - x) + x$ crosses $y = x$ at the point (2, 2), therefore $x = 2$.

Chapter 2 Chapter Test

1. 240 ft
2. 1 sec and 7 sec
3. about 256 ft
4. 8 sec

5. $P(x_1, y_1) = (-2, -5)$, $Q(x_2, y_2) = (7, 8)$

$$x_M = \frac{x_1 + x_2}{2} = \frac{-2+(7)}{2} = \frac{5}{2} \qquad y_M = \frac{y_1 + y_2}{2} = \frac{-5+8}{2} = \frac{3}{2}$$

The midpoint is $\left(\frac{5}{2}, \frac{3}{2}\right)$

6.

x	$y = -\frac{2}{3}x - 2$	y
0	$y = 0 - 2$	-2
-3	$0 = -\frac{2}{3}x - 2$	0

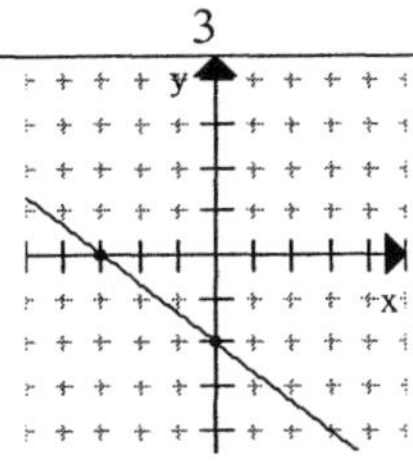

7.

x	$2x - 5y = 10$	y
0	$2(0) - 5y = 10$	-2
5	$2x - 5(0) = 10$	0

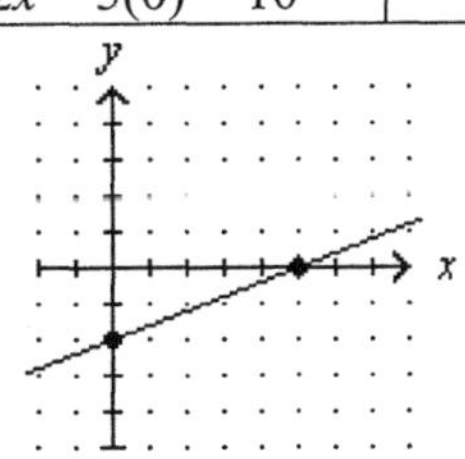

8.

x	$y = -2$	y
3	The value of x	-2
-3	can be any number	-2

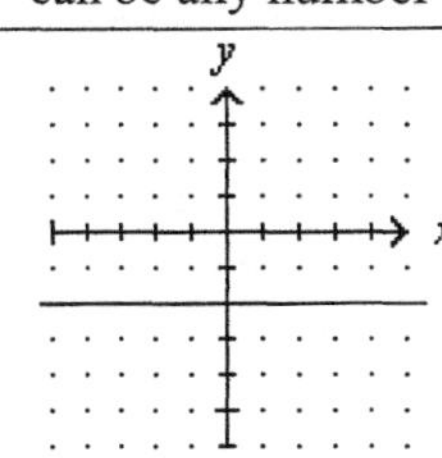

9. Calculate the slope by looking at two points on the line and substituting in slope formula. Sample points: $(-1, -2)$ and $(1, 4)$.

$$\text{slope} = \frac{\text{change in y}}{\text{change in x}} = \frac{y_2 - y_1}{x_2 - x_1} = \frac{4-(-2)}{1-(-1)} = \frac{6}{2} = 3$$

10. Rate of decrease $= \dfrac{\text{change in temperature}}{\text{change in time}} = \dfrac{20-35}{10-0} = \dfrac{-15}{10} = -1.5$ degree/hr.

11. Points: $(x_1, y_1) = (-2, 4)$ and $(x_2, y_2) = (6, 8)$

$$m = \frac{y_2 - y_1}{x_2 - x_1} = \frac{8-4}{6-(-2)} = \frac{4}{8} = \frac{1}{2}$$

12. Find the slope of $2x - 3y = 8$ (solve for y)

$$y = \frac{2}{3}x - \frac{8}{3} \qquad m = \frac{2}{3}$$

13. Since slope $= \dfrac{\text{change in y}}{\text{change in x}}$ and the change in x is always 0, then slope is undefined.

14. Since slope $= \dfrac{\text{change in y}}{\text{change in x}}$ and the change is y is always 0, then slope equals 0.

15. a. $m = \frac{2}{3}$, passing through (4, –5)

Find b:

$$\begin{aligned} y &= mx + b \\ -5 &= \tfrac{2}{3}(4) + b \\ -\frac{15}{3} &= \frac{8}{3} + b \\ -\frac{23}{3} &= b \end{aligned}$$

The equation is: $y = \frac{2}{3}x - \frac{23}{3}$

16. passing through (–2, 6) and (–4, –10)

$$m = \frac{y_2 - y_1}{x_2 - x_1} = \frac{-10-6}{-4-(-2)} = \frac{-16}{-2} = 8$$

Find b:

$$\begin{aligned} y &= mx + b \\ 6 &= 8(-2) + b \\ 6 &= -16 + b \\ 22 &= b \end{aligned}$$

The equation is: $y = 8x + 22$ or $8x - y = -22$

17.

$$\begin{aligned} -2x + 6 &= 6y + 15 \\ -6y &= 2x + 9 \\ \frac{-6y}{-6} &= \frac{2x}{-6} + \frac{9}{-6} \\ y &= -\frac{1}{3}x - \frac{3}{2} \end{aligned}$$

slope $= -\frac{1}{3}$, y-intercept $= \left(0, -\frac{3}{2}\right)$

18. Solve $4x - y = 12$ for y: $y = 4x - 12$. The slopes are reciprocals but not negative reciprocals, therefore the lines are neither parallel nor perpendicular.

19. Find the slope of $y = \frac{3}{2}x - 7$ New equation: $m = \frac{3}{2}$, $(x_1, y_1) = (0, 0)$

$$\begin{aligned} y - y_1 &= m(x - x_1) \\ y - (0) &= \frac{3}{2}[x - (0)] \\ y &= \frac{3}{2}x \end{aligned}$$

20. a. If the x-coordinate represents years and the v -coordinate represents value, the table gives the points (0,4000) and (6,400). The slope through these points is $m = \frac{4000-400}{0-6} = -600$. The v-intercept is 4,000. Using the formula $v = y = mx + b$ gives the equation $v = -600x + 4000$.

b. The v-intercept (0,4000) gives the value ($4,000) of the copier when it is new.

21. No, because there would be two values of y for one value of x.

22. Yes, the table defines a function because for each x value there is no more than one y value.

23. D = the set of all real numbers. R = the set of all nonnegative real numbers.

24. D = the set of all real numbers. R = the set of all real numbers.

25. $f(x) = 3x + 1$; $f(3) = 3(3) + 1 = 9 + 1 = 10$ Replace x with 3.

26. $g(x) = x^2 - 2x - 1$; $g(0) = (0)^2 - 2(0) - 1 = -1$ Replace x with 0.

27. $f(x) = 3x + 1$; $f(\frac{2}{3}) = 3(\frac{2}{3}) + 1 = 2 + 1 = 3$ Replace x with $\frac{2}{3}$.

28. $(x) = x^2 - 2x - 1$; $g(r) = (r)^2 - 2(r) - 1 = r^2 - 2r - 1$ Replace x with r.

29. A function, because every vertical line would intersect the graph only one time.

30. Not a function, because some vertical lines would intersect the graph more than once.

31. $f(x) = x^2 + 3$
Translate the associated function, $f(x) = x^2$, up 3 units.

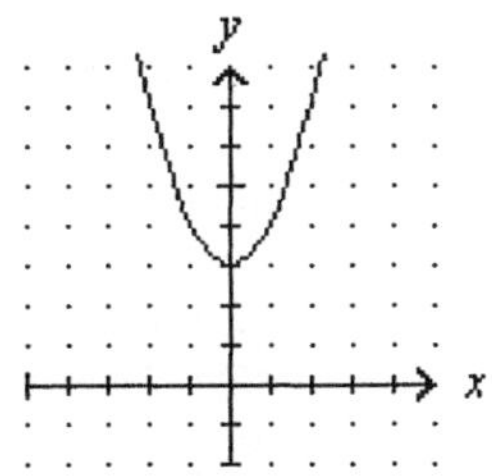

32. $g(x) = -|x + 2|$
Translate the associated function $f(x) = |x|$, left 2 units and reflect about the x–axis.

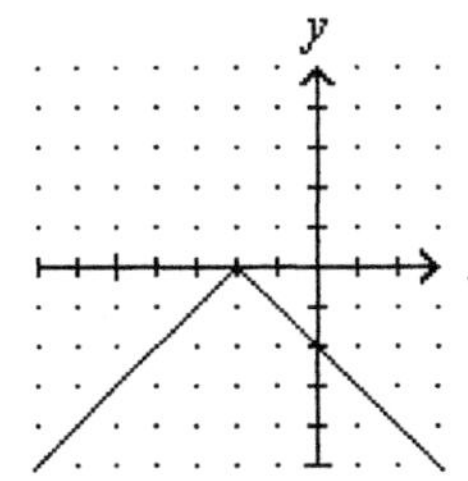

33. Find the x–coordinate of the x–intercept of the graph of $y = 3(x - 2) - 2(-2 + x)$; The solution is $x = 2$.

34. Find the x–coordinate of the point of intersection of the graph of $y = 3(x - 2) - 2(-2 + x)$ and $y = 1$. The solution is $x = 3$.

35. Words, equations, tables, and graphs.

36. Because a circle would fail the vertical line test.

Chapters 1–2 Cumulative Review Exercises

1. 1, 2, 6, 7

2. 0, 1, 2, 6, 7

3. $-2, 0, 1, 2, \frac{13}{12}, 6, 7$

4. $\sqrt{5}, \pi$

5. -2

6. $-2, 0, 1, 2, \frac{13}{12}, 6, 7, \sqrt{5}, \pi$

7. 2, 7

8. 6

9. $-2, 0, 2, 6$

10. 1, 7

11. $-|5| + |-3| = -5 + 3 = -2$

12. $\frac{|-5|+|-3|}{-|4|} = \frac{5+3}{-4} = \frac{8}{-4} = -2$

13. $2 + 4 \bullet 5 = 2 + 20 = 22$

14. $\frac{8-4}{2-4} = \frac{4}{-2} = -2$

15. $-\frac{16}{5} \div \left(-\frac{10}{3}\right) = -\frac{16}{5} \cdot \left(-\frac{3}{10}\right) = \frac{\overset{1}{\cancel{2}} \cdot 8 \cdot 3}{5 \cdot \underset{1}{\cancel{2}} \cdot 5} = \frac{24}{25}$

16. $\frac{(9-8)^4 + 21}{3^3 - (\sqrt{16})^2} = \frac{(1)^4 + 21}{27 - (4)^2} = \frac{1+21}{27-16} = \frac{22}{11} = 2$

17. $-x - 2y = -(2) - 2(-3) = -2 + 6 = 4$ Substitute 2 for x and -3 for y and evaluate.

18. $\frac{x^2 - y^2}{2x + y} = \frac{(2)^2 - (-3)^2}{2(2) + (-3)} = \frac{4-9}{4-3} = \frac{-5}{1} = -5$ Substitute 2 for x and -3 for y and evaluate.

19. associative property of addition

20. distributive property

21. commutative property of addition

22. associative property of multiplication

23. $12y - 17y = (12 - 17)y = -5y$

24. $-7s(-4t)(-1) = (-7 \cdot -4 \cdot -1)(s \cdot t) = -28st$

25. $3x^2 + 2x^2 - 5x^2 = (3 + 2 - 5)x^2 = 0 \cdot x^2 = 0$

26. $-(4 + z) + 2z = -4 - z + 2z = z - 4$

27.
$$\begin{aligned} 2x - 5 &= 11 \\ 2x &= 16 \\ x &= 8 \end{aligned}$$

28.
$$\begin{aligned} \frac{2x-6}{3} &= x+7 \\ 2x - 6 &= 3(x+7) \\ 2x - 6 &= 3x + 21 \\ -x &= 27; \quad x = -27 \end{aligned}$$

29.
$$\begin{aligned} 4(y-3)+4 &= -3(y+5) \\ 4y - 12 + 4 &= -3y - 15 \\ 4y - 8 &= -3y - 15 \\ 7y &= -7 \\ y &= -1 \end{aligned}$$

30.
$$\begin{aligned} 2x - \frac{3(x-2)}{2} &= 7 - \frac{x-3}{3} \\ 6(2x) - 6\left(\frac{3(x-2)}{2}\right) &= 6(7) - 6\left(\frac{x-3}{3}\right) \\ 12x - 3(3x-6) &= 42 - 2(x-3) \\ 12x - 9x + 18 &= 42 - 2x + 6 \\ 3x + 18 &= 48 - 2x \\ 5x &= 30 \\ x &= 6 \end{aligned}$$

31.
$$\begin{aligned} -3 &= -\frac{9}{8}s \\ \left(-\frac{8}{9}\right)(-3) &= \left(-\frac{8}{9}\right)\left(-\frac{9}{8}s\right) \\ \frac{8}{3} &= s \end{aligned}$$

32.
$$\begin{aligned} 0.04(24) + 0.02x &= 0.04(12 + x) \\ 0.96 + 0.02x &= 0.48 + 0.04x \\ -0.02x &= -0.48 \\ x &= 24 \end{aligned}$$

33.
$$\begin{aligned} S &= \frac{n(a+l)}{2} \\ 2S &= na + ln \\ 2S + ln &= na \\ \frac{2S - ln}{n} &= a \end{aligned}$$

34.
$$\begin{aligned} A &= \frac{1}{2}h(b_1 + b_2) \\ 2A &= h(b_1 + b_2) \\ \frac{2A}{b_1 + b_2} &= h \end{aligned}$$

35. **Analysis:** A total of \$20,000 is split between two investments. The interest received will be \$1260 per year.

Equation: Let x represent the number of dollars invested at 6%. Then \$(20,000 – x) is the number of dollars invested at 7%.

	P •	r •	t =	I
First investment	x	0.06	1	$0.06x$
Second investment	$20{,}000 - x$	0.07	1	$0.07(20{,}000 - x)$

Interest earned at 6%	plus	interest earned at 7%	is	total interest earned.
$0.06x$	+	$0.07(20{,}000 - x)$	=	1260

Solve:

$0.06x + 0.07(20{,}000 - x) = 1260$		
$0.06x + 1400 - 0.07x = 1260$		Distribute the multiplication by 0.07.
$-0.01x + 1400 = 1260$		Combine like terms.
$-0.01x = -140$		Subtract 1400 from both sides.
$x = 14{,}000$		Use a calculator to divide both sides by –0.01.

Conclusion: \$14,000 was invested at 6%.

36. **Analysis:** John drove the same distance to the city and when he returned from the city.

Equation: Let r represent John's rate when he drove to the city.

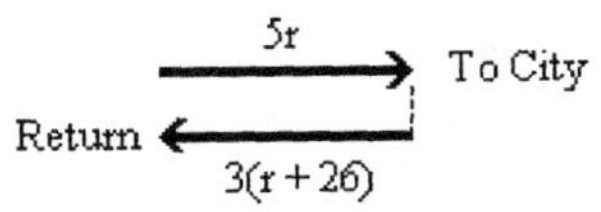

	r	• t	= d
To the city	r	5	$5r$
Return trip	$r+26$	3	$3(r+26)$

To distance equals return distance.

$$5r = 3(r+26)$$

Solve:

$$\begin{aligned} 5r &= 3(r+26) \\ 5r &= 3r+78 && \text{Distribute the multiplication by 3.} \\ 2r &= 78 && \text{Subtract } 3r \text{ from both sides.} \\ r &= 39 && \text{Divide both sides by 2.} \end{aligned}$$

Conclusion: John drove $r = 39$ mph going to the city and $r + 26 = 65$ mph returning.

37.

x	$2x - 3y = 6$	y
0	$2(0) - 3y = 6$	−2
3	$2x - 3(0) = 6$	0

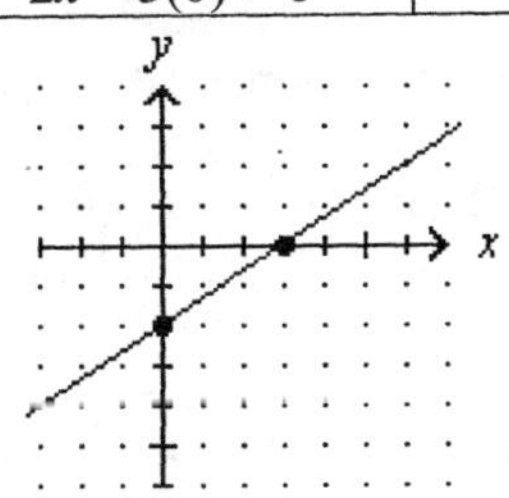

38. Calculate the slope by looking at two points on the line and substituting in slope formula.
Sample points: (−2, 1) and (4, −4).

$$\text{slope} = \frac{\text{change in y}}{\text{change in x}} = \frac{y_2 - y_1}{x_2 - x_1} = \frac{-4-(1)}{4-(-2)} = \frac{-5}{6} = -\frac{5}{6}$$

39. passing through (−2, 5) and (8, −9)

$$m = \frac{y_2 - y_1}{x_2 - x_1} = \frac{-9-5}{8-(-2)} = \frac{-14}{10} = -\frac{7}{5}$$

Find b:

$$\begin{aligned} y &= mx + b \\ 5 &= -\frac{7}{5}(-2) + b \\ \frac{25}{5} &= \frac{14}{5} + b \\ \frac{11}{5} &= b \end{aligned}$$

The equation is: $y = -\frac{7}{5}x + \frac{11}{5}$

40. Find the slope of $3x + y = 8$

$$y = -3x + 8$$

New equation: $m = -3$, $(x_1, y_1) = (-2, 3)$

$$\begin{aligned} y - y_1 &= m(x - x_1) \\ y - (3) &= -3[x - (-2)] \\ y - 3 &= -3(x + 2) \\ y - 3 &= -3x - 6 \end{aligned}$$

The equation is: $y = -3x - 3$

41. $f(x) = 3x^2 + 2;$ $f(-1) = 3(-1)^2 + 2 = 3 + 2 = 5$ Replace x with -1.

42. $g(x) = -2x - 1;$ $g(0) = -2(0) - 1 = -1$ Replace x with 0.

43. $g(x) = -2x - 1;$ $g(-2) = -2(-2) - 1 = 4 - 1 = 3$ Replace x with -2.

44. $f(x) = 3x^2 + 2;$ $f(-r) = 3(-r)^2 + 2 = 3r^2 + 2$ Replace x with $-r$.

45.

x	$f(x) = -x^2 + 1$	$f(x)$
-2	$f(-2) = -(-2)^2 + 1$	-3
0	$f(0) = -(0)^2 + 1$	1
2	$f(2) = -(2)^2 + 1$	-3

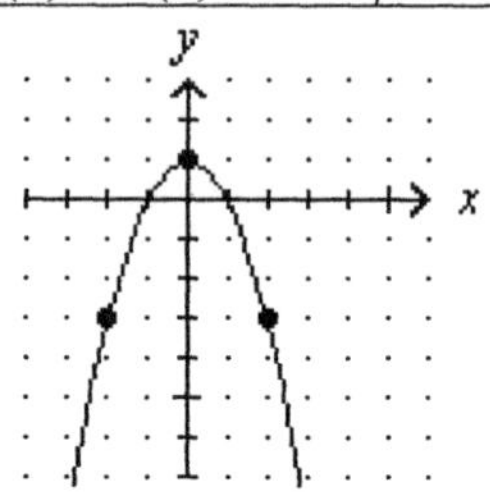

A function.
D = the set of all real numbers,
R = the set of all real numbers less than or equal to 1.

46.

x	$f(x) = \|x - 3\|$	$f(x)$
0	$f(-1) = \|-1 - 3\|$	3
3	$f(0) = \|0 - 3\|$	0
5	$f(2) = \|2 - 3\|$	2

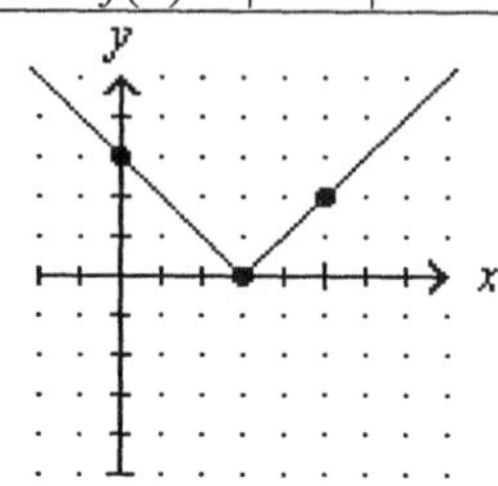

A function.
D = the set of all real numbers,
R = the set of all nonnegative real numbers

47. The points given by (height, range) are represented by the points A(2,20), B(3,40), and C(6,60). Since the slope between points A and B equals 20 and the slope between points B and C equals $\frac{20}{3}$, the relationship is not linear.

48. a. Find the x–coordinate of the x–intercept of the graph of $y = -7 - 5(x - 1) + 4x$. The solution is -2.

b. Find the x–coordinate of the point of intersection of the graph of $y = -7 - 5(x - 1) + 4x$ and $y = -4$. The solution is 2.

Giving up is not falling down;
giving up is staying there once you've hit the ground.

Chapter 3 — Systems of Equations

Study Tip
Is College a full time JOB?

Write the number of credits hours for which you are registered:___________
(These are the number of hours you should be in class per week)
Multiply this number by 2: ___________
(These are the numbers of hours you should study outside of class)
Add these two numbers together: ___________

Example: 16 credits + 16 X 2 study hours = 48 hours

You should also include *extra* study time for a special project or test.

You decide: Is College a full time JOB?

Section 3.1 — Solving Systems by Graphing

VOCABULARY

1. system
3. inconsistent
5. dependent

CONCEPTS

7. a. true b. false c. true d. true

9. a.

x	y	(x, y)
–4	0	(–4, 0)
0	**2**	(0, 2)
2	**3**	(2, 3)

b. (–4, 0); (0, 2)

10. a.

x	y	(x, y)
–1	–2	(–1, 2)
0	1	(0, 1)
2	7	(2, 7)

b. $m = 3$; (0, 1) Compare $y = 3x + 1$ to $y = mx + b$.

11. Answers will vary. Possible solutions follow:

a. If $x = 2$ and $y = 3$, then $x + y = 5$ and $x - y = -1$ gives a system with one solution. $\begin{cases} x + y = 5 \\ x - y = -1 \end{cases}$

b. To make a system with infinite solutions, the equations must be equivalent.
Multiply either equation from a. by some number. For example: $x + y = 5$ by 2. $\begin{cases} x + y = 5 \\ 2x + 2y = 10 \end{cases}$

c. To make a system that has no solutions, the equations must have the same slope but different y–intercepts.
For example change $x - y = 1$ to $x + y = 4$. The system then becomes $\begin{cases} x + y = 5 \\ x + y = 4 \end{cases}$

NOTATION

13. brace {

PRACTICE

15. (1, 2); First equation:

$$\begin{aligned} 2x - y &= 0 \\ 2(1) - (2) &\stackrel{?}{=} 0 \\ 2 - 2 &\stackrel{?}{=} 0 \\ 0 &= 0 \end{aligned}$$

Second equation:

$$\begin{aligned} y &= \frac{1}{2}x + \frac{3}{2} \\ 2 &\stackrel{?}{=} \frac{1}{2}(1) + \frac{3}{2} \\ 2 &\stackrel{?}{=} \frac{4}{2} \\ 2 &= 2 \end{aligned}$$

Yes, the point (1, 2) satisfies both equations.

17. (2, –3); First equation:

$$\begin{aligned} y + 2 &= \frac{1}{2}x \\ (-3) + 2 &\stackrel{?}{=} \frac{1}{2}(2) \\ -1 &\neq 1 \end{aligned}$$

Second equation:

$$\begin{aligned} 3x + 2y &= 0 \\ 3(2) + 2(-3) &\stackrel{?}{=} 0 \\ 6 - 6 &\stackrel{?}{=} 0 \\ 0 &= 0 \end{aligned}$$

No, the point (2, –3) does not satisfy the first equation.

19. Equation #1: $x + y = 6$

x	y
1	5
6	0
3	3

Equation #2: $x - y = 2$

x	y
0	–2
2	0
4	2

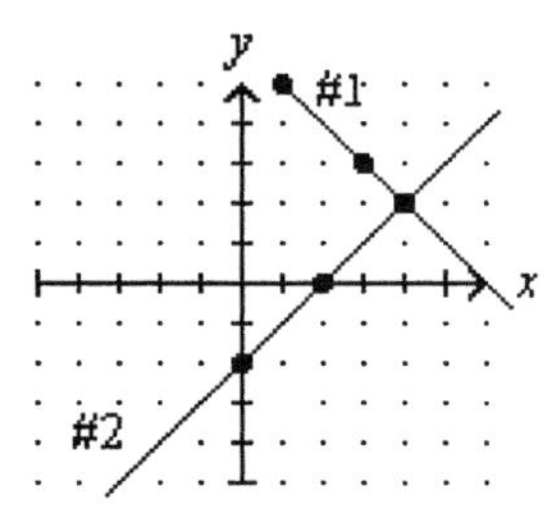

The solution is (4, 2).

21. Equation #1: $2x + y = 1$

x	y
0	1
–2	5
2	–3

Equation #2: $x - 2y = -7$

x	y
1	4
–5	1
–3	2

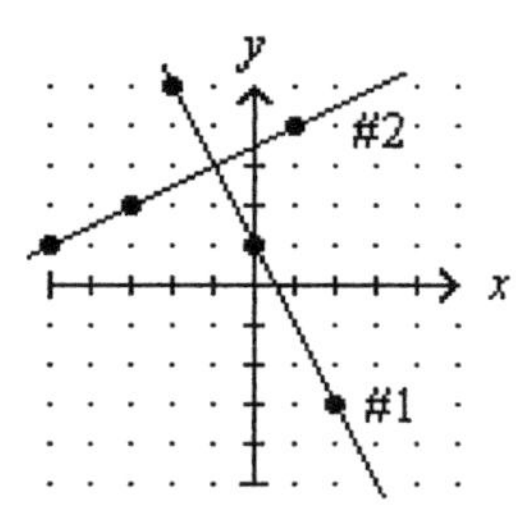

The solution is (–1, 3).

23. Equation #1: $3x - 2y = 0$

x	y
2	3
0	0
-2	-3

Equation #2: $2x + 3y = 0$

x	y
-3	2
0	0
3	-2

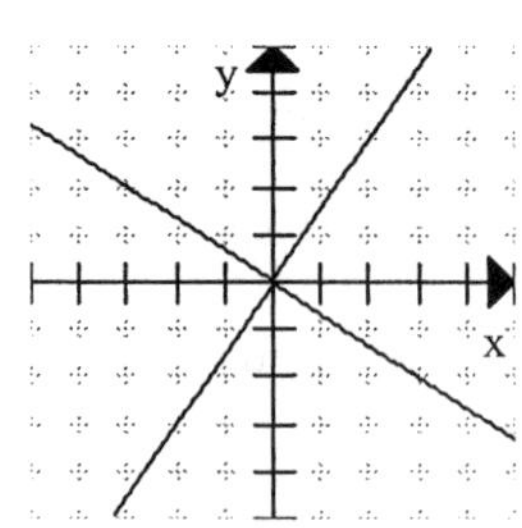

The solution is (0, 0)

25. Equation #1: $4x - 3y = 5$

x	y
–1	–3
2	1
5	5

Equation #2: $2x + y = 0$

x	y
–2	4
0	0
2	–4

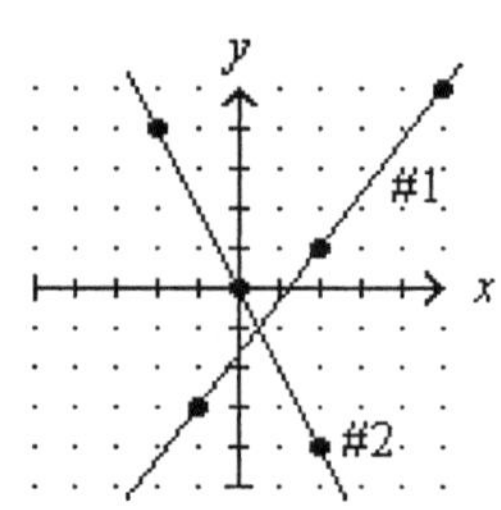

The solution is $\left(\frac{1}{2}, -1\right)$.

27. Equation #1

$3x + y = 3$

x	y
0	3
1	0
2	−3

Equation #2

$3x + 2y = 0$

x	y
−2	3
0	0
2	−3

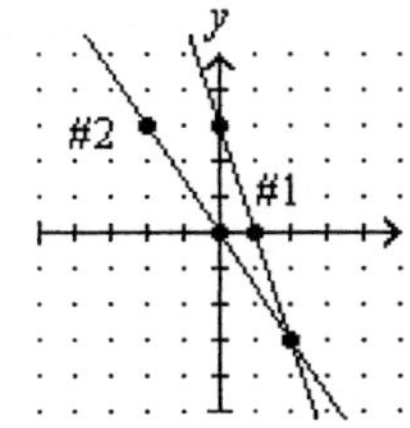

The solution is (2, −3).

29. Equation #1

$x = 13 - 4y$

x	y
1	3
−3	4
5	2

Equation #2

$3x = 4 + 2y$

x	y
−2	−5
0	−2
2	1

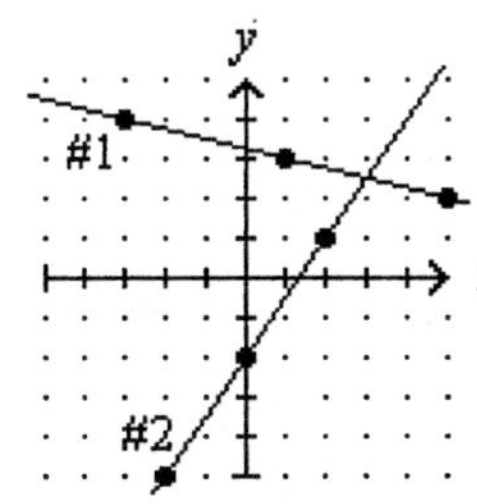

The solution is $\left(3, \frac{5}{2}\right)$.

31. Equation #1

$x = 3 - 2y$

x	y
−3	3
−1	2
3	0

Equation #2

$2x + 4y = 6$

x	y
−3	3
−1	2
3	0

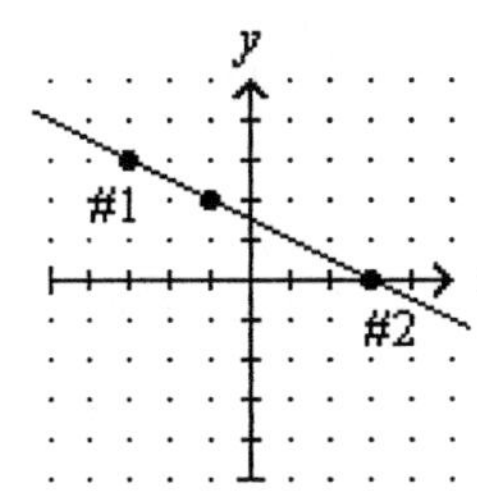

Infinitely many solutions; dependent equations.

33. Equation #1

$x = 2$

x	y
2	−3
2	0
2	3

Equation #2

$y = -\frac{1}{2}x + 2$

x	y
−2	3
0	2
4	0

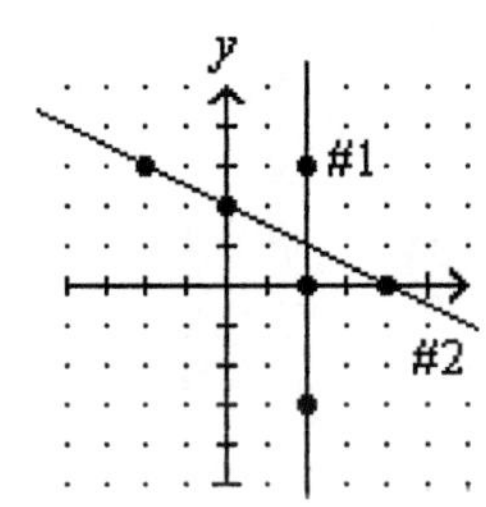

The solution is (2, 1).

35. Equation #1

$y = 3$

x	y
−2	3
0	3
2	3

Equation #2

$x = 2$

x	y
2	−3
2	0
2	3

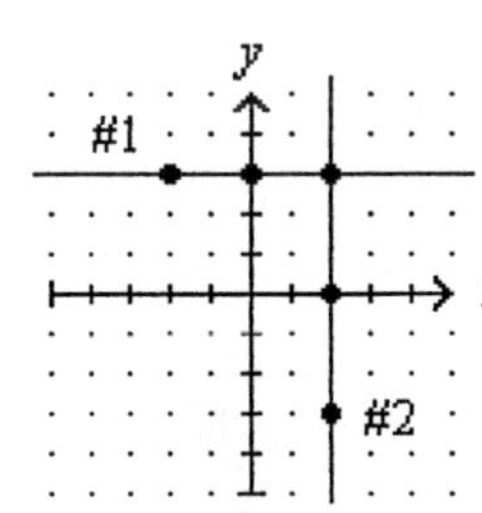

The solution is (2, 3).

37. Equation #1

$$x = \frac{11-2y}{3}$$

x	y
1	4
3	1
5	−2

Equation #2

$$y = \frac{11-6x}{4}$$

x	y
0	$\frac{11}{4} = 2\frac{3}{4}$
1	$\frac{5}{4} = 1\frac{1}{4}$
3	$-\frac{7}{4} = -1\frac{3}{4}$

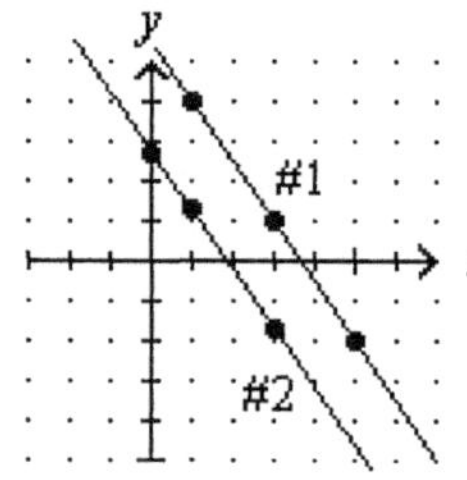

No solution; inconsistent system.

39. Equation #1

$$y = -\frac{5}{2}x + \frac{1}{2}$$

x	y
−1	3
1	−2
3	−7

Equation #2

$$2x - \frac{3}{2}y = 5$$

x	y
−2	−6
1	−2
4	2

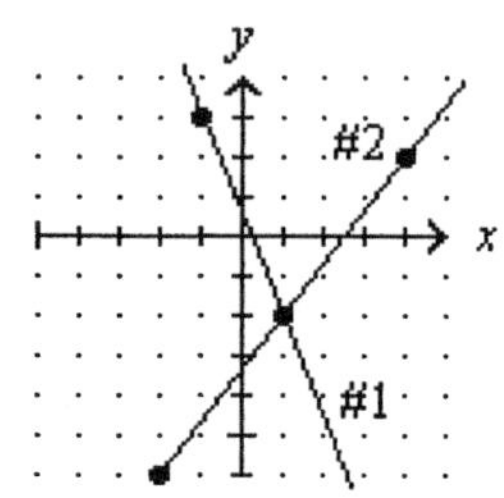

The solution is (1, –2).

41. Equation #1

$$x = \frac{5y-4}{2}$$

x	y
3	2
−2	0
−7	−2

Equation #2

$$x - \frac{5}{3}y + \frac{1}{3} = 0$$

x	y
−2	−1
3	2
8	5

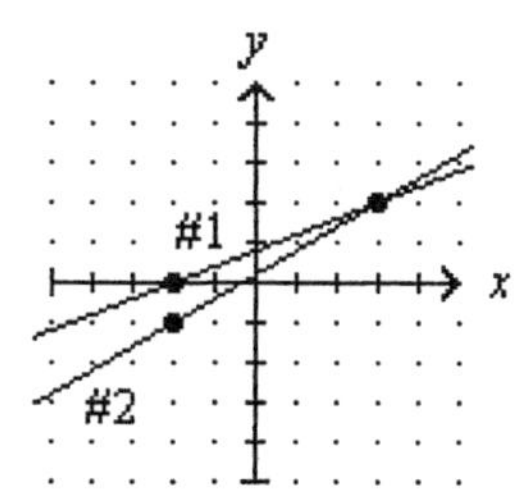

The solution is (3, 2).

43. Equation #1

$$x = -\frac{3}{2}y$$

x	y
−3	2
0	0
3	−2

Equation #2

$$x = \frac{3}{2}y - 2$$

x	y
−2	0
1	2
4	4

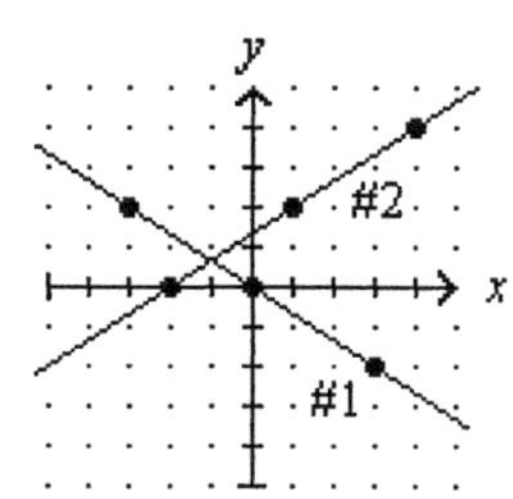

The solution is $\left(-1, \frac{2}{3}\right)$.

In exercises 45 through 48, enter the equations on your calculator and then use the appropriate calculator functions to solve the system. The following graphs are given as starting points to begin the approximation of the solution.

45.

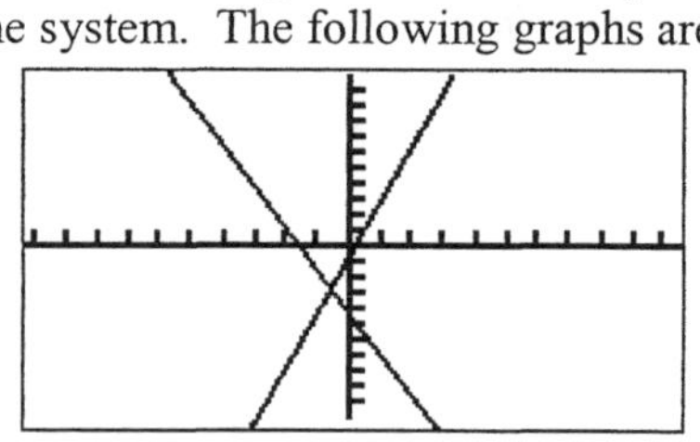

The solution is (–0.37, –2.69)

47. Rewrite $1.7x + 2.3y = 3.2$ as $y = (-1.7x + 3.2) \div 2.3$

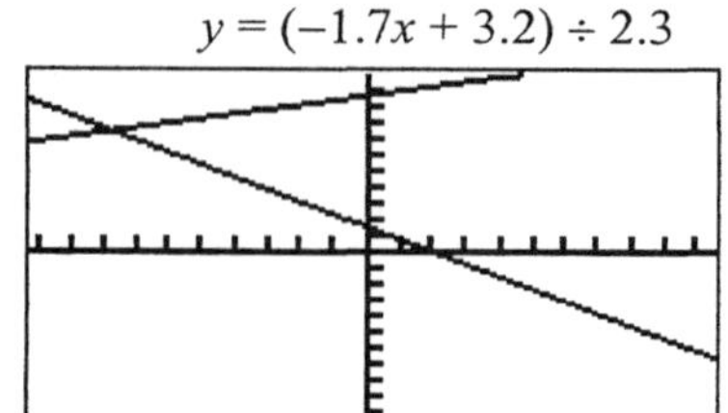

The solution is (–7.64, 7.04)

APPLICATIONS

49. The cities on Interstate 40 are Gallop, Grants, Albuquerque, and Tucumcari.
The cities on Interstate 25 are Las Vegas, Santa Fe, Albuquerque, Socorro, and Las Cruces.
The city on both interstates is Albuquerque.

51. The point where the 2 lines cross would be (2000, 50).

53. a. $S(x) = \frac{25}{4}x - 525$

x	$S(x)$
100	100
160	475
200	725

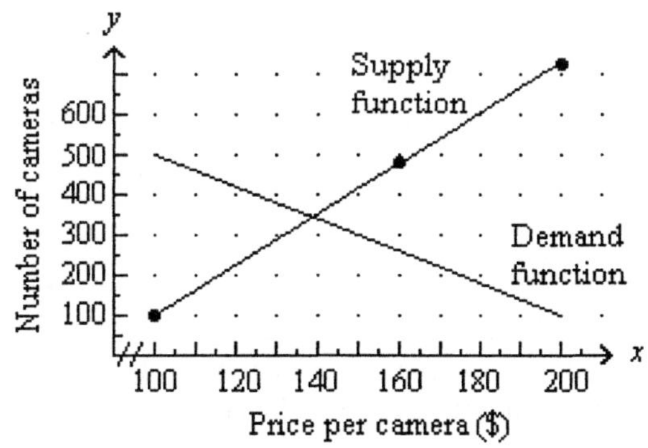

b. $140

c. Supply increases and demand decreases

55. a. First ship

$2x + 3y = 6$

x	y
0	2
3	0
–3	4

Second ship

$y = \frac{2}{3}x - 3$

x	y
–3	–5
0	–3
3	–1

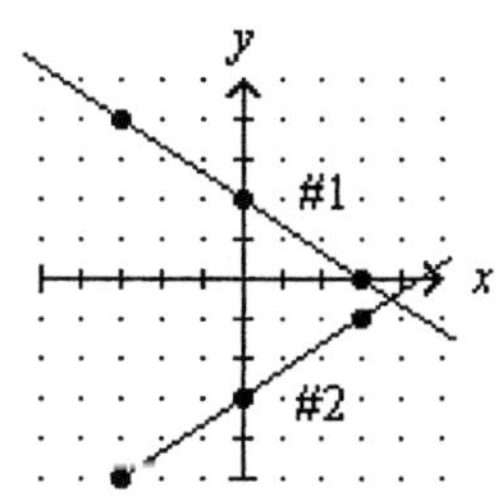

Yes, there is a possibility of a collision.

b. The danger point would be about (3.75, –0.5)

c. No, it is not a certainty because it is not stated which direction the ships are traveling and they might not be at the danger point at the same time.

WRITING

57. Graphs are usually completed in full units. Determining fractional answers would be very difficult to read from such a graph.

REVIEW

59. $f(x) = -x^3 + 2x - 2$; $f(-1) = -(-1)^3 + 2(-1) - 2 = -(-1) - 2 - 2 = 1 - 4 = -3$ Replace x with –1.

61. $g(x) = \frac{2-x}{9+x}$ $g(2) = \frac{2-2}{9+2} = \frac{0}{11} = 0$ Replace x with 2.

63. D = the set of all real numbers. R = the set of all real numbers greater than or equal to –2.

65. The area of the square is $A = s^2 = 81 \text{ cm}^2$. The area of a triangle is $\frac{1}{2}$ the area of a square or rectangle.

Area of triangle $= \frac{1}{2}s^2 = \frac{1}{2}(81) = 40.5 \text{ cm}^2$.

Section 3.2 Solving Systems Algebraically

VOCABULARY

1. general
3. eliminated

CONCEPTS

5. Before adding the two equations together it is necessary for the two coefficients of the variable being eliminated to be opposites.
   a. Option one: Multiply equation 1 by 3 and equation 2 by –4 or
      Option two: Multiply equation 1 by –3 and equation 2 by 4.
   b. Option one: Multiply equation 1 by 2 and equation 2 by –3 or
      Option two: Multiply equation 1 by –2 and equation 2 by 3.

7. The addition method.

9. a. This is a contradiction or a false statement. Lines are parallel. Graph ii.
   b. This is a conditional statement. The lines will intersect. Graph iii.
   c. This is an identity or a true statement. The two equations represent the same line. Graph i.

PRACTICE

11. $\begin{cases} y = x \\ x + y = 4 \end{cases}$

Substitute x for y in Equation 2.

$$\begin{aligned} x + x &= 4 \\ 2x &= 4 \\ x &= 2 \end{aligned}$$

Substitute 2 for x in Equation 1.

$y = 2$ The solution is (2, 2).

13. $\begin{cases} x = 2 + y \\ 2x + y = 13 \end{cases}$

Substitute 2 + y for x in Equation 2.

$$\begin{aligned} 2(2 + y) + y &= 13 \\ 4 + 2y + y &= 13 \\ 4 + 3y &= 13 \\ 3y &= 9 \\ y &= 3 \end{aligned}$$

Substitute 3 for y Equation 1.

$x = 2 + (3) = 5$

The solution is (5, 3).

15. $\begin{cases} x + 2y = 6 \\ 3x - y = -10 \end{cases}$ Solve Equation 1 for x.

$$\begin{aligned} x + 2y &= 6 \\ x &= -2y + 6 \end{aligned}$$

Substitute $-2y + 6$ for x in Equation 2.

$$\begin{aligned} 3(-2y + 6) - y &= -10 \\ -6y + 18 - y &= -10 \\ 18 - 7y &= -10 \\ -7y &= -28 \\ y &= 4 \end{aligned}$$

Substitute 4 for y in Equation 1.

$$\begin{aligned} x + 2(4) &= 6 \\ x + 8 &= 6 \\ x &= -2 \end{aligned}$$

The solution is (–2, 4).

17. $\begin{cases} \frac{3}{2}x + 2 = y \\ 0.6x - 0.4y = -0.4 \end{cases}$

Multiply Equation 2 by 10 to clear decimals.

$6x - 4y = -4$

Substitute $\frac{3}{2}x + 2$ for y in Equation 2.

$$\begin{aligned} 6x - 4\left(\frac{3}{2}x + 2\right) &= -4 \\ 6x - 6x - 8 &= -4 \\ -8 &\neq -4 \end{aligned}$$

There is no solution; inconsistent system.

19. $\begin{cases} x - y = 3 \\ x + y = 7 \end{cases}$

Add the two equations to eliminate y

$$\begin{aligned} x - y &= 3 \\ \underline{x + y} &\underline{= 7} \\ 2x &= 10 \Rightarrow \quad x = 5 \end{aligned}$$

Substitute 5 for x in Equation 2 and solve for y.

$$\begin{aligned} 5 + y &= 7 \\ y &= 2 \end{aligned}$$

The solution is (5, 2).

21. $\begin{cases} 2x + y = -10 \\ 2x - y = -6 \end{cases}$

Add the two equations to eliminate y

$$\begin{aligned} 2x + y &= -10 \\ \underline{2x - y} &\underline{= -6} \\ 4x &= -16 \Rightarrow \quad x = -4 \end{aligned}$$

Substitute −4 for x in Equation 1 and solve for y.

$$\begin{aligned} 2(-4) + y &= -10 \\ -8 + y &= -10 \Rightarrow \quad y = -2 \end{aligned}$$

The solution is (−4, −2).

23. $\begin{cases} 2x + 3y = 8 \\ 3x - 2y = -1 \end{cases}$

Since the coefficients of y have opposite signs, multiply Equation 1 by 2 and Equation 2 by 3 to eliminate y.

$$\begin{cases} 4x + 6y = 16 \\ 9x - 6y = -3 \end{cases}$$

Add the two equations and solve for x.

$$13x = 13 \Rightarrow \quad x = 1$$

Substitute 1 for x in Equation 1 and solve for y.

$$\begin{aligned} 2(1) + 3y &= 8 \\ 2 + 3y &= 8 \\ 3y &= 6 \Rightarrow \quad y = 2 \end{aligned}$$

The solution is (1, 2).

25. $\begin{cases} 4(x - 2) = -9y \\ 2(x - 3y) = -3 \end{cases}$

Remove the parentheses.

$$\begin{cases} 4x - 8 = -9y \\ 2x - 6y = -3 \end{cases}$$

Write the equations in general form.

$$\begin{cases} 4x + 9y = 8 \\ 2x - 6y = -3 \end{cases}$$

Multiply Equation 2 by −2 to eliminate x.

$$\begin{cases} 4x + 9y = 8 \\ -4x + 12y = 6 \end{cases}$$

Add the two equations and solve for y.

$$\begin{aligned} 21y &= 14 \\ y &= \frac{14}{21} = \frac{2}{3} \end{aligned}$$

Substitute $\frac{2}{3}$ for y in Equation 1 and solve for x.

$$\begin{aligned} 4[x - 2] &= -9\left(\frac{2}{3}\right) \\ 4x - 8 &= -6 \\ 4x &= 2 \\ x &= \frac{1}{2} \end{aligned}$$

The solution is $\left(\frac{1}{2}, \frac{2}{3}\right)$.

27. $\begin{cases} 3x - 4y = 9 \\ x + 2y = 8 \end{cases}$

Multiply Equation 2 by 2 to eliminate y.

$\begin{cases} 3x - 4y = 9 \\ 2x + 4y = 16 \end{cases}$

Add the two equations and solve for x.

$$\begin{aligned} 5x &= 25 \\ x &= 5 \end{aligned}$$

Substitute 5 for x in Equation 1 and solve for y.

$$\begin{aligned} 3(5) - 4y &= 9 \\ 15 - 4y &= 9 \\ -4y &= -6 \\ y &= \frac{-6}{-4} = \frac{3}{2} \end{aligned}$$

The solution is $\left(5, \frac{3}{2}\right)$.

29. $\begin{cases} 2(x + y) + 1 = 0 \\ 3x + 4y = 0 \end{cases}$

Remove the parentheses.

$\begin{cases} 2x + 2y + 1 = 0 \\ 3x + 4y = 0 \end{cases}$

Write the equations in general form.

$\begin{cases} 2x + 2y = -1 \\ 3x + 4y = 0 \end{cases}$

Multiply Equation 1 by –2 to eliminate y.

$\begin{cases} -4x - 4y = 2 \\ 3x + 4y = 0 \end{cases}$

Add the two equations and solve for x.

$$\begin{aligned} -x &= 2 \\ x &= -2 \end{aligned}$$

Substitute –2 for x in Equation 2 and solve for y.

$$\begin{aligned} 3(-2) + 4y &= 0 \\ 4y &= 6 \\ y &= \frac{3}{2} \end{aligned}$$

The solution is $\left(-2, \frac{3}{2}\right)$.

31. $\begin{cases} 0.16x - 0.08y = 0.32 \\ 2x - 4 = y \end{cases}$

Multiply Equation 1 by 100.

$\begin{cases} 16x - 8y = 32 \\ 2x - 4 = y \end{cases}$

Write the equations in general form.

$\begin{cases} 16x - 8y = 32 \\ 2x - y = 4 \end{cases}$

Multiply Equation 2 by –8 to eliminate x.

$\begin{cases} 16x - 8y = 32 \\ -16x + 8y = -32 \end{cases}$

Add the two equations and solve for y.

$$0 = 0$$

Infinitely many solutions; dependent equations.

33. $\begin{cases} x = \frac{3}{2}y + 5 \\ 2x - 3y = 8 \end{cases}$

Multiply Equation 1 by the LCD 2.

$\begin{cases} 2x = 3y + 10 \\ 2x - 3y = 8 \end{cases}$

Write the equations in general form.

$\begin{cases} 2x - 3y = 10 \\ 2x - 3y = 8 \end{cases}$

Multiply Equation 2 by –1 to eliminate x.

$\begin{cases} 2x - 3y = 10 \\ -2x + 3y = -8 \end{cases}$

Add the two equations and solve for y.

$$0 = 2$$

No solution, inconsistent systems.

35. $\begin{cases} 0.5x + 0.5y = 6 \\ \dfrac{x}{2} - \dfrac{y}{2} = -2 \end{cases}$

Multiply Equation 1 by 10 and Equation 2 by 2.

$$\begin{cases} 5x + 5y = 60 \\ x - y = -4 \end{cases}$$

Multiply Equation 2 by 5 to eliminate y.

$$\begin{cases} 5x + 5y = 60 \\ 5x - 5y = -20 \end{cases}$$

Add the two equations and solve for x.

$$\begin{aligned} 10x &= 40 \\ x &= 4 \end{aligned}$$

Substitute 4 for x in Equation 1 and solve for y.

$$\begin{aligned} 0.5(4) + 0.5y &= 6 \\ 2 + 0.5y &= 6 \\ 0.5y &= 4 \\ y &= 8 \end{aligned}$$

The solution is (4, 8).

37. $\begin{cases} \dfrac{3}{4}x + \dfrac{2}{3}y = 7 \\ \dfrac{3}{5}x - \dfrac{1}{2}y = 18 \end{cases}$

Multiply Equation 1 by 12 and Equation 2 by 10.

$$\begin{cases} 9x + 8y = 84 \\ 6x - 5y = 180 \end{cases}$$

Multiply Equation 1 by 5 and Equation 2 by 8 to eliminate y.

$$\begin{cases} 45x + 40y = 420 \\ 48x - 40y = 1440 \end{cases}$$ Add the two equations and solve for x

$$93x = 1860 \Rightarrow x = 20$$

Substitute 20 for x in Equation 1 and solve for y.

$$\begin{aligned} \frac{3}{4}(20) + \frac{2}{3}y &= 7 \\ 15 + \frac{2}{3}y &= 7 \\ 45 + 2y &= 21 \\ 2y &= -24 \\ y &= -12 \end{aligned}$$

The solution is (20, –12).

43. $\begin{cases} \dfrac{3}{2}p + \dfrac{1}{3}q = 2 \\ \dfrac{2}{3}p + \dfrac{1}{9}q = 1 \end{cases}$

Multiply Equation 1 by 6 and Equation 2 by 9.

$$\begin{cases} 9p + 2q = 12 \\ 6p + 1q = 9 \end{cases}$$

Multiply Equation 2 by –2 to eliminate q.

$$\begin{cases} 9p + 2q = 12 \\ -12p - 2q = -18 \end{cases}$$

Add the two equations and solve for a.

$$\begin{aligned} -3p &= -6 \\ p &= 2 \end{aligned}$$

Substitute 2 for p in Equation 1 and solve for q.

$$\begin{aligned} \frac{3}{2}(2) + \frac{1}{3}q &= 2 \\ 3 + \frac{1}{3}q &= 2 \\ 9 + q &= 6 \\ q &= -3 \end{aligned}$$

The solution is (2, –3).

45. $\begin{cases} \dfrac{m-n}{5} + \dfrac{m+n}{2} = 6 \\ \dfrac{m-n}{2} - \dfrac{m+n}{4} = 3 \end{cases}$

Multiply Equation 1 by 10 and Equation 2 by 4.

$$\begin{cases} 2(m-n) + 5(m+n) = 60 \\ 2(m-n) - (m+n) = 12 \end{cases}$$

Write the equations in general form.

$$\begin{cases} 7m + 3n = 60 \\ m - 3n = 12 \end{cases}$$

Add the two equations and solve for m.

$$\begin{aligned} 8m &= 72 \\ m &= 9 \end{aligned}$$

Substitute 9 for m in Equation 1 and solve for n.

$$\begin{aligned} \frac{(9)-n}{5} + \frac{(9)+n}{2} &= 6 \\ 2(9-n) + 5(9+n) &= 60 \\ 18 - 2n + 45 + 5n &= 60 \\ 3n &= -3 \\ n &= -1 \end{aligned}$$

The solution is (9, –1).

47. $\begin{cases} \frac{1}{x}+\frac{1}{y}=\frac{5}{6} \\ \frac{1}{x}-\frac{1}{y}=\frac{1}{6} \end{cases}$

Substitute a for $\frac{1}{x}$ and b for $\frac{1}{y}$.

$$\begin{cases} a+b=\frac{5}{6} \\ a-b=\frac{1}{6} \end{cases}$$

Add the two equations and solve for a.

$$\begin{aligned} 2a &= 1 \\ a &= \frac{1}{2} \end{aligned}$$

Substitute $\frac{1}{2}$ for a in Equation 3 and solve for b.

$$\begin{aligned} \frac{1}{2}+b &= \frac{5}{6} \\ 3+6b &= 5 \\ 6b &= 2 \\ b &= \frac{1}{3} \end{aligned}$$

Substitute $\frac{1}{2}$ for a in $a=\frac{1}{x}$ and $\frac{1}{3}$ for b in $b=\frac{1}{y}$.

$$\frac{1}{2} = \frac{1}{x}, \quad \frac{1}{3} = \frac{1}{y}$$
$$x = 2, \quad y = 3$$

The solution is (2, 3).

49. $\begin{cases} \frac{1}{x}+\frac{2}{y}=-1 \\ \frac{2}{x}-\frac{1}{y}=-7 \end{cases}$

Substitute a for $\frac{1}{x}$ and b for $\frac{1}{y}$.

$$\begin{cases} a+2b=-1 \\ 2a-b=-7 \end{cases}$$

Multiply Equation 4 by 2 to eliminate b.

$$\begin{cases} a+2b=-1 \\ 4a-2b=-14 \end{cases}$$

Add the two equations and solve for a.

$$\begin{aligned} 5a &= -15 \\ a &= -3 \end{aligned}$$

Substitute –3 for a in Equation 3 and solve for b.

$$\begin{aligned} -3+2b &= -1 \\ 2b &= 2 \\ b &= 1 \end{aligned}$$

Substitute –3 for a in $a=\frac{1}{x}$ and 1 for b in $b=\frac{1}{y}$.

$$-3 = \frac{1}{x}, \quad 1 = \frac{1}{y}$$
$$x = -\frac{1}{3}, \quad y = 1$$

The solution is $\left(-\frac{1}{3},1\right)$.

APPLICATIONS

51. **Analysis:** Find the cost of a 15–second and a 30–second radio commercial when Plan 1 costs \$6050 and Plan 2 costs \$4775.

Equations: Let x represent the cost of the 15–second spot and y represent the 30–second spot. For Plan 1, the cost of six 15–second spots is 6 • \$$x$ = \$6x ,and the cost of four 30–second spots is 4 • \$$y$ = \$4y. For Plan 2, the cost of five 15–second spots is 5 • \$$x$ = \$5x ,and the cost of three 30–second spots is 3 • \$$y$ = \$3y.

Plan 1:	Cost of six 15–sec spots	plus	cost of four 30–sec spots	is	\$6050.
	$6x$	+	$4y$	=	6050
Plan 2:	Cost of five 15–sec spots	plus	cost of three 30–sec spots	is	\$4775.
	$5x$	+	$3y$	=	4775

Solve: $\begin{cases} 6x+4y=6050 \\ 5x+3y=4775 \end{cases}$ Multiply Equation 1 by 3 and Equation 2 by –4. $\begin{cases} 18x+12y=18{,}150 \\ -20x-12y=-19{,}100 \end{cases}$

Add the equations and solve for x.

$$\begin{aligned} -2x &= -950 \\ x &= 475 \end{aligned}$$

Substitute 475 for x and solve for y.

$$\begin{aligned} 6(475)+4y &= 6050 \\ 2850+4y &= 6050 \\ 4y &= 3200 \\ y &= 800 \end{aligned}$$

Conclusion: The 15–second spot would cost x = \$475 and the 30–second spot would cost y = \$800.

53. **Analysis:** Find the number of cats and dogs in the United States.

Equations: Let c represent the number of cats and d represent the number of dogs. The total number of cats and dogs was $c + d = 126.65$ million. There were 1.85 million more cats than dogs, so $c - d = 1.85$.

Number of cats	plus	number of dogs	is	126.65 million.
c	$+$	d	$=$	126.65
Number of cats	less	number of dogs	is	1.85 million.
c	$-$	d	$=$	1.85

Solve: $\begin{cases} c + d = 126.65 \\ c - d = 1.85 \end{cases}$

Add the equations and solve for c.

$$2c = 128.5$$
$$c = 64.25$$

Substitute 64.25 for c and solve for d.

$$64.25 + d = 126.65$$
$$d = 62.4$$

Conclusion: The number of cats was 64.25 million and the number of dogs was 62.4 million.

55. **Analysis:** Find the dimensions of a field that has a perimeter of 72 meters. When one extra width of fencing is added the fencing used is 88 meters.

Equations: Let x represent the width of the field and y represent the length of the field. Substitute into the formula $P = 2l + 2w$ to calculate the perimeter.

Perimeter:	2 widths	plus	2 lengths	is	72 meters.
	$2x$	$+$	$2y$	$=$	72
Fencing:	3 widths	plus	2 lengths	is	88 meters.
	$3x$	$+$	$2y$	$=$	88

Solve: $\begin{cases} 2x + 2y = 72 \\ 3x + 2y = 88 \end{cases}$ Multiply Equation 2 by -1 $\begin{cases} 2x + 2y = 72 \\ -3x - 2y = -88 \end{cases}$

Add the equations and solve for x.

$$-x = -16$$
$$x = 16$$

Substitute 16 for x and solve for y.

$$2(16) + 2y = 72$$
$$32 + 2y = 72$$
$$2y = 40$$
$$y = 20$$

Conclusion: The dimensions of the field is 16 m by 20 m.

57. **Analysis:** Find the values of x and y in the illustration of the parallelogram. The alternate interior angles are equal to each other and the opposite angles are equal to each other.

Equations: The values of x and y are as represented in the illustration.

Alternate Angles: $(x - y)° = 50°$.

Opposite Angles: $(x + y)° = 100°$.

Solve: $\begin{cases} x - y = 50 \\ x + y = 100 \end{cases}$

Add the equations and solve for x.

$$2x = 150$$
$$x = 75$$

Substitute 75 for x and solve for y.

$$75 + y = 100$$
$$y = 25$$

Conclusion: $x = 75°$ and $y = 25°$.

59. **Analysis:** Find the two amounts invested when part of \$8000 is invested at 10% and the rest is invested at 12%. The annual income from these investments is \$900.

Equations: Let x represent the amount invested at 10% and y represent the amount invested at 12%.

	P •	r •	t =	I
First	x	0.10	1	$0.10x$
Second	y	0.12	1	$0.12y$

Principal: Principal at 10% plus principal at 12% is \$8000.
$x + y = 8000$

Interest Interest at 10% plus interest at 12% is \$900.
$0.10x + 0.12y = 900$

Solve: $\begin{cases} x+y=8000 \\ 0.10x+0.12y=900 \end{cases}$ Multiply Equation 1 by −0.12 $\begin{cases} -0.12x-0.12y=-960 \\ 0.10x+0.12y=900 \end{cases}$

Add the equations and solve for x.
$$-0.02x = -60$$
$$x = 3000$$

Substitute 3000 for x and solve for y.
$$3000 + y = 8000$$
$$y = 5000$$

Conclusion: The investment at 10% was $x = \$3000$ and the investment at 12% was $y = \$5000$.

61. **Analysis:** Find the distance a van traveled in comparison to a helicopter.

Equations: Let x represent the distance the van traveled and y represent the distance the helicopter traveled.

Sum: Van distance plus helicopter distance is 145 miles.
$x + y = 145$

Comparison: Helicopter distance less van distance is 55 miles.
$y - x = 55$

Solve: $\begin{cases} x+y=145 \\ y-x=55 \end{cases}$ Rewrite in general form $\begin{cases} x+y=145 \\ -x+y=55 \end{cases}$

Add the equations and solve for y.
$$2y = 200$$
$$y = 100$$

Substitute 100 for x and solve for x.
$$x + 100 = 145$$
$$x = 45$$

Conclusion: The van traveled $x = 45$ miles to the story.

63. **Analysis:** Find the number of racing bikes and mountain bikes that can be manufactured with a total materials cost of \$26,150 and labor costs of \$31,800.

Equations: Let x represent the number of racing bikes and y represent the number of mountain bikes.

Materials: Cost for racing bikes plus cost for mountain bikes is \$26,150.
$110x + 140y = 26{,}150$

Labor: Cost for racing bikes plus cost for mountain bikes is \$31,800.
$120x + 180y = 31{,}800$

Solve: $\begin{cases} 110x+140y=26{,}150 \\ 120x+180y=31{,}800 \end{cases}$ Multiply Equation 1 by 180 and Equation 2 by −140.

$$\begin{cases} 19{,}800x+25{,}200y=4{,}707{,}000 \\ -16{,}800x-25{,}200y=-4{,}452{,}000 \end{cases}$$

Add the equations and solve for x.
$$3000x = 255{,}000$$
$$x = 85$$

Substitute 85 for x and solve for y.
$$110(85) + 140y = 26{,}150$$
$$140y = 16{,}800$$
$$y = 120$$

Conclusion: The number of racing bikes would be $x = 85$ and the number of mountain bikes would be $y = 120$.

65. **Analysis:** Find the number of sets of CDs that need to be sold to make a profit. We need to find the break point and then add 1 set to that number to make a profit. The break point occurs when the revenues generated equals the costs incurred.

Equations: Let x represent the number of sets of CDs produced.

Cost:	Cost of producing x CDs	plus	original investment	is	total cost.
	$\$18.95x$	$+$	$\$105,000$	$=$	$C(x)$

Revenue:	Revenue from x CDs	is	total revenue.
	$45x$	$=$	$R(x)$

Break point:	Revenue	equals	cost.
	$R(x)$	$=$	$C(x)$

Solve: Substitute the cost and revenue equations into the break point equation.

$$\begin{aligned} 45x &= 18.95x + 105,000 \\ 26.05x &= 105,000 \Rightarrow x = 4031 \end{aligned}$$

Conclusion: If the partners make 4030 sets of CDs they will lose money, but if they make 4031 sets of CDs they will make a profit.

67. **Analysis:** Find the number of books that will cost equal amounts to produce on either of two presses with different setup costs and different unit costs.

Equations: Let x represent the number of books to be produced. The break point occurs when the two costs are the same. The cost on either press is the cost per book times the number of books produced plus the setup costs for that press.

Cost 1:	Cost on press 1	plus	setup cost	is	total cost of press 1.
	$\$5.98x$	$+$	$\$210$	$=$	C_1
Cost 2:	Cost on press 2	plus	setup cost	is	total cost of press 2.
	$\$5.95x$	$+$	$\$350$	$=$	C_2

Break point:	Cost of press 1	equals	cost of press 2.
	C_1	$=$	C_2

Solve: Substitute the two cost equations into the break point equation.

$$\begin{aligned} 5.98x + 210 &= 5.95x + 350 \\ 0.03x &= 140 \\ x &= 4666.67 \end{aligned}$$

Conclusion: The break point occurs when $4666\frac{2}{3}$ books are printed.

69. **Analysis:** Find the number of permanents that need to be given to break even. The break point occurs when the revenues generated equals the costs incurred.

Equations: Let x represent the number of permanents given.

Cost:	Cost of giving x permanents	plus	fixed costs	is	total cost.
	$\$23.60x$	$+$	$\$2101.20$	$=$	$C(x)$

Revenue:	Revenue from x permanents	is	total revenue.
	$44x$	$=$	$R(x)$

Break point:	Revenue	equals	cost.
	$R(x)$	$=$	$C(x)$

Solve: Substitute the cost and revenue equations into the break point equation.

$$\begin{aligned} 44x &= 23.60x + 2101.20 \\ 20.40x &= 2101.20 \\ x &= 103 \end{aligned}$$

Conclusion: The number of permanents that need to be given is 103.

71. **Analysis:** Find the number of grams of each of two face–wash creams that need to be used to provide 185 grams of a 0.3% face wash.

Equations: Let x represent the grams of 0.2% wash and y represent the grams of 0.7% cream.

	percent Triclosan	• total quantity	= quantity of Triclosan
0.2% cream	0.002	x	$0.002x$
0.7% cream	0.007	y	$0.007y$
0.3% cream	0.003	185	$0.003(185) = 0.555$

Total quantity: Grams of 0.2% plus grams of 0.7% is grams of 0.3%

$$x + y = 185$$

Quantity of Triclosan: Triclosan in 0.2% plus Triclosan in 0.7% is Triclosan in 0.3%

$$0.002x + 0.007y = 0.555$$

Solve: $\begin{cases} x + y = 185 \\ 0.002x + 0.007y = 0.555 \end{cases}$ Multiply Equation 1 by -0.007 $\begin{cases} -0.007x - 0.007y = -1.295 \\ 0.002x + 0.007y = 0.555 \end{cases}$

Add the equations and solve for x.

$$\begin{aligned} -0.005x &= -0.74 \\ x &= 148 \end{aligned}$$

Substitute 148 for x and solve for y.

$$\begin{aligned} 148 + y &= 185 \\ y &= 37 \end{aligned}$$

Conclusion: The amount of 0.2% cream would be $x = 148$ g and the amount of 0.7% cream would be $y = 37$ g.

73. **Analysis:** Find the number of water pumps manufactured. The break point occurs when the revenues generated equals the costs incurred.

Equations: Let x represent the number of water pumps manufactured.

Cost A: Cost of making x water pumps plus fixed costs is total cost for process A.

$$\$29x + \$12,390 = C(x)$$

Cost B: Cost of making x water pumps plus fixed costs is total cost for process B.

$$\$17x + \$20,460 = C(x)$$

Revenue: Revenue from x water pumps is total revenue.

$$\$50x = R(x)$$

Break point: Revenue equals cost.

$$R(x) = C(x)$$

Solve: Substitute the cost and revenue equations into the break point equation.

a. $$\begin{aligned} 50x &= 29x + 12,390 \\ 21x &= 12,390 \\ x &= 590 \end{aligned}$$

b. $$\begin{aligned} 50x &= 17x + 20,460 \\ 33x &= 20,460 \\ x &= 620 \end{aligned}$$

Conclusion: a. The break point for process A is 590 water pumps per month.

b. The break point for process B is 620 water pumps per month.

c. Process A should be used because it would result in the smaller loss. Loss is revenue less costs.

Profit From Process A:

$$\begin{aligned} R(x) - C(x) &= R(550) - C(550) \\ &= 50(550) - [29(550) + 12,390] \\ &= 27,500 - (15,950 + 12,390) \\ &= 27,500 - 28,340 \\ &= -840 \end{aligned}$$

Profit from Process B:

$$\begin{aligned} R(x) - C(x) &= R(550) - C(550) \\ &= 50(550) - [17(550) + 20,460] \\ &= 27,500 - (9350 + 20,460) \\ &= 27,500 - 29,810 \\ &= -2310 \end{aligned}$$

WRITING

75. The addition method. None of the variables have a coefficient of 1 or −1. Therefore the substitution method would be cumbersome. Also the coefficients on y have opposite signs.

77. If we only have one equation with two variables there are an infinite number of possible solutions. We need two equations to arrive at a unique solution.

79. Answers will vary.

REVIEW

81. Calculate the slope by finding two points on the line and substituting in slope formula.
Sample points: (−1, 4) and (1, −1). Substitute −1 for y_2, 4 for y_1, 1 for x_2, and −1 for x_1.

$$\text{slope} = \frac{\text{change in y}}{\text{change in x}} = \frac{y_2 - y_1}{x_2 - x_1} = \frac{-1-(4)}{1-(-1)} = \frac{-5}{2} = -\frac{5}{2}$$

83. Points: $(x_1, y_1) = (0, -8)$ and $(x_2, y_2) = (-5, 0)$

$$m = \frac{y_2 - y_1}{x_2 - x_1} = \frac{0-(-8)}{-5-(0)} = \frac{8}{-5} = -\frac{8}{5}$$

85. Find the slope of $4x - 3y = -3$.

$$\begin{aligned} -3y &= -4x - 3 \\ y &= \frac{-4}{-3}x - \frac{3}{-3} \\ y &= \frac{4}{3}x + 1 \end{aligned}$$

$$m = \frac{4}{3}$$

Section 3.3 Systems of Three Equations

VOCABULARY

1. system
3. three
5. dependent

CONCEPTS

7. a. no solution b. no solution

NOTATION

9. $3z - 2y = x + 6$ would become $-x - 2y + 3z = 6$ or $x + 2y - 3z = -6$

PRACTICE

11. Substitute $x = 2$, $y = 1$, and $z = 1$ into each equation in the system.

$$\begin{aligned} x - y + z &= 2 \\ (2) - (1) + (1) &\stackrel{?}{=} 2 \\ 2 &= 2 \end{aligned} \qquad \begin{aligned} 2x + y - z &= 4 \\ 2(2) + (1) - (1) &\stackrel{?}{=} 4 \\ 4 &= 4 \end{aligned} \qquad \begin{aligned} 2x - 3y + z &= 2 \\ 2(2) - 3(1) + (1) &\stackrel{?}{=} 2 \\ 2 &= 2 \end{aligned}$$

Yes, (2, 1, 1) is a solution.

13. $\begin{cases} x+y+z=4 \\ 2x+y-z=1 \\ 2x-3y+z=1 \end{cases}$ Eliminate z.

Add Equations 1 and 2.

$$\begin{array}{lrcl} 1. & x+y+z & = & 4 \\ 2. & \underline{2x+y-z} & \underline{=} & \underline{1} \\ 4. & 3x+2y & = & 5 \end{array}$$

Add Equations 2 and 3.

$$\begin{array}{lrcl} 2. & 2x+\ y-z & = & 1 \\ 3. & \underline{2x-3y+z} & \underline{=} & \underline{1} \\ 5. & 4x-2y & = & 2 \end{array}$$

Add Equations 4 and 5.

$$\begin{array}{lrcl} 4. & 3x+2y & = & 5 \\ 5. & \underline{4x-2y} & \underline{=} & \underline{2} \\ 6. & 7x & = & 7 \\ & x & = & 1 \end{array}$$

Substitute $x = 1$ into Equation 4 and solve for y.

$$\begin{aligned} 3(1)+2y &= 5 \\ 2y &= 2 \\ y &= 1 \end{aligned}$$

Substitute $x = 1$ and $y = 1$ into Equation 1 and solve for z.

$$\begin{aligned} (1)+(1)+z &= 4 \\ z &= 2 \end{aligned}$$

The solution is (1, 1, 2).

15. $\begin{cases} 2x+2y+3z=10 \\ 3x+y-z=0 \\ x+y+2z=6 \end{cases}$ Eliminate y.

Multiply Equation 2 by –2 and add Equations 1 and 2.

$$\begin{array}{lrcl} 1. & 2x+2y+3z & = & 10 \\ 2. & \underline{-6x-2y+2z} & \underline{=} & \underline{0} \\ 4. & -4x \qquad +5z & = & 10 \end{array}$$

Multiply Equation 2 by –1 and add Equations 2 and 3.

$$\begin{array}{lrcl} 2. & -3x-y+\ z & = & 0 \\ 3. & \underline{x+y+2z} & \underline{=} & \underline{6} \\ 5. & -2x \qquad +3z & = & 6 \end{array}$$

Multiply Equation 5 by –2 and add Equations 4 and 5.

$$\begin{array}{lrcl} 4. & -4x+5z & = & 10 \\ 5. & \underline{4x-6z} & \underline{=} & \underline{-12} \\ 6. & -z & = & -2 \\ & z & = & 2 \end{array}$$

Substitute $z = 2$ into Equation 4 and solve for x.

$$\begin{aligned} -4x+5(2) &= 10 \\ -4x &= 0 \\ x &= 0 \end{aligned}$$

Substitute $x = 0$ and $z = 2$ into Equation 1 and solve for y.

$$\begin{aligned} 2(0)+2y+3(2) &= 10 \\ 0+2y+6 &= 10 \\ 2y &= 4 \\ y &= 2 \end{aligned}$$

The solution is (0, 2, 2).

17. $\begin{cases} b+2c=7-a \\ a+c=8-2b \\ 2a+b+c=9 \end{cases}$ Write in general form $\begin{cases} a+b+2c=7 \\ a+2b+c=8 \\ 2a+b+c=9 \end{cases}$ Eliminate c.

Multiply Equation 2 by –2 and add Equations 1 and 2.

$$\begin{array}{lrcl} 1. & a+\ b+2c & = & 7 \\ 2. & \underline{-2a-4b-2c} & \underline{=} & \underline{-16} \\ 4. & -a-3b & = & -9 \end{array}$$

Multiply Equation 3 by –1 and add Equations 2 and 3.

$$\begin{array}{lrcl} 2. & a+2b+c & = & 8 \\ 3. & \underline{-2a-\ b-c} & \underline{=} & \underline{-9} \\ 5. & -a\ +\ b & = & -1 \end{array}$$

Multiply Equation 5 by –1 and add Equations 4 and 5.

$$\begin{array}{lrcl} 4. & -a-3b & = & -9 \\ 5. & \underline{a\ -\ b} & \underline{=} & \underline{1} \\ 6. & -4b & = & -8 \\ & b & = & 2 \end{array}$$

Substitute $b = 2$ into Equation 5 and solve for a.

$$\begin{aligned} -a+(2) &= -1 \\ -a &= -3 \\ a &= 3 \end{aligned}$$

Substitute $a = 3$ and $b = 2$ into Equation 3 and solve for c.

$$\begin{aligned} 2(3)+(2)+c &= 9 \\ 8+c &= 9 \\ c &= 1 \end{aligned}$$

The solution is (3, 2, 1).

19. $\begin{cases} 2x+y-z=1 \\ x+2y+2z=2 \\ 4x+5y+3z=3 \end{cases}$ Eliminate x.

Multiply Equation 2 by –2 and add Equations 1 and 2.

$$\begin{array}{lrcr} 1. & 2x+\ y-\ z &=& 1 \\ 2. & \underline{-2x-4y-4z} &=& \underline{-4} \\ 4. & -3y-5z &=& -3 \end{array}$$

Multiply Equation 2 by –4 and add Equations 2 and 3.

$$\begin{array}{lrcr} 2. & -4x-8y-8z &=& -8 \\ 3. & \underline{4x+5y+3z} &=& \underline{3} \\ 5. & -3y-5z &=& -5 \end{array}$$

Multiply Equation 4 by –1 and add Equations 4 and 5.

$$\begin{array}{lrcr} 4. & -3y-5z &=& -3 \\ 5. & \underline{-3y-5z} &=& \underline{-5} \\ 6. & 0 &\neq& -8 \end{array}$$

Since 0 cannot equal –2, the system is inconsistent and has no solution.

21. $\begin{cases} a+b+c=180 \\ \frac{a}{4}+\frac{b}{2}+\frac{c}{3}=60 \\ 2b+3c-330=0 \end{cases}$ Multiply Equation 2 by 12

$\begin{cases} a+\ b+\ c=180 \\ 3a+6b+4c=720 \\ \quad 2b+3c=330 \end{cases}$ Eliminate a.

Multiply Equation 1 by –3 and add Equations 1 and 2.

$$\begin{array}{lrcr} 1. & -3a-3b-3c &=& -540 \\ 2. & \underline{3a+6b+4c} &=& \underline{720} \\ 4. & 3b\ +\ c &=& 180 \end{array}$$

Multiply Equation 4 by –3 and add Equations 3 and 4.

$$\begin{array}{lrcr} 3. & 2b+3c &=& 330 \\ 4. & \underline{-9b-3c} &=& \underline{-540} \\ 5. & -7b &=& -210 \\ & b &=& 30 \end{array}$$

Substitute $b = 30$ into Equation 3 and solve for c.

$$\begin{array}{rcl} 2(30)+3c &=& 330 \\ 60+3c &=& 330 \\ 3c &=& 270 \\ c &=& 90 \end{array}$$

Substitute $b = 30$, $c = 90$ into Equation 1 and solve for a.

$$\begin{array}{rcl} a+(30)+(90) &=& 180 \\ a+120 &=& 60 \\ a &=& 60 \end{array}$$

The solution is (60, 30, 90).

23. $\begin{cases} 0.5a+0.3b=2.2 \\ 1.2c-8.5b=-24.4 \\ 3.3c+1.3a=29 \end{cases}$ Clear decimals and write in general form

$\begin{cases} 5a+3b \qquad\quad =22 \\ -85b+12c=-244 \\ 13a \qquad +33c=290 \end{cases}$ Eliminate a.

Multiply Equation 1 by - by 5 and add Equations 1

$$\begin{array}{lrcr} 1. & -65a-39b &=& -286 \\ 3. & \underline{65a \qquad +165c} &=& \underline{1450} \\ 4. & -39b+165c &=& 1164 \end{array}$$

Multiply Equation 2 by –165 and Equation 4 by 12 and then add Equations 2 and 4.

$$\begin{array}{lrcr} 2. & 14{,}025b-1980c &=& 40{,}260 \\ 4. & \underline{-468b+1980c} &=& \underline{13{,}968} \\ 5. & 13{,}557b &=& 54{,}228 \\ & b &=& 4 \end{array}$$

Substitute $b = 4$ into Equation 1 and solve for a.

$$\begin{array}{rcl} 0.5a+0.3(4) &=& 2.2 \\ 0.5a-1.2 &=& 2.2 \\ 0.5a &=& 1.0 \\ a &=& 2 \end{array}$$

Substitute $b = 4$ and $a = 2$ into Equation 3 and solve for c.

$$\begin{array}{rcl} 3.3c+1.3(2) &=& 29 \\ 3.3c+2.6 &=& 29 \\ 3.3c &=& 26.4 \\ c &=& 8 \end{array}$$

The solution is (2, 4, 8).

25. $\begin{cases} 2x+3y+4z=6 \\ 2x-3y-4z=-4 \\ 4x+6y+8z=12 \end{cases}$ Eliminate x.

Multiply Equation 1 by –2 and add Equations 1 and 3.

$$\begin{array}{lrcr} 1. & -4x-6y-8z &=& -12 \\ 3. & 4x+6y+8z &=& 12 \\ \hline 4 & 0 &=& 0 \end{array}$$

Since this statement is always true, the system has infinitely many solutions and the equations are dependent.

27. $\begin{cases} x+\frac{1}{3}y+z=13 \\ \frac{1}{2}x-y+\frac{1}{3}z=-2 \\ x+\frac{1}{2}y-\frac{1}{3}z=2 \end{cases}$ Clear fractions by multiplying by LCDs $\begin{cases} 3x+y+3z=39 \\ 3x-6y+2z=-12 \\ 6x+3y-2z=12 \end{cases}$ Eliminate x.

Multiply Equation 2 by –1 and add Equations 1 and 2.

$$\begin{array}{lrcr} 1. & 3x+y+3z &=& 39 \\ 2. & -3x+6y-2z &=& 12 \\ \hline 4. & 7y+z &=& 51 \end{array}$$

Multiply Equation 2 by –2 and add Equations 2 and 3.

$$\begin{array}{lrcr} 2. & -6x+12y-4z &=& 24 \\ 3. & 6x+3y-2z &=& 12 \\ \hline 5. & 15y-6z &=& 36 \end{array}$$

Multiply Equation 4 by 6 and add Equations 4 and 5.

$$\begin{array}{lrcr} 4. & 42y+6z &=& 306 \\ 5. & 15y-6z &=& 36 \\ \hline 6. & 57y &=& 342 \\ & y &=& 6 \end{array}$$

Substitute $y = 6$ into Equation 4 and solve for z.

$$\begin{aligned} 7(6)+z &= 51 \\ 42+z &= 51 \\ z &= 9 \end{aligned}$$

Substitute $y = 6$ and $z = 9$ into Equation 1 and solve for x.

$$\begin{aligned} x+\frac{1}{3}(6)+(9) &= 13 \\ x+2+9 &= 13 \\ x &= 2 \end{aligned}$$

The solution is (2, 6, 9).

APPLICATIONS

29. **Analysis:** Find the number of each of 3 types of statues that should be made to produce \$2100 in monthly revenue when the monthly cost is \$650 for 180 statues.

Equations: Let x represent the number of expensive statues. Let y represent the number of middle–priced statues, and z represent the number of inexpensive statues. The cost of each statue will be the unit cost multiplied by the number of statues and the revenue of each statue will be the unit price multiplied by the number of statues.

	Quantity	Cost	Revenue
Expensive	x	\$5x	\$20x
Middle–priced	y	\$4y	\$12y
Inexpensive	z	\$3z	\$9z
Total	180	\$650	\$2100

Number of expensive	plus	number of middle–priced	plus	number of inexpensive	is	total number of statues.
x	+	y	+	z	=	180
Cost of expensive	plus	cost of middle–priced	plus	cost of inexpensive	is	total cost of statues.
$5x$	+	$4y$	+	$3z$	=	650
Revenue from expensive	plus	revenue from middle–priced	plus	revenue from inexpensive	is	total revenue from statues.
$20x$	+	$12y$	+	$9z$	=	2100

29. cont.

Solve: $\begin{cases} x+y+z=180 \\ 5x+4y+3z=650 \\ 20x+12y+9z=2100 \end{cases}$ Eliminate z.

Multiply Equation 1 by –3 and add Equations 1 and 2.

$$\begin{array}{lrcr} 1. & -3x-3y-3z & = & -540 \\ 2. & 5x+4y+3z & = & 650 \\ \hline 4. & 2x+y & = & 110 \end{array}$$

Multiply Equation 1 by –9 and add Equations 1 and 3.

$$\begin{array}{lrcr} 1. & -9x-9y-9z & = & -1620 \\ 3. & 20x+12y+9z & = & 2100 \\ \hline 5. & 11x+3y & = & 480 \end{array}$$

Multiply Equation 4 by –3 and add Equations 4 and 5.

$$\begin{array}{lrcr} 4. & -6x-3y & = & -330 \\ 5. & 11x+3y & = & 480 \\ \hline 6. & 5x & = & 150 \\ & x & = & 30 \end{array}$$

Substitute $x = 30$ into Equation 4 and solve for y.

$$\begin{aligned} 2(30)+y &= 110 \\ 60+y &= 110 \\ y &= 50 \end{aligned}$$

Substitute $x = 30$ and $y = 50$ into Equation 1 and solve for z.

$$\begin{aligned} (30)+(50)+z &= 180 \\ 80+z &= 180 \\ z &= 100 \end{aligned}$$

Conclusion: The number of expensive statues would be $x = 30$. The number of middle–priced statues would be $y = 50$. The number of inexpensive statues would be $z = 100$.

31. **Analysis:** Find the number of ounces of each of three foods that need to be used to provide 14 grams of fat, 9 grams of carbohydrates, and 9 grams of protein.

Equations: Let A represent the ounces of Food A, B represent the ounces of Food B, and C represent the ounces of Food C.

	Grams from Food A	plus	grams from Food B	plus	grams from Food C	is	total grams.
Fat:	$2A$	+	$3B$	+	$1C$	=	14

	Grams from Food A	plus	grams from Food B	plus	grams from Food C	is	total grams.
Carbohydrates:	$1A$	+	$2B$	+	$1C$	=	9

	Grams from Food A	plus	grams from Food B	plus	grams from Food C	is	total grams.
Protein:	$2A$	+	$1B$	+	$2C$	=	9

Solve: $\begin{cases} 2A+3B+C=14 \\ A+2B+C=9 \\ 2A+B+2C=9 \end{cases}$ Eliminate C.

Multiply Equation 2 by –1 and add Equations 1 and 2.

$$\begin{array}{lrcr} 1. & 2A+3B+C & = & 14 \\ 2. & -A-2B-C & = & -9 \\ \hline 4. & A+B & = & 5 \end{array}$$

Multiply Equation 2 by –2 and add Equations 2 and 3.

$$\begin{array}{lrcr} 2. & -2A-4B-2C & = & -18 \\ 3. & 2A+B+2C & = & 9 \\ \hline 5. & -3B & = & -9 \\ & B & = & 3 \end{array}$$

Substitute $B = 3$ into Equation 4 and solve for A.

$$\begin{aligned} A+(3) &= 5 \\ A &= 2 \end{aligned}$$

Substitute $A = 2$ and $B = 3$ into Equation 2 and solve for C.

$$\begin{aligned} (2)+2(3)+C &= 9 \\ 8+C &= 9 \\ C &= 1 \end{aligned}$$

Conclusion: Use 2 ounces of Food A, 3 ounces of Food B, and 1 ounce of Food C.

33. **Analysis:** Find the number of each of 3 types of statues that should be produced to use 14 hr of carving time, 15 hr of sanding time, and 21 hr of painting time.

Equations: Let x, y, and z respectively represent the number of totem poles, bears, and deer to be produced .

Carving on totem poles	plus	carving on bears	plus	carving on deer	is	total carving on statues.
$2x$	$+$	$2y$	$+$	z	$=$	14

Sanding on totem poles	plus	sanding on bears	plus	sanding on deer	is	total sanding on statues.
$1x$	$+$	$2y$	$+$	$2z$	$=$	15

Painting on totem poles	plus	painting on bears	plus	painting on deer	is	total painting on statues.
$3x$	$+$	$2y$	$+$	$2z$	$=$	21

Solve: $\begin{cases} 2x+2y+z=14 \\ x+2y+2z=15 \\ 3x+2y+2z=21 \end{cases}$ Eliminate y.

Multiply Equation 1 by -1 and add Equations 1 and 2.

$$\begin{array}{lrcr} 1. & -2x-2y-z & = & -14 \\ 2. & x+2y+2z & = & 15 \\ \hline 4. & -x+z & = & 1 \end{array}$$

Multiply Equation 1 by -1 and add Equations 1 and 3.

$$\begin{array}{lrcr} 1. & -2x-2y-z & = & -14 \\ 3. & 3x+2y+2z & = & 21 \\ \hline 5. & x+z & = & 7 \end{array}$$

Add Equations 4 and 5.

$$\begin{array}{lrcr} 4. & -x+z & = & 1 \\ 5. & x+z & = & 7 \\ \hline 6. & 2z & = & 8 \\ & z & = & 4 \end{array}$$

Substitute $z=4$ into Equation 4 and solve for x.

$$\begin{aligned} -x+(4) &= 1 \\ -x &= -3 \\ x &= 3 \end{aligned}$$

Substitute $x=3$ and $z=4$ into Equation 1 and solve for y.

$$\begin{aligned} 2(3)+2y+(4) &= 14 \\ 10+2y &= 14 \\ 2y &= 4 \\ y &= 2 \end{aligned}$$

Conclusion: Produce $x=3$ totem poles, $y=2$ bears, and $z=4$ deer.

35. **Analysis:** Find the percent of the earth's atmosphere that is nitrogen, oxygen and other gases.

Equations: Let x, y, and z respectively represent the percentage of nitrogen, oxygen and other gases.

Percent of nitrogen	plus	percent of oxygen	plus	percent of other gases	is	total percent.
x	$+$	y	$+$	z	$=$	100%

Percent of nitrogen	is	three times the sum of the other two	plus	12%.
x	$=$	$3(y+z)$	$+$	12

Percent of other gases	is	percent of oxygen	minus	20%.
z	$=$	y	$-$	20

Solve: $\begin{cases} x+y+z=100 \\ x=3(y+z)+12 \\ z=y-20 \end{cases}$ Simplify $\begin{cases} x+y+z=100 \\ x=3y+3z+12 \\ z=y-20 \end{cases}$ Rewrite $\begin{cases} x+y+z=100 \\ x-3y-3z=12 \\ y-z=20 \end{cases}$ Eliminate x.

35. cont.

Multiply Equation 1 by –1 and add Equations 1 and 2.

1.	$-x - y - z$	=	-100
2.	$x - 3y - 3z$	=	12
4.	$-4y - 4z$	=	-88

Multiply Equation 3 by 4 and add Equations 4 and 3.

3.	$4y - 4z$	=	80
4.	$-4y - 4z$	=	-88
5.	$-8z$	=	-8
	z	=	1

Substitute $z = 1$ into Equation 3 and solve for y.

$$y - (1) = 20$$
$$y = 21$$

Substitute $y = 21$ and $z = 1$ into Equation 1 and solve for x.

$$x + (21) + (1) = 100$$
$$x = 78$$

Conclusion: The earth's atmosphere is made up of 78% nitrogen, 21% oxygen and 1% other gases.

37. a. The three pages (planes) are intersecting in a straight line (the binding). There are infinitely many solutions.
 b. The shelves are three parallel planes. There is no solution.
 c. The three cards represent planes. Pairs of planes intersect, but the three planes do not all intersect. There is no solution.
 d. Each side of the die represents a plane. The three planes shown intersect at a common corner. There is one solution at that corner.

39. Find the equation of a parabola that fits the points (–2, 5), (2, –3), and (4, –1).

Substitute each point into the equation $y = ax^2 + bx + c$ and simplify.

(–2, 5):	$5 = a(-2)^2 + b(-2) + c$	or	$4a - 2b + c = 5$
(2, –3):	$-3 = a(2)^2 + b(2) + c$	or	$4a + 2b + c = -3$
(4, –1):	$-1 = a(4)^2 + b(4) + c$	or	$16a + 4b + c = -1$

Write as a system and solve.
$$\begin{cases} 4a - 2b + c = 5 \\ 4a + 2b + c = -3 \\ 16a + 4b + c = -1 \end{cases}$$
Eliminate c.

Multiply Equation 2 by –1 and add Equations 1 and 2.

1.	$4a - 2b + c$	=	5
2.	$-4a - 2b - c$	=	3
4.	$-4b$	=	8
	b	=	-2

Multiply Equation 2 by –1 and add Equations 2 and 3.

2.	$-4a - 2b - c$	=	3
3.	$16a + 4b + c$	=	-1
5.	$12a + 2b$	=	2

Substitute $b = -2$ into Equation 5 and solve for a.

$$12a + 2(-2) = 2$$
$$12a - 4 = 2$$
$$12a = 6$$
$$a = \frac{1}{2}$$

Substitute $a = \frac{1}{2}$ and $b = -2$ into Equation 1 and solve for c.

$$4\left(\frac{1}{2}\right) - 2(-2) + c = 5$$
$$6 + c = 5$$
$$c = -1$$

Substitute $a = \frac{1}{2}$, $b = -2$, and $c = -1$ into $y = ax^2 + bx + c$. The equation is $y = \frac{1}{2}x^2 - 2x - 1$.

41. Find the equation of a circle that fits the points (1, 3), (3, 1), and (1, –1).

Substitute each point into the equation $x^2 + y^2 + Cx + Dy + E = 0$ and simplify.

(1, 3):	$(-1)^2 + (-3)^2 + C(1) + D(3) + E = 0$	or	$C + 3D + E = -10$
(3, 1):	$(3)^2 + (1)^2 + C(3) + D(1) + E = 0$	or	$3C + D + E = -10$
(1, –1):	$(1)^2 + (-1)^2 + C(1) + D(-1) + E = 0$	or	$C - D + E = -2$

Write as a system and solve. $\begin{cases} C + 3D + E = -10 \\ 3C + D + E = -10 \\ C - D + E = -2 \end{cases}$ Eliminate E.

Multiply Equation 2 by –1 and add Equations 1 and 2.

$$\begin{array}{lrcl} 1. & C + 3D + E & = & -10 \\ 2. & \underline{-3C - D - E} & \underline{=} & \underline{10} \\ 4. & -2C + 2D & = & 0 \end{array}$$

Multiply Equation 2 by –1 and add Equations 2 and 3.

$$\begin{array}{lrcl} 2. & -3C - D - E & = & 10 \\ 3. & \underline{C - D + E} & \underline{=} & \underline{-2} \\ 5. & -2C - 2D & = & 8 \end{array}$$

Multiply Equation 5 by –1 and add Equations 4 and 5.

$$\begin{array}{lrcl} 4. & -2C + 2D & = & 0 \\ 5. & \underline{2C + 2D} & \underline{=} & \underline{-8} \\ & 4D & = & -8 \\ & D & = & -2 \end{array}$$

Substitute $D = -2$ into Equation 5 and solve for C.

$$\begin{aligned} -2C - 2(-2) &= 8 \\ -2C &= 4 \\ C &= -2 \end{aligned}$$

Substitute $C = -2$,$D = -2$ into Equation 3 and solve for E.

$$\begin{aligned} (-2) - (-2) + E &= -2 \\ E &= -2 \end{aligned}$$

Substitute $C = -2$, $D = -2$, and $E = -2$ into $x^2 + y^2 + Cx + Dy + E = 0$. The equation is $x^2 + y^2 - 2x - 2y - 2 = 0$.

43. **Analysis:** Find the measures of the three angles of a triangle.

Equations: Let A, B, and C respectively represent the measure of the three angles.

Angle A	plus	angle B	plus	angle C	is	180°.
A	$+$	B	$+$	C	$=$	180

Angle A	is	sum of angles B and C	less	100°.
A	$=$	$(B + C)$	$-$	100

Angle C	is	twice angle B	less	40°.
C	$=$	$2B$	$-$	40

Solve: $\begin{cases} A + B + C = 180 \\ A = B + C - 100 \\ C = 2B - 40 \end{cases}$ Rewrite in general form $\begin{cases} A + B + C = 180 \\ A - B - C = -100 \\ -2B + C = -40 \end{cases}$

Add Equations 1 and 2.

$$\begin{array}{lrcl} 1. & A + B + C & = & 180 \\ 2. & \underline{A - B - C} & \underline{=} & \underline{-100} \\ 4. & 2A & = & 80 \\ & A & = & 40 \end{array}$$

Add Equations 2 and 3.

$$\begin{array}{lrcl} 2. & A - B - C & = & -100 \\ 3. & \underline{-2B + C} & \underline{=} & \underline{-40} \\ 5. & A - 3B & = & -140 \end{array}$$

Substitute $A = 40$ into Equation 5, solve for B.

$$\begin{aligned} (40) - 3B &= -140 \\ -3B &= -180 \\ B &= 60 \end{aligned}$$

Substitute $A = 40$ and $B = 60$ into Equation 1, solve for C.

$$\begin{aligned} (40) + (60) + C &= 180 \\ 100 + C &= 180 \\ C &= 80 \end{aligned}$$

Conclusion: The measures of the angles A, B, and C are 40°, 60° and 80° respectively.

45. **Analysis:** Find the three integers that meet the given conditions.

Equations: Let *x*, *y*, and *z* respectively represent the three different integers.

First integer	plus	second integer	plus	third integer	is	sum
x	$+$	y	$+$	z	$=$	48

Double first integer	plus	second integer	plus	third integer	is	second sum
$2x$	$+$	y	$+$	z	$=$	60

First integer	plus	double second integer	plus	third integer	is	third sum
x	$+$	$2y$	$+$	z	$=$	63

Solve: $\begin{cases} x+y+z=48 \\ 2x+y+z=60 \\ x+2y+z=63 \end{cases}$ Eliminate *z*.

Multiply Equation 1 by -1 and add Equations 1 and 2.

$$\begin{array}{lrcr} 1. & -x-y-z & = & -48 \\ 2. & \underline{2x+y+z} & \underline{=} & \underline{60} \\ 4. & x & = & 12 \end{array}$$

Multiply Equation 1 by -1 and add Equations 1 and 3.

$$\begin{array}{lrcr} 1. & -x-\;y-z & = & -48 \\ 3. & \underline{x+2y+z} & \underline{=} & \underline{63} \\ 5. & y & = & 15 \end{array}$$

Substitute $x = 12$ and $y = 15$ into Equation 1 and solve for *z*.

$$\begin{aligned} (12)+(15)+z &= 48 \\ z &= 21 \end{aligned}$$

Conclusion: The three integers are 12, 15, and 21.

WRITING

47. Choose two different pairs of the three equations. Using the multiplication principle and the addition principal, eliminate the same variable from the pairs of equations. This would reduce the original system to a system of two equations with two variables.

REVIEW

49. $f(x) = |x|$
The absolute value function without any translation or reflection

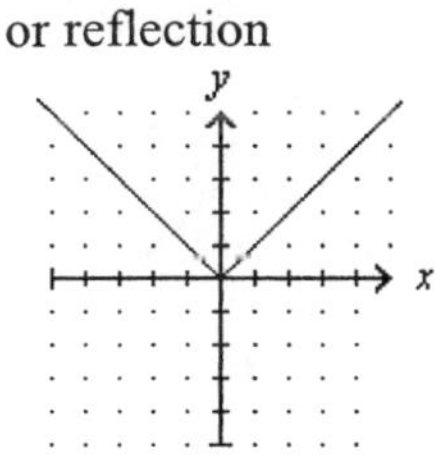

51. $h(x) = x^3$
A cubic function without any translation or reflection.

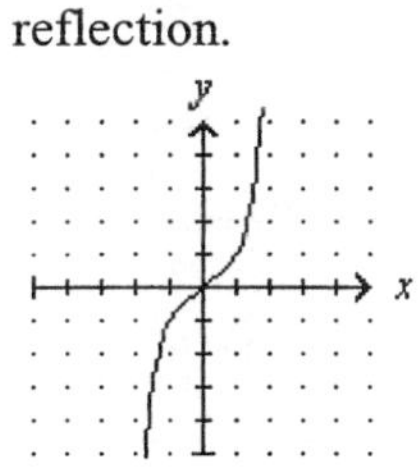

Section 3.4 Solving Systems Using Matrices

VOCABULARY

1. matrix
3. rows; columns
5. augmented

CONCEPTS

7. a. 2×3 b. 3×4

9. Write the resulting system.

$$\begin{cases} x - y = -10 \\ y = 6 \end{cases}$$

Use back substitution to find the solution.

$$x - (6) = -10$$
$$x = -4$$

Solution is (–4, 6)

11. The system is inconsistent and has no solution.

NOTATION

13. a. The notation $\frac{1}{3}R_1$ means multiply row 1 by $\frac{1}{3}$. $\left[\begin{array}{ccc|c} 1 & 2 & -3 & 0 \\ 1 & 5 & -2 & 1 \\ -2 & 2 & -2 & 5 \end{array}\right]$

b. The notation $-R_1 + R_2$ means multiply row 1 by -1 and add this result to row 2. $\left[\begin{array}{ccc|c} 1 & 2 & -3 & 0 \\ 0 & 3 & 1 & 1 \\ -2 & 2 & -2 & 5 \end{array}\right]$

15. Solve $\begin{cases} 4x - y = 14 \\ x + y = 6 \end{cases}$

$$\left[\begin{array}{cc|c} 4 & -1 & 14 \\ 1 & 1 & 6 \end{array}\right]$$

$$\left[\begin{array}{cc|c} 1 & 1 & 6 \\ 4 & -1 & 14 \end{array}\right] \quad R_1 \leftrightarrow R_2$$

$$\left[\begin{array}{cc|c} 1 & 1 & 6 \\ 0 & -5 & -10 \end{array}\right] \quad -4R_1 + R_2$$

$$\left[\begin{array}{cc|c} 1 & 1 & 6 \\ 0 & 1 & 2 \end{array}\right] \quad -\tfrac{1}{5}R_2$$

This matrix represents the system.

$$\begin{cases} x + y = 6 \\ y = 2 \end{cases}$$

The solution is (4, 2).

PRACTICE

17. $\begin{cases} x+y=2 \\ x-y=0 \end{cases}$ $\left[\begin{array}{cc|c} 1 & 1 & 2 \\ 1 & -1 & 0 \end{array}\right]$

Write system as an augmented matrix.

Multiply row 1 by –1 and add to row 2.

$\left[\begin{array}{cc|c} 1 & 1 & 2 \\ 0 & -2 & -2 \end{array}\right]$ $-R_1+R_2$

Multiply row 2 by $-\frac{1}{2}$.

$\left[\begin{array}{cc|c} 1 & 1 & 2 \\ 0 & 1 & 1 \end{array}\right]$ $-\frac{1}{2}R_2$

Write the resulting system.

$\begin{cases} x+y=2 \\ y=1 \end{cases}$

Use back substitution to find the solution.

$x+1=2$

$x=1$

Solution: (1, 1)

$\begin{cases} 2x+y=1 \\ x+2y=-4 \end{cases}$ $\left[\begin{array}{cc|c} 2 & 1 & 1 \\ 1 & 2 & -4 \end{array}\right]$

Write system as an augmented matrix.

Interchange row 1 and row 2.

$\left[\begin{array}{cc|c} 1 & 2 & -4 \\ 2 & 1 & 1 \end{array}\right]$ $R_1 \leftrightarrow R_2$

Multiply row 1 by –2 and add to row 2.

$\left[\begin{array}{cc|c} 1 & 2 & -4 \\ 0 & -3 & 9 \end{array}\right]$ $-2R_1+R_2$

Multiply row 2 by $-\frac{1}{3}$.

$\left[\begin{array}{cc|c} 1 & 2 & -4 \\ 0 & 1 & -3 \end{array}\right]$ $-\frac{1}{3}R_2$

Write the resulting system.

$\begin{cases} x+2y=-4 \\ y=-3 \end{cases}$

Use back substitution to find the solution.

$x+2(-3)=-4$

$x=2$

Solution: (2, –3)

21. $\begin{cases} 2x-y=-1 \\ x-2y=1 \end{cases}$ $\left[\begin{array}{cc|c} 2 & -1 & -1 \\ 1 & -2 & 1 \end{array}\right]$

Write system as an augmented matrix.

Interchange row 1 and row 2.

$\left[\begin{array}{cc|c} 1 & -2 & 1 \\ 2 & -1 & -1 \end{array}\right]$ $R_1 \leftrightarrow R_2$

Multiply row 1 by –2 and add to row 2.

$\left[\begin{array}{cc|c} 1 & -2 & 1 \\ 0 & 3 & -3 \end{array}\right]$ $-2R_1+R_2$

Multiply row 2 by $\frac{1}{3}$.

$\left[\begin{array}{cc|c} 1 & -2 & 1 \\ 0 & 1 & -1 \end{array}\right]$ $\frac{1}{3}R_2$

Write the resulting system.

$\begin{cases} x-2y=1 \\ y=-1 \end{cases}$

Use back substitution to find the solution.

$x-2(-1)=1$

$x=-1$

Solution: (–1, –1)

23. $\begin{cases} 3x+4y=-12 \\ 9x-2y=6 \end{cases}$ $\left[\begin{array}{cc|c} 3 & 4 & -12 \\ 9 & -2 & 6 \end{array}\right]$

Write system as an augmented matrix.

Multiply row 1 by $\frac{1}{3}$.

$\left[\begin{array}{cc|c} 1 & 4/3 & -4 \\ 9 & 2 & 6 \end{array}\right]$ $\frac{1}{3}R_1$

Multiply row 1 by –9 and add to row 2.

$\left[\begin{array}{cc|c} 1 & 4/3 & -4 \\ 0 & -14 & 42 \end{array}\right]$ $-9R_1+R_2$

Multiply row 2 by $-\frac{1}{14}$.

$\left[\begin{array}{cc|c} 1 & 4/3 & -4 \\ 0 & 1 & -3 \end{array}\right]$ $-\frac{1}{14}R_2$

Write the resulting system.

$\begin{cases} x+\frac{4}{3}y=-4 \\ y=-3 \end{cases}$

Use back substitution to find the solution.

$x+\frac{4}{3}(-3)=-4$

$x=0$

Solution: (0, –3)

25. $\begin{cases} x+y+z=6 \\ x+2y+z=8 \\ x+y+2z=9 \end{cases}$ $\left[\begin{array}{ccc|c} 1 & 1 & 1 & 6 \\ 1 & 2 & 1 & 8 \\ 1 & 1 & 2 & 9 \end{array}\right]$

Write system as an augmented matrix.

Multiply row 1 by –1 and add to row 2.
Multiply row 1 by –1 and add to row 3.

$\left[\begin{array}{ccc|c} 1 & 1 & 1 & 6 \\ 0 & 1 & 0 & 2 \\ 0 & 0 & 1 & 3 \end{array}\right]$ $\begin{array}{l} -R_1+R_2 \\ -R_1+R_3 \end{array}$

Write the resulting system.

$\begin{cases} x+y+z=6 \\ y=2 \\ z=3 \end{cases}$

Use back substitution to find the solution.

$x+2+3=6$
$x=1$

Solution: (1, 2, 3)

27. $\begin{cases} 3x+y-3z=5 \\ x-2y+4z=10 \\ x+y+z=13 \end{cases}$ $\left[\begin{array}{ccc|c} 3 & 1 & -3 & 5 \\ 1 & -2 & 4 & 10 \\ 1 & 1 & 1 & 13 \end{array}\right]$

Write system as an augmented matrix.

Interchange row 1 and row 3.

$\left[\begin{array}{ccc|c} 1 & 1 & 1 & 13 \\ 1 & -2 & 4 & 10 \\ 3 & 1 & -3 & 5 \end{array}\right]$ $R_1 \leftrightarrow R_3$

Multiply row 1 by –1 and add to row 2.
Multiply row 1 by –3 and add to row 3.

$\left[\begin{array}{ccc|c} 1 & 1 & 1 & 13 \\ 0 & -3 & 3 & -3 \\ 0 & -2 & -6 & -34 \end{array}\right]$ $\begin{array}{l} -1R_1+R_2 \\ -3R_1+R_3 \end{array}$

Multiply row 2 by $-\frac{1}{3}$.

$\left[\begin{array}{ccc|c} 1 & 1 & 1 & 13 \\ 0 & 1 & -1 & 1 \\ 0 & -2 & -6 & -34 \end{array}\right]$ $-\frac{1}{3}R_2$

Multiply row 2 by 2 and add to row 3.

$\left[\begin{array}{ccc|c} 1 & 1 & 1 & 13 \\ 0 & 1 & -1 & 1 \\ 0 & 0 & -8 & -32 \end{array}\right]$ $2R_2+R_3$

Multiply row 3 by $-\frac{1}{8}$.

$\left[\begin{array}{ccc|c} 1 & 1 & 1 & 13 \\ 0 & 1 & -1 & 1 \\ 0 & 0 & 1 & 4 \end{array}\right]$ $-\frac{1}{8}R_3$

Write the resulting system.

$\begin{cases} x+y+z=13 \\ y-z=1 \\ z=4 \end{cases}$

Use back substitution to find the solution.

$y-4=1$
$y=5$
$x+5+4=13$
$x=4$

Solution: (4, 5, 4)

29. $\begin{cases} 3x-2y+4z=4 \\ x+y+z=3 \\ 6x-2y-3z=10 \end{cases}$ $\left[\begin{array}{ccc|c} 3 & -2 & 4 & 4 \\ 1 & 1 & 1 & 3 \\ 6 & -2 & -3 & 10 \end{array}\right]$

Write system as an augmented matrix.

Interchange row 1 and row 2.

$$\left[\begin{array}{ccc|c} 1 & 1 & 1 & 3 \\ 3 & -2 & 4 & 4 \\ 6 & -2 & -3 & 10 \end{array}\right] \quad R_1 \leftrightarrow R_2$$

Multiply row 1 by –3 and add to row 2.
Multiply row 1 by –6 and add to row 3.

$$\left[\begin{array}{ccc|c} 1 & 1 & 1 & 3 \\ 0 & -5 & 1 & -5 \\ 0 & -8 & -9 & -8 \end{array}\right] \quad \begin{array}{l} -3R_1+R_2 \\ -6R_1+R_3 \end{array}$$

Multiply row 2 by $-\frac{1}{5}$.

$$\left[\begin{array}{ccc|c} 1 & 1 & 1 & 3 \\ 0 & 1 & -1/5 & 1 \\ 0 & -8 & -9 & -8 \end{array}\right] \quad -\tfrac{1}{5}R_2$$

Multiply row 2 by 8 and add to row 3.

$$\left[\begin{array}{ccc|c} 1 & 1 & 1 & 3 \\ 1 & 1 & -1/5 & 1 \\ 0 & 0 & -53/5 & 0 \end{array}\right] \quad 8R_2+R_3$$

Multiply row 3 by $-\frac{5}{53}$.

$$\left[\begin{array}{ccc|c} 1 & 1 & 1 & 3 \\ 0 & 1 & -1/5 & 1 \\ 0 & 0 & 1 & 0 \end{array}\right] \quad -\tfrac{5}{53}R_3$$

Write the resulting system.

$$\begin{cases} x+y+z=3 \\ y-\frac{1}{5}z=1 \\ z=0 \end{cases}$$

Use back substitution to find the solution.

$$y-\tfrac{1}{5}(0)=1$$
$$y=1$$
$$x+1+0=3$$
$$x=2$$

Solution: $(2, 1, 0)$

31. $\begin{cases} 2a+b+3c=3 \\ -2a-b+c=5 \\ 4a-2b+2c=2 \end{cases}$ $\left[\begin{array}{ccc|c} 2 & 1 & 3 & 3 \\ -2 & -1 & 1 & 5 \\ 4 & -2 & 2 & 2 \end{array}\right]$

Write system as an augmented matrix.

Multiply row 1 by $\frac{1}{2}$.

$$\left[\begin{array}{ccc|c} 1 & 1/2 & 3/2 & 3/2 \\ -2 & -1 & 1 & 5 \\ 4 & -2 & 2 & 2 \end{array}\right] \quad \tfrac{1}{2}R_1$$

Multiply row 1 by 2 and add to row 2.
Multiply row 1 by –4 and add to row 3.

$$\left[\begin{array}{ccc|c} 1 & 1/2 & 3/2 & 3/2 \\ 0 & 0 & 4 & 8 \\ 0 & -4 & -4 & -4 \end{array}\right] \quad \begin{array}{l} 2R_1+R_2 \\ -4R_1+R_3 \end{array}$$

Interchange row 2 and row 3.

$$\left[\begin{array}{ccc|c} 1 & 1/2 & 3/2 & 3/2 \\ 0 & -4 & -4 & -4 \\ 0 & 0 & 4 & 8 \end{array}\right] \quad R_2 \leftrightarrow R_3$$

Multiply row 2 by $-\frac{1}{4}$ and row 3 by $\frac{1}{4}$.

$$\left[\begin{array}{ccc|c} 1 & 1/2 & 3/2 & 3/2 \\ 0 & 1 & 1 & 1 \\ 0 & 0 & 1 & 2 \end{array}\right] \quad \begin{array}{l} -\frac{1}{4}R_2 \\ \frac{1}{4}R_3 \end{array}$$

Write the resulting system.

$$\begin{cases} a+\frac{1}{2}b+\frac{3}{2}c=\frac{3}{2} \\ b+c=1 \\ c=2 \end{cases}$$

Use back substitution to find the solution.

$$b+2=1$$
$$b=-1$$
$$a+\tfrac{1}{2}(-1)+\tfrac{3}{2}(2)=\tfrac{3}{2}$$
$$a-\tfrac{1}{2}+3=\tfrac{3}{2}$$
$$a=-1$$

Solution: $(-1, -1, 2)$

33. $\begin{cases} x-3y=9 \\ -2x+6y=18 \end{cases}$

Write system as an augmented matrix.

$$\left[\begin{array}{cc|c} 1 & -3 & 9 \\ -2 & 6 & 18 \end{array}\right]$$

Multiply row 1 by 2 and add to row 2.

$$\left[\begin{array}{cc|c} 1 & -3 & 9 \\ 0 & 0 & 36 \end{array}\right] \quad 2R_1 + R_2$$

The second row of this final matrix states that 0 = 36. This is a false statement and indicates that the system is inconsistent. There is no solution.

35. $\begin{cases} 4x+4y=12 \\ -x-y=-3 \end{cases}$

Write system as an augmented matrix.

$$\left[\begin{array}{cc|c} 4 & 4 & 12 \\ -1 & -1 & -3 \end{array}\right]$$

Multiply row 1 by $\frac{1}{4}$.

$$\left[\begin{array}{cc|c} 1 & 1 & 3 \\ -1 & -1 & -3 \end{array}\right] \quad \tfrac{1}{4}R_1$$

Add row 1 to row 2.

$$\left[\begin{array}{cc|c} 1 & 1 & 3 \\ 0 & 0 & 0 \end{array}\right] \quad R_1 + R_2$$

The second row of this final matrix states that 0 = 0. This is an identity or a statement that is always true. It indicates the system is dependent. There are infinitely many solutions.

37. $\begin{cases} 6x+y-z=-2 \\ x+2y+z=5 \\ 5y-z=2 \end{cases}$ $\left[\begin{array}{ccc|c} 6 & 1 & -1 & -2 \\ 1 & 2 & 1 & 5 \\ 0 & 5 & -1 & 2 \end{array}\right]$

Write system as an augmented matrix.

Interchange row 1 and row 2.

$$\left[\begin{array}{ccc|c} 1 & 2 & 1 & 5 \\ 6 & 1 & -1 & -2 \\ 0 & 5 & -1 & 2 \end{array}\right] \quad R_1 \leftrightarrow R_2$$

Multiply row 1 by –6 and add to row 2.

$$\left[\begin{array}{ccc|c} 1 & 2 & 1 & 5 \\ 0 & -11 & -7 & -32 \\ 0 & 5 & -1 & 2 \end{array}\right] \quad -6R_1 + R_2$$

Multiply row 2 by $-\frac{1}{11}$.

$$\left[\begin{array}{ccc|c} 1 & 2 & 1 & 5 \\ 0 & 1 & 7/11 & 32/11 \\ 0 & 5 & -1 & 2 \end{array}\right] \quad -\tfrac{1}{11}R_2$$

Multiply row 2 by –5 and add to row 3.

39. $\begin{cases} 2x+y-z=1 \\ x+2y+2z=2 \\ 4x+5y+3z=3 \end{cases}$ $\left[\begin{array}{ccc|c} 2 & 1 & -1 & 1 \\ 1 & 2 & 2 & 2 \\ 4 & 5 & 3 & 3 \end{array}\right]$

Write system as an augmented matrix.

Interchange row 1 and row 2.

$$\left[\begin{array}{ccc|c} 1 & 2 & 2 & 2 \\ 2 & 1 & -1 & 1 \\ 4 & 5 & 3 & 3 \end{array}\right] \quad R_1 \leftrightarrow R_2$$

Multiply row 1 by –2 and add to row 2.
Multiply row 1 by –4 and add to row 3.

$$\left[\begin{array}{ccc|c} 1 & 2 & 1 & 5 \\ 0 & 1 & 7/11 & 32/11 \\ 0 & 0 & 1 & 3 \end{array}\right] \quad -\tfrac{11}{46}R_3$$

Write the resulting system.

$$\begin{cases} x+2y+z=5 \\ y+\frac{7}{11}z=\frac{32}{11} \\ z=3 \end{cases}$$

Use back substitution to find the solution.

37. cont.

$$\left[\begin{array}{ccc|c} 1 & 2 & 1 & 5 \\ 0 & 1 & 7/11 & 32/11 \\ 0 & 0 & -46/11 & -138/11 \end{array}\right] \quad -5R_2 + R_3$$

Multiply row 3 by $-\frac{11}{46}$.

$$\left[\begin{array}{ccc|c} 1 & 2 & 2 & 2 \\ 0 & -3 & -5 & -3 \\ 0 & -3 & -5 & -5 \end{array}\right] \quad \begin{array}{l} -2R_1 + R_2 \\ -4R_1 + R_3 \end{array}$$

Multiply row 2 by $-\frac{1}{3}$.

$$\left[\begin{array}{ccc|c} 1 & 2 & 2 & 2 \\ 0 & 1 & 5/3 & 1 \\ 0 & -3 & -5 & -5 \end{array}\right] \quad -\tfrac{1}{3}R_2$$

Multiply row 2 by 3 and add to row 3.

$$\left[\begin{array}{ccc|c} 1 & 2 & 2 & 2 \\ 0 & 1 & 5/3 & 1 \\ 0 & 0 & 0 & -2 \end{array}\right] \quad 3R_2 + R_3$$

The third row of this final matrix states that $0 = -2$. This is a contradiction. It indicates that the system is inconsistent. There is no solution.

39. cont.

$$y + \tfrac{7}{11}(3) = \tfrac{32}{11}$$
$$y + \tfrac{21}{11} = \tfrac{32}{11}$$
$$y = 1$$
$$x + 2(1) + 3 = 5$$
$$x = 0$$

Solution: (0, 1, 3)

41. $\begin{cases} 5x+3y=4 \\ 3y-4z=4 \\ x+z=1 \end{cases}$

Write system as an augmented matrix.

$$\left[\begin{array}{ccc:c} 5 & 3 & 0 & 4 \\ 0 & 3 & -4 & 4 \\ 1 & 0 & 1 & 1 \end{array}\right]$$

Interchange row 1 and row 3.

$$\left[\begin{array}{ccc:c} 1 & 0 & 1 & 1 \\ 0 & 3 & -4 & 4 \\ 5 & 3 & 0 & 4 \end{array}\right] \quad R_1 \leftrightarrow R_3$$

Multiply row 1 by –5 and add to row 3.

$$\left[\begin{array}{ccc:c} 1 & 0 & 1 & 1 \\ 0 & 3 & -4 & 4 \\ 0 & 3 & -5 & -1 \end{array}\right] \quad -5R_1 + R_3$$

Multiply row 2 by $\frac{1}{3}$.

$$\left[\begin{array}{ccc:c} 1 & 0 & 1 & 1 \\ 0 & 1 & -\frac{4}{3} & \frac{4}{3} \\ 0 & 3 & -5 & -1 \end{array}\right] \quad \tfrac{1}{3}R_2$$

Multiply row 2 by –3 and add to row 3.

$$\left[\begin{array}{ccc:c} 1 & 0 & 1 & 1 \\ 0 & 1 & -\frac{4}{3} & \frac{4}{3} \\ 0 & 0 & -1 & -5 \end{array}\right] \quad -3R_2 + R_3$$

Multiply row 3 by –1.

$$\left[\begin{array}{ccc:c} 1 & 0 & 1 & 1 \\ 0 & 1 & -\frac{4}{3} & \frac{4}{3} \\ 0 & 0 & 1 & 5 \end{array}\right] \quad -R_3$$

Write the resulting system.

$$\begin{cases} x+z=1 \\ y-\frac{4}{3}z=\frac{4}{3} \\ z=5 \end{cases}$$

Use back substitution to find the solution.

$$y-\tfrac{4}{3}(5)=\tfrac{4}{3}$$
$$y-\tfrac{20}{3}=\tfrac{4}{3}$$
$$y=8$$
$$x+5=1$$
$$x=-4$$

Solution: (–4, 8, 5)

43. $\begin{cases} x-y=1 \\ 2x-z=0 \\ 2y-z=-2 \end{cases}$

Write system as an augmented matrix.

$$\left[\begin{array}{ccc:c} 1 & -1 & 0 & 1 \\ 2 & 0 & -1 & 0 \\ 0 & 2 & -1 & -2 \end{array}\right]$$

Multiply row 1 by –2 and add to row 2.

$$\left[\begin{array}{ccc:c} 1 & -1 & -0 & 1 \\ 0 & 2 & -1 & -2 \\ 0 & 2 & -1 & -2 \end{array}\right] \quad -2R_1 + R_2$$

The last two rows of this matrix are identical. This indicates that the system is dependent. There will be infinitely many solutions.

APPLICATIONS

45. **Analysis:** Find two angles where one angle is 46° larger than its complement.

Equations: Let x represent the measure of one angle and y represent the measure of the second angle. The sum of the two angles would be 90° since they are complementary.

Sum:	First angle	plus	second angle	is	90°.
	x	$+$	y	$=$	90
Comparison:	First angle	is	second angle	plus	46°
	x	$=$	y	$+$	46

Solve: $\begin{cases} x+y=90 \\ x=y+46 \end{cases}$ Rewrite in general form $\begin{cases} x+y=90 \\ x-y=46 \end{cases}$ Augmented matrix: $\left[\begin{array}{cc|c} 1 & 1 & 90 \\ 1 & -1 & 46 \end{array}\right]$

Multiply row 1 by -1 and add to row 2.

$\left[\begin{array}{cc|c} 1 & 1 & 90 \\ 0 & -2 & -44 \end{array}\right]$ $\quad -1R_1+R_2$

Multiply row 2 by $-\frac{1}{2}$.

$\left[\begin{array}{cc|c} 1 & 1 & 90 \\ 0 & 1 & 22 \end{array}\right]$ $\quad -\frac{1}{2}R_2$

Write the resulting system.

$\begin{cases} x+y=90 \\ \quad\; y=22 \end{cases}$

Use back substitution to find the solution.

$x+22=90$

$x=68$

Conclusion: The measures of the two angles are 22° and 68°.

47. **Analysis:** Find the measures of the three angles of the triangle in the illustration.

Equations: Let A, B, and C respectively represent the measures of the angles of the triangle. The sum of the three angles would be 180° since they represent the interior angles in a triangle.

Sum:	Angle A	plus	angle B	plus	angle C	is	180°.
	A	$+$	B	$+$	C	$=$	180

Comparison:	Angle B	is	angle A	plus	25°.
	B	$=$	A	$+$	25
Comparison:	Angle C	is	twice angle A	less	5°.
	C	$=$	$2A$	$-$	5

Solve: $\begin{cases} A+B+C=180 \\ B=A+25 \\ C=2A-5 \end{cases}$ Rewrite. $\begin{cases} A+B+C=180 \\ -A+B=25 \\ -2A+C=-5 \end{cases}$ Augmented matrix. $\left[\begin{array}{ccc|c} 1 & 1 & 1 & 180 \\ -1 & 1 & 0 & 25 \\ -2 & 0 & 1 & -5 \end{array}\right]$

Multiply row 1 by 1 and add to row 2.
Multiply row 1 by 2 and add to row 3.

$$\left[\begin{array}{ccc|c} 1 & 1 & 1 & 180 \\ 0 & 2 & 1 & 205 \\ 0 & 2 & 3 & 355 \end{array}\right] \quad \begin{array}{l} R_1+R_2 \\ 2R_1+R_3 \end{array}$$

Multiply row 2 by $\frac{1}{2}$.

$$\left[\begin{array}{ccc|c} 1 & 1 & 1 & 180 \\ 0 & 1 & 1/2 & 205/2 \\ 0 & 2 & 3 & 355 \end{array}\right] \quad \tfrac{1}{2}R_2$$

Multiply row 2 by −2 and add to row 3.

$$\left[\begin{array}{ccc|c} 1 & 1 & 1 & 180 \\ 0 & 1 & 1/2 & 205/2 \\ 0 & 0 & 2 & 150 \end{array}\right] \quad -2R_2+R_3$$

Multiply row 3 by $\frac{1}{2}$.

$$\left[\begin{array}{ccc|c} 1 & 1 & 1 & 180 \\ 0 & 1 & 1/2 & 205/2 \\ 0 & 0 & 1 & 75 \end{array}\right] \quad \tfrac{1}{2}R_3$$

Write the resulting system.

$$\begin{cases} A+B+C=180 \\ B+\frac{1}{2}C=\frac{205}{2} \\ C=75 \end{cases}$$

Use back substitution to find the solution.

$$B+\tfrac{1}{2}(75)=\tfrac{205}{2}$$
$$B=\tfrac{130}{2}=65$$
$$A+65+75=180$$
$$A=40$$

Conclusion: The measures of the three angles are 40°, 65° and 75° respectively.

49. **Analysis:** Find two angles where one angle is 28° less than its supplement.

Equations: Let A represent the measure of Angle 1 and B represent the measure of the Angle 2. The sum of the two angles would be 180° since they are supplementary.

Sum:	Angle 1	plus	Angle 2	is	180°.
	A	$+$	B	$=$	180
Comparison:	Angle 1	is	Angle 2	less	28°
	A	$=$	B	$-$	28

Solve: $\begin{cases} A+B=180 \\ A=B-28 \end{cases}$ Rewrite in general form $\begin{cases} A+B=180 \\ A-B=-28 \end{cases}$ Augmented matrix: $\left[\begin{array}{cc|c} 1 & 1 & 180 \\ 1 & -1 & -28 \end{array}\right]$

Multiply row 1 by −1 and add to row 2.

$$\left[\begin{array}{cc|c} 1 & 1 & 180 \\ 0 & -2 & -208 \end{array}\right] \quad -1R_1+R_2$$

Multiply row 2 by $-\frac{1}{2}$.

$$\left[\begin{array}{cc|c} 1 & 1 & 180 \\ 0 & 1 & 104 \end{array}\right] \quad -\tfrac{1}{2}R_2$$

49. cont.

Write the resulting system.

$$\begin{cases} A + B = 180 \\ B = 104 \end{cases}$$

Use back substitution to find the solution.

$$A + 104 = 180$$
$$A = 76$$

Conclusion: The measure of Angle 1 is 76° and the measure of Angle 2 is 104°.

51. **Analysis:** Find the number of seats in each of three sections in an 800–seat theater.

Equations: Let *x*, *y*, and *z* respectively represent the number of Founder's circle seats, the Box seats, and the Promenade seats. The value of the each type of ticket will be the ticket price times the quantity of tickets sold.

	Quantity	Matinee	Evening
Founder's circle	x	\$30x	\$40x
Box seats	y	\$20y	\$30y
Promenade	z	\$10z	\$25z
Total	800	\$13,000	\$23,000

Number of Founder's circle	plus	number of Box seats	plus	number of Promenade	is	total number of seats.
x	$+$	y	$+$	z	$=$	800

Matinee value of Founder's circle	plus	matinee value of Box seats	plus	matinee value of Promenade	is	total value of matinee seats.
$30x$	$+$	$20y$	$+$	$10z$	$=$	13,000

Evening value of Founder's circle	plus	evening value of Box seats	plus	evening value of Promenade	is	total value of Evening seats.
$40x$	$+$	$30y$	$+$	$25z$	$=$	23,000

Solve:
$$\begin{cases} x + y + z = 800 \\ 30x + 20y + 10z = 13{,}000 \\ 40x + 30y + 25z = 23{,}000 \end{cases}$$

Augmented matrix.
$$\left[\begin{array}{ccc|c} 1 & 1 & 1 & 800 \\ 30 & 20 & 10 & 13000 \\ 40 & 30 & 25 & 23{,}000 \end{array}\right]$$

Multiply row 1 by –30 and add to row 2.
Multiply row 1 by –40 and add to row 3.

$$\left[\begin{array}{ccc|c} 1 & 1 & 1 & 800 \\ 0 & -10 & 20 & -11{,}000 \\ 0 & -10 & -15 & -9{,}000 \end{array}\right] \begin{array}{l} \\ -30R_1 + R_2 \\ -40R_1 + R_3 \end{array}$$

Multiply row 2 by $-\frac{1}{10}$.

$$\left[\begin{array}{ccc|c} 1 & 1 & 1 & 800 \\ 0 & 1 & 2 & 1100 \\ 0 & -10 & -15 & -9000 \end{array}\right] \begin{array}{l} \\ -\frac{1}{10}R_2 \\ \end{array}$$

Multiply row 2 by 10 and add to row 3.

$$\left[\begin{array}{ccc|c} 1 & 1 & 1 & 800 \\ 0 & 1 & 2 & 1100 \\ 0 & 0 & 5 & 2000 \end{array}\right] \begin{array}{l} \\ \\ 10R_2 + R_3 \end{array}$$

Multiply row 3 by $\frac{1}{5}$.

$$\left[\begin{array}{ccc|c} 1 & 1 & 1 & 800 \\ 0 & 1 & 2 & 1100 \\ 0 & 0 & 1 & 400 \end{array}\right] \begin{array}{l} \\ \\ \frac{1}{5}R_3 \end{array}$$

Write the resulting system.

$$\begin{cases} x + y + z = 800 \\ y + 2z = 1100 \\ z = 400 \end{cases}$$

Use back substitution to find the solution.

$$y + 2(400) = 1100$$
$$y = 300$$
$$x + 300 + 400 = 800$$
$$x = 100$$

Conclusion: There are 100 Founder's circle seats, 300 Box seats, and 400 Promenade seats.

WRITING

53. "Back substitution" is the method of finding all of the values of the variables in a system of equations after the value for one of the variables has been found. It involves using the equations "above" the current equation and substituting the known values for the variables and solving for the unknown variable.

REVIEW

55. $m = \dfrac{y_2 - y_1}{x_2 - x_1}, \ (x_2 \neq x_1)$

57. $y - y_1 = m(x - x_1)$

Section 3.5 Solving Systems Using Determinants

VOCABULARY

1. determinant
3. minor
5. rows; columns

CONCEPTS

7. dependent; inconsistent
9. $ad - bc$

11. $\begin{vmatrix} 3 & 4 \\ 2 & -3 \end{vmatrix}$

13. $x = \dfrac{D_x}{D} = \dfrac{-7}{-11} = \dfrac{7}{11};\ y = \dfrac{D_y}{D} = \dfrac{5}{-11} = -\dfrac{5}{11}$; The solution is $\left(\dfrac{7}{11}, -\dfrac{5}{11}\right)$.

NOTATION

15. $\begin{vmatrix} 5 & -2 \\ -2 & 6 \end{vmatrix}$

$= 5(\ \mathbf{6}\ ) - (-2)(-2)$

$= \mathbf{30}\ - 4$

$= 26$

PRACTICE

17. $\begin{vmatrix} 2 & 3 \\ -2 & 1 \end{vmatrix} = 2(1) - (3)(-2) = 2 + 6 = 8$

19. $\begin{vmatrix} -1 & 2 \\ 3 & -4 \end{vmatrix} = (-1)(-4) - 2(3) = 4 - 6 = -2$

21. $\begin{vmatrix} 10 & 0 \\ 1 & 20 \end{vmatrix} = 10(20) - 0(1) = 200 - 0 = 200$

23. $\begin{vmatrix} -6 & -2 \\ 15 & 4 \end{vmatrix} = (-6)(4) - (-2)(15) = -24 + 30 = 6$

25. $\begin{vmatrix} 1 & 2 & 0 \\ 0 & 1 & 2 \\ 0 & 0 & 1 \end{vmatrix} = 1\begin{vmatrix} 1 & 2 \\ 0 & 1 \end{vmatrix} - 0\begin{vmatrix} 2 & 0 \\ 0 & 1 \end{vmatrix} + 0\begin{vmatrix} 2 & 0 \\ 1 & 2 \end{vmatrix}$ Expanding on the first column (signs are + – +).

$= 1[1(1) - 2(0)] - 0 + 0 = 1(1 - 0) = 1$

27. $\begin{vmatrix} 1 & -2 & 3 \\ -2 & 1 & 1 \\ -3 & -2 & 1 \end{vmatrix} = 1\begin{vmatrix} 1 & 1 \\ -2 & 1 \end{vmatrix} - (-2)\begin{vmatrix} -2 & 1 \\ -3 & 1 \end{vmatrix} + 3\begin{vmatrix} -2 & 1 \\ -3 & -2 \end{vmatrix}$ Expanding on the first row (signs are + – +).

$= 1[1(1) - 1(-2)] + 2[-2(1) - 1(-3)] + 3[-2(-2) - 1(-3)]$

$= 1(1 + 2) + 2(-2 + 3) + 3(4 + 3) = 1(3) + 2(1) + 3(7) = 3 + 2 + 21 = 26$

29. $\begin{vmatrix} 1 & 0 & 1 \\ 0 & 1 & 0 \\ 1 & 1 & 1 \end{vmatrix} = 1\begin{vmatrix} 1 & 0 \\ 1 & 1 \end{vmatrix} - 0\begin{vmatrix} 0 & 0 \\ 1 & 1 \end{vmatrix} + 1\begin{vmatrix} 0 & 1 \\ 1 & 1 \end{vmatrix}$ Expanding on the first row (signs are + – +).

$= 1[1(1) - 0(1)] - 0 + 1[0(1) - 1(1)] = 1(1) + 1(-1) = 1 - 1 = 0$

31. $\begin{vmatrix} 1 & 2 & 1 \\ -3 & 7 & 3 \\ -4 & 3 & -5 \end{vmatrix} = 1\begin{vmatrix} 7 & 3 \\ 3 & -5 \end{vmatrix} - 2\begin{vmatrix} -3 & 3 \\ -4 & -5 \end{vmatrix} + 1\begin{vmatrix} -3 & 7 \\ -4 & 3 \end{vmatrix}$ Expanding on the first row (signs are + – +).

$= 1[7(-5) - 3(3)] - 2[-3(-5) - 3(-4)] + 1[-3(3) - 7(-4)]$

$= 1(-35 - 9) - 2(15 + 12) + 1(-9 + 28) = 1(-44) - 2(27) + 1(19) = -44 - 54 + 19 = -79$

33. $\begin{cases} x + y = 6 \\ x - y = 2 \end{cases}$ $x = \dfrac{D_x}{D} = \dfrac{\begin{vmatrix} 6 & 1 \\ 2 & -1 \end{vmatrix}}{\begin{vmatrix} 1 & 1 \\ 1 & -1 \end{vmatrix}} = \dfrac{6(-1) - 1(2)}{1(-1) - 1(1)} = \dfrac{-6-2}{-1-1} = \dfrac{-8}{-2} = 4$

$y = \dfrac{D_y}{D} = \dfrac{\begin{vmatrix} 1 & 6 \\ 1 & 2 \end{vmatrix}}{\begin{vmatrix} 1 & 1 \\ 1 & -1 \end{vmatrix}} = \dfrac{1(2) - 6(1)}{-2} = \dfrac{2-6}{-2} = \dfrac{-4}{-2} = 2$ Solution: (4, 2)

35. $\begin{cases} 2x + 3y = 0 \\ 4x - 6y = -4 \end{cases}$ $x = \dfrac{D_x}{D} = \dfrac{\begin{vmatrix} 0 & 3 \\ -4 & -6 \end{vmatrix}}{\begin{vmatrix} 2 & 3 \\ 4 & -6 \end{vmatrix}} = \dfrac{0(-6) - 3(-4)}{2(-6) - 3(4)} = \dfrac{0+12}{-12-12} = \dfrac{12}{-24} = -\dfrac{1}{2}$

$y = \dfrac{D_y}{D} = \dfrac{\begin{vmatrix} 2 & 0 \\ 4 & -4 \end{vmatrix}}{\begin{vmatrix} 2 & 3 \\ 4 & 6 \end{vmatrix}} = \dfrac{2(-4) - 0(4)}{-24} = \dfrac{-8-0}{-24} = \dfrac{-8}{-24} = \dfrac{1}{3}$ Solution: $\left(-\dfrac{1}{2}, \dfrac{1}{3}\right)$

37. $\begin{cases} 3x + 2y = 11 \\ 6x + 4y = 11 \end{cases}$ $x = \dfrac{D_x}{D} = \dfrac{\begin{vmatrix} 11 & 2 \\ 11 & 4 \end{vmatrix}}{\begin{vmatrix} 3 & 2 \\ 6 & 4 \end{vmatrix}} = \dfrac{11(4) - 2(11)}{3(4) - 2(6)} = \dfrac{44-22}{12-12} = \dfrac{22}{0}$ is undefined

When the denominator determinant equals 0, but the numerator determinant is not equal to 0, then the system is inconsistent. There is no solution.

39. $\begin{cases} y = \dfrac{-2x+1}{3} \\ 3x - 2y = 8 \end{cases}$ Rewrite Equation 1 in general form. $\begin{cases} 2x + 3y = 1 \\ 3x - 2y = 8 \end{cases}$

$$x = \frac{D_x}{D} = \frac{\begin{vmatrix} 1 & 3 \\ 8 & -2 \end{vmatrix}}{\begin{vmatrix} 2 & 3 \\ 3 & -2 \end{vmatrix}} = \frac{1(-2)-3(8)}{2(-2)-3(3)} = \frac{-2-24}{-4-9} = \frac{-26}{-13} = 2$$

$$y = \frac{D_y}{D} = \frac{\begin{vmatrix} 2 & 1 \\ 3 & 8 \end{vmatrix}}{\begin{vmatrix} 2 & 3 \\ 3 & -2 \end{vmatrix}} = \frac{2(8)-1(3)}{-13} = \frac{16-3}{-13} = \frac{13}{-13} = -1$$

Solution: (2, −1)

41. $\begin{cases} x + y + z = 4 \\ x + y - z = 0 \\ x - y + z = 2 \end{cases}$

$$x = \frac{D_x}{D} = \frac{\begin{vmatrix} 4 & 1 & 1 \\ 0 & 1 & -1 \\ 2 & -1 & 1 \end{vmatrix}}{\begin{vmatrix} 1 & 1 & 1 \\ 1 & 1 & -1 \\ 1 & -1 & 1 \end{vmatrix}} = \frac{4\begin{vmatrix} 1 & -1 \\ -1 & 1 \end{vmatrix} - 0\begin{vmatrix} 1 & 1 \\ -1 & 1 \end{vmatrix} + 2\begin{vmatrix} 1 & 1 \\ 1 & -1 \end{vmatrix}}{1\begin{vmatrix} 1 & -1 \\ -1 & 1 \end{vmatrix} - 1\begin{vmatrix} 1 & -1 \\ 1 & 1 \end{vmatrix} + 1\begin{vmatrix} 1 & 1 \\ 1 & -1 \end{vmatrix}} = \frac{4(1-1)-0+2(-1-1)}{1(1-1)-1(2)+1(-2)} = \frac{0-0-4}{0-2-2} = \frac{-4}{-4} = 1$$

$$y = \frac{D_y}{D} = \frac{\begin{vmatrix} 1 & 4 & 1 \\ 1 & 0 & -1 \\ 1 & 2 & 1 \end{vmatrix}}{\begin{vmatrix} 1 & 1 & 1 \\ 1 & 1 & -1 \\ 1 & -1 & 1 \end{vmatrix}} = \frac{1\begin{vmatrix} 0 & -1 \\ 2 & 1 \end{vmatrix} - 1\begin{vmatrix} 4 & 1 \\ 2 & 1 \end{vmatrix} + 1\begin{vmatrix} 4 & 1 \\ 0 & -1 \end{vmatrix}}{-4} = \frac{1(0+2)-1(4-2)+1(-4-0)}{-4} = \frac{2-2-4}{-4} = \frac{-4}{-4} = 1$$

$$z = \frac{D_z}{D} = \frac{\begin{vmatrix} 1 & 1 & 4 \\ 1 & 1 & 0 \\ 1 & -1 & 2 \end{vmatrix}}{\begin{vmatrix} 1 & 1 & 1 \\ 1 & 1 & -1 \\ 1 & -1 & 1 \end{vmatrix}} = \frac{1\begin{vmatrix} 1 & 0 \\ -1 & 2 \end{vmatrix} - 1\begin{vmatrix} 1 & 4 \\ -1 & 2 \end{vmatrix} + 1\begin{vmatrix} 1 & 4 \\ 1 & 0 \end{vmatrix}}{-4} = \frac{1(2-0)-1(2+4)+1(0-4)}{-4} = \frac{2-6-4}{-4} = \frac{-8}{-4} = 2$$

Solution: (1, 1, 2)

43. $\begin{cases} x+y+2z=7 \\ x+2y+z=8 \\ 2x+y+z=9 \end{cases}$

$$x=\frac{D_x}{D}=\frac{\begin{vmatrix}7&1&2\\8&2&1\\9&1&1\end{vmatrix}}{\begin{vmatrix}1&1&2\\1&2&1\\2&1&1\end{vmatrix}}=\frac{7\begin{vmatrix}2&1\\1&1\end{vmatrix}-8\begin{vmatrix}1&2\\1&1\end{vmatrix}+9\begin{vmatrix}1&2\\2&1\end{vmatrix}}{1\begin{vmatrix}2&1\\1&1\end{vmatrix}-1\begin{vmatrix}1&2\\1&1\end{vmatrix}+2\begin{vmatrix}1&2\\2&1\end{vmatrix}}=\frac{7(2-1)-8(1-2)+9(1-4)}{1(2-1)-1(1-2)+2(1-4)}=\frac{7+8-27}{1+1-6}=\frac{-12}{-4}=3$$

$$y=\frac{D_y}{D}=\frac{\begin{vmatrix}1&7&2\\1&8&1\\2&9&1\end{vmatrix}}{\begin{vmatrix}1&1&2\\1&2&1\\2&1&1\end{vmatrix}}=\frac{1\begin{vmatrix}8&1\\9&1\end{vmatrix}-1\begin{vmatrix}7&2\\9&1\end{vmatrix}+2\begin{vmatrix}7&2\\8&1\end{vmatrix}}{-4}=\frac{1(8-9)-1(7-18)+2(7-16)}{-4}=\frac{-1+11-18}{-4}=\frac{-8}{-4}=2$$

$$z=\frac{D_z}{D}=\frac{\begin{vmatrix}1&1&7\\1&2&8\\2&1&9\end{vmatrix}}{\begin{vmatrix}1&1&2\\1&2&1\\2&1&1\end{vmatrix}}=\frac{1\begin{vmatrix}2&8\\1&9\end{vmatrix}-1\begin{vmatrix}1&7\\1&9\end{vmatrix}+2\begin{vmatrix}1&7\\2&8\end{vmatrix}}{-4}=\frac{1(18-8)-1(9-7)+2(8-14)}{-4}=\frac{10-2-12}{-4}=\frac{-4}{-4}=1$$

Solution: (3, 2, 1)

45. $\begin{cases} 2x+y+z=5 \\ x-2y+3z=10 \\ x+y-4z=-3 \end{cases}$

$$x=\frac{D_x}{D}=\frac{\begin{vmatrix}5&1&1\\10&-2&3\\-3&1&-4\end{vmatrix}}{\begin{vmatrix}2&1&1\\1&-2&3\\1&1&-4\end{vmatrix}}=\frac{5\begin{vmatrix}-2&3\\1&-4\end{vmatrix}-10\begin{vmatrix}1&1\\1&-4\end{vmatrix}+(-3)\begin{vmatrix}1&1\\-2&3\end{vmatrix}}{2\begin{vmatrix}-2&3\\1&-4\end{vmatrix}-1\begin{vmatrix}1&1\\1&-4\end{vmatrix}+1\begin{vmatrix}1&1\\-2&3\end{vmatrix}}=\frac{5(8-3)-10(-4-1)-3(3+2)}{2(8-3)-1(-4-1)+1(3+2)}=\frac{25+50-15}{10+5+5}=\frac{60}{20}=3$$

$$y=\frac{D_y}{D}=\frac{\begin{vmatrix}2&5&1\\1&10&3\\1&-3&-4\end{vmatrix}}{\begin{vmatrix}2&1&1\\1&-2&3\\1&1&-4\end{vmatrix}}=\frac{2\begin{vmatrix}10&3\\-3&-4\end{vmatrix}-1\begin{vmatrix}5&1\\-3&-4\end{vmatrix}+1\begin{vmatrix}5&1\\10&3\end{vmatrix}}{20}=\frac{2(-40+9)-1(-20+3)+1(15-10)}{20}=\frac{-62+17+5}{20}=\frac{-40}{20}=-2$$

$$z=\frac{D_z}{D}=\frac{\begin{vmatrix}2&1&5\\1&-2&10\\1&1&-3\end{vmatrix}}{\begin{vmatrix}2&1&1\\1&-2&3\\1&1&-4\end{vmatrix}}=\frac{2\begin{vmatrix}-2&10\\1&-3\end{vmatrix}-1\begin{vmatrix}1&5\\1&-3\end{vmatrix}+1\begin{vmatrix}1&5\\-2&10\end{vmatrix}}{20}=\frac{2(6-10)-1(-3-5)+1(10+10)}{20}=\frac{-8+8+20}{20}=\frac{20}{20}=1$$

Solution: (3, −2, 1)

47. $\begin{cases} 4x-3y \quad\quad = 1 \\ 6x \quad\quad -8z = 1 \\ \quad\quad 2y-4z = 0 \end{cases}$

$$x=\frac{D_x}{D}=\frac{\begin{vmatrix}1&-3&0\\1&0&-8\\0&2&-4\end{vmatrix}}{\begin{vmatrix}4&-3&0\\6&0&-8\\0&2&-4\end{vmatrix}}=\frac{1\begin{vmatrix}0&-8\\2&-4\end{vmatrix}-1\begin{vmatrix}-3&0\\2&-4\end{vmatrix}+0\begin{vmatrix}-3&0\\0&-8\end{vmatrix}}{4\begin{vmatrix}0&-8\\2&-4\end{vmatrix}-6\begin{vmatrix}-3&0\\2&-4\end{vmatrix}+0\begin{vmatrix}-3&0\\2&-8\end{vmatrix}}=\frac{1(0+16)-1(12-0)-0}{4(0+16)-6(12-0)+0}=\frac{16-12-0}{64-72+0}=\frac{4}{-8}=-\frac{1}{2}$$

$$y=\frac{D_y}{D}=\frac{\begin{vmatrix}4&1&0\\6&1&-8\\0&0&-4\end{vmatrix}}{\begin{vmatrix}4&-3&0\\6&0&-8\\0&2&-4\end{vmatrix}}=\frac{4\begin{vmatrix}1&-8\\0&-4\end{vmatrix}-6\begin{vmatrix}1&0\\0&-4\end{vmatrix}+0\begin{vmatrix}1&0\\1&-8\end{vmatrix}}{-8}=\frac{4(-4+0)-6(-4+0)+0}{-8}=\frac{-16+24+0}{-8}=\frac{8}{-8}=-1$$

$$z=\frac{D_z}{D}=\frac{\begin{vmatrix}4&-3&1\\6&0&1\\0&2&0\end{vmatrix}}{\begin{vmatrix}4&-3&0\\6&0&-8\\0&2&-4\end{vmatrix}}=\frac{4\begin{vmatrix}0&1\\2&0\end{vmatrix}-6\begin{vmatrix}-3&1\\2&0\end{vmatrix}+0\begin{vmatrix}-3&1\\0&1\end{vmatrix}}{-8}=\frac{4(0-2)-6(0-2)+0}{-8}=\frac{-8+12+0}{-8}=\frac{4}{-8}=-\frac{1}{2}$$

Solution: $\left(-\frac{1}{2},-1,-\frac{1}{2}\right)$

49. $\begin{cases} 2x+3y+4z=6 \\ 2x-3y-4z=-4 \\ 4x+6y+8z=12 \end{cases}$

$$x=\frac{D_x}{D}=\frac{\begin{vmatrix}6&3&4\\-4&-3&-4\\12&6&8\end{vmatrix}}{\begin{vmatrix}2&3&4\\2&-3&-4\\4&6&8\end{vmatrix}}=\frac{6\begin{vmatrix}-3&-4\\6&8\end{vmatrix}-(-4)\begin{vmatrix}3&4\\6&8\end{vmatrix}+12\begin{vmatrix}3&4\\-3&-4\end{vmatrix}}{2\begin{vmatrix}-3&-4\\6&8\end{vmatrix}-2\begin{vmatrix}3&4\\6&8\end{vmatrix}+4\begin{vmatrix}3&4\\-3&-4\end{vmatrix}}=\frac{6(-24+24)+4(24-24)+12(-12+12)}{2(-24+24)-2(24-24)+4(-12+12)}=\frac{0+0+0}{0-0+0}=\frac{0}{0}$$

When both the numerator and denominator determinants equal 0, the system is dependent. There are infinitely many solutions.

51. $\begin{cases} 2x+y-z-1=0 \\ x+2y+2z-2=0 \\ 4x+5y+3z-3=0 \end{cases}$ Rewrite in general form. $\begin{cases} 2x+y-z=1 \\ x+2y+2z=2 \\ 4x+5y+3z=3 \end{cases}$

$$x=\frac{D_x}{D}=\frac{\begin{vmatrix} 1 & 1 & -1 \\ 2 & 2 & 2 \\ 3 & 5 & 3 \end{vmatrix}}{\begin{vmatrix} 2 & 1 & -1 \\ 1 & 2 & 2 \\ 4 & 5 & 3 \end{vmatrix}}=\frac{1\begin{vmatrix} 2 & 2 \\ 5 & 3 \end{vmatrix}-2\begin{vmatrix} 1 & -1 \\ 5 & 3 \end{vmatrix}+3\begin{vmatrix} 1 & -1 \\ 2 & 2 \end{vmatrix}}{2\begin{vmatrix} 2 & 2 \\ 5 & 3 \end{vmatrix}-1\begin{vmatrix} 1 & -1 \\ 5 & 3 \end{vmatrix}+4\begin{vmatrix} 1 & -1 \\ 2 & 2 \end{vmatrix}}=\frac{1(6-10)-2(3+5)+3(2+2)}{2(6-10)-1(3+5)+4(2+2)}=\frac{-4-16+12}{-8-8+16}=\frac{-8}{0}$$

When the denominator determinant equals 0, but the numerator determinant is not equal to 0, then the system is inconsistent. There are no solutions.

53. $\begin{cases} x+y=1 \\ \frac{1}{2}y+z=\frac{5}{2} \\ x-z=-3 \end{cases}$ Rewrite in general form. $\begin{cases} x+y \quad =1 \\ \quad y+2z=5 \\ x \quad -z=-3 \end{cases}$

$$x=\frac{D_x}{D}=\frac{\begin{vmatrix} 1 & 1 & 0 \\ 5 & 1 & 2 \\ -3 & 0 & -1 \end{vmatrix}}{\begin{vmatrix} 1 & 1 & 0 \\ 0 & 1 & 2 \\ 1 & 0 & -1 \end{vmatrix}}=\frac{1\begin{vmatrix} 1 & 2 \\ 0 & -1 \end{vmatrix}-5\begin{vmatrix} 1 & 0 \\ 0 & -1 \end{vmatrix}+(-3)\begin{vmatrix} 1 & 0 \\ 1 & 2 \end{vmatrix}}{1\begin{vmatrix} 1 & 2 \\ 0 & -1 \end{vmatrix}-0\begin{vmatrix} 1 & 0 \\ 0 & -1 \end{vmatrix}+1\begin{vmatrix} 1 & 0 \\ 1 & 2 \end{vmatrix}}=\frac{1(-1-0)-5(-1-0)-3(2-0)}{1(-1-0)-0(-1-0)+1(2-0)}=\frac{-1+5-6}{-1-0+2}=\frac{-2}{1}=-2$$

$$y=\frac{D_y}{D}=\frac{\begin{vmatrix} 1 & 1 & 0 \\ 0 & 5 & 2 \\ 1 & -3 & -1 \end{vmatrix}}{\begin{vmatrix} 1 & 1 & 0 \\ 0 & 1 & 2 \\ 1 & 0 & -1 \end{vmatrix}}=\frac{1\begin{vmatrix} 5 & 2 \\ -3 & -1 \end{vmatrix}-0\begin{vmatrix} 1 & 0 \\ -3 & -1 \end{vmatrix}+1\begin{vmatrix} 1 & 0 \\ 5 & 2 \end{vmatrix}}{1}=\frac{1(-5+6)-0+1(2-0)}{1}=\frac{1-0+2}{1}=\frac{3}{1}=3$$

$$z=\frac{D_z}{D}=\frac{\begin{vmatrix} 1 & 1 & 1 \\ 0 & 1 & 5 \\ 1 & 0 & -3 \end{vmatrix}}{\begin{vmatrix} 1 & 1 & 0 \\ 0 & 1 & 2 \\ 1 & 0 & -1 \end{vmatrix}}=\frac{1\begin{vmatrix} 1 & 5 \\ 0 & -3 \end{vmatrix}-0\begin{vmatrix} 1 & 1 \\ 0 & -3 \end{vmatrix}+1\begin{vmatrix} 1 & 1 \\ 1 & 5 \end{vmatrix}}{1}=\frac{1(-3-0)-0+1(5-1)}{1}=\frac{-3-0+4}{1}=\frac{1}{1}=1$$

Solution: (–2, 3, 1)

APPLICATIONS

55. **Analysis:** Find how many of each model of the two cordless phones were in the warehouse inventory.

Equations: Let x represent the number of cordless phones valued at \$67 and y represent the number of cordless phones valued at \$100. There are 360 cordless phones valued at \$29,400. The value of the phones is the number of phones times the unit value of the phone.

Number of \$67 phones	plus	number of \$100 phones	is	total number of phones.
x	$+$	y	$=$	360

Value of \$67 phones	plus	value of \$100 phones	is	total value of phones.
\$67x	$+$	\$100y	$=$	\$29,400

Solve: $\begin{cases} x + y = 360 \\ 67x + 100y = 29{,}400 \end{cases}$

$$x = \frac{D_x}{D} = \frac{\begin{vmatrix} 360 & 1 \\ 29{,}400 & 100 \end{vmatrix}}{\begin{vmatrix} 1 & 1 \\ 67 & 100 \end{vmatrix}} = \frac{36{,}000 - 29{,}400}{100 - 67} = \frac{6600}{33} = 200$$

$$y = \frac{D_y}{D} = \frac{\begin{vmatrix} 1 & 360 \\ 67 & 29{,}400 \end{vmatrix}}{\begin{vmatrix} 1 & 1 \\ 67 & 100 \end{vmatrix}} = \frac{29{,}400 - 24{,}120}{33} = \frac{5280}{33} = 160$$

Conclusion: There are 200 phones valued at \$67 and 160 phones valued at \$100.

57. **Analysis:** Find how an investment of \$20,000 should be split between 3 stocks to obtain a 6.6% average return.

Equations: Let x, y, and z respectively represent the amount of the investment in HiTech, SaveTel and OilCo. The interest earned would be 20,000(0.066) = \$1320.

	P	• r	• t	$= I$
HiTech	x	0.10	1	$0.10x$
SaveTel	y	0.05	1	$0.05y$
OilCo	z	0.06	1	$0.06z$

Principal at 10%	plus	principal at 5%	plus	principal at 6%	is	\$20,000.
$0.10x$	$+$	$0.05y$	$+$	$0.06z$	$=$	20,000.

Interest at 10%	plus	interest at 5%	plus	interest at 6%	is	\$1320.
$0.10x$	$+$	$0.05y$	$+$	$0.06z$	$=$	1320

Three times principal at 10%	is	principal at 5%	plus	principal at 6%.
$3x$	$=$	y	$+$	z

Solve: $\begin{cases} x + y + z = 20{,}000 \\ 0.10x + 0.05y + 0.06z = 1320 \\ 3x = y + z \end{cases}$ Rewrite in general form. $\begin{cases} x + y + z = 20{,}000 \\ 10x + 5y + 6z = 132{,}000 \\ 3x - y - z = 0 \end{cases}$

57. cont.

$$x=\frac{D_x}{D}=\frac{\begin{vmatrix}20{,}000 & 1 & 1\\ 132{,}000 & 5 & 6\\ 0 & -1 & -1\end{vmatrix}}{\begin{vmatrix}1 & 1 & 1\\ 10 & 5 & 6\\ 3 & -1 & -1\end{vmatrix}}=\frac{20{,}000\begin{vmatrix}5 & 6\\ -1 & -1\end{vmatrix}-132{,}000\begin{vmatrix}1 & 1\\ -1 & -1\end{vmatrix}+0\begin{vmatrix}1 & 1\\ 5 & 6\end{vmatrix}}{1\begin{vmatrix}5 & 6\\ -1 & -1\end{vmatrix}-10\begin{vmatrix}1 & 1\\ -1 & -1\end{vmatrix}+3\begin{vmatrix}1 & 1\\ 5 & 6\end{vmatrix}}$$

$$=\frac{20{,}000(-5+6)-132{,}000(-1+1)+0}{1(-5+6)-10(-1+1)+3(6-5)}=\frac{20{,}000-0+0}{1-0+3}=\frac{20{,}000}{4}=5000$$

$$y=\frac{D_y}{D}=\frac{\begin{vmatrix}1 & 20{,}000 & 1\\ 10 & 132{,}000 & 6\\ 3 & 0 & -1\end{vmatrix}}{\begin{vmatrix}1 & 1 & 1\\ 10 & 5 & 6\\ 3 & -1 & -1\end{vmatrix}}=\frac{-20{,}000\begin{vmatrix}10 & 6\\ 3 & -1\end{vmatrix}+132{,}000\begin{vmatrix}1 & 1\\ 3 & -1\end{vmatrix}-0\begin{vmatrix}1 & 1\\ 10 & 6\end{vmatrix}}{4}$$

$$=\frac{-20{,}000(-10-18)+132{,}000(-1-3)-0}{4}=\frac{560{,}000-528{,}000}{4}=\frac{32{,}000}{4}=8000$$

$$z=\frac{D_z}{D}=\frac{\begin{vmatrix}1 & 1 & 20{,}000\\ 10 & 5 & 132{,}000\\ 3 & -1 & 0\end{vmatrix}}{\begin{vmatrix}1 & 1 & 1\\ 10 & 5 & 6\\ 3 & -1 & -1\end{vmatrix}}=\frac{20{,}000\begin{vmatrix}10 & 5\\ 3 & -1\end{vmatrix}-132{,}000\begin{vmatrix}1 & 1\\ 3 & -1\end{vmatrix}+0\begin{vmatrix}1 & 1\\ 10 & 5\end{vmatrix}}{4}$$

$$=\frac{20{,}000(-10-15)-132{,}000(-1-3)+0}{4}=\frac{-500{,}000+528{,}000+0}{4}=\frac{28{,}000}{4}=7000$$

Conclusion: The investment in HiTech should be $x = \$5000$. The investment in SaveTel should be $y = \$8000$. The investment in OilCo should be $z = \$7000$.

59. Follow the directions from the text or in your calculator manual. The solution is –23.

61. Follow the directions from the text or in your calculator manual. The solution is 26.

WRITING

63. Eliminate the row and the column of the determinant containing the element. The remaining elements make up the minor of the element.

REVIEW

65. Solve $x - 2y = 7$ for y.

$$\begin{aligned} x-2y &= 7\\ -2y &= -x+7\\ y &= \tfrac{-1}{-2}x+\tfrac{7}{-2}\\ y &= \tfrac{1}{2}x-\tfrac{7}{2}\end{aligned}$$

The slopes are reciprocals, but not negative reciprocals, so the lines are not perpendicular.

67. yes

69. x

71. y–intercept

73. The independent variable is x, and the dependent variable is y.

Chapter 3 Key Concept: Systems of Equations

Solutions of a System of Equations

1. Substitute $\left(\frac{1}{4}, -\frac{1}{2}\right)$ in $2x - y = 1$; $\left(\frac{1}{4}\right) - \left(-\frac{1}{2}\right) = \frac{1}{2} + \frac{1}{2} = 1$ True

   Substitute $\left(\frac{1}{4}, -\frac{1}{2}\right)$ in $4x + 2y = 0$; $4\left(\frac{1}{4}\right) + 2\left(-\frac{1}{2}\right) = 1 - 1 = 0$ True

   Yes, the point $\left(\frac{1}{4}, -\frac{1}{2}\right)$ satisfies both equations.

2. Substitute (4, 0, 1) in $2x - y + z = 9$; $2(4) - (0) + (1) = 8 - 0 + 1 = 9$ True
   Substitute (4, 0, 1) in $3x + y - 4z = 8$; $3(4) + (0) - 4(1) = 12 + 0 - 4 = 8$ True
   Substitute (4, 0, 1) in $2x - 7z = -1$; $2(4) - 7(1) = 8 - 7 = 1$ False
   No, the point (4, 0, 1) is not a solution for the system.

Methods of Solving Systems of Linear Equations

3. Equation #1: $2x + 5y = 8$

x	y
−6	4
−1	2
4	0

Equation #2: $y = 3x + 5$

x	y
−2	−1
−1	2
0	5

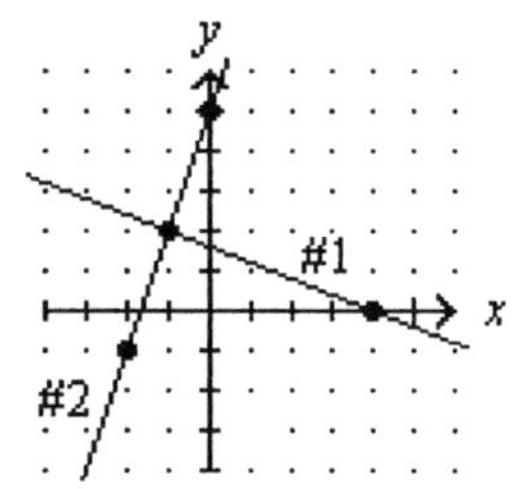

Solution: (−1, 2).

4. $\begin{cases} 9x - 8y = 1 \\ 6x + 12y = 5 \end{cases}$

Multiply Equation 1 by 3 and Equation 2 by 2.

$$\begin{cases} 27x - 24y = 3 \\ 12x + 24y = 10 \end{cases}$$

Add the two equations and solve for x.

$$39x = 13$$
$$x = \frac{13}{39} = \frac{1}{3}$$

Substitute $\frac{1}{3}$ for x in Equation 1 and solve for y.

$$9\left(\tfrac{1}{3}\right) - 8y = 1$$
$$3 - 8y = 1$$
$$-8y = -2$$
$$y = \frac{-2}{-8} = \frac{1}{4}$$

Solution: $\left(\frac{1}{3}, \frac{1}{4}\right)$

5. $\begin{cases} 4x - y - 10 = 0 \\ 3x + 5y = 19 \end{cases}$

Solve Equation 1 for y.

$$-y = -4x + 10$$
$$y = 4x - 10$$

Substitute $4x - 10$ for y in Equation 2.

$$3x + 5(4x - 10) = 19$$
$$3x + 20x - 50 = 19$$
$$23x = 69$$
$$x = 3$$

Substitute 3 for x in Equation 1.

$$4(3) - y - 10 = 0$$
$$12 - y - 10 = 0$$
$$-y = -2$$
$$y = 2$$

Solution: (3, 2)

6. $\begin{cases} -x+3y+2z=5 \\ 3x+2y+z=-1 \\ 2x-y+3z=4 \end{cases}$ Eliminate y.

Multiply Equation 3 by 3 and add Equations 1 and 3.

$$\begin{array}{lrcr} 1. & -x+3y+2z & = & 5 \\ 2. & \underline{6x-3y+9z} & = & \underline{12} \\ 4. & 5x \qquad +11z & = & 17 \end{array}$$

Multiply Equation 3 by 2 and add Equations 2 and 3.

$$\begin{array}{lrcr} 2. & 3x+2y+\ z & = & -1 \\ 3. & \underline{4x-2y+6z} & = & \underline{8} \\ 5. & 7x \qquad +7z & = & 7 \end{array}$$

Multiply Equation 4 by -7 and Equation 5 by 5 and add.

$$\begin{array}{lrcr} 4. & -35x-77z & = & -119 \\ 5. & \underline{35x+35z} & = & \underline{35} \\ 6. & -42z & = & -84 \\ & z & = & 2 \end{array}$$

Substitute $z=2$ into Equation 4 and solve for x.

$$\begin{aligned} 5x+11(2) &= 17 \\ 5x &= -5 \\ x &= -1 \end{aligned}$$

Substitute $x=-1$ and $z=2$ into Equation 1 and solve for y.

$$\begin{aligned} -(-1)+2y+2(2) &= 5 \\ 1+2y+4 &= 5 \\ 2y &= 0 \\ y &= 0 \end{aligned}$$

The solution is $(-1, 0, 2)$.

7. $\begin{cases} x-6y=3 \\ x+3y=21 \end{cases}$

Write system as an augmented matrix.

$$\left[\begin{array}{cc|c} 1 & -6 & 3 \\ 1 & 3 & 21 \end{array}\right]$$

Multiply row 1 by -1 and add to row 2.

$$\left[\begin{array}{cc|c} 1 & -6 & 3 \\ 0 & 9 & 18 \end{array}\right] \quad -R_1+R_2$$

Multiply row 2 by $\frac{1}{9}$.

$$\left[\begin{array}{cc|c} 1 & -6 & 3 \\ 0 & 1 & 2 \end{array}\right] \quad \tfrac{1}{9}R_2$$

Write the resulting system.

$$\begin{aligned} x-6y&=3 \\ y&=2 \end{aligned}$$

Use back substitution to find the solution.

$$\begin{aligned} x-6(2)&=3 \\ x-12&=3 \\ x&=15 \end{aligned}$$

Solution: $(15, 2)$

8. $\begin{cases} x \quad\quad + 2z = 7 \\ 2x - y + 3z = 9 \\ \quad\quad y - z = 1 \end{cases}$

$$x = \frac{D_x}{D} = \frac{\begin{vmatrix} 7 & 0 & 2 \\ 9 & -1 & 3 \\ 1 & 1 & -1 \end{vmatrix}}{\begin{vmatrix} 1 & 0 & 2 \\ 2 & -1 & 3 \\ 0 & 1 & -1 \end{vmatrix}} = \frac{7\begin{vmatrix} -1 & 3 \\ 1 & -1 \end{vmatrix} - 0 + 2\begin{vmatrix} 9 & -1 \\ 1 & 1 \end{vmatrix}}{1\begin{vmatrix} -1 & 3 \\ 1 & -1 \end{vmatrix} - 2\begin{vmatrix} 0 & 2 \\ 1 & -1 \end{vmatrix} + 0} = \frac{7(1-3)+2(9+1)}{1(1-3)-2(0-2)} = \frac{-14+20}{-2+4} = \frac{6}{2} = 3$$

$$y = \frac{D_y}{D} = \frac{\begin{vmatrix} 1 & 7 & 2 \\ 2 & 9 & 3 \\ 0 & 1 & -1 \end{vmatrix}}{\begin{vmatrix} 1 & 0 & 2 \\ 2 & -1 & 3 \\ 0 & 1 & -1 \end{vmatrix}} = \frac{1\begin{vmatrix} 9 & 3 \\ 1 & -1 \end{vmatrix} - 2\begin{vmatrix} 7 & 2 \\ 1 & -1 \end{vmatrix} + 0}{2} = \frac{1(-9-3)-2(-7-2)}{2} = \frac{-12+18}{2} = \frac{6}{2} = 3$$

$$z = \frac{D_z}{D} = \frac{\begin{vmatrix} 1 & 0 & 7 \\ 2 & -1 & 9 \\ 0 & 1 & 1 \end{vmatrix}}{\begin{vmatrix} 1 & 0 & 2 \\ 2 & -1 & 3 \\ 0 & 1 & -1 \end{vmatrix}} = \frac{1\begin{vmatrix} -1 & 9 \\ 1 & 1 \end{vmatrix} - 2\begin{vmatrix} 0 & 7 \\ 1 & 1 \end{vmatrix} + 0}{2} = \frac{1(-1-9)-2(0-7)}{2} = \frac{-10+14}{2} = \frac{4}{2} = 2$$

Solution: (3, 3, 2)

Dependent Equations and Inconsistent Systems

9. The equations of the system are dependent. There are infinitely many solutions.

10. The system is inconsistent. There are no solutions.

Chapter 3 — Chapter Review

Section 3.1

1. Possible answers: (1, 3); (2, 1); (4, –3)   2. Possible answers: (0, –4); (2, –2); (4, 0)   3. (3, –1)

4. President Clinton's job approval and disapproval ratings were the same: approximately 47% in 5/94 and approximately 48% in 5/95.

5. Equation #1

$2x + y = 11$

x	y
2	7
4	3
6	–1

Equation #2

$-x + 2y = 7$

x	y
–3	2
–1	3
5	6

The solution is (3, 5).

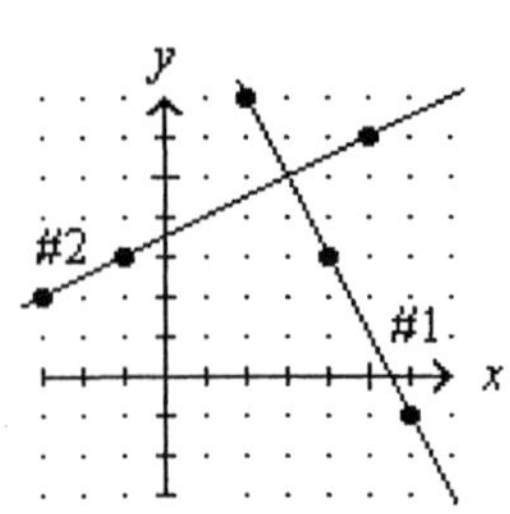

6. Equation #1

$y = -\frac{3}{2}x$

x	y
–4	6
0	0
2	–3

Equation #2

$2x - 3y + 13 = 0$

x	y
–5	1
–2	3
1	5

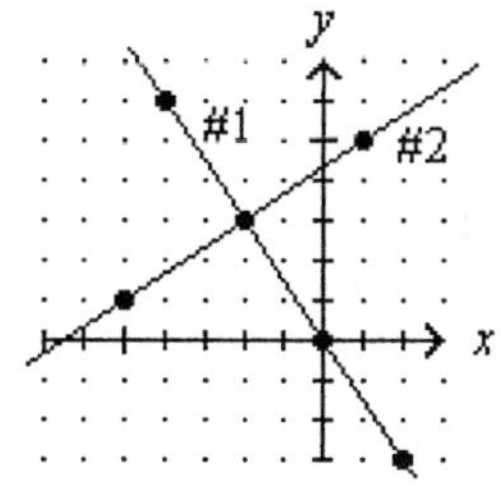

The solution is (–2, 3).

7. Equation #1

$\frac{1}{2}x + \frac{1}{3}y = 2$

x	y
0	6
2	3
4	0

Equation #2

$y = 6 - \frac{3}{2}x$

x	y
0	6
2	3
4	0

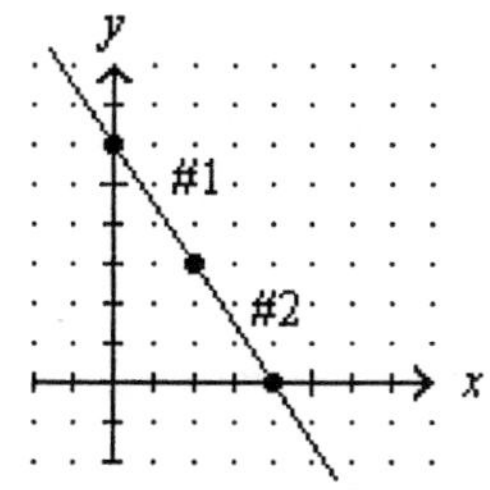

Infinitely many solutions; dependent equations.

8. Equation #1

$\frac{x}{3} - \frac{y}{2} = 1$

x	y
–3	–4
3	-0
–3	–4

Equation #2

$6x - 9y = 3$

x	y
–1	–1
2	1
5	3

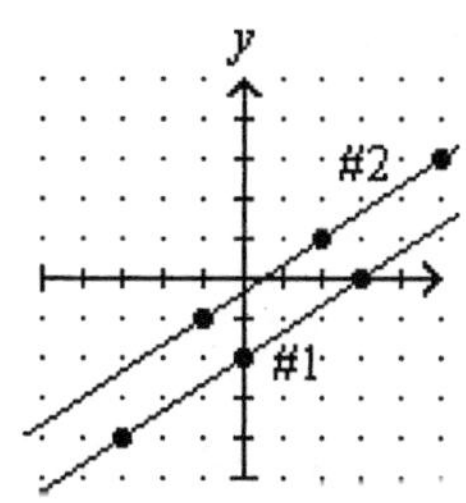

No solution; inconsistent system.

Section 3.2

9. $\begin{cases} x = y - 4 \\ 2x + 3y = 7 \end{cases}$

Substitute $y - 4$ for x in Equation 2.

$$\begin{aligned} 2(y-4) + 3y &= 7 \\ 2y - 8 + 3y &= 7 \\ 5y &= 15 \\ y &= 3 \end{aligned}$$

Substitute 3 for y in Equation 1.

$$\begin{aligned} x &= (3) - 4 \\ x &= -1 \end{aligned}$$

The solution is (–1, 3).

10. $\begin{cases} y = 2x + 5 \\ 3x - 5y = -4 \end{cases}$

Substitute $2x + 5$ for y in Equation 2.

$$\begin{aligned} 3x - 5(2x + 5) &= -4 \\ 3x - 10x - 25 &= -4 \\ -7x &= 21 \\ x &= -3 \end{aligned}$$

Substitute –3 for x in Equation 1.

$$\begin{aligned} y &= 2(-3) + 5 \\ y &= -1 \end{aligned}$$

The solution is (–3,– 1).

11. $\begin{cases} 0.1x + 0.2y = 1.1 \\ 2x - y = 2 \end{cases}$

Multiply Equation 1 by 10 to clear decimals and solve Equation 2 for y.

$$\begin{cases} x + 2y = 11 \\ y = 2x - 2 \end{cases}$$

Substitute $2x - 2$ for y in Equation 1.

$$\begin{aligned} x + 2(2x - 2) &= 11 \\ x + 4x - 4 &= 11 \\ 5x &= 15 \\ x &= 3 \end{aligned}$$

Substitute 3 for x in Equation 2.

$$\begin{aligned} 2(3) - y &= 2 \\ -y &= -4 \\ y &= 4 \end{aligned}$$

The solution is (3, 4).

12. $\begin{cases} x = -2 - 3y \\ -2x - 6y = 4 \end{cases}$

Substitute $-2 - 3y$ for x in Equation 2.

$$\begin{aligned} &-2(-2 - 3y) - 6y \\ &4 + 6y - 6y = 4 \\ &0 = 0 \end{aligned}$$

Infinitely many solutions; dependent equations.

13. $\begin{cases} x + y = -2 \\ 2x + 3y = -3 \end{cases}$

Multiply Equation 1 by –2 to eliminate x.

$$\begin{cases} -2x - 2y = 4 \\ 2x + 3y = -3 \end{cases}$$

Add the two equations and solve for y.

$$y = 1$$

Substitute 1 for y in Equation 1 and solve for x.

$$\begin{aligned} x + (1) &= -2 \\ x &= -3 \end{aligned}$$

The solution is (–3, 1).

14. $\begin{cases} 2x - 3y = 5 \\ 2x - 3y = 8 \end{cases}$

Multiply Equation 1 by –1 to eliminate y.

$$\begin{cases} -2x + 3y = -5 \\ 2x - 3y = 8 \end{cases}$$

Add the two equations and solve for y.

$$0 = 3$$

No solution, inconsistent systems.

15. $\begin{cases} x + \frac{1}{2}y = 7 \\ -2x = 3y - 6 \end{cases}$

Multiply Equation 1 by the LCD 2.

$$\begin{cases} 2x + y = 14 \\ -2x = 3y - 6 \end{cases}$$

Write the equations in general form.

$$\begin{cases} 2x + y = 14 \\ -2x - 3y = -6 \end{cases}$$

Add the two equations and solve for y.

$$\begin{aligned} -2y &= 8 \\ y &= -4 \end{aligned}$$

Substitute –4 for y in Equation 1 and solve for x.

16. $\begin{cases} y = \dfrac{x - 3}{2} \\ x = \dfrac{2y + 7}{2} \end{cases}$

Multiply both equation by the LCD 2.

$$\begin{cases} 2y = x - 3 \\ 2x = 2y + 7 \end{cases}$$

Write the equations in general form.

$$\begin{cases} -x + 2y = -3 \\ 2x - 2y = 7 \end{cases}$$

Add the two equations and solve for x.

$$x = 4$$

Substitute 4 for x in Equation 1 and solve for y.

15. cont.

$$x + \frac{1}{2}(-4) = 7$$
$$x - 2 = 7$$
$$x = 9$$

The solution is (9, –4).

16. cont.

$$y = \frac{(4) - 3}{2} = \frac{1}{2}$$

The solution is $\left(4, \frac{1}{2}\right)$.

17. The computations would be easier using the addition method.

18. $\begin{cases} y = -\frac{2}{3}x \\ 2x - 3y = -4 \end{cases}$

Substitute $-\frac{2}{3}x$ for y in the Equation 2.

$$2x - 3\left(-\frac{2}{3}x\right) = -4$$
$$2x + 2x = -4$$
$$4x = -4$$
$$x = -1$$

Substitute –1 for x in Equation 1.

$$y = -\frac{2}{3}(-1) = \frac{2}{3}$$

The solution is $\left(-1, \frac{2}{3}\right)$.

19. **Analysis:** Find the distance between Austin and Houston and between Austin and San Antonio. The distance between San Antonio and Houston is 197 miles. The round trip distance is 442 miles.

Equation: Let x represent the distance between Austin and Houston. Let y represent the mileage between Austin and San Antonio. Then $x = 2y - 4$ miles.

Austin/Houston	plus	Austin/San Antonio	plus	197 miles	equals	round trip distance.
x	$+$	y	$+$	197	$=$	442

Austin/Houston	equals	two	times	Austin/San Antonio	less	4 miles.
x	$=$	2	•	y	$-$	4

Solve: $\begin{cases} x + y + 197 = 442 \\ x = 2y - 4 \end{cases}$ Collect like terms. $\begin{cases} x + y = 245 \\ x = 2y - 4 \end{cases}$

Substitute $2y - 4$ for x and solve for y.

$$2y - 4 + y = 245$$
$$3y = 249$$
$$y = 83$$

Substitute 83 for y in Equation 2 and solve for x.

$$x = 2(83) - 4$$
$$x = 162$$

Conclusion: The distance from Austin to Houston is 162 miles and the distance from Austin to San Antonio is 83 miles.

20. **Analysis:** Find the speed of a riverboat in still water and the speed of the current of the river. When the riverboat goes downstream the speed of the current will add to the speed of the boat and when the riverboat goes upstream the speed of the current will subtract from the speed of the boat.

Equation: Let x represent the speed of the riverboat and let y represent the speed of the current.

If $rt = d$, then $r = \dfrac{d}{t}$

	r =	d ÷	t
Downstream	$x+y$	30	3
Upstream	$x-y$	30	5

Downstream rate	equals	distance	÷	downstream time.
$x+y$	=	30	÷	3
Upstream rate	equals	distance	÷	upstream time.
$x-y$	=	30	÷	5

Solve: $\begin{cases} x+y=\dfrac{30}{3} \\ x-y=\dfrac{30}{5} \end{cases}$ Do the division on the right sides. $\begin{cases} x+y=10 \\ x-y=6 \end{cases}$

Add the two equations and solve for x.

$$2x = 16$$
$$x = 8$$

Substitute 8 for x in Equation 1 and solve for y

$$(8)+y = 10$$
$$y = 2$$

Conclusion: The speed of the boat is $x = 8$ mph and the speed of the current is $y = 2$ mph.

21. **Analysis:** Find the break even point for a bottling company by comparing the cost of two machines.

Equation: Set x represent the number of bottles processed and use the substitution method to solve the system .

	Setup Cost	plus	Cost for x bottles	=	Cost
Old Machine:	250	+	$0.04x$	=	C
New Machine:	600	+	$0.02x$	=	C

Solve: $250 + 0.04x = 600 + 0.02x$

$$0.02x = 350$$
$$x = 17{,}500$$

Conclusion: The company must process 17,500 bottles in order to break even.

Section 3.3

22. Substitute $(2, -1, 1)$ in $x - y + z = 4$; $(2)-(-1)+(1) = 2+1+1 = 4$ True
Substitute $(2, -1, 1)$ in $x + 2y - z = -1$; $(2)+2(-1)-(1) = 2-2-1 = -1$ True
Substitute $(2, -1, 1)$ in $x + y - 3z = -1$; $(2)+(-1)-3(1) = 2-1-3 = -2$ False
No, the point $(2, -1, 1)$ is not a solution for the system.

23. $\begin{cases} 2x + y + z = -1 \\ 6x - 3y - 2z = 3 \\ 4x - y - z = 4 \end{cases}$ Eliminate z.

Multiply Equation 1 by 2 and add Equations 1 and 2.

$$\begin{array}{lrcr} 1. & 4x + 2y + 2z & = & -2 \\ 2. & 6x - 3y - 2z & = & 3 \\ \hline 4. & 10x - y & = & 1 \end{array}$$

Multiply Equation 3 by –2 and add Equations 2 and 3.

$$\begin{array}{lrcr} 2. & 6x - 3y - 2z & = & 3 \\ 3. & -8x + 2y + 2z & = & -8 \\ \hline 5. & -2x - y & = & -5 \end{array}$$

Multiply Equation 5 by –1 and add Equations 4 and 5.

$$\begin{array}{lrcr} 4. & 10x - y & = & 1 \\ 5. & 2x + y & = & 5 \\ \hline 6. & 12x & = & 6 \\ & x & = & 1/2 \end{array}$$

Substitute $x = \frac{1}{2}$ into Equation 5, solve for y.

$$\begin{aligned} -2\left(\frac{1}{2}\right) - y &= -5 \\ -1 - y &= -5 \\ -y &= -4 \\ y &= 4 \end{aligned}$$

Substitute $x = \frac{1}{2}$ and $y = 4$ into Equation 1, solve for z.

$$\begin{aligned} 2\left(\frac{1}{2}\right) + (4) + z &= -1 \\ 1 + 4 + z &= -1 \\ z &= -6 \end{aligned}$$

The solution is $\left(\frac{1}{2}, 4, -6\right)$.

24. $\begin{cases} 2x + 3y + z = -5 \\ -x + 2y - z = -6 \\ 3x + y + 2z = 4 \end{cases}$ Eliminate z.

Add Equations 1 and 2.

$$\begin{array}{lrcr} 1. & 2x + 3y + z & = & -5 \\ 2. & -x + 2y - z & = & -6 \\ \hline 4. & x + 5y & = & -11 \end{array}$$

Multiply Equation 2 by 2 and add Equations 2 and 3.

$$\begin{array}{lrcr} 2. & -2x + 4y - 2z & = & -12 \\ 3. & 3x + y + 2z & = & 4 \\ \hline 5. & x + 5y & = & -8 \end{array}$$

Multiply Equation 5 by –1 and add Equations 4 and 5.

$$\begin{array}{lrcr} 4. & x + 5y & = & -11 \\ 5. & -x - 5y & = & 8 \\ \hline 6. & 0 & = & -3 \end{array}$$

Since 0 cannot equal –3, the system is inconsistent and has no solution.

25. $\begin{cases} x + y - z = -3 \\ x \qquad + z = 2 \\ 2x - y + 2z = 3 \end{cases}$ Eliminate y.

Add Equations 1 and 3.

$$\begin{array}{lrcr} 1. & x + y - z & = & -3 \\ 3. & 2x - y + 2z & = & 3 \\ \hline 4. & 3x \quad + z & = & 0 \end{array}$$

Multiply Equation 4 by –1 and add Equations 2 and 4.

$$\begin{array}{lrcr} 2. & x + z & = & 2 \\ 4. & -3x - z & = & 0 \\ \hline 5. & -2x & = & 2 \\ & x & = & -1 \end{array}$$

Substitute $x = -1$ into Equation 2, solve for z.

$$\begin{aligned} (-1) + z &= 2 \\ z &= 3 \end{aligned}$$

Substitute $x = -1$ and $z = 3$ into Equation 1, solve for y.

$$(-1) + y - (3) = -3$$

The solution is (–1, 1, 3).

26. $\begin{cases} 3x + 3y + 6z = -6 \\ -x - y - 2z = 2 \\ 2x + 2y + 4z = -4 \end{cases}$ Eliminate z.

Multiply Equation 2 by 3 and add Equations 1 and 2.

$$\begin{array}{lrcr} 1. & 3x + 3y + 6z & = & -6 \\ 2. & -3x - 3y - 6z & = & 6 \\ \hline 4. & 0 & = & 0 \end{array}$$

Since this statement is always true, the system has infinitely many solutions and the equations are dependent.

27. Since the planes intersect in a line, there are infinitely many solutions.

28. **Analysis:** Find the number of pounds of each nut that need to be used to get 50 pounds of a mixture that would sell for \$6 per pound. The total value of the mixture will be \$6(50) = \$300.

Equations: Let x, y and z represent peanuts, cashews and Brazil nuts respectively.

	p	• n	= v
peanuts	\$3	x	$3x$
cashews	\$9	y	$9y$
Brazil nuts	\$9	z	$9z$

Peanut value plus cashew value plus Brazil nut value is value of mixture.
$$3x + 9y + 9z = 300$$

Peanut quantity plus cashew quantity plus Brazil nut quantity is total quantity.
$$x + y + z = 50$$

Cashew quantity is peanut quantity less 15 pounds.
$$y = x - 15$$

Solve: $\begin{cases} 3x+9y+9z=300 \\ x+y+z=50 \\ y=x-15 \end{cases}$ Multiply Equation 2 by –9 $\begin{cases} 3x+9y+9z=300 \\ -9x-9y-9z=-450 \\ y=x-15 \end{cases}$

Add Equation 1 and 2 and solve for x.
$$-6x = -150$$
$$x = 25$$

Substitute 25 for x in Equation 3 and solve for y.
$$y = 25-15$$
$$y = 10$$

Substitute 25 for x and 10 for y in Equation 2.
$$25 + 10 + z = 50$$
$$z = 15$$

Conclusion: It will take 25 pounds of peanuts, 10 pounds of cashews, and 15 pounds of Brazil nuts to make the mixture.

Section 3.4

29. $\begin{cases} 5x+4y=3 \\ x-y=-3 \end{cases}$ $\left[\begin{array}{cc|c} 5 & 4 & 3 \\ 1 & -1 & -3 \end{array}\right]$

Write system as an augmented matrix.

30. $\begin{cases} x+2y+3z=6 \\ x-3y-z=4 \\ 6x+y-2z=-1 \end{cases}$ $\left[\begin{array}{ccc|c} 1 & 2 & 3 & 6 \\ 1 & -3 & -1 & 4 \\ 6 & 1 & -2 & -1 \end{array}\right]$

Write system as an augmented matrix.

31. $\begin{cases} x - y = 4 \\ 3x + 7y = -18 \end{cases}$

Write system as an augmented matrix.

$$\left|\begin{array}{cc:c} 1 & -1 & 4 \\ 3 & 7 & -18 \end{array}\right|$$

Multiply row 1 by –3 and add to row 2.

$$\left|\begin{array}{cc:c} 1 & -1 & 4 \\ 0 & 10 & -30 \end{array}\right| \quad -3R_1 + R_2$$

Multiply row 2 by $\frac{1}{3}$.

$$\left|\begin{array}{cc:c} 1 & -1 & 4 \\ 0 & 1 & -3 \end{array}\right| \quad \tfrac{1}{3}R_2$$

Write the resulting system.

$\begin{cases} x - y = 4 \\ y = -3 \end{cases}$

Use back substitution to find the solution.

$x - (-3) = 4$

$x = 1$ Solution: $(1, -3)$

32. $\begin{cases} x + 2y - 3z = 5 \\ x + y + z = 0 \\ 3x + 4y + 2z = -1 \end{cases}$

Write system as an augmented matrix.

$$\left[\begin{array}{ccc:c} 1 & 2 & -3 & 5 \\ 1 & 1 & 1 & 0 \\ 3 & 4 & 2 & -1 \end{array}\right]$$

Multiply row 1 by –1 and add to row 2.
Multiply row 1 by –3 and add to row 3.

$$\left[\begin{array}{ccc:c} 1 & 2 & -3 & 5 \\ 0 & -1 & 4 & -5 \\ 0 & -2 & 11 & -16 \end{array}\right] \quad \begin{array}{l} -R_1 + R_2 \\ -3R_1 + R_3 \end{array}$$

Multiply row 2 by –1.

$$\left[\begin{array}{ccc:c} 1 & 2 & -3 & 5 \\ 0 & 1 & -4 & 5 \\ 0 & -2 & 11 & -16 \end{array}\right] \quad -R_2$$

Multiply row 2 by 2 and add to row 3.

$$\left[\begin{array}{ccc:c} 1 & 2 & -3 & 5 \\ 0 & 1 & -4 & 5 \\ 0 & 0 & 3 & -6 \end{array}\right] \quad 2R_2 + R_3$$

Multiply row 3 by $\frac{1}{3}$.

$$\left[\begin{array}{ccc:c} 1 & 2 & -3 & 5 \\ 0 & 1 & -4 & 5 \\ 0 & 0 & 1 & -2 \end{array}\right] \quad \tfrac{1}{3}R_3$$

Write the resulting system.

$\begin{cases} x + 2y - 3z = 5 \\ y - 4z = 5 \\ z = -2 \end{cases}$

Use back substitution to find the solution.

$y - 4(-2) = 5$

$y = -3$ Solution: $(5, -3, -2)$

$x + 2(-3) - 3(-2) = 5$

$x = 5$

33. $\begin{cases} 16x-8y=32 \\ -2x+y=-4 \end{cases}$ $\left|\begin{array}{cc|c} 16 & -8 & 32 \\ -2 & 1 & -4 \end{array}\right|$

Write system as an augmented matrix.

Multiply row 1 by $\frac{1}{16}$.

$\left|\begin{array}{cc|c} 1 & -\frac{1}{2} & 2 \\ -2 & 1 & -4 \end{array}\right|$ $\frac{1}{16}R_1$

Multiply row 1 by 2 and add to row 2.

$\left|\begin{array}{cc|c} 1 & -\frac{1}{2} & 2 \\ 0 & 0 & 0 \end{array}\right|$ $2R_1+R_2$

The second row of this matrix states that $0 = 0$. This is an identity or a statement that is always true. It indicates the system is dependent. There are infinitely many solutions.

34. $\begin{cases} x+2y+2z=2 \\ 4x+5y+3z=3 \\ 2x+y-z=1 \end{cases}$ $\left[\begin{array}{ccc|c} 1 & 2 & 2 & 2 \\ 4 & 5 & 3 & 3 \\ 2 & 1 & -1 & 1 \end{array}\right]$

Write system as an augmented matrix.

Multiply row 1 by –4 and add to row 2.
Multiply row 1 by –2 and add to row 3.

$\left[\begin{array}{ccc|c} 1 & 2 & 2 & 2 \\ 0 & -3 & -5 & -5 \\ 0 & -3 & -5 & -3 \end{array}\right]$ $\begin{array}{l} -4R_1+R_2 \\ -2R_1+R_3 \end{array}$

Multiply row 2 by $-\frac{1}{3}$.

$\left[\begin{array}{ccc|c} 1 & 2 & 2 & 2 \\ 0 & 1 & \frac{5}{3} & \frac{5}{3} \\ 0 & -3 & -5 & -3 \end{array}\right]$ $-\frac{1}{3}R_2$

Multiply row 2 by 3 and add to row 3.

$\left[\begin{array}{ccc|c} 1 & 2 & 2 & 2 \\ 0 & 1 & \frac{5}{3} & \frac{5}{3} \\ 0 & 0 & 0 & 2 \end{array}\right]$ $2R_2+R_3$

The third row of this matrix states that $0 = 2$. This is a contradiction. It indicates that the system is inconsistent. There is no solution.

35. **Analysis:** Find how an investment of \$10,000 was split between 2 investments to obtain a return of \$960.

Equations: Let x and y respectively represent the amount of the investment in the mini–mall and the skateboard park.

	P •	r •	t =	I
Mini–mall	x	0.06	1	$0.06x$
Skateboard	y	0.12	1	$0.12y$

Principal at 6%	plus	principal at 12%	is	\$10,000.
$0.06x$	+	$0.12y$	=	10,000.

Profit at 6%	plus	profit at 12%	is	\$960.
$0.06x$	+	$0.12y$	=	960

Solve: $\begin{cases} x+y=10{,}000 \\ 0.06x+0.12y=960 \end{cases}$ Rewrite. $\begin{cases} x+y=10{,}000 \\ 6x+12y=96{,}000 \end{cases}$ Augmented matrix: $\left[\begin{array}{cc|c} 1 & 1 & 10{,}000 \\ 6 & 12 & 96{,}000 \end{array}\right]$

Multiply row 1 by –6 and add to row 2.

$\left[\begin{array}{cc|c} 1 & 1 & 10{,}000 \\ 0 & 6 & 36{,}000 \end{array}\right]$ $-6R_1+R_2$

Multiply row 2 by $\frac{1}{6}$

$\left[\begin{array}{cc|c} 1 & 1 & 10{,}000 \\ 0 & 1 & 6000 \end{array}\right]$ $\frac{1}{6}R_2$

Write the resulting system.

$x+y=10{,}000$

$y=6000$

Use back substitution to find the solution.

$x+6000=10{,}000$

$x=4000$

Conclusion: The investment in the mini–mall was \$4000 and the investment in the skateboard park was \$6000.

Section 3.5

36. $\begin{vmatrix} 2 & 3 \\ -4 & 3 \end{vmatrix} = 2(3) - 3(-4) = 6 + 12 = 18$

37. $\begin{vmatrix} -3 & -4 \\ 5 & -6 \end{vmatrix} = -3(-6) - (-4)(5) = 18 + 20 = 38$

38. $\begin{vmatrix} -1 & 2 & -1 \\ 2 & -1 & 3 \\ 1 & -2 & 2 \end{vmatrix} = -1\begin{vmatrix} -1 & 3 \\ -2 & 2 \end{vmatrix} - 2\begin{vmatrix} 2 & -1 \\ -2 & 2 \end{vmatrix} + 1\begin{vmatrix} 2 & -1 \\ -1 & 3 \end{vmatrix}$ Expanding on the first column (signs are + – +).

$$= -1(-2 + 6) - 2(4 - 2) + 1(6 - 1) = -1(4) - 2(2) + 1(5) = -4 - 4 + 5 = -3$$

39. $\begin{vmatrix} 3 & -2 & 2 \\ 1 & -2 & -2 \\ 2 & 1 & -1 \end{vmatrix} = 3\begin{vmatrix} -2 & -2 \\ 1 & -1 \end{vmatrix} - 1\begin{vmatrix} -2 & 2 \\ 1 & -1 \end{vmatrix} + 2\begin{vmatrix} -2 & 2 \\ -2 & -2 \end{vmatrix}$ Expanding on the first column (signs are + – +).

$$= 3(2 + 2) - 1(2 - 2) + 2(4 + 4) = 3(4) - 1(0) + 2(8) = 12 - 0 + 16 = 28$$

40. $\begin{cases} 3x + 4y = 10 \\ 2x - 3y = 1 \end{cases}$ $x = \dfrac{D_x}{D} = \dfrac{\begin{vmatrix} 10 & 4 \\ 1 & -3 \end{vmatrix}}{\begin{vmatrix} 3 & 4 \\ 2 & -3 \end{vmatrix}} = \dfrac{-30-4}{-9-8} = \dfrac{-34}{-17} = 2$

$y = \dfrac{D_y}{D} = \dfrac{\begin{vmatrix} 3 & 10 \\ 2 & 1 \end{vmatrix}}{\begin{vmatrix} 3 & 4 \\ 2 & -3 \end{vmatrix}} = \dfrac{3-20}{-17} = \dfrac{-17}{-17} = 1$ Solution: (2, 1)

41. $\begin{cases} -6x - 4y = -6 \\ 3x + 2y = 5 \end{cases}$ $x = \dfrac{D_x}{D} = \dfrac{\begin{vmatrix} -6 & -4 \\ 5 & 2 \end{vmatrix}}{\begin{vmatrix} -6 & -4 \\ 3 & 2 \end{vmatrix}} = \dfrac{-12+20}{-12+12} = \dfrac{8}{0}$ is undefined.

When the denominator determinant equals 0, but the numerator determinant is not equal to 0, then the system is inconsistent. There is no solution.

42. $\begin{cases} x+2y+z=0 \\ 2x+y+z=3 \\ x+y+2z=5 \end{cases}$

$$x=\frac{D_x}{D}=\frac{\begin{vmatrix}0&2&1\\3&1&1\\5&1&2\end{vmatrix}}{\begin{vmatrix}1&2&1\\2&1&1\\1&1&2\end{vmatrix}}=\frac{0\begin{vmatrix}1&1\\1&2\end{vmatrix}-3\begin{vmatrix}2&1\\1&2\end{vmatrix}+5\begin{vmatrix}2&1\\1&1\end{vmatrix}}{1\begin{vmatrix}1&1\\1&2\end{vmatrix}-2\begin{vmatrix}2&1\\1&2\end{vmatrix}+1\begin{vmatrix}2&1\\1&1\end{vmatrix}}=\frac{0-3(4-1)+5(2-1)}{1(2-1)-2(4-1)+1(2-1)}=\frac{0-9+5}{1-6+1}=\frac{-4}{-4}=1$$

$$y=\frac{D_y}{D}=\frac{\begin{vmatrix}1&0&1\\2&3&1\\1&5&2\end{vmatrix}}{\begin{vmatrix}1&2&1\\2&1&1\\1&1&2\end{vmatrix}}=\frac{1\begin{vmatrix}3&1\\5&2\end{vmatrix}-2\begin{vmatrix}0&1\\5&2\end{vmatrix}+1\begin{vmatrix}0&1\\3&1\end{vmatrix}}{-4}=\frac{1(6-5)-2(0-5)+1(0-3)}{-4}=\frac{1+10-3}{-4}=\frac{8}{-4}=-2$$

$$z=\frac{D_z}{D}=\frac{\begin{vmatrix}1&2&0\\2&1&3\\1&1&5\end{vmatrix}}{\begin{vmatrix}1&2&1\\2&1&1\\1&1&2\end{vmatrix}}=\frac{1\begin{vmatrix}1&3\\1&5\end{vmatrix}-2\begin{vmatrix}2&0\\1&5\end{vmatrix}+1\begin{vmatrix}2&0\\1&3\end{vmatrix}}{-4}=\frac{1(5-3)-2(10-0)+1(6-0)}{-4}=\frac{2-20+6}{-4}=\frac{-12}{-4}=3$$

Solution: (1, –2, 3)

43. $\begin{cases} 2x+3y+z=2 \\ x+3y+2z=7 \\ x-y-z=-7 \end{cases}$

$$x=\frac{D_x}{D}=\frac{\begin{vmatrix}2&3&1\\7&3&2\\-7&-1&-1\end{vmatrix}}{\begin{vmatrix}2&3&1\\1&3&2\\1&-1&-1\end{vmatrix}}=\frac{2\begin{vmatrix}3&2\\-1&-1\end{vmatrix}-7\begin{vmatrix}3&1\\-1&-1\end{vmatrix}+(-7)\begin{vmatrix}3&1\\3&2\end{vmatrix}}{2\begin{vmatrix}3&2\\-1&-1\end{vmatrix}-1\begin{vmatrix}3&1\\-1&-1\end{vmatrix}+1\begin{vmatrix}3&1\\3&2\end{vmatrix}}=\frac{2(-3+2)-7(-3+1)-7(6-3)}{2(-3+2)-1(-3+1)+1(6-3)}=\frac{-2+14-21}{-2+2+3}=\frac{-9}{3}=-3$$

$$y=\frac{D_y}{D}=\frac{\begin{vmatrix}2&2&1\\1&7&2\\1&-7&-1\end{vmatrix}}{\begin{vmatrix}2&3&1\\1&3&2\\1&-1&-1\end{vmatrix}}=\frac{2\begin{vmatrix}7&2\\-7&-1\end{vmatrix}-1\begin{vmatrix}2&1\\-7&-1\end{vmatrix}+1\begin{vmatrix}2&1\\7&2\end{vmatrix}}{3}=\frac{2(-7+14)-1(-2+7)+1(4-7)}{3}=\frac{14-5-3}{3}=\frac{6}{3}=2$$

43 cont.

$$z=\frac{D_z}{D}=\frac{\begin{vmatrix}2&3&2\\1&3&7\\1&-1&-7\end{vmatrix}}{\begin{vmatrix}2&3&1\\1&3&2\\1&-1&-1\end{vmatrix}}=\frac{2\begin{vmatrix}3&7\\-1&-7\end{vmatrix}-1\begin{vmatrix}3&2\\-1&-7\end{vmatrix}+1\begin{vmatrix}3&2\\3&7\end{vmatrix}}{3}=\frac{2(-21+7)-1(-21+2)+1(21-6)}{3}=\frac{-28+19+15}{3}=\frac{6}{3}=2$$

Solution: (–3, 2, 2)

44. **Analysis:** Find the number of cups of each of three nutrients that need to be used to provide 24 grams of protein, 10 grams of carbohydrates, and 5 grams of fatty acids.

Equations: Let *A* represent the cups of Mix *A*, *B* represent the cups of Mix *B*, and *C* represent the cups of Mix *C*.

	Grams from Mix *A*		grams from Mix *B*		grams from Mix *C*		
Protein:	Grams from Mix *A*	plus	grams from Mix *B*	plus	grams from Mix *C*	is	total grams.
	$5A$	+	$6B$	+	$8C$	=	24
Carbohydrates:	Grams from Mix *A*	plus	grams from Mix *B*	plus	grams from Mix *C*	is	total grams.
	$2A$	+	$3B$	+	$3C$	=	10
Fatty acids:	Grams from Mix *A*	plus	grams from Mix *B*	plus	grams from Mix *C*	is	total grams.
	$1A$	+	$2B$	+	$1C$	=	5

Solve: $\begin{cases}5A+6B+8C=24\\2A+3B+3C=10\\A+2B+C=5\end{cases}$

$$A=\frac{D_A}{D}=\frac{\begin{vmatrix}24&6&8\\10&3&3\\5&2&1\end{vmatrix}}{\begin{vmatrix}5&6&8\\2&3&3\\1&2&1\end{vmatrix}}=\frac{24\begin{vmatrix}3&3\\2&1\end{vmatrix}-10\begin{vmatrix}6&8\\2&1\end{vmatrix}+5\begin{vmatrix}6&8\\3&3\end{vmatrix}}{5\begin{vmatrix}3&3\\2&1\end{vmatrix}-2\begin{vmatrix}6&8\\2&1\end{vmatrix}+1\begin{vmatrix}6&8\\3&3\end{vmatrix}}=\frac{24(3-6)-10(6-16)+5(18-24)}{5(3-6)-2(6-16)+1(18-24)}=\frac{-72+100-30}{-15+20-6}=\frac{-2}{-1}=2$$

$$B=\frac{D_B}{D}=\frac{\begin{vmatrix}5&24&8\\2&10&3\\1&5&1\end{vmatrix}}{\begin{vmatrix}5&6&8\\2&3&3\\1&2&1\end{vmatrix}}=\frac{5\begin{vmatrix}10&3\\5&1\end{vmatrix}-2\begin{vmatrix}24&8\\5&1\end{vmatrix}+1\begin{vmatrix}24&8\\10&3\end{vmatrix}}{-1}=\frac{5(10-15)-2(24-40)+1(72-80)}{-1}=\frac{-25+32-8}{-1}=\frac{-1}{-1}=1$$

$$C=\frac{D_C}{D}=\frac{\begin{vmatrix}5&6&24\\2&3&10\\1&2&5\end{vmatrix}}{\begin{vmatrix}5&6&8\\2&3&3\\1&2&1\end{vmatrix}}=\frac{5\begin{vmatrix}3&10\\2&5\end{vmatrix}-2\begin{vmatrix}6&24\\2&5\end{vmatrix}+1\begin{vmatrix}6&24\\3&10\end{vmatrix}}{-1}=\frac{5(15-20)-2(30-48)+1(60-72)}{-1}=\frac{-25+36-12}{-1}=\frac{-1}{-1}=1$$

Conclusion: Use 2 cups of Mix A, 1 cup of Mix B, and 1 cup of Mix C.

Chapter 3 — Chapter Test

1. Equation #1

$2x + y = 5$

x	y
0	5
2	1
4	−3

Equation #2

$y = 2x - 3$

x	y
−1	−5
1	−1
3	3

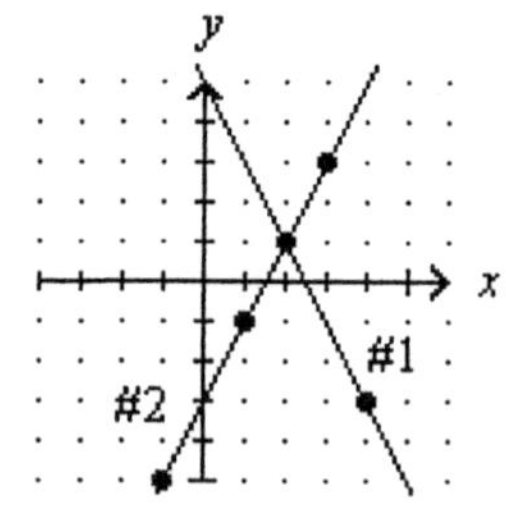

The solution is (2, 1).

2. $\begin{cases} 2x - 4y = 14 \\ x + 2y = 7 \end{cases}$

Solve Equation 2 for x.

$$x = -2y + 7$$

Substitute $-2y + 7$ for x in Equation 1.

$$\begin{aligned} 2(-2y+7) - 4y &= 14 \\ -4y + 14 - 4y &= 14 \\ -8y &= 0 \\ y &= 0 \end{aligned}$$

Substitute 0 for y in Equation 2.

$$\begin{aligned} x &= -2(0) + 7 \\ &= 7 \end{aligned}$$

Solution: (7, 0)

3. $\begin{cases} 2x + 3y = -5 \\ 3x - 2y = 12 \end{cases}$

Multiply Equation 1 by 2 and Equation 2 by 3.

$$\begin{cases} 4x + 6y = -10 \\ 9x - 6y = 36 \end{cases}$$

Add the two equations and solve for x.

$$\begin{aligned} 13x &= 26 \\ x &= 2 \end{aligned}$$

Substitute 2 for x in Equation 1 and solve for y.

$$\begin{aligned} 4 + 3y &= -5 \\ 3y &= -9 \\ y &= -3 \end{aligned}$$

Solution: (2, −3)

4. $\begin{cases} 3(x+y) = x - 3 \\ -y = \dfrac{2x+3}{3} \end{cases}$ Simplify. $\begin{cases} 3x + 3y = x - 3 \\ -3y = 2x + 3 \end{cases}$ Write in general form. $\begin{cases} 2x + 3y = -3 \\ -2x - 3y = 3 \end{cases}$

The sum of the equations is $0 = 0$ which indicates that the equations of the system are dependent.

5. Substitute $x = -1$, $y = -\frac{1}{2}$, and $z = 5$ into each equation in the system.

$$\begin{aligned} x - 2y + z &= 5 \\ (-1) - 2\left(-\tfrac{1}{2}\right) + (5) &\overset{?}{=} 5 \\ -1 + 1 + 5 &\overset{?}{=} 5 \\ 5 &= 5 \end{aligned}$$

$$\begin{aligned} 2x + 4y &= -4 \\ 2(-1) + 4\left(-\tfrac{1}{2}\right) &\overset{?}{=} -4 \\ -2 - 2 &\overset{?}{=} -4 \\ -4 &= -4 \end{aligned}$$

$$\begin{aligned} -6y + 4z &= 22 \\ -6\left(-\tfrac{1}{2}\right) + 4(5) &\overset{?}{=} 22 \\ 3 + 20 &\overset{?}{=} 22 \\ 23 &\neq 22 \end{aligned}$$

No, Equation 3 is not satisfied by $\left(-1, -\frac{1}{2}, 5\right)$

6. $\begin{cases} x+y+z=4 \\ x+y-z=6 \\ 2x-3y+z=-1 \end{cases}$ Eliminate z.

Add Equations 1 and 2.

$$\begin{array}{lrcl} 1. & x+y+z & = & 4 \\ 2. & \underline{x+y-z} & \underline{=} & \underline{6} \\ 4. & 2x+2y & = & 10 \end{array}$$

Add Equations 2 and 3.

$$\begin{array}{lrcl} 2. & x+y-z & = & 6 \\ 3. & \underline{2x-3y+z} & \underline{=} & \underline{-1} \\ 5. & 3x-2y & = & 5 \end{array}$$

Add Equations 4 and 5.

$$\begin{array}{lrcl} 4. & 2x+2y & = & 10 \\ 5. & \underline{3x-2y} & \underline{=} & \underline{5} \\ 6. & 5x & = & 15 \\ & x & = & 3 \end{array}$$

Substitute $x = 3$ into Equation 4 and solve for x.

$$\begin{aligned} 2(3)+2y &= 10 \\ 2y &= 4 \\ y &= 2 \end{aligned}$$

Substitute $x = 3$ and $y = 2$ into Equation 1 and solve for z.

$$\begin{aligned} (3)+(2)+z &= 4 \\ z &= -1 \end{aligned}$$

The solution is $(3, 2, -1)$.

7. **Analysis:** Find the measures of the two angles, x and y, in the illustration.

Equations: Let x represent the measure of two of the angles and $y = x + 15$ represents the measure of the third angle. There are 180° in a triangle.

Sum:	First angle	plus	second angle	plus	third angle	is	180°.
	x	$+$	x	$+$	y	$=$	180

Comparison:	Third angle	is	first angle	plus	15°.
	y	$=$	x	$+$	15

Solve: $\begin{cases} 2x+y=180 \\ y=x+15 \end{cases}$

Substitute $x + 15$ for y into Equation 1.

$$\begin{aligned} 2x+(x+15) &= 180 \\ 3x+15 &= 180 \\ 3x &= 165\ ;\ x = 55 \end{aligned}$$

Substitute 55 for x in Equation 2 and solve for y.

$$\begin{aligned} y &= (55)+15 \\ y &= 70 \end{aligned}$$

Conclusion: The measures of the two angles x are 55° and the measure of the third angle is $y = 70°$.

8. **Analysis:** Find the quantity of 40% antifreeze solution and 80% antifreeze solution needed to obtain 20 gallons of a 50% antifreeze solution.

Equations: Let x represent the quantity of 40% antifreeze solution needed. Let y represent the quantity of 80% antifreeze solution needed. The total quantity of antifreeze needed is 50% of 20 gallons = 10 gallons.

	percent antifreeze •	total quantity =	quantity of antifreeze
40% solution	0.40	x	$0.4x$
80% solution	0.80	y	$0.8y$

Quantity of 40% solution	plus	quantity of 80% solution	is	total quantity of 50% solution.
x	$+$	y	$=$	20

Quantity of antifreeze in 40% solution	plus	quantity of antifreeze in 80% solution	is	quantity of antifreeze in 50% solution.
$0.4x$	$+$	$0.8y$	$=$	10

Solve: $\begin{cases} x+y=20 \\ 0.4x+0.8y=10 \end{cases}$ Solve Equation 1 for y: $y = 20 - x$

Substitute $20 - x$ for y into Equation 2.

$$\begin{aligned} 0.4x+0.8(20-x) &= 10 \\ 0.4x+16-0.8x &= 10 \\ -0.4x &= -6 \\ x &= 15 \end{aligned}$$

Substitute 15 for x in Equation 1 and solve for y.

$$\begin{aligned} y &= 20-(15) \\ y &= 5 \end{aligned}$$

Conclusion: The final antifreeze solution would have 15 gallons of 40% and 5 gallons of 80% antifreeze solutions.

9. $\begin{cases} x+y=4 \\ 2x-y=2 \end{cases}$

Write system as an augmented matrix.

$$\left[\begin{array}{cc|c} 1 & 1 & 4 \\ 2 & -1 & 2 \end{array}\right]$$

Multiply row 1 by –2 and add to row 2.

$$\left[\begin{array}{cc|c} 1 & 1 & 4 \\ 0 & -3 & -6 \end{array}\right] \qquad -2R_1 + R_2$$

Multiply row 2 by $-\frac{1}{3}$.

$$\left[\begin{array}{cc|c} 1 & 1 & 4 \\ 0 & 1 & 2 \end{array}\right] \qquad -\tfrac{1}{3}R_2$$

Write the resulting system.

$$\begin{aligned} x+y&=4 \\ y&=2 \end{aligned}$$

Use back substitution to find the solution.

$$\begin{aligned} x+(2)&=4 \\ x&=2 \end{aligned}$$

Solution: (2, 2)

10. $\begin{cases} x+y+2z=-1 \\ x+3y-6z=7 \\ 2x-y+2z=0 \end{cases}$

Write system as an augmented matrix.

$$\left[\begin{array}{ccc|c} 1 & 1 & 2 & -1 \\ 1 & 3 & -6 & 7 \\ 2 & -1 & 2 & 0 \end{array}\right]$$

Multiply row 1 by –1 and add to row 2.
Multiply row 1 by –2 and add to row 3.

$$\left[\begin{array}{ccc|c} 1 & 1 & 2 & -1 \\ 0 & 2 & -8 & 8 \\ 0 & -3 & -2 & 2 \end{array}\right] \qquad \begin{array}{l} -R_1 + R_2 \\ -2R_1 + R_3 \end{array}$$

Multiply row 2 by $\frac{1}{2}$.

$$\left[\begin{array}{ccc|c} 1 & 1 & 2 & -1 \\ 0 & 1 & -4 & 4 \\ 0 & -3 & -2 & 2 \end{array}\right] \qquad \tfrac{1}{2}R_2$$

Multiply row 2 by 3 and add to row 3.

$$\left[\begin{array}{ccc|c} 1 & 1 & 2 & -1 \\ 0 & 1 & -4 & 4 \\ 0 & 0 & -14 & 14 \end{array}\right] \qquad 3R_2 + R_3$$

Multiply row 3 by $-\frac{1}{14}$.

$$\left[\begin{array}{ccc|c} 1 & 1 & 2 & -1 \\ 0 & 1 & -4 & 4 \\ 0 & 0 & 1 & -1 \end{array}\right] \qquad -\tfrac{1}{14}R_2$$

Write the resulting system.

$$\begin{aligned} x+y+2z&=-1 \\ y-4z&=4 \\ z&=-1 \end{aligned}$$

Use back substitution to find the solution.

$$\begin{aligned} y-4(-1)&=4 \\ y&=0 \\ x+(0)+2(-1)&=-1 \\ x&=1 \end{aligned}$$

Solution: (1, 0, –1)

11. $\begin{vmatrix} 2 & -3 \\ 4 & 5 \end{vmatrix} = 2(5) - (-3)(4) = 10 - (-12) = 10 + 12 = 22$

12. $\begin{vmatrix} 1 & 2 & 0 \\ 2 & 0 & 3 \\ 1 & -2 & 2 \end{vmatrix} = 1\begin{vmatrix} 0 & 3 \\ -2 & 2 \end{vmatrix} - 2\begin{vmatrix} 2 & 3 \\ 1 & 2 \end{vmatrix} + 0 = 1(0+6) - 2(4-3) + 0 = 6 - 2 = 4$ Expanded on first row.

13. $D_x = \begin{vmatrix} -6 & -1 \\ -6 & 1 \end{vmatrix}$

14. $D = \begin{vmatrix} 1 & -1 \\ 3 & 1 \end{vmatrix}$

15. $x = \dfrac{D_x}{D} = \dfrac{\begin{vmatrix} -6 & -1 \\ -6 & 1 \end{vmatrix}}{\begin{vmatrix} 1 & -1 \\ 3 & 1 \end{vmatrix}} = \dfrac{-6-(6)}{1-(-3)} = \dfrac{-12}{4} = -3$

16. $y = \dfrac{D_y}{D} = \dfrac{\begin{vmatrix} 1 & -6 \\ 3 & -6 \end{vmatrix}}{\begin{vmatrix} 1 & -1 \\ 3 & 1 \end{vmatrix}} = \dfrac{-6-(-18)}{1-(-3)} = \dfrac{12}{4} = 3$

17. $\begin{cases} x + y + z = 4 \\ x + y - z = 6 \\ 2x - 3y + z = -1 \end{cases}$

$$z = \frac{D_z}{D} = \frac{\begin{vmatrix} 1 & 1 & 4 \\ 1 & 1 & 6 \\ 2 & -3 & -1 \end{vmatrix}}{\begin{vmatrix} 1 & 1 & 1 \\ 1 & 1 & -1 \\ 2 & -3 & 1 \end{vmatrix}} = \frac{1\begin{vmatrix} 1 & 6 \\ -3 & -1 \end{vmatrix} - 1\begin{vmatrix} 1 & 4 \\ -3 & -1 \end{vmatrix} + 2\begin{vmatrix} 1 & 4 \\ 1 & 6 \end{vmatrix}}{1\begin{vmatrix} 1 & -1 \\ -3 & 1 \end{vmatrix} - 1\begin{vmatrix} 1 & 1 \\ -3 & 1 \end{vmatrix} + 2\begin{vmatrix} 1 & 1 \\ 1 & -1 \end{vmatrix}} = \frac{1(-1+18) - 1(-1+12) + 2(6-4)}{1(1-3) - 1(1+3) + 2(-1-1)} = \frac{17 - 11 + 4}{-2 - 4 - 4} = \frac{10}{-10} = -1$$

Solution: $z = -1$

18. **Analysis:** Find the number of each type of ticket that was sold to produce $410 in receipts when 100 people attended the movie.

Equations: Let x represent the number of children tickets, y represent the number of general admission tickets, and z represent the number of senior tickets.

	Ticket price •	Quantity =	Receipts
Children	\$3	x	\$$3x$
General Admission	\$6	y	\$$6y$
Seniors	\$5	z	\$$5z$

Number of children	plus	number of general admission	plus	number of seniors	is	total number of tickets.
x	$+$	y	$+$	z	$=$	100

Receipts from children	plus	receipts from general admission	plus	receipts from seniors	is	total revenue from tickets.
$3x$	$+$	$6y$	$+$	$5z$	$=$	410

2	times	general admission	equals	children.
2	•	y	$=$	x

Solve: $\begin{cases} x+y+z=100 \\ 3x+6y+5z=410 \\ 2y=x \end{cases}$ Rewrite in general form. $\begin{cases} x+y+z=100 \\ 3x+6y+5z=410 \\ -x+2y=0 \end{cases}$ Eliminate z.

Multiply Equation 1 by –5 and add Equations 1 and 2.

$$\begin{array}{lrcr} 1. & -5x-5y-5z & = & -500 \\ 2. & \underline{3x+6y+5z} & = & \underline{410} \\ 4. & -2x+\ y & = & -90 \end{array}$$

Multiply Equation 4 by –2 and add Equations 4 and 3.

$$\begin{array}{lrcr} 4. & 4x-2y & = & 180 \\ 3. & \underline{-x+2y} & = & \underline{0} \\ 5. & 3x & = & 180 \\ & x & = & 60 \end{array}$$

Substitute $x=60$ into Equation 3 and solve for y.

$$\begin{aligned} -(60)+2y &= 0 \\ 2y &= 60 \\ y &= 30 \end{aligned}$$

Substitute $x=60$ and $y=30$ into Equation 1 and solve for z.

$$\begin{aligned} (60)+(30)+z &= 100 \\ 90+z &= 100 \\ z &= 10 \end{aligned}$$

Conclusion: The number of children's tickets would be $x=60$. The number of general admission tickets would be $y=30$. The number of senior tickets would be $z=10$.

19. Substitution would be the easiest method to use because Equation 2 is already solved for y.

20. The system would have no solutions.

21. Two equations need to be used to solve problems that use two variables because there would be an infinite number of solutions that would satisfy either of the two equations individually. However, there is only one pair that will satisfy both equations unless the equations have the same graph.

22. Estimated point of intersection could be (2035, 25). This would mean that in the year 2035, the percent of the U. S. population that is children under the age of 18 and the percent of the U. S. population that is adults 65 and older will be the same, about 25%.

Chapters 1–3 Cumulative Review Exercises

1.

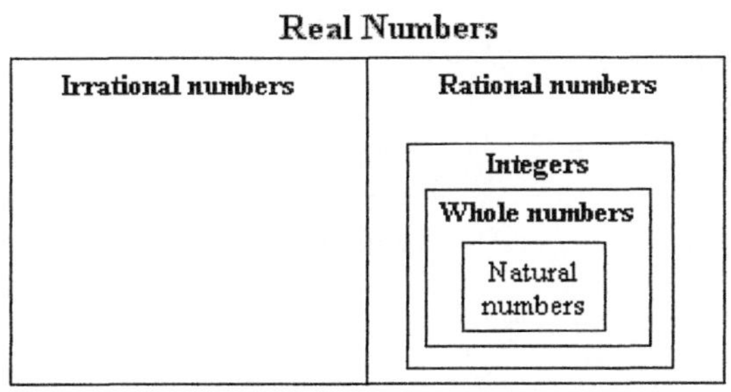

2. **Analysis**: From each dollar of the proposed budget, 21 cents would be spent on Social Security. The total budget was \$2.4 trillion. Find the amount of the total budget that will be spent on Social Security.

Equation: Let x represent the amount of proposed SS spending. \$2.4 trillion can be written as 2.4×10^{12}.

Social Security budget	is	21%	of	total budget.
x	$=$	0.21	$\bullet$	2.4×10^{12}

Solve:
$$x = 0.21(2.4 \times 10^{12})$$
$$= 0.504 \times 10^{12} = 5.04 \times 10^{11}$$

Conclusion: The amount of the proposed budget that would be spent on Social Security would be \$504,000,000,000.

3. $-|b| - ab^2 = -|-5| - (-3)(-5)^2 = -(5) - (-3)(25) = -5 + 75 = 70$

4. $$\frac{14+2[2a-(b-a)]}{-b-2} = \frac{14+2\{2(-3)-[-5-(-3)]\}}{-(-5)-2}$$
$$= \frac{14+2[-6-(-5+3)]}{5-2} = \frac{14+2[-6-(-2)]}{3} = \frac{14+2(-4)}{3} = \frac{14-8}{3} = \frac{6}{3} = 2$$

5. $0.5x^2 - 6(2.1x^2 - x) + 6.7x = 0.5x^2 - 12.6x^2 + 6x + 6.7x = -12.1x^2 + 12.7x$

6. $-(c+2) - (2-c) = -c - 2 - 2 + c = -4$

7. **Analysis**: Find the average speed of a commuter when driving to work.

Equation: Let x represent the rate of the bus. The rate of the car is $x + 10$. The distance is the same for both.

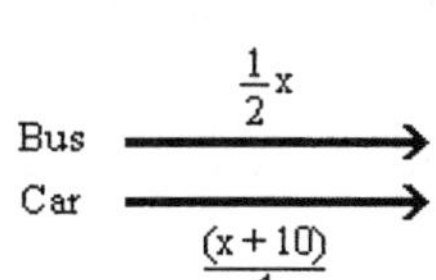

	r	$\bullet$ t	$=$ d
Bus	x	$\frac{1}{2}$	$\frac{1}{2}x$
Car	$x + 10$	$\frac{1}{4}$	$\frac{x+10}{4}$

Distance of the bus	equals	distance of the car.
$\frac{1}{2}x$	$=$	$\frac{x+10}{4}$

Solve:
$$\frac{x}{2} = \frac{x+10}{4}$$
$$2x = x + 10$$
$$x = 10$$

Conclusion: The average speed of the commuter when she drives to work would be $x + 10 = 20$ mph.

8. **Analysis**: Find the number of pounds needed to create a 10–pound mixture that sells for \$4 per pound.

Equation: Let x represent the number of pounds of dried apple slices. Then $10 - x$ will represent the number of pounds of dried banana chips.

	p •	n =	v
Apple slices	4.60	x	$4.60x$
Banana chips	3.40	$10 - x$	$3.40(10 - x)$
Mixture	4.00	10	40

Apple slices value	plus	banana chips value	is	value of mixture.
$4.60x$	+	$3.40(10 - x)$	=	40

Solve:
$$\begin{aligned} 4.60x + 3.40(10 - x) &= 40 \\ 4.60x + 34 - 3.40x &= 40 \\ 1.20x &= 6 \\ x &= 5 \end{aligned}$$

Conclusion: The mixture would need 5 pounds of apple chips and $10 - x = 5$ pounds of banana chips.

9.
$$\begin{aligned} \frac{3}{4}x + 1.5 &= -19.5 \\ 3x + 6 &= -78 \\ 3x &= -84 \\ x &= -28 \end{aligned}$$

10.
$$\begin{aligned} 7 - x - x - x &= 8 \\ 7 - 3x &= 8 \\ -3x &= 1 \\ x &= -\frac{1}{3} \end{aligned}$$

11.
$$\begin{aligned} \frac{x+7}{3} &= \frac{x-2}{5} - \frac{x}{15} + \frac{7}{3} \\ 5(x + 7) &= 3(x - 2) - x + 5(7) \\ 5x + 35 &= 3x - 6 - x + 35 \\ 5x + 35 &= 2x + 29 \\ 3x &= -6 \\ x &= -2 \end{aligned}$$

12.
$$\begin{aligned} 3p - 6 &= 4(p - 2) + 2 - p \\ 3p - 6 &= 4p - 8 + 2 - p \\ 3p - 6 &= 3p - 6 \\ 0 &= 0 \end{aligned}$$
This is an identity.

All real numbers will make this equation true.

13.
$$\begin{aligned} C &= Ax + AB \\ C - Ax &= AB \\ \frac{C - Ax}{A} &= B \end{aligned}$$

14.
$$l = a + (n - 1)d$$
$$l = a + nd - d$$
$$l - a + d = nd$$
$$\frac{l - a + d}{d} = n \quad \text{or} \quad n = \frac{l - a}{d} + 1$$

15.

x	$3x = 4y - 11$	y
-5	$3x = 4(-1) - 11$	-1
-1	$3x = 4(2) - 11$	2
3	$3x = 4(5) - 11$	5

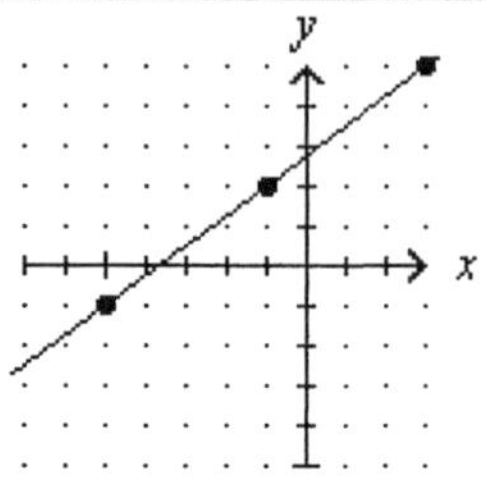

16.

x	$y = -4$	y
-1	The value of x can be any number.	-4
0		-4
3		-4

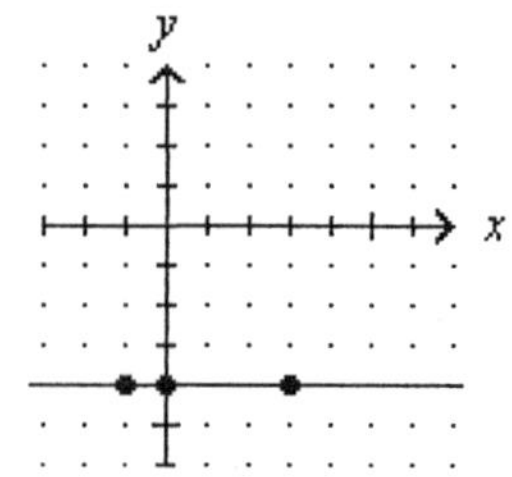

17. Find the slope of $y = -3x$

New equation: $m = -3$, $(x_1, y_1) = (4, 5)$

$$\begin{aligned} y - y_1 &= m(x - x_1) \\ y - (5) &= -3[x - (4)] \\ y - 5 &= -3x + 12 \\ y &= -3x + 17 \end{aligned}$$

18. Using points $(x_1, y_1) = (-5, 0)$ and $(x_2, y_2) = (0, -4)$

$$m = \frac{y_2 - y_1}{x_2 - x_1} = \frac{-4-(0)}{0-(-5)} = \frac{-4}{5} = -\frac{4}{5}$$

19. $f(x) = -x^2 - \dfrac{x}{2}$; $f(10) = -(10)^2 - \dfrac{10}{2} = -100 - 5 = -105$

20. $f(x) = -x^2 - \dfrac{x}{2}$; $f(-10) = -(-10)^2 - \dfrac{-10}{2} = -100 - (-5) = -100 + 5 = -95$

21. Since -27 is $(-3)^3$, one function could be $f(x) = x^3$.

22. The graph is not a function. It does not pass the vertical line test.

23. The equation $x = y^2$ is not a function because for each positive value of x, there are two values for y.

24. Using the points $(0,300) = (0,b)$ and $(20,650)$, write the equation of a line.

$$m = \frac{650-300}{20-0} = 17.5 \qquad \begin{aligned} y &= mx + b \\ y &= 17.5x + 300 \end{aligned}$$

25. $$\begin{cases} y = \dfrac{-2x+1}{3} \\ 3x - 2y = 8 \end{cases}$$

Multiply Equation 1 by 3 and rewrite in standard form.

$$\begin{cases} 2x + 3y = 1 \\ 3x - 2y = 8 \end{cases}$$

Multiply Equation 1 by 2 and equation 2 by 3 to eliminate y.

$$\begin{cases} 4x + 6y = 2 \\ 9x - 6y = 24 \end{cases}$$

Add the two equations and solve for x.

$$\begin{aligned} 13x &= 26 \\ x &= 2 \end{aligned}$$

Substitute 2 for x in Equation 2 and solve for y.

$$\begin{aligned} 3(2) - 2y &= 8 \\ 6 - 2y &= 8 \\ -2y &= 2 \\ y &= -1 \end{aligned}$$

The solution is $(2, -1)$.

26. $$\begin{cases} -x + 3y + 2z = 5 \\ 3x + 2y + z = -1 \\ 2x - y + 3z = 4 \end{cases}$$

Multiply Equation 2 by -2 and add Equations 1 and 2.

$$\begin{aligned} &1. & -x + 3y + 2z &= 5 \\ &2. & \underline{-6x - 4y - 2z} &= \underline{2} \\ &4. & -7x - y \qquad &= 7 \end{aligned}$$

Multiply Equation 2 by -3 and add Equations 2 and 3.

$$\begin{aligned} &2. & -9x - 6y - 3z &= 3 \\ &3. & \underline{2x - y + 3z} &= \underline{4} \\ &5. & -7x - 7y \qquad &= 7 \end{aligned}$$

Multiply Equation 4 by -1 and add Equations 4 and 5.

$$\begin{aligned} &4. & 7x + y &= -7 \\ &5. & \underline{-7x - 7y} &= \underline{7} \\ &6. & -6y &= 0 \\ & & y &= 0 \end{aligned}$$

Substitute $y = 0$ into Equation 4 and solve for x.

$$\begin{aligned} -7x - (0) &= 7 \\ x &= -1 \end{aligned}$$

Substitute $y = 0$ and $x = -1$ in Equation 2 and solve for z.

$$\begin{aligned} 3(-1) + 2(0) + z &= -1 \\ -3 + 0 + z &= -1 \\ z &= 2 \end{aligned}$$

The solution is $(-1, 0, 2)$.

27. $\begin{vmatrix} 5 & -2 \\ -2 & 6 \end{vmatrix} = 5(6) - (-2)(-2) = 30 - 4 = 26$

28. $\begin{vmatrix} 2 & 1 & -3 \\ -2 & 2 & 4 \\ 1 & -2 & 2 \end{vmatrix} = 2\begin{vmatrix} 2 & 4 \\ -2 & 2 \end{vmatrix} - (-2)\begin{vmatrix} 1 & -3 \\ -2 & 2 \end{vmatrix} + 1\begin{vmatrix} 1 & -3 \\ 2 & 4 \end{vmatrix} = 2(4+8) + 2(2-6) + 1(4+6) = 24 - 8 + 10 = 26$

Your math problems can be stepping stones or stumbling blocks.
It's really just a matter of how you view them.

Chapter 4 Inequalities

Study Tip
How do you study for a Test?

You should study for the next test every day, beginning on the first day of class.
Your textbook helps you with this by providing review problems at the end of each problem set.
Do all of these review problems. In addition, do a few problems from previous sections.
Plan on studying previous sections for at least 15 minutes each day.

A few days before your test, begin preparing by doing Chapter Review problems.
Ask your instructor what kinds of problems to expect on the test.
Finally, the day before the test, take the Chapter Test, as if you were taking a test- set a time limit, allow no interruptions, and do not use your notes or examples.
Check the answers in the back and focus your final studying on topics you missed.

Section 4.1 Solving Linear Inequalities

VOCABULARY

1. inequality 3. parenthesis 5. linear 7. is less than; is greater than or equal to

CONCEPTS

9. a. $(4, \infty)$
$\{x \mid x > 4\}$

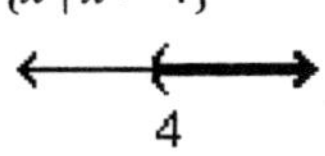

b. $(-\infty, -4)$
$\{x \mid x < -4\}$

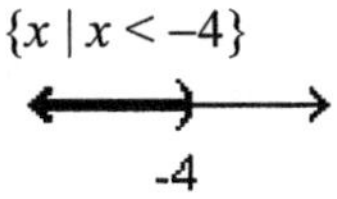

c. $(-\infty, 4]$
$\{x \mid x \leq 4\}$

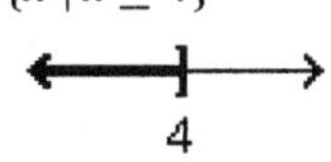

11. a. equation b. expression c. inequality d. expression e. inequality

13. i and ii

15. a. yes; $6 > 0$ b. yes; $0 > -6$ c. yes; $16 > -8$ d. no; -2 is not > 1

17. a. the number of seriously injured ≤ 16 b. the number of references to carpools ≥ 10
c. the age of the car ≥ 25

NOTATION

19. $-5x - 1 \geq -11$

$-5x \geq \mathbf{-10}$

$\dfrac{-5x}{-5} \geq \dfrac{-10}{5}$

$x \leq 2$

Using interval notation, the result is $(-\infty, 2]$.
Using set–builder notation, the result is $\{x \mid \boldsymbol{x \leq 2}\}$.

21. a. $x < -10$ b. $x > \frac{7}{8}$ c. $x \geq 0$

PRACTICE

23. $3x > -9$

$x > -3$ Divide by 3. $(-3, \infty)$

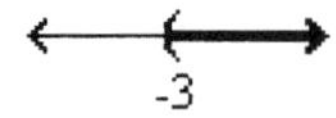

25. $-30y \le -600$

$y \ge 20$ Divide by -30 and reverse the $\le$ symbol. $[20, \infty)$

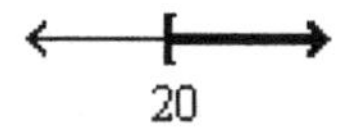

27. $0.6x \ge 36$

$x \ge 60$ Divide by 0.6. $[60, \infty)$

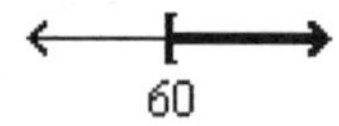

29. $3 > -\frac{9}{10}x$

$30 > -9x$ Multiply both sides by LCD of 10. $\left(-\frac{10}{3}, \infty\right)$

$-\frac{10}{3} = \frac{30}{-9} < x$ Divide both sides by -9, simplify, and reverse the $>$ symbol.

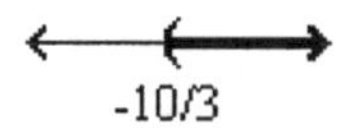

31. $x + 4 < 5$

$x < 1$ Subtract -4 from both sides. $(-\infty, 1)$

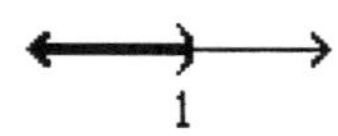

33. $-5t + 3 \le 5$

$-5t \le 2$ Subtract 3 from both sides.

$t \ge -\frac{2}{5}$ Divide both sides by -5 and reverse the $\le$ symbol. $\left[-\frac{2}{5}, \infty\right)$

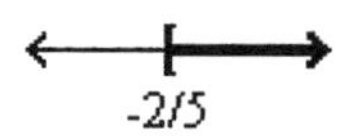

35. $-3x - 1 \le 5$

$-3x \le 6$ Add 1 to both sides.

$x \ge -2$ Divide both sides by -3 and reverse the $\le$ symbol. $[-2, \infty)$

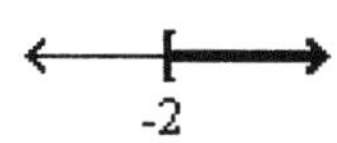

37. $7 < \frac{5}{3}a - 3$

$21 < 5a - 9$ Multiply all the terms by LCD of 3. $(6, \infty)$

$30 < 5a$ Add 9 to both sides.

$6 < a$ Divide both sides by 5.

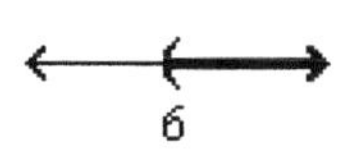

39. $0.4x + 0.4 \le 0.1x + 0.85$

$40x + 40 \le 10x + 85$ Multiply both sides by 100.

$30x \le 45$ Subtract $10x$ and 40 from both sides. $(-\infty, 1.5]$

$x \le 1.5$ Divide both sides by 30.

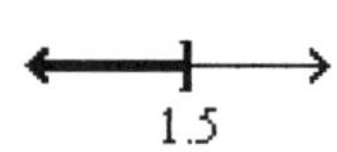

41. $3(z - 2) \le 2(z + 7)$

$3z - 6 \le 2z + 14$ Remove parentheses. $(-\infty, 20]$

$z \le 20$ Subtract $2z$ and add 6 to both sides.

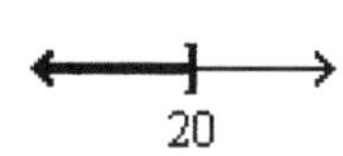

43. $-11(2 - b) < 4(2b + 2)$

$-22 + 11b < 8b + 8)$ Remove parentheses.

$3b < 30$ Subtract $8b$ and add 22 to both sides. $(-\infty, 10)$

$b < 10$ Divide by 3.

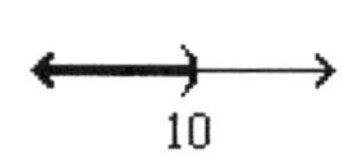

45. $\frac{1}{2}y + 2 \ge \frac{1}{3}y - 4$

$3y + 12 \ge 2y - 24$ Multiply by LCD of 6. $[-36, \infty)$

$y \ge -36$ Subtract $2y$ and 12 from both sides.

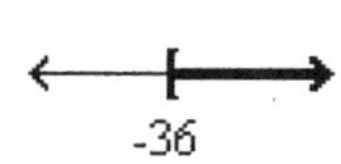

47. $$\begin{aligned} \frac{2}{3}x + \frac{3}{2}(x-5) &\le x \\ 4x + 9(x-5) &\le 6x && \text{Multiply each term by LCD of 6.} \\ 4x + 9x - 45 &\le 6x && \text{Remove parentheses.} \\ 13x - 45 &\le 6x && \text{Combine like terms.} \\ 7x &\le 45 && \text{Subtract } 6x \text{ and add 45 to both sides.} \\ x &\le \frac{45}{7} && \text{Divide both sides by 7.} \end{aligned}$$

$\left(-\infty, \frac{45}{7}\right]$

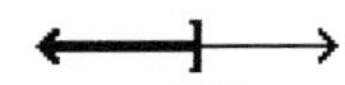

APPLICATIONS

49. The inequality was true for the Midwest and the South.

51. If the two shorter sides are added together, the sum is less than the longer side. $45 + 6 \ngtr 52$

53. **Analysis**: Find how many hours the tank can be rented when the cost must be less or equal to \$185.

Inequality: Let x represent the number of additional hours the tank can be rented. The cost of the additional hours would be \$19.50x.

Cost of first 3 hours	plus	cost of additional hours	is less than or equal to	total cost.
\$85	+	\$19.50x	$\le$	\$185

Solve:
$$\begin{aligned} 85 + 19.50x &\le 185 \\ 19.50x &\le 100 && \text{Subtract 85 from both sides.} \\ x &\le 5.128\ldots && \text{Divide both sides by 19.50.} \end{aligned}$$

Conclusion: The tank could be rented for the first 3 hours plus an additional 5 hours for a total of 8 hours.

55. **Analysis**: Find how many CD–ROMs can be purchased when a computer is purchased for \$1695.95 and the CD–ROMs cost \$19.95 each. The total amount that can be spent is less than or equal to \$2000.

Inequality: Let x represent the number of CD–ROMs to be purchased. The cost of the CD–ROMs would be \$19.95x.

Cost of the computer	plus	cost of CD–ROMs	is less than or equal to	total cost.
\$1695.95	+	\$19.95x	$\le$	\$2000

Solve:
$$\begin{aligned} 1695.95 + 19.95x &\le 2000 \\ 19.95x &\le 304.05 && \text{Subtract 1695.95 from both sides.} \\ x &\le 15.240\ldots && \text{Divide both sides by 19.95.} \end{aligned}$$

Conclusion: The number of CD–ROMS that could be purchased would be 15.

57. **Analysis**: Find how many hours can the student work at the library and earn at least \$175 per week when his hours are limited to no more than 20 hours.

Inequality: Let x represent the number of hours to worked at the library. Then the number of hours worked in construction would be $20 - x$. The amount earned at the library would be \$7x and the amount earned in construction would be \$12$(20 - x)$.

Library earnings	plus	construction earnings	is at least	\$175.
$7x$	+	$12(20 - x)$	$\ge$	175

Solve:
$$\begin{aligned} 7x + 12(20-x) &\ge 175 \\ 7x + 240 - 12x &\ge 175 && \text{Distribute to clear the parentheses.} \\ -5x &\ge -65 && \text{Subtract 240 from both sides and combine like terms.} \\ x &\le 13 && \text{Divide both sides by } -5. \end{aligned}$$

Conclusion: The student could work no more than 13 hours at the library.

59. **Analysis**: Find the size of the hospital bills for which the employee would pay less with Plan 2 than with Plan 1 of the two medical plans presented.

Inequality: Let x represent the amount of the hospital bill. The employee will pay a percentage of the hospital bill plus a deductible that varies with each plan. The percentage is applied to the hospital bill less the amount of the deductible. The percentage that the employee pays will be 100% less the amount that the plan pays.

Plan 2 deductible	plus	20% of bill	is less than	Plan 1 deductible	plus	30% of bill.
200	$+$	$0.20(x-200)$	$<$	100	$+$	$0.30(x-100)$

Solve:

$200 + 0.20(x-200)$	$<$	$100 + 0.30(x-100)$	
$200 + 0.20x - 40$	$<$	$100 + 0.30x - 30$	Remove the parentheses.
$0.20x + 160$	$<$	$70 + 0.30x$	Combine like terms.
$-0.10x$	$<$	-90	Subtract $0.30x$ and 160 and combine like terms.
x	$>$	900	Divide both sides by -0.10 and reverse the $<$ symbol.

Conclusion: Whenever the size of the hospital bill is more than \$900 Plan 2 would pay more of the hospital bill than Plan 1.

61. Use any of the methods described in the text. The answer would be $x < 1$.

63. Use any of the methods described in the text. The answer would be $x \geq -4$.

WRITING

65. The techniques are the same except when multiplying or dividing by a negative value. When this happens, the direction of the inequality must be reversed.

67. Trace the graph of $y = 2x + 1$ below the graph of $y = 3$ for x–values in the interval $(-\infty, 1)$. These values are the solution because in this interval $2x + 1 < 3$.

REVIEW

69. From the illustration, $f(-1) = 4$, $f(0) = 5$, and $f(2) = 3$.

71.

input	$f(x) = x - x^3$	output
-2	$f(x) = (-2) - (-2)^3$ $= -2 - (-8)$	**6**
2	$f(x) = (2) - (2)^3$ $2 - (8)$	**−6**

Section 4.2 Solving Compound Inequalities

VOCABULARY

1. compound
3. interval

CONCEPTS

5. both
7. reversed.

9. a.

$$\frac{x}{3} + 1 \geq 0 \quad \text{and} \quad 2x - 3 < -10$$

$$\frac{-3}{3} + 1 \overset{?}{\geq} 0 \qquad 2(-3) - 3 \overset{?}{<} -10$$

$$-1 + 1 \overset{?}{\geq} 0 \qquad -6 - 3 \overset{?}{<} -10$$

$$0 \geq 0 \qquad -9 \not< -10$$

Since both statements are not true, $x = -3$ is not a solution.

b.

$$2x \leq 0 \quad \text{or} \quad -3x < -5$$

$$2(-3) \overset{?}{\leq} 0 \qquad -3(-3) \overset{?}{<} -5$$

$$-6 \leq 0 \qquad 9 \not< -5$$

Since one statement is true, $x = -3$ is a solution.

11. a. Since $x < -3$ and $x > 3$ do not include common points and both statements must be true in an "and" statement, there is no solution.

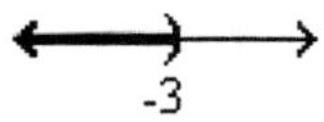 and 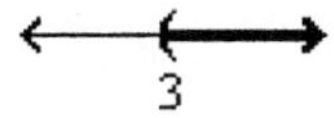 is 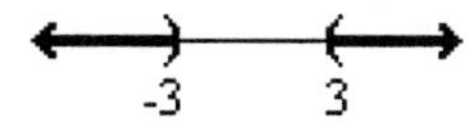

b. Since $x < 3$ and $x > -3$ cover the entire number line and only one statement needs to be true in an "or" statement, all numbers will satisfy. $(-\infty, \infty)$

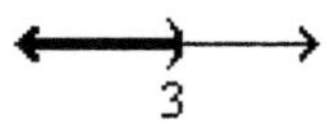 or 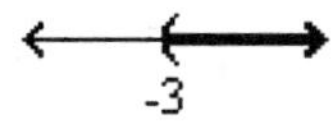 is 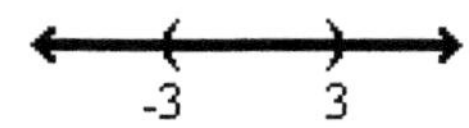

13. a. $[2, 3)$ ii b. $(2, 3)$ iii c. $[2, 3]$ *i*

NOTATION

15. a. "or" statement 

b. "and" statement 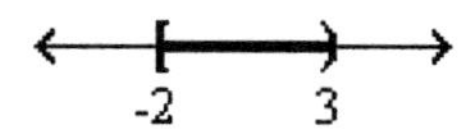

c. "and" statement 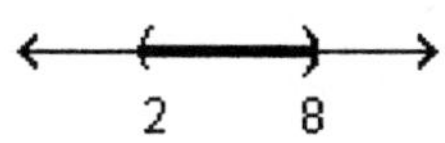

d. "or" statement 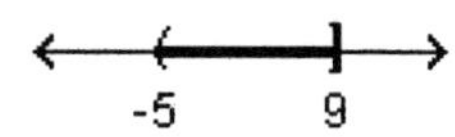

17. If the expression between < and < is removed, $3 \not< -3$

PRACTICE

19. $x > -2$ and $x \leq 5$

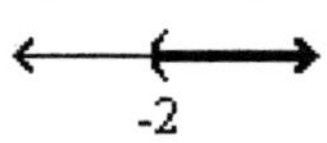 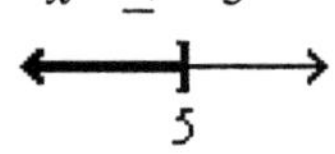 $(-2, 5]$ 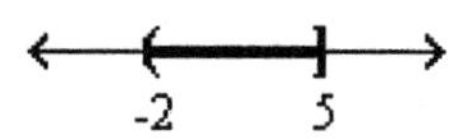

21.
$$\begin{aligned} 2.2x &< -19.8 \\ x &< -9 \end{aligned} \quad \text{and} \quad \begin{aligned} -4x &< 40 \\ x &> -1 \end{aligned}$$

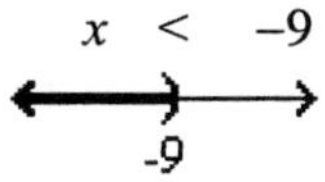 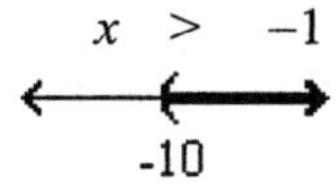 $(-10, -9)$ 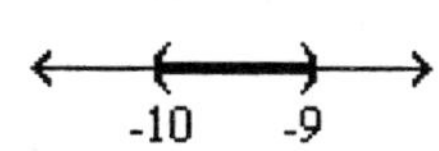

23.
$$\begin{aligned} x + 3 &< 3x - 1 \\ -2x &< -4 \\ x &> 2 \end{aligned} \quad \text{and} \quad \begin{aligned} 4x - 3 &\leq 3x \\ x &\leq 3 \end{aligned}$$

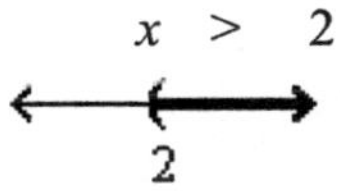 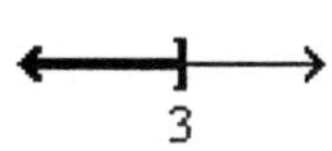 $(2, 3]$ 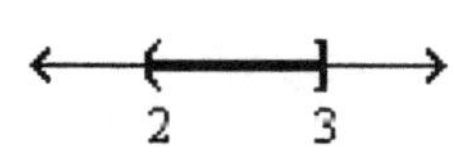

25.
$$\begin{aligned} x + 2 &< -\frac{1}{3}x \\ 3x + 6 &< -x \\ 4x &< -6 \\ x &< -\frac{3}{2} \end{aligned} \quad \text{and} \quad \begin{aligned} -6x &< 9x \\ -15x &< 0 \\ x &> 0 \end{aligned}$$

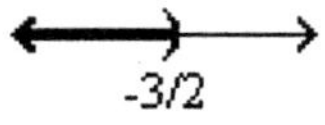 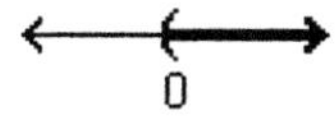

There is no intersection between the two graphs. There is no solution.

27.
$$\begin{array}{rcl} x-1 & \le & 2(x+2) \\ x-1 & < & 2x+4 \\ -x & \le & 5 \\ x & \ge & -5 \end{array} \quad \text{and} \quad \begin{array}{rcl} x & \le & 2x-5 \\ -x & \le & -5 \\ x & \ge & 5 \end{array}$$

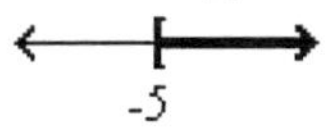
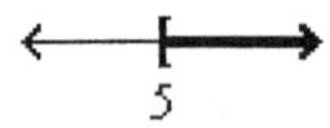
$[5, \infty)$
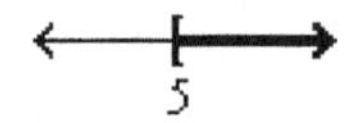

29.
$$\begin{array}{ccccc} 4 & \le & x+3 & \le & 7 \\ 1 & \le & x & \le & 4 \end{array}$$
$[1, 4]$
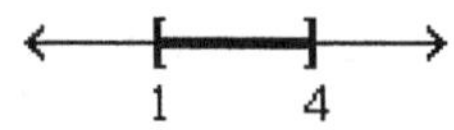

31.
$$\begin{array}{ccccc} 15 & > & 2x-7 & > & 9 \\ 22 & > & 2x & > & 16 \\ 11 & > & x & > & 8 \end{array}$$
$(8, 11)$
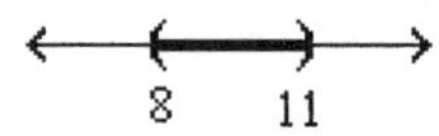

33.
$$\begin{array}{ccccc} -2 & < & -b+3 & < & 5 \\ -5 & < & -b & < & 2 \\ 5 & > & b & > & -2 \end{array}$$
$(-2, 5)$
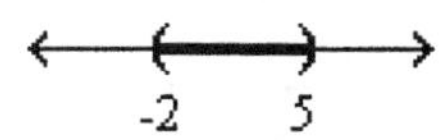

35.
$$\begin{array}{ccccc} -6 & < & -3(x-4) & \le & 24 \\ -6 & < & -3x+12 & \le & 24 \\ -18 & < & -3x & \le & 12 \\ 6 & > & x & \ge & -4 \end{array}$$
$[-4, 6)$
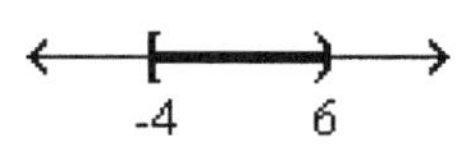

37.
$$\begin{array}{rcl} 2x+1 & \ge & 5 \\ 2x & \ge & 4 \\ x & \ge & 2 \end{array} \quad \text{and} \quad \begin{array}{rcl} -3(x+1) & \ge & -9 \\ -3x-3 & \ge & -9 \\ -3x & \ge & -6 \\ x & \le & 2 \end{array}$$

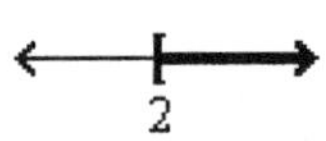
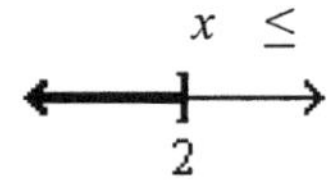
$[2, 2]$
The point only.
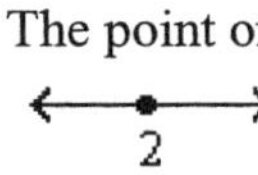

39.
$$\begin{array}{rcl} \dfrac{x}{0.7}+5 & > & 4 \\ x+3.5 & > & 2.8 \\ x & > & -0.7 \end{array} \quad \text{and} \quad \begin{array}{rcl} -4.8 & \le & \dfrac{3x}{-0.125} \\ -0.6 & \le & -3x \\ 0.2 & \ge & x \end{array}$$

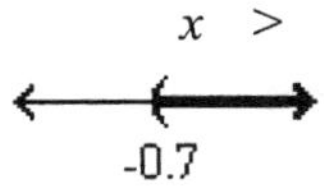
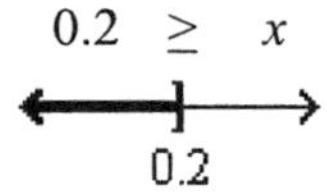
$(-0.7, 0.2]$
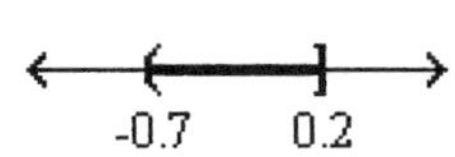

41.
$$\begin{array}{ccccc} -4 & > & \frac{2}{3}x-2 & > & -6 \\ -12 & > & 2x-6 & > & -18 \\ -6 & > & 2x & > & -12 \\ -3 & > & x & > & -6 \end{array}$$
$(-6, -3)$
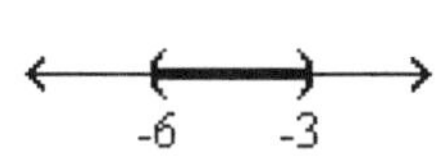

43.
$$\begin{array}{ccccc} 0 & \le & \dfrac{4-x}{3} & \le & 2 \\ 0 & \le & 4-x & \le & 6 \\ -4 & \le & -x & \le & 2 \\ 4 & \ge & x & \ge & -2 \end{array}$$
$[-2, 4]$
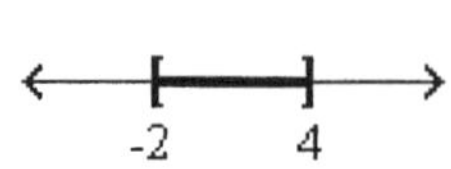

45. $x \le -2$ or $x > 6$

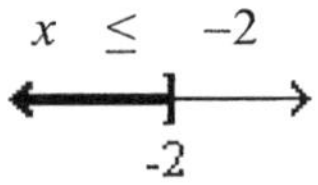
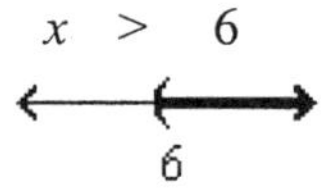
$(-\infty, -2) \cup (6, \infty)$
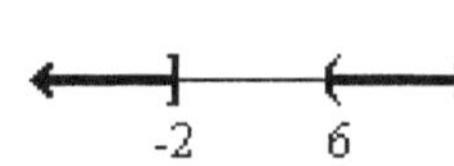

47.

$$\begin{aligned} x-3 &< -4 \\ x &< -1 \end{aligned} \quad \text{or} \quad \begin{aligned} x-2 &> 0 \\ x &> 2 \end{aligned}$$

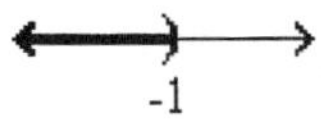
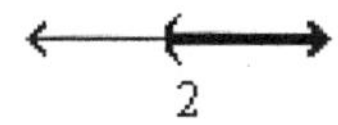

$(-\infty, -1) \cup (2, \infty)$

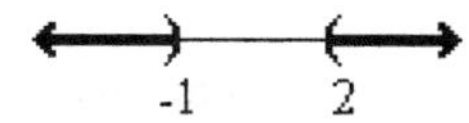

49.

$$\begin{aligned} 3x+2 &< 8 \\ 3x &< 6 \\ x &< 2 \end{aligned} \quad \text{or} \quad \begin{aligned} 2x-3 &> 11 \\ 2x &> 14 \\ x &> 7 \end{aligned}$$

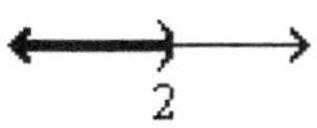
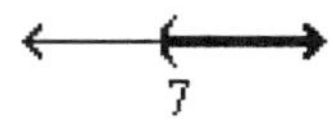

$(-\infty, 2) \cup (7, \infty)$

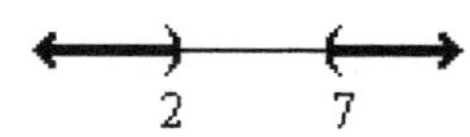

51.

$$x > 3 \quad \text{or} \quad x < 5$$

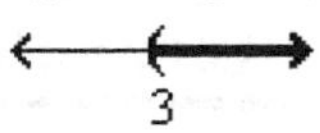
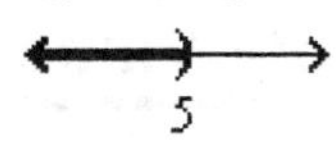

$(-\infty, \infty)$

53.

$$\begin{aligned} -4(x+2) &\geq 12 \\ -4x-8 &\geq 12 \\ -4x &\geq 20 \\ x &\leq -5 \end{aligned} \quad \text{or} \quad \begin{aligned} 3x+8 &< 11 \\ 3x &< 3 \\ x &< 1 \end{aligned}$$

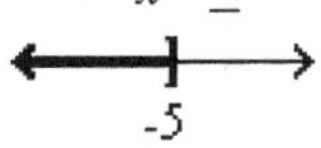
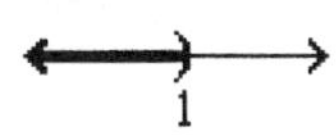

$(-\infty,1)$

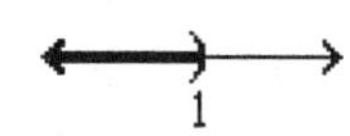

55.

$$\begin{aligned} 4.5x-2 &> 2.5 \\ 4.5x &> 4.5 \\ x &> 1 \end{aligned} \quad \text{or} \quad \begin{aligned} \frac{1}{2}x &\leq 1 \\ x &\leq 2 \end{aligned}$$

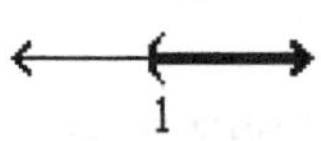
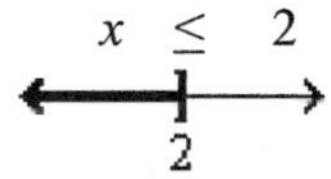

$(-\infty, \infty)$

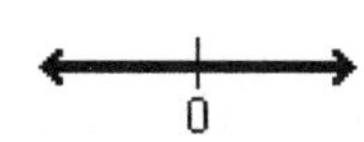

APPLICATIONS

57. a. $128 \leq 4s \leq 192$
    b. $32 \leq s \leq 48$

59. Compare $T = 99.8°$, A = 13 months, S = 3 days = 72 hours to inequalities in flow chart. The child should see the doctor today because his fever has shown no improvement in 72 hours.

61. a. 1999 b. 1998, 1999, 2000, and 2001 c. 1999,2000 d. 1998, 1999, 2000, 2001

WRITING

63. Graph each inequality on the same number line. Find the area "shaded" by both graphs for the union. Find the area common to both graphs for the intersection.

65. The intersection is the set of elements in both sets.

REVIEW

67. The find the mean, we add the values and divide by the number of values, which is 6.

$$\text{mean} = \frac{82+88+94+86+86+78}{6} = \frac{514}{6} = 85.\bar{6} = 85.7$$

To find the median, arrange the values in increasing order: 78, 82, **86, 86**, 88, 94 There are an even number of values, so sum the middle two values and divide by 2.

$$\text{median} = \frac{86+86}{2} = \frac{172}{2} = 86$$

Mode = 86 since it occurs most often (2 times).

69.

	Margin of victory
1st Round:	15
2nd Round	27
Regional Semifinal	26
Regional Final	2
National Semifinal	1
Championship	9
Total	80

$\text{Mean} = \frac{80}{6} = 13.3$ points/game

Section 4.3 Solving Absolute Value Equations and Inequalities

VOCABULARY / CONCEPTS

1. equation 3. isolate 5. 0 7. more than 9. 5

11. a. $|x-1| = 4$; $|-3-1| \stackrel{?}{=} 4$; $|-4| \stackrel{?}{=} 4$; $4 = 4$. Yes, –3 is a solution.

b. $|x-1| > 4$; $|-3-1| \stackrel{?}{>} 4$; $|-4| \stackrel{?}{>} 4$; $4 \not> 4$. No, –3 is not a solution

c. $|x-1| \le 4$; $|-3-1| \stackrel{?}{\le} 4$; $|-4| \stackrel{?}{\le} 4$; $4 \le 4$. Yes, –3 is a solution.

d. $|5-x| = |x+12|$; $|5-(-3)| \stackrel{?}{=} |-3+12|$; $|8| \stackrel{?}{=} |9|$; $8 \neq 9$. No, –3 is not a solution.

NOTATION

13. a. ii b. iii c. *i*

15. $-4 < x < 4$ is the same as $|x| < 4$

17. $x+3 < -6$ or $x+3 > 6$ is the same as $|x+3| > 6$

PRACTICE

19. $|8| = 8$

21. $-|0.02| = -0.02$

23. $-\left|-\frac{31}{16}\right| = -\frac{31}{16}$

25. $|\pi| = \pi$

27. $5|-5| = 5(5) = 25$

29. $-\frac{1}{2}|-4| = -\frac{1}{2}(4) = -2$

31. Rewrite $|x| = 23$ as

$x = 23$ or $x = -23$

33. Rewrite $|x-3.1| = 6$ as

$x - 3.1 = 6$ or $x - 3.1 = -6$

$x = 9.1$ $\quad$ $x = -2.9$

35. Rewrite $|3x+2| = 16$ as

$3x+2 = 16$ or $3x+2 = -16$

$3x = 14$ $\quad$ $3x = -18$

$x = \frac{14}{3}$ $\quad$ $x = -6$

37. $\left|\frac{7}{2}x+3\right| = -5$

An absolute value cannot be negative. This equation has no solution.

39. Rewrite $|3-4x| = 5$ as

$3-4x = 5$ or $3-4x = -5$

$-4x = 2$ $\quad$ $-4x = -8$

$x = -\frac{1}{2}$ $\quad$ $x = 2$

41. Solve $2|3x+24| = 0$ Divide both sides by 2.

$|3x+24| = 0$

$3x+24 = 0$

$x = -8$

43. Rewrite $\left|\frac{3x+48}{3}\right| = 12$ as

$$\begin{aligned} \frac{3x+48}{3} &= 12 \\ 3x+48 &= 36 \\ 3x &= -12 \\ x &= -4 \end{aligned} \quad \text{or} \quad \begin{aligned} \frac{3x+48}{3} &= -12 \\ 3x+48 &= -36 \\ 3x &= -84 \\ x &= -28 \end{aligned}$$

45. If $|x+3|+7=10$, then $|x+3|=3$
Rewrite $|x+3|=3$ as

$$\begin{aligned} x+3 &= 3 \\ x &= 0 \end{aligned} \quad \text{or} \quad \begin{aligned} x+3 &= -3 \\ x &= -6 \end{aligned}$$

47. If $8=-1+|0.3x-3|$, then $9=|0.3x-3|$
Rewrite $9=|0.3x-3|$ as

$$\begin{aligned} 9 &= 0.3x-3 \\ 12 &= 0.3x \\ 40 &= x \end{aligned} \quad \text{or} \quad \begin{aligned} -9 &= 0.3x-3 \\ -6 &= 0.3x \\ -20 &= x \end{aligned}$$

49. Rewrite $|2x+1|=|3x+3|$ as

$$\begin{aligned} 2x+1 &= 3x+3 \\ -x &= 2 \\ x &= -2 \end{aligned} \quad \text{or} \quad \begin{aligned} 2x+1 &= -(3x+3) \\ 2x+1 &= -3x-3 \\ 5x &= -4 \\ x &= -\frac{4}{5} \end{aligned}$$

51. Rewrite $|2-x|=|3x+2|$ as

$$\begin{aligned} 2-x &= 3x+2 \\ -4x &= 0 \\ x &= 0 \end{aligned} \quad \text{or} \quad \begin{aligned} 2-x &= -(3x+2) \\ 2-x &= -3x-2 \\ 2x &= -4 \\ x &= -2 \end{aligned}$$

53. Rewrite $\left|\frac{x}{2}+2\right| = \left|\frac{x}{2}-2\right|$ as

$$\begin{aligned} \frac{x}{2}+2 &= \frac{x}{2}-2 \\ x+4 &= x-4 \\ 0 &= -8 \end{aligned} \quad \text{or} \quad \begin{aligned} \frac{x}{2}+2 &= -\left(\frac{x}{2}-2\right) \\ x+4 &= -x+4 \\ 2x &= 0 \\ x &= 0 \end{aligned}$$

Not possible.

55. Rewrite $\left|x+\frac{1}{3}\right| = |x-3|$ as

$$\begin{aligned} x+\frac{1}{3} &= x-3 \\ 3x+1 &= 3x-9 \\ 0 &= -10 \end{aligned} \quad \text{or} \quad \begin{aligned} x+\frac{1}{3} &= -(x-3) \\ 3x+1 &= -3x+9 \\ 6x &= 8 \\ x &= \frac{8}{6} = \frac{4}{3} \end{aligned}$$

Not possible.

57. Rewrite $|x|<4$ as $-4<x<4$

$(-4, 4)$

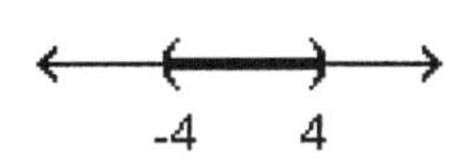

59. Rewrite $|x+9| \le 12$ as

$$\begin{aligned} -12 \le\; & x+9 \le 12 \\ -21 \le\; & x \le 3 \end{aligned}$$

$[-21, 3]$

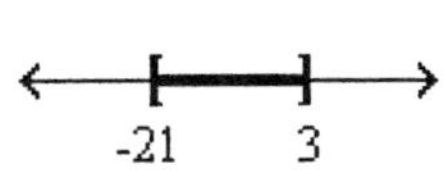

61. Rewrite $|3x-2|<10$ as

$$\begin{aligned} -10 <\; & 3x-2 < 10 \\ -8 <\; & 3x < 12 \\ -\frac{8}{3} <\; & x < 4 \end{aligned}$$

$\left(-\frac{8}{3}, 4\right)$

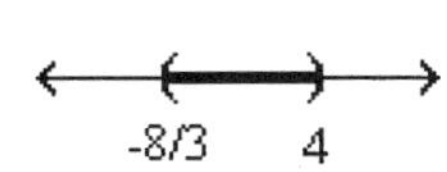

63. $|3x+2| \le -3$

An absolute value cannot be negative. This inequality has no solution.

65. Rewrite $|x|>3$ as
$x<-3$ or $x>3$
$(-\infty, -3) \cup (3, \infty)$

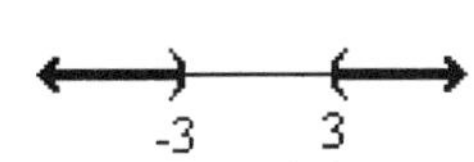

67. Rewrite $|x - 12| > 24$ as

$x - 12 < -24$ or $x - 12 > 24$

$x < -12$ $\qquad$ $x > 36$

$(-\infty, -12) \cup (36, \infty)$

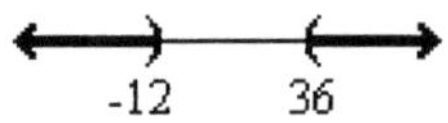

69. Rewrite $|3x + 2| > 14$ as

$3x + 2 < -14$ or $3x + 2 > 14$

$3x < -16$ $\qquad$ $3x > 12$

$x < -\frac{16}{3}$ $\qquad$ $x > 4$

$\left(-\infty, -\frac{16}{3}\right) \cup (4, \infty)$

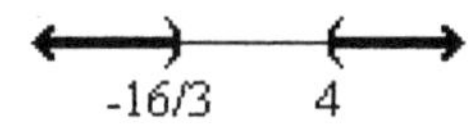

71. Rewrite $|4x + 3| > -5$ as

An absolute value is always positive. This statement is always true. All real numbers will satisfy.

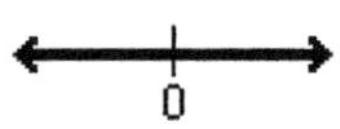

$(-\infty, \infty)$

73. Rewrite $|2 - 3x| \geq 8$ as

$2 - 3x \leq -8$ or $2 - 3x \geq 8$

$-3x \leq -10$ $\qquad$ $-3x \geq 6$

$x \geq \frac{10}{3}$ $\qquad$ $x \leq -2$

$(-\infty, -2] \cup \left[\frac{10}{3}, \infty\right)$

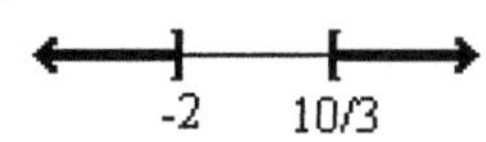

75. If $-|2x - 3| < -7$, then $|2x - 3| > 7$

Rewrite $|2x - 3| > 7$ as

$2x - 3 < -7$ or $2x - 3 > 7$

$2x < -4$ $\qquad$ $2x > 10$

$x < -2$ $\qquad$ $x > 5$

$(-\infty, -2) \cup (5, \infty)$

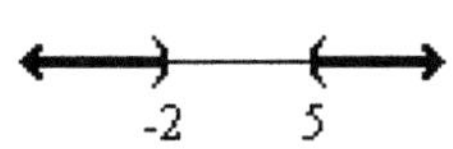

77. Rewrite $\left|\frac{x-2}{3}\right| \leq 4$ as

$$-4 \leq \frac{x-2}{3} \leq 4$$

$$-12 \leq x - 2 \leq 12$$

$$-10 \leq x$$

$[-10, 14]$

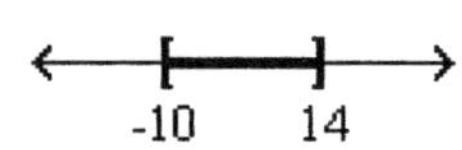

79. If $|3x + 1| + 2 < 6$, then $|3x + 1| < 4$

Rewrite $|3x + 1| < 4$ as

$$-4 < 3x + 1 < 4$$

$$-5 < 3x < 3$$

$$-\frac{5}{3} < x < 1$$

$\left(-\frac{5}{3}, 1\right)$

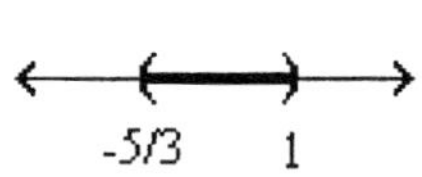

81. If $\left|\frac{1}{3}x + 7\right| + 5 > 6$, then $\left|\frac{1}{3}x + 7\right| > 1$.

Rewrite $\left|\frac{1}{3}x + 7\right| > 1$ as

$\frac{1}{3}x + 7 < -1$ or $\frac{1}{3}x + 7 > 1$

$x + 21 < -3$ $\qquad$ $x + 21 > 3$

$x < -24$ $\qquad$ $x > -18$

$(-\infty, -24) \cup (-18, \infty)$

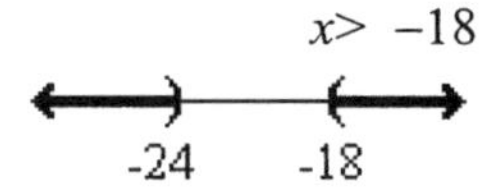

83. If $|0.5x + 1| + 2 \leq 0$, then $|0.5x + 1| \leq -2$.

An absolute value cannot be negative. This inequality has no solution.

APPLICATIONS

85. Rewrite $|t - 78°| \leq 8°$ as

$$-8° \leq t - 78° \leq 8°$$

$$70° \leq t \leq 86°$$

87. a. $|c - 0.6°| \leq 0.5°$

b. Rewrite $|c - 0.6°| \leq 0.5°$ as

$$\begin{aligned} -0.5 &\leq c - 0.6 \leq 0.5 \\ 0.1 &\leq c \leq 1.1 \end{aligned}$$

Interval is [0.1°, 1.1°]

89.a.

Trial #1	$\lvert 22.91 - 25.46\rvert = 2.55$	$\not\leq 1.00$	no
Trial #2	$\lvert 26.45 - 25.46\rvert = 0.99$	≤ 1.00	yes
Trial #3	$\lvert 26.49 - 25.46\rvert = 1.03$	$\not\leq 1.00$	no
Trial #4	$\lvert 24.76 - 25.46\rvert = 0.70$	≤ 1.00	yes

b. error for Trials #2 and #4 is less than 1%.

WRITING

91. If the number is greater than or equal to zero, the absolute value is the number itself. If the number is less than zero, the absolute value is the negative of the number.

93. Parentheses mean that the value next to it is not included in the solution set. Brackets indicate that it is included in the solution set.

95. The inequality $|x - 2| < 3$ will be true for all x–coordinates of points that lie on the graph of $y = |x - 2|$ and below the graph of $y = 3$.

REVIEW

97. **Analysis**: Find the measures of the two angles in the illustration. The angles are supplementary.

Equations: The sum of the measures of the two angles is 180°.

Angle x	plus	angle y	is	180°.
x	$+$	y	$=$	180

Angle y	is	twice angle x	plus	30°.
y	$=$	$2x$	$+$	30

Solve: $\begin{cases} x + y = 180 \\ y = 2x + 30 \end{cases}$

$x + (2x + 30) = 180$		Substitute Equation 2 into Equation 1.
$3x + 30 = 180$		Combine like terms.
$3x = 150$		Subtract 30 from both sides.
$x = 50$		Divide each side by 3.

Conclusion: The measure of angle x is 50° and the measure of angle y is 2(50°) + 30 = 130°.

Section 4.4 Linear Inequalities in Two Variables

VOCABULARY

1. linear; two
3. edge

CONCEPTS

5. a. $3(3) - 2(1) = 9 - 2 = 7 \geq 5$. Yes, (3, 1) is a solution.
   b. $3(0) - 2(3) = 0 - 6 = -6 \not\geq 5$. No, (0, 3) is not a solution.
   c. $3(-1) - 2(-4) = -3 + 8 = 5 \geq 5$. Yes, (−1, −4) is a solution.
   d. $3(1) - 2\left(\frac{1}{2}\right) = 3 - 1 = 2 \not\geq 5$. No, $\left(1, \frac{1}{2}\right)$ is not a solution.

7. Compare $y = 3x - 1$ to $y = mx + b$. slope $= m = 3$; y–intercept $= (0, b) = (0, -1)$

9. no

NOTATION

11. a.

$$\begin{aligned} 2x+4 &\geq 8 \\ 2x &\geq 4 \\ x &\geq 2 \end{aligned}$$

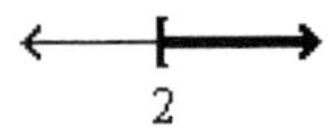

b.

x	$2x+4y \geq 8$ Graph: $2x+4y=8$	y
0	$2(0)+4y=8$	2
4	$2x+4(0)=8$	0
Solid line	Test Point: (0, 0) $2(0)+4(0) \geq 8$ $0 \geq 8$	False

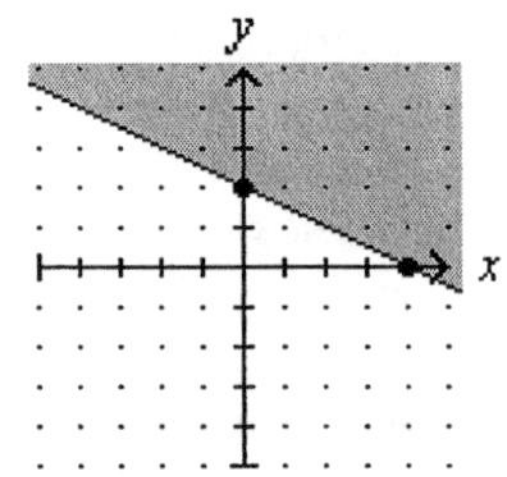

PRACTICE

13.

x	$y > x+1$ Graph: $y=x+1$	y
0	$y=(0)+1$	1
2	$y=(2)+1$	3
Broken line	Test Point: (0, 0) $(0) > (0)+1$ $0 > 1$	False

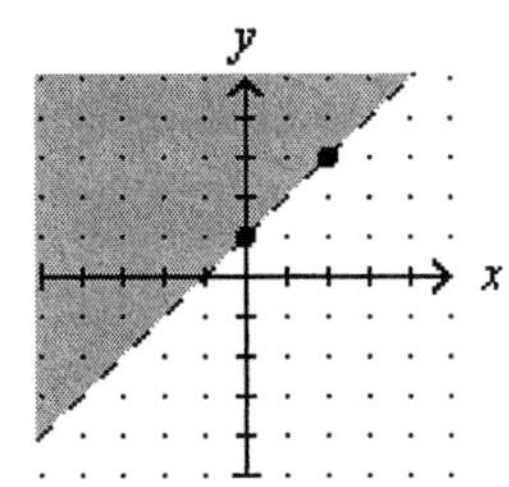

15.

x	$y \geq x$ Graph: $y=x$	y
0	$y=(0)$	0
2	$y=(2)$	2
Solid line	Test Point: (3, 0) $(0) \geq (3)$ $0 \geq 3$	False

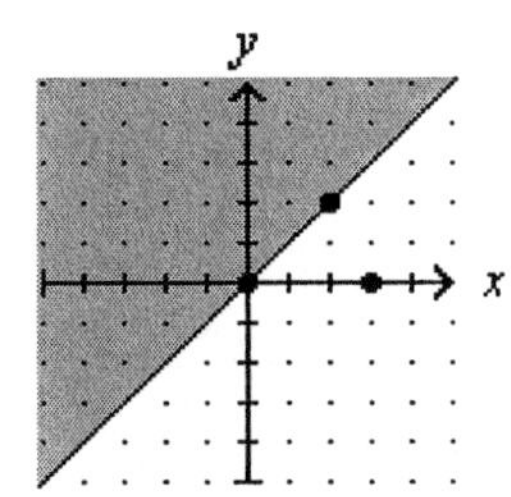

17.

x	$2x+y \leq 6$ Graph: $2x+y=6$	y
0	$2(0)+y=6$	6
3	$2x+(0)=6$	0
Solid line	Test Point: (0, 0) $2(0)+(0) \leq 6$ $0 \leq 6$	True

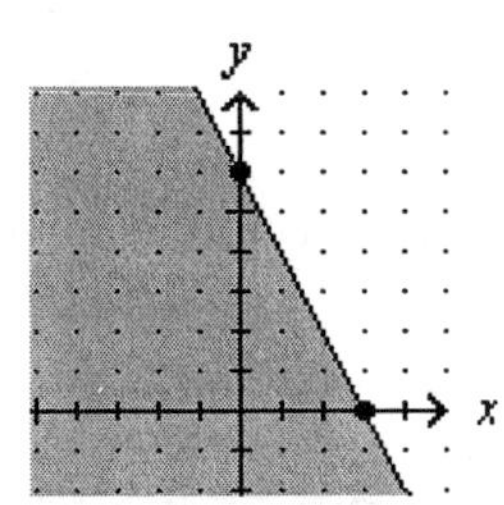

19.

x	$3x \geq -y+3$ Graph: $3x=-y+3$ or $y=-3x+3$	y
0	$y=-3(0)+3$	3
1	$(0)=-3x+3$	0
Solid line	Test Point: (0, 0) $3(0) \geq -(0)+3$ $0 \geq 3$	False

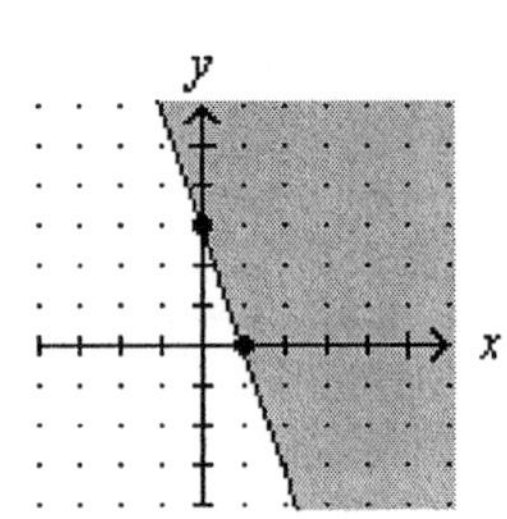

21.

x	$y \ge 1 - \frac{3}{2}x$ Graph: $2y = 2 - 3x$	y
0	$2y = 2 - 3(0)$	1
2	$2y = 2 - 3(2)$	-2
Solid line	Test Point: (0, 0) $0 \ge 1 - \frac{3}{2}(0)$ $0 \ge 1$	False

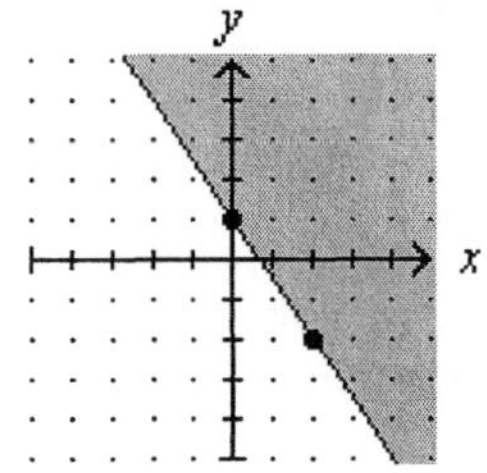

23.

x	$2y < -x$ Graph: $2y = -x$	y
0	$2y = -(0)$	0
4	$2y = -(4)$	-2
Broken line	Test Point: (4, 4) $2(4) < -4$ $8 < -4$	False

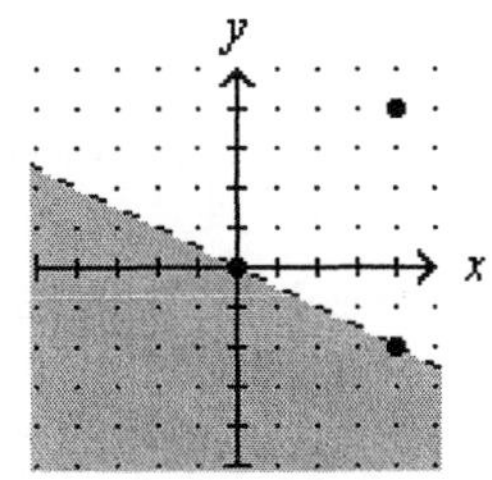

25.

x	$3x + y > 2 + x$ Graph: $y = -2x + 2$	y
0	$y = -2(0) + 2$	2
3	$y = -2(3) + 2$	-4
Broken line	Test Point: (0, 0) $3(0) + (0) > 2 + (0)$ $0 > 2$	False

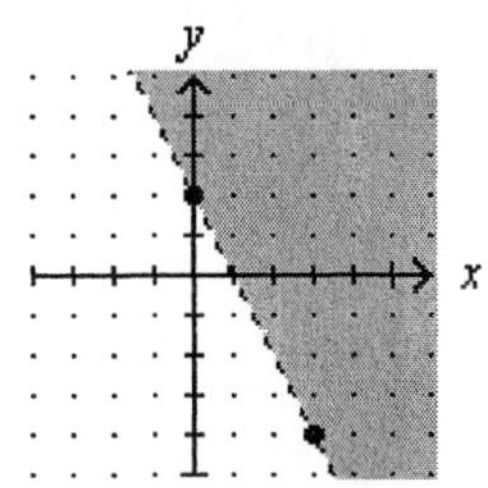

27.

x	$2(x+1) \ge 3\left(y - \frac{4}{3}\right)$ Graph: $2x - 3y = -6$	y
0	$2(0) - 3y = -6$	2
-3	$2x - 3(0) = -6$	0
Solid line	Test Point: (0, 0) $2(0+1) > 3\left(0 - \frac{4}{3}\right)$ $2 > -4$	True

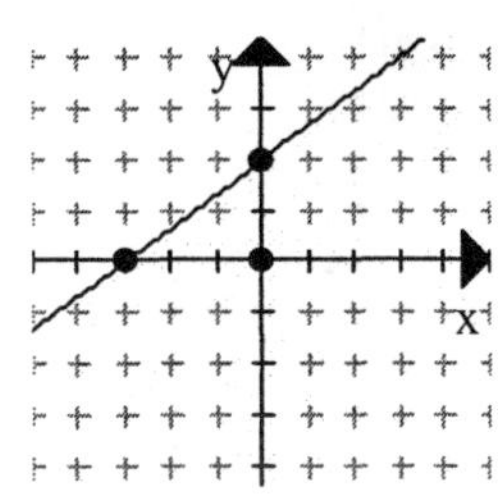

29.

x	$\frac{x}{2} + \frac{y}{2} \le 2$ Graph: $x + y = 4$	y
0	$(0) + y = 4$	4
4	$x + (0) = 4$	0
Solid line	Test Point: (0, 0) $\frac{0}{2} + \frac{0}{2} \le 4$ $0 \le 4$	True

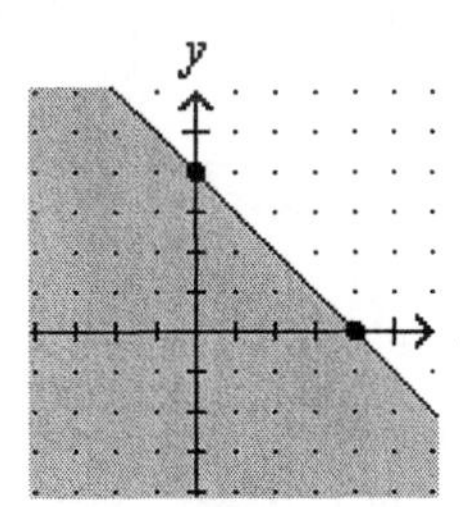

31.

	$x < 4$ Graph: $x = 4$
	A vertical line through $x = 4$.
Broken line	No test point needed. Shade the half–plane to the left of the line.

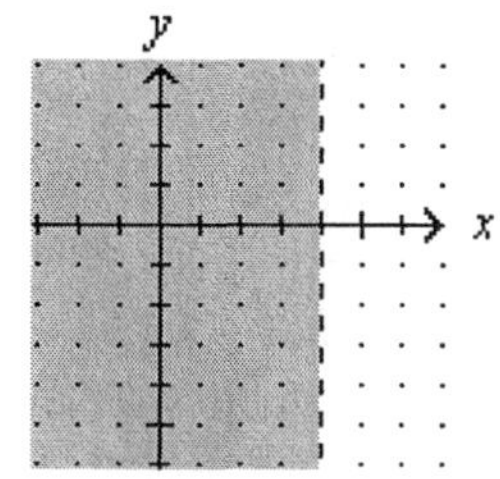

33.

	$y < 0$ Graph: $y = 0$
	A horizontal line through the x–axis.
Broken line	No test point needed. Shade the half–plane under the line.

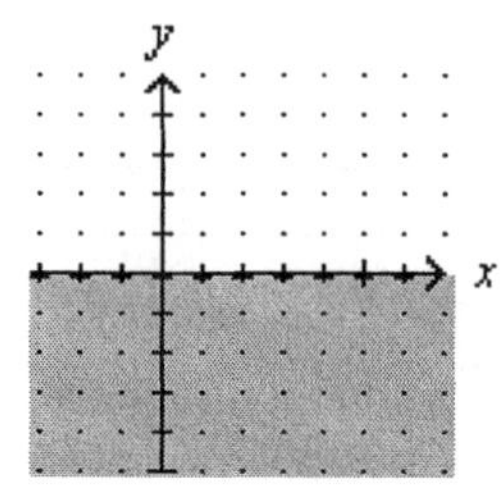

35. Use points (0, 3) and (2, 0). $m = \frac{3-0}{0-2} = -\frac{3}{2}$

Boundary line: $y = -\frac{3}{2}x + 3$ or $3x + 2y = 6$ Inequality: $3x + 2y > 6$

37. Vertical line through $x = 3$ Inequality: $x \leq 3$

39. $y < 0.27x - 1$

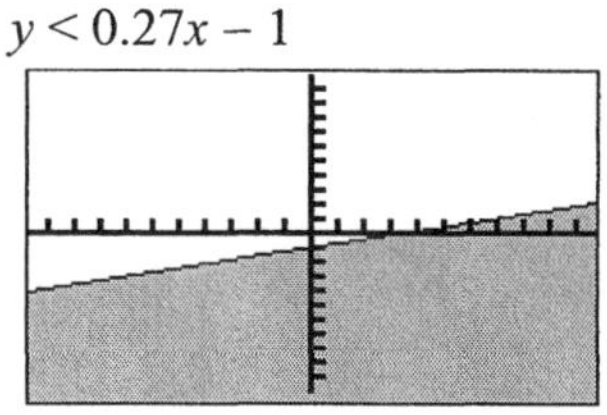

41. $y \geq -2.37x + 1.5$

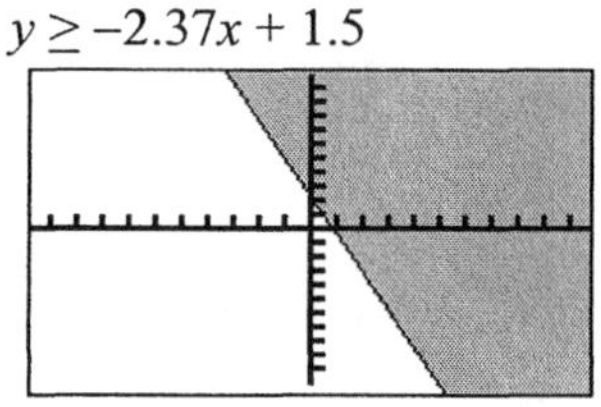

APPLICATIONS

43. a. the Mississippi River
    b. The shaded area is the region of the United States of America west of the Mississippi River.

45. **Analysis**: Find the combination of booths and tables so that a restaurant can seat at most 120 persons.

Inequality: Let x represent the number of booths that seat 4 persons and y represent the number of tables that seat 6 persons. The possible seating from the booths is $4x$ and the possible seating from the tables is $6y$.

Seating from booths	plus	seating from tables	is less than or equal to	120.
$4x$	$+$	$6y$	$\leq$	120

Solve:

x	$4x + 6y \leq 120$ Graph: $4x + 6y = 120$	y
0	$4(0) + 6y = 120$	20
30	$4x + 6(0) = 120$	0
Solid line	Test Point: (0, 0) $4(0) + 6(0) \leq 120$ $0 \leq 120$	True

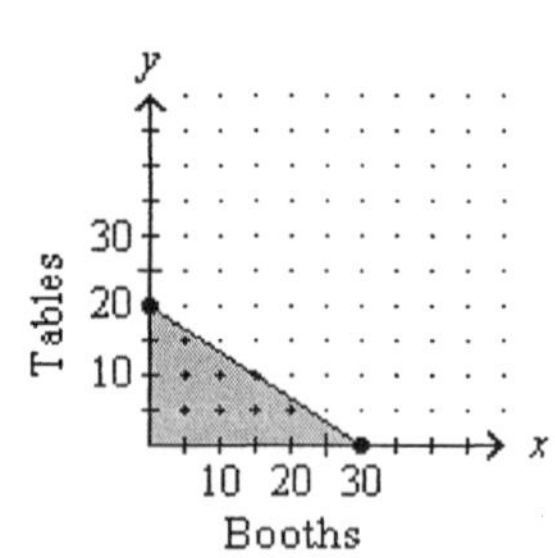

Conclusion: Possible combinations of booths and tables would be: 5 booths and 15 tables, 15 booths and 10 tables, or 20 booths and 5 tables.

47. **Analysis**: Find the possible ways to schedule time to make rods and reels so that at least 1200 units of time per day are used.

Inequality: Let x represent the number of rods that can be made using 10 units of time and y represent the number of reels that can be made using 15 units of time. The time used for rods is $10x$ and the time used for reels is $15y$.

Time used for reels	plus	time used for rods	is at least	1200.
$10x$	$+$	$15y$	$\geq$	1200

Solve:

x	$10x + 15y \geq 1200$ Graph: $10x + 15y = 1200$	y
0	$10(0) + 15y = 1200$	80
120	$10x + 15(0) = 1200$	0
Solid line	Test Point: (0, 0) $10(0) + 15(0) \geq 1200$ $0 \geq 1200$	False

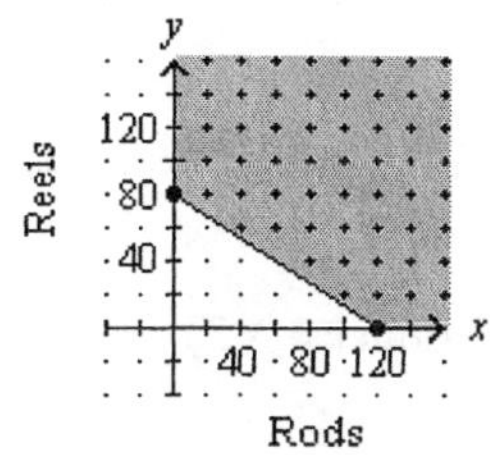

Conclusion: Possible combinations of rods and reels would be: 40 rods and 80 reels, 80 rods and 80 reels, or 120 rods and 40 reels.

WRITING

49. The boundary line is determined by making the inequality an equation and graphing it. If there is an equal sign as part of the inequality the boundary line will be solid. If there is no equal sign as part of the inequality the boundary line will be broken.

REVIEW

51. $4(-4) - (3) = -16 - 3 = -19$ True
$3(-4) + 2(3) = -12 + 6 = -6$ True Yes, the pair (−4, 3) is a solution of the system.

53. $\begin{cases} x + y = 4 \\ x - y = 2 \end{cases}$

Add Equation 1 and Equation 2.

$$2x = 6$$
$$x = 3$$

Substitute 3 for x in Equation 1.

$$(3) + y = 4$$
$$y = 1$$

The solution is (3, 1).

Section 4.5 Systems of Linear Inequalities

VOCABULARY

1. inequalities
3. intersect

CONCEPTS

5. a. $\begin{cases} x+y\le 2 \\ x-3y>10 \end{cases}$ Substitute (2, −3) $\begin{cases} (2)+(-3)\le 2 \\ (2)-3(-3)>10 \end{cases}$ Simplify $\begin{cases} -1\le 2 & \text{True} \\ 11>10 & \text{True} \end{cases}$ Yes

b. $\begin{cases} x+y\le 2 \\ x-3y>10 \end{cases}$ Substitute (12, −1) $\begin{cases} (12)+(-1)\le 2 \\ (12)-3(-1)>10 \end{cases}$ Simplify $\begin{cases} 11\le 2 & \text{False} \\ 16>10 & \text{True} \end{cases}$ No

c. $\begin{cases} x+y\le 2 \\ x-3y>10 \end{cases}$ Substitute (0, −3) $\begin{cases} (0)+(-3)\le 2 \\ (0)-3(-3)>10 \end{cases}$ Simplify $\begin{cases} -3\le 2 & \text{True} \\ 9>10 & \text{False} \end{cases}$ No

d. $\begin{cases} x+y\le 2 \\ x-3y>10 \end{cases}$ Substitute (−0.5, −5) $\begin{cases} (-0.5)+(-5)\le 2 \\ (-0.5)-3(-5)>10 \end{cases}$ Simplify $\begin{cases} -5.5\le 2 & \text{True} \\ 14.5>10 & \text{True} \end{cases}$ Yes

7. a. false b. true c. true d. false e. true f. true

PRACTICE

9. Inequality #1: $y < 3x + 2$

Graph: $y = 3x + 2$

x	y
0	2
−2	−4

Broken line. Test point (0, 0).

$(0) < 3(0) + 2$

$0 < 2$ True

Inequality #2: $y < -2x + 3$

Graph: $y = -2x + 3$

x	y
0	3
2	−1

Broken line. Test point (0, 0).

$(0) < -2(0) + 3$

$0 < 3$ True

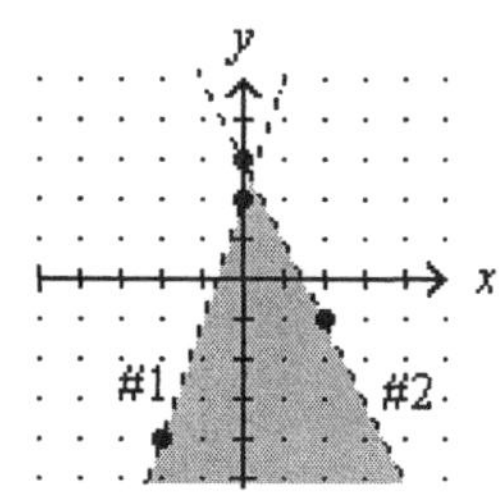

11. Inequality #1: $3x + 2y > 6$

Graph: $3x + 2y = 6$

x	y
0	3
2	0

Broken line. Test point (0, 0).

$3(0) + 2(0) > 6$

$0 > 6$ False

Inequality #2: $x + 3y \le 2$

Graph: $x + 3y = 2$

x	y
−4	2
2	0

Solid line. Test point (0, 0).

$(0) + 3(0) \le 2$

$0 \le 2$ True

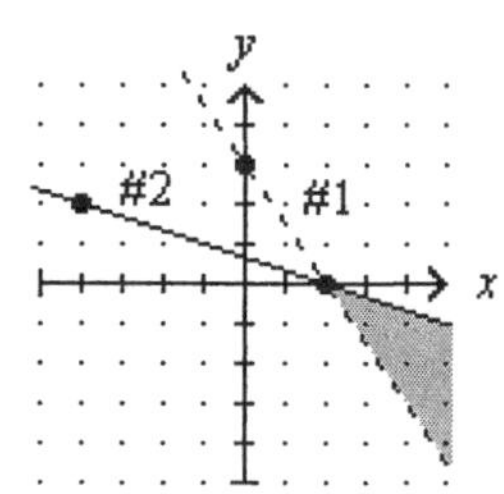

13. Inequality #1: $x + y < 2$

Graph: $x + y = 2$

x	y
0	2
2	0

Broken line. Test point (0, 0).

$(0) + (0) < 2$

$0 < 2$ True

Inequality #2: $x + y < 1$

Graph: $x + y = 1$

x	y
0	1
1	0

Solid line. Test point (0, 0).

$(0) + (0) \le 1$

$0 \le 1$ True

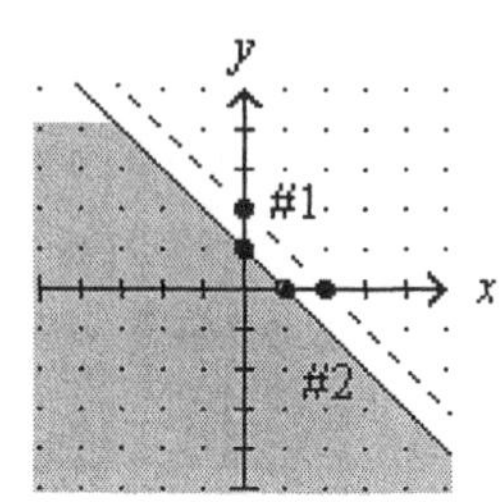

15. Inequality #1: $x > 0$
Graph: $x = 0$
A vertical line through the y–axis.

Broken line. No test point needed. Shade the half–plane to the right of the y–axis.

Inequality #2: $y > 0$
Graph: $y = 0$
A horizontal line through the x–axis.

Broken line. No test point needed. Shade the half–plane over the x–axis.

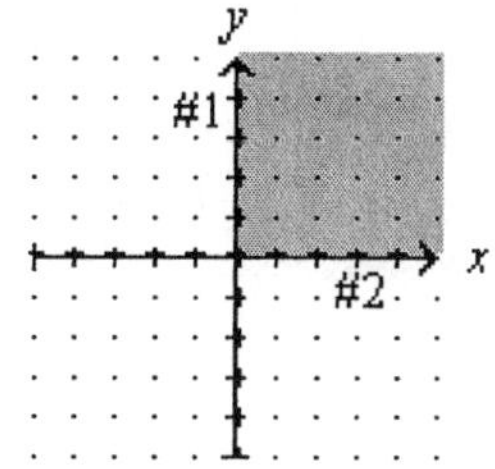

17. Inequality #1: $2x + 3y \le 6$
Graph: $2x + 3y = 6$

x	y
0	2
3	0

Solid line. Test point (0, 0).
$2(0) + 3(0) \le 6$
$0 \le 6$ True

Inequality #2: $3x + y \le 1$
Graph: $3x + y = 1$

x	y
0	1
1	−2

Solid line. Test point (0, 0).
$3(0) + (0) \le 1$
$0 \le 1$ True

Inequality #3: $x \le 0$
Graph: $x = 0$. A vertical line through the y– axis.

#3 Continued. Solid line. No test point needed. Shade the half–plane to the left of the y–axis.

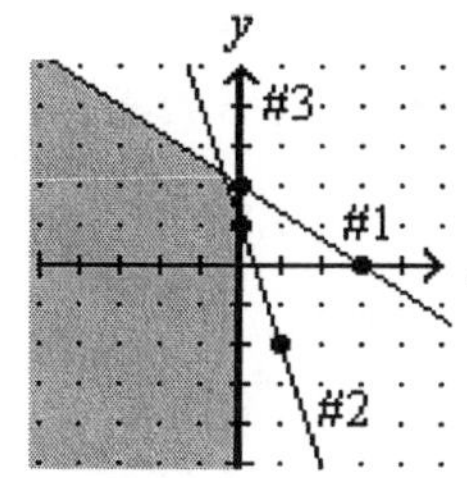

19. Inequality #1: $x - y < 4$
Graph: $x - y = 4$

x	y
0	−4
4	0

Broken line. Test point (0, 0).
$(0) - (0) < 4$
$0 < 4$ True

Inequality #2: $y \le 0$
Graph: $y = 0$
A horizontal line through the x–axis.

Solid line. No test point needed. Shade the half–plane under the x–axis.

Inequality #3: $x \ge 0$
Graph: $x = 0$. A vertical line through the y– axis.

#3 Continued. Solid line. No test point needed. Shade the half–plane to the right of the y–axis.

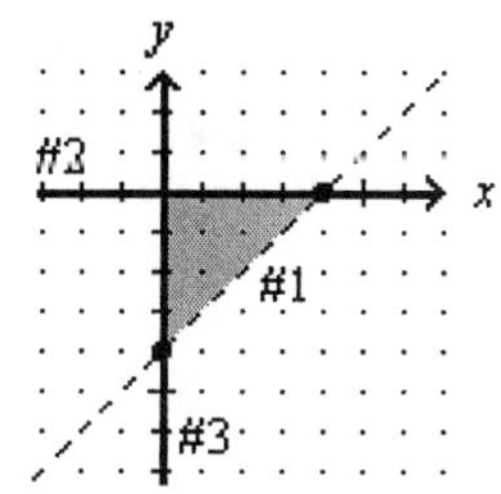

21. Inequality: $2 \le x < 0$
Rewrite as #1: $-2 \le x$
Graph: $x = -2$

A vertical line through $x = -2$.

Solid line. No test point needed. Shade the half–plane to the right of the line $x = -2$.

and #2: $x < 0$
Graph: $x = 0$
A vertical line through the y–axis.
Broken line. No test point needed. Shade the half–plane to the left of the y–axis.

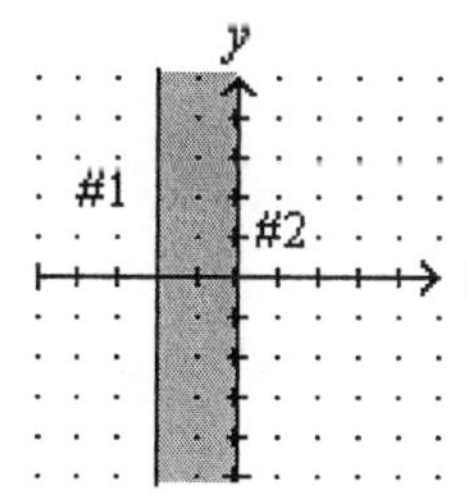

23. Inequality #1 $y < -2$
Graph: $y = -2$

A horizontal line through $y = -2$.
Broken line. No test point needed. Shade the half–plane under the line $y = -2$.

or #2: $y > 3$
Graph: $y = 3$

A horizontal line through $y = 3$.
Broken line. No test point needed. Shade the half–plane over the line $y = 3$.

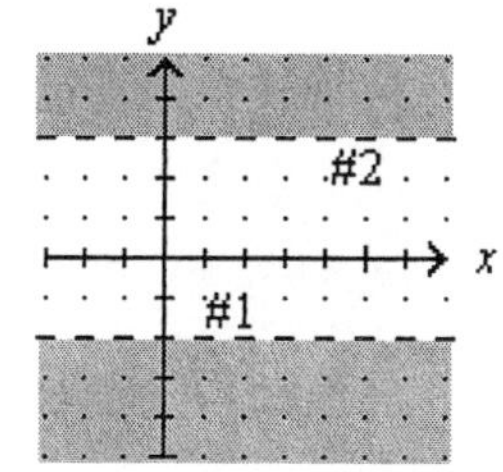

25. 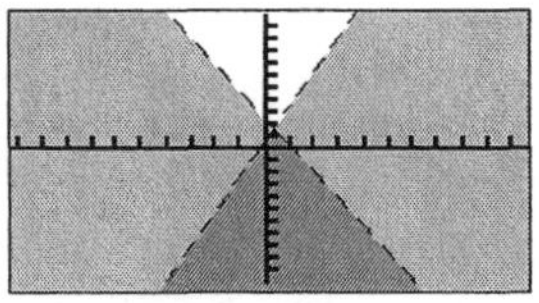

27. 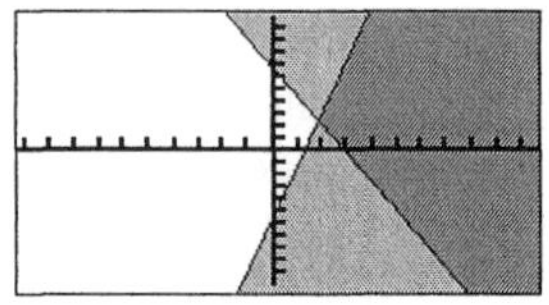

APPLICATIONS

29. Shade the area between "*G*" and 20 on the end of the field labeled BRONCOS.

31. Label the lines with the appropriate parallel. Shade the part of Iraq that is above the 36th parallel and below the 33rd parallel.

33. **Analysis**: Find the possible ways to spend at least \$30 but no more than \$60 on CDs that cost either \$10 or \$15.

Inequality: Let x represent the number of CDs that cost \$10 and y represent the number of CDs that cost \$15. The total cost of \$10 CDs is \$10x and the total cost of \$15 CDs is \$15y. Only a nonnegative number of CDs can be purchased, so $x \geq 0$ and $y \geq 0$.

Cost of \$10 CDs	plus	cost of \$15 CDS	is at least	\$30.
$10x$	$+$	$15y$	$\geq$	30

Cost of \$10 CDs	plus	cost of \$15 CDS	is no more than	\$60.
$10x$	$+$	$15y$	$\leq$	60

Solve:

Inequality #1: $10x + 15y \geq 30$
Graph: $10x + 15y = 30$

x	y
0	2
3	0

Solid line. Test point (0, 0).
$10(0) + 15(0) \geq 30$
$0 \geq 30$
False

Inequality #2: $10x + 15y \leq 60$
Graph: $10x + 15y = 60$

x	y
0	4
6	0

Solid line. Test point (0, 0).
$10(0) + 15(0) \leq 60$
$0 \leq 60$
True

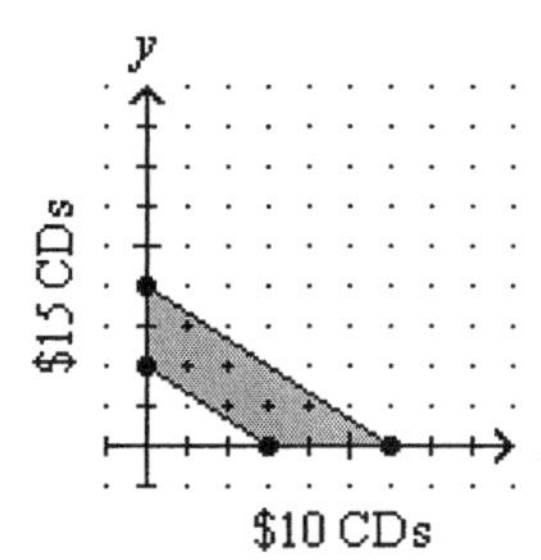

The graph is valid only in Quadrant I because $x \geq 0$ and $y \geq 0$.

Conclusion: Two possible combinations of CD purchases would be one \$10 CD and two \$15 CDs or four \$10 CDs and one \$15 CD.

35. **Analysis**: Find the possible ways to order no more than \$900 worth of chairs when desk chairs cost \$150 and side chairs cost \$100. There needs to be more side chairs than desk chairs.

Inequality: Let x represent the number of desk chairs at \$150 and y represent the number of side chairs at \$100. The cost of the desk chairs is \$150x and the cost of the side chairs is \$100y. Only a nonnegative number of chairs can be ordered, so $x \geq 0$ and $y \geq 0$.

Cost of desk chairs	plus	cost of side chairs	is no more than	900.
$150x$	$+$	$100y$	$\leq$	900

Number of side chairs	is more than	number of desk chairs.
y	>	x

Solve:

Inequality #1: $150x + 100y \leq 900$
Graph: $150x + 100y = 900$

x	y
0	9
6	0

Solid line. Test point (0, 0).

$$150(0) + 100(0) \leq 900$$
$$0 \leq 900 \quad \text{True}$$

Inequality #2: $y > x$
Graph: $y = x$

x	y
0	0
6	6

Broken line. Test point (4, 1).

$$(1) > (4)$$
$$1 > 4 \quad \text{False}$$

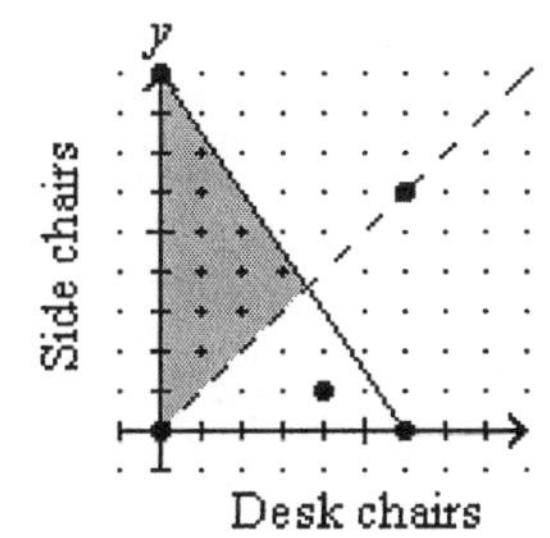

The graph is valid only in Quadrant I because $x \geq 0$ and $y \geq 0$.

Conclusion: Two possible combinations of desk chairs and side chairs would be 2 desk chairs and 4 side chairs or 1 desk chair and 5 side chairs.

WRITING

37. Graph each of the linear inequalities on the same coordinate system. Lightly shade the part of the coordinate system where the points will satisfy each of the inequalities individually. The solution of the system is the part of the coordinate system where all the inequalities are satisfied.

REVIEW

39. Quadrant IV

41. Quadrant II

Chapter 4 Key Concept: Inequalities

Types of Inequalities

1. a. compound inequality
   b. system of linear inequalities
   c. absolute value inequality
   d. linear inequality in two variables.
   e. linear inequality in one variable
   f. double linear inequality
   g. compound inequality
   h. absolute value inequality
   i. linear inequality in two variables

Solutions of Inequalities

2. a.

$$x - 3 < -4 \quad \text{or} \quad x - 2 > 0$$
$$(-2) - 3 \overset{?}{<} -4 \qquad (-2) - 2 \overset{?}{>} 0$$
$$-5 < -4 \qquad -4 \not> 0$$

Since at least one statement is true, $x = -2$ is a solution.

b.

$$\begin{cases} y < 3x + 2 \\ y < -2x + 3 \end{cases} \quad \text{Substitute } (-1, 3)$$
$$\begin{cases} (3) < 3(-1) + 2 \\ (3) < -2(-1) + 3 \end{cases} \quad \text{Simplify.}$$
$$\begin{cases} 3 < -1 \\ 3 < 5 \end{cases}$$

Since both statements are not true, (−1, 3) is not a solution.

2. cont.

c. $|x-8| \le 12$

$|-2-8| \overset{?}{\le} 12$

$|-10| \overset{?}{\le} 12$

$10 \le 12$

Yes, $x = -2$ is a solution.

d. $y < \frac{x}{3} - 1$

$(3) \overset{?}{<} \frac{-1}{3} - 1$

$9 \overset{?}{<} -1 - 3$

$9 \not< -4$

No, $(-1, 3)$ is not a solution.

e. $\frac{1}{2}x + 2 \ge \frac{1}{3}x - 4$

$\frac{1}{2}(-2) + 2 \overset{?}{\ge} \frac{1}{3}(-2) - 4$

$-1 + 2 \overset{?}{\ge} -\frac{2}{3} - 4$

$1 \ge -4\frac{2}{3}$

Yes, $x = -2$ is a solution.

f. $-6 < -3(x-4) \le 24$

$-6 \overset{?}{<} -3(-2-4) \overset{?}{\le} 24$

$-6 \overset{?}{<} -3(-6) \overset{?}{\le} 24$

$-6 < 18 \le 24$

Yes, $x = -2$ is a solution.

g. $5(x-2) \ge 0$ and $-3x < 9$

$5(-2-2) \overset{?}{\ge} 0$ $\qquad -3(-2) \overset{?}{<} 9$

$5(-4) \overset{?}{\ge} 0$ $\qquad 6 < 9$

$-20 \not\ge 0$

Since both statements are not true,
$x = -2$ is not a solution.

h. $|-1-2x| > 5$

$|-1-2(-2)| \overset{?}{>} 5$

$|3| \overset{?}{>} 5$

$3 \not> 5$

No, $x = -2$ is not a solution.

i. $y > -x$

$3 \overset{?}{>} -(-1)$

$3 > 1$

Yes, -2 is a solution.

Graphs of Inequalities

3. $2x + 1 > 4$

$2x > 3$

$x > \frac{3}{2}$

4.

x	$2x + y \ge 4$ Graph: $2x + y = 4$	y
0	$2(0) + y = 4$	4
2	$2x + (0) = 4$	0
Solid line	Test Point: (0, 0) $2(0) + (0) \ge 4$ $0 \ge 4$	False

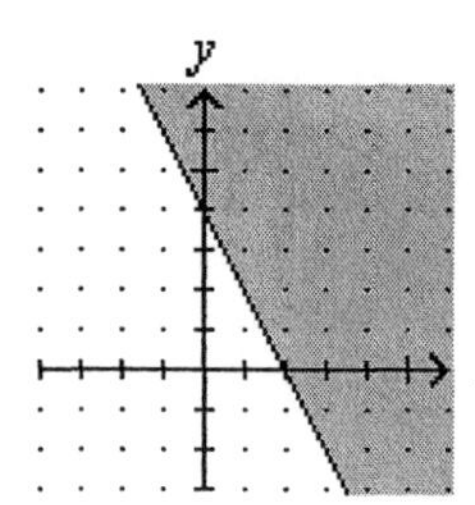

Chapter 4 — Chapter Review

Section 4.1

1.
$5(x-2) \le 5$
$5x - 10 \le 5$ — Remove the parentheses.
$5x \le 15$ — Add 10 to both sides.
$x \le 3$ — Divide both sides by 5.

$(-\infty, 3]$

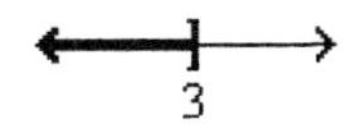

2.
$0.3x - 0.4 \ge 1.2 - 0.1x$
$0.4x \ge 1.6$ — Add $0.1x$ and 0.4 to both sides.
$x \ge 4$ — Divide both sides by 0.4

$[4, \infty)$

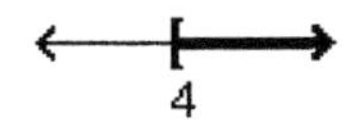

3.
$-16 < -\frac{4}{5}x$
$-80 < -4x$ — Multiply both sides by LCD 5.
$20 > x$ — Divide both sides by –4 and reverse the direction of the inequality symbol.

$(-\infty, 20)$

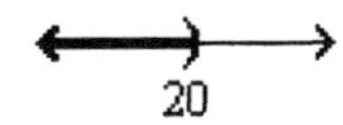

4.
$\frac{7}{4}(x+3) < \frac{3}{8}(x-3)$
$14(x+3) < 3(x-3)$ — Multiply both sides by LCD 8.
$14x + 42 < 3x - 9$ — Remove the parentheses.
$11x < -51$ — Subtract $3x$ and 42 from both sides.
$x < -\frac{51}{11}$ — Divide both sides by 11.

$\left(-\infty, -\frac{51}{11}\right)$

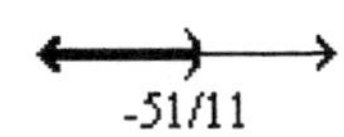

5. $[2, \infty)$

6. **Analysis**: Find how much additional investment over $10,000 must be made so that the annual income at 7% is at least $2000.

Inequality: Let x represent the amount of additional investment.

	P •	r •	t =	I
Investment at 6%	10,000	0.06	1	600
Investment at 7%	x	0.07	1	$0.07x$

Income at 6%	plus	income at 7%	is at least	$2000.
$600	+	$0.07x$	$\ge$	2000.

Solve:
$600 + 0.07x \ge 2000$
$0.07x \ge 1400$ — Subtract 600 from both sides.
$x \ge 20{,}000$ — Divide both sides by 0.07.

Conclusion: The additional investment would need to be $20,000 or more.

Section 4.2

7.
$x < 0$ and $x > -5$
$-4 < 0$ True; $-4 > -5$ True
Yes, $x = -4$ is a solution.

8.
$x + 3 < -3x - 1$ and $4x - 3 > 3x$
$(-4) + 3 < -3(-4) - 1$; $4(-4) - 3 > 3(-4)$
$-1 < 12 - 1$; $-16 - 3 > -12$
$-1 < 11$ True; $-19 > -12$ False
No, $x = -4$ is not a solution because both inequalities need to be true in an "and" statement.

9. $\begin{aligned} -2x &> 8 \\ x &< -4 \end{aligned}$ and $\begin{aligned} x+4 &\geq -6 \\ x &\geq -10 \end{aligned}$

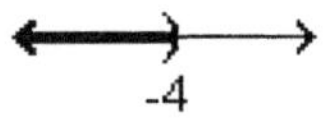

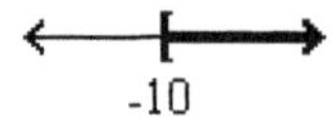

$[-10, -4)$

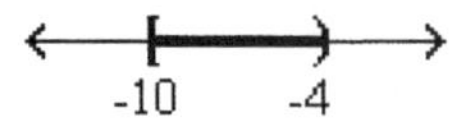

10. $\begin{aligned} 5(x+2) &\leq 4(x+1) \\ 5x+10 &\leq 4x+4 \\ x &\leq -6 \end{aligned}$ and $\begin{aligned} 11+x &< 0 \\ x &< -11 \end{aligned}$

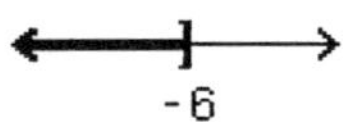

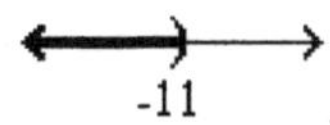

$[-\infty, -11)$

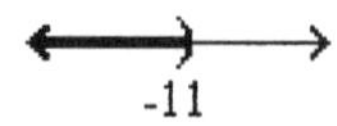

11. $\begin{aligned} 3 &< 3x+4 < 10 \\ -1 &< 3x < 6 \\ -\frac{1}{3} &< x < 2 \end{aligned}$ $\left(-\frac{1}{3}, 2\right)$

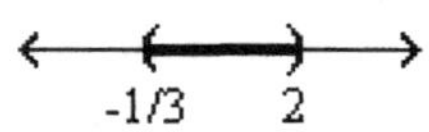

12. $\begin{aligned} -2 &\leq \frac{5-x}{2} \leq 2 \\ -4 &\leq 5-x \leq 4 \\ -9 &\leq -x \leq -1 \\ 9 &\geq x \geq 1 \end{aligned}$ $[1, 9]$

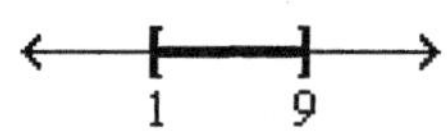

13. $\begin{aligned} x &< 1.6 \\ (-4) &\ ?_< 1.6 \end{aligned}$ or $\begin{aligned} x &> -3.9 \\ (-4) &\ ?_> -3.9 \end{aligned}$

$3(-4)$

$-4 < 1.6$ $\quad 6-3 \ ?_> -3.9$

$3 > -10$

Since at least one statement is true, $x = -4$ is a solution.

14. $\begin{aligned} x+1 &< 2x-1 \\ (-4)+1 &\ ?_< 2(-4)-1 \\ -4+1 &\ ?_< -8-1 \\ -3 &\not< -9 \end{aligned}$ or $\begin{aligned} 4x-3 &> 3x \\ 4(-4)-3 &\ ?_> \\ -16-3 &\ ?_> -12 \\ -19 &\not> -12 \end{aligned}$

Since neither statement is true, $x = -4$ is not a solution.

15. $\begin{aligned} x+1 &< -4 \\ x &< -5 \end{aligned}$ or $\begin{aligned} x-4 &> 0 \\ x &> 4 \end{aligned}$

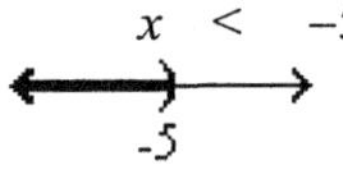

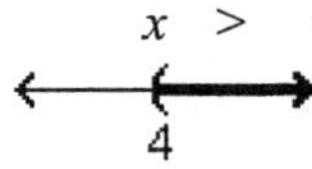

$(-\infty, -5) \cup (4, \infty)$

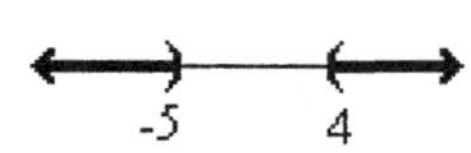

16. $\begin{aligned} \frac{x}{2}+3 &> -2 \\ x+6 &> -4 \\ x &> -10 \end{aligned}$ or $\begin{aligned} 4-x &> 4 \\ -x &> 0 \\ x &< 0 \end{aligned}$

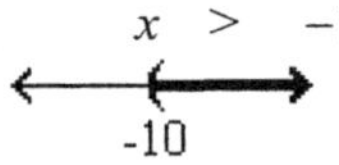

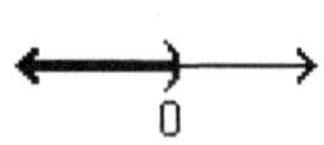

$(-\infty, \infty)$

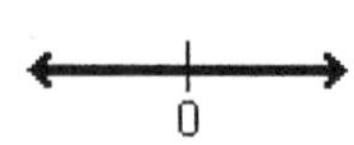

17. The area of the rug is $A = 4l$. This area varies from 17 ft^2 to 25 ft^2. The double inequality is: $17 \leq 4l \leq 25$. Solving for l: $4.25 \text{ ft} \leq l \leq 6.25 \text{ ft}$.

18. a. ii, iv b. i, iii

Section 4.3

19. $|-7| = 7$

20. $\left|\frac{5}{16}\right| = \frac{5}{16}$

21. $-|71.05| = -71.05$

22. $-|-12| = -(12) = -12$

23. Rewrite $|4x| = 8$ as

$$4x = -8 \quad\text{or}\quad 4x = 8$$
$$x = -2 \qquad\qquad x = 2$$

24. If $2|3x + 1| = 20$, then $|3x + 1| = 10$.
Rewrite $|3x + 1| = 10$ as

$$\begin{aligned} 3x + 1 &= -10 \\ 3x &= -11 \\ x &= -\frac{11}{3} \end{aligned} \quad\text{or}\quad \begin{aligned} 3x + 1 &= 10 \\ 3x &= 9 \\ x &= 3 \end{aligned}$$

25. If $\left|\frac{3}{2}x - 4\right| - 10 = -1$, then $\left|\frac{3}{2}x - 4\right| = 9$.

Rewrite $\left|\frac{3}{2}x - 4\right| = 9$ as

$$\begin{aligned} \frac{3}{2}x - 4 &= -9 \\ 3x - 8 &= -18 \\ 3x &= -10 \\ x &= -\frac{10}{3} \end{aligned} \quad\text{or}\quad \begin{aligned} \frac{3}{2}x - 4 &= 9 \\ 3x - 8 &= 18 \\ 3x &= 26 \\ x &= \frac{26}{3} \end{aligned}$$

26. Rewrite $\left|\frac{2 - x}{3}\right| = 4$ as

$$\begin{aligned} \frac{2 - x}{3} &= -4 \\ 2 - x &= -12 \\ -x &= -14 \\ x &= 14 \end{aligned} \quad\text{or}\quad \begin{aligned} \frac{2 - x}{3} &= 4 \\ 2 - x &= 12 \\ -x &= 10 \\ x &= -10 \end{aligned}$$

27. Rewrite $|3x + 2| = |2x - 3|$ as

$$\begin{aligned} 3x + 2 &= 2x - 3 \\ x &= -5 \end{aligned} \quad\text{or}\quad \begin{aligned} 3x + 2 &= -(2x - 3) \\ 3x + 2 &= -2x + 3 \\ 5x &= 1 \\ x &= \frac{1}{5} \end{aligned}$$

28. Rewrite $\left|\frac{3 - 2x}{2}\right| = \left|\frac{3x - 2}{3}\right|$ as

$$\begin{aligned} \frac{3 - 2x}{2} &= \frac{3x - 2}{3} \\ 3(3 - 2x) &= 2(3x - 2) \\ 9 - 6x &= 6x - 4 \\ -12x &= -13 \\ x &= \frac{13}{12} \end{aligned} \quad\text{or}\quad \begin{aligned} \frac{3 - 2x}{2} &= -\left(\frac{3x - 2}{3}\right) \\ 3(3 - 2x) &= -2(3x - 2) \\ 9 - 6x &= -6x + 4 \\ 9 &= 4 \end{aligned}$$

Not possible.

29. Rewrite $|x| \le 3$ as
$$-3 \le x \le 3$$

$[-3, 3]$

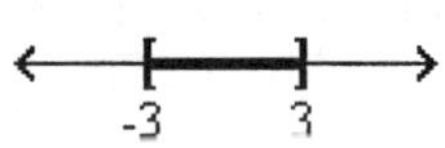

30. Rewrite $|2x + 7| < 3$ as
$$\begin{aligned} -3 &< 2x + 7 < 3 \\ -10 &< 2x < -4 \\ -5 &< x < -2 \end{aligned}$$

$(-5, -2)$

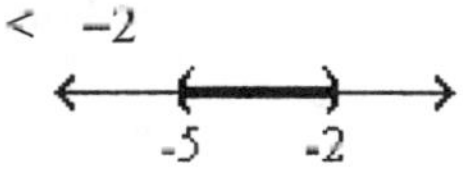

31. Rewrite $|5 - 3x| \le 14$ as

$$\begin{aligned} -14 &\le 5 - 3x \le 14 \\ -19 &\le -3x \le 9 \\ \frac{19}{3} &\ge x \ge -3 \end{aligned}$$

$\left[-3, \frac{19}{3}\right]$

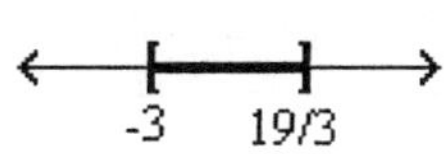

32. $\left|\frac{2}{3}x + 14\right| < 0$

An absolute value cannot be negative. This inequality has no solution.

33. Rewrite $|x| > 1$ as

$x < -1$ or $x > 1$

$(-\infty, -1) \cup (1, \infty)$

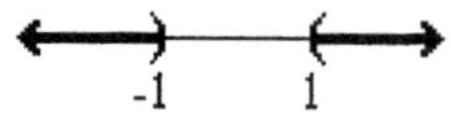

34. Rewrite $\left|\dfrac{1-5x}{3}\right| \geq 7$ as

$$\begin{aligned} \frac{1-5x}{3} &\leq -7 \\ 1-5x &\leq -21 \\ -5x &\leq -22 \\ x &\geq \frac{22}{5} \end{aligned} \quad \text{or} \quad \begin{aligned} \frac{1-5x}{3} &\geq 7 \\ 1-5x &\geq 21 \\ -5x &\geq 20 \\ x &\leq -4 \end{aligned}$$

$(-\infty, -4] \cup \left[\dfrac{22}{5}, \infty\right)$

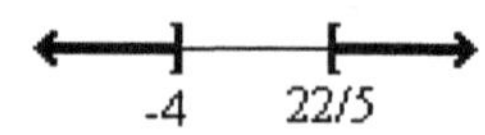

35. Rewrite $|3x - 8| > 4$ as

$$\begin{aligned} 3x-8 &< -4 \\ 3x &< 4 \\ x &< \frac{4}{3} \end{aligned} \quad \text{or} \quad \begin{aligned} 3x-8 &> 4 \\ 3x &> 12 \\ x &> 4 \end{aligned}$$

$\left(-\infty, \dfrac{4}{3}\right) \cup (4, \infty)$

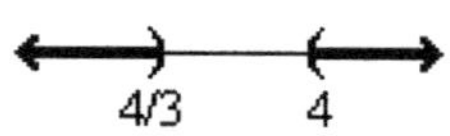

36. $\left|\dfrac{3}{2}x - 14\right| \geq 0$

Since an absolute value is always positive, this inequality is always true.

$(-\infty, \infty)$

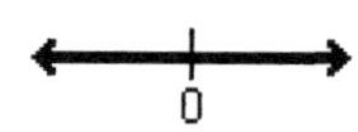

37. a. The weight is two ounces on either side of 8 ounces, $|w - 8| \leq 2$.

b. Rewrite $|w - 8| \leq 2$ as $-2 \leq w - 8 \leq 2$. Then add 8 to all three parts to obtain $6 \leq w \leq 10$. In interval notation this is expressed as [6, 10].

38. We must isolate the variable before we split into two separate equations.

Section 4.4

39.

x	$2x + 3y > 6$ Graph: $2x + 3y = 6$	y
0	$2(0) + 3y = 6$	2
3	$2x + 3(0) = 6$	0
Broken line	Test Point: (0, 0) $2(0) + 3(0) > 6$ $0 > 6$	False

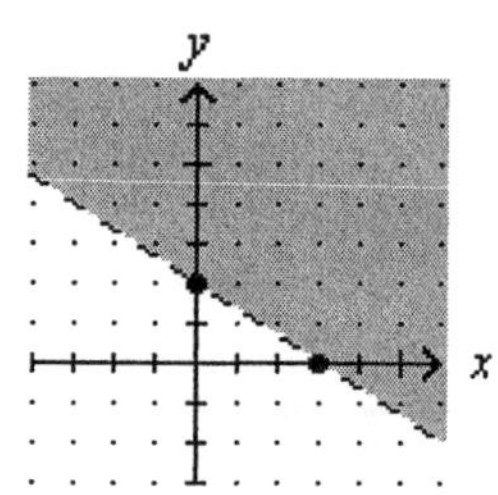

40.

x	$y \leq 4 - x$ Graph: $y = 4 - x$	y
0	$y = 4 - (0)$	4
4	$(0) = 4 - x$	0
Solid line	Test Point: (0, 0) $(0) \leq 4 - (0)$ $0 \leq 4$	True

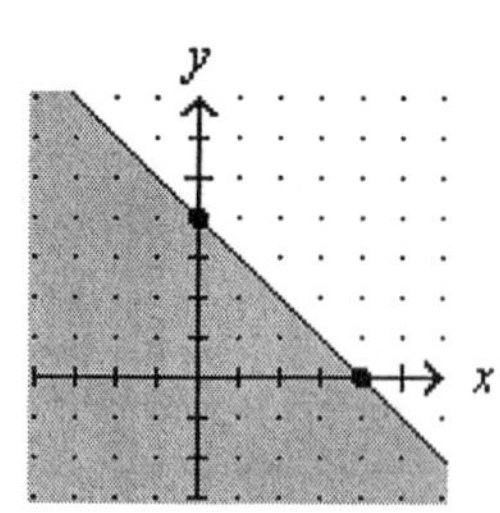

41.

x	$y < \frac{1}{2}x$ Graph: $2y = x$	y
0	$2y = (0)$	0
4	$2(2) = x$	2
Broken line	Test Point: (4, 0) $2(0) < \frac{1}{2}(4)$ $0 < 2$	True

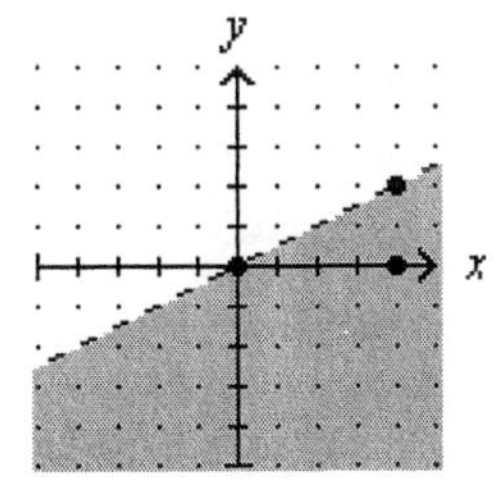

42.

	$x \geq -\frac{3}{2}$ Graph: $x = -\frac{3}{2}$
	A vertical line through $x = -\frac{3}{2}$.
Solid line	No test point needed. Shade the half–plane to the right of the line.

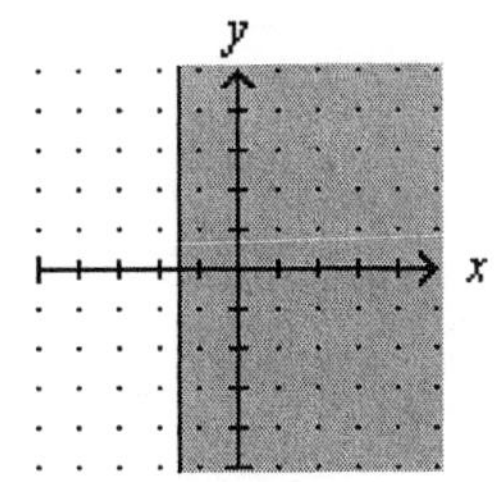

43. **Analysis**: Find the possible ways to sell \$6 reserved seats and \$4 general admission tickets and generate receipts of at least \$10,200.

Inequality: Let x represent the number of reserved seats and y represent the number of general admission tickets. The receipts from the reserved seats will be \$$6x$ and the receipts from the general admission tickets will be \$$4y$. Only a nonnegative number of tickets can be sold.

Receipts from reserved seats	plus	receipts from general admissions	is at least	\$10,200.
$6x$	$+$	$4y$	$\geq$	10,200

Solve:

x	$6x + 4y \geq 10{,}200$ Graph: $6x + 4y = 10{,}200$	y
0	$6(0) + 4y = 10{,}200$	2550
1700	$6x + 4(0) = 10{,}200$	0
Solid line	Test Point: (0, 0) $6(0) + 4(0) \geq 10{,}200$ $0 \geq 10{,}200$	False

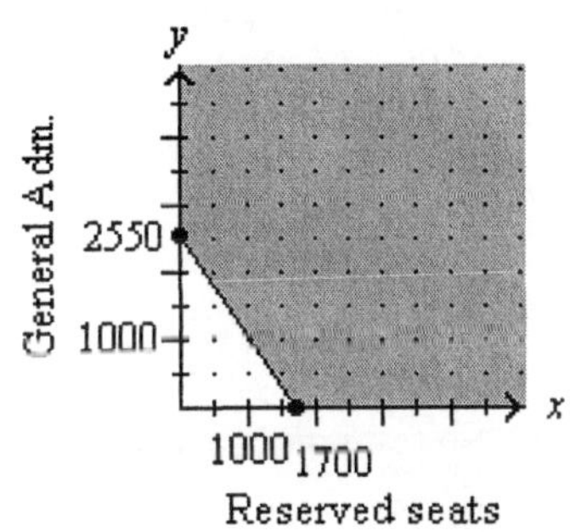

Conclusion: Possible combinations of reserved seats and general admission tickets are 1800 reserved and no general admission, 1000 reserved and 1500 general admission, and 2000 reserved and 2000 general admission.

44. **Analysis:** Use the points (4,0) and (0,–3) to write the equation of the dashed line.

$$m = \frac{0-(-3)}{4-0} = \frac{3}{4} \;;\; b = -3 \;;\; \text{use } y = mx + b \text{ to obtain } \; y = \frac{3}{4}x - 3$$

Since shading is under the dashed line the inequality is: $y < \frac{3}{4}x - 3$. Now multiply by 4 and rearrange.

Conclusion: The inequality is $3x - 4y > 12$

Section 4.5

45. Inequality #1: $y \geq x + 1$

Graph: $y = x + 1$

x	y
0	1
−1	0

Solid line. Test point (0, 0).

$(0) \geq (0) + 1$

$0 \geq 1$ False

Inequality #2: $3x + 2y < 6$

Graph: $3x + 2y = 6$

x	y
0	3
2	0

Broken line. Test point (0, 0).

$3(0) + 2(0) < 6$

$0 < 6$ True

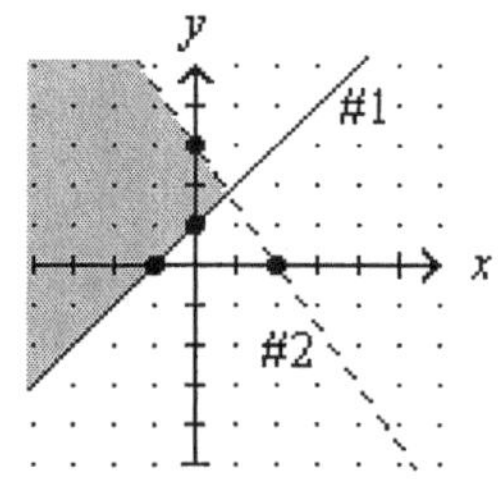

46. Inequality #1: $x - y < 3$

Graph: $x - y = 3$

x	y
0	−3
3	0

Broken line. Test point (0, 0).

$(0) - (0) < 3$

$0 < 3$ True

Inequality #2: $y \leq 0$

Graph: $y = 0$

A horizontal line through the x–axis.

Solid line. No test point needed. Shade the half–plane under the x–axis.

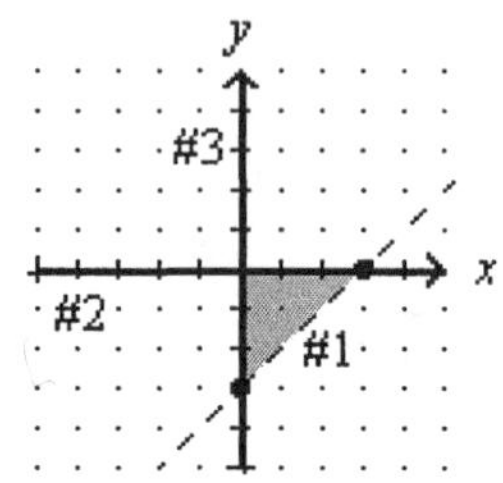

Inequality #3: $x \geq 0$

Graph: $x = 0$. A vertical line through the y– axis.

#3 Continued. Solid line. No test point needed. Shade the half–plane to the right of the y–axis.

47. Inequality: $-2 < x < 4$

Rewrite as: $-2 < x$ and $x < 4$

Graph: $x = -2$

A vertical line through $x = -2$

Broken line. No test point needed. Shade the half–plane to the right of the line $x = -2$.

Graph: $x = 4$

A vertical line through $x = 4$.

Broken line. No test point needed. Shade the half–plane to the left of the line $x = 4$.

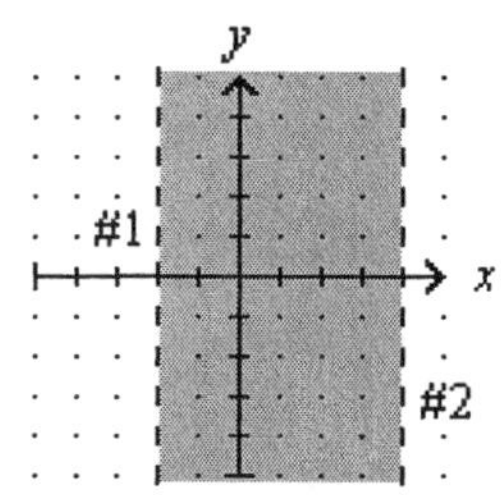

48. Inequality: $y \leq -2$ or $y > 1$

Graph: $y = -2$

A horizontal line through $y = -2$.

Solid line. No test point needed. Shade the half–plane under the line $y = -2$.

Graph: $y = 1$

A horizontal line through $y = 1$.

Broken line. No test point needed. Shade the half–plane over the line $y = 1$.

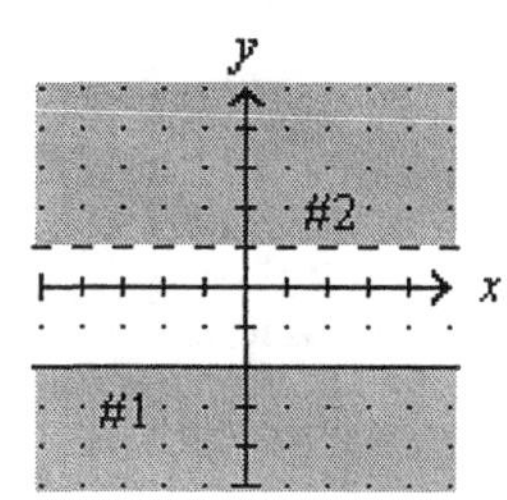

49. The line on the left is $x = 35$ and the area to the right is shaded so $x \geq 35$.
The line on the right is $x = 130$ and the area to the left is shaded so $x \leq 130$.
The line at the bottom is $y = -56x + 280$ and the area above that line is shaded so $y \geq -56x + 280$.
The line at the top is $y = -18x + 90$ and the area below that line is shaded so $y \leq -18x + 90$.

Chapter 4 — Chapter Test

1. The statement is false because –5.67 is less than –5.

2.
$$\begin{aligned} 3(x-2) &\le 2(x+7) \\ 3[(-2)-2] &\overset{?}{\le} 2[((-2)+7] \\ 3(-4) &\overset{?}{\le} 2(5) \\ -12 &\le 10 \end{aligned}$$
True. Yes, $x = -2$ is a solution.

3. $(-\infty, -5)$. x is in the interval between $-\infty$ and -5 and -5 is not included.

4. $c \le d$

5.
$$\begin{aligned} 7 &< \frac{2}{3}t - 1 \\ 21 &< 2t - 3 && \text{Multiply both sides by the LCD 3.} \\ 24 &< 2t && \text{Add 3 to both sides.} \\ 12 &< t && \text{Divide both sides by 2.} \end{aligned}$$
$(12, \infty)$

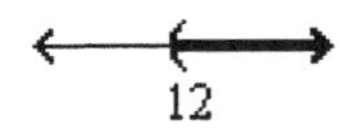

6.
$$\begin{aligned} -2(2x+3) &\ge 14 \\ -4x - 6 &\ge 14 && \text{Clear the parentheses.} \\ -4x &\ge 20 && \text{Add 6 to both sides.} \\ x &\le -5 && \text{Divide by –4 and reverse the symbol.} \end{aligned}$$
$(-\infty, -5]$

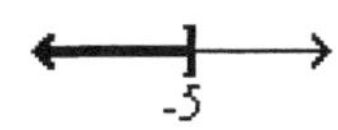

7. Let x represent the score from Exam 5. Calculate the mean by adding all the scores and dividing by the number of exams which is 5.

Exam average is more than 80, where the exam average is $\frac{70+79+85+88+x}{5} = \frac{322+x}{5}$

$$\frac{322+x}{5} > 80$$

Solve for x:
$$\begin{aligned} \frac{322+x}{5} &> 80 \\ 322 + x &> 400 \\ x &> 78 \end{aligned}$$
The Exam 5 score needs to be more than 78.

8.
$$\begin{aligned} 3x &\ge -2x + 5 \\ 5x &\ge 5 \\ x &\ge 1 \end{aligned} \qquad \text{and} \qquad \begin{aligned} 7 &\ge 4x - 2 \\ 9 &\ge 4x \\ \frac{9}{4} &\ge x \end{aligned}$$

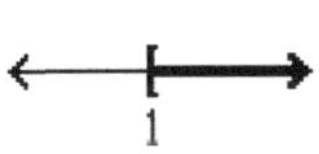

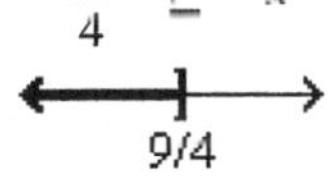

$\left[1, \frac{9}{4}\right]$

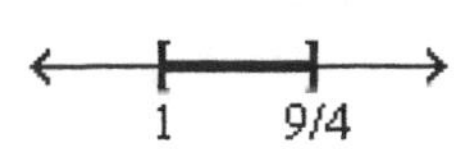

9.
$$\begin{aligned} 3x &< -9 \\ x &< -3 \end{aligned} \qquad \text{or} \qquad \begin{aligned} -\frac{x}{4} &< -2 \\ -x &< -8 \\ x &> 8 \end{aligned}$$

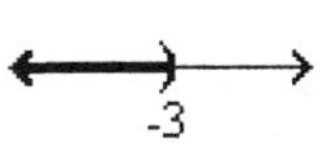

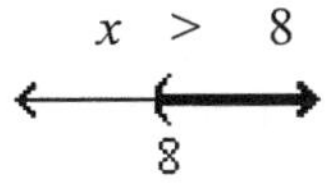

$(-\infty, -3) \cup (8, \infty)$

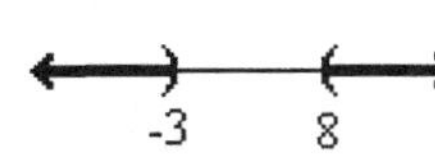

10.
$$\begin{aligned} -2 &< \frac{x-4}{3} < 4 \\ -6 &< x - 4 < 12 \\ -2 &< x < 16 \end{aligned}$$
$(-2, 16)$

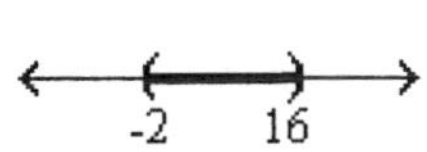

11. $|8| = 8$

12. $-|-4.75| = -(4.75) = -4.75$

13. Rewrite $|4 - 3x| = 19$ as

$$\begin{aligned} 4 - 3x &= 19 & \text{or} \quad 4 - 3x &= -19 \\ -3x &= 15 & -3x &= -23 \\ x &= -5 & x &= \frac{23}{3} \end{aligned}$$

14. Rewrite $|3x + 4| = |x + 12|$ as

$$\begin{aligned} 3x + 4 &= -(x + 12) & \text{or} \quad 3x + 4 &= x + 12 \\ 3x + 4 &= -x - 12 & 2x &= 8 \\ 4x &= -16 & x &= 4 \\ x &= -4 \end{aligned}$$

15. Rewrite $|x + 3| \leq 4$ as

$$\begin{aligned} -4 \leq x + 3 &\leq 4 \\ -7 \leq x &\leq 1 \end{aligned}$$

$[-7, 1]$

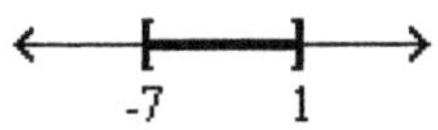

16. Rewrite $|2x - 4| > 22$ as

$$\begin{aligned} 2x - 4 &< -22 & \text{or} \quad 2x - 4 &> 22 \\ 2x &< -18 & 2x &> 26 \\ x &< -9 & x &> 13 \end{aligned}$$

$(-\infty, -9) \cup (13, \infty)$

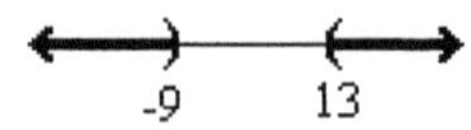

17. If $|4 - 2x| + 1 > 3$, then $|4 - 2x| > 2$
Rewrite $|4 - 2x| > 2$ as

$$\begin{aligned} 4 - 2x &< -2 & \text{or} \quad 4 - 2x &> 2 \\ -2x &< -6 & -2x &> -2 \\ x &> 3 & x &< 1 \end{aligned}$$

$(-\infty, 1) \cup (3, \infty)$

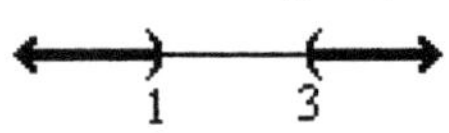

18. If $2|2x - 4| \leq 4$, then $|2x - 4| \leq 2$
Rewrite $|2x - 4| \leq 2$ as

$$\begin{aligned} -2 \leq 2x - 4 &\leq 2 \\ 2 \leq 2x &\leq 6 \\ 1 \leq x &\leq 3 \end{aligned}$$

$[1, 3]$

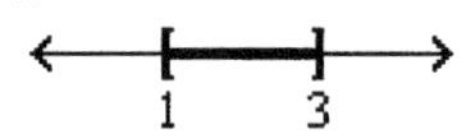

19.

x	$3x + 2y \geq 6$ Graph: $3x + 2y = 6$	y
0	$3(0) + 2y = 6$	3
2	$3x + 2(0) = 6$	0
Solid line	Test Point: (0, 0) $3(0) + 2(0) \geq 6$ $0 \geq 6$	False

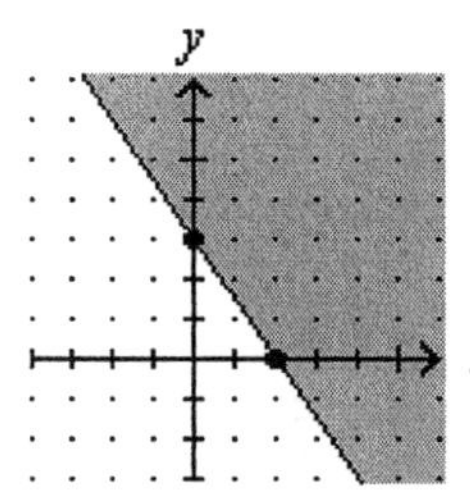

20.

x	$y < x$ Graph: $y = x$	y
0	$y = (0)$	0
2	$y = (2)$	2
Broken line	Test Point: (3, 0) $(0) < (3)$ $0 < 3$	True

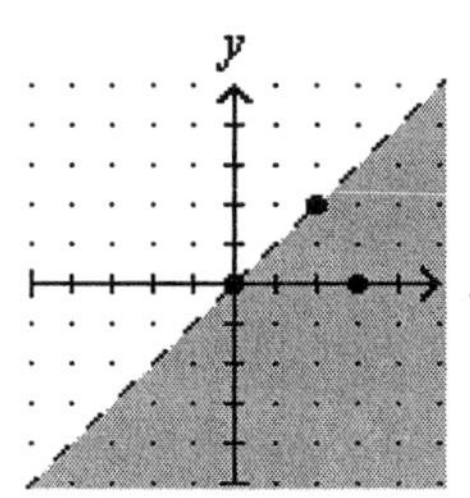

21. Inequality #1: $2x - 3y \geq 6$
Graph: $2x - 3y = 6$

x	y
0	−2
3	0

Solid line. Test point (0, 0).

$$\begin{aligned} 2(0) - 3(0) &\geq 6 \\ 0 &\geq 6 \quad \text{False} \end{aligned}$$

Inequality #2: $y \leq -x + 1$
Graph: $y = -x + 1$

x	y
0	1
5	−4

Solid line. Test point (0, 0).

$$\begin{aligned} (0) &\leq -(0) + 1 \\ 0 &\leq 1 \quad \text{True} \end{aligned}$$

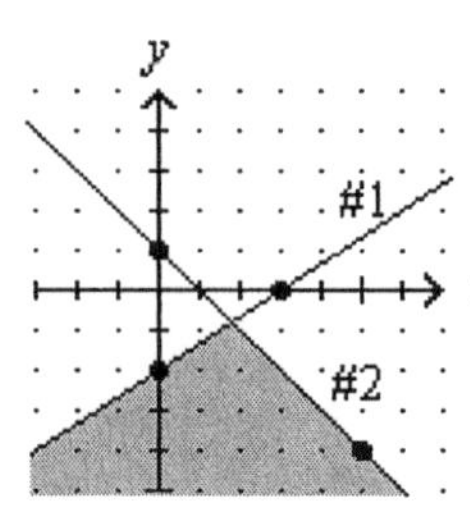

22. $-2 \le y < 5$

Rewrite as #1: $y \ge -2$ and #2: $y < 5$

#1	#2
Graph: $y = -2$	Graph: $y = 5$
A horizontal line through $y = -2$.	A horizontal line through $y = 5$.
Solid line. No test point needed. Shade the half–plane above the line $y = -2$.	Broken line. No test point needed. Shade the half–plane under the line $y = 5$.

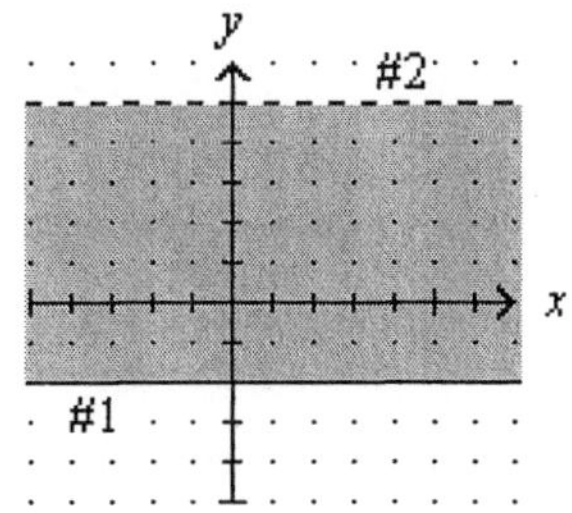

23. **Analysis**: Find the possible combinations of simple tax returns taking 1 hour to complete and complicated returns taking 3 hours to complete that an accountant can complete in less than 9 hours per day.

Inequality: Let x represent the number of simple returns and y represent the number of complicated returns The possible hours from simple returns is $1x$ and the possible hours from complicated returns is $3y$. Only a nonnegative number of returns can be prepared.

Hours from simple returns	plus	hours from complicated returns	is less than	9 hours.
$1x$	$+$	$3y$	$<$	9

Solve:

x	$x + 3y < 9$ Graph: $x + 3y = 9$	y
0	$(0) + 3y = 9$	3
9	$x + 3(0) = 9$	0
Broken line	Test Point: (0, 0) $(0) + 3(0) < 9$ $0 < 9$	True

The graph is valid only in Quadrant I because $x \ge 0$ and $y \ge 0$.

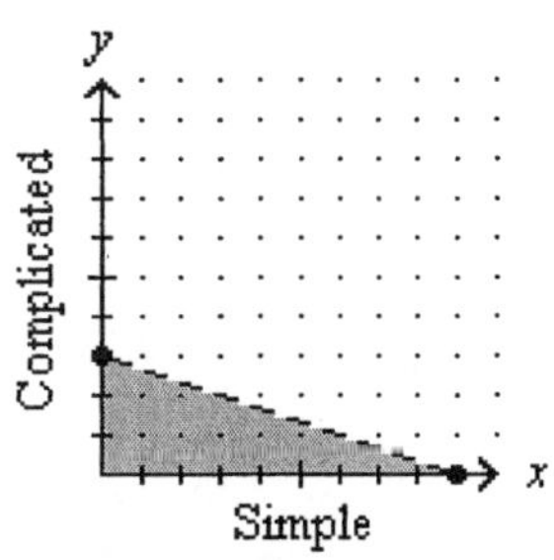

Conclusion: Possible combinations of returns would be 1 simple and 1 complicated, 2 simple and 1 complicated, or 2 simple and 2 complicated.

24. Yes, it can be determined from the graph that (3, –4) is a solution because it is within the blue shaded area. It is not a solution of the system because it is not in the area shaded by both inequalities.

25. The line at the top is $y = 60$ and the area under that line is shaded so $y \le 60$.
The line at the bottom is $y = 27$ and the area above that line is shaded so $y \ge 27$.
The line on the left is $y = -11x + 852$ and the area to the right and above that line is shaded so $y \ge -11x + 852$.
The line on the right is $y = -5x + 445$ and the area to the left and below that line is shaded so $y \le -5x + 445$.

26. Answers vary- possible answer: Multiplying by a negative number will make –2 become +2 and +5 become –5. The inequality must be switched in order to be true.

Chapter 1–4 Cumulative Review Exercises

1. Rational numbers: terminating and repeating decimals
Irrational numbers: nonterminating, nonrepeating decimals

2. $p = 0.125,\ \frac{p}{2} = \frac{0.125}{2} = 0.0625,\ \frac{p}{4} = \frac{0.125}{4} = 0.03125,\ 0.433p = 0.433(0.125) = 0.054125$

3. $|x| - xy = |2| - (2)(-4) = 2 + 8 = 10$

4. $\frac{x^2 - y^2}{3x + y} = \frac{(2)^2 - (-4)^2}{3(2) + (-4)} = \frac{4 - 16}{6 - 4} = \frac{-12}{2} = -6$

5. $3p^2 - 6(5p^2 + p) + p^2 = 3p^2 - 30p^2 - 6p + p^2 = (3 - 30 + 1)p^2 - 6p = -26p^2 - 6p$

6. $-(a + 2) - (a - b) = -a - 2 - a + b = -2a + b - 2$

7. Convert $11\frac{3}{4}$ inches to feet. $11\frac{3}{4}\text{ in.}\left(\dfrac{1\text{ ft}}{12\text{ in,}}\right) = 0.9791\overline{6}\text{ ft.}$

Multiply the converted $11\frac{3}{4}$ in. by 205 ft. $(205\text{ ft}) \bullet (0.9791666) = 200.72916\text{ ft}^2 \approx 201\text{ ft}^2$.

8. Find P when $r = 8\frac{7}{8}\% = 0.08875$, A = \$1775, and $t = 1$ year. Use the formula $A = Pr\,t$.

$1775 = P(0.08875)(1)$; $\$20{,}000 = P$

9.
$$\begin{aligned} 3x - 6 &= 20 \\ 3x &= 26;\ x = \frac{26}{3} \end{aligned}$$

10.
$$\begin{aligned} 6(x-1) &= 2(x+3) \\ 6x - 6 &= 2x + 6 \\ 4x &= 12;\ x = 3 \end{aligned}$$

11.
$$\begin{aligned} \frac{5b}{2} - 10 &= \frac{b}{3} + 3 \\ 15b - 60 &= 2b = 18 \\ 13b &= 78 \\ b &= 6 \end{aligned}$$

12.
$$\begin{aligned} 2a - 5 &= -2a + 4(a-2) + 1 \\ 2a - 5 &= -2a + 4a - 8 + 1 \\ 2a - 5 &= 2a - 7 \\ 0 &= -2 \end{aligned}$$
There is no solution.

13. Solve each equation for y and then compare to the slope–intercept form, $y = mx + b$.

$$\begin{aligned} 3x + 2y &= 12 \\ 2y &= -3x + 12 \\ y &= -\frac{3}{2} + 6,\ m = -\frac{3}{2} \end{aligned} \qquad \begin{aligned} 2x - 3y &= 5 \\ -3y &= -2x + 5 \\ y &= \frac{2}{3}x - \frac{5}{3},\ m = \frac{2}{3} \end{aligned}$$

The slopes are negative reciprocals, therefore the lines are perpendicular.

14. Solve each equation for y and then compare to the slope–intercept form, $y = mx + b$.

$$\begin{aligned} 3x &= y + 4 \\ y &= 3x - 4,\ m = 3 \end{aligned} \qquad \begin{aligned} y &= 3(x-4) - 1 \\ y &= 3x - 12 - 1 \\ y &= 3x - 13,\ m = 3 \end{aligned}$$

The slopes are the same, therefore the lines are parallel.

15. Rewrite the given equation in slope–intercept form. $3x + y = 8$ becomes $y = -3x + 8$. The new slope is $m = \frac{1}{3}$.

Substitute the point $(-2, 3)$ into the point–slope form.
$$\begin{aligned} y - y_1 &= m(x - x_1) \\ y - (3) &= \frac{1}{3}[x - (-2)] \\ y - 3 &= \frac{1}{3}x + \frac{2}{3} \\ y &= \frac{1}{3}x + \frac{11}{3} \end{aligned}$$

16. $(x_1, y_1) = (0, -8)$ and $(x_2, y_2) = (-5, 0)$; $m = \dfrac{y_2 - y_1}{x_2 - x_1} = \dfrac{0-(-8)}{-5-0} = \dfrac{8}{-5} = -\dfrac{8}{5}$

17. Rate of change $= \dfrac{1975\text{ prisoners} - 1970\text{ prisoners}}{1975 - 1970} = \dfrac{241{,}000 - 196{,}000}{5} = \dfrac{45{,}000}{5} = 9{,}000$ prisoners/yr.

18. The greatest rate of change would be in the period 1990–1995.

Rate of change $= \dfrac{1995\text{ prisoners} - 1990\text{ prisoners}}{1995 - 1990} = \dfrac{1{,}100{,}000 - 739{,}000}{5} = \dfrac{361{,}000}{5} = 72{,}200$ prisoners/yr

19. $f(x) = 3x^2 - x;\quad f(2) = 3(2)^2 - (2) = 3(4) - 2 = 12 - 2 = 10$

20. $f(x) = 3x^2 - x;\quad f(-2) = 3(-2)^2 - (-2) = 3(4) + 2 = 12 + 2 = 14$

21. Equation #1

$2x + y = 5$

x	y
0	5
2	1
4	−3

Equation #2

$x - 2y = 0$

x	y
0	0
−4	−2
4	2

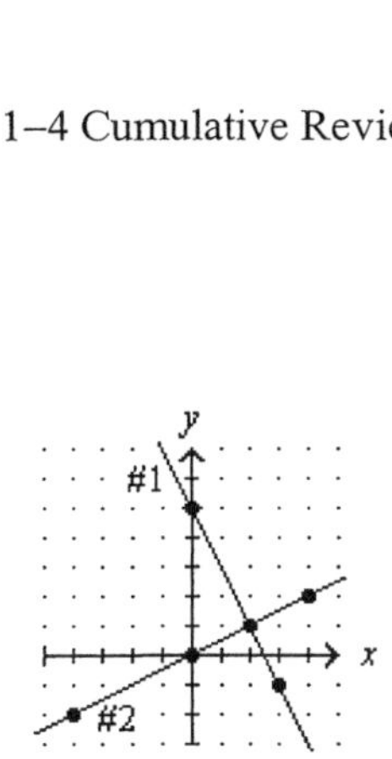

The solution is (2, 1).

22. $$\begin{cases} \dfrac{x}{10} + \dfrac{y}{5} = \dfrac{1}{2} \\ \dfrac{x}{2} - \dfrac{y}{5} = \dfrac{13}{10} \end{cases}$$

Multiply both equation by LCD 10.

$$\begin{cases} x + 2y = 5 \\ 5x - 2y = 13 \end{cases}$$

Add the two equations and solve for x.

$$\begin{aligned} 6x &= 18 \\ x &= 3 \end{aligned}$$

Substitute 3 for x in Equation 3.

$$\begin{aligned} (3) + 2y &= 5 \\ 2y &= 2 \\ y &= 1 \end{aligned}$$

The solution is (3, 1).

23. $$\begin{cases} y = 4 - 3x \\ 2x - 3y = -1 \end{cases}$$

Substitute $4 - 3x$ for y in Equation 2.

$$\begin{aligned} 2x - 3(4 - 3x) &= -1 \\ 2x - 12 + 9x &= -1 \\ 11x - 12 &= -1 \\ 11x &= 11 \\ x &= 1 \end{aligned}$$

Substitute 1 for x in Equation 1 and solve for y.

$$y = 4 - 3(1) = 4 - 3 = 1$$

The solution is (1, 1).

24. $$\begin{cases} x + y + z = 1 \\ 2x - y - z = -4 \\ x - 2y + z = 4 \end{cases}$$ Eliminate z.

Add Equations 1 and 2.

1.	$x + y + z$	=	1
2.	$2x - y - z$	=	−4
4.	$3x$	=	−3
	x	=	−1

Add Equations 2 and 3.

2.	$2x - y - z$	=	−4
3.	$x - 2y + z$	=	4
5.	$3x - 3y$	=	0

Substitute $x = -1$ into Equation 5, solve for y.

$$\begin{aligned} 3(-1) - 3y &= 0 \\ -3y &= 3 \\ y &= -1 \end{aligned}$$

Substitute $x = -1$ and $y = -1$ into Equation 1, solve for z.

$$\begin{aligned} (-1) + (-1) + z &= 1 \\ z &= 3 \end{aligned}$$

The solution is (−1, −1, 3).

25. $$\begin{cases} 4x - 3y = -1 \\ 3x + 4y = -7 \end{cases}$$

$$x = \frac{D_x}{D} = \frac{\begin{vmatrix} -1 & -3 \\ -7 & 4 \end{vmatrix}}{\begin{vmatrix} 4 & -3 \\ 3 & 4 \end{vmatrix}} = \frac{-1(4) - (-3)(-7)}{4(4) - (-3)(3)} = \frac{-4 - 21}{16 + 9} = \frac{-25}{25} = -1$$

$$y = \frac{D_y}{D} = \frac{\begin{vmatrix} 4 & -1 \\ 3 & -7 \end{vmatrix}}{\begin{vmatrix} 4 & -3 \\ 3 & 4 \end{vmatrix}} = \frac{4(-7) - (-1)(3)}{25} = \frac{-28 + 3}{25} = \frac{-25}{25} = -1$$

Solution: (−1, −1)

26. $\begin{cases} x-2y-z=-2 \\ 3x+y-z=6 \\ 2x-y+z=-1 \end{cases}$ write the augmented matrix: $\left[\begin{array}{ccc|c} 1 & -2 & -1 & -2 \\ 3 & 1 & -1 & 6 \\ 2 & -1 & 1 & -1 \end{array}\right]$ $\begin{array}{l} -3R_1+R_2 \\ -2R_1+R_3 \end{array}$

$\left[\begin{array}{ccc|c} 1 & -2 & -1 & -2 \\ 0 & 7 & 2 & 12 \\ 0 & 3 & 3 & 3 \end{array}\right]$ switch R_2 with R_3 $\left[\begin{array}{ccc|c} 1 & -2 & -1 & -2 \\ 0 & 3 & 3 & 3 \\ 0 & 7 & 2 & 12 \end{array}\right]$ $\frac{1}{3}R_2$ $\left[\begin{array}{ccc|c} 1 & -2 & -1 & -2 \\ 0 & 1 & 1 & 1 \\ 0 & 7 & 2 & 12 \end{array}\right]$ $\begin{array}{l} 2R_2+R_1 \\ -7R_2+R_3 \end{array}$

$\left[\begin{array}{ccc|c} 1 & 0 & 1 & 0 \\ 0 & 1 & 1 & 1 \\ 0 & 0 & -5 & 5 \end{array}\right]$ $-\frac{1}{5}R_3$ $\left[\begin{array}{ccc|c} 1 & 0 & 1 & 0 \\ 0 & 1 & 1 & 1 \\ 0 & 0 & 1 & -1 \end{array}\right]$ Yields the system: $\begin{cases} x \quad +z=0 \\ y \ +z=1 \\ z=-1 \end{cases}$

Plug in $z=-1$ to get $y+(-1)=1$ to obtain $y=2$ and $x+(-1)=0$ to obtain $x=1$
Solution is $(1, 2, -1)$

27. The first point of intersection is about (7, 23). In 1907, the percent of U.S. workers in white–collar and farming jobs was the same (23%). The second point is (45, 42). In 1945, the percent of U.S. workers in white–collar and blue–collar jobs was the same (42%).

28. a. Find the slope: $(x_1, y_1) = (20, 7)$ and $(x_2, y_2) = (70, 4)$. $m = \dfrac{y_2-y_1}{x_2-x_1} = \dfrac{4-7}{70-20} = \dfrac{-3}{50} = -0.06$

Substitute the point (20, 7) into the point–slope form.

$$\begin{aligned} y-y_1 &= m(x-x_1) \\ y-(7) &= -0.06[x-(20)] \\ y-7 &= -0.06x+1.2 \\ y &= -0.06x+8.2 \end{aligned}$$

b. Let $x=90$, then $y=2.8$ L/min.

29. **Analysis:** Find the number of pieces of computer software that need to be sold to break even . The break point occurs when the revenues generated equals the costs incurred.

Equations: Let x represent the number of pieces of computer software to be sold. The cost of making the x pieces of software is 5.45x$ and the revenue from x pieces of software is 29.95x$.

Cost:	Cost of making x pieces of software 5.45x$	plus +	fixed costs $18,375	is =	total cost for software. $C(x)$
Revenue:	Revenue from x pieces of software 29.95x$	is =	total revenue. $R(x)$		
Break point:	Revenue $R(x)$	equals =	cost. $C(x)$		

Solve: Substitute the cost and revenue equations into the break point equation.

$$\begin{aligned} 29.95x &= 5.45x + 18{,}375 \\ 24.50x &= 18{,}375 \\ x &= 750 \end{aligned}$$

Conclusion: The number of pieces of software that needs to be sold to break even is 750.

30. **Analysis:** Find the number of each kind of ticket that were sold to have receipts of \$2625. There were 750 tickets sold. Twice as many \$5 tickets were sold as \$2 tickets.

Equations: Let x, y, and z represent the number of \$5, \$3, and \$2 tickets sold, respectively. Then the total value of the tickets were \$5x, \$3y, and \$2z, respectively.

Number of \$5 tickets	plus	number of \$3 tickets	plus	number of \$2 tickets	is	total number tickets.
x	$+$	y	$+$	z	$=$	750
Value of \$5 tickets	plus	value of \$3 tickets	plus	value of \$2 tickets	is	total value of tickets.
$5x$	$+$	$3y$	$+$	$2z$	$=$	2625

Twice the number of \$5 tickets	is	the number of \$2 tickets.
$2x$	$=$	z

Solve: $\begin{cases} x+y+z=750 \\ 5x+3y+2z=2625 \\ x=2z \end{cases}$ Rewrite $\begin{cases} x+y+z=750 \\ 5x+3y+2z=2625 \\ x-2z=0 \end{cases}$ Eliminate y.

Multiply Equation 1 by –3 and add Equations 1 and 2.

$$\begin{aligned} &1. & -3x-3y-3z &= -2250 \\ &2. & 5x+3y+2z &= 2625 \\ \hline &4. & 2x \quad -z &= 375 \end{aligned}$$

Multiply Equation 3 by –2 and add Equations 4 and 3.

$$\begin{aligned} &3. & -2x+4z &= 0 \\ &4. & 2x-\ z &= 375 \\ \hline &5. & 3z &= 375 \\ && z &= 125 \end{aligned}$$

Substitute $z = 125$ into Equation 3, solve for x.

$$\begin{aligned} x-2(125) &= 0 \\ x &= 250 \end{aligned}$$

Substitute $x = 250$ and $z = 125$ into Equation 1, solve for y.

$$\begin{aligned} (250)+y+(125) &= 750 \\ y &= 375 \end{aligned}$$

Conclusion: There were $x = 250$ \$5 tickets, $y = 375$ \$3 tickets, and $z = 125$ \$2 tickets sold.

31. Rewrite $|4x-3|=9$ as

$$\begin{aligned} 4x-3 &= -9 \\ 4x &= -6 \\ x &= -\frac{6}{4} = -\frac{3}{2} \end{aligned} \quad \text{or} \quad \begin{aligned} 4x-3 &= 9 \\ 4x &= 12 \\ x &= 3 \end{aligned}$$

32. Rewrite $|2x-1|=|3x+4|$ as

$2x-1 = 3x+$
4

$$\begin{aligned} -x &= 5 \\ x &= -5 \end{aligned}$$

or

$$\begin{aligned} 2x-1 &= -(3x+4) \\ 2x-1 &= -3x-4 \\ 5x &= -3 \\ x &= -\frac{3}{5} \end{aligned}$$

33.

$$\begin{aligned} -3(x-4) &\ge x-32 \\ -3x+12 &\ge x-32 \\ -4x &\ge -44 \\ x &\le 11 \end{aligned}$$

Reverse the direction.

$(-\infty, 11]$

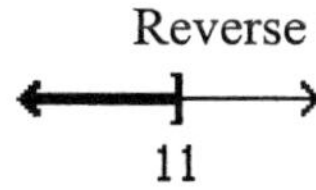

34.

$$\begin{aligned} -8 <\ & -3x+1 < 10 \\ -9 <\ & -3x < 9 \\ 3 >\ & x > -3 \end{aligned}$$

$(-3, 3)$

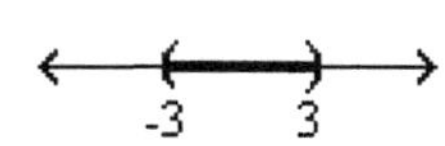

35. Rewrite $|3x-2|\le 4$ as

$$\begin{aligned} -4 \le\ & 3x-2 \le 4 \\ -2 \le\ & 3x \le 6 \\ -\frac{2}{3} \le\ & x \le 2 \end{aligned}$$

$\left[-\frac{2}{3}, 2\right]$

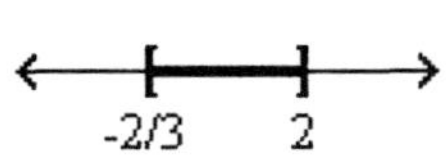

36. If $|2x+3|-1>4$, then $|2x+3|>5$

Rewrite $|2x+3|>5$ as

$$\begin{aligned} 2x+3 &< -5 \\ 2x &< -8 \\ x &< -4 \end{aligned} \quad \text{or} \quad \begin{aligned} 2x+3 &> 5 \\ 2x &> 2 \\ x &> 1 \end{aligned}$$

$(-\infty, -4) \cup (1, \infty)$

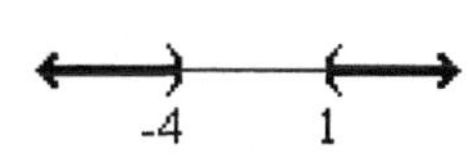

37.

x	$2x - 3y \le 12$ Graph: $2x - 3y = 12$	y
0	$2(0) - 3y = 12$	-4
6	$2x - 3(0) = 12$	0
Solid line	Test Point: (0, 0) $2(0) - 3(0) \le 12$ $0 \le 12$	True

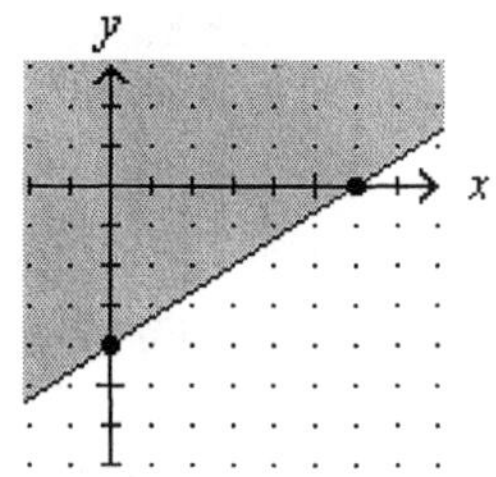

38. Inequality #1: $y < x + 2$
Graph: $y = x + 2$

x	y
0	2
-2	0

Broken line. Test point (0, 0).

$(0) < (0) + 2$

$0 < 2$ True

Inequality #2: $3x + y \le 6$
Graph: $3x + y = 6$

x	y
1	3
2	0

Solid line. Test point (0, 0).

$3(0) + (0) \le 6$

$0 \le 6$ True

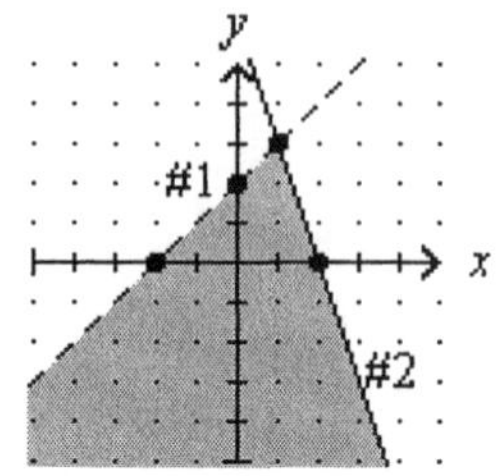

Feeling defeated is often a temporary condition.
Giving up is what makes it permanent.

Chapter 5 Exponents, Polynomials, and Polynomial Functions

Study Tip
When do you study?

Here are some suggestions

- Right after class, take 5 minutes to look over your notes. Circle things you don't understand and be sure to visit your instructor to get those points clarified.
- Do half of your homework the same day as your class. Then do the other half the following day.
- Don't study in large chunks of time. Study for 30 minutes, take a break, then study again. Make sure that between class periods, you have studied for a total of at least two hours.
- Do not study late at night. You think best when you are not tired. Get a good night's rest, then study again in the morning
- Don't study late the night before a test. You perform better when you aren't tired.

Section 5.1 Exponents

VOCABULARY

1. power
3. product; quotient

CONCEPTS

5. $x^m x^n = x^{m+n}$
7. $(xy)^n = x^n y^n$
9. $x^0 = 1$
11. $\frac{x^m}{x^n} = x^{m-n}$

13. Because $x^{-n} \bullet x^n = x^{-n+n} = x^0 = 1$ and $\frac{1}{x^n} \bullet x^n = 1$, x^{-n} is defined to be the reciprocal of x^n.

For example: $4^{-2} = \frac{1}{4^2}$

NOTATION

17. $\frac{x^5x^4}{x^{-2}} = \frac{x^{\ 9}}{x^{-2}} = x^{9-(-2)} = x^{11}$

PRACTICE

19. base is 5, exponent is 3
21. base is x, exponent is 5
23. base is b, exponent is 6
25. base is $\frac{n}{4}$, exponent is 3
27. $3^2 = (3)(3) = 9$
29. $-3^2 = -(3)(3) = -9$
31. $(-3)^2 = (-3)(-3) = 9$
33. $5^{-2} = \frac{1}{5^2} = \frac{1}{25}$
35. $-5^{-2} = -\frac{1}{5^2} = -\frac{1}{25}$
37. $(-5)^{-2} = \frac{1}{(-5)^2} = \frac{1}{25}$
39. $8^0 = 1$
41. $(-8)^0 = 1$
43. $(-2x)^5 = (-2)^5(x)^5 = -32x^5$
45. $x^2x^3 = x^{2+3} = x^5$
47. $x^2x^3x^5 = x^{2+3+5} = x^{10}$

49. $k^0k^7 = 1(k^7) = k^7$

51. $2aba^3b^4 = 2a^{1+3}b^{1+4} = 2a^4b^5$

53. $p^9pp^0 = p^{9+1+0} = p^{10}$

55. $(-x)^2y^4x^3 = x^2x^3y^4 = x^{2+3}y^4 = x^5y^4$

57. $(b^{-8})^9 = b^{-8(9)} = b^{-72} = \dfrac{1}{b^{72}}$

59. $(x^4)^7 = x^{4(7)} = x^{28}$

61. $(r^{-3}s)^3 = r^{-3(3)}s^{1(3)} = r^{-9}s^3 = \dfrac{s^3}{r^9}$

63. $(a^2a^3)^4 = (a^{2+3})^4 = (a^5)^4 = a^{5(4)} = a^{20}$

65. $(-d^2)^3(d^{-3})^3 = -d^{2(3)}(d^{-3(3)}) = -d^6(d^{-9}) = -d^{6-9} = -d^{-3} = -\dfrac{1}{d^3}$

67. $(3x^3y^4)^3 = 3^3x^{3(3)}y^{4(3)} = 27x^9y^{12}$

69. $\left(-\dfrac{1}{3}mn^2\right)^6 = \left(-\dfrac{1}{3}\right)^6 m^6n^{2(6)} = \dfrac{1}{729}m^6n^{12}$

71. $\left(\dfrac{a^3}{b^2}\right)^5 = \dfrac{a^{3(5)}}{b^{2(5)}} = \dfrac{a^{15}}{b^{10}}$

73. $\left(\dfrac{a^{-3}}{b^{-2}}\right)^{-2} = \dfrac{a^{-3(-2)}}{b^{-2(-2)}} = \dfrac{a^6}{b^4}$

75. $\dfrac{a^8}{a^3} = a^{8-3} = a^5$

77. $\dfrac{c^{12}c^5}{c^{10}} = \dfrac{c^{12+5}}{c^{10}} = \dfrac{c^{17}}{c^{10}} = c^{17-10} = c^7$

79. $\left(\dfrac{2}{3}\right)^{-2} = \dfrac{2^{-2}}{3^{-2}} = \left(\dfrac{1}{2^2}\right)\left(3^2\right) = \left(\dfrac{1}{4}\right)(9) = \dfrac{9}{4}$

81. $\dfrac{1}{a^{-4}} = a^4$

83. $\dfrac{(3x^2)^{-2}}{x^3x^{-4}x^0} = \dfrac{3^{-2}x^{2(-2)}}{x^{3-4+0}} = \dfrac{3^{-2}x^{-4}}{x^{-1}} = \dfrac{x^{-4-(-1)}}{3^2} = \dfrac{x^{-3}}{9} = \dfrac{1}{9x^3}$

85. $\left(\dfrac{4a^{-2}b}{3ab^{-3}}\right)^3 = \left(\dfrac{4}{3}a^{-2-1}b^{1-(-3)}\right)^3 = \left(\dfrac{4}{3}a^{-3}b^4\right)^3 = \left(\dfrac{4b^4}{3a^3}\right)^3 = \dfrac{64b^{12}}{27a^9}$

87. $\left(\dfrac{3a^{-2}b^2}{17a^2b^3}\right)^0 = 1$

89. $\left(\dfrac{-2a^4b}{a^{-3}b^2}\right)^3 = \left(-2a^{4-(-3)}b^{1-2}\right)^3 = \left(-2a^7b^{-1}\right)^3 = -8a^{21}b^{-3} = -\dfrac{8a^{21}}{b^3}$

91. $\left(-\dfrac{2a^3b^2}{3a^{-3}b^2}\right)^{-3} = \left(-\dfrac{2}{3}a^{3-(-3)}b^{2-2}\right)^{-3} = \left(-\dfrac{2}{3}a^6b^0\right)^{-3} = -\dfrac{2^{-3}a^{6(-3)}}{3^{-3}} = -\dfrac{3^3a^{-18}}{2^3} = -\dfrac{27}{8a^{18}}$

93. $\left(\dfrac{-3pqr^{-4}}{2p^2q^{-3}r^2}\right)^{-2} = \left(\dfrac{-3q^4}{2pr^6}\right)^{-2} = \left(\dfrac{2pr^6}{-3q^4}\right)^2 = \dfrac{4p^2r^{12}}{9q^8}$

95. $(2x^{-4}y^3)^3(3x^2y^{-1})^{-2} = (8x^{-12}y^9)(\frac{1}{9}x^{-4}y^2) = \dfrac{8}{9}x^{-16}y^{11} = \dfrac{8y^{11}}{9x^{16}}$

97. $\dfrac{(3x^2y^{-4})^{-2}}{(2x^3y^2)^{-3}} = \dfrac{(2x^3y^2)^3}{(3x^2y^{-4})^2} = \dfrac{8x^9y^6}{9x^4y^{-8}} = \dfrac{8x^5y^{14}}{9}$

99. $1.23^6 = 3.462825992$

101. $-6.25^3 = -244.140625$

103. Left side: $(3.68)^0 = 1$ Right side: $1 = 1$

105. Left side: $(7.2)^2(2.7)^2 = (51.84)(7.29) = 377.9136$ Right side: $[(7.2)(2.7)]^2 = (19.44)^2 = 377.9136$

107. Left side: $(3.2)^2(3.2)^{-2} = (10.24)(0.09765625) = 1$ Right sides: $1 = 1$

APPLICATIONS

109. Starting at the bottom moving upward: 10^{-2}, 10^{-3}, 10^{-4}, 10^{-5}, 10^{-6}, 10^{-7}, 10^{-8}, 10^{-9}.

111. $10 \cdot 10 \cdot 10 \cdot 26 \cdot 26 \cdot 26 = 10^3(26^3) = 17{,}576{,}000$

113 a. Area $= s^2 = (x^3\text{ ft})^2 = (x^3)^2(\text{ft})^2 = x^{3(2)}\text{ ft}^2 = x^6\text{ ft}^2$. b. Volume $= s^3 = (x^3\text{ ft})^3 = (x^3)^3(\text{ft})^3 = x^{3(3)}\text{ ft}^3 = x^9\text{ ft}^3$.

WRITING

115. Place the base of the negative exponent in the denominator of a fraction whose numerator is one; change the exponent on the base from negative to positive. For example: $4^{-2} = \dfrac{1}{4^2}$.

117. Only b is the base of the negative exponent. The negative in front of the 8 refers to its position on the number line.

REVIEW

119.

$$\begin{aligned} a + 5 &< 6 \\ a &< 1 \end{aligned}$$

Subtract 5 from both sides. $(-\infty, 1)$

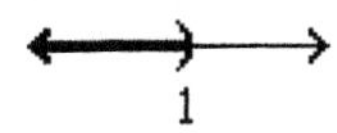

121.

$$\begin{aligned} 6(t-2) &\le 4(t+7) \\ 6t - 12 &\le 4t + 28 \\ 2t &\le 40 \\ t &\le 20 \end{aligned}$$

Remove the parentheses.
Subtract $4t$ and add 12 to both sides.
Divide both sides by 2.

$(-\infty, 20]$

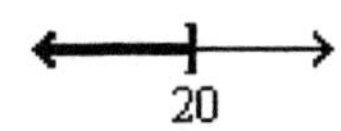

Section 5.2 Scientific Notation

VOCABULARY/ CONCEPTS

1. scientific 3. 10^n; integer 5.left

NOTATION

7. This is not scientific notation because 60.22 is not between 1 and 10.

PRACTICE

9. $3900 = 3.9 \times 1000 = 3.9 \times 10^3$ move 3 places left.

11. $0.0078 = 7.8 \times 0.001 = 7.8 \times 10^{-3}$ move 3 places right.

13. $173{,}000{,}000{,}000{,}000 = 1.73 \times 10^{14}$ move 14 places left.

15. $0.0000096 = 9.6 \times 10^{-6}$ move 6 places right.

17. $323 \times 10^5 = (3.23 \times 10^2) \times 10^5 = 3.23 \times (10^2 \times 10^5) = 3.23 \times 10^7$

19. $6000 \times 10^{-7} = (6.0 \times 10^3) \times 10^{-7} = 6.0 \times (10^3 \times 10^{-7}) = 6.0 \times 10^{-4}$

21. $0.0527 \times 10^5 = (5.27 \times 10^{-2}) \times 10^5 = 5.27 \times (10^{-2} \times 10^5) = 5.27 \times 10^3$

23. $0.0317 \times 10^{-2} = (3.17 \times 10^{-2}) \times 10^{-2} = 3.17 \times (10^{-2} \times 10^{-2}) = 3.17 \times 10^{-4}$

25. $2.7 \times 10^2 = 2.7 \times 100 = 270$ move 2 places right.

27. $3.23 \times 10^{-3} = 3.23 \times 0.001 = 0.00323$ move 3 places left.

29. $7.96 \times 10^5 = 7.96 \times 100{,}000 = 796{,}000$ move 5 places right.

31. $3.7 \times 10^{-4} = 0.00037$ move 4 places left.

33. $5.23 \times 10^{0} = 5.23$ move 0 places.

35. $23.65 \times 10^6 = (2.365 \times 10^1) \times 10^6 = 2.365 \times (10^1 \times 10^6) = 2.365 \times 10^7 = 23{,}650{,}000$

37. $(7.9 \times 10^5)(2.3 \times 10^6) = (7.9 \cdot 2.3)(10^5 \cdot 10^6) = 18.17 \times 10^{11} = (1.817 \times 10^1) \times 10^{11} = 1.817 \times 10^{12}$

39. $(9.1 \times 10^{-5})(5.5 \times 10^{12}) = (9.1 \cdot 5.5)(10^{-5} \cdot 10^{12}) = 50.05 \times 10^7 = (5.005 \times 10^1) \times 10^7 = 5.005 \times 10^8$

41. $\dfrac{4.2\times10^{-12}}{8.4\times10^{-5}} = \dfrac{4.2}{8.4}\times\dfrac{10^{-12}}{10^{-5}} = 0.5\times10^{-12-(-5)} = 0.5\times10^{-7} = (5\times10^{-1})\times10^{-7} = 5\times10^{-8}$

43. $\dfrac{(3.9\times10^{-9})(9.5\times10^{-4})}{1.96\times10^{-2}} = \dfrac{(3.9)(9.5)}{1.95}\times\dfrac{(10^{-9})(10^{-4})}{10^{-2}} = 19\times10^{(^-9+^-4-^-2)} = 19\times10^{-11} = 1.9\times10^{-10}$

45. $(89{,}000{,}000{,}000)(4{,}500{,}000{,}000) = (8.9 \times 10^{10})(4.5 \times 10^9) = (8.9 \cdot 4.5)(10^{10} \cdot 10^9) = 40.05 \times 10^{19}$

$= 4.005 \times 10^{20}$ or $400{,}500{,}000{,}000{,}000{,}000{,}000$

47. $\dfrac{0.00000129}{0.0003} = \dfrac{1.29\times10^{-6}}{3\times10^{-4}} = \dfrac{1.29}{3}\times\dfrac{10^{-6}}{10^{-4}} = 0.43\times10^{-6-(-4)} = 0.43\times10^{-2} = 4.3\times10^{-3}$ or 0.0043

49. $\dfrac{(220{,}000)(0.000009)}{0.00033} = \dfrac{(2.2\times10^5)(9\times10^{-6})}{3.3\times10^{-4}} = \dfrac{(2.2)(9)}{3.3}\times\dfrac{10^5 10^{-6}}{10^{-4}} = \dfrac{19.8}{3.3}\times10^{5-6-(-4)} = 6\times10^3$ or 6000

51. $\dfrac{(0.00024)(96{,}000{,}000)}{640{,}000{,}000} = \dfrac{(2.4\times10^{-4})(9.6\times10^7)}{6.4\times10^8} = \dfrac{(2.4)(9.6)}{6.4}\times10^{-4+7-8} = \dfrac{23.04}{6.4}\times10^{-5} = 3.6\times10^{-5}$ or 0.000036

APPLICATIONS

53. 2.6×10^6 to 1= 2,600,000 to 1

55. \$1.7 trillion = \$1,700,000,000,000 = \$$1.7 \times 10^{12}$; \$3.9 billion = \$3,900,000,000 = \$$3.9 \times 10^9$;
\$275 million = \$275,000,000 = \$$2.75 \times 10^8$; \$312 million = \$312,000,000 = \$$3.12 \times 10^8$

57. The mass of an electron $= \left(\dfrac{1}{2000}\right)$(the mass of a proton)

$= \left(\dfrac{1}{2\times10^3}\right)(1.7 \times 10^{-24}) = \dfrac{1.7}{2} \times 10^{-24-3} = 0.85 \times 10^{-27}$ gm. $= 8.5 \times 10^{-28}$ grams

59. 1 minute = 60 seconds;
1 hour = 60 minutes = 60(60) seconds = 3600 seconds;
1 day = 24 hours = 24(3600) seconds = 86,400 seconds;
1 year = 365 days = 365(86,400)seconds = 31,536,000 seconds.
1 light year = (300,000,000 m/sec)(31,536,000 sec)
$= (3 \times 10^8)(3.1536 \times 10^7) = (3)(3.1536) \times 10^{8+7} = 9.4708 \times 10^{15}$ or about 9.5×10^{15} meters

61. a. $t = \dfrac{\text{distance from Merak to earth}}{\text{rate that light travels}} = \dfrac{4.65\times10^{14}\text{ miles}}{1.86\times10^5\text{ miles/sec}} = \dfrac{4.65}{1.86}\times10^{14-5} = 2.5\times10^9$ sec or 2,500,000,000 sec

b. From Exercise 57 we know that 1 year is 3.1536×10^7 sec.

$\dfrac{\text{number of seconds for light from Merak}}{\text{number of seconds per year}} = \dfrac{2.5\times10^9}{3.1536\times10^7} = \dfrac{2.5}{3.1536}\times10^{9-7} = 0.7927\times10^2$ or about 79 years.

63. Distance in miles = (Number of astronomical units)(miles in one AU)
$= (1.3)(9.3 \times 10^7) = 12.09 \times 10^7 = 1.209 \times 10^8$ miles

65. 1 million million billion $= 1.0 \times 10^6 \cdot 10^6 \cdot 10^9 = 1.0 \times 10^{21}$

WRITING

67. Every positive number written in scientific notation is the product of a number between 1 (including 1) and 10 and an integer power of 10. To change a number from standard notation to scientific notation place the decimal after the first nonzero integer and then count the places from that new decimal to the previous decimal. If the original number was larger than 1 then the exponent will be positive. If the original number was between 0 and 1, then the exponent will be negative.

69. If n is negative then 10^n is less than 1 but is always positive. The largest value for 10^n would be $10^{-1} = 0.1$.Then $9.99 \times 10^{-1} = 9.99 \times (0.1) = .999$ which is less than 1. Therefore $0 < 9.99 \times 10^n < 1$.

REVIEW

71. $4x \geq -x + 5$ and $6 \geq 4x - 3$

$5x \geq 5$ $\quad 9 \geq 4x$

$x \geq 1$ $\quad \frac{9}{4} \geq x$

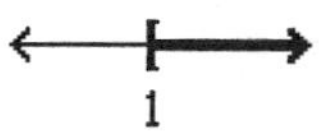

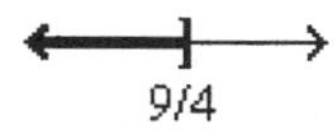

$\left[1, \frac{9}{4}\right]$

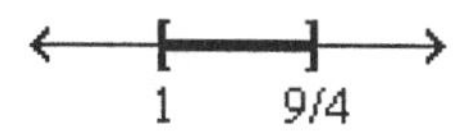

73. $3x + 2 < 8$ or $2x - 3 > 11$

$3x < 6$ $\quad 2x > 14$

$x < 2$ $\quad x > 7$

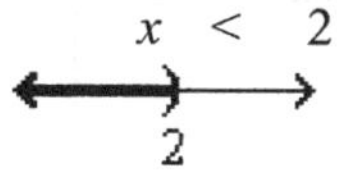

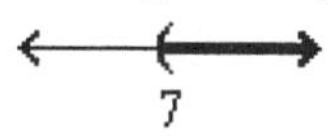

$(-\infty, 2) \cup (7, \infty)$

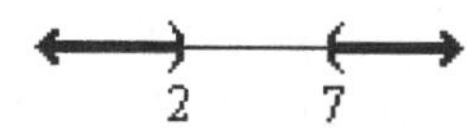

Section 5.3 Polynomials and Polynomial Functions

VOCABULARY

1. polynomial 3. degree 5. cubic 7. like

CONCEPTS

9. $3x^2$ is a monomial with a degree of 2.

11. $3x^2y - 2x + 3y$ is a trinomial with a degree of 3.

13. $x^2 - y^2$ is a binomial with a degree of 2.

15. 5 is a monomial with a degree of 0.

17. $9x^2y^4 - x - y^{10} + 1$ is a none of these with a degree of 10.

19. $4x^9 + 3x^2y^4$ is a binomial with a degree of 9.

21. a. $3x - 2x^4 + 7 - 5x^2 = -2x^4 - 5x^2 + 3x + 7$ in descending order of x.
    b. $a^2x - ax^3 + 7a^3x^5 - 5a^3x^2 = 7a^3x^5 - ax^3 - 5a^3x^2 + a^2x$ in descending order of x.

23. like terms: $3x + 7x = 10x$

25. unlike terms

27. like terms: $3r^2t^3 - 8r^2t^3 = -5r^2t^3$

29. unlike terms

31.
$$\begin{aligned} P &= (2x^2 + 3x + 1) + (3x^2 + x - 1) + (4x^2 - x - 2) && \text{Add the sides of the triangle.} \\ &= (2x^2 + 3x^2 + 4x^2) + (3x + x - x) + (1 - 1 - 2) && \text{Rearrange the terms so like terms are together.} \\ &= 9x^2 + 3x - 2 && \text{Combine like terms.} \end{aligned}$$

NOTATION

33. If $f(x) = 2x^2 + x + 2$, find $f(-1)$.

$$\begin{aligned} f(-1) &= 2(\ \mathbf{-1}\ )^2 + (\ \mathbf{-1}\ ) + 2 \\ &= 2(\ \mathbf{1}\ ) + (-1) + 2 \\ &= \mathbf{3} \end{aligned}$$

PRACTICE

35.

x	$f(x) = 2x^2 - 4x + 2$	$f(x)$
-1	$f(-1) = 2(-1)^2 - 4(-1) + 2$	**8**
0	$f(0) = 2(0)^2 - 4(0) + 2$	**2**
1	$f(1) = 2(1)^2 - 4(1) + 2$	**0**
2	$f(2) = 2(2)^2 - 4(2) + 2$	**2**
3	$f(3) = 2(3)^2 - 4(3) + 2$	**8**

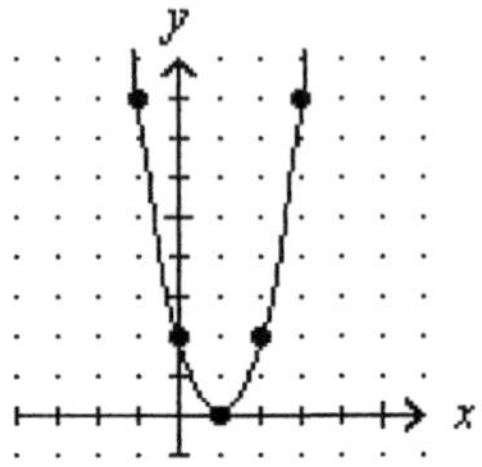

37.

x	$f(x) = 2x^3 - 3x^2 - 11x + 6$	$f(x)$
-3	$f(-3) = 2(-3)^3 - 3(-3)^2 - 11(-3) + 6$	**−42**
-2	$f(-2) = 2(-2)^3 - 3(-2)^2 - 11(-2) + 6$	**0**
-1	$f(-1) = 2(-1)^3 - 3(-1)^2 - 11(-1) + 6$	**12**
0	$f(0) = 2(0)^3 - 3(0)^2 - 11(0) + 6$	**6**
1	$f(1) = 2(1)^3 - 3(1)^2 - 11(1) + 6$	**−6**
2	$f(2) = 2(2)^3 - 3(2)^2 - 11(2) + 6$	**−12**
3	$f(3) = 2(3)^3 - 3(3)^2 - 11(3) + 6$	**0**
4	$f(4) = 2(4)^3 - 3(4)^2 - 11(4) + 6$	**42**

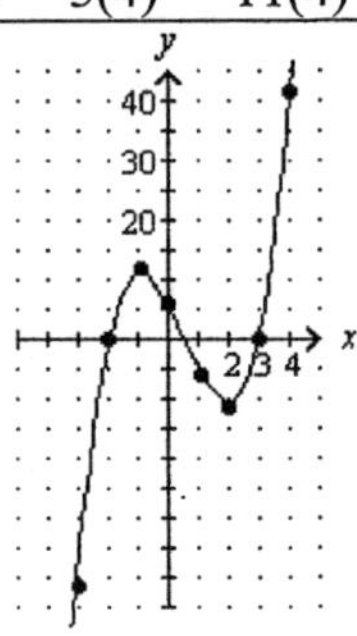

39. $f(x) = 2.75x^2 - 4.7x + 1.5$

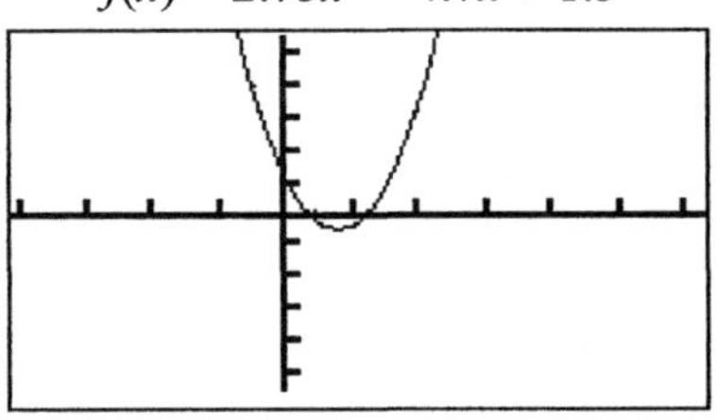

41. $(3x^2 + 2x + 1) + (-2x^2 - 7x + 5)$

$= 3x^2 + 2x + 1 - 2x^2 - 7x + 5$ Remove parentheses.

$= 3x^2 - 2x^2 + 2x - 7x + 1 + 5$ Rearrange terms so that like terms are together.

$= x^2 - 5x + 6$ Combine like terms.

43. $(-a^2 + 2a + 3) - (4a^2 - 2a - 1)$

$= -a^2 + 2a + 3 - 4a^2 + 2a + 1$ Remove parentheses. Change signs of the second polynomial.

$= -a^2 - 4a^2 + 2a + 2a + 3 + 1$ Rearrange terms so that like terms are together.

$= -5a^2 + 4a + 4$ Combine like terms.

45. $(2a^2 + 4a - 7) + (3a^2 - a - 2)$

$= 2a^2 + 4a - 7 + 3a^2 - a - 2$ Remove parentheses.

$= 2a^2 + 3a^2 + 4a - a - 7 - 2$ Rearrange terms so that like terms are together.

$= 5a^2 + 3a - 9$ Combine like terms.

47. $(7y^3 + 4y^2 + y + 3) + (-8y^3 - y + 3)$

$= 7y^3 + 4y^2 + y + 3 - 8y^3 - y + 3$	Remove parentheses.
$= 7y^3 - 8y^3 + 4y^2 + y - y + 3 + 3$	Rearrange terms so that like terms are together.
$= -y^3 + 4y^2 + 6$	Combine like terms.

49. $(3pq + p - q) + (-pq - p - q)$

$= 3pq + p - q - pq - p - q$	Remove parentheses.
$= 3pq - pq + p - p - q - q$	Rearrange terms so that like terms are together.
$= 2pq - 2q$	Combine like terms.

51. $(-2x^2y^3 + 6xy + 5y^2) - (-4x^2y^3 - 7xy + 2y^2)$

$= -2x^2y^3 + 6xy + 5y^2 + 4x^2y^3 + 7xy - 2y^2$	Remove parentheses. Change signs of the second polynomial.
$= -2x^2y^3 + 4x^2y^3 + 6xy + 7xy + 5y^2 - 2y^2$	Rearrange terms so that like terms are together.
$= 2x^2y^3 + 13\,xy + 3y^2$	Combine like terms.

53. $(3x^2 + 4x - 3) + (2x^2 - 3x - 1) - (x^2 + x + 7)$

$= 3x^2 + 4x - 3 + 2x^2 - 3x - 1 - x^2 - x - 7$	Remove parentheses. Change signs of the third polynomial.
$= 3x^2 + 2x^2 - x^2 + 4x - 3x - x - 3 - 1 - 7$	Rearrange terms so that like terms are together.
$= 4x^2 - 11$	Combine like terms.

55.
$$\begin{array}{rr} & 3x^3 - 2x^2 + 4x - 3 \\ + & -2x^3 + 3x^2 + 3x - 2 \\ & \underline{5x^3 - 7x^2 + 7x - 12} \\ & 6x^3 - 6x^2 + 14x - 17 \end{array}$$

57.
$$\begin{array}{rrcrr} & 3x^2 - 4x + 17 & & & 3x^2 - 4x + 17 \\ - & \underline{2x^2 + 4x - 5} & \Rightarrow & + & \underline{-2x^2 - 4x + 5} \\ & & & & x^2 - 8x + 22 \end{array}$$

59.
$$\begin{array}{rrcrr} & -5y^3 + 4y^2 - 11y + 3 & & & -5y^3 + 4y^2 - 11y + 3 \\ - & \underline{-2y^3 - 14y^2 + 17y - 32} & \Rightarrow & + & \underline{2y^3 + 14y^2 - 17y + 32} \\ & & & & -3y^3 + 18y^2 - 28y + 35 \end{array}$$

61.
$$\begin{array}{rrcrr} & 4x^2 + 0x & & & 4x^2 + 0x \\ - & \underline{4x^2 - 2x} & \Rightarrow & + & \underline{-4x^2 + 2x} \\ & & & & 2x \end{array}$$

63. $(4x^2 + 5x - 3) + (7x^2 + 2x - 10) - (-x^2 - 2x + 1)$

$= 4x^2 + 5x - 3 + 7x^2 + 2x - 10 + x^2 + 2x - 1$

$= 4x^2 + 7x^2 + x^2 + 5x + 2x + 2x - 3 - 10 - 1$

$= 12x^2 + 9x - 14$

APPLICATIONS

65. $f(t) = -16t^2 + 32t + 4$
$f(1) = -16(1)^2 + 32(1) + 4 = 20$ ft

67. $V(r) = 4.2r^3 + 37.7r^2$
$V(4) = 4.2(4)^3 + 37.7(4)^2 = 268.8 + 603.2 = 872$ ft^3

69. $f(x) = -240x^2 + 1440x$
$f(3) = -240(3)^2 + 1440(3) = -2160 + 4320 = 2160$ in.^{3}

71. a. $V(x) = R(x) + D(x) = (1100x + 125{,}000) + (1400x + 150{,}000) = 2500x + 275{,}000$
b. $V(20) = 2500(20) + 275{,}000 = 50{,}000 + 275{,}000 = \$325{,}000$

73. a. 4 terms
b. ascending order
c. $-\frac{1}{720}$ is the coefficient of the fourth term.
d. fourth degree
e. sixth degree

WRITING

75. Both terms contain the same variables, but the exponents on those variables differ.

77. The polynomials were subtracted in the wrong order.

79. The student was descending the ladder. The student was ascending the ladder.

REVIEW

81. Rewrite $|x| \le 5$ as
$-5 \le x \le 5$ $[-5, 5]$

83. Rewrite $|x - 4| < 5$ as
$-5 < x - 4 < 5$
$-1 < x < 9$

Section 5.4 Multiplying Polynomials

VOCABULARY

1. product
3. terms

CONCEPTS

5. factors
7. term
9. $x^2 + 2xy + y^2$
11. $x^2 - y^2$

13. $A = lw = (x - 2)(x + 4) = x(x + 4) - 2(x + 4) = x^2 + 4x - 2x - 8 = x^2 + 2x - 8$

15. $A = s^2 = (4a + 3)^2 = (4a)^2 + 2(4a \cdot 3) + 3^2 = 16a^2 + 24a + 9$

17. a. $2x, 4x$ b. $2x, -3$ c. $4, 4x$ d. $4, -3$

PRACTICE

19. $(2a^2)(-3ab) = 2(-3)a^2 \cdot a \cdot b = -6a^3b$ Multiply the numerical factors and then the variable factors.

21. $(-3ab^2c)(5ac^2)$
$= (-3 \cdot 5) \cdot a \cdot a \cdot b^2 \cdot c \cdot c^2$ Rearrange the factors.
$= -15a^2b^2c^3$ Multiply the numerical factors and then the variable factors.

23. $(4a^2b)(-5a^3b^2)(6a^4)$
$= 4(-5)(6) \bullet a^2 \bullet a^3 \bullet a^4 \bullet b \bullet b^2$ Rearrange the factors.
$= -120a^9b^3$ Multiply the numerical factors and then the variable factors.

25. $(-5xx^2)(-3xy)^4$
$= (-5x^3)(81x^4y^4)$ Multiply the exponents in the second factor.
$= -5 \cdot 81 \cdot x^3 \bullet x^4 \bullet y^4$ Rearrange the factors.
$= -405x^7y^4$ Multiply the numerical factors and then the variable factors.

27. $3(x + 2) = 3x + 6$ Distribute the 3.

29. $3x(x^2 + 3x)$
$= 3x \bullet x^2 + 3x \bullet 3x$ Distribute the $3x$.
$= 3x^3 + 9x^2$ Do the multiplications.

31. $-2x(3x^2 - 3x + 2)$
$= -2x(3x^2) - 2x(-3x) - 2x(2)$ Distribute the $-2x$.
$= -6x^3 + 6x^2 - 4x$ Do the multiplications.

33. $7rst(r^2 + s^2 - t^2)$
$= 7rst(r^2) + 7rst(s^2) + 7rst(-t^2)$ Distribute the $7rst$.
$= 7r^3st + 7rs^3t - 7rst^3$ Do the multiplications.

35. $4m^2n(-3mn)(m + n)$
$= -12m^3n^2(m + n)$ Multiply the first two factors.
$= -12m^3n^2 \bullet m - 12m^3n^2 \bullet n$ Distribute the $-12m^3n^2$.
$= -12m^4n^2 - 12m^3n^3$ So the multiplications.

37. $(x + 2)(x + 3)$

$= x(x + 3) + 2(x + 3)$ Multiply each term of the first factor by the second factor.
$= x^2 + 3x + 2x + 6$ Distribute the x and the 2.
$= x^2 + 5x + 6$ Combine like terms.

39. $(3t - 2)(2t + 3)$

$= 3t(2t + 3) - 2(2t + 3)$ Multiply each term of the first factor by the second factor.
$= 6t^2 + 9t - 4t - 6$ Distribute the $3t$ and the -2.
$= 6t^2 + 5t - 6$ Combine like terms.

41. $(3y - z)(2y - z)$

$= 3y(2y - z) - z(2y - z)$ Multiply each term of the first factor by the second factor.
$= 6y^2 - 3yz - 2yz + z^2$ Distribute the $3y$ and the $-z$.
$= 6y^2 - 5yz + z^2$ Combine like terms.

43. $(\frac{1}{2}b + 8)(4b + 6)$

$= \frac{1}{2}b(4b + 6) + 8(4b + 6)$ Multiply each term of the first factor by the second factor.
$= 2b^2 + 3b + 32b + 48$ Distribute the $\frac{1}{2}b$ and the 8.
$= 2b^2 + 35b + 48$ Combine like terms.

45. $(0.4t - 3)(0.5t - 3)$

$= 0.4t(0.5t - 3) - 3(0.5t - 3)$ Multiply each term of the first factor by the second factor.
$= 0.2t^2 - 1.2t - 1.5t + 9$ Distribute the $0.4t$ and the -3.
$= 0.2t^2 - 2.7t + 9$ Combine like terms.

47. $(b^3 - 1)(b + 1)$

$= b^3(b + 1) - 1(b + 1)$ Multiply each term of the first factor by the second factor.
$= b^4 + b^3 - b - 1$ Distribute the b and the -1.

49. $(3tu - 1)(-2tu + 3)$

$= 3tu(-2tu + 3) - 1(-2tu + 3)$ Multiply each term of the first factor by the second factor.
$= -6t^2u^2 + 9tu + 2tu - 3$ Distribute the $3tu$ and the -1.
$= -6t^2u^2 + 11tu - 3$ Combine like terms.

51. $(9b^3 - c)(3b^2 - c)$

$= 9b^3(3b^2 - c) - c(3b^2 - c)$ Multiply each term of the first factor by the second factor.
$= 27b^5 - 9b^3c - 3b^2c + c^2$ Distribute the $9b^3$ and the $-c$.

53. $(11m^2 + 3n^3)(5m + 2n^2)$

$= 11m^2(5m + 2n^2) + 3n^3(5m + 2n^2)$ Multiply each term of the first factor by the second factor.
$= 55m^3 + 22m^2n^2 + 15mn^3 + 6n^5$ Distribute the $11m^2$ and the $3n^3$.

55. $6p^2(3p - 4)(p + 3)$

$= 6p^2(3p^2 + 9p - 4p - 12)$ Use FOIL to multiply the second and third factors.
$= 6p^2(3p^2 + 5p - 12)$ Combine like terms.
$= 18p^4 + 30p^3 - 72p^2$ Distribute the $6p^2$.

57. $(3m - y)(4my)(2m - y)$

$= 4my(3m - y)(2m - y)$ Rearrange the factors.
$= 4my(6m^2 - 3my - 2my + y^2)$ Use FOIL to multiply the second and third factors..
$= 4my(6m^2 - 5my + y^2)$ Combine like terms.
$= 24m^3y - 20m^2y^2 + 4my^3$ Distribute the $4my$.

59. $(x + 2)^2$

$= (x)^2 + 2(x)(2) + (2)^2$ Use the special product formula for squaring a binomial.
$= x^2 + 4x + 4$ Simplify.

61. $(a-4)^2$

$= (a)^2 + 2(a)(-4) + (-4)^2$	Use the special product formula for squaring a binomial.
$= a^2 - 8a + 16$	Simplify.

63. $(2a+b)^2$

$= (2a)^2 + 2(2a)(b) + (b)^2$	Use the special product formula for squaring a binomial.
$= 4a^2 + 4ab + b^2$	Simplify.

65. $(5r^2+6)^2$

$= (5r^2)^2 + 2(5r^2)(6) + (6)^2$	Use the special product formula for squaring a binomial.
$= 25r^4 + 60r^2 + 36$	Simplify.
$= 36p^4 - 36p^2 + 9$	Simplify.

67. $(9ab-4)^2$

$= (9ab)^2 + 2(9ab)(-4) + (-4)^2$	Use the special product formula for squaring a binomial.
$= 81a^2b^2 - 72ab + 16$	Simplify.

69. $(\frac{1}{4}b+2)^2$

$= (\frac{1}{4}b)^2 + 2(\frac{1}{4}b)(2) + (2)^2$	Use the special product formula for squaring a binomial.
$= \frac{1}{16}b^2 + b + 4$	Simplify.

71. $(4k-1.3)^2$

$= (4k)^2 + 2(4k)(-1.3) + (-1.3)^2$	Use the special product formula for squaring a binomial.
$= 16k^2 - 10.4k + 1.69$	Simplify.

73. $(x+2)(x-2)$

$= (x)^2 - (2)^2$	Use the formula for the product of a sum and difference.
$= x^2 - 4$	Simplify.

75. $(y^3+2)(y^3-2)$

$= (y^3)^2 - (2)^2$	Use the formula for the product of a sum and difference.
$= y^6 - 4$	Simplify.

77. $(xy-6)(xy+6)$

$= (xy)^2 - (6)^2$	Use the formula for the product of a sum and difference.
$= x^2y^2 - 36$	Simplify.

79. $(\frac{1}{2}x-16)(\frac{1}{2}x+16)$

$= (\frac{1}{2}x)^2 - (16)^2$	Use the formula for the product of a sum and difference.
$= \frac{1}{4}x^2 - 256$	Simplify.

81. $(2.4+y)(2.4-y)$

$= (2.4)^2 - (y)^2$	Use the formula for the product of a sum and difference.
$= 5.76 - y^2$	Simplify.

83. $(x-y)(x^2+xy+y^2)$

$= x(x^2+xy+y^2) - y(x^2+xy+y^2)$	Multiply each term of the first factor by the second factor.
$= x^3 + x^2y + xy^2 - x^2y - x y^2 - y^3$	Distribute the x and the $-y$.
$= x^3 - y^3$	Combine like terms.

85. $(3y + 1)(2y^2 + 3y + 2)$
$= 3y(2y^2 + 3y + 2) + 1(2y^2 + 3y + 2)$ Multiply each term of the first factor by the second factor.
$= 6y^3 + 9y^2 + 6y + 2y^2 + 3y + 2$ Distribute the $3y$ and the 1.
$= 6y^3 + 11y^2 + 9y + 2$ Combine like terms.

87. $(2a - b)(4a^2 + 2ab + b^2)$
$= 2a(4a^2 + 2ab + b^2) - b(4a^2 + 2ab + b^2)$ Multiply each term of the first factor by the second factor.
$= 8a^3 + 4a^2b + 2ab^2 - 4a^2b - 2ab^2 - b^3$ Distribute the $2a$ and the b.
$= 8a^3 - b^3$ Combine like terms.

89. $(a + b)(a - b)(a - 3b)$
$= (a^2 - b^2)(a - 3b)$ Use special formula to multiply the first and second factors..
$= a^2(a - 3b) - b^2(a - 3b)$ Multiply each term of the first factor by the second factor.
$= a^3 - 3a^2b - ab^2 + 3b^3$ Distribute the a^2 and the $-b^2$.

91. $(a + b + c)(2a - b - 2c)$
$= a(2a - b - 2c) + b(2a - b - 2c) + c(2a - b - 2c)$ Multiply each term of the first factor by second factor.
$= 2a^2 - ab - 2ac + 2ab - b^2 - 2bc + 2ac - bc - 2c^2$ Distribute the a, b, and c.
$= 2a^2 + ab - b^2 - 3bc - 2c^2$ Combine like terms.

93. $(r + s)^2(r - s)^2$
$= (r + s)(r - s)(r + s)(r - s)$ Rearrange the factors.
$= (r^2 - s^2)(r^2 - s^2)$ Multiply the first two factors and the last two factors.
$= (r^2)^2 + 2(r^2)(-s^2) + (-s^2)^2$ Use the formula for squaring a binomial.
$= r^4 - 2r^2s^2 + s^4$ Simplify.

95. $3x(2x + 4) - 3x^2$
$= 6x^2 + 12x - 3x^2$ Distribute the $3x$.
$= 3x^2 + 12x$ Combine like terms.

97. $3pq - p(p - q)$
$= 3pq - p^2 + pq$ Distribute the $-p$.
$= -p^2 + 4pq$ Combine like terms.

99. $(x + 3)(x - 3) + (2x - 1)(x + 2)$
$= (x^2 - 9) + (2x^2 + 4x - x - 2)$ Multiply the first two factors and the last two factors.
$= 3x^2 + 3x - 11$ Remove parentheses and combine like terms.

101. $(3x - 4)^2 - (2x + 3)^2$
$= [9x^2 - 2(12x) + 16] - [4x^2 + 2(6x) + 9]$ Use the formula for squaring a binomial twice.
$= 9x^2 - 24x + 16 - 4x^2 - 12x - 9$ Remove the parentheses.
$= 5x^2 - 36x + 7$ Combine like terms.

103. $(3.21x - 7.85)(2.87x + 4.59)$
$= 9.2127x^2 + 14.7339x - 22.5295x - 36.0315$ Use FOIL to multiply.
$= 9.2127x^2 - 7.7956x - 36.0315$ Combine like terms.

105. $(-17.3y + 4.35)^2$
$= 299.29y^2 - 2(75.255y) + 18.9225$ Use the formula for squaring a binomial.
$= 299.29y^2 - 150.51y + 18.9225$ Remove the parentheses.

APPLICATIONS

107. a. The ads for the movers are the two in the left–most column. That area is represented by $(x + y)(x - y)$.
b. The area of the ad for Budget Moving Co. is represented $x(x - y)$. Multiplied this is $x^2 - xy$.
c. The area of the ad for Snyder Movers is represented by $y(x - y)$. Multiplied this is $xy - y^2$.
d. They represent the same area. $(x + y)(x - y) = x^2 - y^2$

109. The volume of the box is represented by

$$\begin{aligned} x(12-2x)(12-2x) \text{ in.}^3 &= x[12^2 + 2(12)(-2x) + (-2x)^2] \\ &= x(144 - 48x + 4x^2) \\ &= (144x - 48x^2 + 4x^3) \text{ in.}^3 \end{aligned}$$

WRITING

111. The FOIL method is only used when multiplying two binomials. Multiply the first terms of each binomial (**first**); multiply the first term of the first binomial with the last term of the second binomial (**outer**) and add this to the product of the second term of the first binomial and the first term of the second binomial (**inner**); multiply the second term of each binomial (**last**).

113. The student did not find the middle terms for the binomial that was squared.

REVIEW

115.

x	$2x + y \le 2$ Graph: $2x + y = 2$	y
0	$2(0) + y = 2$	2
1	$2x + (0) = 2$	0
Solid line	Test Point: (0, 0) $2(0) + (0) \le 2$ $0 \le 2$	True

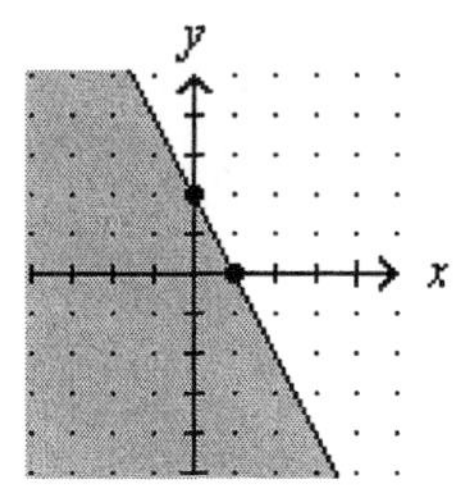

117. Inequality #1: $y - 2 < 3x$

Graph: $y = 3x + 2$

x	y
0	2
−2	−4

Broken line. Test point (0, 0).

$(0) - 2 < 3(0)$

$-2 < 0$ True

Inequality #2: $y + 2x < 3$

Graph: $y + 2x = 3$

x	y
0	3
2	−1

Broken line. Test point (0, 0).

$(0) + 2(0) < 3$

$0 < 3$ True

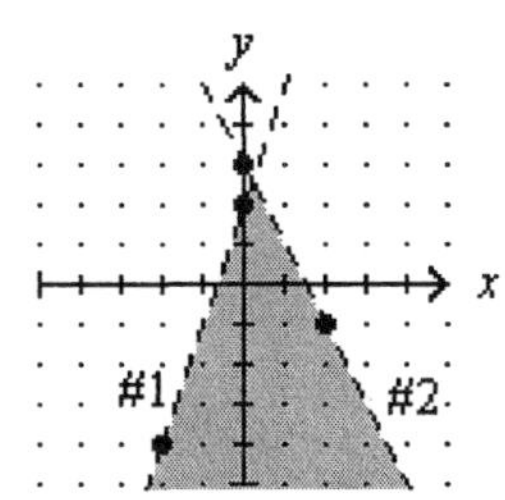

Section 5.5 The Greatest Common Factor and Factoring by Grouping

VOCABULARY

1. factored
3. factor
5. greatest common factor

CONCEPTS

7. GCF is $2 \cdot 3 \cdot x \cdot y \cdot y = 6xy^2$

9. a. The terms inside the parentheses have a common factor of 2.
   b. The terms inside the parentheses have a common factor of t.

NOTATION

11. $3a - 12 = 3(a - \mathbf{4})$

13. $x^3 - x^2 + 2x - 2 = \mathbf{x^2}(x-1) + \mathbf{2}(x-1) = (x-1)(x^2+2)$

PRACTICE

15. $6 = 2 \cdot 3$

17. $135 = 5 \cdot 27 = 5 \cdot 3 \cdot 3 \cdot 3 = 3^3 \cdot 5$

19. $\begin{aligned} 128 &= 4 \cdot 32 \\ &= 4 \cdot 4 \cdot 8 \\ &= 2 \cdot 2 \cdot 2 \cdot 2 \cdot 2 \cdot 2 \cdot 2 = 2^7 \end{aligned}$

21. $\begin{aligned} 325 &= 25 \cdot 13 \\ &= 5 \cdot 5 \cdot 13 \\ &= 5^2 \cdot 13 \end{aligned}$

23. $36 = \mathbf{2 \cdot 2 \cdot 3} \cdot 3$
$48 = \mathbf{2 \cdot 2} \cdot 2 \cdot 2 \cdot \mathbf{3}$

GCF is $2 \cdot 2 \cdot 3 = 12$

25. $42 = \mathbf{2} \cdot 3 \cdot 7$
$36 = \mathbf{2} \cdot 2 \cdot 3 \cdot 3$

$\cdot\, c$

$98 = \mathbf{2} \cdot 7 \cdot 7$
GCF is 2

27. $4a^2 b = \mathbf{2 \cdot 2 \cdot a \cdot a} \cdot b$
$8a^3c = \mathbf{2 \cdot 2} \cdot 2 \cdot \mathbf{a \cdot a} \cdot a$

GCF is $2 \cdot 2 \cdot a \cdot a = 4a^2$

29. $18x^4y^3z^2 = \mathbf{2 \cdot 3} \cdot 3 \cdot \mathbf{x} \cdot x \cdot x \cdot x \cdot \mathbf{y \cdot y} \cdot y \cdot \mathbf{z \cdot z}$
$-12xy^2z^3 = -1 \cdot \mathbf{2} \cdot 2 \cdot \mathbf{3 \cdot x \cdot y \cdot y \cdot z \cdot z} \cdot z$
GCF is $2 \cdot 3 \cdot x \cdot y \cdot y \cdot z \cdot z = 6xy^2z^2$

31. $2x + 8 = 2 \cdot x + 2 \cdot 4 = 2(x + 4)$

33. $2x^2 - 6x = 2x \cdot x - 2x \cdot 3 = 2x(x - 3)$

35. $5xy + 12ab^2$ is prime because there are no common factors.

37. $15x^2y - 10x^2y^2 = 5x^2 y \cdot 3 - 5x^2y \cdot 2y = 5x^2y(3 - 2y)$

39. $14r^2s^3 + 15t^6$ is prime because there are no common factors.

41. $27z^3 + 12z^2 + 3z = 3z \cdot 9z^2 + 3z \cdot 4z + 3z \cdot 1 = 3z(9z^2 + 4z + 1)$

43. $45x^{10}y^3 - 63x^7y^7 + 81x^{10}y^{10} = 9x^7y^3 \cdot 5x^3 - 9x^7y^3 \cdot 7y^4 + 9x^7y^3 \cdot 9x^3y^7 = 9x^7y^3(5x^3 - 7y^4 + 9x^3y^7)$

45. $-3a - 6 = -3 \cdot a - 3 \cdot 2 = -3(a + 2)$

47. $-3x^2 - x = -x \cdot 3x - x \cdot 1 = -x(3x + 1)$

49. $-6x^2 - 3xy = -3x \cdot 2x - 3x \cdot y = -3x(2x + y)$

51. $-18a^2b - 12ab^2 = -6ab \cdot 3a - 6ab \cdot 2b = -6ab(3a + 2b)$

53. $-63u^3v^6z^9 + 28u^2v^7z^2 - 21u^3v^3z^4 = -7u^2v^3z^2 \cdot 9uv^3z^7 - 7u^2v^3z^2 \cdot (-4v^4) - 7u^2v^3z^2 \cdot 3uz^2$
$= -7u^2v^3z^2(9uv^3z^7 - 4v^4 + 3uz^2)$

55. $4(x + y) + t(x + y) = (x + y)(4 + t)$

57. $(a - b)r - (a - b)s = (a - b)(r - s)$

59. $3(m + n + p) + x(m + n + p) = (m + n + p)(3 + x)$

61. $(u + v)^2 - (u + v) = (u + v)(u + v) - 1(u + v) = (u + v)(u + v - 1)$

63. $-a(x + y) + b(x + y) = (x + y)(-a + b)$ or $-(x + y)(a - b)$

65. $ax + bx + ay + by = (ax + bx) + (ay + by) = x(a + b) + y(a + b) = (a + b)(x + y)$

67. $x^2 + yx + 2x + 2y = (x^2 + yx) + (2x + 2y) = x(x + y) + 2(x + y) = (x + y)(x + 2)$

69. $3c - cd + 3d - c^2 = 3c - c^2 + 3d - cd = c(3 - c) + d(3 - c) = (3 - c)(c + d)$ Rearrange terms first.

71. $a^2 - 4b + ab - 4a = a^2 - 4a + ab - 4b = a(a - 4) + b(a - 4) = (a - 4)(a + b)$ Rearrange terms first.

73. $ax + bx - a - b = x(a + b) - 1(a + b) = (a + b)(x - 1)$

75. $x^2 + xy + xz + xy + y^2 + zy = x(x + y + z) + y(x + y + z) = (x + y + z)(x + y)$

77. $mpx + mqx + npx + nqx = x(mp + mq + np + nq) = x[m(p + q) + n(p + q)] = x(p + q)(m + n)$

79. $x^2y + xy^2 + 2xyz + xy^2 + y^3 + 2y^2z = y(x^2 + xy + 2xz + xy + y^2 + 2yz)$
$= y[x(x + y + 2z) + y(x + y + 2z)] = y(x + y)(x + y + 2z)$

81. $2n^4p - 2n^2 - n^3p^2 + np + 2mn^3p - 2mn = n(2n^3p - 2n - n^2p^2 + p + 2mn^2p - 2m)$
$= n(2n^3p - n^2p^2 + 2mn^2p - 2n + p - 2m) = n[n^2p(2n - p + 2m) - 1(2n - p + 2m)] = n(n^2p - 1)(2n - p + 2m)$

83.

$r_1r_2 = rr_2 + rr_1$	Solving for r_1.
$r_1r_2 - rr_1 = rr_2$	Subtract rr_1 from both sides.
$r_1(r_2 - r) = rr_2$	Factor out the common factor r_1.
$r_1 = \dfrac{rr_2}{r_2 - r}$	Divide both sides by $r_2 - r$.

85.
$$\begin{aligned} d_1d_2 &= fd_2 + fd_1 && \text{Solving for } f. \\ d_1d_2 &= f(d_2 + d_1) && \text{Factor out the common factor } f. \\ \frac{d_1d_2}{d_2 + d_1} &= f && \text{Divide both sides by } d_2 + d_1. \end{aligned}$$

87.
$$\begin{aligned} b^2x^2 + a^2y^2 &= a^2b^2 && \text{Solving for } a^2. \\ b^2x^2 &= a^2b^2 - a^2y^2 && \text{Subtract } a^2y^2 \text{ from both sides.} \\ b^2x^2 &= a^2(b^2 - y^2) && \text{Factor out the common factor } a^2. \\ \frac{b^2x^2}{b^2 - y^2} &= a^2 && \text{Divide both sides by } b^2 - y^2. \end{aligned}$$

89.
$$\begin{aligned} S(1 - r) &= a - lr && \text{Solving for } r. \\ S - Sr &= a - lr && \text{Distribute the S.} \\ S - a &= Sr - lr && \text{Subtract a and Sr from both sides.} \\ S - a &= r(S - l) && \text{Factor out the common factor } r. \\ \frac{S - a}{S - l} &= r && \text{Divide both sides by S} - l. \end{aligned}$$

APPLICATIONS

91. a. The red shaded area is a triangle, so $A_{\text{red}} = \frac{1}{2}b_1h$. b. The blue shaded area is a triangle, so $A_{\text{blue}} = \frac{1}{2}b_2h$.

c. $A_{\text{Total}} = A_{\text{red}} + A_{\text{blue}} = \frac{1}{2}b_1h. + \frac{1}{2}b_2h = \frac{1}{2}h(b_1 + b_2)$. This is the formula for the area of a trapezoid.

93. $4r^2 - \pi r^2 = r^2(4 - \pi)$

WRITING

95. When we distribute, we multiply a sum or difference by a common factor. When we factor, we remove a common factor from sums or differences of terms.

97. The factor of five is only present in the first term, $5x^2$. The remaining terms do not contain a factor of 5.

REVIEW

99. a line 101. $-x > 3$; $x < -3$; $(-\infty, -3)$ 103. The slopes are the same.

Section 5.6 The Difference of Two Squares; The Sum and Difference of Two Cubes

VOCABULARY

1. squares

CONCEPTS

3. $1^2 = 1, 2^2 = 4, 3^2 = 9, 4^2 = 16, 5^2 = 25, 6^2 = 36, 7^2 = 49, 8^2 = 64, 9^2 = 81, 10^2 = 100$

5. $(x + 5)(x + 5) = x^2 + 5x + 5x + 25 = x^2 + 10x + 25$ not $x^2 + 25$.

7. a. A common factor of 2 can be factored out of each binomial.
   b. $(1 - t^4)$ can be factored as the difference of two squares.

9. a. $5p^2 + 20 = 5(p^2 + 4)$
   b. $5p^2 - 20 = 5(p^2 - 4) = 5(p^2 - 2^2) = 5(p + 2)(p - 2)$

NOTATION

11. $p^3 + q^3 = (p + q)(\ \mathbf{p^2 - pq + q^2}\ )$ 13. $p^2 - q^2 = (p + q)(\ \mathbf{p - q}\ )$

15. $36y^2 - 49m^2 = (\ \mathbf{6y}\ )^2 - (7m)^2 = (6y\ +\ 7m)(6y - \ \mathbf{7m}\ )$

PRACTICE

17. $x^2 - 4 = x^2 - 2^2 = (x + 2)(x - 2)$ Use $F^2 - L^2 = (F + L)(F - L)$ for these problems.

19. $9y^2 - 64 = (3y)^2 - 8^2 = (3y + 8)(3y - 8)$

21. $x^2 + 25$ is prime The **sum** of two squares is not factorable.

23. $625a^2 - 169b^4 = (25a)^2 - (13b^2) = (25a + 13b^2)(25a - 13b^2)$

25. $81a^4 - 49b^2 = (9a^2)^2 - (7b)^2 = (9a^2 + 7b)(9a^2 - 7b)$

27. $36x^4y^2 - 49z^4 = (6x^2y)^2 - (7z^2)^2 = (6x^2y + 7z^2)(6x^2y - 7z^2)$

29. $(x + y)^2 - z^2 = [(x + y) + z][(x + y) - z] = (x + y + z)(x + y - z)$

31. $(a - b)^2 - c^2 = [(a - b) + c][(a - b) - c] = (a - b + c)(a - b - c)$

33. $x^4 - y^4 = (x^2)^2 - (y^2)^2 = (x^2 + y^2)(x^2 - y^2)$
$= (x^2 + y^2)[(x + y)(x - y)] = (x^2 + y^2)(x + y)(x - y)$ Use the formula twice.

35. $256x^4y^4 - z^8 = (16x^2y^2)^2 - z^4 = (16x^2y^2 + z^4)(16x^2y^2 - z^4)$
$= (16x^2y^2 + z^4)[(4xy)^2 - (z^2)^2] = (16x^2y^2 + z^4)(4xy + z^2)(4xy - z^2)$

37. $2x^2 - 288 = 2(x^2 - 144) = 2(x^2 - 12^2) = 2(x + 12)(x - 12)$ Factor out the common factor first.

39. $2x^3 - 32x = 2x(x^2 - 16) = 2x(x^2 - 4^2) = 2x(x + 4)(x - 4)$

41. $5x^3 - 125x = 5x(x^2 - 25) = =5x(x^2 - 5^2) = 5x(x + 5)(x - 5)$

43. $r^2s^2t^2 - t^2x^4y^2 = t^2(r^2s^2 - x^4y^2) = t^2[(rs)^2 - (x^2y)^2] = t^2(rs + x^2y)(rs - x^2y)$

45. $a^2 - b^2 + a + b = (a + b)(a - b) + 1(a + b) = (a + b)(a - b + 1)$

47. $a^2 - b^2 + 2a - 2b = (a + b)(a - b) + 2(a - b) = (a - b)(a + b + 2)$

49. $2x + y + 4x^2 - y^2 = 1(2x + y) + (2x + y)(2x - y) = (2x + y)(1 + 2x - y)$

51. $r^3 + s^3 = (r + s)(r^2 - rs + s^2)$ Use $F^3 + L^3 = (F + L)(F^2 - FL + L^2)$

53. $x^3 - 8y^3 = x^3 - (2y)^3 = (x - 2y)(x^2 + 2xy + 4y^2)$

55. $64a^3 - 125b^6 = (4a)^3 - (5b^2)^3 = (4a - 5b^2)(16a^2 + 20ab^2 + 25b^4)$

57. $125x^3y^6 + 216z^9 = (5xy^2)^3 + (6z^3)^3 = (5xy^2 + 6z^3)(25x^2y^4 - 30xy^2z^3 + 36z^6)$

59. $x^6 + y^6 = (x^2)^3 + (y^2)^3 = (x^2 + y^2)(x^4 - x^2y^2 + y^4)$

61. $5x^3 + 625 = 5(x^3 + 125) = 5(x^3 + 5^3) = 5(x + 5)(x^2 - 5x + 25)$

63. $4x^5 - 256x^2 = 4x^2(x^3 - 64) = 4x^2(x^3 - 4^3) = 4x^2(x - 4)(x^2 + 4x + 16)$

65. $128u^2v^3 - 2t^3u^2 = 2u^2(64v^3 - t^3) = 2u^2[(4v)^3 - t^3)] = 2u^2(4v - t)(16v^2 + 4tv + t^2)$

67. $(a + b)x^3 + 27(a + b) = (a + b)(x^3 + 27) = (a + b)(x^3 + 3^3) = (a + b)(x + 3)(x^2 - 3x + 9)$

APPLICATIONS

69. $V = \frac{4}{3}\pi r_1^3 - \frac{4}{3}\pi r_2^3 = \frac{4}{3}\pi(r_1^3 - r_2^3) = \frac{4}{3}\pi(r_1 - r_2)(r_1^2 + r_1r_2 + r_2^2)$

WRITING

71. The sum and difference of the factors which make the squares are the factors. Example: $x^2 - y^2 = (x + y)(x - y)$.

REVIEW

73. Point (–2, –1); Slope $= -\frac{2}{3} = \frac{\text{rise}}{\text{run}}$

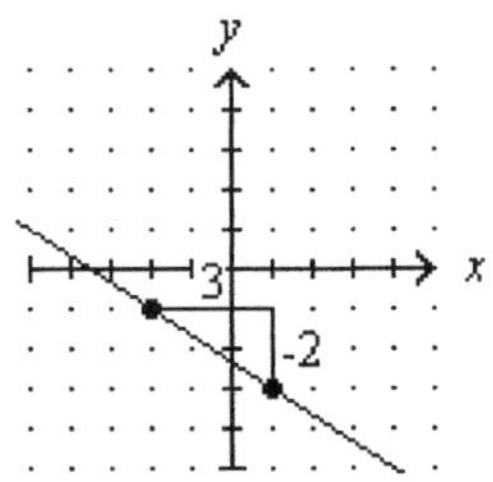

75. Horizontal; y–intercept (0, –2)

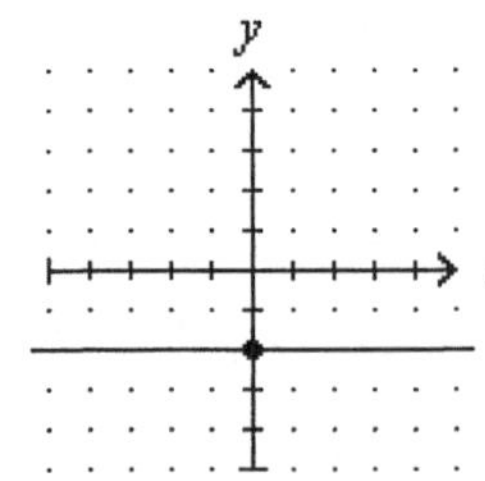

77. The line goes through (0, –4) and is horizontal. The equation would be $y = -4$.

Section 5.7 Factoring Trinomials

VOCABULARY

1. trinomial
3. lead

CONCEPTS

5. a. positive b. negative c. positive
7. integers

NOTATION

9. $(x+y)(x+y) = x^2 + \mathbf{2xy + y^2}$
11. $(x+y)(x-y) = \mathbf{x^2 - y^2}$
13. $a = 4, b = -4, c = 1$

PRACTICE

15. $x^2 + 5x + 6 = (x+3)\ \mathbf{(x+2)}$ If one factor of 6 is 3, the other is 2. Also, the sum of 3 and 2 is 5.

17. $x^2 + 2x - 15 = (x+5)\ \mathbf{(x-3)}$ If one factor of –15 is 5, the other is –3. Also the sum of 5 and –3 is 2.

19. $2a^2 + 9a + 4 = \mathbf{(2a+1)}\ (a+4)$ Since the first term of the trinomial is $2a^2$, the first terms of the binomials must be $2a$ and a. The last terms of the binomial must multiply to 4, so the last terms are 1 and 4. The outer and inner products, $8a$ and a, add to $9a$.

21. $x^2 + 2x + 1 = x^2 + 2(x)(1) + 1^2 = (x+1)(x+1) = (x+1)^2$

23. $a^2 - 18a + 81 = a^2 - 2(a)(9) + 9^2 = (a-9)(a-9) = (a-9)^2$

25. $4y^2 + 4y + 1 = (2y)^2 + 2(2y)(1) + 1^2 = (2y+1)(2y+1) = (2y+1)^2$

27. $9b^2 - 12b + 4 = (3b)^2 - 2(3b)(2) + 2^2 = (3b-2)(3b-2) = (3b-2)^2$

29. Test: $b^2 - 4ac = (-5)^2 - 4(1)(6) = 25 - 24 = 1 = 1^2$ It is factorable.
Negative factors of 6 are (–1)(–6), (–2)(–3). Factors that add to –5 are –2 and –3.
$x^2 - 5x + 6 = (x-2)(x-3)$

31. Test: $b^2 - 4ac = (-7)^2 - 4(1)(10) = 49 - 40 = 9 = 3^2$ It is factorable.
Negative factors of 10 are (–1)(–10), (–2)(–5). Factors that add to –7 are –2 and –5.
$x^2 - 7x + 10 = (x-2)(x-5)$

33. Test: $b^2 - 4ac = (8)^2 - 4(1)(18) = 64 - 72 = -8$ It is not factorable. $b^2 + 8b + 18$ is prime.

35. Test: $b^2 - 4ac = (-1)^2 - 4(1)(-30) = 1 - (-120) = 121 = 11^2$ It is factorable.
Positive and negative factors are (–1)(30), (–2)(15), (–3)(10), (–5)(6) as well as reversing the signs.
Factors that add to –1 are –6 and 5.
$x^2 - x - 30 = (x-6)(x+5)$

37. Test: $b^2 - 4ac = (5)^2 - 4(1)(-50) = 25 - (-200) = 225 = 15^2$ It is factorable.
Positive and negative factors of 50 are $(-1)(50)$, $(-2)(25)$, $(-5)(10)$ as well as reversing the signs.
Factors that add to 5 are 10 and -5. $a^2 + 5a - 50 = (a + 10)(a - 5)$

39. Test: $b^2 - 4ac = (-4)^2 - 4(1)(-21) = 16 - (-84) = 100 = 10^2$ It is factorable.
Positive and negative factors of 21 are $(-1)(21)$, $(-3)(7)$. Factors that add to -4 are -7 and 3.
$x^2 - 4xy - 21y^2 = (x - 7y)(x + 3y)$

Factoring by grouping in used in the following exercises when appropriate.

41. $3x^2 + 12x - 63 = 3(x^2 + 4x - 21) = 3(x^2 + 7x - 3x - 21) = 3[x(x + 7) - 3(x + 7)] = 3(x + 7)(x - 3)$

43. $b^2x^2 - 12bx^2 + 35x^2 = x^2(b^2 - 12b + 35) = x^2[b^2 - 5b - 7b + 35] = x^2[b(b - 5) - 7(b - 5)] = x^2(b - 5)(b - 7)$

45. $-a^2 + 4a + 32 = -(a^2 - 4a - 32) = -(a^2 - 8a + 4a - 32) = -[a(a - 8) + 4(a - 8)] = -(a - 8)(a + 4)$

47. $-3x^2 + 15x - 18 = -3(x^2 - 5x + 6) = -3(x^2 - 2x - 3x + 6) = -3[x(x - 2) - 3(x - 2)] = -3(x - 2)(x - 3)$

49. $-2p^2 - 2pq + 4q^2 = -2(p^2 + pq - 2q^2)$
$= -2(p^2 + 2pq - pq - 2q^2) = -2[p(p + 2q) - q(p + 2q)] = -2(p + 2q)(p - q)$

51. Key number $= ac = (6)(2) = 12$. Use factors $(3)(4) = 12$ because $3 + 4 = 7 = b$.
$6y^2 + 7y + 2 = 6y^2 + 3y + 4y + 2 = 3y(2y + 1) + 2(2y + 1) = (2y + 1)(3y + 2)$

53. Key number $= ac = (8)(-9) = -72$. Use factors $(-6)(12)$ because $-6 + 12 = 6 = b$.
$8a^2 + 6a - 9 = 8a^2 - 6a + 12a - 9 = 2a(4a - 3) + 3(4a - 3) = (4a - 3)(2a + 3)$

55. Key number $= ac = (6)(-4) = -24$. Use factors $(-8)(3)$ because $-8 + 3 = -5 = b$.
$6x^2 - 5xy - 4y^2 = 6x^2 - 8xy + 3xy - 4y^2 = 2x(3x - 4y) + y(3x - 4y) = (3x - 4y)(2x + y)$

57. Key number $= ac = (5)(1) = 5$. There are no factors that add to 4. Also, $b^2 - 4ac = (4)^2 - 4(1)(5) = 16 - 20 = -4$ which is not a perfect square. $5x^2 + 4x + 1$ is prime.

59. Key number $= ac = (8)(3) = 24$. Use factors $(-4)(-6)$ because $-4 + (-6) = -10 = b$.
$8x^2 - 10x + 3 = 8x^2 - 4x - 6x + 3 = 4x(2x - 1) - 3(2x - 1) = (2x - 1)(4x - 3)$

61. Key number $= ac = (1)(-4) = -4$. Use factors $(-4)(1)$ because $-4 + 1 = -3 = b$.
$a^2 - 3ab - 4b^2 = a^2 - 4ab + ab - 4b^2 = a(a - 4b) + b(a - 4b) = (a - 4b)(a + b)$

63. Key number $= ac = (3)(3) = 9$. Use factors $(-1)(-9)$ because $-1 + (-9) = -10 = b$.
$3x^3 - 10x^2 + 3x = x(3x^2 - 10x + 3) = x(3x^2 - x - 9x + 3) = x[x(3x - 1) - 3(3x - 1)] = x(3x - 1)(x - 3)$

65. Key number $= ac = (3)(-2) = -6$. Use factors $(-3)(2)$ because $-3 + 2 = -1 = b$.
$-3a^2 + ab + 2b^2 = -(3a^2 - ab - 2b^2) = -(3a^2 - 3ab + 2ab - 2b^2)$
$= -[3a(a - b) + 2b(a - b)] = -(a - b)(3a + 2b)$

67. Key number $= ac = (1)(9) = 9$. Use factors $(-3)(-3)$ because $-3 + (-3) = -6 = b$.
$5a^2 + 45b^2 - 30ab = 5a^2 - 30ab + 45b^2 = 5(a^2 - 6ab + 9b^2) = 5(a^2 - 3ab - 3ab + 9b^2)$
$= 5[a(a - 3b) - 3b(3a - 3b)] = 5(a - 3b)(a - 3b) = 5(a - 3b)^2$

69. Key number $= ac = (21)(16) = 336$. Use factors $(-24)(14)$ because $-24 + 14 = -10 = b$.
$21x^4 - 10x^3 - 16x^2 = x^2(21x^2 - 10x - 16) = x^2(21x^2 - 24x + 14x - 16)$
$= x^2[3x(7x - 8) + 2(7x - 8)] = x^2(7x - 8)(3x + 2)$

71. Key number $= ac = (1)(15) = 15$. Use factors $(3)(5)$ because $3 + 5 = 8 = b$.
$x^4 + 8x^2 + 15 = x^4 + 3x^2 + 5x^2 + 15 = x^2(x^2 + 3) + 5(x^2 + 3) = (x^2 + 3)(x^2 + 5)$

73. Key number $= ac = (1)(30) = 30$. Use factors $(-10)(-3)$ because $-10 + (-3) = -13 = b$.
$y^4 - 13y^2 + 30 = y^4 - 10y^2 - 3y^2 + 30 = y^2(y^2 - 10) - 3(y^2 - 10) = (y^2 - 10)(y^2 - 3)$

75. Key number $= ac = (1)(36) = 36$. Use factors $(-4)(-9)$ because $-4 + (-9) = -13 = b$.
$a^4 - 13a^2 + 36 = a^4 - 4a^2 - 9a^2 + 36 = a^2(a^2 - 4) - 9(a^2 - 4) = (a^2 - 4)(a^2 - 9)$
$= (a^2 - 2^2)(a^2 - 3^2) = (a + 2)(a - 2)(a + 3)(a - 3)$ Special product formula.

77. Rewrite $(x+a)^2+2(x+a)+1$ as z^2+2z+1 where $z=(x+a)$.

$z^2+2z+1 = (z+1)^2$ Special product formula.

$(x+a)^2+2(x+a)+1 = (x+a+1)^2$ Replace z with $x+a$.

79. Rewrite $3(a+b)^2-14(a+b)-24$ as $3z^2-14z-24$ where $z=(a+b)$.

$3z^2-14z-24 = (3z+4)(z-6)$

$(a+b)^2-2(a+b)-24 = (3a+3b+4)(a+b-6)$ Replace z with $a+b$.

81. $x^2+4x+4-y^2 = (x^2+4x+4)-y^2$ Group first three terms.

$= (x+2)(x+2)-y^2 = (x+2)^2-y^2$ Special product formula.

$= (x+2+y)(x+2-y)$ Difference of two squares.

83. $x^2+2x+1-9z^2 = (x^2+2x+1)-9z^2$ Group first three terms.

$= (x+1)(x+1)-9z^2 = (x+1)^2-(3z)^2$ Special product formula.

$= (x+1+3z)(x+1-3z)$ Difference of two squares.

85. $c^2-4a^2+4ab-b^2 = c^2-(4a^2-4ab+b^2)$ Group last three terms.

$= c^2-(2a-b)(2a-b) = c^2-(2a-b)^2$ Special product formula.

$= (c+2a-b)[c-(2a-b)]$ Difference of two squares.

$= (c+2a-b)(c-2a+b)$ Simplify.

87. Key number $= ac = (2)(16) = 32$. Use factors $(-1)(-32)$ because $-1+(-32) = -33 = b$.

$2a^2-33a+16 = 2a^2-a-32a+16 = a(2a-1)-16(2a-1) = (2a-1)(a-16)$

89. Key number $= ac = (2)(3) = 6$. Use factors $(2)(3)$ because $2+3 = 5 = b$.

$2u^2+5u+3 = 2u^2+2u+3u+3 = 2u(u+1)+3(u+1) = (u+1)(2u+3)$

91. Key number $= ac = (20)(-6) = -120$. Use factors $(-15)(8)$ because $-15+8 = -7 = b$.

$20r^2-7rs-6s^2 = 20r^2-15rs+8rs-6s^2 = 5r(4r-3s)+2s(4r-3s) = (4r-3s)(5r+2s)$

APPLICATIONS

93. Since a cube has 6 sides, the surface area is 6 times the area of one side.
The area of one side of the cube is the length of the side squared.

In this case, the surface area is $6x^2+36x+54$. Factored this is $6(x^2+6x+9) = 6(x+3)^2$.

The factored form can be interpreted as (6 sides of the cube) times (length of one side squared).
The length of the edge would be $x+3$.

WRITING

95. Place a negative in front of a set of parentheses, then change the signs of the terms of the trinomial and write the trinomial inside the parentheses.

REVIEW

97. $f(x) = |2x-1|$ $\quad f(-2) = |2(-2)-1| = |-4-1| = |-5| = 5$

99.
$$-3 = -\frac{9}{8}s$$
$$-24 = -9s$$
$$\frac{24}{9} = s$$
$$s = \frac{8}{3}$$

101.
$$3p^2-6(5p^2+p)+p^2$$
$$= 3p^2-30p^2-6p+p^2$$
$$= -26p^2-6p$$

Section 5.8 Summary of Factoring Techniques

VOCABULARY

1. factoring
3. cubes

CONCEPTS

5. common
7. trinomial
9. Multiply the factors to see if the product is the polynomial.

NOTATION

11. $18a^3b + 3a^2b^2 - 6ab^3 = \mathbf{3ab}\ (6a^2 + ab - 2b^2) = 3ab(3a + \mathbf{2b})(\mathbf{2a} - b)$

PRACTICE

13. $x^2 + 16 + 8x = x^2 + 8x + 16 = x^2 + 2(4)(x) + 4^2 = (x + 4)^2$ — Rearrange, then perfect square trinomial.

15. $8x^3y^3 - 27 = (2xy)^3 - 3^3 = (2xy - 3)(4x^2y^2 + 6xy + 9)$ — Difference of two cubes.

17. $xy - ty + xs - ts = y(x - t) + s(x - t) = (x - t)(y + s)$ — Four terms, use grouping.

19. $25x^2 - 16y^2 = (5x)^2 - (4y)^2 = (5x + 4y)(5x - 4y)$ — Difference of two squares.

21. $12x^2 + 52x + 35 = 12x^2 + 10x + 42x + 35$
$= 2x(6x + 5) + 7(6x + 5) = (6x + 5)(2x + 7)$ — Trinomial, use grouping technique.

23. $6x^2 - 14x + 8 = 2(3x^2 - 7x + 4)$
$= 2(3x^2 - 3x - 4x + 4) = 2[3x(x - 1) - 4(x - 1)] = 2(x - 1)(3x - 4)$ — Common factor, then trinomial.

25. $4x^2y^2 + 4xy^2 + y^2 = y^2(4x^2 + 4x + 1)$
$= y^2[(2x)^2 + 2(1)(2x) + 1^2] = y^2(2x + 1)^2$ — Common factor, then perfect square trinomial.

27. $x^3 + (a^2y)^3 = (x + a^2y)(x^2 - a^2xy + a^4y^2)$ — Sum of two cubes.

29. $2x^3 - 54 = 2(x^3 - 27) =$
$= 2(x^3 - 3^3) = 2(x - 3)(x^2 + 3x + 9)$ — Common factor, then difference of two cubes.

31. $ae + bf + af + be = ae + af + bf + be$
$= a(e + f) + b(f + e) = (e + f)(a + b)$ — Rearrange, then four terms for grouping.

33. Rewrite $2(x + y)^2 + (x + y) - 3$ as $2z^2 + z - 3$ where $z = (x + y)$
Factor: $2z^2 + z - 3 = 2z^2 + 3z - 2z - 3 = z(2z + 3) - 1(2z + 3) = (2z + 3)(z - 1)$
$2(x + y)^2 + (x + y) - 3 = [2(x + y) + 3][(x + y) - 1] = (2x + 2y + 3)(x + y - 1)$ — Replace z with $(x + y)$

35. $625x^4 - 256y^4 = (25x^2)^2 - (16y^2)^2 = (25x^2 + 16y^2)(25x^2 - 16y^2)$
$= (25x^2 + 16y^2)[(5x)^2 - (4y)^2] = (25x^2 + 16y^2)(5x + 4y)(5x - 4y)$ — Difference of two squares, twice.

37. $36x^4 - 36 = 36(x^4 - 1) = 36[(x^2)^2 - 1^2]$
$= 36(x^2 + 1)(x^2 - 1) = 36(x^2 + 1)(x + 1)(x - 1)$ — Common factor, then difference of two squares, twice.

39. $a^4 - 13a^2 + 36 = a^4 - 4a^2 - 9a^2 + 36 = a^2(a^2 - 4) - 9(a^2 - 4)$
$= (a^2 - 4)(a^2 - 9) = (a + 2)(a - 2)(a + 3)(a - 3)$ — Trinomial, then two difference of squares.

41. $x^2 + 6x + 9 - y^2 = [x^2 + 2(x)(3) + 3^2] - y^2 = (x + 3)^2 - y^2$
$= (x + 3 + y)(x + 3 - y)$ — Group the first three terms, for perfect square, then difference of squares

43. $4x^2 + 4x + 1 - 4y^2 = [(2x)^2 + 2(1)(2x) + 1^2] - 4y^2$
$= (2x + 1)^2 - (2y)^2 = (2x + 1 + 2y)(2x + 1 - 2y)$ — Group the first three terms for perfect square, then difference of squares.

45. $x^2 - y^2 - 2y - 1 = x^2 - (y^2 + 2y + 1) = x^2 - (y + 1)^2$
$= (x + y + 1)[x - (y + 1)] = (x + y + 1)(x - y - 1)$ — Group the last three terms for perfect square, then difference of squares.

WRITING

47. Factor out all common factors.
 Determine the size of the remaining polynomial. If it is a
 binomial, look for
 the difference of two perfect squares,
 the sum of two perfect cubes, or
 the difference of two perfect cubes.
 trinomial, look for
 perfect square trinomial, otherwise
 use grouping technique to factor into binomials.
 four or more term polynomial,
 group the terms and apply all of the above strategies.
 Check if all the factors are prime, if not, repeat above strategies on each non–prime factor.
 Check the results by multiplying.

REVIEW

49. Solve $x + y = 2$ for y: $y = -x + 2$, so slope is -1. The slope of $y = x + 5$ is 1. The slopes are negative reciprocals of each other so the lines are perpendicular.

51. $\begin{vmatrix} 1 & 15 \\ 15 & 0 \end{vmatrix} = 1(0) - 15(15) = 0 - 225 = -225$

Section 5.9 Solving Equations by Factoring

VOCABULARY

1. quadratic

CONCEPTS

3. At least one of the numbers is 0.

5. a. yes b. no c. yes d. no

7. The graph of the equation crosses the x–axis ($y = 0$) when $x = -1$ and 3.

NOTATION

9. Solve: $y^2 - 3y - 54 = 0$

$(y - 9)(\mathbf{y + 6}) = 0$

$\mathbf{y - 9} = 0$ or $y + 6 = 0$

$y = 9$ $\quad y = \mathbf{-6}$

PRACTICE

11. $4x^2 + 8x = 0$

$4x(x + 2) = 0$

$4x = 0$ or $x + 2 = 0$

$x = 0$ $\quad x = -2$

13. $y^2 - 16 = 0$

$(y + 4)(y - 4) = 0$

$y + 4 = 0$ or $y - 4 = 0$

$y = -4$ $\quad y = 4$

15. $x^2 + x = 0$

$x(x + 1) = 0$

$x = 0$ or $x + 1 = 0$

$x = -1$

17. $5y^2 - 25y = 0$

$5y(y - 5) = 0$

$5y = 0$ or $y - 5 = 0$

$y = 0$ $\quad y = 5$

19.
$$\begin{aligned} z^2 + 8z + 15 &= 0 \\ (z+3)(z+5) &= 0 \end{aligned}$$
$$z + 3 = 0 \quad \text{or} \quad z + 5 = 0$$
$$z = -3 \qquad\qquad z = -5$$

21.
$$\begin{aligned} x^2 + 6x + 8 &= 0 \\ (x+2)(x+4) &= 0 \end{aligned}$$
$$x + 2 = 0 \quad \text{or} \quad x + 4 = 0$$
$$x = -2 \qquad\qquad x = -4$$

23.
$$\begin{aligned} 3m^2 + 10m + 3 &= 0 \\ 3m^2 + 9m + 1m + 3 &= 0 \\ 3m(m+3) + 1(m+3) &= 0 \\ (3m+1)(m+3) &= 0 \end{aligned}$$
$$3m + 1 = 0 \quad \text{or} \quad m + 3 = 0$$
$$3m = -1 \qquad\qquad m = -3$$
$$m = -\tfrac{1}{3}$$

25.
$$\begin{aligned} 2y^2 - 5y + 2 &= 0 \\ 2y^2 - 4y - y + 2 &= 0 \\ 2y(y-2) - 1(y-2) &= 0 \\ (y-2)(2y-1) &= 0 \end{aligned}$$
$$y - 2 = 0 \quad \text{or} \quad 2y - 1 = 0$$
$$y = 2 \qquad\qquad 2y = 1$$
$$y = \tfrac{1}{2}$$

27.
$$\begin{aligned} 2x^2 - x - 1 &= 0 \\ 2x^2 - 2x + x - 1 &= 0 \\ 2x(x-1) + 1(x-1) &= 0 \\ (x-1)(2x+1) &= 0 \end{aligned}$$
$$x - 1 = 0 \quad \text{or} \quad 2x + 1 = 0$$
$$x = 1 \qquad\qquad 2x = -1$$
$$x = -\tfrac{1}{2}$$

29.
$$\begin{aligned} x(x-6) + 9 &= 0 \\ x^2 - 6x + 9 &= 0 \\ (x-3)(x-3) &= 0 \end{aligned}$$
$$x - 3 = 0 \quad \text{or} \quad x - 3 = 0$$
$$x = 3 \qquad\qquad x = 3$$

31.
$$\begin{aligned} 8a^2 &= 3 - 10a \\ 8a^2 + 10a - 3 &= 0 \\ 8a^2 - 2a + 12a - 3 &= 0 \\ 2a(4a-1) + 3(4a-1) &= 0 \\ (4a-1)(2a+3) &= 0 \end{aligned}$$
$$4a - 1 = 0 \quad \text{or} \quad 2a + 3 = 0$$
$$4a = 1 \qquad\qquad 2a = -3$$
$$a = \tfrac{1}{4} \qquad\qquad a = -\tfrac{3}{2}$$

33.
$$\begin{aligned} b(6b-7) &= 10 \\ 6b^2 - 7b - 10 &= 0 \\ 6b^2 - 12b + 5b - 10 &= 0 \\ 6b(b-2) + 5(b-2) &= 0 \\ (b-2)(6b+5) &= 0 \end{aligned}$$
$$b - 2 = 0 \quad \text{or} \quad 6b + 5 = 0$$
$$b = 2 \qquad\qquad 6b = -5$$
$$b = -\tfrac{5}{6}$$

35.
$$\begin{aligned} \frac{3a^2}{2} &= \frac{1}{2} - a \\ 3a^2 &= 1 - 2a \\ 3a^2 + 2a - 1 &= 0 \\ (3a-1)(a+1) &= 0 \end{aligned}$$
$$3a - 1 = 0 \quad \text{or} \quad a + 1 = 0$$
$$3a = 1 \qquad\qquad a = -1$$
$$a = \tfrac{1}{3}$$

37.
$$\begin{aligned} x^2 + 1 &= \frac{5}{2}x \\ 2x^2 + 2 &= 5x \\ 2x^2 - 5x + 2 &= 0 \\ (2x-1)(x-2) &= 0 \end{aligned}$$
$$2x - 1 = 0 \quad \text{or} \quad x - 2 = 0$$
$$2x = 1 \qquad\qquad x = 2$$
$$x = \tfrac{1}{2}$$

39.
$$\begin{aligned} x\left(3x + \frac{22}{5}\right) &= 1 \\ 3x^2 + \frac{22}{5}x &= 1 \\ 15x^2 + 22x &= 5 \\ 15x^2 + 22x - 5 &= 0 \\ 15x^2 + 25x - 3x - 5 &= 0 \\ 5x(3x+5) - 1(3x+5) &= 0 \\ (3x+5)(5x-1) &= 0 \end{aligned}$$
$$3x + 5 = 0 \quad \text{or} \quad 5x - 1 = 0$$
$$3x = -5 \qquad\qquad 5x = 1$$
$$x = -\tfrac{5}{3} \qquad\qquad x = \tfrac{1}{5}$$

41.
$$\begin{aligned} x^3 + x^2 &= 0 \\ x^2(x+1) &= 0 \\ x \bullet x(x+1) &= 0 \end{aligned}$$
$$x = 0 \quad \text{or} \quad x = 0 \quad \text{or} \quad x + 1 = 0$$
$$x = -1$$

43.

$$\begin{aligned} y^3 - 49y &= 0 \\ y(y^2 - 49) &= 0 \\ y(y+7)(y-7) &= 0 \end{aligned}$$

$y = 0$ or $y + 7 = 0$, $y = -7$ or $y - 7 = 0$, $y = 7$

45.

$$\begin{aligned} x^3 - 4x^2 - 21x &= 0 \\ x(x^2 - 4x - 21) &= 0 \\ x(x-7)(x+3) &= 0 \end{aligned}$$

$x = 0$ or $x - 7 = 0$, $x = 7$ or $x + 3 = 0$, $x = -3$

47.

$$\begin{aligned} z^4 - 13z^2 + 36 &= 0 \\ (z^2 - 4)(z^2 - 9) &= 0 \\ (z+2)(z-2)(z+3)(z-3) &= 0 \end{aligned}$$

$z + 2 = 0$, $z = -2$ or $z - 2 = 0$, $z = 2$ or $z + 3 = 0$, $z = -3$ or $z - 3 = 0$, $z = 3$

49.

$$\begin{aligned} 3a(a^2 + 5a) &= -18a \\ 3a^3 + 15a^2 + 18a &= 0 \\ 3a(a^2 + 5a + 6) &= 0 \\ 3a(a+2)(a+3) &= 0 \end{aligned}$$

$3a = 0$, $a = 0$ or $a + 2 = 0$, $a = -2$ or $a + 3 = 0$, $a = -3$

51.

$$\begin{aligned} \frac{x^2(6x+37)}{35} &= x \\ x^2(6x + 37) &= 35x \\ 6x^3 + 37x^2 - 35x &= 0 \\ x(6x^2 + 37x - 35) &= 0 \\ x(6x^2 + 42x - 5x - 35) &= 0 \\ x[6x(x+7) - 5(x+7)] &= 0 \\ x(x+7)(6x-5) &= 0 \end{aligned}$$

$x = 0$ or $x + 7 = 0$, $x = -7$ or $6x - 5 = 0$, $6x = 5$, $x = \frac{5}{6}$

53. Let x represent the first even integer, then $x + 2$ represents the second even integer. The product is $x(x + 2) = 288$.

Solve:

$$\begin{aligned} x(x+2) &= 288 \\ x^2 + 2x &= 288 \\ x^2 + 2x - 288 &= 0 \\ x^2 + 18x - 16x - 288 &= 0 \\ x(x+18) - 16(x+18) &= 0 \\ (x+18)(x-16) &= 0 \end{aligned}$$

$x + 18 = 0$, $x = -18$ or $x - 16 = 0$, $x = 16$

Two values were found for x so there are two pairs of integers: $x = -18$, $x + 2 = -16$ and $x = 16$, $x + 2 = 18$.

55. Follow the directions for your calculator. The answers will be 2.78 and 0.72.

57. Follow the directions for your calculator. The answer will be 1.

APPLICATIONS

59. **Analysis**: Find the length and width of a rectangular griddle that has an area of 160 in.^{2}

Equation: Let w represent the width of the griddle, then $w + 6$ will represent the length.

Width of griddle	times	length of griddle	is	160 in.^{2}
w	$\bullet$	$w + 6$	$=$	160

Solve:

$$\begin{aligned} w(w+6) &= 160 \\ w^2 + 6w - 160 &= 0 \\ (w+16)(w-10) &= 0 \end{aligned}$$

$$w + 16 = 0 \quad \text{or} \quad w - 10 = 0$$
$$w = -16 \qquad\qquad w = 10$$

Conclusion: Possible solutions for the width are -16 and 10. Since a width cannot be negative, discard the -16. Thus, the width is $w = 10$ in. and the length is $w + 6 = 16$ in.

61. **Analysis**: Find the width of a walkway that surrounds a swimming pool. The swimming pool has an area of 1500 ft^2 and the uniform–width walkway has an area of 516 ft^2. The total area is $1500 + 516 = 2016$ ft^2. This is a two–part problem, because the length and width of the pool must be found first. Then the width of the walkway can be found based upon that length and width.

Part 1: Find the length and width of the pool.

Equation: Let x represent the width of the pool, then $2x - 10$ will represent the length.

Width of pool	times	length of pool	is	1500 ft^2
x	$\bullet$	$2x - 10$	$=$	1500

Solve:

$$\begin{aligned} x(2x-10) &= 1500 \\ 2x^2 - 10x - 1500 &= 0 \\ x^2 - 5x - 750 &= 0 \\ (x-30)(x+25) &= 0 \end{aligned}$$

Divide both sides by 2.

$$x - 30 = 0 \quad \text{or} \quad x + 25 = 0$$
$$x = 30 \qquad\qquad x = -25$$

Discard the negative answer. The width of the pool is 30 ft and the length is $2x - 10 = 50$ ft.

Part 2: Find the width of the walkway.

Equation: Let w represent the width of the walkway, then $2w + 30$ will represent the width of the total area and $2w + 50$ will represent the length.

Width of total area	times	length of total area	is	2016 ft^2
$2w + 30$	$\bullet$	$2w + 50$	$=$	2016

Solve:

$$\begin{aligned} (2w+30)(2w+50) &= 2016 \\ 4w^2 + 160w + 1500 &= 2016 \\ 4w^2 + 160w - 516 &= 0 \\ w^2 + 40w - 129 &= 0 \\ (w+43)(w-3) &= 0 \end{aligned}$$

$$w + 43 = 0 \quad \text{or} \quad w - 3 = 0 \quad \text{or}$$
$$w = -43 \qquad\qquad w = 3$$

Conclusion: Possible solutions for the width are -43 and 3. Since a width cannot be negative, discard the -43. Thus, the width of the walkway is $w = 3$ ft.

63. **Analysis**: Find the length and width of a rectangular room. The area of the larger part of the room is 560 ft^2.

Equation: Let x represent the width of the room, then $2x$ will represent the length. The length of the larger part of the room is $2x - 12$.

Width of large part	times	length of large part	is	560 ft^2
x	•	$2x - 12$	=	560

Solve:

$$\begin{aligned} x(2x-12) &= 560 \\ 2x^2 - 12x - 560 &= 0 \\ x^2 - 6x - 280 &= 0 \\ (x+14)(x-20) &= 0 \end{aligned}$$

$$x + 14 = 0 \quad \text{or} \quad x - 20 = 0$$
$$x = -14 \qquad\qquad x = 20$$

Conclusion: Possible solutions for the width are -14 and 20. Since a width cannot be negative, discard the -14. Thus, the width is $x = 20$ ft. and the length is $2x = 40$ ft.

65. **Analysis**: Find the time that it will take a cannonball to be at a height of 3344 ft when it is fired vertically at a velocity of $v = 480$ ft/sec.

Equation: Substitute $h = 3344$ and $v = 480$ into the formula given in Example 7 in the text.

$$\begin{aligned} h &= vt - 16t^2 \\ 3344 &= (480)t - 16t^2 \end{aligned}$$

Solve:

$$\begin{aligned} 3344 &= 480t - 16t^2 \\ 16t^2 - 480t + 3344 &= 0 \\ t^2 - 30t + 209 &= 0 \quad \text{Divide both sides by 16.} \\ (t-19)(t-11) &= 0 \end{aligned}$$

$$t - 19 = 0 \quad \text{or} \quad t - 11 = 0$$
$$t = 19 \qquad\qquad t = 11$$

Conclusion: The times when the cannonball will be 3344 feet above the ground are 19 seconds and 11 seconds.

67. Substitute 148 for h and solve for t in the formula $h = -16t^2 + 212$.

Solve:

$$\begin{aligned} 148 &= -16t^2 + 212 \\ 16t^2 - 64 &= 0 \\ t^2 - 4 &= 0 \\ (t+2)(t-2) &= 0 \end{aligned}$$

$$t + 2 = 0 \quad \text{or} \quad t - 2 = 0$$
$$t = -2 \qquad\qquad t = 2$$

Conclusion:

Discard the negative answer. It would take 2 seconds to reach the point where the bungee cord starts to stretch.

69. Substitute 3 for m and 54 for E and then solve for v in the formula $E = \frac{1}{2}mv^2$.

Solve:

$$\begin{aligned} 54 &= \tfrac{1}{2}(3)v^2 \\ 108 &= 3v^2 \\ 3v^2 - 108 &= 0 \\ v^2 - 36 &= 0 \\ (v+6)(v-6) &= 0 \end{aligned}$$

$$v + 6 = 0 \quad \text{or} \quad v - 6 = 0$$
$$t = -6 \qquad\qquad v = 6$$

Discard the negative answer. The velocity of the club at impact would be 6 m/sec.

71. **Analysis**: Find the number of guitars that must be sold so that the cost equals the revenue.

Equation: Set the cost function equal to the revenue function and solve for x the number of guitars that must be sold so that the cost equals the revenue.

$C(x)$	equals	$R(x)$
$\frac{1}{8}x^2 - x + 6$	$=$	$\frac{1}{4}x^2$

Solve:

$$\begin{aligned} \tfrac{1}{8}x^2 - x + 6 &= \tfrac{1}{4}x^2 \\ x^2 - 8x + 48 &= 2x^2 \quad \text{Multiply both sides by the LCD 8.} \\ -x^2 - 8x + 48 &= 0 \\ x^2 + 8x - 48 &= 0 \\ (x+12)(x-4) &= 0 \end{aligned}$$

$$x + 12 = 0 \quad \text{or} \quad x - 4 = 0$$
$$x = -12 \qquad\qquad x = 4$$

Conclusion: Discard the negative solution. The number of guitars to be sold would be 4.

WRITING

73. The zero–factor property states that if the product of two quantities is zero, then at least one of the quantities must equal 0.

75. Both sides of the equation were divided by x which caused the solution of $x = 0$ to be lost.

77. The two graphs are identical because $2x^3 - 8x$ factors as $2x(x+2)(x-2)$.

REVIEW

79. First, both dimensions of the aluminum foil on the roll must be converted to feet.

$$8\frac{1}{3}\text{ yd}\left(\frac{3\text{ ft}}{1\text{ yd}}\right) = \frac{25}{3}(3) = 25\text{ ft} \qquad 12\text{ in.}\left(\frac{1\text{ ft}}{12\text{ in.}}\right) = 1\text{ ft}$$

Area = (25ft)(1 ft) = 25 ft^2

Chapter 5 Key Concept: Polynomials

Operations with Polynomials

1. $(-2x^2 - 5x - 7) + (-3x^2 + 7x + 1) = -2x^2 - 3x^2 - 5x + 7x - 7 + 1 = -5x^2 + 2x - 6$
2. $(6s^3 + 3s - 2) - (2s^3 + 3s^2 + 5) = 6s^3 + 3s - 2 - 2s^3 - 3s^2 - 5) = 4s^3 - 3s^2 + 3s - 7$
3. $(3m - 4)(m + 3) = 3m^2 + 9m - 4m - 12 = 3m^2 + 5m - 12$
4. $3r^2st(r^2 - 2s + 3t^2) = 3r^4st - 6r^2s^2t + 9r^2st^3$
5. $(a - 2d)^2 = a^2 + 2(a)(-2d) + (-2d)^2 = a^2 - 4ad + 4d^2$
6. $(x - 3y)(x^2 + 3xy + 9y^2) = x(x^2 + 3xy + 9y^2) - 3y(x^2 + 3xy + 9y^2)$
$= x^3 + 3x^2y + 9xy^2 - 3x^2y - 9xy^2 - 27y^3 = x^3 - 27y^3$
7. $(3b + 1)(2b^2 + 3b + 2) = 3b(2b^2 + 3b + 2) + 1(2b^2 + 3b + 2)$
$= 6b^3 + 9b^2 + 6b + 2b^2 + 3b + 2 = 6b^3 + 11b^2 + 9b + 2$
8. $(2y + 3)(y - 1) - (y + 2)(3y - 1) = (2y^2 - 2y + 3y - 3) - (3y^2 - y + 6y - 2)$
$= 2y^2 + y - 3 - 3y^2 - 5y + 2 = -y^2 - 4y - 1$

Polynomial Functions

9. $d(t) = -16t^2 + 576,\ d(6) = -16(6)^2 + 576 = -576 + 576 = 0$
This means that after 6 seconds, the squeegee will strike the ground.

10. $V(x) = 2x(4x-1)(2x+2) = 2x(8x^2+8x-2x-2) = 2x(8x^2+6x-2) = 16x^3+12x^2-4x$
$V(3) = 16(3)^3 + 12(3)^2 - 4(3) = 432 + 108 - 12 = 528$

Solving Equations by Factoring

11.
$$\begin{aligned} x^2 - 81 &= 0 \\ (x+9)(x-9) &= 0 \end{aligned}$$
$x + 9 = 0$ or $x - 9 = 0$
$x = -9$ $\quad x = 9$

12.
$$\begin{aligned} 5x^2 - 25x &= 0 \\ 5x(x-5) &= 0 \end{aligned}$$
$5x = 0$ or $x - 5 = 0$
$x = 0$ $\quad x = 5$

13.
$$\begin{aligned} z^2 + 8z + 15 &= 0 \\ (z+3)(z+5) &= 0 \end{aligned}$$
$z + 3 = 0$ or $z + 5 = 0$
$z = -3$ $\quad z = -5$

14.
$$\begin{aligned} 2r^2 + 5r + 3 &= 0 \\ 2r^2 + 2r + 3r + 3 &= 0 \\ 2r(r+1) + 3(r+1) &= 0 \\ (r+1)(2r+3) &= 0 \end{aligned}$$
$r + 1 = 0$ or $2r + 3 = 0$
$r = -1$ $\quad 2r = -3$
$r = -\frac{3}{2}$

15.
$$\begin{aligned} \frac{3t^2}{2} + t &= \frac{1}{2} \\ 3t^2 + 2t &= 1 \\ 3t^2 + 2t - 1 &= 0 \\ (3t-1)(t+1) &= 0 \end{aligned}$$
$3t - 1 = 0$ or $t + 1 = 0$
$3t = 1$ $\quad t = -1$
$t = \frac{1}{3}$

16.
$$\begin{aligned} m^3 &= 9m - 8m^2 \\ m^3 + 8m^2 - 9m &= 0 \\ m(m^2 + 8m - 9) &= 0 \\ m(m+9)(m-1) &= 0 \end{aligned}$$
$m = 0$ or $m + 9 = 0$ or $m - 1 = 0$
$m = -9$ $\quad m = 1$

Chapter 5 — Chapter Review

Section 5.1

1. $3^6 = 3 \bullet 3 \bullet 3 \bullet 3 \bullet 3 \bullet 3 = 729$
2. $-2^5 = -2 \bullet 2 \bullet 2 \bullet 2 \bullet 2 = -32$
3. $(-4)^3 = (-4)(-4)(-4) = -64$
4. $15^1 = 15$
5. $x^4 \bullet x^2 = x^{4+2} = x^6$
6. $a^3b^5a^2b = a^{3+2}b^{5+1} = a^5b^6$
7. $(m^6)^3 = m^{6(3)} = m^{18}$
8.. $(-t^2)^2(t^3)^3 = t^4t^9 = t^{13}$
9. $(3x^2y^3)^2 = 3^2(x^2)^2(y^3)^2 = 9x^4y^6$
10. $\left(\frac{x^4}{b}\right)^4 = \frac{(x^4)^4}{b^4} = \frac{x^{16}}{b^4}$
11. $-3x^0 = -3(1) = -3$
12. $(x^2)^{-5} = x^{2(-5)} = x^{-10} = \frac{1}{x^{10}}$
13. $-5^{-4} = -\frac{1}{5^4} = -\frac{1}{625}$
14. $\frac{70}{x^{-4}} = 70x^4$
15. $(3x^{-3})^{-2} = 3^{-2}x^{-3(-2)} = \frac{1}{3^2}x^6 = \frac{x^6}{9}$
16. $2x^{-4}x^3 = 2x^{-4+3} = 2x^{-1} = \frac{2}{x}$

17. $-\left(\dfrac{c^{-3}}{c^{-5}}\right)^5 = -(c^{-3-(-5)})^5 = -(c^2)^5 = -c^{10}$

18. $\left(\dfrac{4}{5}\right)^{-2} = \left(\dfrac{5}{4}\right)^2 = \dfrac{25}{16}$

19. $\dfrac{y^{-3}}{y^4 y} = y^{-3-4-1} = y^{-8} = \dfrac{1}{y^8}$

20. $\left(\dfrac{-2a^4b}{a^{-3}b^2}\right)^{-3} = \left(-2a^{4-(-3)}b^{1-2}\right)^{-3} = \left(-2a^7b^{-1}\right)^{-3} = (-2)^{-3}a^{7(-3)}b^{-1(-3)} = \dfrac{1}{(-2)^3}a^{-21}b^3 = -\dfrac{b^3}{8a^{21}}$

Section 5.2

21. $19{,}300{,}000{,}000 = 1.93 \times 10^{10}$

22. $0.00000002735 = 2.735 \times 10^{-8}$

23. a. $7.277 \times 10^7 = 72{,}770{,}000$

24. $8.3 \times 10^{-9} = 0.0000000083$

25. $\dfrac{228{,}000{,}000}{300{,}000} = \dfrac{2.28\times10^8}{3\times10^5} = \dfrac{2.28}{3}\times10^{8-5} = 0.76\times10^3 = \left(7.6\times10^{-1}\right)\times10^3 = 7.6\times10^2$ seconds

26. $(1.67248 \times 10^{-24})(1 \times 10^6) = 1.67248 \times 10^{-24+6} = 1.67248 \times 10^{-18}$ g

27. $\dfrac{(616{,}000{,}000)(0.000009)}{0.00066} = \dfrac{(6.16\times10^8)(9\times10^{-6})}{6.6\times10^{-4}} = \dfrac{(6.16)(9)}{6.6}\times10^{8-6-(-4)} = 8.4\times10^6$

Section 5.3

28. No, there is a variable in the denominator.

29. Yes.

30. Yes.

31. No, The exponents are not whole numbers.

32. $x^2 - 8$ is a binomial with a degree of 2.

33. $-15a^3b$ is a monomial with a degree of 4

34. $x^4 + x^3 - x^2 + x - 4$ is none of these with a degree of 4.

35. $9x^2y + 13x^3y^2 + 8x^4y^4$ is a trinomial with a degree of 8.

36. $V(r) = 4.19r^3 + 25.12r^2$; $V(2) = 4.19(2)^3 + 25.12(2)^2 = 33.52 + 100.48 = 134$ in.^{3}

37.

x	$f(x) = x^2 - 2x$	$f(x)$
-1	$f(-1) = (-1)^2 - 2(-1)$	**3**
0	$f(0) = (0)^2 - 2(0)$	**0**
1	$f(1) = (1)^2 - 2(1)$	**−1**
2	$f(2) = (2)^2 - 2(2)$	**0**
3	$f(3) = (3)^2 - 2(3)$	**3**

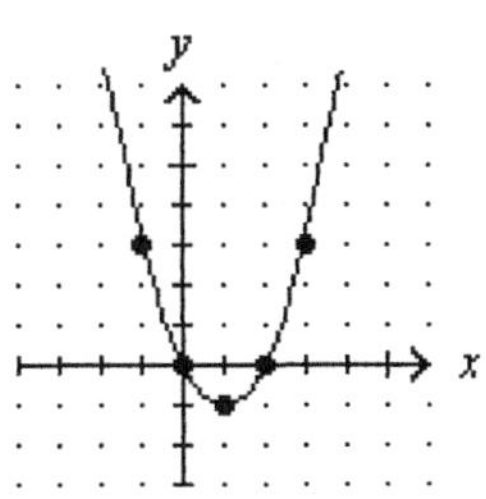

38.

x	$f(x) = x^3 - 3x^2 + 4$	$f(x)$
-2	$f(-2) = (-2)^3 - 3(-2)^2 + 4$	**−16**
-1	$f(-1) = (-1)^3 - 3(-1)^2 + 4$	**0**
0	$f(0) = (0)^3 - 3(0)^2 + 4$	**4**
1	$f(1) = (1)^3 - 3(1)^2 + 4$	**2**
2	$f(2) = (2)^3 - 3(2)^2 + 4$	**0**
3	$f(3) = (3)^3 - 3(3)^2 + 4$	**4**
4	$f(4) = (4)^3 - 3(4)^2 + 4$	**20**

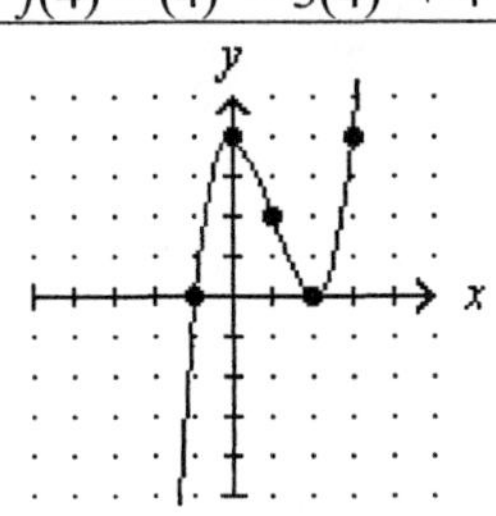

39. $(3x^2 + 4x + 9) + (2x^2 - 2x + 7) = 3x^2 + 4x + 9 + 2x^2 - 2x + 7 = 5x^2 + 2x + 16$

40. $(2x^2y^3 - 5x^2y + 9y) + (x^2y^3 - 3x^2y - y) = 2x^2y^3 - 5x^2y + 9y + x^2y^3 - 3x^2y - y = 3x^2y^3 - 8x^2y + 8y$

41. The opposite of $x^2 - 3x$ is $-x^2 + 3x$

42. The opposite of $-2c^3d - 3c^2d + 1$ is $2c^3d + 3c^2d - 1$

43. $(4x^3 + 4x^2 + 7) - (-2x^3 - x - 2) = 4x^3 + 4x^2 + 7 + 2x^3 + x + 2 = 6x^3 + 4x^2 + x + 9$

44. $\begin{array}{rl} & -10k^4 - 4k^3 + 5k^2 - k + 1 \\ - & \underline{-16k^4 + 2k^3 - 4k^2 - k + 3} \end{array} \Rightarrow \begin{array}{r} -10k^4 - 4k^3 + 5k^2 - k + 1 \\ \underline{16k^4 - 2k^3 + 4k^2 + k - 3} \\ 6k^4 - 6k^3 + 9k^2 \qquad - 2 \end{array}$

Section 5.4

45. $(8a^2)\left(-\frac{1}{2}a\right) = 8\left(-\frac{1}{2}\right) \bullet a^2 \bullet a = -4a^{2+1} = -4a^3$

46. $(-3xy^2z)(-2xz^3)(xz) = (-3)(-2) \bullet x \bullet x \bullet x \bullet y^2 \bullet z \bullet z^3 \bullet z = 6x^3y^2z^5$

47. $2xy^2(x^3y - 4xy^5) = 2xy^2(x^3y) - 2xy^2(4xy^5) = 2x^4y^3 - 8x^2y^7$

48. $-a^2b(-a^2 - 2ab + b^2) = -a^2b(-a^2) - a^2b(-2ab) - a^2b(b^2) = a^4b + 2a^3b^2 - a^2b^3$

49. $(8x - 5)(2x + 3) = (8x)(2x) + (8x)(3) + (-5)(2x) + (-5)(3) = 16x^2 + 24x - 10x - 15 = 16x^2 + 14x - 15$

50. $(3x^2 + 2)(2x - 4) = (3x^2)(2x) + (3x^2)(-4) + 2(2x) + 2(-4) = 6x^3 - 12x^2 + 4x - 8$

51. $(5a - 6)^2 = (5a)^2 + 2(5a)(-6) + (-6)^2 = 25a^2 - 60a + 36$

52. $(0.7c^2 - d)(0.7c^2 + d) = (0.7c)^2 - d^2 = 0.49c^4 - d^2$

53. $\left(\frac{1}{3}a^3 - 1\right)^2 = \left(\frac{1}{3}a^3\right)^2 + 2\left(\frac{1}{3}a^3\right)(-1) + (-1)^2 = \frac{1}{9}a^6 - \frac{2}{3}a^3 + 1$

54. $(ab + 1)(ab + 3) = ab(ab) + 3(ab) + ab + 3 = a^2b^2 + 4ab + 3$

55. $(5x^2 - 4x)(3x^2 - 2x + 10) = 5x^2(3x^2 - 2x + 10) - 4x(3x^2 - 2x + 10)$
$= 15x^4 - 10x^3 + 50x^2 - 12x^3 + 8x^2 - 40x = 15x^4 - 22x^3 + 58x^2 - 40x$

56. $(r + s)(r - s)(r - 3s) = (r^2 - s^2)(r - 3s) = r^2(r - 3s) - s^2(r - 3s) = r^3 - 3r^2s - rs^2 + 3s^3$

57. Divide the razor blade into three sections by drawing lines vertically from the corner of the notches at the top to the corner of the notches at the bottom. The two end sections will each have an area of $x(5x + 1)$. The center section will have an area of $(16x + 1)(5x + 1 + 2x + 2x) = (16x + 1)(9x + 1)$. Add the three sections together and simplify to find the total area.

$$\begin{aligned} A &= 2[x(5x + 1)] + (16x + 1)(9x + 1) \\ &= 2(5x^2 + x) + 144x^2 + 16x + 9x + 1 \\ &= 10x^2 + 2x + 144x^2 + 25x + 1 \\ &= 154x^2 + 27x + 1 \end{aligned}$$

Section 5.5

59. $350 = 10 \bullet 35 = 2 \bullet 5 \bullet 5 \bullet 7 = 2 \bullet 5^2 \bullet 7$

59. $42 = 2 \bullet 3 \bullet 7$
$36 = 2 \bullet 2 \bullet 3 \bullet 3$
$54 = 2 \bullet 3 \bullet 3 \bullet 3$
$\text{GCF} = 2 \bullet 3 = 6$

60. $6x^2y^5 = 2 \bullet 3 \bullet x \bullet x \bullet y \bullet y \bullet y \bullet y \bullet y$
$15xy^3 = 3 \bullet 5 \bullet x \bullet y \bullet y \bullet y$
$3xy^3 = 3 \bullet x \bullet y \bullet y \bullet y$
$\text{GCF} = 3 \bullet x \bullet y \bullet y \bullet y = 3xy^3$

61. $4x + 8 = 4 \bullet x + 4 \bullet 2 = 4(x + 2)$

62. $3x^3 - 6x^2 + 9x = 3x \bullet x^2 - 3x \bullet 2x + 3x \bullet 3 = 3x(x^2 - 2x + 3)$

63. $5x^2y^3 - 11mn^2$ is prime. There are no common factors.

64. $7a^4b^2 + 49a^3b = 7a^3b \bullet ab + 7a^3b \bullet 7 = 7a^3b(ab + 7)$

65. $5x^2(x + y) - 15x^3(x + y) = (x + y)(5x^2 - 15x^3) = 5x^2(x + y)(1 - 3x)$ Common factor after the first factoring.

66. $27x^3y^3z^3 + 81x^4y^5z^2 - 90x^2y^3z^7 = 9x^2y^3z^2(3xz + 9x^2y^2 - 10z^5)$

67. $-7b + 14 = -7(b) - 7(-2) = -7(b - 2)$

68. $-49a^3b^2(a - b)^4 + 63a^2b^4(a - b)^3 = -7a^2b^2(a - b)^3[7a(a - b) - 9b^2] = -7a^2b^2(a - b)^3(7a^2 - 7ab - 9b^2)$

69. $xy + 2y + 4x + 8 = [xy + 2y] + [4x + 8] = y(x + 2) + 4(x + 2) = (x + 2)(y + 4)$

70. $r^2y - ar - ry + a = [r^2y - ar] - [ry - a] = r(ry - a) - 1(ry - a) = (ry - a)(r - 1)$

71.
$$\begin{aligned} m_1m_2 &= mm_2 + mm_1 \quad \text{Solve for } m_1. \\ m_1m_2 - mm_1 &= mm_2 \\ m_1(m_2 - m) &= mm_2 \\ m_1 &= \frac{mm_2}{m_2 - m} \end{aligned}$$

Section 5.6

72. $z^2 - 16 = z^2 - 4^2 = (z + 4)(z - 4)$

73. $y^2 - 121 = y^2 - 11^2 = (y + 11)(y - 11)$

74. $x^2y^4 - 64z^6 = (xy^2)^2 - (8z^3)^2 = (xy^2 + 8z^3)(xy^2 - 8z^3)$

75. $a^2b^2 + c^2$ is prime. Sums of squares are not factorable.

76. $c^2 - (a + b)^2 = [c + (a + b)][c - (a + b)] = (c + a + b)(c - a - b)$

77. $3x^6 - 300x^2 = 3x^2(x^4 - 100) = 3x^2[(x^2)^2 - 10^2] = 3x^2(x^2 + 10)(x^2 - 10)$

78. $t^3 + 64 = t^3 + 4^3 = (t + 4)(t^2 - 4t + 16)$

79. $2x^3y - 54yz^3 = 2y(x^3 - 27z^3) = 2y[x^3 - (3z)^3] = 2y(x - 3z)(x^2 + 3xz + 9z^2)$

Section 5.7

80. $x^2 + 10x + 25 = x^2 + 2(x)(5) + 5^2 = (x + 5)(x + 5) = (x + 5)^2$

81. $a^2 - 14a + 49 = a^2 - 2(a)(7) + 7^2 = (a - 7)(a - 7) = (a - 7)^2$

82. $y^2 + 21y + 20 = y^2 + 20y + y + 20 = y(y + 20) + 1(y + 20) = (y + 20)(y + 1)$

83. $z^2 - 11z + 30 = z^2 - 5z - 6z + 30 = z(z - 5) - 6(z - 5) = (z - 5)(z - 6)$

84. $-x^2 - 3x + 28 = -1(x^2 + 3x - 28) = -1(x^2 + 7x - 4x - 28) = -1[x(x + 7) - 4(x + 7) = -1(x + 7)(x - 4)$

85. $a^2 - 24b^2 - 5ab = a^2 - 5ab - 24b^2 = a^2 - 8ab + 3ab - 24b^2 = a(a - 8b) + 3b(a - 8b) = (a - 8b)(a + 3b)$

86. $4a^2 - 5a + 1 = 4a^2 - 4a - a + 1 = 4a(a - 1) - 1(a - 1) = (a - 1)(4a - 1)$

87. $3b^2 + 2b + 1$ is prime because there are no factors of $ac = 3$ that add to $b = 2$.

88. $y^3 + y^2 - 2y = y(y^2 + y - 2) = y(y^2 + 2y - y - 2) = y[y(y + 2) - 1(y + 2)] = y(y + 2)(y - 1)$

89. $15x^2 - 57xy - 12y^2 \; 3(5x^2 - 19xy - 4y^2) = 3(5x^2 - 20xy + xy - 4y^2)$
$= 3[5x(x - 4y) + y(x - 4y)] = 3(x - 4y)(5x + y)$

90. $2a^4 + 4a^3 - 6a^2 = 2a^2(a^2 + 2a - 3) = 2a^2(a^2 + 3a - a - 3) = 2a^2[a(a + 3) - 1(a + 3)] = 2a^2(a + 3)(a - 1)$

91. $v^4 - 13v^2 + 42 = v^4 - 6v^2 - 7v^2 + 42 = v^2(v^2 - 6) - 7(v^2 - 6) = (v^2 - 6)(v^2 - 7)$

92. Rewrite $(s+t)^2 - 2(s+t) + 1$ as $z^2 - 2z + 1$ where $z = (s+t)$.

$z^2 - 2z + 1 = (z-1)^2$ Special product formula.

$(s+t)^2 - 2(s+t) + 1 = (s+t-1)^2$ Replace z with $s + t$.

93. $k^2 + 2k + 1 - 9m^2 = (k^2 + 2k + 1) - 9m^2 = (k+1)^2 - (3m)^2 = (k+1+3m)(k+1-3m)$

Section 5.8

94. $x^3 + 5x^2 - 6x = x(x^2 + 5x - 6)$ Common factor, then trinomial.

$= x(x^2 - x + 6x - 6) = x[x(x-1) + 6(x-1)] = x(x-1)(x+6)$

95. $3x^2y - 12xy - 63y = 3y(x^2 - 4x - 21)$ Common factor, then trinomial.

$= 3y(x^2 - 7x + 3x - 21) = 3y[x(x-7) + 3(x-7)] = 3y(x-7)(x+3)$

96. $z^2 - 4 + zx - 2x = (z+2)(z-2) + x(z-2)$ Four terms, use grouping with difference of squares.

$= (z-2)[(z+2) + x] = (z-2)(z+x+2)$

97. $x^2 + 2x + 1 - p^2 = [x^2 + 2(x)(1) + 1^2] - p^2 = (x+1)^2 - p^2$ Group the first three terms, for perfect square, then difference of squares

$= (x+1+p)(x+1-p)$

98. $x^2 + 4x + 4 - 4p^4 = [x^2 + 2(x)(2) + 2^2] - 4p^4$ Group the first three terms, for perfect square, then difference of squares

$= (x+2)^2 - (2p^2)^2 = (x+2+2p^2)(x+2-2p^2)$

99. $y^2 + 3y + 2 + 2x + xy = (y+1)(y+2) + x(2+y)$ Group first three terms for trinomial, then factor by grouping.

$= (y+2)[(y+1) + x] = (y+2)(y+1+x)$

100. $4a^3b^3 + 256 = 4(a^3b^3 + 64) =$ Common factor, then sum of two cubes.

$= 4[(ab)^3 + 4^3) = 4(ab+4)(a^2b^2 - 4ab + 16)$

101. $36z^4 - 36 = 36(z^4 - 1) = 36[(z^2)^2 - 1^2]$ Common factor, then difference of two squares, twice.

$= 36(z^2+1)(z^2-1) = 36(z^2+1)(z+1)(z-1)$

102. $V = \frac{\pi}{2}r_1^2h - \frac{\pi}{2}r_2^2h = \frac{\pi}{2}h(r_1^2 - r_2^2) = \frac{\pi}{2}h(r_1+r_2)(r_1-r_2)$

Section 5.9

103.

$$4x^2 - 3x = 0$$
$$x(4x-3) = 0$$
$x = 0$ or $4x - 3 = 0$, $4x = 3$, $x = \frac{3}{4}$

104.

$$x^2 - 36 = 0$$
$$(x+6)(x-6) = 0$$
$x + 6 = 0$, $x = -6$ or $x - 6 = 0$, $x = 6$

105.

$$12x^2 = 5 - 4x$$
$$12x^2 + 4x - 5 = 0$$
$$12x^2 - 6x + 10x - 5 = 0$$
$$6x(2x-1) + 5(2x-1) = 0$$
$$(2x-1)(6x+5) = 0$$
$2x - 1 = 0$, $2x = 1$, $x = \frac{1}{2}$ or $6x + 5 = 0$, $6x = -5$, $x = -\frac{5}{6}$

106.

$$7y^2 - 37y + 10 = 0$$
$$7y^2 - 35y - 2y + 10 = 0$$
$$7y(y-5) - 2(y-5) = 0$$
$$(y-5)(7y-2) = 0$$
$y - 5 = 0$, $y = 5$ or $7y - 2 = 0$, $7y = 2$, $y = \frac{2}{7}$

107.

$$\begin{aligned}
t^2(15t-2) &= 8t\\
15t^3-2t^2-8t &= 0\\
t(15t^2-2t-8) &= 0\\
t(15t^2-12t+10t-8) &= 0\\
t(3t(5t-4)+2(5t-4) &= 0\\
t(5t-4)(3t+2) &= 0
\end{aligned}$$

$t = 0$ or $5t-4 = 0$, $5t = 4$, $t = \frac{4}{5}$ or $3t+2 = 0$, $3t = -2$, $t = -\frac{2}{3}$

108.

$$\begin{aligned}
u^3 &= \frac{u}{3}(19u+14)\\
3u^3 &= u(19u+14)\\
3u^3-19u^2-14u &= 0\\
u(3u^2-19u-14) &= 0\\
u(3u^2-21u+2u-14) &= 0\\
u[3u(u-7)+2(u-7) &= 0\\
u(u-7)(3u+2) &= 0
\end{aligned}$$

$u = 0$ or $u-7 = 0$, $u = 7$ or $3u+2 = 0$, $3u = -2$, $u = -\frac{2}{3}$

109. **Analysis**: Find the dimensions of the rectangular base of a pyramid when the volume is 1020 m^3. One edge of the base is 3 meters longer than the other and the height is 9 meters.

Equation: Let x represent the length of one edge and $x + 3$ the other edge of the rectangular base. The area of the base is $x(x + 3)$. Substitute $V = 1020$, $B = x(x + 3)$, and $h = 9$ into the given formula for the volume.

$$V = \frac{Bh}{3};\quad 1020 = \frac{x(x+3)(9)}{3} = x(x+3)(3) = 3x(x+3) = 3x^2+9x$$

Solve:

$$\begin{aligned}
3x^2+9x &= 1020\\
x^2+3x &= 340\\
x^2+3x-340 &= 0\\
x^2+20x-17x-340 &= 0\\
x(x+20)-17(x+20) &= 0\\
(x+20)(x-17) &= 0
\end{aligned}$$

$x+20 = 0$, $x = -20$ or $x-17 = 0$, $x = 17$

Conclusion: Possible solutions for the height are –20 and 17. Since length cannot be negative, discard the –20. Thus, the one edge of the base is $x = 17$ m and the other edge is $x + 3 = 20$ m.

110. Estimate the solutions as $x = 1$ and $x = -\frac{1}{2}$. Substitute these solutions into $y = 2x^2 - x - 1$ to check that $y = 0$.

$$\begin{aligned}
2(1)^2-(1)-1 &\stackrel{?}{=} 0\\
2-1-1 &= 0
\end{aligned}$$

$$\begin{aligned}
2\left(-\tfrac{1}{2}\right)^2-\left(-\tfrac{1}{2}\right)-1 &\stackrel{?}{=} 0\\
2\left(\tfrac{1}{4}\right)+\tfrac{1}{2}-1 &\stackrel{?}{=} 0\\
\tfrac{1}{2}+\tfrac{1}{2}-1 &= 0
\end{aligned}$$

Both solutions check.

Chapter 5 Chapter Test

1 $x^3x^5x = x^{3+5+1} = x^9$

2. $(-2x^2y^3)^3 = (-2)^3x^{2(3)}y^{3(3)} = -8x^6y^9$

3. $m^3(m^{-4})^2 = m^3m^{-4(2)} = m^3m^{-8} = m^{-5} = \dfrac{1}{m^5}$

4. $\left(\dfrac{3m^2n^3}{m^4n^{-2}}\right)^{-2} = \left(3m^{2-4}n^{3-(-2)}\right)^{-2} = \left(3m^{-2}n^5\right)^{-2} = 3^{-2}m^{-2(-2)}n^{5(-2)} = \dfrac{1}{3^2}m^4n^{-10} = \dfrac{m^4}{9n^{10}}$

5. $4{,}706{,}000{,}000{,}000 = 4.706 \times 10^{12}$

6. $2.45 \times 10^{-4} = 0.000245$

7. $\dfrac{3.19\times10^{15}}{2.2\times10^{-4}} = \dfrac{3.19}{2.2}\times10^{15-(-4)} = 1.45\times10^{19}$

8. $(1.86 \times 10^5)(60) = 111.6 \times 10^5 = 1.116 \times 10^7$ mi/min

9. The degree is 5.

10. The degree is 13.

11. $h(t) = -16t^2 + 80t + 10$

$$\begin{aligned} h(2.5) &= -16(2.5)^2 + 80(2.5) + 10 \\ &= -16(6.25) + 200 + 10 \\ &= -100 + 210 \\ &= 110 \text{ ft} \end{aligned}$$

12. Use $A = lw$

$$\begin{aligned} A &= (3x-2)(x+2) \\ &= 3x^2 + 6x - 2x - 4 \\ &= 3x^2 + 4x - 4 \end{aligned}$$

13.

x	$f(x) = x^2 + 2x$	$f(x)$
−3	$f(-3) = (-3)^2 + 2(-3)$	**3**
−2	$f(-2) = (-2)^2 + 2(-2)$	**0**
−1	$f(-1) = (-1)^2 + 2(-1)$	**−1**
0	$f(0) = (0)^2 + 2(0)$	**0**
1	$f(1) = (1)^2 + 2(1)$	**3**

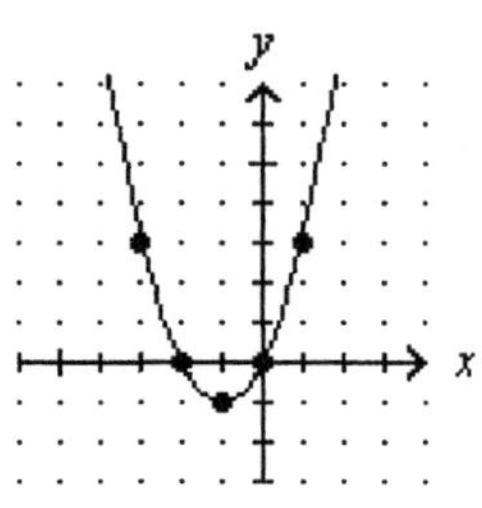

14. a.

x	$f(x) = x^3 + 4x^2 + 4x$	$f(x)$
−3	$f(-3) = (-3)^3 + 4(-3)^2 + 4(-3)$	**−3**
−2	$f(-2) = (-2)^3 + 4(-2)^2 + 4(-2)$	**0**
−1	$f(-1) = (-1)^3 + 4(-1)^2 + 4(-1)$	**−1**
0	$f(0) = (0)^3 + 4(0)^2 + 4(0)$	**0**
1	$f(1) = (1)^3 + 4(1)^2 + 4(1)$	**9**

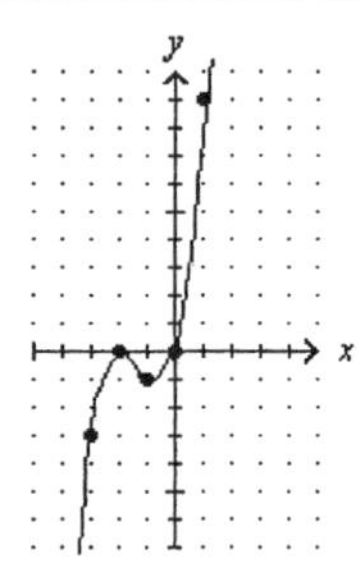

b. -2, 0

15. $(2y^2 + 4y + 3) + (3y^2 - 3y - 4) = 5y^2 + y - 1$

16. $(-3u^2 + 2u - 7) - (u^2 + 7) = -3u^2 + 2u - 7 - u^2 - 7 = -4u^2 + 2u - 14$

17. $(3x^3y^2z)(-2xyz^3) = -6x^{3+1}y^{2+1}z^{1+3} = -6x^4y^3z^4$

18. $-5a^2b(3ab^3 - 2ab^4) = -5a^2b(3ab^3) - 5a^2b(-2ab^4) = -15a^3b^4 + 10a^3b^5$

19. $(z^3 + 0.4)(z^3 - 0.4) = z^6 + 0.4z - 0.4z - 0.16 = z^6 - 0.16$

20. $\left(\dfrac{3}{4}st + 8\right)^2 = \left(\dfrac{3}{4}st\right)^2 + 2\left(\dfrac{3}{4}st\right)(8) + 8^2 = \dfrac{9}{16}s^2t^2 + 12st + 64$

21. $(4t^2 - 9y)^2 = (4t^2)^2 + 2(4t)(-9y) + (-9y)^2 = 16t^4 - 72t^2y + 81y^2$

22. $2s(s - t)(s + t) = 2s(s^2 - t^2) = 2s^3 - 2st^2$

23. $15x^5 + 6x^7 = 3x^5(5 + 2x^2)$

24. $12a^3b^2c - 3a^2b^2c^2 + 6abc^3 = 3abc(4a^2b - abc + 2c^2)$

25. $(u - v)r + (u - v)s = (u - v)(r + s)$

26. $ax - xy + ay - y^2 = x(a - y) + y(a - y) = (a - y)(x + y)$

27. $a^2x^2 - 49 = (ax)^2 - 7^2 = (ax + 7)(ax - 7)$

28. $4y^4 - 64 = 4(y^4 - 16) = 4[(y^2)^2 - 4^2] = 4(y^2 + 4)(y^2 - 4) = 4(y^2 + 4)(y + 2)(y - 2)$

29. $81b^2 + 25$ is prime. Sums of squares are not factorable.

30. $b^3 + 125 = b^3 + 5^3 = (b + 5)(b^2 - 5b + 25)$

31. $3u^3 - 24 = 3(u^3 - 8) = 3(u^3 - 2^3) = 3(u - 2)(u^2 + 2u + 4)$

32. $(b+c)^2 - 5(b+c) - 6 = z^2 - 5z - 6 = (z-6)(z+1) = (b+c-6)(b+c+1)$ where $z = b+c$

33. $6b^2 + bc - 2c^2 = 6b^2 - 3bc + 4bc - 2c^2 = 3b(2b - c) + 2c(2b - 2) = (2b - c)(3b + 2c)$

34. $x^2 + 6x + 9 - y^2 = (x^2 + 6x + 9) - y^2 = (x + 3)^2 - y^2 = (x + 3 + y)(x + 3 - y)$

35.
$$\begin{aligned} 2x(4x + 3) &= 9 \\ 8x^2 + 6x &= 9 \\ 8x^2 + 6x - 9 &= 0 \\ 8x^2 - 6x + 12x - 9 &= 0 \\ 2x(4x - 3) + 3(4x - 3) &= 0 \\ (4x - 3)(2x + 3) &= 0 \end{aligned}$$
$4x - 3 = 0$ or $2x + 3 = 0$

$4x = 3$ $\quad 2x = -3$

$x = \frac{3}{4}$ $\quad x = \frac{3}{2}$

36.
$$\begin{aligned} &x^2 + 4x = 0 \\ &x(x + 4) = 0 \\ &x = 0 \text{ or } x + 4 = 0 \\ &x = 0 \text{ or } x = -4 \end{aligned}$$

37.
$$\begin{aligned} &a^3 - 9a = -8a^2 \\ &a^3 + 8a^2 - 9a = 0 \\ &a(a^2 + 8a - 9) = 0 \\ &a(a + 9)(a - 1) = 0 \\ &a = 0,\ a + 9 = 0, \text{ or } a - 1 = 0 \\ &a = 0, -9, 1 \end{aligned}$$

38.
$$\begin{aligned} v_1v_3 - v_3v &= v_1v \\ v_1v_3 &= v_1v + v_3v \\ v_1v_3 &= v(v_1 + v_3) \\ \frac{v_1v_3}{v_3 + v_1} &= v \end{aligned}$$

39. **Analysis**: Find the dimensions of a slab of concrete that is twice as long as it is wide. The slab area is surrounded by a border that is 1 foot wide and has an area of 70 ft^2.

Equation: Let w represent the width of the slab, then $2w$ will represent the length. The area of the slab would be represented by $2(2w) = 2w^2$ ft^2. The width of the slab plus the border would be $w + 2$ ft and the length would be $2w + 2$ ft. The area of the slab plus the border would be $2w^2 + 70$ ft^2.

Width of slab and border	times	length of slab and border	is	area of slab	plus	border area
$w + 2$	•	$2w + 2$	=	$2w^2$	+	70

Solve:
$$\begin{aligned} (w + 2)(2w + 2) &= 2w^2 + 70 \\ 2w^2 + 6w + 4 &= 2w^2 + 70 \\ 6w - 66 &= 0 \\ 6w &= 66 \\ w &= 11 \end{aligned}$$

Conclusion: The width of the slab will be $w = 11$ ft and the length of the slab will be $2w = 22$ ft.

40. Once a polynomial has been factored, re–multiplying the factors should give the polynomial you started with.
Factoring: $x^2 + 7x + 12 = (x + 4)(x + 3)$
Multiplying: $(x + 4)(x + 3) = x^2 + 3x + 4x + 12 = x^2 + 7x + 12$

1. a. true b. false c. false d. true

2. $-3|(3^2 \cdot 5 - 2^3 \cdot 6)^2| = -3|(9 \cdot 5 - 8 \cdot 6)^2| = -3|(45 - 48)^2| = -3|(-3)^2| = -3|9| = -3(9) = -27$

3.
$$\begin{aligned} \frac{x+2}{5} - 4x &= \frac{8}{5} - \frac{x+9}{2} \\ 2(x+2) - 10(4x) &= 2(8) - 5(x+9) \\ 2x + 4 - 40x &= 16 - 5x - 45 \\ -38x + 4 &= -5x - 29 \\ -33x &= -33 \\ x &= 1 \end{aligned}$$

4.
$$\begin{aligned} F &= \frac{9}{5}C + 32 \\ F - 32 &= \frac{9}{5}C \\ \frac{5}{9}(\text{F} - 32) &= C \qquad \text{Multiply by } \frac{5}{9}. \end{aligned}$$

5. Find the volume of the block using the formula $V = lwh$ and then subtract the volume of the cylinder using the formula $V = \pi r^2 h$ where $l = 12$ cm, $w = 12$ cm, $h = 18$ cm, and $r = 4$ cm.

$$\begin{aligned} V &= l\,w\,h - \pi r^2 h \\ &= (12)(12)(18) - \pi(4)^2(18) \\ &= 2592 - 904.8 \\ &= 1687.22 \text{ cm}^3 \end{aligned}$$

6. **Analysis**: Find the amount of water to add to the bucket to make a 10% solution.

Equation: Let x represent the number of ounces of water added. Then $24 + x$ will represent the number of ounces of the 10% solution.

	percent alcohol •	total quantity =	quantity of alcohol
15% solution	0.15	24	3.6
Water	0.00	x	0
10% solution	0.10	$24 + x$	$0.10(24 + x)$

Quantity of Spic–and–Span in 15% solution	plus	quantity of Spic–and–Span in water	is	quantity of Spic–and–Span in 10% solution.
3.6	+	0	=	$0.10(24 + x)$

Solve:
$$\begin{aligned} 3.6 + 0 &= 0.10(24 + x) \\ 3.6 &= 2.4 + 0.10x \\ 1.2 &= 0.10x \\ 12 &= x \end{aligned}$$

Conclusion: It would be necessary to add 12 ounces of water to the bucket to obtain a 10% solution.

7. Original slope is $m = -\frac{1}{8}$. New slope is the negative reciprocal: $m = 8$. The point is: $(x_1, y_1) = (-2, 5)$

$$\begin{aligned} y - y_1 &= m(x - x_1) \\ y - (5) &= 8[x - (-2)] \\ y - 5 &= 8(x + 2) \\ y - 5 &= 8x + 16 \\ y &= 8x + 21 \end{aligned}$$

8. Calculate the slope through (–5, 4) and (8 –6): $m = \dfrac{-6-4}{8-(-5)} = \dfrac{-10}{13}$

$$\begin{aligned} y - y_1 &= m(x - x_1) \\ y - (4) &= -\frac{10}{13}[x - (-5)] \\ y - 4 &= -\frac{10}{13}(x + 5) \\ y - 4 &= -\frac{10}{13}x - \frac{50}{13} \\ y &= -\frac{10}{13}x + \frac{2}{13} \end{aligned}$$

9. rate of change $= \dfrac{\text{change in percent}}{\text{change in time}} = \dfrac{65.6\% - 74\%}{2000 - 1993} = \dfrac{-8.4}{7} = -1.2\%$ per year

10. $h(x) = x^4 - x^3 - x^2 - x - 1;\ h(1) = (1)^4 - (1)^3 - (1)^2 - (1) - 1 = -3$

11. $f(x) = 0.025x + 95$, where x represents the total number of copies made during the month.

12.

x	$f(x) = x^2 + 2$	$f(x)$
–2	$f(-2) = (-2)^2 + 2$	6
–1	$f(-1) = (-1)^2 + 2$	3
0	$f(0) = (0)^2 + 2$	2
1	$f(1) = (1)^2 + 2$	3
2	$f(2) = (2)^2 + 2$	6

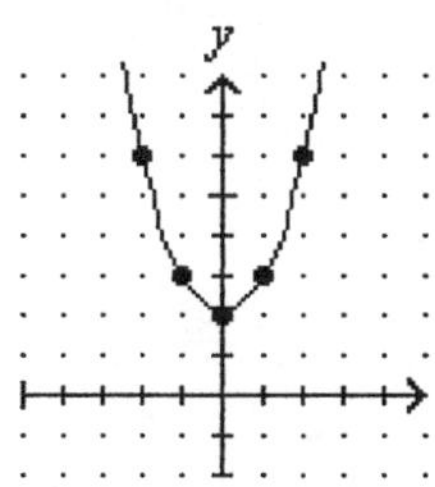

D = the set of real numbers. R = the set of all real numbers greater than or equal to 2.

13. Equation #1

$y = -\dfrac{5}{2}x + \dfrac{1}{2}$

x	y
–1	3
1	–2
3	–7

Equation #2

$2x - \dfrac{3}{2}y = 5$

x	y
–2	–6
1	–2
4	2

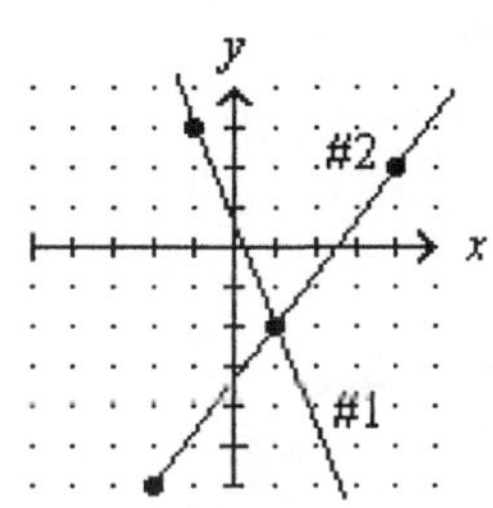

The solution is (1, –2).

14. $\begin{cases} 2(2x+3y) = 5 \\ 8x = 3(1+3y) \end{cases}$ Rewrite in general form $\begin{cases} 4x + 6y = 5 \\ 8x - 9y = 3 \end{cases}$

Multiply Equation 1 by –2 to eliminate x.

$\begin{cases} -8x - 12y = -10 \\ 8x - 9y = 3 \end{cases}$

Add the two equations and solve for y.

$-21y = -7$

$y = \dfrac{1}{3}$

Substitute $\dfrac{1}{3}$ for y in Equation 2.

$$\begin{aligned} 8x &= 3\left[1 + 3\left(\frac{1}{3}\right)\right] = 3(1 + 1) = 6 \\ x &= \frac{6}{8} = \frac{3}{4} \end{aligned}$$

The solution is $\left(\dfrac{3}{4}, \dfrac{1}{3}\right)$.

15. $\begin{cases} 3x+2y-z=-8 \\ 2x-y+7z=10 \\ 2x+2y-3z=-10 \end{cases}$ Eliminate y.

Multiply Equation 2 by 2 and add Equations 1 and 2.

$$\begin{array}{lrcl} 1. & 3x+2y-\quad z & = & -8 \\ 2. & 4x-2y+14z & = & 20 \\ \hline 4. & 7x\qquad +13z & = & 12 \end{array}$$

Multiply Equation 2 by 2 and add Equations 2 and 3.

$$\begin{array}{lrcl} 2. & 4x-2y+14z & = & 20 \\ 3. & 2x+2y-\ 3z & = & -10 \\ \hline 5. & 6x\qquad +11z & = & 10 \end{array}$$

Multiply Equation 4 by 6 and Equation 5 by –7 and add.

$$\begin{array}{lrcl} 4. & 42x+78z & = & 72 \\ 5. & -42x-77z & = & -70 \\ \hline 6. & z & = & 2 \end{array}$$

Substitute $z=2$ into Equation 4 and solve for x.

$$\begin{aligned} 7x+13(2) &= 12 \\ 7x &= -14 \\ x &= -2 \end{aligned}$$

Substitute $x=-2$ and $z=2$ into Equation 1 and solve for y.

$$\begin{aligned} 3(-2)+2y-(2) &= -8 \\ -6+2y-2 &= -8 \\ 2y &= 0 \\ y &= 0 \end{aligned}$$

The solution is (–2, 0, 2).

16. $\begin{vmatrix} 2 & -3 & 4 \\ -1 & 2 & 4 \\ 3 & -3 & 1 \end{vmatrix} = 2\begin{vmatrix} 2 & 4 \\ -3 & 1 \end{vmatrix} - (-1)\begin{vmatrix} -3 & 4 \\ -3 & 1 \end{vmatrix} + 3\begin{vmatrix} -3 & 4 \\ 2 & 4 \end{vmatrix} = 2(2+12)+1(-3+12)+3(-12-8) = 28+9-60 = -23$

17.
$$\begin{aligned} -9(t-3)+2t &\le 8(4-t) \\ -9t+27+2t &\le 32-8t \\ -7t+27 &\le 32-8t \\ t &\le 5 \end{aligned}$$

$(-\infty, 5]$

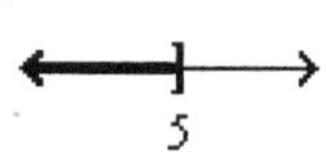

18.
$$\begin{aligned} -6 \le \tfrac{1}{3}h+1 &< 0 \\ -18 \le h+3 &< 0 \\ -21 \le h &< -3 \end{aligned}$$

$[-21, -3)$

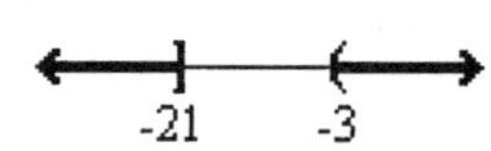

19. Rewrite $|m+5| \ge 7$ as

$$\begin{aligned} m+5 &\le -7 \\ m &\le -12 \end{aligned} \qquad \text{or} \qquad \begin{aligned} m+5 &\ge 7 \\ m &\ge 2 \end{aligned}$$

$(-\infty, -12] \cup [2, \infty)$

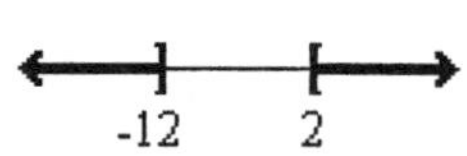

20.
$$\begin{aligned} 4.5x-1 &< -10 \\ 4.5x &< -9 \\ x &< -2 \end{aligned} \qquad \text{or} \qquad \begin{aligned} 6-2x &\ge 12 \\ -2x &\ge 6 \\ x &\le -3 \end{aligned}$$

$(-\infty, -2)$

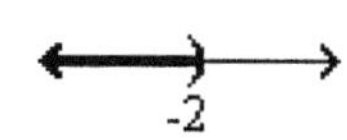

21. iii.

22. Inequality #1: $x-y<4$

Graph: $x-y=4$

x	y
0	–4
4	0

Broken line. Test point (0, 0).

$$\begin{aligned} (0)-(0) &< 4 \\ 0 &< 4 \quad \text{True} \end{aligned}$$

Inequality #2: $y \le 0$

Graph: $y=0$

A horizontal line through the x–axis.

Solid line. No test point needed. Shade the half–plane under the x–axis.

Inequality #3: $x \ge 0$

Graph: $x=0$. A vertical line through the y– axis.

#3 Continued. Solid line. No test point needed. Shade the half–plane to the right of the y–axis.

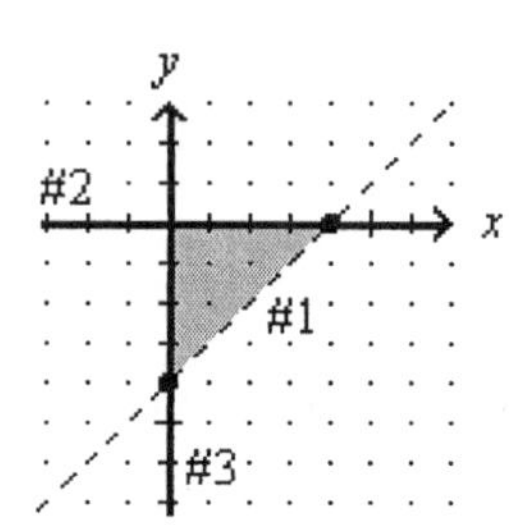

23. $\left(\dfrac{-3a^4b^2}{-9a^5b^{-2}}\right)^{-2} = \left(\dfrac{-3}{-9}a^{4-5}b^{2-(-2)}\right)^{-2} = \left(\dfrac{1}{3}a^{-1}b^4\right)^{-2} = \dfrac{1}{3^{-2}}a^2b^{-8} = \dfrac{3^2a^2}{b^8} = \dfrac{9a^2}{b^8}$

24. $9.0895 \times 10^{-8} = 0.000000090895$

25. $(2a - b)(4a^2 + 2ab + b^2) = 2a(4a^2 + 2ab + b^2) - b(4a^2 + 2ab + b^2)$
$= 8a^3 + 4a^2b + 2ab^2 - 4a^2b - 2ab^2 - b^3 = 8a^3 - b^3$

26. $(3k + 1)^2 + (2k - 4)^2 = 9k^2 + 6k + 1 + 4k^2 - 16k + 16 = 13k^2 - 10k + 17$

27. $x^2 + 4y - xy - 4x = x^2 - xy - 4x + 4y = x(x - y) - 4(x - y) = (x - y)(x - 4)$

28. $6s^4 - 216s^2 = 6s^2(s^2 - 36) = 6s^2(s + 6)(s - 6)$

29. $8x^6 + 125y^3 = (2x^2)^3 + (5y)^3 = (2x^2 + 5y)(4x^4 - 10x^2y + 25y^2)$

30. $-3a^2 + ab + 2b^2 = -(3a^2 - ab - 2b^2)$
$= -(3a^2 - 3ab + 2ab - 2b^2) = -[3a(a - b) + 2b(a - b)] = -(a - b)(3a + 2b)$

31.
$$\begin{aligned} x^2 &= \tfrac{1}{2}(x + 1) \\ 2x^2 &= x + 1 \\ 2x^2 - x - 1 &= 0 \\ (2x + 1)(x - 1) &= 0 \end{aligned}$$
$2x + 1 = 0$ or $x - 1 = 0$
$2x = -1$ $\quad x = 1$
$x = -\frac{1}{2}$

32.
$$\begin{aligned} m_1m_2 &= m_2g + m_1g \\ m_1m_2 - m_1g &= m_2g \\ m_1(m_2 - g) &= m_2g \\ m_1 &= \frac{m_2g}{m_2 - g} \end{aligned}$$

Believing in yourself is the first secret of success with Math.
Hold yourself to high standards and build an image of yourself as an achiever.

Chapter 6 Rational Expressions, Equations, and Functions

Study Tip
Why did you have to buy that textbook anyway?

You know that you need the exercises at the end of each section, because your instructor probably assigns homework from them. But if that was all there was to a textbook, you could just buy a list of problems. Here are some ideas of how you can use the rest of the book!

- Before class, read the section titles and skim the section you will be covering that day so you know what your instructor will be talking about in class.
- Read the examples before you start doing your homework. After reading them, test yourself by doing the self-check problems in the columns.
- Information in a blue box is important for you to know- be sure to read and understand!
- Special comments give you warnings about common errors, or a short explanation of an important idea; read these and ask your instructor if you have more questions.

Section 6.1 Rational Functions and Simplifying Rational Expressions

VOCABULARY

1. rational
3. asymptote

CONCEPTS

5. a. $f(1) = 1$ b. $f(2) = 0.5$ c. $f(4) = 0.25$

7. a. $\frac{x+8}{x+8} = 1$ b. $\frac{x+8}{8+x} = \frac{x+8}{x+8} = 1$ c. $\frac{x-8}{x-8} = 1$

   d. $\frac{x-8}{8-x} = \frac{-1(-x+8)}{-x+8} = -1$

9. Graph 1: The average cost decreases, then steadily increases.
   Graph 2: The average cost increases, then decreases for a while, and then increases steadily.
   Graph 3: The average cost increases steadily.
   Graph 4. The average cost decreases approaching a cost of $2.00 per unit.

NOTATION

11. a. True, because $-\frac{x-4}{x+4} = \frac{-1(x-4)}{x+4} = \frac{-x+4}{x+4} = \frac{4-x}{x+4}$ b. True, because $\frac{a-3b}{2b-a} = \frac{-1(-a+3b)}{-1(-2b+a)} = \frac{3b-a}{a-2b}$

PRACTICE

13. $t = \frac{600}{r} = \frac{600\text{ miles}}{30\text{ mph}} = 20\text{ hr}$

15. $t = \frac{600}{r} = \frac{600\text{ miles}}{50\text{ mph}} = 12\text{ hr}$

17.

x	$f(x) = \frac{6}{x}$	$f(x)$
1	$f(1) = \frac{6}{1}$	**6**
2	$f(2) = \frac{6}{2}$	**3**
4	$f(4) = \frac{6}{4}$	**1.5**
6	$f(6) = \frac{6}{6}$	**1**
8	$f(8) = \frac{6}{8}$	**0.75**
10	$f(10) = \frac{6}{10}$	**0.6**
12	$f(12) = \frac{6}{12}$	**0.5**

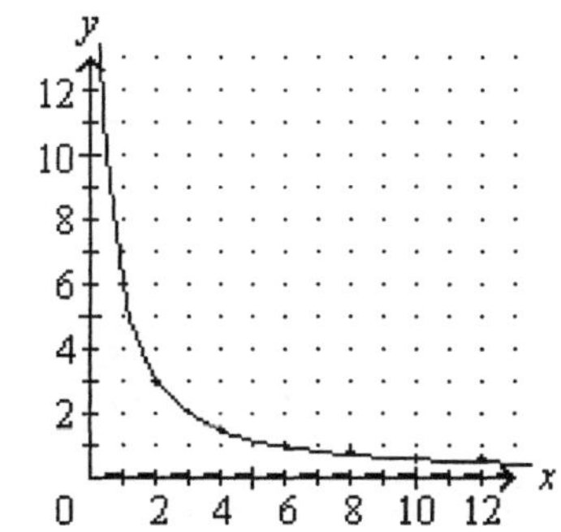

Horizontal Asymptote is $y = 0$
Vertical Asymptote is x=0

19.

x	$f(x) = \frac{x+2}{x}$	$f(x)$
1	$f(1) = \frac{1+2}{1}$	**3**
2	$f(2) = \frac{2+2}{2}$	**2**
4	$f(4) = \frac{4+2}{4}$	**1.5**
6	$f(6) = \frac{6+2}{6}$	**1.33**
8	$f(8) = \frac{8+2}{8}$	**1.25**
10	$f(10) = \frac{10+2}{10}$	**1.2**
12	$f(12) = \frac{12+2}{12}$	**1.17**

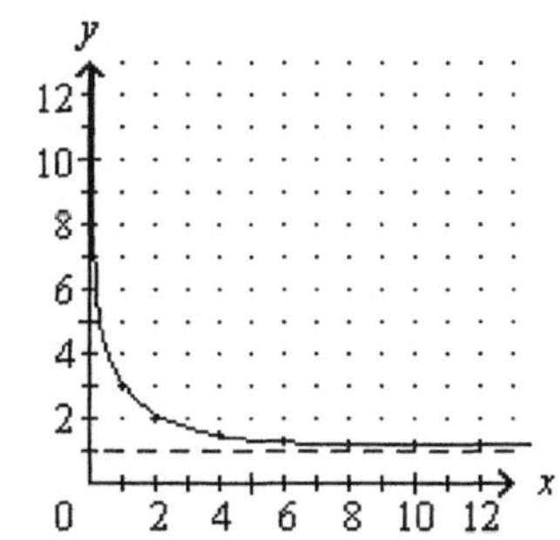

Horizontal Asymptote is $y = 1$
Vertical Asymptote is x=0

In problems 21 – 27, determine the values which cause the denominator to be zeros. These values must be eliminated from the set of real numbers to determine the domain of the function.

21. $f(x) = \frac{2}{x}$

$x \neq 0$

Domain: $(-\infty, 0) \cup (0, \infty)$

23. $f(x) = \frac{2}{x+2}$

$x + 2 \neq 0$

$x \neq -2$

Domain: $(-\infty, -2) \cup (-2, \infty)$

25. $f(x) = \frac{2}{x - x^2}$

$x(1 - x) \neq 0$

$x \neq 0$ or $1 - x \neq 0$

$1 \neq x$

Domain: $(-\infty, 0) \cup (0, 1) \cup (1, \infty)$

27. $f(x) = \frac{2}{x^2 - x - 56}$

$(x - 8)(x + 7) \neq 0$

$x - 8 \neq 0$ or $x + 7 \neq 0$

$x \neq 8$ $\quad$ $x \neq -7$

Domain: $(-\infty, -7) \cup (-7, 8) \cup (8, \infty)$

Note: Cancellation marks have not been used in the following exercises because they make the factors difficult to read.

29. $\frac{12}{18}=\frac{6\cdot 2}{6\cdot 3}=\frac{2}{3}$

31. $-\frac{112}{36}=-\frac{4\cdot 28}{4\cdot 9}=-\frac{28}{9}$

33. $\frac{12x^3}{3x}=\frac{3\cdot 4}{3}x^{3-1}=4x^2$

35. $\frac{-24x^3y^4}{18x^4y^3}=\frac{-6\cdot 4}{6\cdot 3}x^{3-4}y^{4-3}=-\frac{4}{3}x^{-1}y^1=-\frac{4y}{3x}$

37. $-\frac{11x(x-y)}{22(x-y)}=-\frac{11\cdot 1\cdot x}{11\cdot 2}\cdot\frac{(x-y)}{(x-y)}=-\frac{x}{2}$

39. $\frac{(a-b)(d-c)}{(c-d)(a-b)}=\frac{(a-b)}{(a-b)}\cdot\frac{-1(-d+c)}{(c-d)}=-1$

41. $\frac{y+x}{x^2-y^2}=\frac{(y+x)}{(x+y)(x-y)}=\frac{1}{x-y}$

43. $\frac{5x-10}{x^2-4x+4}=\frac{5(x-2)}{(x-2)(x-2)}=\frac{5}{x-2}$

45. $\frac{12-3x^2}{x^2-x-2}=\frac{-3(x^2-4)}{(x+1)(x-2)}=\frac{-3(x+2)(x-2)}{(x+1)(x-2)}=-\frac{3(x+2)}{x+1}$

47. $\frac{x^2+y^2}{x+y}$ is in lowest terms. (x^2+y^2 cannot be factored.)

49. $\frac{x^3+8}{x^2-2x+4}=\frac{(x+2)(x^2-2x+4)}{(x^2-2x+4)}=x+2$

51. $\frac{x^2+2x+1}{x^2+4x+3}=\frac{(x+1)(x+1)}{(x+1)(x+3)}=\frac{x+1}{x+3}$

53. $\frac{sx+4s-3x-12}{sx+4s+6x+24}=\frac{s(x+4)-3(x+4)}{s(x+4)+6(x+4)}=\frac{(x+4)(s-3)}{(x+4)(s+6)}=\frac{s-3}{s+6}$

55. $\frac{4x^2+24x+32}{16x^2+8x-48}=\frac{4(x^2+6x+8)}{8(2x^2+x-6)}=\frac{4((x+2)(x+4)}{8(x+2)(2x-3)}=\frac{x+4}{2(2x-3)}$

57. $\frac{3x^2-3y^2}{x^2+2y+2x+yx}=\frac{3(x^2-y^2)}{yx+x^2+2y+2x}=\frac{3(x+y)(x-y)}{x(y+x)+2(y+x)}=\frac{3(x+y)(x-y)}{(y+x)(x+2)}=\frac{3(x-y)}{x+2}$

59. $\frac{4x^2+8x+3}{6+x-2x^2}=\frac{4x^2+8x+3}{-1(2x^2-x-6)}=\frac{(2x+1)(2x+3)}{-1(2x+3)(x-2)}=-\frac{2x+1}{x-2}$ or $\frac{2x+1}{2-x}$

61. $\frac{a^3+27}{4a^2-36}=\frac{(a+3)(a^2-3a+9)}{4(a^2-9)}=\frac{(a+3)(a^2-3a+9)}{4(a+3)(a-3)}=\frac{a^2-3a+9}{4(a-3)}$

63. $\frac{2x^2-3x-9}{2x^2+3x-9}=\frac{(2x+3)(x-3)}{(2x-3)(x+3)}$ is in lowest terms. No factors will cancel.

65. $\frac{(m+n)^3}{m^2+2mn+n^2}=\frac{(m+n)^3}{(m+n)(m+n)}=\frac{(m+n)^3}{(m+n)^2}=(m+n)^{3-2}=m+n$

67. $\frac{m^3-mn^2}{mn^2+m^2n-2m^3}=\frac{m(m^2-n^2)}{m(n^2+mn-2m^2)}=\frac{m(m^2-n^2)}{-m(2m^2-mn-n^2)}=\frac{m(m+n)(m-n)}{-m(2m+n)(m-n)}=-\frac{m+n}{2m+n}$

69. $\frac{x^4-y^4}{(x^2+2xy+y^2)(x^2+y^2)}=\frac{(x^2+y^2)(x^2-y^2)}{(x+y)^2(x^2+y^2)}=\frac{x^2-y^2}{(x+y)^2}=\frac{(x+y)(x-y)}{(x+y)^2}=\frac{x-y}{x+y}$

71. $\dfrac{6xy-4x-9y+6}{6y^2-13y+6}=\dfrac{2x(3y-2)-3(3y-2)}{(3y-2)(2y-3)}=\dfrac{(3y-2)(2x-3)}{(3y-2)(2y-3)}=\dfrac{2x-3}{2y-3}$

73. $\dfrac{(2x^2+3xy+y^2)(3a+b)}{(x+y)(2xy+2bx+y^2+by)}=\dfrac{(2x+y)(x+y)(3a+b)}{(x+y)[2x(y+b)+y(y+b)]}=\dfrac{(2x+y)(x+y)(3a+b)}{(x+y)(y+b)(2x+y)}=\dfrac{3a+b}{y+b}$

75. $\dfrac{(x^2+2x+1)(x^2-2x+1)}{(x^2-1)^2}=\dfrac{(x+1)^2(x-1)^2}{[(x+1)(x-1)]^2}=\dfrac{(x+1)^2(x-1)^2}{(x+1)^2(x-1)^2}=1$

77. Vertical asymptote at $x=2$. Domain: $(-\infty, 2)\cup(2, \infty)$
Horizontal asymptote at $y=1$. Range: $(-\infty, 1)\cup(1, \infty)$

79. Vertical asymptotes at $x=2$ and $x=-2$. Domain: $(-\infty, -2)\cup(-2, 2)\cup(2, \infty)$
Range: $(-\infty, \infty)$ It appears that there is an asymptote at $y=0$ but there is not, since when $x=-1$, $y=0$.

APPLICATIONS

81. $f(p)=\dfrac{50{,}000p}{100-p}$

a. $f(50)=\dfrac{50{,}000(50)}{100-(50)}=\dfrac{2{,}500{,}000}{50}=\$50{,}000$

b. $f(80)=\dfrac{50{,}000(80)}{100-(80)}=\dfrac{4{,}000{,}000}{80}=\$200{,}000$

83. Let n represent the number of kilowatt hours used.

a. $c(n)=0.09n+7.50$

b. To find the average cost, divide by the number of hours used. $A(n)=\dfrac{0.09n+7.50}{n}$.

c. $A(775)=\dfrac{0.09(775)+7.50}{775}=\dfrac{77.25}{775}=0.0996\ldots\approx 10$ cents

85. a. If $t+3$ represents the smaller pipe, then $t+3=7$ and $t=4$, the time for the larger pipe.

$f(t)=\dfrac{t^2+3t}{2t+3}$, $\quad f(4)=\dfrac{(4)^2+3(4)}{2(4)+3}=\dfrac{28}{11}=2.\overline{54}$ or 2.6 hours

b. $t=8$ for the larger pipe, $f(8)=\dfrac{(8)^2+3(8)}{2(8)+3}=\dfrac{88}{19}=4.631\ldots$ or 4.6 hours

WRITING

87. Factor the denominator of $f(x)=\dfrac{4x+2}{x^2-25}=\dfrac{4x+2}{(x+5)(x-5)}$. Then if x is replaced with either 5 or -5 the denominator would be zero which makes $f(x)$ undefined.

REVIEW

89. $3x^2-9x = 3x\bullet x-3x\bullet 3 = 3x(x-3)$

91. $27x^6+64y^3 = (3x^2)^3+(4y)^3 = (3x^2+4y)(9x^4-12x^2y+16y^2)$

Section 6.2 Multiplying and Dividing Rational Expressions

VOCABULARY

1. rational
3. numerator; denominator

CONCEPTS

5. multiply ; $\frac{ac}{bd}$
7. 0

NOTATION

9. $$\frac{x^2+3x}{5x-25}\cdot\frac{x-5}{x+3}=\frac{(x^2+3x)\ (x-5)}{(5x-25)\ (x+3)}=\frac{x(x+3)\ (x-5)}{5(x-5)\ (x+3)}=\frac{x}{5}$$

11. Answer 1: yes;
Answer 2: no (numerators are opposites, denominators are the same);
Answer 3: yes [(2 – x) is the same as –(x – 2)]

PRACTICE

13. $\frac{3}{4}\cdot\frac{5}{3}=\frac{3\cdot5}{4\cdot3}=\frac{1\cdot5}{4\cdot1}=\frac{5}{4}$

15. $-\frac{6}{11}\div\frac{36}{55}=-\frac{6}{11}\cdot\frac{55}{36}=\frac{-6\cdot5\cdot11}{11\cdot6\cdot6}=\frac{-1\cdot5\cdot1}{1\cdot1\cdot6}=\frac{-5}{6}=-\frac{5}{6}$

17. $\frac{2x^2y^2}{cd}\cdot\frac{cd^2}{4c^2x}=\frac{2x^2y^2cd^2}{4c^3dx}=\frac{1}{2}x^{2-1}y^2c^{1-3}d^{2-1}=x^1y^2c^{-2}d^1=\frac{xy^2d}{2c^2}$

19. $\frac{-x^3y^3}{y^2}\div\frac{y^3}{x^7}=\frac{-x^3y^3}{y^2}\cdot\frac{x^7}{y^3}=\frac{-x^{10}y^3}{y^5}=-\frac{x^{10}}{y^{5-3}}=-\frac{x^{10}}{y^2}$

21. $\frac{x^2+2x+1}{x}\cdot\frac{x^2-x}{x^2-1}=\frac{(x+1)(x+1)(x)(x-1)}{x(x+1)(x-1)}=\frac{x+1}{1}=x+1$

23. $\frac{2x^2-x-3}{x^2-1}\cdot\frac{x^2+x-2}{2x^2+x-6}=\frac{(2x-3)(x+1)(x-1)(x+2)}{(x+1)(x-1)(2x-3)(x+2)}=1$

25. $\frac{x^2-16}{x^2-25}\div\frac{x+4}{x-5}=\frac{x^2-16}{x^2-25}\cdot\frac{x-5}{x+4}=\frac{(x+4)(x-4)(x-5)}{(x+5)(x-5)(x+4)}=\frac{x-4}{x+5}$

27. $-\frac{a^2+2a-35}{12x}\div\frac{ax-3x}{a^2+4a-21}=-\frac{a^2+2a-35}{12x}\cdot\frac{a^2+4a-21}{ax-3x}$

$=-\frac{(a+7)(a-5)(a+7)(a-3)}{12x(x)(a-3)}=-\frac{(a+7)^2(a-5)}{12x^2}$

29. $\frac{3t^2-t-2}{6t^2-5t-6}\cdot\frac{4t^2-9}{2t^2+5t+3}=\frac{(3t+2)(t-1)(2t+3)(2t-3)}{(3t+2)(2t-3)(2t+3)(t+1)}=\frac{t-1}{t+1}$

31. $\dfrac{3n^2+5n-2}{12n^2-13n+3} \div \dfrac{n^2+3n+2}{4n^2+5n-6} = \dfrac{3n^2+5n-2}{12n^2-13n+3} \cdot \dfrac{4n^2+5n-6}{n^2+3n+2} = \dfrac{(3n-1)(n+2)(4n-3)(n+2)}{(3n-1)(4n-3)(n+1)(n+2)} = \dfrac{n+2}{n+1}$

33. $\dfrac{2x^2+5x+3}{3x^2-5x+2} \div \dfrac{2x^2+x-3}{-3x^2+5x-2} = \dfrac{2x^2+5x+3}{3x^2-5x+2} \cdot \dfrac{-(3x^2-5x+2)}{2x^2+x-3} = \dfrac{-(2x+3)(x+1)(3x-2)(x-1)}{(3x-2)(x-1)(2x+3)(x-1)} = -\dfrac{x+1}{x-1}$

35. $(x+1) \cdot \dfrac{1}{x^2+2x+1} = \dfrac{x+1}{1} \cdot \dfrac{1}{x^2+2x+1} = \dfrac{x+1}{(x+1)(x+1)} = \dfrac{1}{x+1}$

37. $(x^2-x-2) \cdot \dfrac{x^2+3x+2}{x^2-4} = \dfrac{x^2-x-2}{1} \cdot \dfrac{x^2+3x+2}{x^2-4} = \dfrac{(x-2)(x+1)(x+2)(x+1)}{(x+2)(x-2)} = \dfrac{(x+1)(x+1)}{1} = (x+1)^2$

39. $$\dfrac{7b^5-7b^2}{5b^5-5b} \div \dfrac{14b^4+14b^3+14b^2}{20b^4-20b^2} = \dfrac{7b^5-7b^2}{5b^5-5b} \cdot \dfrac{20b^4-20b^2}{14b^4+14b^3+14b^2} = \dfrac{7b^2(b^3-1)}{5b(b^4-1)} \cdot \dfrac{20b^2(b^2-1)}{14b^2(b^2+b+1)}$$
$$= \dfrac{140b^4(b^3-1)(b^2-1)}{70b^3(b^4-1)(b^2+b+1)} = \dfrac{2b(b-1)(b^2+b+1)(b+1)(b-1)}{(b^2+1)(b+1)(b-1)(b^2+b+1)} = \dfrac{2b(b-1)}{b^2+1}$$

41. $\dfrac{x^3+y^3}{x^3-y^3} \div \dfrac{x^2-xy+y^2}{x^2+xy+y^2} = \dfrac{x^3+y^3}{x^3-y^3} \cdot \dfrac{x^2+xy+y^2}{x^2-xy+y^2} = \dfrac{(x+y)(x^2-xy+y^2)(x^2+xy+y^2)}{(x-y)(x^2+xy+y^2)(x^2-xy+y^2)} = \dfrac{x+y}{x-y}$

43. $$\dfrac{ax+ay+bx+by}{x^3-27} \div \dfrac{xc+xd+yc+yd}{x^2+3x+9} = \dfrac{ax+ay+bx+by}{x^3-27} \cdot \dfrac{x^2+3x+9}{xc+xd+yc+yd}$$
$$= \dfrac{(a+b)(x+y)(x^2+3x+9)}{(x-3)(x^2+3x+9)(c+d)(x+y)} = \dfrac{a+b}{(x-3)(c+d)}$$

45. $\dfrac{x^2-x-6}{x^2-4} \cdot \dfrac{x^2-x-2}{9-x^2} = \dfrac{(x-3)(x+2)(x-2)(x+1)}{(x+2)(x-2)(3-x)(3+x)} = \dfrac{x-3}{3-x} \cdot \dfrac{x+1}{3+x} = -\dfrac{x+1}{3+x}$ or $\dfrac{-x-1}{x+3}$

47. $(4x+12) \cdot \dfrac{x^2}{2x-6} \div \dfrac{2}{x-3} = \dfrac{4(x+3)}{1} \cdot \dfrac{x^2}{2(x-3)} \cdot \dfrac{x-3}{2} = \dfrac{4x^2(x+3)(x-3)}{4(x-3)} = x^2(x+3)$

49. $(x^2-x-6) \div (x-3) \div (x-2) = \dfrac{(x-3)(x+2)}{1} \cdot \dfrac{1}{x-3} \cdot \dfrac{1}{x-2} = \dfrac{(x-3)(x+2)}{(x-3)(x-2)} = \dfrac{x+2}{x-2}$

51. $$\dfrac{2x^2-2x-4}{x^2+2x-8} \cdot \dfrac{3x^2+15x}{x+1} \div \dfrac{4x^2-100}{x^2-x-20} = \dfrac{2(x^2-x-2)}{x^2+2x-8} \cdot \dfrac{3x(x+5)}{x+1} \cdot \dfrac{x^2-x-20}{4(x^2-25)}$$
$$= \dfrac{2(3x)(x-2)(x+1)(x+5)(x-5)(x+4)}{4(x+4)(x-2)(x+1)(x+5)(x-5)} = \dfrac{3x}{2}$$

53. $$\dfrac{x^2-x-12}{x^2+x-2} \div \dfrac{x^2-6x+8}{x^2-3x-10} \cdot \dfrac{x^2-3x+2}{x^2-2x-15} = \dfrac{x^2-x-12}{x^2+x-2} \cdot \dfrac{x^2-3x-10}{x^2-6x+8} \cdot \dfrac{x^2-3x+2}{x^2-2x-15}$$
$$= \dfrac{(x-4)(x+3)(x-5)(x+2)(x-2)(x-1)}{(x+2)(x-1)(x-2)(x-4)(x-5)(x+3)} = 1$$

55. $\left(\dfrac{x-3}{x^3+4}\right)^2 = \dfrac{(x-3)(x-3)}{(x^3+4)(x^3+4)} = \dfrac{x^2-6x+9}{x^6+8x^3+16}$

57. $\left(\frac{2m^2-m-3}{x^2-1}\right)^2=\frac{(2m^2-m-3)(2m^2-m-3)}{(x^2-1)(x^2-1)}=\frac{4m^4-4m^3-11m^2+6m+9}{x^4-2x^2+1}$

APPLICATIONS

59. Trial 1: Distance can be found by multiplying rate and time.

$$\frac{k_1^2+3k_1+2}{k_1-3}\cdot\frac{k_1^2-3k_1}{k_1+1}=\frac{(k_1+2)(k_1+1)(k_1)(k_1-3)}{(k_1-3)(k_1+1)}=k_1(k_1+2)$$

Trial 2: Time can be found by dividing distance by rate.

$$(k_2^2+11k_2+30)\div\frac{k_2^2+6k_2+5}{k_2+1}=\frac{k_2^2+11k_2+30}{1}\cdot\frac{k_2+1}{k_2^2+6k_2+5}=\frac{(k_2+5)(k_2+6)(k_2+1)}{(k_2+1)(k_2+5)}=k_2+6$$

WRITING

61. Begin by factoring all numerators and denominators. Cancel common factors which exist in the numerator and in the denominator. Multiply the remaining numerators together to get the new numerator and multiply the remaining denominators together to get the new denominator.

REVIEW

63. $-2a^2(3a^3-a^2) = -6a^5+2a^4$

65. $(2g-n)(3g-n) = 6g^2-2gn-3gn+n^2 = 6g^2-5gn+n^2$

Section 6.3 Adding and Subtracting Rational Expressions

VOCABULARY

1. denominator

3. LCD

CONCEPTS

5. subtract; keep; $\frac{a-c}{b}$

7. same

9. a. ii . b. adding or subtracting rational expression c. simplifying a rational expression

NOTATION

11. $\frac{6x-1}{3x-1}+\frac{3x-2}{3x-1}=\frac{6x-1+3x-2}{3x-1}$

$=\frac{9x-3}{3x-1}$

$=\frac{3(3x-1)}{3x-1}$

$=3$

PRACTICE

13. $\frac{3}{4}+\frac{7}{4}=\frac{3+7}{4}=\frac{10}{4}=\frac{2\cdot5}{2\cdot2}=\frac{5}{2}$

15. $\frac{3}{4y}+\frac{8}{4y}=\frac{3+8}{4y}=\frac{11}{4y}$

17. $\frac{3x}{2x+2}+\frac{x+4}{2x+2}=\frac{3x+x+4}{2x+2}=\frac{4x+4}{2x+2}=\frac{4(x+1)}{2(x+1)}=2$

19. $\frac{3x}{x-3}-\frac{9}{x-3}=\frac{3x-9}{x-3}=\frac{3(x-3)}{x-3}=3$

21. $\frac{5x}{x+1}+\frac{3}{x+1}-\frac{2x}{x+1}=\frac{5x+3-2x}{x+1}=\frac{3x+3}{x+1}=\frac{3(x+1)}{x+1}=3$

23. $\dfrac{3(x^2+x)}{x^2-5x+6}+\dfrac{-3(x^2-x)}{x^2-5x+6}=\dfrac{3x^2+3x-3x^2+3x}{x^2-5x+6}=\dfrac{6x}{(x-2)(x-3)}$

25. $\left.\begin{cases}12x=2\cdot 2\cdot 3\cdot x=2^2\cdot 3\cdot x\\ 18x^2=2\cdot 3\cdot 3\cdot x\cdot x=2\cdot 3^2\cdot x^2\end{cases}\right\}\text{LCD}=2^2\cdot 3^2\cdot x^2=4(9)=36x^2$

27. $\left.\begin{cases}x^2+3x=x(x+3)\\ x^2-9=(x+3)(x-3)\end{cases}\right\}\text{LCD}=x(x+3)(x-3)$

29. $\left.\begin{cases}x^3+27=(x+3)(x^2-3x+9)\\ x^2+6x+9=(x+3)^2\end{cases}\right\}\text{LCD}=(x+3)^2(x^2-3x+9)$

31. $\left.\begin{cases}2x^2+5x+3=(2x+3)(x+1)\\ 4x^2+12x+9=(2x+3)^2\\ x^2+2x+1=(x+1)^2\end{cases}\right\}\text{LCD}=(2x+3)^2(x+1)^2$

33. LCD = 2 • 3 = 6 $\quad \dfrac{1}{2}+\dfrac{1}{3}=\dfrac{1\cdot 3}{2\cdot 3}+\dfrac{1\cdot 2}{3\cdot 2}=\dfrac{3}{6}+\dfrac{2}{6}=\dfrac{3+2}{6}=\dfrac{5}{6}$

35. LCD = 2 • 7 = 14 $\quad \dfrac{3a}{2}-\dfrac{4b}{7}=\dfrac{3a\cdot 7}{2\cdot 7}-\dfrac{4b\cdot 2}{7\cdot 2}=\dfrac{21a}{14}-\dfrac{8b}{14}=\dfrac{21a-8b}{14}$

37. LCD = 3 • 4 • x = 12x $\quad \dfrac{3}{4x}+\dfrac{2}{3x}=\dfrac{3\cdot 3}{4x\cdot 3}+\dfrac{2\cdot 4}{3x\cdot 4}=\dfrac{9}{12x}+\dfrac{8}{12x}=\dfrac{9+8}{12x}=\dfrac{17}{12x}$

39. LCD = 2 • 3 • a • b = 6ab $\quad \dfrac{3a}{2b}-\dfrac{2b}{3a}=\dfrac{3a\cdot 3a}{2b\cdot 3a}-\dfrac{2b\cdot 2b}{3a\cdot 2b}=\dfrac{9a^2-4b^2}{6ab}$

41. LCD = a^2b^2 $\quad \dfrac{3}{ab^2}-\dfrac{5}{a^2b}=\dfrac{3\cdot a}{ab^2\cdot a}-\dfrac{5\cdot b}{a^2b\cdot b}=\dfrac{3a-5b}{a^2b^2}$

43. $\{4b^2=2^2\bullet b^2;\ 6b=2\bullet 3\bullet b\}$ LCD = $2^2\bullet 3\bullet b^2=12b^2$ $\quad \dfrac{r}{4b^2}+\dfrac{s}{6b}=\dfrac{r\cdot 3}{4b^2\cdot 3}+\dfrac{s\cdot 2b}{6b\cdot 2b}=\dfrac{3r+2bs}{12b^2}$

45. LCD = 3 • 7 = 21

$$\frac{a+b}{3}+\frac{a-b}{7}=\frac{(a+b)7}{3\cdot 7}+\frac{(a-b)3}{7\cdot 3}=\frac{7a+7b+3a-3b}{21}=\frac{10a+4b}{21}=\frac{2(5a+2b)}{21}$$

47. LCD = $(x+2)(x-4)$

$$\frac{3}{x+2}+\frac{5}{x-4}=\frac{3(x-4)}{(x+2)(x-4)}+\frac{5(x+2)}{(x-4)(x+2)}=\frac{3x-12+5x+10}{(x+2)(x-4)}=\frac{8x-2}{(x+2)(x-4)}=\frac{2(4x-1)}{(x+2)(x-4)}$$

49. LCD = $(x+5)(x+7)$

$$\frac{x+2}{x+5}-\frac{x-3}{x+7}=\frac{(x+2)(x+7)}{(x+5)(x+7)}-\frac{(x-3)(x+5)}{(x+7)(x+5)}=\frac{(x^2+9x+14)-(x^2+2x-15)}{(x+5)(x+7)}$$

$$=\frac{x^2+9x+14-x^2-2x+15}{(x+5)(x+7)}=\frac{7x+29}{(x+5)(x+7)}$$

51. LCD $= x$ $\quad 4+\frac{1}{x}=\frac{4\cdot x}{x}+\frac{1}{x}=\frac{4x+1}{x}$

53. LCD $= (x-3)$ Multiply the numerator and denominator of the last fraction by -1.

$$\frac{x+8}{x-3}-\frac{x-14}{3-x}=\frac{x+8}{x-3}-\frac{-1(x-14)}{-1(3-x)}=\frac{(x+8)-(-1)(x-14)}{x-3}=\frac{x+8+(x-14)}{x-3}=\frac{2x-6}{x-3}=\frac{2(x-3)}{x-3}=2$$

55. LCD $= (3a-2)$ Multiply the numerator and denominator of the last fraction by -1.

$$\frac{2a+1}{3a-2}-\frac{a-4}{2-3a}=\frac{2a+1}{3a-2}-\frac{-1(a-4)}{-1(2-3a)}=\frac{(2a+1)-(-1)(a-4)}{3a-2}=\frac{2a+1+(a-4)}{3a-2}=\frac{3a-3}{3a-2}=\frac{3(a-1)}{3a-2}$$

57. $\{x^2+5x+6=(x+3)(x+2);\ x^2-4=(x+2)(x-2)\}$ LCD $= (x+3)(x+2)(x-2)$

$$\frac{x}{x^2+5x+6}+\frac{x}{x^2-4}=\frac{x(x-2)}{(x+3)(x+2)(x-2)}+\frac{x(x+3)}{(x+2)(x-2)(x+3)}=\frac{x^2-2x+x^2+3x}{(x+2)(x-2)(x+3)}$$
$$=\frac{2x^2+x}{(x+2)(x-2)(x+3)}=\frac{x(2x+1)}{(x+2)(x-2)(x+3)}$$

59. $\{x^2-2x-3=(x-3)(x+1);\ 3x^2-7x-6=(3x+2)(x-3)\}$ LCD $= (x-3)(x+1)(3x+2)$

$$\frac{4}{x^2-2x-3}-\frac{x}{3x^2-7x-6}=\frac{4(3x+2)}{(x-3)(x+1)(3x+2)}-\frac{x(x+1)}{(x-3)(x+1)(3x+2)}=\frac{12x+8-(x^2+x)}{(x-3)(x+1)(3x+2)}$$
$$=\frac{12x+8-x^2-x}{(x-3)(x+1)(3x+2)}=\frac{-x^2+11x+8}{(x-3)(x+1)(3x+2)}$$

61. LCD $= (x+1)$ $\quad \frac{2x+3}{1}+\frac{1}{x+1}=\frac{(2x+3)(x+1)}{x+1}+\frac{1}{x+1}=\frac{2x^2+5x+3+1}{x+1}=\frac{2x^2+5x+4}{x+1}$

63. LCD $= (x-5)$ $\quad \frac{1+x}{1}-\frac{x}{x-5}=\frac{(x+1)(x-5)}{x-5}-\frac{x}{x-5}=\frac{x^2-4x-5-x}{x-5}=\frac{x^2-5x-5}{x-5}$

65. $\{x^2-9=(x+3)(x-3);\ (x-3);\ x\}$ LCD $= x(x+3)(x-3)$

$$\frac{8}{x^2-9}+\frac{2}{x-3}-\frac{6}{x}=\frac{8\cdot x}{(x+3)(x-3)\cdot x}+\frac{2x(x+3)}{x(x+3)(x-3)}-\frac{6(x+3)(x-3)}{x(x+3)(x-3)}=\frac{8x+(2x^2+6x)-6(x^2-9)}{x(x+3)(x-3)}$$
$$=\frac{8x+2x^2+6x-6x^2+54}{x(x+3)(x-3)}=\frac{-4x^2+14x+54}{x(x+3)(x-3)}=\frac{-2(2x^2-7x-27)}{x(x+3)(x-3)}$$

67. $\{(2x-1);\ (3x+2);\ 6x^3+x^2-2x=x(3x+2)(2x-1)\}$ LCD $= x(2x-1)(3x+2)$

$$\frac{3x}{2x-1}+\frac{x+1}{3x+2}-\frac{2x}{6x^3+x^2-2x}=\frac{3x(x)(3x+2)}{x(2x-1)(3x+2)}+\frac{(x+1)(x)(2x-1)}{x(2x-1)(3x+2)}-\frac{2x}{x(2x-1)(3x+2)}$$
$$=\frac{9x^3+6x^2+2x^3+x^2-x-2x}{x(2x-1)(3x+2)}=\frac{11x^3+7x^2-3x}{x(2x-1)(3x+2)}$$
$$=\frac{x(11x^2+7x-3)}{x(2x-1)(3x+2)}=\frac{11x^2+7x-3}{(2x-1)(3x+2)}$$

69. $\{(x+1);\ (x-1);\ x^2-1=(x+1)(x-1)\}$ LCD $=(x+1)(x-1)$

$$\frac{3}{x+1}-\frac{2}{x-1}+\frac{x+3}{x^2-1}=\frac{3(x-1)}{(x+1)(x-1)}-\frac{2(x+1)}{(x+1)(x-1)}+\frac{x+3}{(x+1)(x-1)}=\frac{3x-3-(2x+2)+x+3}{(x+1)(x-1)}$$
$$=\frac{3x-3-2x-2+x+3}{(x+1)(x-1)}=\frac{2x-2}{(x+1)(x-1)}=\frac{2(x-1)}{(x+1)(x-1)}=\frac{2}{x+1}$$

71. $\{(a-b);\ (a+b);\ b^2-a^2=-1(a^2-b^2)=-1(a+b)(a-b)\}$ LCD $=(a+b)(a-b)$

$$\frac{a}{a-b}+\frac{b}{a+b}+\frac{a^2+b^2}{b^2-a^2}=\frac{a(a+b)}{(a-b)(a+b)}+\frac{b(a-b)}{(a-b)(a+b)}+\frac{-1(a^2+b^2)}{-1(b^2-a^2)}=\frac{a^2+ab+ab-b^2-a^2-b^2}{(a-b)(a+b)}$$
$$=\frac{2ab-2b^2}{(a-b)(a+b)}=\frac{2b(a-b)}{(a-b)(a+b)}=\frac{2b}{a+b}$$

73. $\{(m-n);\ (n-m)=-1(m-n);\ m^2-2mn+n^2=(m-n)^2\}$ LCD $=(m-n)^2$
Multiply the numerator and denominator of the middle fraction by -1.

$$\frac{7n^2}{m-n}+\frac{3m}{n-m}-\frac{3m^2-n}{m^2-2mn+n^2}=\frac{7n^2(m-n)}{(m-n)^2}+\frac{-1(3m)(m-n)}{-1(n-m)(m-n)}-\frac{3m^2-n}{(m-n)^2}$$
$$=\frac{7mn^2-7n^3-3m^2+3mn-(3m^2-n)}{(m-n)^2}=\frac{7mn^2-7n^3-6m^2+3mn+n}{(m-n)^2}$$

75. Simplify:

$$\frac{m+1}{m^2+2m+1}+\frac{m-1}{m^2-2m+1}+\frac{2}{m^2-1}=\frac{m+1}{(m+1)^2}+\frac{m-1}{(m-1)^2}+\frac{2}{(m+1)(m-1)}=\frac{1}{m+1}+\frac{1}{m-1}+\frac{2}{(m+1)(m-1)}$$

LCD $=(m+1)(m-1)$

$$\frac{1}{m+1}+\frac{1}{m-1}+\frac{2}{(m+1)(m-1)}=\frac{m-1}{(m+1)(m-1)}+\frac{m+1}{(m+1)(m-1)}+\frac{2}{(m+1)(m-1)}$$
$$=\frac{m-1+m+1+2}{(m+1)(m-1)}=\frac{2m+2}{(m+1)(m-1)}=\frac{2(m+1)}{(m+1)(m-1)}=\frac{2}{m-1}$$

APPLICATIONS

77. The perimeter of the 45°–45°–90° triangle is: $\frac{10}{r}+\frac{10}{r}+10=\frac{10}{r}+\frac{10}{r}+\frac{10r}{r}=\frac{20+10r}{r}$ or $\frac{10r+20}{r}$

The perimeter of the 30°–60°–90° triangle is:

$$3+\frac{6}{t}+\left(\frac{1}{2}\right)\left(\frac{6}{t}\right)=3+\frac{6}{t}+\frac{3}{t}=\frac{3(t)}{t}+\frac{6}{t}+\frac{3}{t}=\frac{3t+6+3}{t}=\frac{3t+9}{t}$$

WRITING

79. Factor each denominator. List all the different factors. Use the factors the most number of times they occur as a factor in any **one** of the original fractions.

81. When multiplying rational expressions a common denominator is not necessary. Answer should be $\frac{3}{2x}$

REVIEW

83.
$$\begin{aligned} a(a-6) &= -9 \\ a^2-6a+9 &= 0 \\ (a-3)(a-3) &= 0 \end{aligned}$$
$a-3=0$ or $a-3=0$
$a=3$ $\quad$ $a=3$

85.
$$\begin{aligned} y^3+y^2 &= 0 \\ y^2(y+1) &= 0 \\ y(y)(y+1) &= 0 \end{aligned}$$
$y=0$ or $y=0$ or $y+1=0$
$y=-1$

Section 6.4 Complex Fractions

VOCABULARY/ CONCEPTS

1. complex

3. $\dfrac{t^2}{t^2}$

NOTATION

5. $$\frac{\frac{5m^2}{6}}{\frac{25m}{3}} = \frac{5m^2}{6} \div \frac{25m}{3} = \frac{5m^2}{6} \cdot \frac{3}{25m} = \frac{5 \cdot m \cdot m \cdot 3}{2 \cdot 3 \cdot 5 \cdot 5 \cdot m} = \frac{m}{10}$$

7. $\dfrac{a}{b} \div \dfrac{c}{d}$

PRACTICE

9. Divide the fractions and simplify.

$$\frac{\frac{1}{2}}{\frac{3}{4}} = \frac{1}{2} \div \frac{3}{4} = \frac{1}{2} \cdot \frac{4}{3} = \frac{2}{3}$$

11. Multiply by LCD = 6 and simplify.

$$\frac{\frac{1}{2} - \frac{2}{3}}{\frac{2}{3} + \frac{1}{2}} = \frac{6\left(\frac{1}{2} - \frac{2}{3}\right)}{6\left(\frac{2}{3} + \frac{1}{2}\right)} = \frac{3-4}{4+3} = -\frac{1}{7}$$

13. Divide the fractions and simplify.

$$\frac{\frac{4x}{y}}{\frac{6xz}{y^2}} = \frac{4x}{y} \div \frac{6xz}{y^2} = \frac{4x}{y} \cdot \frac{y^2}{6xz} = \frac{2y}{3z}$$

15. Write the numerator over 1, divide and simplify.

$$\frac{5ab^2}{\frac{ab}{25}} = \frac{\frac{5ab^2}{1}}{\frac{ab}{25}} = \frac{5ab^2}{1} \div \frac{ab}{25} = \frac{5ab^2}{1} \cdot \frac{25}{ab} = \frac{125b}{1} = 125b$$

17. Divide the fractions and simplify.

$$\frac{\frac{x-y}{xy}}{\frac{y-x}{x}} = \frac{x-y}{xy} \div \frac{y-x}{x} = \frac{x-y}{xy} \cdot \frac{x}{y-x} = \frac{x-y}{y(y-x)} = -\frac{1}{y}$$

19. Multiply by LCD $= xy$ and simplify.

$$\frac{\frac{1}{x} - \frac{1}{y}}{xy} = \frac{xy\left(\frac{1}{x} - \frac{1}{y}\right)}{xy(xy)} = \frac{y-x}{x^2y^2}$$

21. Multiply by LCD $= ab$ and simplify.

$$\frac{\frac{1}{a} + \frac{1}{b}}{\frac{1}{a}} = \frac{ab\left(\frac{1}{a} + \frac{1}{b}\right)}{ab\left(\frac{1}{a}\right)} = \frac{b+a}{b}$$

23. Multiply by LCD = y and simplify.

$$\frac{1+\frac{x}{y}}{1-\frac{x}{y}}=\frac{y\left(1+\frac{x}{y}\right)}{y\left(1-\frac{x}{y}\right)}=\frac{y+x}{y-x}$$

25. Multiply by LCD = xy and simplify.

$$\frac{\frac{y}{x}-\frac{x}{y}}{\frac{1}{x}+\frac{1}{y}}=\frac{xy\left(\frac{y}{x}-\frac{x}{y}\right)}{xy\left(\frac{1}{x}+\frac{1}{y}\right)}=\frac{y^2-x^2}{y+x}=\frac{(y+x)(y-x)}{y+x}=y-x$$

27. Multiply by LCD = ab and simplify.

$$\frac{\frac{1}{a}-\frac{1}{b}}{\frac{a}{b}-\frac{b}{a}}=\frac{ab\left(\frac{1}{a}-\frac{1}{b}\right)}{ab\left(\frac{a}{b}-\frac{b}{a}\right)}=\frac{b-a}{a^2-b^2}=\frac{b-a}{(a-b)(a+b)}=-\frac{1}{a+b}$$

29. Multiply by the LCD = x and simplify.

$$\frac{x+1-\frac{6}{x}}{\frac{1}{x}}=\frac{x\left(x+1-\frac{6}{x}\right)}{x\left(\frac{1}{x}\right)}=x^2+x-6$$

31. Multiply by LCD = xy and simplify.

$$\frac{5xy}{1+\frac{1}{xy}}=\frac{xy(5xy)}{xy\left(1+\frac{1}{xy}\right)}=\frac{5x^2y^2}{xy+1}$$

33. Multiply by the LCD = a and simplify.

$$\frac{a-4+\frac{1}{a}}{-\frac{1}{a}-a+4}=\frac{a\left(a-4+\frac{1}{a}\right)}{a\left(-\frac{1}{a}-a+4\right)}=\frac{a^2-4a+1}{-1-a^2+4a}=\frac{a^2-4a+1}{a^2+4a-1}=\frac{a^2-4a+1}{-1(a^2-4a+1)}=-1$$

35. Multiply by the LCD = x^2 and simplify.

$$\frac{1+\frac{6}{x}+\frac{8}{x^2}}{1+\frac{1}{x}-\frac{12}{x^2}}=\frac{x^2\left(1+\frac{6}{x}+\frac{8}{x^2}\right)}{x^2\left(1+\frac{1}{x}-\frac{12}{x^2}\right)}=\frac{x^2+6x+8}{x^2+x-12}=\frac{(x+2)(x+4)}{(x+4)(x-3)}=\frac{x+2}{x-3}$$

37. Multiply by the LCD = $(a+1)(a-1)$ and simplify.

$$\frac{\frac{1}{a+1}+1}{\frac{3}{a-1}+1}=\frac{(a+1)(a-1)\left(\frac{1}{a+1}+1\right)}{(a+1)(a-1)\left(\frac{3}{a-1}+1\right)}=\frac{(a-1)+(a+1)(a-1)}{3(a+1)+(a+1)(a-1)}=\frac{a-1+a^2-1}{3a+3+a^2-1}$$

$$=\frac{a^2+a-2}{a^2+3a+2}=\frac{(a+2)(a-1)}{(a+2)(a+1)}=\frac{a-1}{a+1}$$

39. Multiply by the LCD = $x(x+1)$ and simplify.

$$\frac{2+\frac{3}{x+1}}{\frac{1}{x}+x}=\frac{x(x+1)\left(2+\frac{3}{x+1}\right)}{x(x+1)\left(\frac{1}{x}+x\right)}=\frac{2x(x+1)+3x}{(x+1)+x^2(x+1)}=\frac{2x^2+2x+3x}{x+1+x^3+x^2}=\frac{2x^2+5x}{x^3+x^2+x+1}$$

41. Write the fraction without negative exponents, multiply by the LCD = xy and simplify.

$$\frac{y}{x^{-1}-y^{-1}}=\frac{y}{\frac{1}{x}-\frac{1}{y}}=\frac{xy(y)}{xy\left(\frac{1}{x}-\frac{1}{y}\right)}=\frac{xy^2}{y-x}$$

43. Write the fraction without negative exponents, multiply by the LCD = x^2y^2 and simplify.

$$\frac{x-y^{-2}}{y-x^{-2}}=\frac{x-\frac{1}{y^2}}{y-\frac{1}{x^2}}=\frac{x^2y^2\left(x-\frac{1}{y^2}\right)}{x^2y^2\left(y-\frac{1}{x^2}\right)}=\frac{x^3y^2-x^2}{x^2y^3-y^2}=\frac{x^2(xy^2-1)}{y^2(x^2y-1)}$$

45. Divide the fractions and simplify.

$$\frac{\frac{t}{x^2-y^2}}{\frac{t}{x+y}}=\frac{t}{x^2-y^2}\div\frac{t}{x+y}=\frac{t}{x^2-y^2}\cdot\frac{x+y}{t}=\frac{x+y}{(x+y)(x-y)}=\frac{1}{x-y}$$

47. Multiply by the LCD = $x^2-9=(x+3)(x-3)$ and simplify.

$$\frac{\frac{2}{x+3}-\frac{1}{x-3}}{\frac{3}{x^2-9}}=\frac{(x+3)(x-3)\left(\frac{2}{x+3}-\frac{1}{x-3}\right)}{(x+3)(x-3)\left(\frac{3}{(x+3)(x-3)}\right)}=\frac{2(x-3)-(x+3)}{3}=\frac{2x-6-x-3}{3}=\frac{x-9}{3}$$

49. Multiply by the LCD = $h^2+3h+2=(h+2)(h+1)$ and simplify.

$$\frac{\frac{h}{h^2+3h+2}}{\frac{4}{h+2}-\frac{4}{h+1}}=\frac{(h+2)(h+1)\left(\frac{h}{(h+2)(h+1)}\right)}{(h+2)(h+1)\left(\frac{4}{(h+2)}-\frac{4}{h+1}\right)}=\frac{h}{4(h+1)-4(h+2)}=\frac{h}{4h+4-4h-8}=\frac{h}{-4}=-\frac{h}{4}$$

51. Multiply the second expression by the LCD = $a+1$ and then simplify.

$$a+\frac{a}{1+\frac{a}{a+1}}=a+\frac{(a+1)a}{(a+1)\left(1+\frac{a}{a+1}\right)}=a+\frac{a^2+a}{(a+1)+a}=a+\frac{a^2+a}{2a+1}$$

Write the a in the first expression over 1 and then multiply by the LCD = $2a+1$ and add the expressions.

$$=\frac{a}{1}+\frac{a^2+a}{2a+1}=\frac{a(2a+1)}{2a+1}+\frac{a^2+a}{2a+1}=\frac{2a^2+a+a^2+a}{2a+1}=\frac{3a^2+2a}{2a+1}$$

53. Multiply the fraction in the numerator by the LCD = 2 and the denominator by the LCD = 3 and then simplify.

$$\frac{x-\dfrac{1}{1-\dfrac{x}{2}}}{\dfrac{3}{x+\dfrac{2}{3}}-x}=\frac{x-\dfrac{2(1)}{2\left(1-\dfrac{x}{2}\right)}}{\dfrac{3(3)}{3\left(x+\dfrac{2}{3}\right)}-x}=\frac{x-\dfrac{2}{2-x}}{\dfrac{9}{3x+2}-x}$$

Write the x's in the expression over 1 and then add the numerator using the LCD = 2 – x and add the denominator using the LCD = $3x + 2$.

$$=\frac{\dfrac{x}{1}-\dfrac{2}{2-x}}{\dfrac{9}{3x+2}-\dfrac{x}{1}}=\frac{\dfrac{x(2-x)}{2-x}-\dfrac{2}{2-x}}{\dfrac{9}{3x+2}-\dfrac{x(3x+2)}{3x+2}}=\frac{\dfrac{2x-x^2-2}{2-x}}{\dfrac{9-3x^2-2x}{3x+2}}=\frac{\dfrac{-x^2+2x-2}{2-x}}{\dfrac{-3x^2-2x+9}{3x+2}}$$

Divide the fraction and simplify.

$$=\frac{-x^2+2x-2}{2-x}\div\frac{-3x^2-2x+9}{3x+2}=\frac{-x^2+2x-2}{2-x}\cdot\frac{3x+2}{-3x^2-2x+9}=\frac{(-x^2+2x-2)(3x+2)}{(2-x)(-3x^2-2x+9)}$$

APPLICATIONS

55. $k=\dfrac{1}{\dfrac{1}{k_1}+\dfrac{1}{k_2}}=\dfrac{k_1k_2(1)}{k_1k_2\left(\dfrac{1}{k_1}+\dfrac{1}{k_2}\right)}=\dfrac{k_1k_2}{k_2+k_1}$; Multiply by the LCD = k_1k_2 and simplify.

57. $\dfrac{\text{opening of tongs}}{\text{opening of handles}}=\dfrac{8-\dfrac{2}{d}}{6-\dfrac{2}{d}}=\dfrac{d\left(8-\dfrac{2}{d}\right)}{d\left(6-\dfrac{2}{d}\right)}=\dfrac{8d-2}{6d-2}=\dfrac{2(4d-1)}{2(3d-1)}=\dfrac{4d-1}{3d-1}$; Multiply by LCD = d and simplify.

WRITING

59. A complex fraction is a fraction that has a fraction in its numerator or denominator or both.

REVIEW

61.
$$\begin{aligned}\frac{8(a-5)}{3}&=2(a-4)\\8(a-5)&=6(a-4)\\8a-40&=6a-24\\2a&=16\\a&=8\end{aligned}$$

63.
$$\begin{aligned}a^4-13a^2+36&=0\\(a^2-4)(a^2-9)&=0\\(a+2)(a-2)(a+3)(a-3)&=0\end{aligned}$$

$a+2=0$ or $a-2=0$ or $a+3=0$ or $a-3=0$

$a=-2$ $\qquad a=2$ $\qquad a=-3$ $\qquad a=3$.

Section 6.5 Equations Containing Rational Expressions

VOCABULARY

1. rational
3. clear

CONCEPTS

5. a. Substitute $x = 2$. $\frac{x+2}{x+3} + \frac{1}{x^2+2x-3} = \frac{(2)+2}{(2)+3} + \frac{1}{(2)^2+2(2)-3} = \frac{4}{5} + \frac{1}{5} = \frac{5}{5} = 1$ Yes, 2 is a solution.

   b. $x = 2$ is not a solution because the first denominator would be 0.

7. If $r \bullet t = d$, then $t = \frac{d}{r}$

	r	• t	= d
Running	x	$\frac{12}{x}$	12
Bicycling	$x + 15$	$\frac{12}{x+15}$	12

9. a. 3, 0 b. 3, 0 c. 3, 0

NOTATION

11.
$$\frac{10}{3y} - \frac{7}{30} = \frac{9}{2y}$$
$$30y\left(\frac{10}{3y} - \frac{7}{30}\right) = 30y\left(\frac{9}{2y}\right)$$
$$30y\left(\frac{10}{3y}\right) - 30y\left(\frac{7}{30}\right) = 30y\left(\frac{9}{2y}\right)$$
$$100 - 7y = 135$$
$$-7y = 35$$
$$y = -5$$

PRACTICE

13. Multiply both sides by LCD = $4x$.
$$\frac{1}{4} + \frac{9}{x} = 1$$
$$4x\left(\frac{1}{4}\right) + 4x\left(\frac{9}{x}\right) = 4x(1)$$
$$x + 36 = 4x$$
$$-3x = -36$$
$$x = 12$$

15. Multiply both sides by LCD = $20x$.
$$\frac{34}{x} - \frac{3}{2} = -\frac{13}{20}$$
$$20x\left(\frac{34}{x}\right) - 20x\left(\frac{3}{2}\right) = 20x\left(-\frac{13}{20}\right)$$
$$20(34) - 10x(3) = x(-13)$$
$$680 - 30x = -13x$$
$$-17x = -680$$
$$x = 40$$

17. Multiply both sides by LCD = $2y$.

$$\frac{3}{y}+\frac{7}{2y}=13$$
$$2y\left(\frac{3}{y}\right)+2y\left(\frac{7}{2y}\right)=2y(13)$$
$$6+7=26y$$
$$13=26y$$
$$\frac{13}{26}=y \quad \text{or} \quad y=\frac{1}{2}$$

19. Multiply both sides by LCD = x.

$$\frac{5x-1}{x-4}=\frac{6x-5}{x-4}$$
$$(x-4)\,\frac{5x-1}{x-4}=\frac{6x-5}{x-4}\,(x-4)$$
$$5x-1=6x-5$$
$$-x=-4$$
$$x=4$$

$x=4$ would make the denominator 0, so it is extraneous.; no solution

21. Multiply both sides by LCD = $30x$.

$$\frac{7}{5x}-\frac{1}{2}=\frac{5}{6x}+\frac{1}{3}$$
$$30x\left(\frac{7}{5x}\right)-30x\left(\frac{1}{2}\right)=30x\left(\frac{5}{6x}\right)+30x\left(\frac{1}{3}\right)$$
$$6(7)-15x(1)=5(5)+10x(1)$$
$$42-15x=25+10x$$
$$-25x=-17$$
$$x=\frac{17}{25}$$

23. Multiply both sides by LCD = $(2+y)(2-y)$.

$$\frac{3-5y}{2+y}=\frac{3+5y}{2-y}$$
$$(2+y)(2-y)\left(\frac{3-5y}{2+y}\right)=(2+y)(2-y)\left(\frac{3+5y}{2-y}\right)$$
$$(2-y)(3-5y)=(2+y)(3+5y)$$
$$6-13y+5y^2=6+13y+5y^2$$
$$-26y=0$$
$$y=0$$

25. Multiply both sides by LCD = $(a+1)(a-3)$.

$$\frac{a+2}{a+1}-\frac{a-4}{a-3}=0$$
$$(a+1)(a-3)\left(\frac{a+2}{a+1}\right)-(a+1)(a-3)\left(\frac{a-4}{a-3}\right)=0$$
$$(a-3)(a+2)-(a+1)(a-4)=0$$
$$(a^2-a-6)-(a^2-3a-4)=0$$
$$a^2-a-6-a^2+3a+4=0$$
$$2a-2=0$$
$$2a=2$$
$$a=1$$

27. Multiply the fraction on the right side by 1 in the form $\frac{-1}{-1}$ and rearrange. Then multiply both sides by $(x + 3)(x - 1)$.

$$\frac{x+2}{x+3}-1=\frac{1}{3-2x-x^2}$$

$$\frac{x+2}{x+3}-1=\frac{-1}{x^2+2x-3}$$

$$(x+3)(x-1)\left(\frac{x+2}{x+3}\right)-(x+3)(x-1)(1)=(x+3)(x-1)\left(\frac{-1}{(x+3)(x-1)}\right)$$

$$(x-1)(x+2)-(x^2+2x-3)=-1$$

$$x^2+x-2-x^2-2x+3=-1$$

$$-x+1=-1$$

$$-x=-2$$

$$x=2$$

29. Multiply both sides by $(x + 2)^2$.

$$\frac{x}{x+2}=1-\frac{3x+2}{x^2+4x+4}$$

$$(x+2)^2\left(\frac{x}{x+2}\right)=(x+2)^2(1)-(x+2)^2\left(\frac{3x+2}{(x+2)(x+2)}\right)$$

$$x(x+2)=(x+2)^2-(3x+2)$$

$$x^2+2x=x^2+4x+4-3x-2$$

$$x^2+2x=x^2+x+2$$

$$x=2$$

31. Multiply by the LCD $= (x - 2)(x + 1)$.

$$\frac{2}{x-2}+\frac{1}{x+1}=\frac{1}{x^2-x-2}$$

$$(x-2)(x+1)\left(\frac{2}{x-2}\right)+(x-2)(x+1)\left(\frac{1}{x+1}\right)=(x-2)(x+1)\left(\frac{1}{(x-2)(x+1)}\right)$$

$$2(x+1)+1(x-2)=1$$

$$2x+2+x-2=1$$

$$3x=1$$

$$x=\frac{1}{3}$$

33. Multiply the second fraction on the left side by 1 in the form $\frac{-1}{-1}$. Then multiply both sides by $(a + 3)(a - 1)$.

$$\frac{a-1}{a+3}-\frac{1-2a}{3-a}=\frac{2-a}{a-3}$$
$$\frac{a-1}{a+3}-\frac{2a-1}{a-3}=\frac{2-a}{a-3}$$
$$(a+3)(a-3)\left(\frac{a-1}{a+3}\right)-(a+3)(a-3)\left(\frac{2a-1}{a-3}\right)=(a+3)(a-3)\left(\frac{2-a}{a-3}\right)$$
$$(a-3)(a-1)-(a+3)(2a-1)=(a+3)(2-a)$$
$$(a^2-4a+3)-(2a^2+5a-3)=2a-a^2+6-3a$$
$$a^2-4a+3-2a^2-5a+3=-a^2-a+6$$
$$-a^2-9a+6=-a^2-a+6$$
$$-8a=0$$
$$a=0$$

35. Multiply both sides by LCD $= x + 4$.

$$\frac{5}{x+4}+\frac{1}{x+4}=x-1$$
$$(x+4)\left(\frac{5}{x+4}\right)+(x+4)\left(\frac{1}{x+4}\right)=(x+4)(x-1)$$
$$5+1=x^2+3x-4$$
$$0=x^2+3x-10$$
$$0=(x+5)(x-2)$$

$x + 5 = 0$ or $x - 2 = 0$

$x = -5$ $\qquad$ $x = 2$

37. Multiply both sides by LCD $= 2(x + 1)$.

$$\frac{3}{x+1}-\frac{x-2}{2}=\frac{x-2}{x+1}$$
$$2(x+1)\left(\frac{3}{x+1}\right)-2(x+1)\left(\frac{x-2}{2}\right)=2(x+1)\left(\frac{x-2}{x+1}\right)$$
$$2(3)-(x+1)(x-2)=2(x-2)$$
$$6-(x^2-x-2)=2x-4$$
$$6-x^2+x+2=2x-4$$
$$0=x^2+x-12$$
$$0=(x+4)(x-3)$$

$x + 4 = 0$ or $x - 3 = 0$

$x = -4$ $\qquad$ $x = 3$

39. Multiply both sides by LCD $= 4x(x - 3)$.

$$\frac{2}{x-3}+\frac{3}{4}=\frac{17}{2x}$$
$$4x(x-3)\left(\frac{2}{x-3}\right)+4x(x-3)\left(\frac{3}{4}\right)=4x(x-3)\left(\frac{17}{2x}\right)$$
$$2(4x)+3x(x-3)=2(x-3)(17)$$
$$8x+3x^2-9x=34x-102$$
$$3x^2-35x+102=0$$
$$(3x-17)(x-6)=0$$

$3x - 17 = 0$ or $x - 6 = 0$

$3x = 17$ $\qquad$ $x = 6$

$x = \frac{17}{3}$

41. Multiply both sides by the LCD = $8(x+7)(x+3)$.

$$\frac{x+4}{x+7}-\frac{x}{x+3}=\frac{3}{8}$$

$$8(x+7)(x+3)\left(\frac{x+4}{x+7}\right)-8(x+7)(x+3)\left(\frac{x}{x+3}\right)=8(x+7)(x+3)\left(\frac{3}{8}\right)$$

$$8(x+3)(x+4)-8x(x+7)=3(x+7)(x+3)$$

$$8(x^2+7x+12)-(8x^2+56x)=3(x^2+10x+21)$$

$$8x^2+56x+96-8x^2-56x=3x^2+30x+63$$

$$96=3x^2+30x+63$$

$$0=3x^2+30x-33$$

$$0=x^2+10x-11$$

$$0=(x-1)(x+11)$$

$$x-1=0 \quad \text{or} \quad x+11=0$$

$$x=1 \qquad\qquad x=-11$$

43. Multiply both sides by LCD = $(a-3)(a+3)$

$$\frac{2a}{a-3}+\frac{24}{a^2-9}=-\frac{4}{a+3}$$

$$\frac{2a}{a-3}+\frac{24}{(a+3)(a-3)}=-\frac{4}{a+3}$$

$$(a-3)(a+3)\frac{2a}{a-3}+\frac{24}{(a+3)(a-3)}=-\frac{4}{a+3}(a-3)(a+3)$$

$$(a+3)(2a)+24=-4(a-3)$$

$$2a^2+6a+24=-4a+12$$

$$2a^2+10a+12=0$$

$$2(a^2+5a+6)=0$$

$$2(a+3)(a+2)=0$$

$$a+3=0 \quad or \quad a+2=0$$

$$a=-3,\ -2$$

solution is $a=-2$; -3 is an extraneous solution

45.

$I=\dfrac{E}{R_L+r}$	Solve for r.
$I(R_L+r)=E$	Multiply both sides by R_l+r.
$IR_L+Ir=E$	Distribute I.
$Ir=E-IR_L$	Subtract IR_l from both sides.
$r=\dfrac{E-IR_L}{I}$	Divide both sides by I.

47.

$S=\dfrac{a-lr}{1-r}$	Solve for r.
$S(1-r)=a-lr$	Multiply both sides by $1-r$.
$S-Sr=a-lr$	Distribute S.
$lr-Sr=a-S$	Add lr and subtract S from both sides.
$r(l-S)=a-S$	Factor.
$r=\dfrac{a-S}{l-S}$ or $r=\dfrac{S-a}{S-l}$	Divide both sides by $l-S$.

49.

$P=\dfrac{Q_1}{Q_2-Q_1}$	Solve for Q_1.
$P(Q_2-Q_1)=Q_1$	Multiply both sides by Q_2-Q_1.
$PQ_2-PQ_1=Q_1$	Distribute P.
$PQ_2=Q_1+PQ_1$	Add PQ_1 to both sides.
$PQ_2=Q_1(1+P)$	Factor.
$\dfrac{PQ_2}{1+P}=Q_1$	Divide both sides by $1+P$.

51.

$\dfrac{1}{R}=\dfrac{1}{R_1}+\dfrac{1}{R_2}+\dfrac{1}{R_3}$	Solve for R.
$RR_1R_2R_3\left(\dfrac{1}{R}\right)=RR_1R_2R_3\left(\dfrac{1}{R_1}\right)+RR_1R_2R_3\left(\dfrac{1}{R_2}\right)+RR_1R_2R_3\left(\dfrac{1}{R_3}\right)$	Multiply both sides by $RR_1R_2R_3$.
$R_1R_2R_3=RR_2R_3+RR_1R_3+RR_1R_2$	Do the multiplication.
$R_1R_2R_3=R(R_2R_3+R_1R_3+R_1R_2)$	Factor.
$\dfrac{R_1R_2R_3}{R_2R_3+R_1R_3+R_1R_2}=R$	Divide.

APPLICATIONS

53.

$\dfrac{1}{f}=\dfrac{1}{s_1}+\dfrac{1}{s_2}$	Substitute $s_1=5$ in. and $s_2=5\text{ ft}\bullet\dfrac{12\text{ in.}}{1\text{ ft}}=60$ in.
$\dfrac{1}{f}=\dfrac{1}{5}+\dfrac{1}{60}$	Solve for f.
$60f\left(\dfrac{1}{f}\right)=60f\left(\dfrac{1}{5}\right)+60f\left(\dfrac{1}{60}\right)$	Multiply by LCD $=60f$.
$60=12f+f$	
$60=13f$	
$\dfrac{60}{13}=f$ or $f=4\dfrac{8}{13}$ in.	

55. Solve for L:

$$V = C - \left(\frac{C-S}{L}\right)N$$
$$L(V) = L(C) - L\left(\frac{C-S}{L}\right)N$$
$$LV = LC - N(C-S)$$
$$LV = LC - CN + SN$$
$$LV - LC = -CN + SN$$
$$L(V-C) = SN - CN$$
$$L = \frac{SN - CN}{V - C}$$

Substitute $V = \$13{,}000$, $N = 4$, $C = \$25{,}000$, $S = \$1{,}000$.

$$L = \frac{SN - CN}{V - C}$$
$$L = \frac{1{,}000(4) - 25{,}000(4)}{13{,}000 - 25{,}000}$$
$$L = \frac{4000 - 100{,}000}{-12{,}000}$$
$$L = 8 \text{ years}$$

57. **Analysis**: Find the amount of time it will take for the painters to do the job together.

Equation: Let x represent the amount of time it will take the two paint contractors working together to finish the job.

Work Santos can do in one day	plus	work Mays can do in one day	is	work they can do together in one day.
$\frac{1}{3}$	$+$	$\frac{1}{5}$	$=$	$\frac{1}{x}$

Solve:

$$\frac{1}{3} + \frac{1}{5} = \frac{1}{x}$$
$$15x\left(\frac{1}{3}\right) + 15x\left(\frac{1}{5}\right) = 15x\left(\frac{1}{x}\right)$$
$$5x + 3x = 15$$
$$8x = 15 \qquad x = \frac{15}{8} = 1\frac{7}{8}$$

Conclusion: **a.** It will take $1\frac{7}{8}$ days to paint the house working together.

b. Santos charges \$220/day so $\$220 \cdot 1\frac{7}{8} = \$412.50$.

Mays charges \$200 a day so $\$200 \cdot 1\frac{7}{8} = \$375$.

59. **Analysis**: Find the amount of time it will take to open 100 oysters working together.

Equation: Let x represent the amount of time it will take the two oyster openers working together to finish the job. The novice will take $8\frac{1}{2}$ minutes or 510 seconds to open 100 oysters.

Part of job Racz can do in one second	plus	part of job the novice can do in one second	is	part of the job they can do together in one second.
$\frac{1}{140}$	$+$	$\frac{1}{510}$	$=$	$\frac{1}{x}$

Solve:

$$\frac{1}{140} + \frac{1}{510} = \frac{1}{x}$$
$$7140x\left(\frac{1}{140}\right) + 7140x\left(\frac{1}{510}\right) = 7140x\left(\frac{1}{x}\right)$$
$$51x + 14x = 7140$$
$$65x = 7140$$
$$x = 109.846...$$

Conclusion: It will take about 110 seconds to open 100 oysters working together.

61. **Analysis**: Find the amount of time it will take the two pipes to fill the pond together.

Equation: Let x represent the amount of time it will take the two pipes working together to finish the job. Evaporation is taking place so this will need to be deducted from the filling process. The time worked in given in weeks.

Amount that pipe 1 fills	plus	amount that pipe 2 fills	less	amount evaporated	is	amount filled together in one week.
$\frac{1}{3}$	$+$	$\frac{1}{5}$	$-$	$\frac{1}{10}$	$=$	$\frac{1}{x}$

Solve:

$$\frac{1}{3}+\frac{1}{5}-\frac{1}{10}=\frac{1}{x}$$

$$30x\left(\frac{1}{3}\right)+30x\left(\frac{1}{5}\right)-30x\left(\frac{1}{10}\right)=30x\left(\frac{1}{x}\right)$$

$$10x+6x-3x=30$$

$$13x=30$$

$$x=\frac{30}{13}=2\frac{4}{13}$$

Conclusion: It will take $2\frac{4}{13}$ weeks to fill the pond working together.

63. **Analysis**: Find how fast the boxer jogs as part of his workout.

Equation: Let x represent the jogging rate for the boxer. Then $x + 6$ represents the bicycling rate of the boxer.

If $r \bullet t = d$, then $t = \frac{d}{r}$

	r	$\bullet\ t$	$= d$
Jogging	x	$\frac{8}{x}$	8
Bicycling	$x+6$	$\frac{8}{x+6}$	8

Jogging time	plus	bicycling time	is	2 hours.
$\frac{8}{x}$	$+$	$\frac{8}{x+6}$	$=$	2

Solve:

$$\frac{8}{x}+\frac{8}{x+6}=2$$

$$x(x+6)\left(\frac{8}{x}\right)+x(x+6)\left(\frac{8}{x+6}\right)=2x(x+6)$$ Multiply both sides by LCD = $x(x + 6)$.

$$8(x+6)+8x=2x^2+12x$$

$$8x+48+8x=2x^2+12x$$

$$0=2x^2-4x-48$$

Divide both sides by 2.

$$0=x^2-2x-24$$

$$0=(x-6)(x+4)$$

$$x-6=0 \quad \text{or} \quad x+4=0$$

$$x=6 \qquad\qquad x=-4$$ Discard the negative solution.

Conclusion: The boxer jogs at a rate of 6 mph.

65. **Analysis**: Find the speed of two trains that make the same trip of 315 miles.

Equation: Let x represent the speed of the slow train. Then $x + 10$ represents the speed of the fast train.

If $r \cdot t = d$, then $t = \frac{d}{r}$

	r	• t	= d
Slow train	x	$\frac{315}{x}$	315
Fast train	$x + 10$	$\frac{315}{x+10}$	315

Fast train time plus 2 hours is slow train time.

$$\frac{315}{x+10} + 2 = \frac{315}{x}$$

Solve:

$$\frac{315}{x+10} + 2 = \frac{315}{x}$$

$$x(x+10)\left(\frac{315}{x+10}\right) + 2x(x+10) = x(x+10)\left(\frac{315}{x}\right)$$ Multiply both sides by LCD = $x(x + 10)$.

$$315x + 2x^2 + 20x = 315x + 3150$$

$$2x^2 + 335x = 315x + 3150$$

$$2x^2 + 20x - 3150 = 0$$

Divide both sides by 2.

$$x^2 - 10x - 1575 = 0$$

$$(x - 35)(x + 45) = 0$$

$x - 35 = 0$ or $x + 45 = 0$

$x = 35$ $\quad$ $x = -45$ $\quad$ Discard the negative solution.

Conclusion: The speed of the slow train was $x = 35$ mph and the speed of the fast train was $x + 10 = 45$ mph.

67. **Analysis**: Find the speed of the wind.

Equation: Let x represent the speed of the wind. Then $45 + x$ represents the speed of the helicopter going downwind (with the wind) and $45 - x$ represents the speed of the helicopter going upwind (against the wind).

If $r \cdot t = d$, then $t = \frac{d}{r}$

	r	• t	= d
Downwind	$45 + x$	$\frac{0.5}{45+x}$	0.5
Upwind	$45 - x$	$\frac{0.4}{45-x}$	0.4

Downwind time equals upwind time.

$$\frac{0.5}{45+x} = \frac{0.4}{45-x}$$

Solve:

$$\frac{0.5}{45+x} = \frac{0.4}{45-x}$$

Multiply the means and extremes.

$$0.5(45 - x) = 0.4(45 + x)$$

$$22.5 - 0.5x = 18 + 0.4x$$

$$-0.9x = -4.5$$

$$x = 5$$

Conclusion: The speed of the wind was $x = 5$ mph.

69. **Analysis**: Find the number of microwave ovens purchased the first month when there was a change in unit costs.

Equation: Let x represent the number of microwave ovens purchased the first month, then $x - 1$ represents the number of microwaves purchased the second month. The total cost ($1800) divided by the number of microwaves purchased will give the unit cost.

First month's unit cost	is	second month's unit cost	less	$25.
$\frac{1800}{x}$	$=$	$\frac{1800}{x-1}$	$-$	25

Solve:

$$\frac{1800}{x} = \frac{1800}{x-1} - 25$$

$$x(x-1)\left(\frac{1800}{x}\right) = x(x-1)\left(\frac{1800}{x-1}\right) - 25x(x-1)$$ Multiply both sides by LCD = $x(x - 1)$.

$$1800x - 1800 = 1800x - 25x^2 + 25x$$

$$25x^2 - 25x - 1800 = 0$$

$$x^2 - x - 72 = 0$$ Divide both sides by 25.

$$(x-9)(x+8) = 0$$

$x - 9 = 0$ or $x + 8 = 0$ Discard the negative solution.

$x = 9$ $\quad$ $x = -8$

Conclusion: There were $x = 9$ microwaves purchased the first month.

WRITING

71. It is possible that the solution found will cause a denominator to be equal to 0.

73. Find the point of intersection of the two graphs by using the INTERSECT feature. In this illustration the graphs intersect at approximately (–3, 2) so the solution would be $x = -3$.

REVIEW

75. $9,000,000,000 = 9.0×10^9

77. 0.0000000000000000000044 = 4.4×10^{-22}

Section 6.6 — Dividing Polynomials

VOCABULARY

1. dividend; divisor; quotient

3. placeholder

CONCEPTS

5. $\frac{a+c}{b} = \frac{a}{b} + \frac{c}{b}$

7. $(2x - 1)(x^2 + 3x - 4) = 2x(x^2 + 3x - 4) - 1(x^2 + 3x - 4) = 2x^3 + 6x^2 - 8x - x^2 - 3x + 4 = 2x^3 + 5x^2 - 11x + 4$

NOTATION

9.
$$\begin{array}{r|l} & 2x \quad +1 \\ x+4 & 2x^2 + 9x + 4 \\ & \underline{2x^2 + 8x} \\ & \qquad x + 4 \\ & \qquad \underline{x + 4} \\ & \qquad\quad 0 \end{array}$$

11. $3a^2 + 5 + \frac{6}{3a - 2}$

13. $\frac{x^2 - x - 12}{x - 4}$, $x - 4\overline{)x^2 - x - 12}$, and $(x^2 - x - 12) \div (x - 4)$

PRACTICE

15. $\dfrac{4x^2y^3}{8x^5y^2}=\dfrac{4}{8}x^{2-5}y^{3-2}=\dfrac{1}{2}x^{-3}y^1=\dfrac{y}{2x^3}$

17. $-\dfrac{33a^2b^2}{44a^4b^2}=-\dfrac{33}{44}a^{2-4}b^{2-2}=-\dfrac{3}{4}a^{-2}b^0=-\dfrac{3}{4a^2}$

19. $\dfrac{4x+6}{2}=\dfrac{4x}{2}+\dfrac{6}{2}=2x+3$

21. $\dfrac{4x^2-x^3}{-6x}=\dfrac{4x^2}{-6x}-\dfrac{x^3}{-6x}=-\dfrac{2x}{3}+\dfrac{x^2}{6}$

23. $\dfrac{12x^2y^3+x^3y^2}{6xy}=\dfrac{12x^2y^3}{6xy}+\dfrac{x^3y^2}{6xy}=2xy^2+\dfrac{x^2y}{6}$

25. $\dfrac{24x^6y^7-12x^5y^{12}+36xy}{48x^2y^3}=\dfrac{24x^6y^7}{48x^2y^3}-\dfrac{12x^5y^{12}}{48x^2y^3}+\dfrac{36xy}{48x^2y^3}=\dfrac{x^4y^4}{2}-\dfrac{x^3y^9}{4}+\dfrac{3}{4xy^2}$

27.
$$\begin{array}{r|l} & x+2 \\ x+3 & x^2+5x+6 \\ & \underline{x^2+3x} \\ & \quad 2x+6 \\ & \quad \underline{2x+6} \\ & \qquad 0 \end{array}$$

Solution: $x+2$

29.
$$\begin{array}{r|l} & x+7 \\ x+3 & x^2+10x+21 \\ & \underline{x^2+3x} \\ & \quad 7x+21 \\ & \quad \underline{7x+21} \\ & \qquad 0 \end{array}$$

Solution: $x+7$

31.
$$\begin{array}{r|l} & 3x-5 \\ 2x+3 & 6x^2-x-12 \\ & \underline{6x^2+9x} \\ & \quad -10x-12 \\ & \quad \underline{-10x-15} \\ & \qquad 3 \end{array}$$

Solution: $3x-5+\dfrac{3}{2x+3}$

33.
$$\begin{array}{r|l} & 3x^2+x+2 \\ x-1 & 3x^3-2x^2+x-6 \\ & \underline{3x^3-3x^2} \\ & \quad x^2+x \\ & \quad \underline{x^2-x} \\ & \qquad 2x-6 \\ & \qquad \underline{2x-2} \\ & \qquad -4 \end{array}$$

Solution: $3x^2+x+2-\dfrac{4}{x-1}$

35.
$$\begin{array}{r|l} & 2x^2+5x+3 \\ 3x-2 & 6x^3+11x^2-x-2 \\ & \underline{6x^3-4x^2} \\ & \quad 15x^2-x \\ & \quad \underline{15x^2-10x} \\ & \qquad 9x-2 \\ & \qquad \underline{9x-6} \\ & \qquad\quad 4 \end{array}$$

Solution: $2x^2+5x+3+\dfrac{4}{3x-2}$

37.
$$\begin{array}{r|l} & 3x^2+4x+3 \\ 2x-3 & 6x^3-x^2-6x-9 \\ & \underline{6x^3-9x^2} \\ & \quad 8x^2-6x \\ & \quad \underline{8x^2-12x} \\ & \qquad 6x-9 \\ & \qquad \underline{6x-9} \\ & \qquad\quad 0 \end{array}$$

Solution: $3x^2+4x+3$

39. Rearrange the terms before dividing.

$$\begin{array}{r} a+1 \\ a+1\overline{)a^2+2a+1} \\ \underline{a^2+a} \\ a+1 \\ \underline{a+1} \\ 0 \end{array}$$

Solution: $a+1$

41. Rearrange the terms before dividing.

$$\begin{array}{r} 2y+2 \\ 5y-2\overline{)10y^2+6y-4} \\ \underline{10y^2-4y} \\ 10y-4 \\ \underline{10y-4} \\ 0 \end{array}$$

Solution: $2y+2$

43. Rearrange the terms before dividing.

$$\begin{array}{r} 6x-12 \\ x-1\overline{)6x^2-18x+12} \\ \underline{6x^2-6x} \\ -12x+12 \\ \underline{-12x+12} \\ 0 \end{array}$$

Solution: $6x-12$

45. Rearrange the terms and add $0x^3$ before dividing.

$$\begin{array}{r} 4x^3-3x^2+3x+1 \\ 4x+3\overline{)16x^4+0x^3+3x^2+13x+3} \\ \underline{16x^4+12x^3} \\ -12x^3+3x^2 \\ \underline{-12x^3-9x^2} \\ 12x^2+13x \\ \underline{12x^2+9x} \\ 4x+3 \\ \underline{4x+3} \\ 0 \end{array}$$

Solution: $4x^3-3x^2+3x+1$

47. Add $0a^2$ and $0a$ before dividing.

$$\begin{array}{r} a^2+a+1 \\ a-1\overline{)a^3+0a^2+0a+1} \\ \underline{a^3-a^2} \\ a^2+0a \\ \underline{a^2-a} \\ a+1 \\ \underline{a-1} \\ 2 \end{array}$$

Solution: $a^2+a+1+\frac{2}{a-1}$

49. Add 0a before dividing.

$$\begin{array}{r} 5a^2-3a-4 \\ 3a-4\overline{)15a^3-29a^2+0a+16} \\ \underline{15a^3-20a^2} \\ -9a^2+0a \\ \underline{-9a^2+12a} \\ -12a+16 \\ \underline{-12a+16} \\ 0 \end{array}$$

Solution: $5a^2-3a-4$

51. Rearrange terms before dividing.

$$\begin{array}{r} 6y-12 \\ y-2\overline{)6y^2-24y+24} \\ \underline{6y^2-12y} \\ -12y+24 \\ \underline{-12y+24} \\ 0 \end{array}$$

Solution: $6y-12$

53. Extra terms are not needed in this problem.

$$\begin{array}{r} x^4+x^2+4 \\ x^2-2\overline{)x^6-x^4+2x^2-8} \\ \underline{x^6-2x^4} \\ x^4+2x^2 \\ \underline{x^4-2x^2} \\ 4x^2-8 \\ \underline{4x^2-8} \\ 0 \end{array}$$

Solution: x^4+x^2+4

55.

$$\begin{array}{r} x^2+x+1 \\ x^2+x+2\overline{)x^4+2x^3+4x^2+3x+2} \\ \underline{x^4+x^3+2x^2} \\ x^3+2x^2+3x \\ \underline{x^3+x^2+2x} \\ x^2+x+2 \\ \underline{x^2+x+2} \\ 0 \end{array}$$

Solution: x^2+x+1

57. Rearrange terms before dividing.

$$\begin{array}{r} x^2 + x + 2 \\ x^2+3 \overline{\big) x^4 + x^3 + 5x^2 + 3x + 6} \\ \underline{x^4 \quad + 3x^2} \\ x^3 + 2x^2 + 3x \\ \underline{x^3 \quad + 3x} \\ 2x^2 \quad + 6 \\ \underline{2x^2 \quad + 6} \\ 0 \end{array}$$

Solution: $x^2 + x + 2$

59.

$$\begin{array}{r} 9.8x + 16.4 \\ x-2 \overline{\big) 9.8x^2 - 3.2x - 69.3} \\ \underline{9.8x^2 - 19.6x} \\ 16.4x - 69.3 \\ \underline{16.4x - 32.8} \\ -36.5 \end{array}$$

Solution: $9.8x + 16.4 - \dfrac{36.5}{x-2}$

61.

2	6	1	−23	2
		12	26	6
	6	13	3	8

63.

1	1	1	−2
		1	2
	1	2	0

Solution: $x + 2$

65. Rearrange the terms before dividing.

−4	1	6	8
		−4	−8
	1	2	0

Solution: $x + 2$

67.

−2	1	−5	14
		−2	14
	1	−7	28

Solution: $x - 7 + \dfrac{28}{x+2}$

69.

3	3	−10	5	−6
		9	−3	6
	3	−1	2	0

Solution: $3x^2 - x + 2$

71. Add $0x^2$ before dividing.

2	2	0	−5	−6
		4	8	6
	2	4	3	0

Solution: $2x^2 + 4x + 3$

73. Rearrange the terms and add $0x$ before dividing.

−1	6	5	0	4
		−6	1	−1
	6	−1	1	3

Solution: $6x^2 - x + 1 + \dfrac{3}{x+1}$

75.

0.2	7.2	−2.1	0.5
		1.44	−0.132
	7.2	−0.66	0.368

Solution: $7.2x - 0.66 + \dfrac{0.368}{x-0.2}$

77.

−1.7	2.7	1	−5.2
		−4.59	6.103
	2.7	−3.59	0.903

Solution: $2.7x - 3.59 + \dfrac{0.903}{x+1.7}$

79.

$$\begin{aligned} P(x) &= 2x^3 - 4x^2 + 2x - 1 \\ P(1) &= 2(1)^3 - 4(1)^2 + 2(1) - 1 \\ &= 2 - 4 + 2 - 1 \\ &= -1 \end{aligned}$$

Divide $P(x)$ by $x - 1$.

1	2	−4	2	−1
		2	−2	0
	2	−2	0	−1

Remainder is -1 so $P(1) = -1$.

81. $\begin{aligned} P(x) &= 2x^3 - 4x^2 + 2x - 1 \\ P(-2) &= 2(-2)^3 - 4(-2)^2 + 2(-2) - 1 \\ &= -16 - 16 - 4 - 1 \\ &= -37 \end{aligned}$

Divide $P(x)$ by $x + 2$.

$$\begin{array}{r|rrrr} -2 & 2 & -4 & 2 & -1 \\ & & -4 & 16 & -36 \\ \hline & 2 & -8 & 18 & -37 \end{array}$$

Remainder is -37 so $P(-2) = -37$.

83. $\begin{aligned} Q(x) &= x^4 - 3x^3 + 2x^2 + x - 3 \\ Q(-1) &= (-1)^4 - 3(-1)^3 + 2(-1)^2 + (-1) - 3 \\ &= 1 + 3 + 2 - 1 - 3 \\ &= 2 \end{aligned}$

Divide $Q(x)$ by $x + 1$.

$$\begin{array}{r|rrrrr} -1 & 1 & -3 & 2 & 1 & -3 \\ & & -1 & 4 & -6 & 5 \\ \hline & 1 & -4 & 6 & -5 & 2 \end{array}$$

Remainder is 2 so $Q(-1) = 2$.

85. $\begin{aligned} Q(x) &= x^4 - 3x^3 + 2x^2 + x - 3 \\ Q(2) &= (2)^4 - 3(2)^3 + 2(2)^2 + (2) - 3 \\ &= 16 - 24 + 8 + 2 - 3 \\ &= -1 \end{aligned}$

Divide $Q(x)$ by $x - 2$.

$$\begin{array}{r|rrrrr} 2 & 1 & -3 & 2 & 1 & -3 \\ & & 2 & -2 & 0 & 2 \\ \hline & 1 & -1 & 0 & 1 & -1 \end{array}$$

Remainder is -1 so $Q(2) = -1$.

87. Divide $P(x) = x^3 - 4x^2 + x - 2$ by $x - 2$.

$$\begin{array}{r|rrrr} 2 & 1 & -4 & 1 & -2 \\ & & 2 & -4 & -6 \\ \hline & 1 & -2 & -3 & -8 \end{array}$$

Remainder is -8 so $P(2) = -8$.

89. Divide $P(x) = 2x^3 + x + 2$ by $x - 3$.

$$\begin{array}{r|rrrr} 3 & 2 & 0 & 1 & 2 \\ & & 6 & 18 & 57 \\ \hline & 2 & 6 & 19 & 59 \end{array}$$

Remainder is 59 so $P(3) = 59$.

91. Divide $P(x) = x^4 - 2x^3 + x^2 - 3x + 2$ by $x + 2$.

$$\begin{array}{r|rrrrr} -2 & 1 & -2 & 1 & -3 & 2 \\ & & -2 & 8 & -18 & 42 \\ \hline & 1 & -4 & 9 & -21 & 44 \end{array}$$

Remainder is 44 so $P(-2) = 44$.

93. Divide $P(x) = 3x^5 + 1$ by $x + \frac{1}{2}$.

$$\begin{array}{r|rrrrrr} -\frac{1}{2} & 3 & 0 & 0 & 0 & 0 & 1 \\ & & -\frac{3}{2} & \frac{3}{4} & -\frac{3}{8} & \frac{3}{16} & -\frac{3}{32} \\ \hline & 3 & -\frac{3}{2} & \frac{3}{4} & -\frac{3}{8} & \frac{3}{16} & \frac{29}{32} \end{array}$$

Remainder is $\frac{29}{32}$ so $P(-\frac{1}{2}) = \frac{29}{32}$.

APPLICATIONS

95. Divide $x^3 - 4x^2 + x + 6$ by $x + 1$. Using long division:

$$\begin{array}{r} x^2 - 5x + 6 \\ x+1 \overline{\smash{\big)}\, x^3 - 4x^2 + x + 6} \\ \underline{x^3 + x^2} \\ -5x^2 + x \\ \underline{-5x^2 - 5x} \\ 6x + 6 \\ \underline{6x + 6} \\ 0 \end{array}$$

The length of the longer side is $x^2 - 5x + 6$.

Or, using synthetic division:

$$\begin{array}{r|rrrr} -1 & 1 & -4 & 1 & 6 \\ & & -1 & 5 & -6 \\ \hline & 1 & -5 & 6 & 0 \end{array}$$

Solution: $x^2 - 5x + 6$

97. **Dog sled**

Divide $12x^2 + 13x - 14$ by $4x + 7$ using long division. Synthetic division does not work in this situation.

$$\begin{array}{r} 3x-2 \\ 4x+7\overline{)12x^2+13x-14} \\ \underline{12x^2+21x} \\ -8x-14 \\ \underline{-8x-14} \\ 0 \end{array}$$

The rate by dog sled is $3x - 2$ mph.

Snowshoes

Divide $3x^2 + 19x + 20$ by $3x + 4$ using long division. Synthetic division does not work in this situation.

$$\begin{array}{r} x+5 \\ 3x+4\overline{)3x^2+19x+20} \\ \underline{3x^2+4x} \\ 15x+20 \\ \underline{15x+20} \\ 0 \end{array}$$

The time by snowshoes is $x + 5$ hours.

WRITING

99. Reduce the coefficients. Subtract the exponents on like variable bases.

101. Divide $P(x)$ by $x - a$. The remainder is $P(a)$.

REVIEW

103. $2(x^2 + 4x - 1) + 3(2x^2 - 2x + 2) = 2x^2 + 8x - 2 + 6x^2 - 6x + 6 = 8x^2 + 2x + 4$

105. $-2(3y^3 - 2y + 7) - (y^2 + 2y - 4) + 4(y^3 + 2y - 1) = -6y^3 + 4y - 14 - y^2 - 2y + 4 + 4y^3 + 8y - 4$
$= -2y^3 - y^2 + 10y - 14$

Section 6.7 Proportion and Variation

VOCABULARY

1. ratio
3. extremes; means
5. direct; inverse
7. Inverse

CONCEPTS

9. direct variation

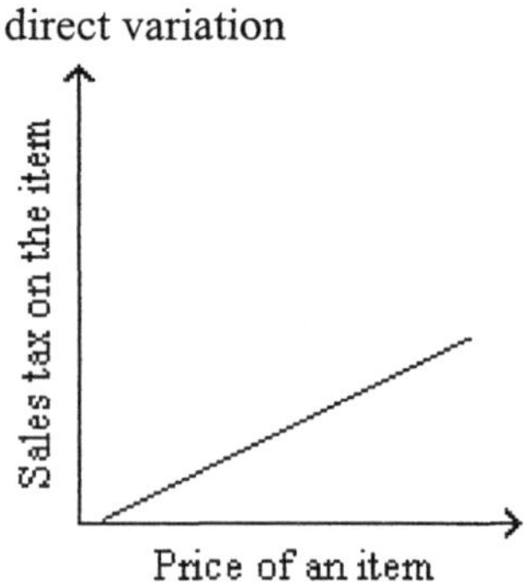

11. inverse variation

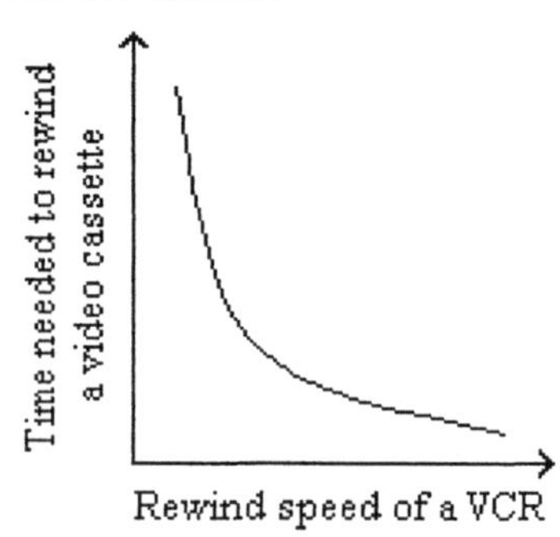

NOTATION

13.
$$\begin{aligned} \frac{-7}{6} &= \frac{x+3}{12} \\ \mathbf{-7}\,(12) &= \mathbf{6}\,(x+3) \\ -84 &= 6x + \mathbf{18} \\ \mathbf{-102} &= 6x \\ -17 &= x \end{aligned}$$

PRACTICE

15.
$$\begin{aligned} \frac{x}{5} &= \frac{15}{25} \\ 25x &= 5(15) \\ 25x &= 75 \\ x &= 3 \end{aligned}$$

17.
$$\begin{aligned} \frac{r-2}{3} &= \frac{r}{5} \\ 5(r-2) &= 3r \\ 5r-10 &= 3r \\ 2r &= 10 \\ r &= 5 \end{aligned}$$

19.
$$\begin{aligned} \frac{5}{5z+3} &= \frac{2z}{2z^2+6} \\ 5(2z^2+6) &= 2z(5z+3) \\ 10z^2+30 &= 10z^2+6z \\ 30 &= 6z \\ 5 &= z \end{aligned}$$

21.
$$\begin{aligned} \frac{2(y+3)}{3} &= \frac{4(y-4)}{5} \\ 5(2)(y+3) &= 3(4)(y-4) \\ 10y+30 &= 12y-48 \\ -2y &= -78 \\ y &= 39 \end{aligned}$$

23.
$$\begin{aligned} \frac{2}{3x} &= \frac{6x}{36} \\ 2(36) &= 3x(6x) \\ 72 &= 18x^2 \\ 4 &= x^2 \\ x^2-4 &= 0 \\ (x+2)(x-2) &= 0 \end{aligned}$$
$x+2 = 0$ or $x-2 = 0$
$x = -2$ $\quad x = 2$

25.
$$\begin{aligned} \frac{2}{c} &= \frac{c-3}{2} \\ c(c-3) &= 4 \\ c^2-3c &= 4 \\ c^2-3c-4 &= 0 \\ (c-4)(c+1) &= 0 \end{aligned}$$
$c-4 = 0$ or $c+1 = 0$
$c = 4$ $\quad c = -1$

27.
$$\begin{aligned} \frac{1}{x+3} &= \frac{-2x}{x+5} \\ 1(x+5) &= -2x(x+3) \\ x+5 &= -2x^2-6x \\ 2x^2+7x+5 &= 0 \\ (2x+5)(x+1) &= 0 \end{aligned}$$
$2x+5 = 0$ or $x+1 = 0$
$2x = -5$ $\quad x = -1$
$x = -\frac{5}{2}$

29.
$$\begin{aligned} \frac{9z+6}{z(z+3)} &= \frac{7}{z+3} \\ 7z(z+3) &= (9z+6)(z+3) \\ 7z^2+21z &= 9z^2+33z+18 \\ 2z^2+12z+18 &= 0 \\ z^2+6z+9 &= 0 \\ (z+3)(z+3) &= 0 \\ z+3 &= 0 \\ z &= -3 \end{aligned}$$
Denominator would be zero. There is no solution.

31. $A = kp^2$

33. $v = \frac{k}{r^2}$

35. $P = \frac{ku^2}{j^3}$

37. L varies jointly with m and n.

39. R varies directly with L and inversely with d^2.

APPLICATIONS

41. Use the ratio $\frac{\text{ounces of soft drink}}{\text{mg of caffeine}}$ to determine the amount of caffeine. Let x represent the number of milligrams of caffeine in 44 ounces of drink and set up a proportion.

Mountain Dew
$$\begin{aligned} \frac{12}{55} &= \frac{44}{x} \\ 12x &= 44(55) \\ 12x &= 2420 \\ x &= 202 \text{ mg} \end{aligned}$$

Coca–Cola Classic
$$\begin{aligned} \frac{12}{47} &= \frac{44}{x} \\ 12x &= 44(47) \\ 12x &= 2068 \\ x &= 172 \text{ mg} \end{aligned}$$

Pepsi
$$\begin{aligned} \frac{12}{37} &= \frac{44}{x} \\ 12x &= 44(37) \\ 12x &= 1628 \\ x &= 136 \text{ mg} \end{aligned}$$

43. Let x represent the number of gallons of wall paper adhesive.

$$\frac{\text{gallons of adhesive}}{\text{square feet of wallpaper}} = \frac{1/2}{140} = \frac{x}{500}$$
$$140x = 250$$
$$x = 1.785\ldots \text{ or about 2 gallons of adhesive.}$$

45. Use the ratio $\frac{\text{height of option}}{\text{height of person}}$ to determine the height of each option. Let x represent the height of the option for a person who is 5ft 11 in. tall. (5 ft 11 in. is 71 in.)

Eye height
$$\frac{48.5}{69} = \frac{x}{71}$$
$$69x = 71(48.5)$$
$$69x = 3443.5$$
$$x = 49.9 \text{ in.}$$

Elbow height
$$\frac{27.0}{69} = \frac{x}{71}$$
$$69x \quad 71(27.0)$$
$$69x = 1917$$
$$x = 27.8 \text{ in.}$$

Seat height
$$\frac{17.1}{69} = \frac{x}{71}$$
$$69x = 71(17.1)$$
$$69x = 1214.1$$
$$x = 17.6 \text{ in.}$$

47. $\frac{\text{width}}{\text{length}} = \frac{3 \text{ in.}}{5 \text{ in.}} = \frac{7.5 \text{ in.}}{x}$, where x represents the length of the enlargement.
$$3x = 5(7.5)$$
$$3x = 37.5$$
$$x = 12.5 \text{ in.}$$

49. $\frac{\text{actual height}}{\text{length of shadow}} = \frac{x}{166\frac{1}{2}\text{ ft}} = \frac{5\text{ ft}}{1\frac{1}{2}\text{ ft}}$, where x represents the actual height of the Washington Monument.
$$1.5x = 5(166.5)$$
$$1.5x = 832.5$$
$$x = 555 \text{ ft.}$$

51. $\frac{\text{distance from common point}}{\text{distance perpendicular to river}} = \frac{75\text{ ft}}{w} = \frac{32\text{ ft}}{20\text{ ft}}$, where w represents the width of the river.
$$32w = 20(75)$$
$$32w = 1500$$
$$w = 46.875 \text{ ft} \quad \text{or } 46\tfrac{7}{8} \text{ ft.}$$

53. $\frac{\text{fall distance}}{\text{run distance}} = \frac{100\text{ ft}}{300\text{ ft}} = \frac{x}{2640\text{ ft}}$, where x represents the height of the hill. ($\frac{1}{2}$ mi $= \frac{1}{2}$(5280 ft) = 2640 ft)
$$300x = 100(2640)$$
$$300x = 264{,}000$$
$$x = 880 \text{ ft}$$

55. Direct variation model
$$s = kt^2$$

Determine k.
$$1024 = k(8)^2$$
$$1024 = 64k$$
$$k = 16$$

Solve for s when $t = 10$ sec.
$$s = 16t^2$$
$$s = 16(10)^2$$
$$s = 1600 \text{ ft}$$

57. Let t represent the time the bushels of corn will feed the cattle and c the number of cows fed.

Inverse variation model
$$t = \frac{k}{c}$$

Determine k.
$$10 = \frac{k}{25}$$
$$250 = k$$

Solve for t when $c = 10$ cows
$$t = \frac{250}{c}$$
$$t = \frac{250}{10} = 25 \text{ days}$$

59. Let V represent the volume of the gas and p the pressure applied.

Inverse variation model
$$V = \frac{k}{p}$$

Determine k.
$$20 = \frac{k}{6}$$
$$120 = k$$

Solve for V when $p = 10$ lbs
$$V = \frac{120}{p}$$
$$V = \frac{120}{10} = 12 \text{ in.}^3$$

61. Let C represent costs, t represent the number of trucks, and h the number of hours used.

Joint variation model
$$C = kth$$

Determine k.
$$1800 = k(4)(6)$$
$$1800 = 24k$$
$$75 = k$$

Solve for C when $t = 10$ and $h = 12$.
$$C = 75th$$
$$C = 75(10)(12)$$
$$C = \$9000$$

63. Let V represent the voltage and c the current flowing through the resistor.

Direct variation model
$$V = kc$$

Determine k, the resistance.
$$6 = k(2)$$
$$3 \text{ ohms} = k$$

65. Let D represent the deflection of a beam, w represent the beam's width, and d represent the beam's depth.

Inverse variation model
$$D = \frac{k}{wd^3}$$

Determine k.
$$1.1 = \frac{k}{4(4)^3}$$
$$281.6 = k$$

Solve for D when $w = 2$ in. and d = 8 in.
$$D = \frac{281.6}{w(d)^3}$$
$$D = \frac{281.6}{2(8)^3} = 0.275 \text{ in.}$$

67. Let T represent the tension, s the speed of the ball, and r the radius of the circle.

Combined variation model
$$T = \frac{ks^2}{r}$$

Determine k.
$$6 = \frac{k(6)^2}{3}$$
$$18 = 36k$$
$$0.5 = k$$

Solve for T when $s = 8$ ft/sec and $r = 2.5$ ft
$$T = \frac{0.5s^2}{r}$$
$$T = \frac{0.5(8)^2}{2.5} = 12.8 \text{ lb}$$

WRITING

69. A ratio is the relationship between quantities. A proportion is a relationship between ratios.

REVIEW

71. $(x^2x^3)^2 = (x^{2+3})^2 = (x^5)^2 = x^{5(2)} = x^{10}$

73. $\dfrac{b^0 - 2b^0}{b^0} = \dfrac{1-2(1)}{1} = 1 - 2 = -1$

Chapter 6 Key Concept: Expressions and Equations

Rational Expressions

1. a. $\frac{6x^2+x-2}{8x^2+2x-3}=\frac{(3x+2)(2x-1)}{(4x+3)(2x-1)}=\frac{3x+2}{4x+3}$ b. $2x-1$

2. a. $\frac{3d^2-d-2}{6d^2-5d-6}\cdot\frac{4d^2-9}{2d^2+5d+3}=\frac{(3d+2)(d-1)}{(3d+2)(2d-3)}\cdot\frac{(2d+3)(2d-3)}{(2d+3)(d+1)}=\frac{d-1}{d+1}$ b. $3d+2$, $2d-3$, $2d-3$

3. a. LCD = $(x+2)(x-4)$

$$\frac{3}{x+2}+\frac{5}{x-4}=\frac{3(x-4)}{(x+2)(x-4)}+\frac{5(x+2)}{(x-4)(x+2)}=\frac{3x-12+5x+10}{(x+2)(x-4)}=\frac{8x-2}{(x+2)(x-4)}=\frac{2(4x-1)}{(x+2)(x-4)}$$

b. $\frac{x-4}{x-4}$; $\frac{x+2}{x+2}$

4. a. $\frac{n-1-\frac{2}{n}}{\frac{n}{3}}=\frac{3n\left(n-1-\frac{2}{n}\right)}{3n\left(\frac{n}{3}\right)}=\frac{3n^2-3n-6}{n^2}=\frac{3(n^2-n-2)}{n^2}$ b. $3n$

Rational Equations

5. a.

$$\frac{t-3}{t-2}-\frac{t-3}{t}=\frac{1}{t}$$

$$t(t-2)\left(\frac{t-3}{t-2}\right)-t(t-2)\left(\frac{t-3}{t}\right)=t(t-2)\left(\frac{1}{t}\right)$$

$$t(t-3)-(t-2)(t-3)=1(t-2)$$

$$t^2-3t-(t^2-5t+6)=t-2$$

$$t^2-3t-t^2+5t-6=t-2$$

$$2t-6=t-2$$

$$t=4$$

b. Multiply both sides by LCD = $t(t-2)$.

6. a.

$$\frac{5}{2x^2+x-3}-\frac{x+1}{x-1}=\frac{2}{2x+3}-1$$

$$(2x+3)(x-1)\left(\frac{5}{(2x+3)(x-1)}\right)-(2x+3)(x-1)\left(\frac{x+1}{x-1}\right)=(2x+3)(x-1)\left(\frac{2}{2x+3}\right)-(1)(2x+3)(x-1)$$

$$5-(2x+3)(x+1)=2(x-1)-(2x^2+x-3)$$

$$5-(2x^2+5x+3)=2x-2-2x^2-x+3$$

$$5-2x^2-5x-3=-2x^2+x+1$$

$$-6x=-1$$

$$x=\frac{1}{6}$$

b. Multiply both sides by LCD = $(2x+3)(x-1)$.

7. a.

$$\frac{x+1}{x+2}=\frac{x-3}{x-4}$$
$$(x+2)(x-4)\left(\frac{x+1}{x+2}\right)=(x+2)(x-4)\left(\frac{x-3}{x-4}\right)$$
$$(x-4)(x+1)=(x+2)(x-3)$$
$$x^2-3x-4=x^2-x-6$$
$$-2x=-1$$
$$x=1$$

b. Multiply both sides by LCD = $(x+2)(x-4)$.

8. a.

$$\frac{x^2}{a^2}-\frac{y^2}{b^2}=1$$
$$a^2b^2\left(\frac{x^2}{a^2}\right)-a^2b^2\left(\frac{y^2}{b^2}\right)=(1)a^2b^2$$
$$b^2x^2-a^2y^2=a^2b^2$$
$$b^2x^2=a^2b^2+a^2y^2$$
$$b^2x^2=a^2(b^2+y^2)$$
$$\frac{b^2x^2}{b^2+y^2}=a^2$$

b. Multiply both sides by LCD = a^2b^2.

Chapter 6 Chapter Review

Section 6.1

1.

x	$f(x)=4/x$	$f(x)$
$1/2$	$f(x)=\frac{4}{\frac{1}{2}}=\frac{4}{1}\left(\frac{2}{1}\right)$	**8**
1	$f(x)=4/1$	**4**
2	$f(x)=4/2$	**2**
3	$f(x)=4/3$	**1.33**
4	$f(x)=4/4$	**1**
5	$f(x)=4/5$	**0.8**
6	$f(x)=4/6$	**0.67**
7	$f(x)=4/7$	**0.57**
8	$f(x)=4/8$	**0.5**

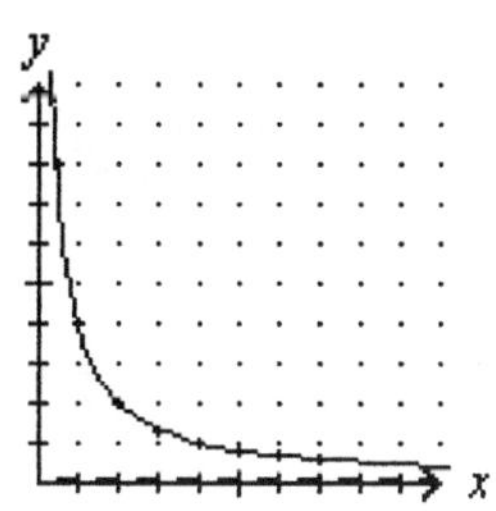

Horizontal asymptote is $y=0$

2. Vertical asymptote at $x=0$. D: $(-\infty, 0)\cup(0,\infty)$
Horizontal asymptote at $y=3$. R: $(-\infty, 3)\cup(3,\infty)$

3. $$\frac{248x^2y}{576xy^2}=\frac{8(31)}{8(72)}x^{2-1}y^{1-2}=\frac{31}{72}x^1y^{-1}=\frac{31x}{72y}$$

4. $$\frac{x^2-49}{x^2+14x+49}=\frac{(x+7)(x-7)}{(x+7)(x+7)}=\frac{x-7}{x+7}$$

5. $$\frac{x^2-2x+4}{2x^3+16}=\frac{x^2-2x+4}{2(x^3+8)}=\frac{x^2-2x+4}{2(x+2)(x^2-2x+4)}=\frac{1}{2(x+2)}$$

6. $\dfrac{x^2+6x+36}{x^3-216}=\dfrac{x^2+6x+36}{(x-6)(x^2+6x+36)}=\dfrac{1}{x-6}$

7. $\dfrac{ac-ad+bc-bd}{d^2-c^2}=\dfrac{a(c-d)+b(c-d)}{(d+c)(d-c)}=\dfrac{(c-d)((a+b)}{(d+c)(d-c)}=\dfrac{c-d}{d-c}\bullet\dfrac{a+b}{c+d}=-1\dfrac{a+b}{c+d}=\dfrac{-a-b}{c+d}$

8. $\dfrac{m^3+m^2n-2mn^2}{mn^2+m^2n-2m^3}=\dfrac{m(m^2+mn-2n^2)}{m(n^2+mn-2m^2)}=\dfrac{m(m+2n)(m-n)}{m(n+2m)((n-m)}=-1\dfrac{m+2n}{n+2m}=\dfrac{-m-2n}{2m+n}$

9. $\dfrac{x-y}{y-x}=\dfrac{-1(-x+y)}{y-x}=-1$

10. $\dfrac{2m-2n}{n-m}=\dfrac{2(m-n)}{n-m}=2(-1)=-2$

Section 6.2

11. $\dfrac{x^2+4x+4}{x^2-x-6}\bullet\dfrac{9-x^2}{x^2+5x+6}=\dfrac{(x+2)(x+2)(3-x)(3+x)}{(x-3)(x+2)(x+3)(x+2)}=\dfrac{3-x}{x-3}=-1$

12. $\dfrac{2a^2-5a-3}{a^2-9}\div\dfrac{2a^2+5a+2}{2a^2+5a-3}=\dfrac{2a^2-5a-3}{a^2-9}\bullet\dfrac{2a^2+5a-3}{2a^2+5a+2}=\dfrac{(2a+1)(a-3)(2a-1)(a+3)}{(a+3)(a-3)(2a+1)(a+2)}=\dfrac{2a-1}{a+2}$

13. $\left(\dfrac{h-2}{h^3+4}\right)^2=\dfrac{(h-2)(h-2)}{(h^3+4)(h^3+4)}=\dfrac{h^2-4h+4}{h^6+8h^3+16}$

14. $\dfrac{t^2-4}{t}\div(t+2)=\dfrac{t^2-4}{t}\bullet\dfrac{1}{t+2}=\dfrac{(t+2)(t-2)}{t(t+2)}=\dfrac{t-2}{t}$

15.
$$\dfrac{x^2+3x+2}{x^2-x-6}\bullet\dfrac{3x^2-3x}{x^2-3x-4}\div\dfrac{x^2+3x+2}{x^2-2x-8}=\dfrac{x^2+3x+2}{x^2-x-6}\bullet\dfrac{3x^2-3x}{x^2-3x-4}\bullet\dfrac{x^2-2x-8}{x^2+3x+2}$$
$$=\dfrac{(x+2)(x+1)(3x)(x-1)(x-4)(x+2)}{(x-3)(x+2)(x-4)(x+1)(x+2)(x+1)}=\dfrac{3x(x-1)}{(x-3)(x+1)}$$

Section 6.3

16. $\dfrac{5y}{x-y}-\dfrac{3}{x-y}=\dfrac{5y-3}{x-y}$

17. $\dfrac{3x-1}{x^2+2}+\dfrac{3(x-2)}{x^2+2}=\dfrac{3x-1+3x-6}{x^2+2}=\dfrac{6x-7}{x^2+2}$

18. $\dfrac{4}{t-3}+\dfrac{6}{3-t}=\dfrac{4}{t-3}+\dfrac{-1(6)}{-1(3-t)}=\dfrac{4-6}{t-3}=-\dfrac{2}{t-3}$

19. $\dfrac{p+3}{p^2+13p+12}-\dfrac{2p+4}{p^2+13p+12}=\dfrac{p+3-2p-4}{p^2+13p+12}=\dfrac{-p-1}{(p+12)(p+1)}=\dfrac{-(p+1)}{(p+12)(p+1)}=-\dfrac{1}{p+12}$

20. $\left.\begin{cases}15a^2h=3\cdot5\cdot a\cdot a\cdot h\\ 20ah^3=2\cdot2\cdot5\cdot a\cdot h\cdot h\cdot h\end{cases}\right\}\ \text{LCD}=2^2\cdot3\cdot5\cdot a^2\cdot h^3=60a^2h^3$

21. $\left.\begin{cases}ab^2-ab=ab(b-1)\\ ab^2=a\cdot b^2\\ b^2-b=b(b-1)\end{cases}\right\}\ \text{LCD}=a\cdot b^2(b-1)=ab^2(b-1)$

22. $\left.\begin{cases} x^2 - 4x - 5 = (x-5)(x+1) \\ x^2 - 25 = (x+5)(x-5) \end{cases}\right\}$ LCD = $(x-5)(x+5)(x+1)$

23. $\left.\begin{cases} m^2 - 4m + 4 = (m-2)(m-2) \\ m^3 - 8 = (m-2)(m^2+2m+4) \end{cases}\right\}$ LCD = $(m-2)^2(m^2+2m+4)$

24. LCD = (a + 1) $\quad 9 - \dfrac{1}{a+1} = \dfrac{9(a+1)}{a+1} - \dfrac{1}{a+1} = \dfrac{9a+9-1}{a+1} = \dfrac{9a+8}{a+1}$

25. LCD = 2 • 2 • 3 • z^2 = $12z^2$ $\quad \dfrac{x}{4z^2} + \dfrac{y}{6z} = \dfrac{x(3)}{4z^2(3)} + \dfrac{y(2z)}{6z(2z)} = \dfrac{3x}{12z^2} + \dfrac{2yz}{12z^2} = \dfrac{3x+2yz}{12z^2}$

26. LCD = $(x-4)(x+3)$

$$\frac{4x}{x-4} - \frac{3}{x+3} = \frac{4x(x+3)}{(x-4)(x+3)} - \frac{3(x-4)}{(x+3)(x-4)} = \frac{4x^2+12x-(3x-12)}{(x-4)(x+3)} = \frac{4x^2+12x-3x+12}{(x-4)(x+3)} = \frac{4x^2+9x+12}{(x-4)(x+3)}$$

27.. $\{3xy - 6y = 3y(x-2);\ 10 - 5x = 5(2-x)\}$ LCD = $3y(5)(x-2) = 15y(x-2)$

Multiply the numerator and denominator of the last fraction by –1.

$$\frac{4}{3xy-6y} - \frac{4}{10-5x} = \frac{4(5)}{3y(x-2)(5)} - \frac{-1(4)(3y)}{(-1)(5)(2-x)(3y)} = \frac{20+12y}{15y(x-2)} = \frac{4(5+3y)}{15y(x-2)}$$

28. $\{x^2 + 6x + 9 = (x+3)^2;\ x^2 - 6x + 9 = (x-3)^2;\ x^2 - 9 = (x+3)(x-3)\}$ LCD = $(x+3)(x-3)^2$

Cancel one $(x+3)$ from the first fraction before multiplying.

$$\frac{-2(3+x)}{x^2+6x+9} + \frac{3(x+2)}{x^2-6x+9} - \frac{1}{x^2-9} = \frac{-2(x+3)(x-3)^2}{(x+3)^2(x-3)^2} + \frac{(3x+6)(x+3)}{(x-3)^2(x+3)} - \frac{1(x-3)}{(x+3)(x-3)^2}$$

$$= \frac{-2(x^2-6x+9)}{(x+3)(x-3)^2} + \frac{3x^2+15x+18}{(x-3)^2(x+3)} - \frac{x-3}{(x+3)(x-3)^2}$$

$$= \frac{-2x^2+12x-18+3x^2+15x+18-x+3}{(x+3)(x-3)^2} = \frac{x^2+26x+3}{(x+3)(x-3)^2}$$

Section 6.4

29. Multiply by LCD = $6pt$ and simplify.

$$\frac{\dfrac{p^2-9}{6pt}}{\dfrac{p^2+5p+6}{3pt}} = \frac{6pt\left(\dfrac{p^2-9}{6pt}\right)}{6pt\left(\dfrac{p^2+5p+6}{3pt}\right)} = \frac{p^2-9}{2(p^2+5p+6)} = \frac{(p+3)(p-3)}{2(p+3)(p+2)} = \frac{p-3}{2(p+2)}$$

30. Multiply by LCD = ab and simplify.

$$\frac{\dfrac{1}{a}+\dfrac{2}{b}}{\dfrac{2}{a}-\dfrac{1}{b}} = \frac{ab\left(\dfrac{1}{a}+\dfrac{2}{b}\right)}{ab\left(\dfrac{2}{a}-\dfrac{1}{b}\right)} = \frac{b+2a}{2b-a}$$

31. Multiply by LCD = x^2 and simplify.

$$\frac{1-\frac{1}{x}-\frac{2}{x^2}}{1+\frac{4}{x}+\frac{3}{x^2}}=\frac{x^2\left(1-\frac{1}{x}-\frac{2}{x^2}\right)}{x^2\left(1+\frac{4}{x}+\frac{3}{x^2}\right)}=\frac{x^2-x-2}{x^2+4x+3}=\frac{(x-2)(x+1)}{(x+3)(x+1)}=\frac{x-2}{x+3}$$

32. Write the fraction without negative exponents, multiply by LCD = $x^2y^2(x-y)^2$ and simplify.

$$\frac{(x-y)^{-2}}{x^{-2}-y^{-2}}=\frac{\frac{1}{(x-y)^2}}{\frac{1}{x^2}-\frac{1}{y^2}}=\frac{x^2y^2(x-y)^2\left(\frac{1}{(x-y)^2}\right)}{x^2y^2(x-y)^2\left(\frac{1}{x^2}-\frac{1}{y^2}\right)}=\frac{x^2y^2}{y^2(x-y)^2-x^2(x-y)^2}=\frac{x^2y^2}{(x-y)^2(y^2-x^2)}$$

33. Multiply the second fraction in the numerator by LCD = 2 and then by $\frac{-1}{-1}$.

$$\frac{3x-\frac{1}{3-\frac{x}{2}}}{\frac{3}{x-6}+x}=\frac{3x-\frac{2(1)}{2\left(3-\frac{x}{2}\right)}}{\frac{3}{x-6}+x}=\frac{3x-\frac{2}{6-x}}{\frac{3}{x-6}+x}=\frac{3x+\frac{2}{x-6}}{\frac{3}{x-6}+x}$$

Next, multiply by the LCD = $(x-6)$ and simplify.

$$=\frac{3x+\frac{2}{x-6}}{\frac{3}{x-6}+x}=\frac{(x-6)\left(3x+\frac{2}{x-6}\right)}{(x-6)\left(\frac{3}{x-6}+x\right)}=\frac{3x(x-6)+2}{3+x(x-6)}=\frac{3x^2-18x+2}{3+x^2-6x}=\frac{3x^2-18x+2}{x^2-6x+3}$$

Section 6.5

34. Multiply both sides by LCD = $10x$.

$$\begin{aligned}\frac{4}{x}-\frac{1}{10}&=\frac{7}{2x}\\10x\left(\frac{4}{x}\right)-10x\left(\frac{1}{10}\right)&=10x\left(\frac{7}{2x}\right)\\10(4)-x(1)&=5(7)\\40-x&=35\\-x&=-5\\x&=5\end{aligned}$$

35. Multiply both sides by LCD = $t+7$.

$$\begin{aligned}\frac{5}{7+t}-1&=\frac{-4}{t+7}\\(t+7)\left(\frac{5}{7+t}\right)-1(t+7)&=(t+7)\left(\frac{-4}{t+7}\right)\\5-t-7&=-4\\-t&=-2\\t&=2\end{aligned}$$

36. Multiply both sides by LCD = $6(q + 4)$.

$$\frac{q-3}{6} = \frac{10}{q+4}$$
$$6(q+4)\left(\frac{q-3}{6}\right) = 6(q+4)\left(\frac{10}{q+4}\right)$$
$$(q+4)(q-3) = 6(10)$$
$$q^2 + q - 12 = 60$$
$$q^2 + q - 72 = 0$$
$$(q+9)(q-8) = 0$$

$$q+9 = 0 \quad \text{or} \quad q-8 = 0$$
$$q = -9 \qquad\qquad q = 8$$

37. Multiply both sides by the LCD = $6(x + 5)(x + 4)$.

$$\frac{2}{x+5} - \frac{1}{6} = \frac{1}{x+4}$$
$$6(x+5)(x+4)\left(\frac{2}{x+5}\right) - 6(x+5)(x+4)\left(\frac{1}{6}\right) = 6(x+5)(x+4)\left(\frac{1}{x+4}\right)$$
$$6(2)(x+4) - (x+5)(x+4) = 6(1)(x+5)$$
$$12x + 48 - (x^2 + 9x + 20) = 6x + 30$$
$$12x + 48 - x^2 - 9x - 20 = 6x + 30$$
$$0 = x^2 + 3x + 2$$
$$0 = (x+1)(x+2)$$
$$x+1 = 0 \quad \text{or} \quad x+2 = 0$$
$$x = -1 \qquad\qquad x = -2$$

38. Multiply the fraction on the right side by 1 in the form $\frac{-1}{-1}$ and rearrange. Then multiply both sides by the LCD = $(x + 2)(x - 2)$.

$$\frac{2(x-5)}{x-2} = \frac{6x+12}{4-x^2}$$
$$\frac{2(x-5)}{x-2} = \frac{-1(6x+12)}{-1(4-x^2)}$$
$$(x+2)(x-2)\left(\frac{2x-10}{x-2}\right) = (x+2)(x-2)\left(\frac{-6x-12}{x^2-4}\right)$$
$$(x+2)(2x-10) = -6x - 12$$
$$2x^2 - 6x - 20 = -6x - 12$$
$$2x^2 - 8 = 0$$
$$x^2 - 4 = 0$$
$$(x+2)(x-2) = 0$$
$$x+2 = 0 \quad \text{or} \quad x-2 = 0$$
$$x = -2 \qquad\qquad x = 2$$

Both $x = -2$ and $x = 2$ would make a denominator zero so they are extraneous and there is no solution.

39. Multiply both sides by the LCD = $x^2 - 7x + 10 = (x-2)(x-5)$.

$$\frac{x+3}{x-5}+\frac{6+2x^2}{x^2-7x+10}=\frac{3x}{x-2}$$

$$(x-2)(x-5)\left(\frac{x+3}{x-5}\right)+(x-2)(x-5)\left(\frac{6+2x^2}{x^2-7x+10}\right)=(x-2)(x-5)\left(\frac{3x}{x-2}\right)$$

$$(x-2)(x+3)+(6+2x^2)=(x-5)(3x)$$

$$x^2+x-6+6+2x^2=3x^2-15x$$

$$16x=0$$

$$x=0$$

40.

$\frac{x^2}{a^2}-\frac{y^2}{b^2}=1$ Solve for y^2.

$a^2b^2\left(\frac{x^2}{a^2}\right)-a^2b^2\left(\frac{y^2}{b^2}\right)=a^2b^2(1)$ Multiply both sides by a^2b^2.

$b^2x^2-a^2y^2=a^2b^2$ Do the multiplication.

$b^2x^2-a^2b^2=a^2y^2$ Subtract a^2b^2 and add a^2y^2 to both sides.

$\frac{b^2x^2-a^2b^2}{a^2}=y^2$ Divide both sides by a^2

41.

$H=\frac{2ab}{a+b}$ Solve for b.

$H(a+b)=2ab$ Multiply both sides by $a+b$.

$aH+bH=2ab$ Distribute H.

$bH-2ab=-aH$ Subtract $2ab$ and aH from both sides.

$b(H-2a)=aH$ Factor.

$b=\frac{-aH}{H-2a}$ or $b=\frac{Ha}{2a-H}$ Divide both sides by $H-2a$.

42. **Analysis**: Find the rate of the wind.

Equation: Let x represent the rate of the wind. Then $75+x$ represents the speed of the plane going downwind (with the wind) and $75-x$ represents the speed of the plane going upwind (against the wind).

If $r \bullet t = d$, then $t=\frac{d}{r}$

	r •	t =	d
Downwind	$75+x$	$\frac{40}{75+x}$	40
Upwind	$75-x$	$\frac{35}{75-x}$	35

Downwind time equals upwind time.

$$\frac{40}{75+x}=\frac{35}{75-x}$$

42. cont.

Solve:

$$\frac{40}{75+x}=\frac{35}{75-x}$$
$$(75+x)(75-x)\left(\frac{40}{75+x}\right)=(75+x)(75-x)\left(\frac{35}{75-x}\right)$$
$$40(75-x)=35(75+x)$$
$$3000-40x=2625+35x$$
$$-75x=-375$$
$$x=5$$

Conclusion: The rate of the wind was $x = 5$ mph.

43. **Analysis**: Find the a driver's usual speed for a trip of 200 miles.

Equation: Let x represent the driver's usual speed of the slow train. Then $x - 10$ represents the reduced speed.

If $r \bullet t = d$, then $t = \dfrac{d}{r}$

	r	• t	$= d$
Usual speed	x	$\frac{200}{x}$	200
Reduced speed	$x - 10$	$\frac{200}{x-10}$	200

Usual time	plus	1 hour	is	reduced speed time.
$\frac{200}{x}$	$+$	1	$=$	$\frac{200}{x-10}$

Solve:

$$\frac{200}{x}+1=\frac{200}{x-10}$$
$$x(x-10)\left(\frac{200}{x}\right)+1x(x-10)=x(x-10)\left(\frac{200}{x-10}\right)$$ Multiply both sides by LCD = $x(x - 10)$.
$$200(x-10)+(x^2-10x)=200x$$
$$200x-2000+x^2-10x=200x$$
$$x^2-10x-2000=0$$
$$(x-50)(x+40)=0$$
$$x-50=0 \quad \text{or} \quad x+40=0$$
$$x=50 \qquad\qquad x=-40$$ Discard the negative solution.

Conclusion: The driver's usual speed is $x = 50$ mph.

44. **Analysis**: Find the amount of time it will take the two pipes to drain a tank together.

Equation: Let x represent the amount of time it will take the two pipes working together to finish the job. The time worked in given in hours.

Amount that pipe 1 drains	plus	amount that pipe 2 drains	is	amount drained together in one hour.
$\frac{1}{24}$	$+$	$\frac{1}{36}$	$=$	$\frac{1}{x}$

Solve:

$$\frac{1}{24}+\frac{1}{36}=\frac{1}{x}$$
$$72x\left(\frac{1}{24}\right)+72x\left(\frac{1}{36}\right)=72x\left(\frac{1}{x}\right)$$
$$3x+2x=72$$
$$5x=72$$
$$x=\frac{72}{5}=14\frac{2}{5}$$

Conclusion: It will take $14\frac{2}{5}$ hours to drain the pond working together.

45. **Analysis**: Find the amount of time it will take for the one man to side the house alone when the man who got sick could do the job alone in 14 days.

Equation: Let x represent the amount of time it will take the one man working alone. It would have taken the two men 8 days to do the job together

Work man can do in one day	plus	work sick man can do in one day	is	work they can do together in one day.
$\frac{1}{x}$	$+$	$\frac{1}{14}$	$=$	$\frac{1}{8}$

Solve:

$$\frac{1}{x}+\frac{1}{14}=\frac{1}{8}$$
$$56x\left(\frac{1}{x}\right)+56x\left(\frac{1}{14}\right)=56x\left(\frac{1}{8}\right)$$
$$56+4x=7x$$
$$-3x=-56$$
$$x=\frac{-56}{-3}=18\frac{2}{3}$$

Conclusion: It will take the one man $18\frac{2}{3}$ days to side the house working alone.

46 Substitute $k = 1{,}900{,}000$ and $k_2 = 4{,}200{,}000$.

$$k = \frac{1}{\dfrac{1}{k_1} + \dfrac{1}{k_2}}$$

$$1{,}900{,}000 = \frac{1}{\dfrac{1}{k_1} + \dfrac{1}{4{,}200{,}000}}$$

$$1{,}900{,}000\left(\frac{1}{k_1} + \frac{1}{4{,}200{,}000}\right) = 1$$ Multiply the means and the extremes.

$$\frac{1{,}900{,}000}{k_1} + \frac{1{,}900{,}000}{4{,}200{,}000} = 1$$

$$\frac{1{,}900{,}000}{k_1} = \frac{4{,}200{,}000 - 1{,}900{,}000}{4{,}200{,}000} = \frac{2{,}300{,}000}{4{,}200{,}000} = \frac{23}{42}$$

$$23k_1 = 42(1{,}900{,}000) = 79{,}800{,}000$$

$$k_1 = 3{,}469{,}565 \text{ or about } 3{,}500{,}000 \text{ lb/rad}$$

Section 6.6

47. $\dfrac{25h^4k^7}{5hk^9} = \dfrac{25}{5}h^{4-1}k^{7-9} = 5h^3k^{-2} = \dfrac{5h^3}{k^2}$

48. $(-5x^6y^3) \div (10x^3y^6) = \dfrac{-5x^6y^3}{10x^3y^6} = \dfrac{-5}{10}x^{6-3}y^{3-6} = -\dfrac{1}{2}x^3y^{-3} = -\dfrac{x^3}{2y^3}$

49. $\dfrac{36a+32}{6} = \dfrac{36a}{6} + \dfrac{32}{6} = \dfrac{6 \cdot 6a}{6} + \dfrac{2 \cdot 16}{2 \cdot 3} = 6a + \dfrac{16}{3}$

50. $\dfrac{30x^3y^2 - 15x^2y - 10xy^2}{-10xy} = \dfrac{30x^3y^2}{-10xy} + \dfrac{-15x^2y}{-10xy} + \dfrac{-10xy^2}{-10xy} = -3x^2y + \dfrac{3x}{2} + y$

51.

$$\begin{array}{r} b+4 \\ b+5\overline{)\,b^2+9b+20} \\ \underline{b^2+5b} \\ 4b+20 \\ \underline{4b+20} \\ 0 \end{array}$$

Solution: $b+4$

52. Rearrange the terms before dividing.

$$\begin{array}{r} v^2-3v-10 \\ 3v+1\overline{)\,3v^3-8v^2-33v-10} \\ \underline{3v^3+v^2} \\ -9v^2-33v \\ \underline{-9v^2-3v} \\ -30v-10 \\ \underline{-30v-10} \\ 0 \end{array}$$

Solution: $v^2 - 3v - 10$

53. Add $0x^2$ and $0x$ before dividing.

$$\begin{array}{r|l} & x^2 - 2x + 4 \\ \hline x+2 & x^3 + 0x^2 + 0x + 8 \\ & \underline{x^3 + 2x^2} \\ & -2x^2 + 0x \\ & \underline{-2x^2 - 4x} \\ & 4x + 8 \\ & \underline{4x + 8} \\ & 0 \end{array}$$

Solution: $x^2 - 2x + 4$

54. Rearrange the terms before dividing.

$$\begin{array}{r|l} & x^2 + 2x - 1 \\ \hline 2x+3 & 2x^3 + 7x^2 + 4x + 3 \\ & \underline{2x^3 + 3x^2} \\ & 4x^2 + 4x \\ & \underline{4x^2 + 6x} \\ & -2x + 3 \\ & \underline{-2x - 3} \\ & 6 \end{array}$$

Solution: $x^2 + 2x - 1 + \dfrac{6}{2x+3}$

55. Add $0x^2$ before dividing.

$$\begin{array}{r|rrrr} 4 & 1 & 0 & -13 & -12 \\ & & 4 & 16 & 12 \\ \hline & 1 & 4 & 3 & 0 \end{array}$$

Solution: $x^2 + 4x + 3$

56. Rearrange the terms and add $0x^3$ and $0x$ before dividing.

$$\begin{array}{r|rrrrr} -1 & 1 & 0 & 1 & 0 & 1 \\ & & -1 & 1 & -2 & 2 \\ \hline & 1 & -1 & 2 & -2 & 3 \end{array}$$

Solution: $x^3 - x^2 + 2x - 2 + \dfrac{3}{x+1}$

57. Divide $P(x) = 3x^2 - 2x + 3$ by $x - (-1)$.

$$\begin{array}{r|rrr} -1 & 3 & -2 & 3 \\ & & -3 & 5 \\ \hline & 3 & -5 & 8 \end{array}$$

Remainder is 8 so $P(-1) = 8$.

Section 6.7

58.

$$\begin{aligned} \frac{x+1}{8} &= \frac{4x-2}{24} \\ 24(x+1) &= 8(4x-2) \\ 24x + 24 &= 32x - 16 \\ -8x &= -40 \\ x &= 5 \end{aligned}$$

59.

$$\begin{aligned} \frac{1}{x+6} &= \frac{x+10}{12} \\ (x+6)(x+10) &= 12 \\ x^2 + 16x + 60 &= 12 \\ x^2 + 16x + 48 &= 0 \\ (x+12)(x+4) &= 0 \end{aligned}$$

$x + 12 = 0$ or $x + 4 = 0$

$x = -12$ $\qquad$ $x = -4$

60. $\dfrac{\text{actual height}}{\text{length of shadow}} = \dfrac{x}{44\text{ ft}} = \dfrac{4\text{ ft}}{2\frac{1}{2}\text{ ft}}$, where x represents the actual height of the tree.

$$\begin{aligned} 2.5x &= 4(44) \\ 2.5x &= 176 \\ x &= 70.4\text{ ft.} \end{aligned}$$

61. Let T represent the property tax and v the assessed value.

Direct variation model

$T = kv$

Determine k.

$$\begin{aligned} 1575 &= k(90{,}000) \\ k &= 0.0175 \end{aligned}$$

Solve for T when $v = \$312{,}000$.

$$\begin{aligned} T &= 0.0175v \\ T &= 0.0175(312{,}000) \\ T &= \$5460 \end{aligned}$$

62. Let C represent the current and r the resistance in the circuit.

Inverse variation model
$$C = \frac{k}{r}$$

Determine k.
$$2.5 = \frac{k}{150}$$
$$375 = k$$

Solve for C when $r = 2(150) = 300$ ohms
$$C = \frac{375}{r}$$
$$C = \frac{375}{300} = 1.25 \text{ amps}$$

63. Let F represent the wind force, A represent the surface area, and v the velocity of the wind. The area is $lw = (3 \text{ ft})(1.5\text{ft}) = 4.5 \text{ ft}^2$.

Joint variation model
$$F = kAv^2$$

Determine k.
$$1.98 = k(4.5)(10)^2$$
$$1.98 = 450k$$
$$0.0044 = k$$

Solve for F when $A = 4.5$, and $v = 80$.
$$F = 0.0044Av^2$$
$$F = 0.0044(4.5)(80)^2$$
$$F = 126.72 \text{ lb}$$

64. inverse variation

65. Combined variation model
$$x_1 = \frac{kt^3}{x_2}$$

Determine k.
$$1.6 = \frac{k(8)^3}{64}$$
$$102.4 = 512k$$
$$0.2 = k$$

Chapter 6 Chapter Test

1. $\dfrac{-12x^2y^3z^2}{18x^3y^4z^2} = \dfrac{-12}{18}x^{2-3}y^{3-4}z^{2-2} = -\dfrac{2}{3}x^{-1}y^{-1}z^0 = -\dfrac{2}{3xy}$

2. $\dfrac{2x+4}{x^2-4} = \dfrac{2(x+2)}{(x+2)(x-2)} = \dfrac{2}{x-2}$

3. $\dfrac{3y-6z}{2z-y} = \dfrac{3(y-2z)}{2z-y} = 3\cdot\dfrac{y-2z}{2z-y} = 3(-1) = -3$

4. $\dfrac{2x^2+7x+3}{4x+12} = \dfrac{(2x+1)(x+3)}{4(x+3)} = \dfrac{2x+1}{4}$

5. $\dfrac{x^2}{x^3z^2y^2}\cdot\dfrac{x^2z^4}{y^2z} = \dfrac{x^2x^2z^4}{x^3z^2y^2y^2z} = x^{2+2-3}y^{-2-2}z^{4-2-1} = x^1y^{-4}z^1 = \dfrac{xz}{y^4}$

6. $\dfrac{(x+1)(x+2)}{10}\cdot\dfrac{5}{x+2} = \dfrac{5(x+1)(x+2)}{2\cdot 5(x+2)} = \dfrac{x+1}{2}$

7. $\dfrac{u^2+5u+6}{u^2-4}\cdot\dfrac{u^2-5u+6}{u^2-9} = \dfrac{(u+3)(u+2)(u-2)(u-3)}{(u+2)(u-2)(u+3)(u-3)} = 1$

8. $\dfrac{x^3+y^3}{4}\div\dfrac{x^2-xy+y^2}{2x+2y} = \dfrac{x^3+y^3}{4}\cdot\dfrac{2x+2y}{x^2-xy+y^2} = \dfrac{(x+y)(x^2-xy+y^2)(2)(x+y)}{2(2)(x^2-xy+y^2)} = \dfrac{(x+y)^2}{2}$

9. $\dfrac{xu+2u+3x+6}{u^2-9}\cdot\dfrac{2u-6}{x^2+3x+2} = \dfrac{[u(x+2)+3(x+2)](2)(u-3)}{(u+3)(u-3)(x+2)(x+1)} = \dfrac{2(x+2)(u+3)(u-3)}{(u+3)(u-3)(x+2)(x+1)} = \dfrac{2}{x+1}$

10. $\dfrac{a^2+7a+12}{a+3}\div\dfrac{16-a^2}{a-4} = \dfrac{a^2+7a+12}{a+3}\cdot\dfrac{a-4}{16-a^2} = \dfrac{(a+3)(a+4)(a-4)}{(a+3)(4+a)(4-a)} = \dfrac{a-4}{4-a} = -1$

11. $\dfrac{-3t+4}{t^2+t-20}+\dfrac{6+5t}{t^2+t-20}=\dfrac{-3t+4+6+5t}{t^2+t-20}=\dfrac{2t+10}{(t+5)(t-4)}=\dfrac{2(t+5)}{(t+5)(t-4)}=\dfrac{2}{t-4}$

12. LCD = $w-5$ Multiply the numerator and denominator of the last fraction by -1.

$$\frac{3w}{w-5}+\frac{w+10}{5-w}=\frac{3w}{w-5}+\frac{-1(w+10)}{-1(5-w)}=\frac{3w-w-10}{w-5}=\frac{2w-10}{w-5}=\frac{2(w-5)}{w-5}=2$$

13. LCD = $3b+1$ $\quad 8b-5+\dfrac{5b+4}{3b+1}=\left(\dfrac{8b-5}{1}\right)\left(\dfrac{3b+1}{3b+1}\right)+\dfrac{5b+4}{3b+1}=\dfrac{24b^2-7b-5+5b+4}{3b+1}=\dfrac{24b^2-2b-1}{3b+1}$

14. LCD = $(x+1)(x+2)$

$$\frac{x+2}{x+1}-\frac{x+1}{x+2}=\frac{(x+2)(x+2)}{(x+1)(x+2)}-\frac{(x+1)(x+1)}{(x+2)(x+1)}=\frac{x^2+4x+4}{(x+1)(x+2)}-\frac{x^2+2x+1}{(x+1)(x+2)}$$

$$=\frac{x^2+4x+4-x^2-2x-1}{(x+1)(x+2)}=\frac{2x+3}{(x+1)(x+2)}$$

15. Multiply by LCD = uv^2

$$\frac{\dfrac{2u^2w^3}{v^2}}{\dfrac{4uw^4}{uv}}=\frac{uv^2\left(\dfrac{2u^2w^3}{v^2}\right)}{uv^2\left(\dfrac{4uw^4}{uv}\right)}=\frac{u(2u^2w^3)}{v(4uw^4)}=\frac{2u^3w^3}{4uvw^4}=\frac{2}{4}u^{3-1}v^{-1}w^{3-4}=\frac{1}{2}u^2v^{-1}w^{-1}=\frac{u^2}{2vw}$$

16. Multiply by LCD = $3k(k+1)$

$$\frac{\dfrac{4}{3k}+\dfrac{k}{k+1}}{\dfrac{k}{k+1}-\dfrac{3}{k}}=\frac{3k(k+1)\left(\dfrac{4}{3k}+\dfrac{k}{k+1}\right)}{3k(k+1)\left(\dfrac{k}{k+1}-\dfrac{3}{k}\right)}=\frac{4(k+1)+3k(k)}{3k(k)-3(3)(k+1)}=\frac{4k+4+3k^2}{3k^2-9k-9}=\frac{3k^2+4k+4}{3k^2-9k-9}$$

17. Multiply both sides by LCD = $20x$.

$$\frac{34}{x}+\frac{13}{20}=\frac{3}{2}$$
$$20x\left(\frac{34}{x}\right)+20x\left(\frac{13}{20}\right)=20x\left(\frac{3}{2}\right)$$
$$20(34)+13x=3(10x)$$
$$680+13x=30x$$
$$-17x=-680$$
$$x=40$$

18. Multiply the last fraction by 1 in the form $\frac{-1}{-1}$ and then multiply both sides by LCD = $u-3$.

$$\frac{u-2}{u-3}+3=u+\frac{u-4}{3-u}$$
$$\frac{u-2}{u-3}+3=u+\frac{-1(u-4)}{-1(3-u)}$$
$$(u-3)\left(\frac{u-2}{u-3}\right)+3(u-3)=u(u-3)+(u-3)\left(\frac{-u+4}{u-3}\right)$$
$$u-2+3u-9=u^2-3u-u+4$$
$$4u-11=u^2-4u+4$$
$$0=u^2-8u+15$$
$$0=(u-5)(u-3)$$

$u-5=0$ or $u-3=0$

$u=5$ $\qquad u=3$

$x=3$ would make the denominator 0, so it is extraneous.

19. Multiply both sides by LCD = $2x(x-2)$.

$$\frac{3}{x-2}=\frac{x+3}{2x}$$

$$2x(x-2)\left(\frac{3}{x-2}\right)=2x(x-2)\left(\frac{x+3}{2x}\right)$$

$$3(2x)=(x-2)(x+3)$$

$$6x=x^2+x-6$$

$$0=x^2-5x-6$$

$$0=(x-6)(x+1)$$

$$x-6=0 \quad \text{or} \quad x+1=0$$

$$x=6 \qquad\qquad x=-1$$

20. Multiply both sides by the LCD = $(m+3)(m-3)(m-4)$.

$$\frac{4}{m^2-9}+\frac{5}{m^2-m-12}=\frac{7}{m^2-7m+12}$$

$$(m+3)(m-3)(m-4)\left(\frac{4}{(m+3)(m-3)}+\frac{5}{(m+3)(m-4)}\right)=(m+3)(m-3)(m-4)\left(\frac{7}{(m-4)(m-3)}\right)$$

$$4(m-4)+5(m-3)=7(m+3)$$

$$4m-16+5m-15=7m+21$$

$$9m-31=7m+21$$

$$2m=52$$

$$m=26$$

21.

$$\frac{x^2}{a^2}+\frac{y^2}{b^2}=1$$ Solve for a^2.

$$a^2b^2\left(\frac{x^2}{a^2}\right)+a^2b^2\left(\frac{y^2}{b^2}\right)=a^2b^2(1)$$ Multiply both sides by a^2b^2.

$$b^2x^2+a^2y^2=a^2b^2$$ Do the multiplication.

$$a^2y^2-a^2b^2=-b^2x^2$$ Subtract a^2b^2 and b^2x^2 from both sides.

$$a^2(y^2-b^2)=-b^2x^2$$ Factor.

$$a^2=\frac{-b^2x^2}{y^2-b^2}=\frac{b^2x^2}{b^2-y^2}$$ Divide both sides by y^2-b^2. Then multiply by $\frac{-1}{-1}$.

22.

$$\frac{1}{r}=\frac{1}{r_1}+\frac{1}{r_2}$$ Solve for r_2.

$$rr_1r_2\left(\frac{1}{r}\right)=rr_1r_2\left(\frac{1}{r_1}\right)+rr_1r_2\left(\frac{1}{r_2}\right)$$ Multiply both sides by rr_1r_2.

$$r_1r_2=rr_2+rr_1$$ Do the multiplication.

$$r_1r_2-rr_2=rr_1$$ Subtract rr_2 from both sides.

$$r_2(r_1-r)=rr_1$$ Factor.

$$r_2=\frac{rr_1}{r_1-r}$$ Divide both sides by r_1-r.

23. **Analysis**: Find the amount of time it will take for the roofing crews to do the job together. If that time is over 5 hours, then subtract 5 hours from that time to see how long they need to work in the rain.

Equation: Let x represent the amount of time it will take the two roofing crews working together to finish the job.

Work one crew can do in one hour	plus	work other crew can do in one hour	is	work they can do together in one hour.
$\frac{1}{10}$	$+$	$\frac{1}{12}$	$=$	$\frac{1}{x}$

Solve:

$$\frac{1}{10}+\frac{1}{12}=\frac{1}{x}$$
$$60x\left(\frac{1}{10}\right)+60x\left(\frac{1}{12}\right)=60x\left(\frac{1}{x}\right)$$
$$6x+5x=60$$
$$11x=60$$
$$x=\frac{60}{11}=5\frac{5}{11}$$

Conclusion: It will take $5\frac{5}{11}$ hours to roof the house working together. They will need to work $\frac{5}{11}$ of an hour in the rain.

24. **Analysis**: Find how fast the man walks.

Equation: Let x represent the walking rate for the man. Then $x + 5$ represents the bicycling rate of the man.

If $r \bullet t = d$, then $t = \frac{d}{r}$

	r	$\bullet\ t$	$= d$
Walking	x	$\frac{24}{x}$	24
Bicycling	$x+5$	$\frac{24}{x+5}$	24

Walking time	plus	bicycling time	is	11 hours.
$\frac{24}{x}$	$+$	$\frac{24}{x+5}$	$=$	11

Solve:

$$\frac{24}{x}+\frac{24}{x+5}=11$$
$$x(x+5)\left(\frac{24}{x}\right)+x(x+5)\left(\frac{24}{x+5}\right)=11x(x+5)$$ Multiply both sides by LCD $= x(x + 5)$.
$$24x+120+24x=11x^2+55x$$
$$0=11x^2+7x-120$$
$$0=(11x+40)(x-3)$$

$11x+40=0$ or $x-3=0$

$x=-40$ $\quad$ $x=3$ $\quad$ Discard the negative solution.

$x=-\frac{40}{11}$

Conclusion: The man walks at a rate of 3 mph.

25. $\dfrac{18x^2y^3 - 12x^3y^2 + 9xy}{-3xy^4} = \dfrac{18x^2y^3}{-3xy^4} - \dfrac{12x^3y^2}{-3xy^4} + \dfrac{9xy}{-3xy^4} = -\dfrac{6x}{y} + \dfrac{4x^2}{y^2} - \dfrac{3}{y^3}$

26. Add $0y^2$ and $0y$ before dividing.

$$\begin{array}{r}
y^2 - 2y + 4 \\
y+2 \overline{\big) y^3 + 0y^2 + 0y - 48} \\
\underline{y^3 + 2y^2} \\
-2y^2 + 0y \\
\underline{-2y^2 - 4y} \\
4y - 48 \\
\underline{4y + 8} \\
-56
\end{array}$$

Solution: $y^2 - 2y + 4 - \dfrac{56}{y+2}$

27. Divide $a^3 + 0a^2 + 0a + 1$ by $a - 1$.

1⌋	1	0	0	1
		1	1	1
	1	1	1	2

Solution is $a^2 + a + 1 + \dfrac{2}{a-1}$

28. Divide $4x^3 + 3x^2 + 2x - 1$ by $x - 2$.

2⌋	4	3	2	−1
		8	22	48
	4	11	24	47

The remainder is 47.

29. In the first expression the quantities are being multiplied. In the second expression the quantities are being added. "Dividing out" can only occur when the original operation is multiplication.

30. When two quantities vary inversely one quantity will get larger in a predictable way when the other quantity gets smaller. Examples of this will vary.

31. $\dfrac{\text{cost of a dozen roses}}{12 \text{ roses}} = \dfrac{57.99}{12} = \dfrac{x}{16}$ where x represents the cost for 16 roses.

$$\begin{aligned} 12x &= 927.84 \\ x &= 77.32 \end{aligned}$$ the cost for 16 roses is $77.32.

32. Let I represent the intensity of the sound and d the distance from the source.

Inverse variation model

$$I = \frac{k}{d^2}$$

Determine k.

$$\begin{aligned} 100 &= \frac{k}{(30)^2} \\ 90{,}000 &= k \end{aligned}$$

Solve for I when $d = 60$ ft

$$\begin{aligned} I &= \frac{90{,}000}{d^2} \\ V &= \frac{90{,}000}{(60)^2} = 25 \text{ decibels} \end{aligned}$$

33.

x	$f(x)=\frac{2}{x}$	$f(x)$
$\frac{1}{4}$	$f(x)=\frac{2}{\frac{1}{4}}=\frac{2}{1}\left(\frac{4}{1}\right)$	8
$\frac{1}{2}$	$f(x)=\frac{2}{\frac{1}{2}}=\frac{2}{1}\left(\frac{2}{1}\right)$	4
1	$f(x)=\frac{2}{1}$	2
2	$f(x)=\frac{2}{2}$	1
4	$f(x)=\frac{2}{4}$	0.5
8	$f(x)=\frac{2}{8}$	0.25

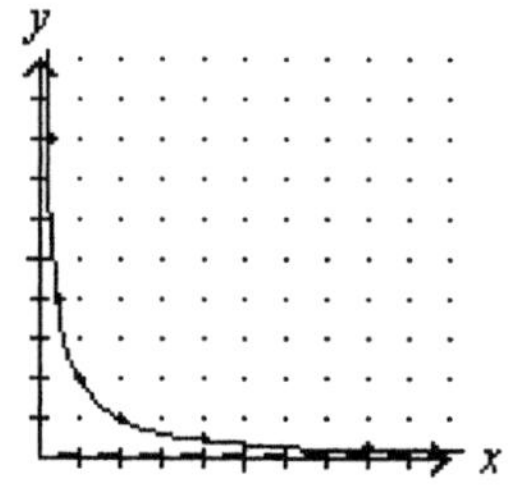

Horizontal asymptote is $y = 0$

34.

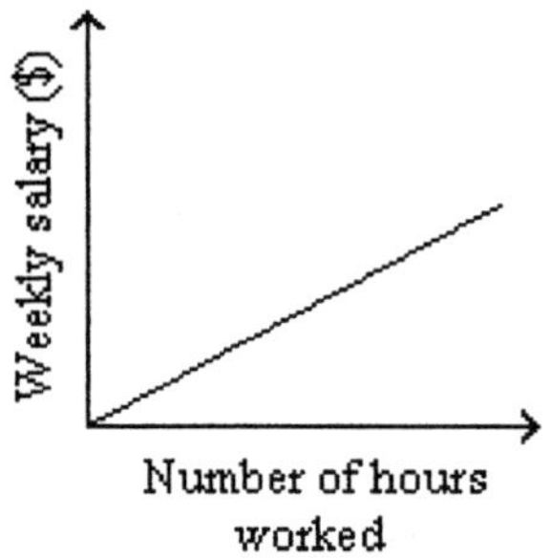

Chapters 1–6 Cumulative Review Exercises

1.

$$\begin{aligned} -3 &= -\frac{9}{8}t \\ -24 &= -9t \\ \frac{-24}{-9} &= t \text{ or } t = \frac{8}{3} \end{aligned}$$

2.

$$\frac{3x-4}{6}-\frac{x-2}{2}=\frac{-2x-3}{3}$$
$$6\left(\frac{3x-4}{6}\right)-6\left(\frac{x-2}{2}\right)=6\left(\frac{-2x-3}{3}\right)$$
$$(3x-4)-3(x-2)=2(-2x-3)$$
$$3x-4-3x+6=-4x-6$$
$$+2=-4x-6$$
$$8=-4x$$
$$-2=x$$

3. The number of green escorts that should be ordered can be found by multiplying 17.5% (0.175) by the total number of cars to be ordered, 80.

   $0.175(80) = 14$ green cars

4. To determine the rate of change, find the slope of the line connecting the two data points (2000, 78.8) and (2050, 83.8).

$$m = \frac{83.8-78.8}{2050-2000}=\frac{5}{50}=0.1$$

The life expectancy will increase 0.1 year each year during this period.

5. $m = -7,\ (x_1, x_2) = (7, 5)$

$$\begin{aligned} y - y_1 &= m(x - x_1) \\ y - 5 &= -7(x - 7) \\ y - 5 &= -7x + 49 \\ y &= -7x + 54 \end{aligned}$$

6. $(x_1, y_1) = (-4, 5)$ and $(x_2, y_2) = (2, -6)$

$$m = \frac{-6-5}{2-(-4)} = \frac{-11}{6}$$

$$\begin{aligned} y - y_1 &= m(x - x_1) \\ y - 5 &= -\frac{11}{6}[x - (-4)] \\ y - 5 &= -\frac{11}{6}x - \frac{44}{6} \\ y &= -\frac{11}{6}x - \frac{7}{3} \end{aligned}$$

7.

x	$f(x) = \frac{2}{3}x - 2$	y
0	$f(x) = \frac{2}{3}(0) - 2$	-2
3	$f(x) = \frac{2}{3}(3) - 2$	0

D: $(-\infty, \infty)$
R: $(-\infty, \infty)$

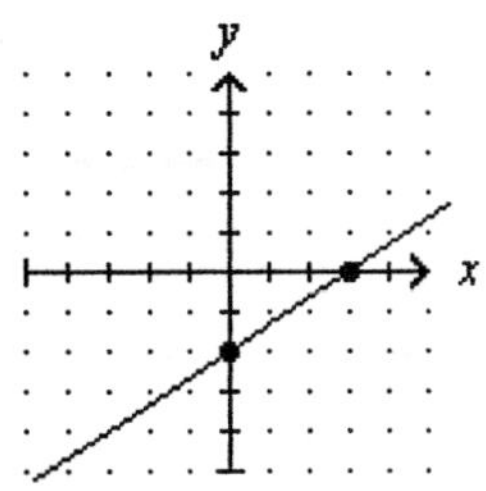

8. $\begin{cases} 0.6T_1 - 0.8T_2 = 0 \\ 0.8T_1 + 0.6T_2 = 100 \end{cases}$

Multiply each equation by 10

$$\begin{cases} 6T_1 - 8T_2 = 0 \\ 8T_1 + 6T_2 = 1000 \end{cases}$$

Multiply Equation 1 by 6 and Equation 2 by 8 to eliminate T_2.

$$\begin{cases} 36T_1 - 48T_2 = 0 \\ 64T_1 + 48T_2 = 8000 \end{cases}$$

Add the equations

$$\begin{aligned} 100T_1 &= 8000 \\ T_1 &= 80 \end{aligned}$$

Substitute $T_1 = 80$ into Equation 1 to solve for T_2.

$$\begin{aligned} 0.6(80) - 0.8T_2 &= 0 \\ 48 - 0.8T_2 &= 0 \\ -0.8T_2 &= -48 \\ T_2 &= 60 \end{aligned}$$

9. $$\begin{aligned} -2x &\le -5 \\ x &\ge \frac{5}{2} \end{aligned}$$

Divide by -2 and reverse the $\le$ symbol.

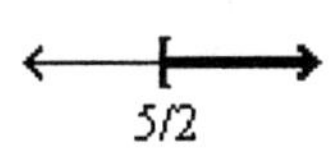

10. If $\left|\frac{3a}{5}-2\right|+1 \geq \frac{6}{5}$ then $\left|\frac{3a}{5}-2\right| \geq \frac{1}{5}$. Rewrite this as

$$\frac{3a}{5}-2 \leq -\frac{1}{5} \quad \text{or} \quad \frac{3a}{5}-2 \geq \frac{1}{5}$$

$$3a-10 \leq -1 \qquad 3a-10 \geq 1$$

$$3a \leq 9 \qquad 3a \geq -11$$

$$a \leq 3 \qquad a \geq \frac{11}{3}$$

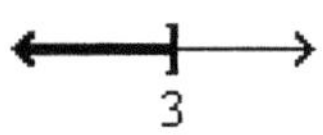

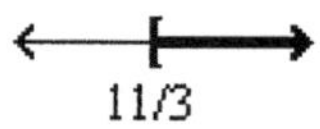

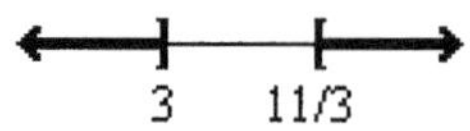

11.
$$5(x+2) \leq 4(x+1) \quad \text{and} \quad 11+x < 0$$

$$5x+10 \leq 4x+4 \qquad x < -11$$

$$x \leq -6$$

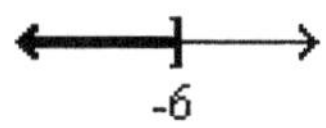

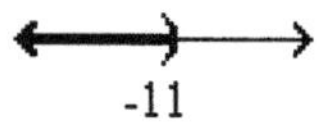

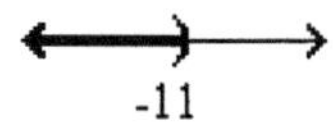

12.
$$x+1 < -4 \quad \text{or} \quad x-4 > 0$$

$$x < -5 \qquad x > 4$$

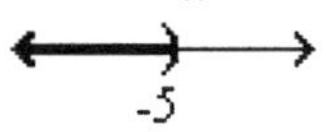

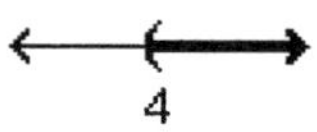

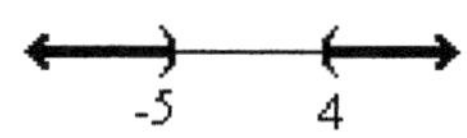

13. $a^3b^2a^5b^2 = a^{3+5}b^{2+2} = a^8b^4$

14. $\frac{a^3b^6}{a^7b^2} = a^{3-7}b^{6-2} = a^{-4}b^4 = \frac{b^4}{a^4}$

15. $\frac{1}{3^{-4}} = 3^4 = 3(3)(3)(3) = 81$

16. $\left(\frac{2x^{-2}y^3}{x^2x^3y^4}\right)^{-3} = \left(2x^{-2-2-3}y^{3-4}\right)^{-3} = \left(2x^{-7}y^{-1}\right)^{-3} = 2^{-3}x^{-7(-3)}y^{-1(-3)} = \frac{1}{2^3}x^{21}y^3 = \frac{x^{21}y^3}{8}$

17. $4.25 \times 10^4 = 42{,}500$

18. $7.12 \times 10^{-4} = 0.000712$

19. $y = \frac{kxz}{r}$

20. $\begin{vmatrix} 3 & -2 \\ -2 & 4 \end{vmatrix} = 3(4)-(-2)(-2) = 12-4 = 8$

21.

x	$y < 4 - x$ Graph: $y = 4 - x$	y
0	$y = 4-(0)$	4
4	$y = 4-(4)$	0
Broken line	Test Point: (0, 0) $0 < 4-(0)$ $0 < 4$	True

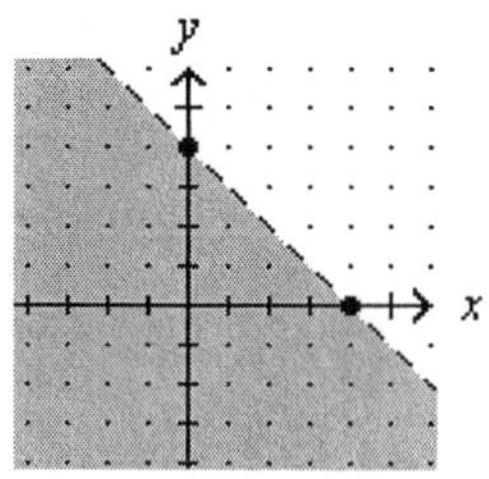

22.

x	$f(x) = 2x^2 - 3$	$f(x)$
−2	$f(x) = 2(-2)^2 - 3$	5
−1	$f(x) = 2(-1)^2 - 3$	−1
0	$f(x) = 2(0)^2 - 3$	−3
1	$f(x) = 2(1)^2 - 3$	−1
2	$f(x) = 2(2)^2 - 3$	5

D = $(-\infty, \infty)$ R = $[-3, \infty)$

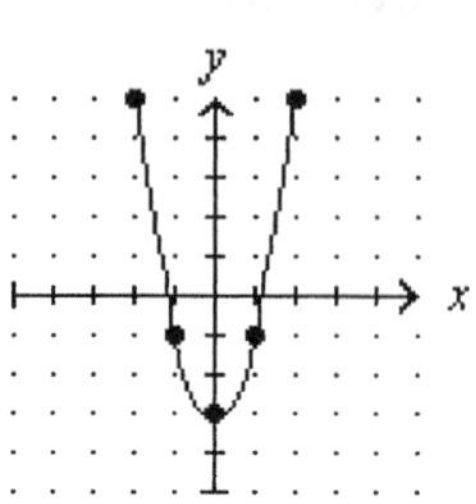

23. $g(x) = -3x^3 + x - 4$; $g(-2) = -3(-2)^3 + (-2) - 4 = 24 - 2 - 4 = 18$

24. The degree is the largest sum of the exponents in any one term. In this case, the degree is 7

25. a. x–intercept is $(-3, 0)$; y–intercept is $(0, -7)$ b. slope is $\frac{\text{rise}}{\text{run}} = \frac{\text{down } 7}{\text{over } 3} = \frac{-7}{3} = -\frac{7}{3}$

c. Quadrant I d. Use slope–intercept form: $y = mx + b$. $y = -\frac{7}{3}x - 7$

e. Substitute -42 for y and 15 for x.

$$\begin{aligned} -42 &\stackrel{?}{=} -\frac{7}{3}(15) - 7 \\ -42 &\stackrel{?}{=} -7(5) - 7 \\ -42 &\stackrel{?}{=} -35 - 7 \\ -42 &= -42 \end{aligned}$$

Yes, the line passes through $(15, -42)$

26. penny: $\$0.01 = \$10^{-2}$; dime: $\$0.10 = \$10^{-1}$; one dollar bill: $\$1.00 = \$10^0$; one hundred thousand dollar bill: $\$100{,}000 = \$10^5$

27. $(x^3 + 3x^2 - 2x + 7) + (x^3 - 2x^2 + 2x + 5) = 2x^3 + x^2 + 12$

28. $(-5x^2 + 3x + 4) - (-2x^2 + 3x + 7) = -5x^2 + 3x + 4 + 2x^2 - 3x - 7 = -3x^2 - 3$

29. $(3x + 4)(2x - 5) = 3x(2x) + 3x(-5) + 4(2x) + 4(-5) = 6x^2 - 15x + 8x - 20 = 6x^2 - 7x - 20$

30. $(2x^3 - 1)^2 = (2x^3)^2 - 2(2x^3)(1) + (-1)^2 = 4x^6 - 4x^3 + 1$

31. Yes, there is only one y value for each x value.

32. This reading would occur during month 10. The maximum number of minutes ahead would be about 17.

33. This reading would occur during month 2. The maximum number of minutes behind would be about 14.

34. four times

35. $3r^2s^3 - 6rs^4 = 3rs^3(r - 2s)$

36. $5(x - y) - a(x - y) = (x - y)(5 - a)$

37. $xu + yv + xv + yu = xu + xv + yv + yu = x(u + v) + y(v + u) = (u + v)(x + y)$

38. $81x^4 - 16y^4 = (9x^2)^2 - (4y^2)^2 = (9x^2 + 4y^2)(9x^2 - 4y^2) = (9x^2 + 4y^2)[(3x)^2 - (2y)^2] = (9x^2 - 4y^2)(3x + 2y)(3x - 2y)$

39. $8x^3 - 27y^6 = (2x)^3 - (3y^2)^3 = (2x - 3y^2)(4x^2 + 6xy^2 + 9y^4)$

40. $6x^2 + 5x - 6 = 6x^2 + 9x - 4x - 6 = 3x(2x + 3) - 2(2x + 3) = (2x + 3)(3x - 2)$

41. $9x^2 - 30x + 25 = (3x)^2 - 2(3x)(5) + 5^2 = (3x - 5)^2$

42. $15x^2 - x - 6 = 15x^2 - 10x + 9x - 6 = 5x(3x - 2) + 3(3x - 2) = (3x - 2)(5x + 3)$

43.

$$\begin{aligned} 6x^2 + 7 &= -23x \\ 6x^2 + 23x + 7 &= 0 \\ 6x^2 + 21x + 2x + 7 &= 0 \\ 3x(2x + 7) + 1(2x + 7) &= 0 \\ (2x + 7)(3x + 1) &= 0 \end{aligned}$$

$$\begin{aligned} 2x + 7 &= 0 \\ 2x &= -7 \\ x &= -\tfrac{7}{2} \end{aligned} \quad \text{or} \quad \begin{aligned} 3x + 1 &= 0 \\ 3x &= -1 \\ x &= -\tfrac{1}{3} \end{aligned}$$

44.

$$\begin{aligned} x^3 - 4x &= 0 \\ x(x^2 - 4) &= 0 \\ x(x + 2)(x - 2) &= 0 \end{aligned}$$

$$x = 0 \quad \text{or} \quad \begin{aligned} x + 2 &= 0 \\ x &= -2 \end{aligned} \quad \text{or} \quad \begin{aligned} x - 2 &= 0 \\ x &= 2 \end{aligned}$$

45.
$$\begin{aligned} b^2x^2 + a^2y^2 &= a^2b^2 \\ b^2x^2 - a^2b^2 &= -a^2y^2 \\ b^2(x^2 - a^2) &= -a^2y^2 \\ b^2 &= -\frac{a^2y^2}{x^2 - a^2} \quad \text{or} \quad \frac{a^2y^2}{a^2 - x^2} \end{aligned}$$

46. Let x represent the width of the surface, then $x + 3$ represents the length of the surface. Use $A = lw = 108 \text{ in.}^2$.

$$\begin{aligned} A = x(x+3) &= 108 \\ x^2 + 3x &= 108 \\ x^2 + 3x - 108 &= 0 \\ (x+12)(x-9) &= 0 \end{aligned}$$

$$x + 12 = 0 \quad \text{or} \quad x - 9 = 0$$

$$x = -12 \qquad\qquad x = 9 \qquad\qquad \text{Discard the negative solution.}$$

The cooking surface would have a width of $x = 9$ in. and a length of $x + 3 = 12$ in.

47. $$\frac{2x^2y + xy - 6y}{3x^2y + 5xy - 2y} = \frac{y(2x^2 + x - 6)}{y(3x^2 + 5x - 2)} = \frac{y(2x-3)(x+2)}{y(3x-1)(x+2)} = \frac{2x-3}{3x-1}$$

48. $$\frac{p^3 - q^3}{q^2 - p^2} \cdot \frac{q^2 + pq}{p^3 + p^2q + pq^2} = \frac{(p-q)(p^2 + pq + q^2)(q)(q+p)}{(q+p)(q-p)(p)(p^2 + pq + q^2)} = \frac{q(p-q)}{p(q-p)} = -\frac{q}{p}$$

49. LCD $= (x+y)(x-y)$

$$\begin{aligned} \frac{2}{x+y} + \frac{3}{x-y} - \frac{x-3y}{x^2-y^2} &= \frac{2(x-y)}{(x+y)(x-y)} + \frac{3(x+y)}{(x-y)(x+y)} - \frac{x-3y}{(x+y)(x-y)} \\ &= \frac{2x - 2y + 3x + 3y - (x-3y)}{(x+y)(x-y)} = \frac{2x - 2y + 3x + 3y - x + 3y}{(x+y)(x-y)} \\ &= \frac{4x+4y}{(x+y)(x-y)} = \frac{4(x+y)}{(x+y)(x-y)} = \frac{4}{x-y} \end{aligned}$$

50. LCD $= ab$

$$\frac{\frac{a}{b} + b}{a - \frac{b}{a}} = \frac{ab\left(\frac{a}{b} + b\right)}{ab\left(a - \frac{b}{a}\right)} = \frac{a^2 + ab^2}{a^2b - b^2}$$

51.
$$\begin{aligned} \frac{5x-3}{x+2} &= \frac{5x+3}{x-2} \\ (x+2)(x-2)\left(\frac{5x-3}{x+2}\right) &= (x+2)(x-2)\left(\frac{5x+3}{x-2}\right) \\ (x-2)(5x-3) &= (x+2)(5x+3) \\ 5x^2 - 13x + 6 &= 5x^2 + 13x + 6 \\ -26x &= 0 \\ x &= 0 \end{aligned}$$

52.

$$\frac{3}{x-2}+\frac{x^2}{(x+3)(x-2)}=\frac{x+4}{x+3}$$

$$(x-2)(x+3)\left(\frac{3}{x-2}\right)+(x-2)(x+3)\left(\frac{x^2}{(x+3)(x-2)}\right)=(x-2)(x+3)\left(\frac{x+4}{x+3}\right)$$

$$3(x+3)+x^2=(x-2)(x+4)$$

$$3x+9+x^2=x^2+2x-8$$

$$x=-17$$

53.

$$\begin{array}{r}
x\ +4 \\
x+5\overline{)x^2+9x+20} \\
\underline{x^2+5x} \\
4x+20 \\
\underline{4x+20} \\
0
\end{array}$$

Solution: $x+4$

54. Rearrange the terms before dividing.

$$\begin{array}{r}
-x^2+\ x+\ 5 \\
x-1\overline{)-x^3+2x^2+4x+3} \\
\underline{-x^3+\ x^2} \\
x^2+4x \\
\underline{x^2-\ x} \\
5x+3 \\
\underline{5x-5} \\
8
\end{array}$$

Solution: $-x^2+x+5+\dfrac{8}{x-1}$

55. The inflation rate will rise sharply.

56. The inflation rate has dropped.

The teachers whom we remember with fondness and appreciation are those who pushed us, not those who pampered us.

Chapter 7 Radical Expressions, Equations, and Functions

Study Tip
My family motto: "A strong mind in a strong body"

Studying hard and consistently is a great way to build a strong mind, but what about your body?

Eat healthy foods- at meal time and for snack
Get lots of exercise- a brisk walk can clear your mind
Get plenty of rest – you cannot perform if you are too tired to think!

If your body is not healthy, your mind cannot work at its full potential.

Section 7.1 Radical Expressions and Radical Functions

VOCABULARY

1. square root
3. radical
5. odd
7. simplified

CONCEPTS

9. $b^2 = a$
11. two; positive
13. x
15. 3; up

17.

x	$f(x) = -\sqrt{x}$	$f(x)$
0	$f(0) = -\sqrt{0}$	**0**
1	$f(1) = -\sqrt{1}$	**–1**
4	$f(4) = -\sqrt{4}$	**–2**
9	$f(9) = -\sqrt{9}$	**–3**
16	$f(16) = -\sqrt{16}$	**–4**

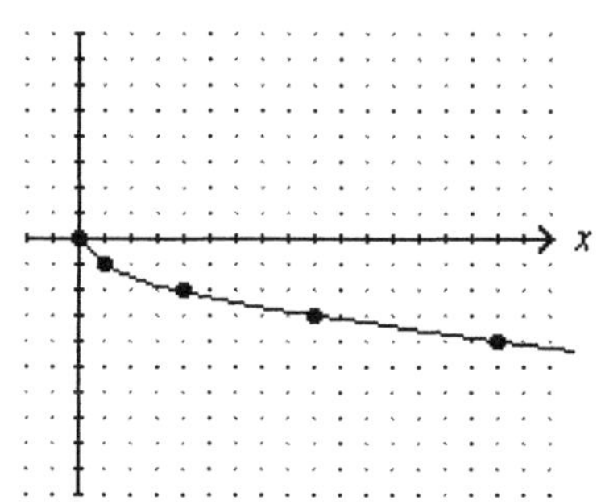

NOTATION

19. $\sqrt{x^2} = |x|$
21. $f(x) = \sqrt{x-5}$

PRACTICE

23. $\sqrt{121} = \sqrt{(11)^2} = 11$
25. $-\sqrt{64} = -\sqrt{(8)^2} = -8$
27. $\sqrt{\frac{1}{9}} = \sqrt{\left(\frac{1}{3}\right)^2} = \frac{1}{3}$
29. $\sqrt{0.25} = \sqrt{(0.5)^2} = 0.5$
31. $\sqrt{-25}$ is not a real number
33. $\sqrt{(-4)^2} = \sqrt{16} = \sqrt{(4)^2} = 4$
35. $\sqrt{12} = 3.4641$
37. $\sqrt{679.25} = 26.0624$

39. $\sqrt{4x^2} = \sqrt{(2x)^2} = |2x| = 2|x|$

41. $\sqrt{(t+5)^2} = |t+5|$

43. $\sqrt{(-5b)^2} = \sqrt{25b^2} = \sqrt{(5b)^2} = |5b| = 5|b|$

45. $\sqrt{a^2+6a+9} = \sqrt{(a+3)^2} = |a+3|$

47. $f(x) = \sqrt{x-4}$; $f(4) = \sqrt{4-4} = \sqrt{0} = 0$

49. $f(x) = \sqrt{x-4}$; $f(20) = \sqrt{20-4} = \sqrt{16} = 4$

51. $g(x) = \sqrt[3]{x-4}$; $g(12) = \sqrt[3]{12-4} = \sqrt[3]{8} = \sqrt[3]{2^3} = 2$

53. $g(x) = \sqrt[3]{x-4}$; $g(-996) = \sqrt[3]{-996-4} = \sqrt[3]{-1000} = \sqrt[3]{(-10)^3} = -10$

55. $f(x) = \sqrt{x^2+4}$; $f(4) = \sqrt{4^2+4} = \sqrt{16+4} = \sqrt{20} \approx 4.4721$

57. $g(x) = \sqrt[3]{x^2+1}$; $g(6) = \sqrt[3]{6^2+1} = \sqrt[3]{36+1} = \sqrt[3]{37} \approx 3.3322$

59.

x	$f(x) = \sqrt{x+4}$	$f(x)$
-4	$f(-4) = \sqrt{-4+4} = \sqrt{0}$	0
-3	$f(-3) = \sqrt{-3+4} = \sqrt{1}$	1
0	$f(0) = \sqrt{0+4} = \sqrt{4}$	2
5	$f(5) = \sqrt{5+4} = \sqrt{9}$	3
12	$f(12) = \sqrt{12+4} = \sqrt{16}$	4

D: $[-4, \infty)$; R: $[0, \infty)$

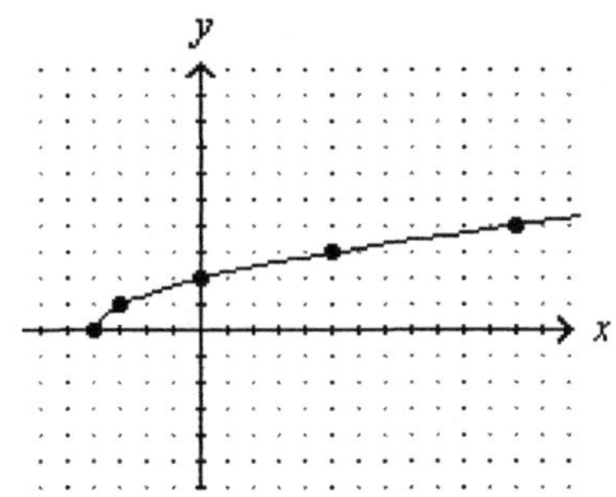

61.

x	$f(x) = -\sqrt[3]{x} - 3$	$f(x)$
-8	$f(-8) = -\sqrt[3]{-8} - 3$ $= -\sqrt[3]{(-2)^3} - 3 = -(-2) - 3$	-1
-1	$f(-1) = -\sqrt[3]{-1} - 3$ $= -\sqrt[3]{(-1)^3} - 3 = -(-1) - 3$	-2
0	$f(0) = -\sqrt[3]{0} - 3 = -0 - 3$	-3
1	$f(1) = -\sqrt[3]{1} - 3$ $= -\sqrt[3]{(1)^3} - 3 = -(1) - 3$	-4
8	$f(8) = -\sqrt[3]{8} - 3$ $= -\sqrt[3]{(2)^3} - 3 = -(2) - 3$	-5

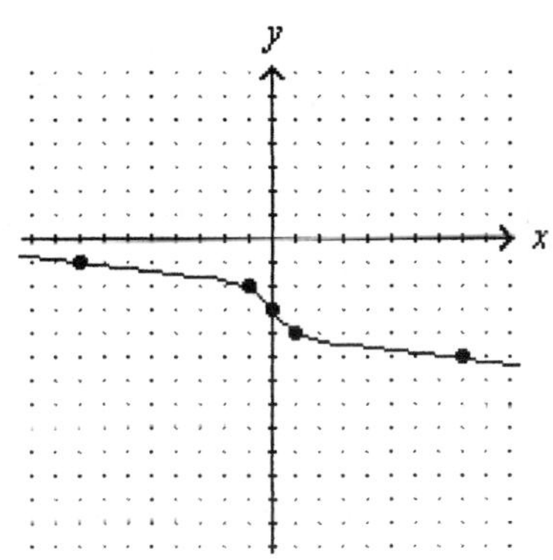

D: $(-\infty, \infty)$; R: $(-\infty, \infty)$

63. $\sqrt[3]{1} = \sqrt[3]{1^3} = 1$

65. $\sqrt[3]{-125} = \sqrt[3]{(-5)^3} = -5$

67. $\sqrt[3]{-\frac{8}{27}} = \sqrt[3]{\left(-\frac{2}{3}\right)^3} = -\frac{2}{3}$

69. $\sqrt[3]{0.064} = \sqrt[3]{(0.4)^3} = 0.4$

71. $\sqrt[3]{8a^3} = \sqrt[3]{(2a)^3} = 2a$

73. $\sqrt[3]{-1000p^3q^3} = \sqrt[3]{(-10pq)^3} = -10pq$

75. $\sqrt[3]{-\frac{1}{8}m^6n^3} = \sqrt[3]{\left(-\frac{1}{2}m^2n\right)^3} = -\frac{1}{2}m^2n$

77. $\sqrt[3]{-0.064s^9t^6} = \sqrt[3]{(-0.4s^3t^2)^3} = -0.4s^3t^2$

79. $\sqrt[4]{81}=\sqrt[4]{3^4}=3$

81. $-\sqrt[5]{243}=-\sqrt[5]{3^5}=-3$

83. $\sqrt[4]{-256}$ is not a real number, (even root of a negative number).

85. $\sqrt[4]{\dfrac{16}{625}}=-\sqrt[4]{\left(\dfrac{2}{5}\right)^4}=\dfrac{2}{5}$

87. $-\sqrt[5]{-\dfrac{1}{32}}=-\sqrt[5]{\left(-\dfrac{1}{2}\right)^5}=-\left(-\dfrac{1}{2}\right)=\dfrac{1}{2}$

89. $\sqrt[5]{32a^5}=\sqrt[5]{(2a)^5}=2a$

91. $\sqrt[4]{16a^4}=\sqrt[4]{(2a)^4}=2a$

93. $\sqrt[4]{k^{12}}=\sqrt[4]{\left(k^3\right)^4}=k^3$

95. $\sqrt[4]{\dfrac{1}{16}m^4}=\sqrt[4]{\left(\dfrac{1}{2}m\right)^4}=\dfrac{1}{2}m$

97. $\sqrt[25]{(x+2)^{25}}=x+2$

APPLICATIONS

99. Substitute $A = 38.5$ in.^{2} $r=\sqrt{\dfrac{A}{\pi}}=\sqrt{\dfrac{38.5}{\pi}}=\sqrt{12.2549...}=3.5007..$ The diameter is $2r = 2(3.5) = 7$ in.

101. Convert $t = 7$ ft. $8\frac{1}{2}$ in. to inches. 7ft (12 in./ft) + $8\frac{1}{2}$ in. = 84 in. + $8\frac{1}{2}$ in. = $92\frac{1}{2}$ in. $= t$

$$p(t)=\frac{590}{\sqrt{t}},\quad p\left(92\tfrac{1}{2}\right)=\frac{590}{\sqrt{92\frac{1}{2}}}=\frac{590}{9.617...}=61.3 \text{ beats/min}$$

103. Since there are 5 rats and each rat requires 125 cu. ft of space, the volume of the hemisphere needs to be 5(125) or 625 cu. ft.

$$d(V)=\sqrt[3]{12\left(\frac{V}{\pi}\right)},\quad d((625)=\sqrt[3]{12\left(\frac{625}{\pi}\right)}=\sqrt[3]{2387.324...}=13.365... \text{ or } 13.4 \text{ ft.}$$

105. Substitute P = \$800, A = \$950, and n = 5 years.

$$r=\sqrt[n]{\frac{A}{P}}-1=\sqrt[5]{\frac{950}{800}}-1=\sqrt[5]{1.1875}-1=1.0349...-1=0.0349 \text{ or about } 3.5\%.$$

WRITING

107. There are two integers that when squared will give 36 as an answer, but $\sqrt{36}$ asks us to find the *principal* or positive value that when squared will give 36 as an answer.

109. Since x cannot be a negative number, the portion of the graph to the left of 0 is incorrect.

REVIEW

111. $\dfrac{x^2-x-6}{x^2-2x-3}\cdot\dfrac{x^2-1}{x^2+x-2}=\dfrac{(x-3)(x+2)(x+1)(x-1)}{(x-3)(x+1)(x+2)(x-1)}=1$

113. $\dfrac{3}{m+1}+\dfrac{3m}{m-1}=\dfrac{3(m-1)}{(m+1)(m-1)}+\dfrac{3m(m+1)}{(m+1)(m-1)}=\dfrac{3m-3+3m^2+3m}{(m+1)(m-1)}=\dfrac{3m^2+6m-3}{(m+1)(m-1)}=\dfrac{3(m^2+2m-1)}{(m+1)(m-1)}$

Section 7.2 Simplifying and Combining Radical Expressions

VOCABULARY

1. like
3. factor

CONCEPTS

5. $\sqrt[n]{a}\sqrt[n]{b}$; product; roots
7. a. $\sqrt{4\cdot 5}$ b. $\sqrt{4}\sqrt{5}$ c. $\sqrt{4\cdot 5} = \sqrt{4}\sqrt{5}$
9. a. $\sqrt{5}$, $\sqrt[3]{5}$ (answers may vary), The expressions cannot be added.
   b. $\sqrt{5}$, $\sqrt{6}$ (answers may vary), The expressions cannot be added.

NOTATION

11. $\sqrt[3]{32k^4} = \sqrt[3]{8k^3 \cdot 4k}$
    $= \sqrt[3]{8k^3}\sqrt[3]{4k}$
    $= 2k\sqrt[3]{4k}$

PRACTICE

13. $\sqrt{6}\sqrt{6} = \sqrt{6\cdot 6} = \sqrt{6^2} = 6$

15. $\sqrt{t}\sqrt{t} = \sqrt{t\cdot t} = \sqrt{t^2} = t$

17. $\sqrt[3]{5x^2}\sqrt[3]{25x} = \sqrt[3]{5x^2\cdot 25x} = \sqrt[3]{5^3x^3} = 5x$

19. $\dfrac{\sqrt{500}}{\sqrt{5}} = \sqrt{\dfrac{500}{5}} = \sqrt{100} = 10$

21. $\dfrac{\sqrt{98x^3}}{\sqrt{2x}} = \sqrt{\dfrac{98x^3}{2x}} = \sqrt{49x^2} = 7x$

23. $\dfrac{\sqrt{180ab^4}}{\sqrt{5ab^2}} = \sqrt{\dfrac{180ab^4}{5ab^2}} = \sqrt{36b^2} = 6b$

25. $\dfrac{\sqrt[3]{48}}{\sqrt[3]{6}} = \sqrt[3]{\dfrac{48}{6}} = \sqrt[3]{8} = 2$

27. $\dfrac{\sqrt[3]{189a^4}}{\sqrt[3]{7a}} = \sqrt[3]{\dfrac{189a^4}{7a}} = \sqrt[3]{27a^3} = 3a$

29. $\sqrt{20} = \sqrt{4\cdot 5} = \sqrt{4}\sqrt{5} = 2\sqrt{5}$

31. $-\sqrt{200} = -\sqrt{100\cdot 2} = -\sqrt{100}\sqrt{2} = -10\sqrt{2}$

33. $\sqrt[3]{80} = \sqrt[3]{8\cdot 10} = \sqrt[3]{8}\sqrt[3]{10} = 2\sqrt[3]{10}$

35. $\sqrt[3]{81} = \sqrt[3]{-27\cdot 3} = \sqrt[3]{-27}\sqrt[3]{3} = -3\sqrt[3]{3}$

37. $\sqrt[4]{32} = \sqrt[4]{16\cdot 2} = \sqrt[4]{16}\sqrt[4]{2} = 2\sqrt[4]{2}$

39. $\sqrt[5]{96} = \sqrt[5]{32\cdot 3} = \sqrt[5]{32}\sqrt[5]{3} = 2\sqrt[5]{3}$

41. $\sqrt{\dfrac{7}{9}} = \dfrac{\sqrt{7}}{\sqrt{9}} = \dfrac{\sqrt{7}}{3}$

43. $\sqrt[3]{\dfrac{7}{64}} = \dfrac{\sqrt[3]{7}}{\sqrt[3]{64}} = \dfrac{\sqrt[3]{7}}{4}$

45. $\sqrt[4]{\dfrac{3}{10{,}000}} = \dfrac{\sqrt[4]{3}}{\sqrt[4]{10^4}} = \dfrac{\sqrt[4]{3}}{10}$

47. $\sqrt[5]{\dfrac{3}{32}} = \dfrac{\sqrt[5]{3}}{\sqrt[5]{2^5}} = \dfrac{\sqrt[5]{3}}{2}$

49. $\sqrt{50x^2} = \sqrt{25x^2\cdot 2} = \sqrt{25x^2}\sqrt{2} = 5x\sqrt{2}$

51. $\sqrt{32b} = \sqrt{16\cdot 2b} = \sqrt{16}\sqrt{2b} = 4\sqrt{2b}$

53. $-\sqrt{112a^3} = -\sqrt{16a^2}\sqrt{7a} = -4a\sqrt{7a}$

55. $\sqrt{175a^2b^3} = \sqrt{25a^2b^2}\sqrt{7b} = 5ab\sqrt{7b}$

57. $-\sqrt{300xy} = -\sqrt{100}\sqrt{3xy} = -10\sqrt{3xy}$

59. $\sqrt[3]{-54x^6} = \sqrt[3]{-27x^6}\sqrt[3]{2} = -3x^2\sqrt[3]{2}$

61. $\sqrt[3]{16x^{12}y^4} = \sqrt[3]{8x^{12}y^3}\,\sqrt[3]{2y} = 2x^4y\sqrt[3]{2y}$

63. $\sqrt[4]{32x^{12}y^5} = \sqrt[4]{16x^{12}y^4}\,\sqrt[4]{2y} = 2x^3y\sqrt[4]{2y}$

64. $\sqrt[5]{64x^{10}y^6} = \sqrt[5]{32x^{10}y^5 \cdot 2y} = \sqrt[5]{2^5x^{10}y^5}\,\sqrt[5]{2y} = 2x^2y\sqrt[5]{2y}$

65. $\sqrt{\dfrac{z^2}{16x^2}} = \dfrac{\sqrt{z^2}}{\sqrt{16x^2}} = \dfrac{z}{4x}$

67. $\sqrt[4]{\dfrac{5x}{16z^4}} = \dfrac{\sqrt[4]{5x}}{\sqrt[4]{16z^4}} = \dfrac{\sqrt[4]{5x}}{2z}$

69. $4\sqrt{2x} + 6\sqrt{2x} = (4+6)\sqrt{2x} = 10\sqrt{2x}$

71. $8\sqrt[5]{7a^2} - 7\sqrt[5]{7a^2} = (8-7)\sqrt[5]{7a^2} = \sqrt[5]{7a^2}$

73. $\sqrt{2} - \sqrt{8} = \sqrt{2} - \sqrt{4 \cdot 2} = \sqrt{2} - 2\sqrt{2} = -\sqrt{2}$

75. $\sqrt{98} - \sqrt{50} = \sqrt{49 \cdot 2} - \sqrt{25 \cdot 2} = 7\sqrt{2} - 5\sqrt{2} = 2\sqrt{2}$

77. $3\sqrt{24} + \sqrt{54} = 3\sqrt{4 \cdot 6} + \sqrt{9 \cdot 6} = 3 \cdot 2\sqrt{6} + 3\sqrt{6} = 6\sqrt{6} + 3\sqrt{6} = 9\sqrt{6}$

79. $\sqrt[3]{24x} + \sqrt[3]{3x} = \sqrt[3]{8 \cdot 3x} + \sqrt[3]{3x} = 2\sqrt[3]{3x} + \sqrt[3]{3x} = 3\sqrt[3]{3x}$

81. $\sqrt[3]{32} - \sqrt[3]{108} = \sqrt[3]{8 \cdot 4} - \sqrt[3]{27 \cdot 4} = 2\sqrt[3]{4} - 3\sqrt[3]{4} = -\sqrt[3]{4}$

83. $2\sqrt[3]{125} - 5\sqrt[3]{64} = 2(5) - 5(4) = 10 - 20 = -10$

85. $14\sqrt[4]{32} - 15\sqrt[4]{162} = 14\sqrt[4]{16 \cdot 2} - 15\sqrt[4]{81 \cdot 2} = 14 \cdot 2\sqrt[4]{2} - 15 \cdot 3\sqrt[4]{2} = 28\sqrt[4]{2} - 45\sqrt[4]{2} = -17\sqrt[4]{2}$

87. $3\sqrt[4]{512} + 2\sqrt[4]{32} = 3\sqrt[4]{256 \cdot 2} + 2\sqrt[4]{16 \cdot 2} = 3 \cdot 4\sqrt[4]{2} + 2 \cdot 2\sqrt[4]{2} = 12\sqrt[4]{2} + 4\sqrt[4]{2} = 16\sqrt[4]{2}$

89. $\sqrt{98} - \sqrt{50} - \sqrt{72} = \sqrt{49 \cdot 2} - \sqrt{25 \cdot 2} - \sqrt{36 \cdot 2} = 7\sqrt{2} - 5\sqrt{2} - 6\sqrt{2} = -4\sqrt{2}$

91. $\sqrt{18t} + \sqrt{300t} - \sqrt{243t} = \sqrt{9 \cdot 2t} + \sqrt{100 \cdot 3t} - \sqrt{81 \cdot 3t} = 3\sqrt{2t} + 10\sqrt{3t} - 9\sqrt{3t} = 3\sqrt{2t} + \sqrt{3t}$

93. $2\sqrt[3]{16} - \sqrt[3]{54} - 3\sqrt[3]{128} = 2\sqrt[3]{8 \cdot 2} - \sqrt[3]{27 \cdot 2} - 3\sqrt[3]{64 \cdot 2} = 2 \cdot 2\sqrt[3]{2} - 3\sqrt[3]{2} - 3 \cdot 4\sqrt[3]{2} = 4\sqrt[3]{2} - 3\sqrt[3]{2} - 12\sqrt[3]{2} = -11\sqrt[3]{2}$

95. $\sqrt{25y^2z} - \sqrt{16y^2z} = \sqrt{25y^2}\,\sqrt{z} - \sqrt{16y^2}\,\sqrt{z} = 5y\sqrt{z} - 4y\sqrt{z} = y\sqrt{z}$

97. $\sqrt{36xy^2} + \sqrt{49xy^2} = \sqrt{36y^2}\,\sqrt{x} + \sqrt{49y^2}\,\sqrt{x} = 6y\sqrt{x} + 7y\sqrt{x} = 13y\sqrt{x}$

99. $2\sqrt[3]{64a} + 2\sqrt[3]{8a} = 2\sqrt[3]{64}\sqrt[3]{a} + 2\sqrt[3]{8}\sqrt[3]{a} = 2 \cdot 4\sqrt[3]{a} + 2 \cdot 2\sqrt[3]{a} = 8\sqrt[3]{a} + 4\sqrt[3]{a} = 12\sqrt[3]{a}$

101. $\sqrt{y^5} - \sqrt{9y^5} - \sqrt{25y^5} = \sqrt{y^4 \cdot y} - \sqrt{9y^4 \cdot y} - \sqrt{25y^4y} = y^2\sqrt{y} - 3y^2\sqrt{y} - 5y^2\sqrt{y} = -7y^2\sqrt{y}$

103. $\sqrt[5]{x^6y^2} + \sqrt[5]{32x^6y^2} + \sqrt[5]{x^6y^2} = \sqrt[5]{x^5 \cdot xy^2} + \sqrt[5]{32x^5 \cdot xy^2} + \sqrt[5]{x^5 \cdot xy^2} = x\sqrt[5]{xy^2} + 2x\sqrt[5]{xy^2} + x\sqrt[5]{xy^2} = 4x\sqrt[5]{xy^2}$

APPLICATIONS

105. Substitute $r = 4$ ft and $h = 2$ ft.

$S = \pi r\sqrt{r^2 + h^2} = \pi(4)\sqrt{(4)^2 + 2^2} = 4\pi\sqrt{16+4} = 4\pi\sqrt{20} = 4\pi\sqrt{4 \bullet 5} = 4\pi\left(2\sqrt{5}\right) = 8\pi\sqrt{5} \approx 56.198\ldots \approx 56.2\text{ ft}^2$

107. Substitute $P = 1200$ watt and $R = 16$ ohms.

$$I = \sqrt{\frac{P}{R}} = \sqrt{\frac{1200}{16}} = \sqrt{75} = \sqrt{25 \cdot 3} = 5\sqrt{3} \approx 8.660\ldots \approx 8.7\text{ amps}$$

109. Sum the lengths of each side for the total length.

$$\begin{aligned}\text{Total length} &= \sqrt{80}+\sqrt{20}+\sqrt{80}+\sqrt{20}+\sqrt{45}+\sqrt{75}+\sqrt{80}+\sqrt{75}+\sqrt{80}+\sqrt{45}\\ &= 4\sqrt{80}+2\sqrt{20}+2\sqrt{45}+2\sqrt{75}\\ &= 4\sqrt{16\cdot 5}+2\sqrt{4\cdot 5}+2\sqrt{9\cdot 5}+2\sqrt{25\cdot 3}\\ &= 4\cdot 4\sqrt{5}+2\cdot 2\sqrt{5}+2\cdot 3\sqrt{5}+2\cdot 5\sqrt{3}\\ &= 16\sqrt{5}+4\sqrt{5}+6\sqrt{5}+10\sqrt{3}\\ &= 26\sqrt{5}+10\sqrt{3}\\ &\approx 58.137\ldots+17.320\ldots \approx 75.458\ldots \approx 75.5 \text{ in.}\end{aligned}$$

WRITING

111. $\sqrt[3]{9x^4}$ is not simplified because a perfect cube factor, x^3, is present under the radical sign.

113. Since the two graphs are the same, the right side and the left side of the equations are equal.

REVIEW

115. $3x^2y^3(-5x^3y^{-4}) = 3(-5)x^{2+3}y^{3-4} = -15x^5y^{-1} = -\dfrac{15x^5}{y}$

117.
$$\begin{array}{r l} & 3p + 4 \\ 2p-5 \overline{)} & 6p^2 - 7p - 25 \\ & 6p^2 - 15p \\ & \quad\; 8p - 25 \\ & \quad\; 8p - 20 \\ & \qquad\; -5 \end{array}$$

Solution: $3p+4-\dfrac{5}{2p-5}$

Section 7.3 Multiplying and Dividing Radical Expressions

VOCABULARY

1. FOIL
3. irrational
5. perfect

CONCEPTS

7. a. $4\sqrt{6}+2\sqrt{6} = (4+2)\sqrt{6} = 6\sqrt{6}$

   b. $4\sqrt{6}\left(2\sqrt{6}\right) = 4\cdot 2\sqrt{6}\sqrt{6} = 8\sqrt{36} = 8\cdot 6 = 48$

   c. $3\sqrt{2}-2\sqrt{3}$ cannot be simplified

   d. $3\sqrt{2}\left(-2\sqrt{3}\right) = 3(-2)\sqrt{2}\sqrt{3} = -6\sqrt{6}$

9. When the numerator and the denominator of a fraction are multiplied by the same nonzero number, the value of the fraction is not changed.

11. A radical appears in the denominator.

NOTATION

13. Multiply $5\sqrt{8}\cdot 7\sqrt{6}$.

$$5\sqrt{8}\cdot 7\sqrt{6} = 5(7)\sqrt{8}\ \sqrt{6}$$
$$= 35\sqrt{\ 48\ }$$
$$= 35\sqrt{\ 16\cdot 3\ }$$
$$= 35(\ 4\ )\sqrt{3}$$
$$= 140\sqrt{3}$$

PRACTICE

15. $\sqrt{11}\sqrt{11} = \sqrt{11^2} = 11$

17. $\left(\sqrt{7}\right)^2 = \sqrt{7}\sqrt{7} = \sqrt{7^2} = 7$

19. $\sqrt{2}\sqrt{8} = \sqrt{16} = 4$

21. $\sqrt{5}\sqrt{10} = \sqrt{50} = \sqrt{25}\sqrt{2} = 5\sqrt{2}$

23. $2\sqrt{3}\sqrt{6} = 2\sqrt{18} = 2\sqrt{9}\sqrt{2} = 2\cdot 3\sqrt{2} = 6\sqrt{2}$

25. $\sqrt[3]{5}\sqrt[3]{25} = \sqrt[3]{125} = 5$

27. $\left(3\sqrt{2}\right)^2 = 3^2\left(\sqrt{2}\right)^2 = 9\cdot 2 = 18$

29. $\left(-2\sqrt{2}\right)^2 = (-2)^2\left(\sqrt{2}\right)^2 = 4\cdot 2 = 8$

31. $\left(3\sqrt[3]{9}\right)\left(2\sqrt[3]{3}\right) = 3\cdot 2\sqrt[3]{9\cdot 3} = 6\sqrt[3]{27} = 6\cdot 3 = 18$

33. $\sqrt[3]{2}\sqrt[3]{12} = \sqrt[3]{2\cdot 12} = \sqrt[3]{24} = \sqrt[3]{8\cdot 3} = 2\sqrt[3]{3}$

35. $\sqrt{ab^3}\sqrt{ab} = \sqrt{a^2b^4} = ab^2$

37. $\sqrt{5ab}\sqrt{5a} = \sqrt{25a^2b} = 5a\sqrt{b}$

39. $-4\sqrt[3]{5r^2s}\left(5\sqrt[3]{2r}\right) = -4\cdot 5\sqrt[3]{(5r^2s)(2r)} = -20\sqrt[3]{10r^3s} = -20r\sqrt[3]{10s}$

41. $\sqrt{x(x+3)}\ \sqrt{x^3(x+3)} = \sqrt{x^4(x+3)^2} = x^2(x+3)$

43. $3\sqrt{5}\left(4-\sqrt{5}\right) = 3\sqrt{5}\cdot 4 - 3\sqrt{5}\cdot\sqrt{5} = 12\sqrt{5} - 3\sqrt{25} = 12\sqrt{5} - 3\cdot 5 = 12\sqrt{5} - 15$

45. $3\sqrt{2}\left(4\sqrt{6} + 2\sqrt{7}\right) = 3\sqrt{2}\cdot 4\sqrt{6} + 3\sqrt{2}\cdot 2\sqrt{7} = 12\sqrt{12} + 6\sqrt{14}$

$= 12\sqrt{4\cdot 3} + 6\sqrt{14} = 12\cdot 2\sqrt{3} + 6\sqrt{14} = 24\sqrt{3} + 6\sqrt{14}$

47. $-2\sqrt{5x}\left(4\sqrt{2x} - 3\sqrt{3}\right) = -8\sqrt{10x^2} + 6\sqrt{15x} = -8x\sqrt{10} + 6\sqrt{15x}$

49. $\left(\sqrt{2}+1\right)\left(\sqrt{2}-3\right) = \sqrt{2^2} - 3\sqrt{2} + 1\sqrt{2} - 3 = 2 - 2\sqrt{2} - 3 = -1 - 2\sqrt{2}$

51. $\left(\sqrt{5z}+\sqrt{3}\right)\left(\sqrt{5z}+\sqrt{3}\right) = \sqrt{(5z)^2} + \sqrt{5z}\sqrt{3} + \sqrt{5z}\sqrt{3} + \sqrt{3^2} = 5z + \sqrt{15z} + \sqrt{15z} + 3 = 5z + 2\sqrt{15z} + 3$

53. $\left(\sqrt{3x}-\sqrt{2y}\right)\left(\sqrt{3x}+\sqrt{2y}\right) = \sqrt{(3x)^2} + \sqrt{3x}\sqrt{2y} - \sqrt{3x}\sqrt{2y} - \sqrt{(2y)^2} = 3x + \sqrt{6xy} - \sqrt{6xy} - 2y = 3x - 2y$

55. $\left(2\sqrt{3a}-\sqrt{b}\right)\left(\sqrt{3a}+3\sqrt{b}\right) = 2\sqrt{(3a)^2} + 6\sqrt{3ab} - \sqrt{3ab} - 3\sqrt{b^2} = 2(3a) + 5\sqrt{3ab} - 3b = 6a + 5\sqrt{3ab} - 3b$

57. $\left(3\sqrt{2r}-2\right)^2 = \left(3\sqrt{2r}-2\right)\left(3\sqrt{2r}-2\right) = 9\sqrt{(2r)^2} - 6\sqrt{2r} - 6\sqrt{2r} + 4 = 9(2r) - 12\sqrt{2r} + 4 = 18r - 12\sqrt{2r} + 4$

59. $-2\left(\sqrt{3x}+\sqrt{3}\right)^2 = -2\left(\sqrt{3x}+\sqrt{3}\right)\left(\sqrt{3x}+\sqrt{3}\right) = -2\left[\sqrt{(3x)^2} + 2\sqrt{9x} + \sqrt{3^2}\right]$

$= -2\left(3x + 2\cdot 3\sqrt{x} + 3\right) = -6x - 12\sqrt{x} - 6$

61. $\sqrt{\frac{1}{7}} = \frac{\sqrt{1}}{\sqrt{7}} = \frac{\sqrt{1}\cdot\sqrt{7}}{\sqrt{7}\cdot\sqrt{7}} = \frac{\sqrt{7}}{\sqrt{49}} = \frac{\sqrt{7}}{7}$

63. $\dfrac{6}{\sqrt{30}}=\dfrac{6\cdot\sqrt{30}}{\sqrt{30}\cdot\sqrt{30}}=\dfrac{6\sqrt{30}}{30}=\dfrac{6\sqrt{30}}{6\cdot 5}=\dfrac{\sqrt{30}}{5}$

65. $\dfrac{\sqrt{5}}{\sqrt{8}}=\dfrac{\sqrt{5}\cdot\sqrt{2}}{\sqrt{8}\cdot\sqrt{2}}=\dfrac{\sqrt{10}}{\sqrt{16}}=\dfrac{\sqrt{10}}{4}$

67. $\dfrac{\sqrt{8}}{\sqrt{2}}=\dfrac{\sqrt{8}\cdot\sqrt{2}}{\sqrt{2}\cdot\sqrt{2}}=\dfrac{\sqrt{16}}{\sqrt{4}}=\dfrac{4}{2}=2$

69. $\dfrac{1}{\sqrt[3]{2}}=\dfrac{\sqrt[3]{4}}{\sqrt[3]{2}\cdot\sqrt[3]{4}}=\dfrac{\sqrt[3]{4}}{\sqrt[3]{8}}=\dfrac{\sqrt[3]{4}}{2}$

71. $\dfrac{3}{\sqrt[3]{9}}=\dfrac{3\cdot\sqrt[3]{3}}{\sqrt[3]{9}\cdot\sqrt[3]{3}}=\dfrac{3\sqrt[3]{3}}{\sqrt[3]{27}}=\dfrac{3\sqrt[3]{3}}{3}=\sqrt[3]{3}$

73. $\dfrac{\sqrt[3]{2}}{\sqrt[3]{9}}=\dfrac{\sqrt[3]{2}\cdot\sqrt[3]{3}}{\sqrt[3]{9}\cdot\sqrt[3]{3}}=\dfrac{\sqrt[3]{6}}{\sqrt[3]{27}}=\dfrac{\sqrt[3]{6}}{3}$

75. $\dfrac{\sqrt{8}}{\sqrt{xy}}=\dfrac{\sqrt{8}\cdot\sqrt{xy}}{\sqrt{xy}\cdot\sqrt{xy}}=\dfrac{\sqrt{8xy}}{\sqrt{x^2y^2}}=\dfrac{\sqrt{4\cdot 2xy}}{xy}=\dfrac{2\sqrt{2xy}}{xy}$

77. $\dfrac{\sqrt{10xy^2}}{\sqrt{2xy^3}}=\sqrt{\dfrac{10xy^2}{2xy^3}}=\sqrt{\dfrac{5}{y}}=\dfrac{\sqrt{5\cdot y}}{\sqrt{y\cdot y}}=\dfrac{\sqrt{5y}}{\sqrt{y^2}}=\dfrac{\sqrt{5y}}{y}$

79. $\dfrac{\sqrt[3]{4a^2}}{\sqrt[3]{2ab}}=\sqrt[3]{\dfrac{4a^2}{2ab}}=\sqrt[3]{\dfrac{2a}{b}}=\sqrt[3]{\dfrac{2a\cdot b^2}{b\cdot b^2}}=\dfrac{\sqrt[3]{2ab^2}}{b}$

81. $\dfrac{1}{\sqrt[4]{4}}=\dfrac{\sqrt[4]{4}}{\sqrt[4]{4}\cdot\sqrt[4]{4}}=\dfrac{\sqrt[4]{4}}{\sqrt[4]{16}}=\dfrac{\sqrt[4]{4}}{2}$

83. $\dfrac{1}{\sqrt[5]{16}}=\dfrac{\sqrt[5]{2}}{\sqrt[5]{16}\cdot\sqrt[5]{2}}=\dfrac{\sqrt[5]{2}}{\sqrt[5]{32}}=\dfrac{\sqrt[5]{2}}{2}$

85. $\dfrac{1}{\sqrt{2}-1}=\dfrac{1\left(\sqrt{2}+1\right)}{\left(\sqrt{2}-1\right)\left(\sqrt{2}+1\right)}=\dfrac{\sqrt{2}+1}{2-1}=\sqrt{2}+1$

87. $\dfrac{\sqrt{2}}{\sqrt{5}+3}=\dfrac{\sqrt{2}\left(\sqrt{5}-3\right)}{\left(\sqrt{5}+3\right)\left(\sqrt{5}-3\right)}=\dfrac{\sqrt{10}-3\sqrt{2}}{5-9}=\dfrac{\sqrt{10}-3\sqrt{2}}{-4}=\dfrac{3\sqrt{2}-\sqrt{10}}{4}$

89. $\dfrac{\sqrt{3}+1}{\sqrt{3}-1}=\dfrac{\left(\sqrt{3}+1\right)\left(\sqrt{3}+1\right)}{\left(\sqrt{3}-1\right)\left(\sqrt{3}+1\right)}=\dfrac{\sqrt{9}+2\sqrt{3}+1}{\sqrt{9}-1}=\dfrac{3+2\sqrt{3}+1}{3-1}=\dfrac{4+2\sqrt{3}}{2}=\dfrac{2\left(2+\sqrt{3}\right)}{2}=2+\sqrt{3}$

91. $\dfrac{\sqrt{7}-\sqrt{2}}{\sqrt{2}+\sqrt{7}}=\dfrac{\left(\sqrt{7}-\sqrt{2}\right)\left(\sqrt{2}-\sqrt{7}\right)}{\left(\sqrt{2}+\sqrt{7}\right)\left(\sqrt{2}-\sqrt{7}\right)}=\dfrac{\sqrt{14}-\sqrt{49}-\sqrt{4}+\sqrt{14}}{\sqrt{4}-\sqrt{49}}=\dfrac{2\sqrt{14}-7-2}{2-7}=\dfrac{2\sqrt{14}-9}{-5}=\dfrac{9-2\sqrt{14}}{5}$

93. $\dfrac{2}{\sqrt{x}+1}=\dfrac{2\left(\sqrt{x}-1\right)}{\left(\sqrt{x}+1\right)\left(\sqrt{x}-1\right)}=\dfrac{2\left(\sqrt{x}-1\right)}{\sqrt{x^2}-1}=\dfrac{2\left(\sqrt{x}-1\right)}{x-1}$

95. $\dfrac{2z-1}{\sqrt{2z}-1}=\dfrac{(2z-1)\left(\sqrt{2z}+1\right)}{\left(\sqrt{2z}-1\right)\left(\sqrt{2z}+1\right)}=\dfrac{(2z-1)\left(\sqrt{2z}+1\right)}{2z-1}=\sqrt{2z}+1$

97. $\dfrac{\sqrt{x}-\sqrt{y}}{\sqrt{x}+\sqrt{y}}=\dfrac{\left(\sqrt{x}-\sqrt{y}\right)\left(\sqrt{x}-\sqrt{y}\right)}{\left(\sqrt{x}+\sqrt{y}\right)\left(\sqrt{x}-\sqrt{y}\right)}=\dfrac{x-\sqrt{xy}-\sqrt{xy}+y}{\sqrt{x^2}-\sqrt{y^2}}=\dfrac{x-2\sqrt{xy}+y}{x-y}$

APPLICATIONS

99. $\dfrac{1}{\sigma\sqrt{2\pi}}=\dfrac{1\cdot\sqrt{2\pi}}{\sigma\sqrt{2\pi}\cdot\sqrt{2\pi}}=\dfrac{\sqrt{2\pi}}{\sigma(2\pi)}=\dfrac{\sqrt{2\pi}}{2\pi\sigma}$

101. $\dfrac{\text{length of side } AC}{\text{length of side } AB}=\dfrac{1}{\sqrt{2}}=\dfrac{1\cdot\sqrt{2}}{\sqrt{2}\cdot\sqrt{2}}=\dfrac{\sqrt{2}}{2}$

103. The new area of the aperture is $\frac{9\pi}{4}$. Find the diameter using $A=\pi r^2=\pi\left(\frac{d}{2}\right)^2=\pi\left(\frac{d^2}{4}\right)$

$$\frac{9\pi}{4}=\pi\left(\frac{d^2}{4}\right)$$
$$9\pi=\pi d^2$$
$$9=d^2$$
$$d=3$$

$$f\text{-number}=\frac{\text{focal length}}{\text{diameter}}=\frac{f}{d}=\frac{12}{3}=4$$

WRITING

105. m times m is a perfect square, it is not a perfect cube.

REVIEW

107.
$$\frac{2}{3-a}=1$$
$$(3-a)\left(\frac{2}{3-a}\right)=1(3-a)$$
$$2=3-a$$
$$a=1$$

109.
$$\frac{8}{b-2}+\frac{3}{2-b}=-\frac{1}{b}$$
$$b(b-2)\left(\frac{8}{b-2}\right)+b(b-2)\left(\frac{-3}{b-2}\right)=b(b-2)\left(-\frac{1}{b}\right)$$
$$8b-3b=-(b-2)$$
$$5b=-b+2$$
$$6b=2$$
$$b=\frac{2}{6}=\frac{1}{3}$$

Section 7.4 Radical Equations

VOCABULARY

1. radical
3. extraneous

CONCEPTS

5. a. square both sides b. cube both sides

7. a. $\left(\sqrt{x}\right)^2=x$ b. $\left(\sqrt{x-5}\right)^2=x-5$

c. $\left(4\sqrt{2x}\right)^2=4^2\left(\sqrt{2x}\right)^2=16(2x)=32x$ d. $\left(-\sqrt{x+3}\right)^2=(-1)^2\left(\sqrt{x+3}\right)^2=x+3$

9. Only one side of the equation was squared.

11. If we raise two equal quantities to the same power, the results are equal quantities.

NOTATION

13.
$$2\sqrt{x-2}=4$$
$$\left(2\sqrt{x-2}\right)^2=4^2$$
$$4(x-2)=16$$
$$4x-8=16$$
$$4x=24$$
$$x=6$$

PRACTICE

15. $\sqrt{5x-6}=2$

$\left(\sqrt{5x-6}\right)^2=2^2$	Square both sides.	Check:
$5x-6=4$	Do the squaring.	$\sqrt{5(2)-6}\ \underset{=}{?}\ 2$
$5x=10$	Add 6 to both sides.	$\sqrt{4}=2$
$x=2$	Divide both sides by 5.	

17. $\sqrt{6x+1}+2=7$

$\sqrt{6x+1}=5$	Subtract 2 from both sides.	Check:
$\left(\sqrt{6x+1}\right)^2=5^2$	Square both sides.	$\sqrt{6(4)+1}+2\ \underset{=}{?}\ 7$
$6x+1=25$	Do the squaring.	$\sqrt{25}+2\ \underset{=}{?}\ 7$
$6x=24$	Subtract 1 from both sides.	$5+2=7$
$x=4$	Divide both sides by 6.	

19. $2\sqrt{4x+1}=\sqrt{x+4}$

$\left(2\sqrt{4x+1}\right)^2=\left(\sqrt{x+4}\right)^2$	Square both sides.	Check:
$4(4x+1)=x+4$	Do the squaring.	$2\sqrt{4(0)+1}\ \underset{=}{?}\ \sqrt{0+4}$
$16x+4=x+4$	Remove the parentheses.	$2\sqrt{1}\ \underset{=}{?}\ \sqrt{4}$
$15x=0$	Subtract x and 4 from both sides.	$2=2$
$x=0$	Divide both sides by 15.	

21. $\sqrt[3]{7n-1}=3$

$\left(\sqrt[3]{7n-1}\right)^3=3^3$	Cube both sides.	Check:
$7n-1=27$	Do the cubing.	$\sqrt[3]{7(4)-1}\ \underset{=}{?}\ 3$
$7n=28$	Add 1 to both sides.	$\sqrt[3]{27}\ \underset{=}{?}\ 3$
$n=4$	Divide both sides by 7.	$3=3$

23. $\sqrt[4]{10p+1}=\sqrt[4]{11p-7}$

$\left(\sqrt[4]{10p+1}\right)^4=\left(\sqrt[4]{11p-7}\right)^4$	Raise both sides to the 4th power.	Check:
$10p+1=11p-7$	Simplify.	$\sqrt[4]{10(8)+1}\ \underset{=}{?}\ \sqrt[4]{11(8)-7}$
$-p=-8$	Subtract $11p$ and 1 from both sides.	$\sqrt[4]{81}\ \underset{=}{?}\ \sqrt[4]{81}$
$p=8$	Multiply both sides by -1.	$3=3$

25.

$$x = \frac{\sqrt{12x-5}}{2}$$

$2x = \sqrt{12x-5}$ Multiply both sides by 2.

$(2x)^2 = \left(\sqrt{12x-5}\right)^2$ Square both sides.

$4x^2 = 12x-5$ Do the squaring.

$4x^2 - 12x + 5 = 0$ Subtract $12x$ and add 5.

$(2x-5)(2x-1) = 0$ Factor.

$2x-5 = 0$ or $2x-1 = 0$

$2x = 5$ $\quad 2x = 1$

$x = \frac{5}{2}$ $\quad x = \frac{1}{2}$

Check:

$$\frac{5}{2} \stackrel{?}{=} \frac{\sqrt{12\left(\frac{5}{2}\right)-5}}{2}$$

$$\frac{5}{2} \stackrel{?}{=} \frac{\sqrt{25}}{2} = \frac{5}{2}$$

and

$$\frac{1}{2} \stackrel{?}{=} \frac{\sqrt{12\left(\frac{1}{2}\right)-5}}{2}$$

$$\frac{1}{2} \stackrel{?}{=} \frac{\sqrt{1}}{2} = \frac{1}{2}$$

27. $\sqrt{x+2} - \sqrt{4-x} = 0$

$\sqrt{x+2} = \sqrt{4-x}$ Add $\sqrt{4-x}$ to both sides.

$\left(\sqrt{x+2}\right)^2 = \left(\sqrt{4-x}\right)^2$ Square both sides.

$x+2 = 4-x$ Do the squaring.

$2x = 2$ Add x and subtract 2 from both sides.

$x = 1$ Divide both sides by 2.

Check:

$$\sqrt{1+2} - \sqrt{4-1} \stackrel{?}{=} 0$$

$$\sqrt{3} - \sqrt{3} \stackrel{?}{=} 0$$

$$0 = 0$$

29. $2\sqrt{x} = \sqrt{5x-16}$

$\left(2\sqrt{x}\right)^2 = \left(\sqrt{5x-16}\right)^2$ Square both sides.

$4x = 5x - 16$ Do the squaring.

$-x = -16$ Subtract $5x$ from both sides.

$x = 16$ Multiply both sides by -1.

Check:

$$2\sqrt{16} \stackrel{?}{=} \sqrt{5(16)-16}$$

$$2(4) \stackrel{?}{=} \sqrt{64}$$

$$8 = 8$$

31.

$r - 9 = \sqrt{2r-3}$

$(r-9)^2 = \left(\sqrt{2r-3}\right)^2$ Square both sides.

$r^2 - 18r + 81 = 2r - 3$ Do the squaring.

$r^2 - 20r + 84 = 0$ Subtract $2r$ and add 3.

$(r-14)(r-6) = 0$ Factor.

$r - 14 = 0$ or $r - 6 = 0$

$r = 14$ $\quad r = 6$ is an extraneous solution.

Check:

$$14 - 9 \stackrel{?}{=} \sqrt{2(14)-3}$$

$$5 = \sqrt{25}$$

and

$$6 - 9 \stackrel{?}{=} \sqrt{2(6)-3}$$

$$-3 \stackrel{?}{=} \sqrt{9}$$

$$-3 \neq 3$$

33. $\sqrt{-5x+24} = 6-x$

$\left(\sqrt{-5x+24}\right)^2 = (6-x)^2$ Square both sides.

$-5x + 24 = 36 - 12x + x^2$ Do the squaring.

$0 = x^2 - 7x + 12$ Subtract 24 and add $5x$.

$0 = (x-4)(x-3)$ Factor.

$x - 4 = 0$ or $x - 3 = 0$

$x = 4$ $\quad x = 3$

Check:

$$\sqrt{-5(4)+24} \stackrel{?}{=} 6-4$$

$$\sqrt{4} = 2$$

and

$$\sqrt{-5(3)+24} \stackrel{?}{=} 6-3$$

$$\sqrt{9} = 3$$

35. $\sqrt{y+2}=4-y$ Check:

$\left(\sqrt{y+2}\right)^2=(4-y)^2$	Square both sides.	$\sqrt{(2)+2}\stackrel{?}{=}4-2$
$y+2=16-8y+y^2$	Do the squaring.	$\sqrt{4}=2$
$0=y^2-9y+14$	Subtract y and 2.	and
$0=(y-2)(y-7)$	Factor.	$\sqrt{(7)+2}\stackrel{?}{=}4-7$
		$\sqrt{9}\neq-3$

$y-2=0$ or $y-7=0$

$y=2$ $\quad y=7$ is an extraneous solution.

37. $\sqrt[3]{x^3-7}=x-1$ Check:

$\left(\sqrt[3]{x^3-7}\right)^3=(x-1)^3=(x-1)(x^2-2x+1)$	Cube both sides.	$\sqrt[3]{(2)^3-7}\stackrel{?}{=}(2)-1$
$x^3-7=x^3-3x^2+3x-1$	Do the multiplication.	$\sqrt[3]{1}=1$
$0=-3x^2+3x+6$	Subtract x^3 and add 7.	and
$0=x^2-x-2$	Divide by –3.	$\sqrt[3]{(-1)^3-7}\stackrel{?}{=}(-1)-1$
$0=(x-2)(x+1)$	Factor.	$\sqrt[3]{-8}=-2$

$x-2=0$ or $x+1=0$

$x=2$ $\quad x=-1$

39. $\sqrt[4]{x^4+4x^2-4}=-x$ Check:

$\left(\sqrt[4]{x^4+4x^2-4}\right)^4=(-x)^4$	Raise to the 4th power.	$\sqrt[4]{(-1)^4+4(-1)^2-4}\stackrel{?}{=}-(-1)$
$x^4+4x^2-4=x^4$	Simplify.	$\sqrt[4]{1}=1$
$4x^2-4=0$	Subtract x^4.	and
$x^2-1=0$	Divide by 4.	$\sqrt[4]{(1)^4+4(1)^2-4}\stackrel{?}{=}-(1)$
$(x+1)(x-1)=0$	Factor.	$\sqrt[4]{1}\neq-1$

$x+1=0$ or $x-1=0$

$x=-1$ $\quad x=1$ is an extraneous solution.

41. $\sqrt[4]{12t+4}+2=0$ Check:

$\sqrt[4]{12t+4}=-2$	Subtract 2 from both sides.	$\sqrt[4]{12(1)+4}+2\stackrel{?}{=}0$
$\left(\sqrt[4]{12t+4}\right)^4=(-2)^4$	Raise both sides to the 4th power.	$\sqrt[4]{16}+2\stackrel{?}{=}0$
$12t+4=16$	Simplify.	$2+2\neq0$
$12t=12$	Subtract 4 from both sides.	
$t=1$	Divide both sides by 12.	

$t=1$ is an extraneous solution, so there are no solutions.

43.

$\sqrt{2y+1}=1-2\sqrt{y}$	
$\left(\sqrt{2y+1}\right)^2=\left(1-2\sqrt{y}\right)^2$	Square both sides.
$2y+1=1-4\sqrt{y}+4y$	Do the squaring.
$-2y=-4\sqrt{y}$	Subtract 1 and 4y.
$y=2\sqrt{y}$	Divide both sides by –2.
$y^2=\left(2\sqrt{y}\right)^2$	Square both sides.
$y^2=4y$	Do the squaring.
$y^2-4y=0$	Subtract 4*y*.
$y(y-4)=0$	Factor.

$y=0$ or $y-4=0$

$y=4$ is an extraneous solution.

Check:

$\sqrt{2(0)+1}\overset{?}{=}1-2\sqrt{0}$

$\sqrt{1}\overset{?}{=}1-0$

$1=1$

and

$\sqrt{2(4)+1}\overset{?}{=}1-2\sqrt{4}$

$\sqrt{9}\overset{?}{=}1-2(2)$

$3\neq-3$

45.

$\sqrt{y+7}+3=\sqrt{y+4}$	
$\left(\sqrt{y+7}+3\right)^2=\left(\sqrt{y+4}\right)^2$	Square both sides.
$(y+7)+6\sqrt{y+7}+9=y+4$	Do the squaring.
$y+16+6\sqrt{y+7}=y+4$	Simplify.
$6\sqrt{y+7}=-12$	Subtract 16 and y.
$\sqrt{y+7}=-2$	Divide both sides by 6.
$\left(\sqrt{y+7}\right)^2=(-2)^2$	Square both sides.
$y+7=4$	Do the squaring.
$y=-3$	Subtract 7.

$y=-3$ is an extraneous solution, so there are no solutions.

Check:

$\sqrt{(-3)+7}+3\overset{?}{=}\sqrt{(-3)+4}$

$\sqrt{4}+3\overset{?}{=}\sqrt{1}$

$2+3\neq1$

47.

$2+\sqrt{u}=\sqrt{2u+7}$	
$\left(2+\sqrt{u}\right)^2=\left(\sqrt{2u+7}\right)^2$	Square both sides.
$4+4\sqrt{u}+u=2u+7$	Do the squaring.
$4\sqrt{u}=u+3$	Subtract 4 and *u*.
$\left(4\sqrt{u}\right)^2=(u+3)^2$	Square both sides.
$16u=u^2+6u+9$	Do the squaring.
$0=u^2-10u+9$	Subtract 16*u*.
$0=(u-9)(u-1)$	Factor.

$u-9=0$ or $u-1=0$

$u=9$ $\quad$ $u=1$

Check:

$2+\sqrt{9}\overset{?}{=}\sqrt{2(9)+7}$

$2+3\overset{?}{=}\sqrt{25}$

$5=5$

and

$2+\sqrt{1}\overset{?}{=}\sqrt{2(1)+7}$

$2+1\overset{?}{=}\sqrt{9}$

$3=3$

49. $\sqrt{6t+1}-3\sqrt{t}=-1$

$\sqrt{6t+1}=3\sqrt{t}-1$	Add $3\sqrt{t}$ to both sides.
$\left(\sqrt{6t+1}\right)^2=\left(3\sqrt{t}-1\right)^2$	Square both sides.
$6t+1=9t-6\sqrt{t}+1$	Do the squaring.
$-3t=-6\sqrt{t}$	Subtract 1 and 6t.
$t=2\sqrt{t}$	Divide both sides by –3.
$t^2=\left(2\sqrt{t}\right)^2$	Square both sides.
$t^2=4t$	Do the squaring.
$t^2-4t=0$	Subtract 4t.
$t(t-4)=0$	Factor.

$t-4=0$ or $t=0$ is an extraneous solution.
$t=4$
$s-6=0$ or $s=0$ is an extraneous solution.
$s=6$

Check:

$\sqrt{6(4)+1}-3\sqrt{4}\overset{?}{=}-1$
$\sqrt{25}-3(2)\overset{?}{=}-1$
$5-6=-1$

and

$\sqrt{6(0)+1}-3\sqrt{0}\overset{?}{=}-1$
$\sqrt{1}-0\overset{?}{=}-1$
$1\neq-1$

51. $\sqrt{2x+5}+\sqrt{x+2}=5$

$\sqrt{2x+5}=5-\sqrt{x+2}$	Subtract $\sqrt{x+2}$.
$\left(\sqrt{2x+5}\right)^2=\left(5-\sqrt{x+2}\right)^2$	Square both sides.
$2x+5=25-10\sqrt{x+2}+x+2$	Do the squaring.
$x-22=-10\sqrt{x+2}$	Subtract x and 27.
$(x-22)^2=\left(-10\sqrt{x+2}\right)^2$	Square both sides.
$x^2-44x+484=100(x+2)$	Do the squaring.
$x^2-44x+484=100x+200$	Remove the parentheses.
$x^2-144x+284=0$	Subtract 100x and 200.
$(x-2)(x-142)=0$	Factor.

$x-2=0$ or $x-142=0$
$x=2$ $x=142$ is an extraneous solution.

Check:

$\sqrt{2(2)+5}+\sqrt{2+2}\overset{?}{=}5$
$\sqrt{9}+\sqrt{4}\overset{?}{=}5$
$3+2=5$

and

$\sqrt{2(142)+5}+\sqrt{(142)+2}\overset{?}{=}5$
$\sqrt{289}+\sqrt{144}\overset{?}{=}5$
$17+12\neq5$

53. $\sqrt{x-5}-\sqrt{x+3}=4$

$\sqrt{x-5}=4+\sqrt{x+3}$	Add $\sqrt{x+3}$.
$\left(\sqrt{x-5}\right)^2=\left(4+\sqrt{x+3}\right)^2$	Square both sides.
$x-5=16+8\sqrt{x+3}+x+3$	Do the squaring.
$-24=8\sqrt{x+3}$	Subtract x and 19.
$-3=\sqrt{x+3}$	Divide by 8.
$(-3)^2=\left(\sqrt{x+3}\right)^2$	Square both sides.
$9=x+3$	Do the squaring.
$6=x$	Subtract 3.

$x=6$ is an extraneous solution, so there are no solutions.

Check:

$\sqrt{(6)-5}-\sqrt{(6)+3}\overset{?}{=}4$
$\sqrt{1}-\sqrt{9}\overset{?}{=}4$
$1-3\neq4$

55.

$v = \sqrt{2gh}$	Solve for h.
$v^2 = \left(\sqrt{2gh}\right)^2$	Square both sides.
$v^2 = 2gh$	Do the squaring.
$\dfrac{v^2}{2g} = h$	Divide both sides by $2g$.

57.

$T = 2\pi\sqrt{\dfrac{l}{32}}$	Solve for l.
$\dfrac{T}{2\pi} = \sqrt{\dfrac{l}{32}}$	Divide by 2π.
$\left(\dfrac{T}{2\pi}\right)^2 = \left(\sqrt{\dfrac{l}{32}}\right)^2$	Square both sides.
$\dfrac{T^2}{4\pi^2} = \dfrac{l}{32}$	Do the squaring.
$32\left(\dfrac{T^2}{4\pi^2}\right) = 32\left(\dfrac{l}{32}\right)$	Multiply by LCD 32.
$\dfrac{8T^2}{\pi^2} = l$	Do the multiplication.

59.

$r = \sqrt[3]{\dfrac{A}{P}} - 1$	Solve for A.
$r + 1 = \sqrt[3]{\dfrac{A}{P}}$	Add 1 to both sides.
$(r+1)^3 = \left(\sqrt[3]{\dfrac{A}{P}}\right)^3$	Cube both sides.
$(r+1)^3 = \dfrac{A}{P}$	Do the cubing.
$P(r+1)^3 = A$	Multiply both sides by P.

61.

$$L_A = L_B\sqrt{1-\frac{v^2}{c^2}}$$ Solve for v^2.

$$\frac{L_A}{L_B} = \sqrt{1-\frac{v^2}{c^2}}$$ Divide both sides by L_B.

$$\left(\frac{L_A}{L_B}\right)^2 = \left(\sqrt{1-\frac{v^2}{c^2}}\right)^2$$ Square both sides.

$$\frac{L_A^{\,2}}{L_B^{\,2}} = 1-\frac{v^2}{c^2}$$ Do the squaring.

$$\frac{L_A^{\,2}}{L_B^{\,2}} - 1 = -\frac{v^2}{c^2}$$ Subtract 1 from both sides.

$$-c^2\left(\frac{L_A^{\,2}}{L_B^{\,2}} - 1\right) = -c^2\left(-\frac{v^2}{c^2}\right)$$ Multiply both sides by $-c^2$.

$$c^2\left(1-\frac{L_A^{\,2}}{L_B^{\,2}}\right) = v^2$$ Do the multiplication

APPLICATIONS

63. Find r when $s = 40$ mph.

$$s = 3\sqrt{r}$$
$$40 = 3\sqrt{r}$$
$$\frac{40}{3} = \sqrt{r}$$
$$\left(\frac{40}{3}\right)^2 = \left(\sqrt{r}\right)^2$$
$$\left(\frac{40}{3}\right)^2 = r$$
$$r = 177.\overline{7} \quad \text{or} \quad 178 \text{ ft.}$$

65. Find P when $v = 29$ mph.

$$v = \sqrt[3]{\frac{P}{0.02}}$$
$$29 = \sqrt[3]{\frac{P}{0.02}}$$
$$29^3 = \left(\sqrt[3]{\frac{P}{0.02}}\right)^3$$
$$29^3 = \frac{P}{0.02}$$
$$0.02(29^3) = P$$
$$P = 487.78 \quad \text{or about 488 watts}$$

67. Find w_1 when $w_2 = 12.5$ lb and $w_3 = 7.5$ lb.

$$w_2 = \sqrt{w_1^2 + w_3^2}$$
$$12.5 = \sqrt{w_1^2 + (7.5)^2}$$
$$(12..5)^2 = \left(\sqrt{w_1^2 + (7.5)^2}\right)^2$$
$$12.5^2 = w_1^2 + 7.5^2$$
$$12.5^2 - 7.5^2 = w_1^2$$
$$\sqrt{12.5^2 - 7.5^2} = w_1$$
$$w_1 = \sqrt{100} = 10 \text{ lb}$$

69. Find x when supply equals demand

$$\text{supply} = \text{demand}$$
$$s = d$$
$$\sqrt{5x} = \sqrt{100 - 3x^2}$$
$$\left(\sqrt{5x}\right)^2 = \left(\sqrt{100 - 3x^2}\right)^2$$
$$5x = 100 - 3x^2$$
$$3x^2 + 5x - 100 = 0$$
$$(3x + 20)(x - 5) = 0$$

$$3x + 20 = 0 \quad \text{or} \quad x - 5 = 0$$
$$3x = -20 \qquad x = \$5$$

Discard, x is negative.

WRITING

71. Raising both sides of an equation to the same power can change the signs of an expression which will produce solutions that don't satisfy the original equation.

73. Look for the same Y–value in each table.
for $x = 7$, $Y_1 = 3$ and $Y_2 = 3$ so, $x = 7$ is a solution.
for $x = 3$, $Y_1 = 1$ and $Y_2 = 1$, so $x = 3$ is a solution

REVIEW

75. Let I represent the intensity of the light and d the distance from the bulb.

Inverse variation model

$$I = \frac{k}{d^2}$$

Determine k.

$$40 = \frac{k}{5^2}$$
$$40(5^2) = k$$
$$1000 = k$$

Solve for V when $p = 10$ lbs

$$I = \frac{1000}{d^2}$$
$$I = \frac{1000}{20^2} = 2.5 \text{ foot–candles}$$

77. Let x represent the size of 30–point type and set up a proportion.

$\frac{12}{0.166044} = \frac{30}{x}$	The proportion.
$12x = 30(0.166044)$	Multiply the means and the extremes.
$12x = 4.98132$	Do the multiplication.
$x = 0.41511$ in.	Divide by 12.

Section 7.5 — Rational Exponents

VOCABULARY

1. rational (or fractional)
3. index; radicand

CONCEPTS

5.

Radical form	Exponential form
$\sqrt[5]{25}$	$25^{1/5}$
$(\sqrt[3]{-27})^2$	$(-27)^{2/3}$
$(\sqrt[4]{16})^{-3}$	$16^{-3/4}$
$(\sqrt{81})^3$	$81^{3/2}$
$-\sqrt{\frac{9}{64}}$	$-\left(\frac{9}{64}\right)^{1/2}$

7. $8^{2/3} = 4$; $(-125)^{1/3} = -5$; $-16^{-1/4} = -1/2$; $4^{3/2} = 8$; $-(9/100)^{-1/2} = -10/3$

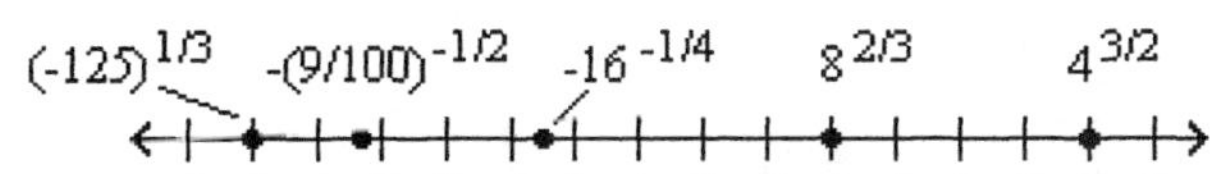

9. $x^m x^n = x^{m+n}$

11. $\dfrac{x^m}{x^n} = x^{m-n}$

13. $x^{1/n} = \sqrt[n]{x}$

15.

x	$f(x) = x^{1/2}$	$f(x)$
0	$f(0) = 0^{1/2} = \sqrt{0}$	0
1	$f(1) = 1^{1/2} = \sqrt{1}$	1
4	$f(4) = 4^{1/2} = \sqrt{4}$	2
9	$f(9) = 9^{1/2} = \sqrt{9}$	3
16	$f(16) = 16^{1/2} = \sqrt{16}$	4

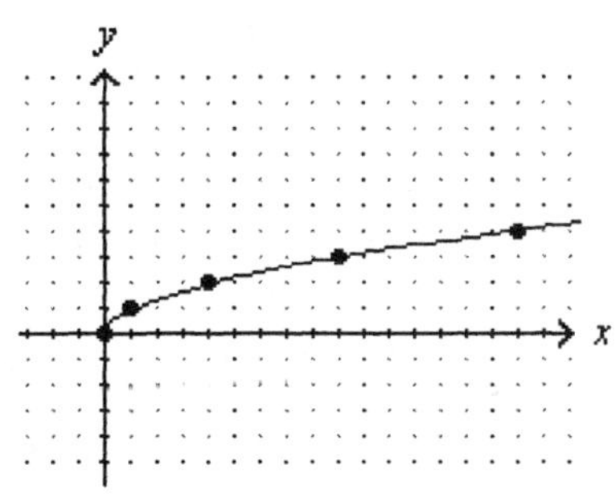

NOTATION

17.
$$\left(100a^4\right)^{3/2} = \left(\sqrt{100a^4}\right)^3 = \left(10a^2\right)^3 = 1{,}000a^6$$

PRACTICE

19. $x^{1/3} = \sqrt[3]{x}$

21. $(3x)^{1/4} = \sqrt[4]{3x}$

23. $\left(17x^3y\right)^{1/4} = \sqrt[4]{17x^3y}$

25. $\left(x^2+y^2\right)^{1/2} = \sqrt{x^2+y^2}$

27. $\sqrt{m} = m^{1/2}$

29. $\sqrt[4]{3a} = (3a)^{1/4}$

31. $\sqrt[6]{\frac{1}{7}abc} = \left(\frac{1}{7}abc\right)^{1/6}$

33. $\sqrt[3]{a^2-b^2} = \left(a^2-b^2\right)^{1/3}$

35. $4^{1/2} = \sqrt{4} = 2$

37. $125^{1/3} = \sqrt[3]{125} = 5$

39. $16^{1/4} = \sqrt[4]{16} = 2$

41. $32^{1/5} = \sqrt[5]{32} = 2$

43. $\left(\frac{1}{16}\right)^{1/2} = \sqrt{\frac{1}{16}} = \frac{1}{4}$

45. $-16^{1/4} = -\sqrt[4]{16} = -2$

47. $(-64)^{1/2}$ is not real because no real number raised to the 2nd power is -64.

49. $(-27)^{1/3} = \sqrt[3]{-27} = -3$

51. $\left(25y^2\right)^{1/2} = \sqrt{25y^2} = 5|y|$

53. $\left(16x^4\right)^{1/4} = \sqrt[4]{16x^4} = 2\,|\,x\,|$

55. $\left(243x^5\right)^{1/5} = \sqrt[5]{243x^5} = 3x$

57. $\left(-64x^8\right)^{1/4}$ is not real.

59. $36^{3/2} = \left(\sqrt{36}\right)^3 = 6^3 = 216$

61 $81^{3/4} = \left(\sqrt[4]{81}\right)^3 = 3^3 = 27$

63. $144^{3/2} = \left(\sqrt{144}\right)^3 = 12^3 = 1728$

65. $\left(\frac{1}{8}\right)^{2/3} = \left(\sqrt[3]{\frac{1}{8}}\right)^2 = \left(\frac{1}{2}\right)^2 = \frac{1}{4}$

67. $\left(25x^4\right)^{3/2} = \left(\sqrt{25x^4}\right)^3 = \left(5x^2\right)^3 = 125x^6$

69. $\left(\frac{8x^3}{27}\right)^{2/3} = \left(\sqrt[3]{\frac{8x^3}{27}}\right)^2 = \left(\frac{2x}{3}\right)^2 = \frac{4x^2}{9}$

71. $4^{-1/2} = \frac{1}{4^{1/2}} = \frac{1}{\sqrt{4}} = \frac{1}{2}$

73. $4^{-3/2} = \frac{1}{4^{3/2}} = \frac{1}{\left(\sqrt{4}\right)^3} = \left(\frac{1}{2}\right)^3 = \frac{1}{8}$

75. $\left(16x^2\right)^{-3/2} = \frac{1}{\left(16x^2\right)^{3/2}} = \frac{1}{\left(\sqrt{16x^2}\right)^3} = \left(\frac{1}{4x}\right)^3 = \frac{1}{64x^3}$

77. $\left(-27y^3\right)^{-2/3} = \frac{1}{\left(-27y^3\right)^{2/3}} = \frac{1}{\left(\sqrt[3]{-27y^3}\right)^2} = \frac{1}{(-3y)^2} = \frac{1}{9y^2}$

79. $\left(\frac{27}{8}\right)^{-4/3} = \frac{27^{-4/3}}{8^{-4/3}} = \frac{8^{4/3}}{27^{4/3}} = \frac{\left(\sqrt[3]{8}\right)^4}{\left(\sqrt[3]{27}\right)^4} = \frac{2^4}{3^4} = \frac{16}{81}$

81. $\left(-\frac{8x^3}{27}\right)^{-1/3} = \frac{\left(-8x^3\right)^{-1/3}}{27^{-1/3}} = \frac{27^{1/3}}{(-8x^3)^{1/3}} = \frac{\sqrt[3]{27}}{\sqrt[3]{-8x^3}} = \frac{3}{-2x} = -\frac{3}{2x}$

83. $\sqrt[3]{15} = 2.47$

85. $\sqrt[5]{1.045} = 1.01$

87. $5^{3/7}5^{2/7} = 5^{\frac{3}{7}+\frac{2}{7}} = 5^{5/7}$

89. $\left(4^{1/5}\right)^3 = 4^{\left(\frac{1}{5}\right)(3)} = 4^{3/5}$

91. $\frac{9^{4/5}}{9^{3/5}} = 9^{\frac{4}{5}-\frac{3}{5}} = 9^{1/5}$

93. $6^{-2/3}6^{-4/3} = 6^{\frac{-2}{3}-\frac{4}{3}} = 6^{-6/3} = 6^{-2} = \frac{1}{6^2} = \frac{1}{36}$

95. $\frac{3^{4/3}3^{1/3}}{3^{2/3}} = 3^{\frac{4}{3}+\frac{1}{3}-\frac{2}{3}} = 3^{3/3} = 3$

97. $a^{2/3}a^{1/3} = a^{\frac{2}{3}+\frac{1}{3}} = a^{3/3} = a$

99. $\left(a^{2/3}\right)^{1/3} = a^{\left(\frac{2}{3}\right)\left(\frac{1}{3}\right)} = a^{2/9}$

101. $\left(a^{1/2}b^{1/3}\right)^{3/2} = a^{\left(\frac{1}{2}\right)\left(\frac{3}{2}\right)}b^{\left(\frac{1}{3}\right)\left(\frac{3}{2}\right)} = a^{3/4}b^{1/2}$

103. $\frac{\left(4x^3y\right)^{1/2}}{(9xy)^{1/2}} = \frac{4^{1/2}x^{3/2}y^{1/2}}{9^{1/2}x^{1/2}y^{1/2}} = \frac{\sqrt{4}}{\sqrt{9}}x^{\frac{3}{2}-\frac{1}{2}}y^{\frac{1}{2}-\frac{1}{2}} = \frac{2}{3}x^{2/2}y^0 = \frac{2x}{3}$

105. $\left(27x^{-3}\right)^{-1/3} = 27^{-1/3}x^{-3\left(\frac{-1}{3}\right)} = \frac{1}{27^{1/3}}x^1 = \frac{x}{3}$ or $\frac{1}{3}x$

107. $(2x^2y^{-1/4})^3(8y^{-2})^{2/3} = (8x^6y^{-3/4})\left(4y^{-4/3}\right) = 32x^6y^{-\frac{3}{4}-\frac{4}{3}} = 32x^6y^{-\frac{25}{12}} = \frac{32x^6}{y^{25/12}}$

109. $y^{1/3}\left(y^{2/3}+y^{5/3}\right)=y^{\frac{1}{3}+\frac{2}{3}}+y^{\frac{1}{3}+\frac{5}{3}}=y^{3/3}+y^{6/3}=y+y^2$

111. $x^{3/5}\left(x^{7/5}-x^{2/5}+1\right)=x^{\frac{3}{5}+\frac{7}{5}}-x^{\frac{3}{5}+\frac{2}{5}}+x^{3/5}=x^{10/5}-x^{5/5}+x^{3/5}=x^2-x+x^{3/5}$

113. $\sqrt[6]{p^3}=p^{3/6}=p^{1/2}=\sqrt{p}$

115. $\sqrt[4]{25b^2}=\left(25b^2\right)^{1/4}=\left(5^2b^2\right)^{1/4}=5^{2/4}b^{2/4}=5^{1/2}b^{1/2}=\sqrt{5b}$

APPLICATIONS

117. Find v when $m = 0.0625$ lb, $M = 6.0$ lb, $g = 32$, and $h = 0.9$ ft.

$$v=\frac{m+M}{m}(2gh)^{1/2}$$
$$v=\frac{0.0625+6.0}{0.0625}[(2(32)(0.9)]^{1/2}$$
$$=736 \quad \text{ft/sec.}$$

119. Find m when $m_0 = 1$ unit, $v = 160{,}000$ mi/sec, and $c = 186{,}000$ mi/sec.

$$m=m_0\left(1-\frac{v^2}{c^2}\right)^{-1/2}$$
$$m=1(1-\left(\frac{160{,}000^2}{186{,}000^2}\right)^{-1/2}$$
$$=1.96 \quad \text{units}$$

121. First, find the floor space lost by 1 cubicle.

$$A(x)=V^{2/3}$$
$$A(x)=(4096)^{2/3}=256 \quad \text{in.}^2$$

Next, find the floor space lost by 18 cubicles.

$$18(256)=4608 \text{ in.}^2$$

Then, convert from in.^{2} to ft^2.

There are 144 in.^{2} in 1 ft^2.

$$4608 \div 144 = 32 \text{ ft}^2.$$

WRITING

123. A rational exponent is an exponent that is a fraction. An example is $4^{1/2}$, where 4 is the base and $\frac{1}{2}$ is the exponent.

REVIEW

125.
$$5x-4 < 11$$
$$5x < 15$$
$$x < 3$$
$$(-\infty,3)$$

127.
$$\frac{4}{5}(r-3)>\frac{2}{3}(r+2)$$
$$15\left[\frac{4}{5}(r-3)\right]>15\left[\frac{2}{3}(r+2)\right]$$
$$12(r-3)>10(r+2)$$
$$12r-36>10r+20$$
$$2r>56$$
$$r>28 \qquad (28,\infty)$$

Section 7.6 Geometric Applications of Radicals

VOCABULARY

1. hypotenuse
3. Pythagorean

CONCEPTS

5. $a^2 + b^2 = c^2$
7. $\sqrt{2}$
9. $\sqrt{3}$
11. 30°; 60°

13. a. Take the positive square root of both sides.
b. Subtract 25 from both sides.

NOTATION

15.
$$\sqrt{(-1-3)^2 + [2-(-4)]^2} = \sqrt{(-4)^2 + [\ 6\ ]^2}$$
$$= \sqrt{\ 52\ }$$
$$= \sqrt{\ 4 \cdot 13}$$
$$= 2\sqrt{13}$$
$$\approx 7.21$$

PRACTICE

17. $a = 6$ ft and $b = 8$ft. Find c.

$$c^2 = a^2 + b^2$$
$$c^2 = 6^2 + 8^2$$
$$c^2 = 36 + 64$$
$$c^2 = 100$$
$$\sqrt{c^2} = \sqrt{100}$$
$$c = 10 \text{ ft}$$

19. $b = 18$ m and $c = 82$ m. Find a.

$$c^2 = a^2 + b^2$$
$$82^2 = a^2 + 18^2$$
$$6724 = a^2 + 324$$
$$6400 = a^2$$
$$\sqrt{6400} = \sqrt{a^2}$$
$$a = 80 \text{ m}$$

21. 45°–45°–90° triangle.

Legs are the same length.
$x = 2$.

Hypotenuse is length of leg times $\sqrt{2}$.
$h = 2\sqrt{2} \approx 2.83$

23. 30°–60°–90° triangle.

Hypotenuse is twice as long as shorter leg.
$h = 2(5) = 10$

Longer leg is shorter leg times $\sqrt{3}$.
$x = 5\sqrt{3} \approx 8.66$

25. 30°–60°–90° triangle.

Hypotenuse is twice as long as shorter leg.
$h = 2x$
$9.37 = 2x$
$4.69 = x$

Longer leg is shorter leg times $\sqrt{3}$.
$y = 4.69\sqrt{3} = 8.11$

27. 45°–45°–90° triangle.

Hypotenuse is length of leg times $\sqrt{2}$.
$h = x\sqrt{2}$
$17.12 = x\sqrt{2}$
$12.11 = x$

Legs are the same length.
$y = x = 12.11$

29. 45°–45°–90° triangle, because legs are the same length, 7 cm.

Diagonal is hypotenuse which is length of leg times $\sqrt{2}$.
diagonal of face $= 7\sqrt{2}$ cm

31. $P(x_1, y_1) = P(3, -4),\ Q(x_2, y_2) = (0, 0)$

$$\begin{aligned} d(PQ) &= \sqrt{(x_2 - x_1)^2 + (y_2 - y_1)^2} \\ &= \sqrt{(0-3)^2 + [0-(-4)]^2} \\ &= \sqrt{3^2 + (4)^2} \\ &= \sqrt{9+16} \\ &= \sqrt{25} \\ &= 5 \end{aligned}$$

33. $P(x_1, y_1) = P(-2, -8),\ Q(x_2, y_2) = (3, 4)$

$$\begin{aligned} d(PQ) &= \sqrt{(x_2 - x_1)^2 + (y_2 - y_1)^2} \\ &= \sqrt{[3-(-2)]^2 + [4-(-8)]^2} \\ &= \sqrt{5^2 + (12)^2} \\ &= \sqrt{25+144} \\ &= \sqrt{169} \\ &= 13 \end{aligned}$$

35. $P(x_1, y_1) = P(6, 8),\ Q(x_2, y_2) = (12,$ 16)

$$\begin{aligned} d(PQ) &= \sqrt{(x_2 - x_1)^2 + (y_2 - y_1)^2} \\ &= \sqrt{(12-6)^2 + (16-8)^2} \\ &= \sqrt{6^2 + 8^2} \\ &= \sqrt{36+64} \\ &= \sqrt{100} \\ &= 10 \end{aligned}$$

37. $P(x_1, y_1) = P(-5, -5),\ Q(x_2, y_2) = (-3, 5)$

$$\begin{aligned} d(PQ) &= \sqrt{(x_2 - x_1)^2 + (y_2 - y_1)^2} \\ &= \sqrt{[-3-(-5)]^2 + [5-(-5)]^2} \\ &= \sqrt{2^2 + 10^2} \\ &= \sqrt{4+100} \\ &= \sqrt{104} \\ &= 2\sqrt{26} \end{aligned}$$

39. (–2, 4) and (2, 8)

$$\begin{aligned} d &= \sqrt{(x_2 - x_1)^2 + (y_2 - y_1)^2} \\ &= \sqrt{[2-(-2)]^2 + (8-4)^2} \\ &= \sqrt{4^2 + 4^2} \\ &= \sqrt{16+16} \\ &= \sqrt{32} \\ &= 4\sqrt{2} \end{aligned}$$

(2, 8) and (6, 4)

$$\begin{aligned} d &= \sqrt{(x_2 - x_1)^2 + (y_2 - y_1)^2} \\ &= \sqrt{(6-2)^2 + (4-8)^2} \\ &= \sqrt{4^2 + (-4)^2} \\ &= 0\sqrt{16+16} \\ &= \sqrt{32} \\ &= 4\sqrt{2} \end{aligned}$$

(–2, 4) and (6, 4)

$$\begin{aligned} d &= \sqrt{(x_2 \quad x_1)^2 + (y_2 - y_1)^2} \\ &= \sqrt{[6-(-2)]^2 + (4-4)^2} \\ &= \sqrt{8^2 + 0^2} \\ &= 0\sqrt{64+0} \\ &= \sqrt{64} \\ &= 8 \end{aligned}$$

Since two of the legs have the same length, the triangle is isosceles.

APPLICATIONS

41. The area of the square is 100 mi^2, so the length of each side is $\sqrt{100} = 10$ mi. The x– and y–axes together with each of the fours sides, form 4 isosceles right triangles. The "side" of the square is the hypotenuse of the right triangle. Let a represent the length along the x–axis from the origin to the corner of the square. Then a will also represent the length along the y–axis from the origin to the corner of the square.

The length of the hypotenuse is the length of leg a times $\sqrt{2}$.

$$a\sqrt{2} = h$$

$$a\sqrt{2} = 10 \qquad a = \frac{10}{\sqrt{2}} = \frac{10\sqrt{2}}{\sqrt{2}\cdot\sqrt{2}} = \frac{10\sqrt{2}}{2} = 5\sqrt{2}$$

The coordinates of the 4 corners are ($5\sqrt{2}$, 0); (0, $5\sqrt{2}$); ($-5\sqrt{2}$, 0); (0, $-5\sqrt{2}$).
The approximation is (7.07, 0); (0, 7.07); (–7.07, 0); (0, – 7.07).

43. The altitude of the triangle shown divides the flat side of the nut in half. So the base of the 30°–60°–90° triangle is 5 mm. The longer side of the 30°–60°–90° triangle is then $5\sqrt{3}$. The height of the nut is 2 times the longer leg .

$$h = 2(5\sqrt{3}) = 10\sqrt{3}\ \text{mm} \approx 17.32\ \text{mm}$$

45. "10 feet directly behind third base" means that the ball was fielded at a distance of 100 ft from home plate. First base is 90 ft from home plate. The third base line and the first base line form a right angle at home plate. Thus the ball is thrown along the hypotenuse of a right triangle having legs of 90 ft and 100 ft.

$$c^2 = a^2 + b^2$$
$$c^2 = 90^2 + 100^2$$
$$c^2 = 18{,}100$$
$$\sqrt{c^2} = \sqrt{18{,}100}$$
$$c = 10\sqrt{181} \approx 134.54\ \text{ft}$$

47. The pants hanging on the line creates a right triangle with half of the clothesline being the hypotenuse. The distance the pants pulls the line down and the original position of the clothesline form the legs. The legs are 1 ft and $\frac{15}{2} = 7.5$ ft respectively.

$$c^2 = a^2 + b^2$$
$$c^2 = 1^2 + 7.5^2$$
$$c^2 = 1 + 56.25$$
$$c^2 = 57.25$$
$$\sqrt{c^2} = \sqrt{57.25}$$
$$c = \sqrt{57.25} \approx 7.57\ \text{ft}$$

To find the amount the clothesline was stretched, use twice the length of the hypotenuse and then subtract the original length. $2\sqrt{57.25} - 15 \approx 0.13$ ft.

49. a. Foot is at (5, 0) and eye is at (8, 21)

$$d = \sqrt{(x_2 - x_1)^2 + (y_2 - y_1)^2}$$
$$= \sqrt{(8-5)^2 + (21-0)^2}$$
$$= \sqrt{3^2 + 21^2}$$
$$= \sqrt{450} \approx 21.21\ \text{units}$$

b. Belt is at (10, 13) and hand is at (2, 11)

$$d = \sqrt{(x_2 - x_1)^2 + (y_2 - y_1)^2}$$
$$= \sqrt{(2-10)^2 + (11-13)^2}$$
$$= \sqrt{(-8)^2 + (-2)^2}$$
$$= \sqrt{68} \approx 8.25\ \text{units}$$

c. Shoulder is at (7, 19) and symbol is at (12, 7)

$$d = \sqrt{(x_2 - x_1)^2 + (y_2 - y_1)^2}$$
$$= \sqrt{(12-7)^2 + (7-19)^2}$$
$$= \sqrt{5^2 + (-12)^2}$$
$$= \sqrt{169} = 13.00\ \text{units}$$

51. Find the diagonal of the box d when $a = 24$ in., $b = 24$ in., and $c = 4$ in.

$$d = \sqrt{a^2 + b^2 + c^2} = \sqrt{24^2 + 24^2 + 4^2} = \sqrt{1168} \approx 34.18\ \text{in.}$$ Yes, the bone will fit.

WRITING

53. The Pythagorean Theorem says that the sum of the squares of the legs of a right triangle equals the square of the hypotenuse.

REVIEW

55. **Analysis**: Find the number of original motors purchased.

Equation: Let x represent the represent the number of motors originally purchased, then $x + 1$ represents the number of motors purchased the second time. The total cost ($224) divided by the number of motors purchased will give the unit cost.

Original unit cost	less	$4	is	second purchase unit cost.
$\frac{224}{x}$	$-$	4	$=$	$\frac{224}{x+1}$

Solve:

$$\frac{224}{x} - 4 = \frac{224}{x+1}$$

$$x(x+1)\left(\frac{224}{x}\right) - x(x+1)(4) = x(x+1)\left(\frac{224}{x+1}\right)$$ Multiply both sides by LCD = $x(x + 1)$.

$$224(x+1) - 4x(x+1) = 224x$$

$$224x + 224 - 4x^2 - 4x = 224x$$

$$-4x^2 - 4x + 224 = 0$$ Divide both sides by –4.

$$x^2 + x - 56 = 0$$

$$(x-7)(x+8) = 0$$

$x - 7 = 0$ or $x + 8 = 0$ Discard the negative solution.

$x = 7$ $\quad$ $x = -8$

Conclusion: There were $x = 7$ motors purchased originally.

57. $\text{mean} = \frac{\text{sum of values}}{\text{number of values}} = \frac{16+6+10+4+5+13}{6} = \frac{54}{6} = 9$

Chapter 7 Key Concept: Radicals

Expressions Containing Radicals

1. $\sqrt[3]{-54h^6} = \sqrt[3]{-27h^6} \cdot \sqrt[3]{2} = -3h^2\sqrt[3]{2}$

2. $2\sqrt[3]{64e} + 3\sqrt[3]{8e} = 2\sqrt[3]{4^3 e} + 3\sqrt[3]{2^3 e} = 2 \cdot 4\sqrt[3]{e} + 3 \cdot 2\sqrt[3]{e} = 8\sqrt[3]{e} + 6\sqrt[3]{e} = 14\sqrt[3]{e}$

3. $\sqrt{72} - \sqrt{200} = \sqrt{36 \cdot 2} - \sqrt{100 \cdot 2} = 6\sqrt{2} - 10\sqrt{2} = (6-10)\sqrt{2} = -4\sqrt{2}$

4. $-4\sqrt[3]{5r^2s}\left(5\sqrt[3]{2r}\right) = -4 \cdot 5\sqrt[3]{5r^2s \cdot 2r} = -20\sqrt[3]{10r^3s} = -20r\sqrt[3]{10s}$

5. $\left(\sqrt{3s} - \sqrt{2t}\right)\left(\sqrt{3s} + \sqrt{2t}\right) = \left(\sqrt{3s}\right)^2 - \left(\sqrt{2t}\right)^2 = 3s - 2t$

6. $-\sqrt{3}\left(\sqrt{7} - \sqrt{5}\right) = -\sqrt{3 \cdot 7} + \sqrt{3 \cdot 5} = -\sqrt{21} + \sqrt{15}$

7. $\left(3\sqrt{2n} - 2\right)^2 = \left(3\sqrt{2n}\right)^2 - 2\left(3\sqrt{2n}\right)(2) + 2^2 = 9(2n) - 12\sqrt{2n} + 4 = 18n - 12\sqrt{2n} + 4$

8. $\frac{\sqrt[3]{9j}}{\sqrt[3]{3jk}} = \sqrt[3]{\frac{9j}{3jk}} = \sqrt[3]{\frac{3}{k}} = \frac{\sqrt[3]{3} \cdot \sqrt[3]{k^2}}{\sqrt[3]{k}\sqrt[3]{k^2}} = \frac{\sqrt[3]{3k^2}}{\sqrt[3]{k^3}} = \frac{\sqrt[3]{3k^2}}{k}$

Equations Containing Radicals

9. $\sqrt{1-2g} = \sqrt{g+10}$

$\left(\sqrt{1-2g}\right)^2 = \left(\sqrt{g+10}\right)^2$ Square both sides.

$1-2g = g+10$ Do the squaring.

$-3g = 9$ Subtract g and 1 from both sides.

$g = -3$ Divide both sides by –3.

Check:

$\sqrt{1-2(-3)} \stackrel{?}{=} \sqrt{(-3)+10}$

$\sqrt{7} = \sqrt{7}$

10. $4-\sqrt[3]{4+12x} = 0$

$4 = \sqrt[3]{4+12x}$ Add $\sqrt[3]{4+12x}$.

$4^3 = \left(\sqrt[3]{4+12x}\right)^3$ Cube both sides.

$64 = 4+12x$ Do the cubing.

$60 = 12x$ Subtract 4.

$5 = x$ Divide by 12.

Check:

$4-\sqrt[3]{4+12(5)} \stackrel{?}{=} 0$

$4-\sqrt[3]{64} \stackrel{?}{=} 0$

$4-4 = 0$

11. $\sqrt{y+2} - 4 = -y$

$\sqrt{y+2} = 4-y$ Add 4.

$\left(\sqrt{y+2}\right)^2 = (4-y)^2$ Square both sides.

$y+2 = 16-8y+y^2$ Do the squaring.

$0 = y^2-9y+14$ Subtract y and 2.

$0 = (y-2)(y-7)$ Factor.

$y-2 = 0$ or $y-7 = 0$

$y = 2$ $\quad$ $y = 7$ is an extraneous solution.

Check:

$\sqrt{(2)+2} - 4 \stackrel{?}{=} -(2)$

$\sqrt{4} - 4 \stackrel{?}{=} -2$

$2-4 = -2$

and

$\sqrt{(7)+2} - 4 \stackrel{?}{=} -(7)$

$\sqrt{9} - 4 \stackrel{?}{=} -7$

$3-4 \neq -7$

12. $\sqrt[4]{12t+4} + 2 = 0$

$\sqrt[4]{12t+4} = -2$ Subtract 2.

$\left(\sqrt[4]{12t+4}\right)^4 = (-2)^4$ Raise to the 4th power.

$12t+4 = 16$ Simplify.

$12t = 12$ Subtract 4.

$t = 1$ Divide by 12.

$t = 1$ is an extraneous solution, so there are no solutions.

Check:

$\sqrt[4]{12(1)+4} + 2 \stackrel{?}{=} 0$

$\sqrt[4]{16} + 2 \stackrel{?}{=} 0$

$2+2 \neq 0$

Radicals and Rational Exponents

13. $\sqrt[3]{3} = 3^{1/3}$

14. $5a^{2/5} = 5\sqrt[5]{a^2}$

Chapter 7 — Chapter Review

Section 7.1

1. $\sqrt{49}=\sqrt{7^2}=7$

2. $-\sqrt{121}=-\sqrt{11^2}=-11$

3. $\sqrt{\dfrac{225}{49}}=\sqrt{\left(\dfrac{15}{7}\right)^2}=\dfrac{15}{7}$

4. $\sqrt{-4}$ is not a real number

5. $\sqrt{0.01}=\sqrt{0.1^2}=0.1$

6. $\sqrt{25x^2}=\sqrt{5^2x^2}=|5x|=5|x|$

7. $\sqrt{x^8}=\sqrt{(x^4)^2}=x^4$

8. $\sqrt{x^2+4x+4}=\sqrt{(x+2)^2}=|x+2|$

9. $\sqrt[3]{-27}=\sqrt[3]{(-3)^3}=-3$

10. $-\sqrt[3]{216}=-\sqrt[3]{6^3}=-6$

11. $\sqrt[3]{64a^6b^3}=\sqrt[3]{\left(4a^2b\right)^3}=4a^2b$

12. $\sqrt[3]{\dfrac{s^9}{125}}=\sqrt[3]{\left(\dfrac{s^3}{5}\right)^3}=\dfrac{s^3}{5}$

13. $\sqrt[4]{625}=\sqrt[4]{5^4}=5$

14. $\sqrt[5]{-32}=\sqrt[5]{(-2)^5}=-2$

15. $\sqrt[4]{256x^8y^4}=\sqrt[4]{(4x^2y)^4}=|4x^2y|=4x^2|y|$

16. $\sqrt{(-22y)^2}=|-22y|=22|y|$

17. $-\sqrt[4]{\dfrac{1}{16}}=-\sqrt[4]{\left(\dfrac{1}{2}\right)^4}=-\dfrac{1}{2}$

18. $\sqrt[6]{-1}$ is not a real number

19. $\sqrt{0}=0$

20. $\sqrt[3]{0}=0$

21. If $s(A)=\sqrt{A}$, then $s(144)=\sqrt{144}=12$ ft is the length of the side.
The perimeter will be $P=4(12)=48$ ft

22. If $A(V)=6\sqrt[3]{V^2}$, then $A(8)=6\sqrt[3]{8^2}=6\sqrt[3]{64}=6(4)=24\ \text{cm}^2$

23.

x	$f(x)=\sqrt{x+2}$	$f(x)$
−2	$f(-2)=\sqrt{-2+2}=\sqrt{0}$	0
−1	$f(-1)=\sqrt{-1+2}=\sqrt{1}$	1
2	$f(2)=\sqrt{2+2}=\sqrt{4}$	2
7	$f(7)=\sqrt{7+2}=\sqrt{9}$	3
14	$f(14)=\sqrt{14+2}=\sqrt{16}$	4

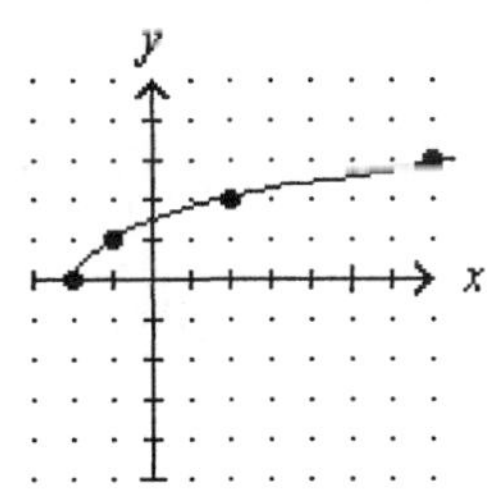

D: $[-2,\infty)$; R: $[0,\infty)$

24.

x	$f(x) = -\sqrt[3]{x} + 3$	$f(x)$
−8	$f(-8) = -\sqrt[3]{-8} + 3$ $= \sqrt[3]{(-2)^3} + 3 = -(-2) + 3$	5
−1	$f(-1) = -\sqrt[3]{-1} + 3$ $= \sqrt[3]{(-1)^3} + 3 = -(-1) + 3$	4
0	$f(0) = -\sqrt[3]{0} + 3 = -0 + 3$	3
1	$f(1) = -\sqrt[3]{1} + 3$ $= -\sqrt[3]{1^3} + 3 = -1 + 3$	2
8	$f(8) = -\sqrt[3]{8} + 3$ $= -\sqrt[3]{2^3} + 3 = -2 + 3$	1

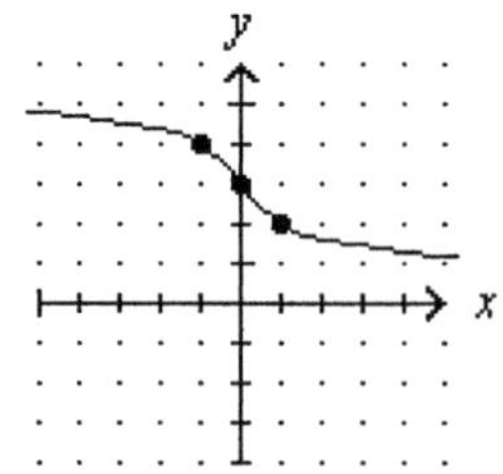

D: $(-\infty, \infty)$; R: $(-\infty, \infty)$

Section 7.2

25. $\sqrt{240} = \sqrt{16}\sqrt{15} = 4\sqrt{15}$

26. $\sqrt[3]{54} = \sqrt[3]{27}\sqrt[3]{2} = 3\sqrt[3]{2}$

27. $\sqrt[4]{32} = \sqrt[4]{16}\sqrt[4]{2} = 2\sqrt[4]{2}$

28. $-2\sqrt[5]{-96} = -2\sqrt[5]{-32}\sqrt[5]{3} = -2(-2)\sqrt[5]{3} = 4\sqrt[5]{3}$

29. $\sqrt{8x^5} = \sqrt{4x^4}\sqrt{2x} = 2x^2\sqrt{2x}$

30. $\sqrt[3]{r^{17}} = \sqrt[3]{r^{15}}\sqrt[3]{r^2} = \sqrt[3]{(r^5)^3}\sqrt[3]{r^2} = r^5\sqrt[3]{r^2}$

31. $\sqrt[3]{16x^5y^4} = \sqrt[3]{8x^3y^3}\sqrt[3]{2x^2y} = 2xy\sqrt[3]{2x^2y}$

32. $3\sqrt[3]{27j^7k} = 3\sqrt[3]{27j^6}\sqrt[3]{jk} = 3(3j^2)\sqrt[3]{jk} = 9j^2\sqrt[3]{jk}$

33. $\dfrac{\sqrt{32x^3}}{\sqrt{2x}} = \sqrt{\dfrac{32x^3}{2x}} = \sqrt{\dfrac{2x \cdot 16x^2}{2x}} = \sqrt{16x^2} = 4x$

34. $\sqrt{\dfrac{17xy}{64a^4}} = \dfrac{\sqrt{17xy}}{\sqrt{64a^4}} = \dfrac{\sqrt{17xy}}{8a^2}$

35. $\sqrt{2} + 2\sqrt{2} = (1+2)\sqrt{2} = 3\sqrt{2}$

36. $6\sqrt{20} - \sqrt{5} = 6\sqrt{4 \cdot 5} - \sqrt{5} = 6 \cdot 2\sqrt{5} - \sqrt{5} = 12\sqrt{5} - \sqrt{5} = 11\sqrt{5}$

37. $2\sqrt[3]{3} - \sqrt[3]{24} = 2\sqrt[3]{3} - \sqrt[3]{8 \cdot 3} = 2\sqrt[3]{3} - 2\sqrt[3]{3} = 0$

38. $-\sqrt[4]{32} - 2\sqrt[4]{162} = -\sqrt[4]{16 \cdot 2} - 2\sqrt[4]{81 \cdot 2} = -2\sqrt[4]{2} - 2 \cdot 3\sqrt[4]{2} = -2\sqrt[4]{2} - 6\sqrt[4]{2} = -8\sqrt[4]{2}$

39. $2x\sqrt{8} + 2\sqrt{200x^2} + \sqrt{50x^2} = 2x \cdot 2\sqrt{2} + 2 \cdot 10x\sqrt{2} + 5x\sqrt{2} = 4x\sqrt{2} + 20x\sqrt{2} + 5x\sqrt{2} = 29x\sqrt{2}$

40. $\sqrt[3]{54} - 3\sqrt[3]{16} + 4\sqrt[3]{128} = \sqrt[3]{27 \cdot 2} - 3\sqrt[3]{8 \cdot 2} + 4\sqrt[3]{64 \cdot 2} = 3\sqrt[3]{2} - 3 \cdot 2\sqrt[3]{2} + 4 \cdot 4\sqrt[3]{2} = 3\sqrt[3]{2} - 6\sqrt[3]{2} + 16\sqrt[3]{2} = 13\sqrt[3]{2}$

41. Sum the lengths of each side for the total length.
Total length $= \sqrt{40} + \sqrt{8} + \sqrt{32} = \sqrt{4 \cdot 10} + \sqrt{4 \cdot 2} + \sqrt{16 \cdot 2} = 2\sqrt{10} + 2\sqrt{2} + 4\sqrt{2} = \left(2\sqrt{10} + 6\sqrt{2}\right)$ in. $\approx$ 14.8 in.

Section 7.3

42.. $\sqrt{7}\sqrt{7} = \sqrt{7^2} = 7$

43. $\left(2\sqrt{5}\right)\left(3\sqrt{2}\right) = 2 \cdot 3\sqrt{5 \cdot 2} = 6\sqrt{10}$

44. $\left(-2\sqrt{8}\right)^2 = (-2)^2\left(\sqrt{8}\right)^2 = 4(8) = 32$

45. $2\sqrt{6}\sqrt{216} = 2\sqrt{6 \cdot 216} = 2\sqrt{36^2} = 2(36) = 72$

46. $\sqrt{9x}\sqrt{x} = \sqrt{9x^2} = \sqrt{(3x)^2} = 3x$

47. $\left(\sqrt{33}\right)^2 = 33$

48. $-\sqrt[3]{2x^2}\sqrt[3]{4x} = -\sqrt[3]{8x^3} = -\sqrt[3]{(2x)^3} = -2x$

49. $4\sqrt[3]{9}\,\sqrt[3]{9} = 4\sqrt[3]{81} = 4\sqrt[3]{3^3 \cdot 3} = 4 \cdot 3\sqrt[3]{3} = 12\sqrt[3]{3}$

50. $\sqrt{2}\left(\sqrt{8} - 3\right) = \sqrt{2} \cdot \sqrt{8} - \sqrt{2} \cdot 3 = \sqrt{16} - 3\sqrt{2} = 4 - 3\sqrt{2}$

51.
$$-\sqrt[4]{256x^5y^{11}}\,\sqrt[4]{625x^9y^3} = -\sqrt[4]{4^4 \cdot 5^4 x^{14} y^{14}} = -4 \cdot 5\sqrt[4]{\left(x^3y^3\right)^4 \cdot x^2y^2}$$
$$= -20x^3y^3\sqrt[4]{x^2y^2} = -20x^3y^3\left(x^{2/4}y^{2/4}\right) = -20x^3y^3\left(x^{1/2}y^{1/2}\right) = -20x^3y^3\sqrt{xy}$$

52. $\left(\sqrt{3b} + \sqrt{3}\right)^2 = \left(\sqrt{3b}\right)^2 + 2\sqrt{3b} \cdot \sqrt{3} + \left(\sqrt{3}\right)^2 = 3b + 2\sqrt{9b} + 3 = 3b + 2 \cdot 3\sqrt{b} + 3 = 3b + 6\sqrt{b} + 3$

53. $\left(2\sqrt{u} + 3\right)\left(3\sqrt{u} - 4\right) = 6\sqrt{u^2} - 8\sqrt{u} + 9\sqrt{u} - 12 = 6u + \sqrt{u} - 12$

54. $\dfrac{10}{\sqrt{3}} = \dfrac{10 \cdot \sqrt{3}}{\sqrt{3} \cdot \sqrt{3}} = \dfrac{10\sqrt{3}}{3}$

55. $\sqrt{\dfrac{3}{5}} = \sqrt{\dfrac{3 \cdot 5}{5 \cdot 5}} = \dfrac{\sqrt{15}}{5}$

56. $\dfrac{x}{\sqrt{xy}} = \dfrac{x \cdot \sqrt{xy}}{\sqrt{xy} \cdot \sqrt{xy}} = \dfrac{x\sqrt{xy}}{xy} = \dfrac{\sqrt{xy}}{y}$

57. $\dfrac{\sqrt[3]{uv}}{\sqrt[3]{u^5v^7}} = \dfrac{\sqrt[3]{uv} \cdot \sqrt[3]{uv^2}}{\sqrt[3]{u^5v^7} \cdot \sqrt[3]{uv^2}} = \dfrac{\sqrt[3]{u^2v^3}}{\sqrt[3]{u^6v^9}} = \dfrac{v\sqrt[3]{u^2}}{u^2v^3} = \dfrac{\sqrt[3]{u^2}}{u^2v^2}$

58. $\dfrac{2}{\sqrt{2} - 1} = \dfrac{2\left(\sqrt{2} + 1\right)}{\left(\sqrt{2} - 1\right)\left(\sqrt{2} + 1\right)} = \dfrac{2\sqrt{2} + 2}{2 - 1} = \dfrac{2\sqrt{2} + 2}{1} = 2\left(\sqrt{2} + 1\right)$

59. $\dfrac{\sqrt{a} + 1}{\sqrt{a} - 1} = \dfrac{\left(\sqrt{a} + 1\right)\left(\sqrt{a} + 1\right)}{\left(\sqrt{a} - 1\right)\left(\sqrt{a} + 1\right)} = \dfrac{a + 2\sqrt{a} + 1}{a - 1}$

60. $\dfrac{3 - \sqrt{x}}{2} = \dfrac{\left(3 - \sqrt{x}\right)\left(3 + \sqrt{x}\right)}{2\left(3 + \sqrt{x}\right)} = \dfrac{9 - x}{2\left(3 + \sqrt{x}\right)}$

61. $\dfrac{\sqrt{a} - \sqrt{b}}{\sqrt{a}} = \dfrac{\left(\sqrt{a} - \sqrt{b}\right)\left(\sqrt{a} + \sqrt{b}\right)}{\sqrt{a}\left(\sqrt{a} + \sqrt{b}\right)} = \dfrac{u - b}{a + \sqrt{ab}}$

62. $r = \sqrt[3]{\dfrac{3V}{4\pi}} = \dfrac{\sqrt[3]{3V} \cdot \sqrt[3]{2\pi^2}}{\sqrt[3]{4\pi} \cdot \sqrt[3]{2\pi^2}} = \dfrac{\sqrt[3]{6\pi^2 V}}{\sqrt[3]{8\pi^3}} = \dfrac{\sqrt[3]{6\pi^2 V}}{2\pi}$

Section 7.4

63.

$\sqrt{7x - 10} - 1 = 11$		Check:
$\sqrt{7x - 10} = 12$	Add 1 to both sides.	
$\left(\sqrt{7x - 10}\right)^2 = 12^2$	Square both sides.	$\sqrt{7(22) - 10} - 1 \overset{?}{=} 11$
$7x - 10 = 144$	Do the squaring.	$\sqrt{144} - 1 \overset{?}{=} 11$
$7x = 154$	Add 10 to both sides.	$12 - 1 = 11$
$x = 22$	Divide both sides by 7.	

64.

$u = \sqrt{25u - 144}$		Check:
$u^2 = \left(\sqrt{25u - 144}\right)^2$	Square both sides.	$16 \overset{?}{=} \sqrt{25(16) - 144}$
$u^2 = 25u - 144$	Do the squaring.	$16 = \sqrt{256}$
$u^2 - 25u + 144 = 0$	Subtract $25u$ and add 144.	and
$(u - 16)(u - 9) = 0$	Factor.	$9 \overset{?}{=} \sqrt{25(9) - 144}$
$u - 16 = 0$ or $u - 9 = 0$		$9 = \sqrt{81}$
$u = 16$ $u = 9$		

65.

$$2\sqrt{y-3} = \sqrt{2y+1}$$
$$\left(2\sqrt{y-3}\right)^2 = \left(\sqrt{2y+1}\right)^2 \quad \text{Square both sides.}$$
$$4(y-3) = 2y+1 \quad \text{Do the squaring.}$$
$$4y-12 = 2y+1 \quad \text{Remove the parentheses.}$$
$$2y = 13 \quad \text{Subtract } 2y \text{ and add 12.}$$
$$y = \frac{13}{2} \quad \text{Divide both sides by 2.}$$

Check:

$$2\sqrt{\tfrac{13}{2}-3} \overset{?}{=} \sqrt{2\left(\tfrac{13}{2}\right)+1}$$
$$2\sqrt{\tfrac{7}{2}} \overset{?}{=} \sqrt{14}$$
$$2\tfrac{\sqrt{14}}{2} = \sqrt{14}$$

66.

$$\sqrt{z+1} + \sqrt{z} = 2$$
$$\sqrt{z+1} = 2-\sqrt{z} \quad \text{Subtract } \sqrt{z} \text{ from both sides.}$$
$$\left(\sqrt{z+1}\right)^2 = \left(2-\sqrt{z}\right)^2 \quad \text{Square both sides.}$$
$$z+1 = 4-4\sqrt{z}+z \quad \text{Do the squaring.}$$
$$-3 = -4\sqrt{z} \quad \text{Subtract 4 and } z.$$
$$(-3)^2 = \left(-4\sqrt{z}\right)^2 \quad \text{Square both sides.}$$
$$9 = 16z \quad \text{Do the squaring.}$$
$$\frac{9}{16} = z \quad \text{Divide by 16.}$$

Check:

$$\sqrt{\tfrac{9}{16}+1} + \sqrt{\tfrac{9}{16}} \overset{?}{=} 2$$
$$\sqrt{\tfrac{25}{16}} + \sqrt{\tfrac{9}{16}} \overset{?}{=} 2$$
$$\tfrac{5}{4} + \tfrac{3}{4} = 2$$

67.

$$\sqrt[3]{x^3+56} - 2 = x$$
$$\sqrt[3]{x^3+56} = x+2 \quad \text{Add 2 to both sides.}$$
$$\left(\sqrt[3]{x^3+56}\right)^3 = (x+2)^3 = (x+2)(x^2+4x+4) \quad \text{Cube both sides.}$$
$$x^3+56 = x^3+6x^2+12x+8 \quad \text{Do the multiplication.}$$
$$0 = 6x^2+12x-48 \quad \text{Subtract } x^3 \text{ and 56.}$$
$$0 = x^2+2x-8 \quad \text{Divide by 6.}$$
$$0 = (x-2)(x+4) \quad \text{Factor.}$$

$x-2$	$=$	0	or	$x+4$	$=$	0
x	$=$	2		x	$=$	-4

Check:

$$\sqrt[3]{2^3+56} - 2 \overset{?}{=} 2$$
$$\sqrt[3]{64} - 2 = 2$$

and

$$\sqrt[3]{(-4)^3+56} - 2 \overset{?}{=} -4$$
$$\sqrt[3]{-8} - 2 = -4$$

68.

$$\sqrt[4]{8x-8} + 2 = 0$$
$$\sqrt[4]{8x-8} = -2 \quad \text{Subtract 2 from both sides.}$$
$$\left(\sqrt[4]{8x-8}\right)^4 = (-2)^4 \quad \text{Raise to the 4}^{\text{th}} \text{ power.}$$
$$8x-8 = 16$$
$$8x = 24 \quad \text{Add 8 to both sides..}$$
$$x = 3 \quad \text{Divide by 8.}$$

$x = 3$ is an extraneous solution, so there is no solution.

Check:

$$\sqrt[4]{8(3)-8} + 2 \overset{?}{=} 0$$
$$\sqrt[4]{16} + 2 \overset{?}{=} 0$$
$$2+2 \neq 0$$

69. The solution is the x–value where the graphs cross. The estimated value of x is 2. Check by substitution.

$$\sqrt{2x-3} = -2x+5$$
$$\sqrt{2(2)-3} \stackrel{?}{=} -2(2)+5$$
$$\sqrt{4-3} \stackrel{?}{=} -4+5$$
$$\sqrt{1} = 1$$

70.

$r = \sqrt{\dfrac{A}{P}} - 1$ Solve for P.

$r+1 = \sqrt{\dfrac{A}{P}}$ Add 1 to both sides.

$(r+1)^2 = \left(\sqrt{\dfrac{A}{P}}\right)^2$ Square both sides.

$(r+1)^2 = \dfrac{A}{P}$ Do the squaring.

$P(r+1)^2 = A$ Multiply both sides by P.

$P = \dfrac{A}{(r+1)^2}$ Divide both sides by $(r+1)^2$.

71.

$h = \sqrt[3]{\dfrac{12I}{b}}$ Solve for I.

$h^3 = \left(\sqrt[3]{\dfrac{12I}{b}}\right)^3$ Cube both sides.

$h^3 = \dfrac{12I}{b}$ Do the cubing.

$\left(\dfrac{b}{12}\right)h^3 = \left(\dfrac{b}{12}\right)\left(\dfrac{12I}{b}\right)$ Multiply by reciprocal $\dfrac{b}{12}$.

$\dfrac{bh^3}{12} = I$ Do the multiplication.

72. Find R when $P = 980$ watts and $I = 7.4$ amps.

$$I = \sqrt{\frac{P}{R}}$$
$$7.4 = \sqrt{\frac{980}{R}}$$
$$(7.4)^2 = \left(\sqrt{\frac{980}{R}}\right)^2$$
$$54.76 = \frac{980}{R}$$
$$54.76R = 980$$
$$R \approx 17.9 \text{ ohms}$$

Section 7.5

73. $t^{1/2} = \sqrt{t}$

74. $\left(5xy^3\right)^{1/4} = \sqrt[4]{5xy^3}$

75. $25^{1/2} = \sqrt{25} = 5$

76. $-36^{1/2} = -\sqrt{36} = -6$

77. $(-36)^{1/2}$ is not a real number

78. $1^{1/2} = \sqrt{1} = 1$

79. $\left(\dfrac{9}{x^2}\right)^{1/2} = \sqrt{\dfrac{9}{x^2}} = \dfrac{3}{x}$

80. $\left(\dfrac{1}{27}\right)^{1/3} = \sqrt[3]{\dfrac{1}{27}} = \dfrac{1}{3}$

81. $(-8)^{1/3} = \sqrt[3]{-8} = -2$

82. $625^{1/4} = \sqrt[4]{625} = 5$

83. $(27a^3b)^{1/3}=\sqrt[3]{27a^3b}=\sqrt[3]{27a^3}\cdot\sqrt[3]{b}=3a\sqrt[3]{b}=3ab^{1/3}$

84. $(81c^4d^4)^{1/4}=\sqrt[4]{3^4c^4d^4}=3cd$

85. $9^{3/2}=(\sqrt{9})^3=3^3=27$

86. $8^{-2/3}=\dfrac{1}{8^{2/3}}=\dfrac{1}{(\sqrt[3]{8})^2}=\dfrac{1}{2^2}=\dfrac{1}{4}$

87. $-49^{5/2}=-(\sqrt{49})^5=-7^5=-16{,}807$

88. $\dfrac{1}{100^{-1/2}}=100^{1/2}=\sqrt{100}=10$

89. $\left(\dfrac{4}{9}\right)^{-3/2}=\dfrac{4^{-3/2}}{9^{-3/2}}=\dfrac{9^{3/2}}{4^{3/2}}=\dfrac{(\sqrt{9})^3}{(\sqrt{4})^3}=\dfrac{3^3}{2^3}=\dfrac{27}{8}$

90. $\dfrac{1}{25^{5/2}}=\dfrac{1}{(\sqrt{25})^5}=\dfrac{1}{5^5}=\dfrac{1}{3{,}125}$

91. $(25x^2y^4)^{3/2}=\left(\sqrt{25x^2y^4}\right)^3=(5xy^2)^3=125x^3y^6$

92. $(8u^6v^3)^{-2/3}=\dfrac{1}{(8u^6v^3)^{2/3}}=\dfrac{1}{\left(\sqrt[3]{8u^6v^3}\right)^2}=\dfrac{1}{(2u^2v)^2}=\dfrac{1}{4u^4v^2}$

93. $5^{1/4}5^{1/2}=5^{\frac{1}{4}+\frac{1}{2}}=5^{\frac{1}{4}+\frac{2}{4}}=5^{3/4}$

94. $a^{3/7}a^{-2/7}=a^{\frac{3}{7}-\frac{2}{7}}=a^{1/7}$

95. $(k^{4/5})^{10}=k^{\left(\frac{4}{5}\right)(10)}=k^8$

96. $\dfrac{(4g^3h)^{1/2}}{(9gh^{-1})^{1/2}}=\dfrac{4^{1/2}g^{3/2}h^{1/2}}{9^{1/2}g^{1/2}h^{-1/2}}=\dfrac{\sqrt{4}}{\sqrt{9}}g^{\frac{3}{2}-\frac{1}{2}}h^{\frac{1}{2}-\left(\frac{-1}{2}\right)}=\dfrac{2}{3}g^{2/2}h^{2/2}=\dfrac{2gh}{3}$

97. $u^{1/2}(u^{1/2}-u^{-1/2})=u^{1/2}u^{1/2}-u^{1/2}u^{-1/2}=u^{\frac{1}{2}+\frac{1}{2}}-u^{\frac{1}{2}-\frac{1}{2}}=u^1-u^0=u-1$

98. $v^{2/3}(v^{1/3}+v^{4/3})=v^{2/3}v^{1/3}+v^{2/3}v^{4/3}=v^{\frac{2}{3}+\frac{1}{3}}+v^{\frac{2}{3}+\frac{4}{3}}=v^{3/3}+v^{6/3}=v+v^2$

99. $\sqrt[4]{\dfrac{a^2}{25b^2}}=\left(\dfrac{a^2}{5^2b^2}\right)^{1/4}=\dfrac{a^{2/4}}{5^{2/4}b^{2/4}}=\dfrac{a^{1/2}}{5^{1/2}b^{1/2}}=\sqrt{\dfrac{a}{5b}}=\dfrac{\sqrt{5ab}}{5b}$

100. $(x,y)=(-64,64)$: $x^{2/3}+y^{2/3}=(-64)^{2/3}+64^{2/3}=(\sqrt[3]{-64})^2+(\sqrt[3]{64})^2=(-4)^2+4^2=16+16=32$

$(x,y)=(64,64)$: $x^{2/3}+y^{2/3}=(64)^{2/3}+64^{2/3}=(\sqrt[3]{64})^2+(\sqrt[3]{64})^2=4^2+4^2=16+16=32$

Section 7.6

101. The roof line is the hypotenuse of a right triangle whose legs are 8 ft and 15 ft long. Find h.

$$h^2=8^2+15^2$$
$$h^2=64+225=289$$
$$\sqrt{h^2}=\sqrt{289}$$
$$h=17\text{ ft}$$

102. Half of the distance, d, is a leg of a right triangle with the length of the other leg equal to 117 yards and the length of the hypotenuse equal to 125 yards.

$$\left(\frac{d}{2}\right)^2+117^2=125^2$$
$$\frac{d^2}{4}=125^2-117^2$$
$$d^2=4(125^2-117^2)$$
$$\sqrt{d^2}=\sqrt{4(125^2-117^2)}$$
$$d=\sqrt{7744}=88\text{ yd}$$

103. An isosceles right triangle is a 45°–45°–90° triangle. The given legs are 7 m.

Hypotenuse is length of a leg times $\sqrt{2}$.

$$h = 7\sqrt{2} \text{ m}$$

104. 30°–60°–90° triangle.

Hypotenuse is twice as long as shorter leg.

$$2x = h$$
$$2x = 12\sqrt{3}$$
$$x = 6\sqrt{3} \text{ cm}$$

Longer leg is shorter leg times $\sqrt{3}$.

$$6\sqrt{3}\left(\sqrt{3}\right) = 6\sqrt{9} = 6 \bullet 3 = 18 \text{ cm}$$

105. 45°–45°–90° triangle.

Hypotenuse is length of leg times $\sqrt{2}$.

$$h = 5\sqrt{2} \approx 7.07 \text{ in.}$$

106. 30°–60°–90° triangle.

Hypotenuse is twice as long as shorter leg, so shorter leg is $10 \div 2 = 5$ cm..

Longer leg is shorter leg times $\sqrt{3}$.

$$x = 5\sqrt{3} = 8.66 \text{ cm}$$

107. $P(x_1, y_1) = P(0, 0),\ Q(x_2, y_2) = (5, -12)$

$$\begin{aligned} d(PQ) &= \sqrt{(x_2 - x_1)^2 + (y_2 - y_1)^2} \\ &= \sqrt{(5-0)^2 + (-12-0)^2} \\ &= \sqrt{5^2 + (-12)^2} \\ &= \sqrt{25 + 144} \\ &= \sqrt{169} \\ &= 13 \end{aligned}$$

108. $P(x_1, y_1) = P(-4, 6),\ Q(x_2, y_2) = (-2, 8)$

$$\begin{aligned} d(PQ) &= \sqrt{(x_2 - x_1)^2 + (y_2 - y_1)^2} \\ &= \sqrt{[-2-(-4)]^2 + (8-6)^2} \\ &= \sqrt{2^2 + 2^2} \\ &= \sqrt{4 + 4} \\ &= \sqrt{8} \\ &= 2\sqrt{2} \end{aligned}$$

Chapter 7 Chapter Test

1.

x	$f(x) = \sqrt{x-1}$	$f(x)$
1	$f(x) = \sqrt{1-1}$	**0**
2	$f(x) = \sqrt{2-1}$	**1**
3	$f(x) = \sqrt{3-1}$	**1.41**
5	$f(x) = \sqrt{5-1}$	**2**
10	$f(x) = \sqrt{10-1}$	**3**
12	$f(x) = \sqrt{12-1}$	**3.32**
17	$f(x) = \sqrt{17-1}$	**4**

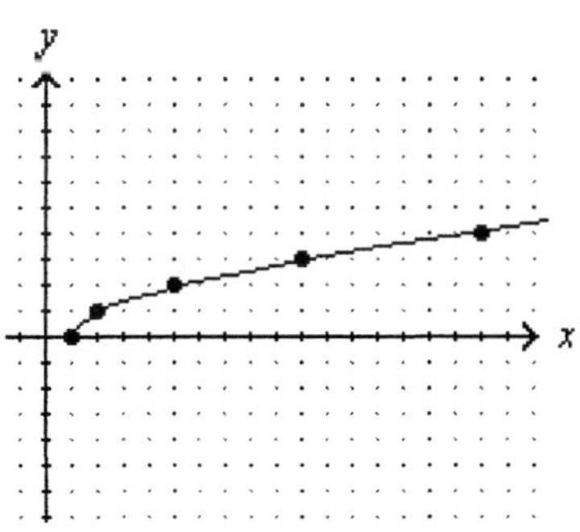

2. If d(h) = $d(h) = 1.4\sqrt{h}$, then $d(4.7) = 1.4\sqrt{4.7} = 3.0$ mi

3. $\sqrt{x^2} = |x|$

4. $\sqrt{8x^2} = \sqrt{(2x)^2 \cdot 2} = 2|x|\sqrt{2}$

5. $\sqrt[3]{54x^5} = \sqrt[3]{27x^3}\,\sqrt[3]{2x^2} = 3x\sqrt[3]{2x^2}$

6. $\sqrt{18x^4y^8} = \sqrt{\left(3x^2y^4\right)^2 \cdot 2} = 3x^2y^4\sqrt{2}$

7. $\sqrt[3]{-64x^3y^6} = -4xy^2$

8. $\sqrt{\dfrac{4a^2}{9}} = \sqrt{\left(\dfrac{2a}{3}\right)^2} = \dfrac{2a}{3}$

9. $\sqrt[4]{-16}$ is not a real number.

10. $\sqrt[5]{(t+8)^5} = t+8$

11. $\sqrt{48} = \sqrt{16\cdot 3} = 4\sqrt{3}$

12. $\sqrt{250x^3y^5} = \sqrt{25x^2y^4\cdot 10xy} = 5xy^2\sqrt{10xy}$

13. $\dfrac{\sqrt[3]{24x^{15}y^4}}{\sqrt[3]{y}} = \sqrt[3]{\dfrac{24x^{15}y^4}{y}} = \sqrt[3]{8\cdot 3x^{15}y^3} = 2x^5y\sqrt[3]{3}$

14. $\sqrt{\dfrac{3a^5}{48a^7}} = \sqrt{\dfrac{1}{16a^2}} = \dfrac{1}{4a}$

15. $\sqrt{12} - \sqrt{27} = \sqrt{4\cdot 3} - \sqrt{9\cdot 3} = 2\sqrt{3} - 3\sqrt{3} = -\sqrt{3}$

16. $2\sqrt[3]{40} - \sqrt[3]{5000} + 4\sqrt[3]{625} = 2\sqrt[3]{8\cdot 5} - \sqrt[3]{1000\cdot 5} + 4\sqrt[3]{125\cdot 5}$
$= 2\cdot 2\sqrt[3]{5} - 10\sqrt[3]{5} + 4\cdot 5\sqrt[3]{5} = 4\sqrt[3]{5} - 10\sqrt[3]{5} + 20\sqrt[3]{5} = 14\sqrt[3]{5}$

17. $2\sqrt{48y^5} - 3y\sqrt{12y^3} = 2\sqrt{16y^4\cdot 3y} - 3y\sqrt{4y^2\cdot 3y} = 2\left(4y^2\sqrt{3y}\right) - 3y\left(2y\sqrt{3y}\right) = 8y^2\sqrt{3y} - 6y^2\sqrt{3y} = 2y^2\sqrt{3y}$

18. $\sqrt[4]{768z^5} + z\sqrt[4]{48z} = \sqrt[4]{256z^4\cdot 3z} + z\sqrt[4]{16\cdot 3z} = 4z\sqrt[4]{3z} + 2z\sqrt[4]{3z} = 6z\sqrt[4]{3z}$

19. $-2\sqrt{xy}\left(3\sqrt{x} + \sqrt{xy^3}\right) = -2\sqrt{xy}\left(3\sqrt{x}\right) - 2\sqrt{xy}\left(\sqrt{xy^3}\right) = -6\sqrt{x^2y} - 2\sqrt{x^2y^4} = -6x\sqrt{y} - 2xy^2$

20. $\left(3\sqrt{2} + \sqrt{3}\right)\left(2\sqrt{2} - 3\sqrt{3}\right) = 3\cdot 2\sqrt{2\cdot 2} - 3\cdot 3\sqrt{2\cdot 3} + 2\sqrt{2\cdot 3} - 3\sqrt{3\cdot 3}$
$= 6\sqrt{4} - 9\sqrt{6} + 2\sqrt{6} - 3\sqrt{9} \doteq 6\cdot 2 - 7\sqrt{6} - 3\cdot 3 = 12 - 7\sqrt{6} - 9 = 3 - 7\sqrt{6}$

21. $\left(\sqrt[3]{2a} + 9\right)^2 = \sqrt[3]{4a^2} + 18\sqrt[3]{2a} + 81$

21. $\dfrac{1}{\sqrt{5}} = \dfrac{1\cdot\sqrt{5}}{\sqrt{5}\cdot\sqrt{5}} = \dfrac{\sqrt{5}}{5}$

23. $\dfrac{3t-1}{\sqrt{3t}-1} = \dfrac{(3t-1)\left(\sqrt{3t}+1\right)}{\left(\sqrt{3t}-1\right)\left(\sqrt{3t}+1\right)} = \dfrac{(3t-1)\left(\sqrt{3t}+1\right)}{3t-1} = \sqrt{3t}+1$

24. $\sqrt[3]{\dfrac{9}{4a^2}} = \sqrt[3]{\dfrac{9}{4a^2}}\sqrt[3]{\dfrac{4^2a}{4^2a}} = \dfrac{\sqrt[3]{144a}}{\sqrt[3]{4^3a^3}} = \dfrac{\sqrt[3]{2^3 18a}}{4a} = \dfrac{2\sqrt[3]{18a}}{4a} = \dfrac{\sqrt[3]{18a}}{2a}$

25.

		Check:
$2\sqrt{x} = \sqrt{x+1}$		
$\left(2\sqrt{x}\right)^2 = \left(\sqrt{x+1}\right)^2$	Square both sides.	$2\sqrt{\frac{1}{3}} \stackrel{?}{=} \sqrt{\frac{1}{3}+1}$
$4x = x+1$	Do the squaring.	
$3x = 1$	Subtract x and x from both sides.	$2\sqrt{\frac{1}{3}} \stackrel{?}{=} \sqrt{\frac{4}{3}}$
$x = \dfrac{1}{3}$	Divide both sides by 3.	$2\sqrt{\frac{1}{3}} = 2\sqrt{\frac{1}{3}}$

26. $\sqrt[3]{6n+4}-4=0$

		Check:
$\sqrt[3]{6n+4}=4$	Add 4 to both sides.	$\sqrt[3]{6(10)+4}-4 \stackrel{?}{=} 0$
$\left(\sqrt[3]{6n+4}\right)^3=4^3$	Cube both sides.	$\sqrt[3]{64}-4 \stackrel{?}{=} 0$
$6n+4=64$	Do the cubing.	$4-4=0$
$6n=60$	Subtract 4 from both sides.	
$n=10$	Divide both sides by 6.	

27. $1-\sqrt{u}=\sqrt{u-3}$

		Check:
$\left(1-\sqrt{u}\right)^2=\left(\sqrt{u-3}\right)^2$	Square both sides.	$1-\sqrt{4} \stackrel{?}{=} \sqrt{4-3}$
$1-2\sqrt{u}+u=u-3$	Do the squaring.	$1-2 \stackrel{?}{=} \sqrt{1}$
$-2\sqrt{u}=-4$	Subtract u and 1.	$-1\neq 1$
$\sqrt{u}=2$	Divide both sides by –2.	
$\left(\sqrt{u}\right)^2=2^2$	Square both sides.	
$u=4$	Do the squaring.	

$u = 4$ is an extraneous solution, so there is no solution.

28.

$r=\sqrt[3]{\dfrac{GMt^2}{4\pi^2}}$	Solve for G.
$r^3=\left(\sqrt[3]{\dfrac{GMt^2}{4\pi^2}}\right)^3$	Cube both sides.
$r^3=\dfrac{GMt^2}{4\pi^2}$	Do the cubing.
$\left(\dfrac{4\pi^2}{Mt^2}\right)r^3=\left(\dfrac{4\pi^2}{Mt^2}\right)\left(\dfrac{GMt^2}{4\pi^2}\right)$	Multiply by reciprocal $\dfrac{\pi}{12}$.
$\dfrac{4\pi^2r^3}{Mt^2}=G$	Do the multiplication.

29. $16^{1/4}=\sqrt[4]{16}=2$

30. $27^{2/3}=\left(\sqrt[3]{27}\right)^2=3^2=9$

31. $36^{-3/2}=\dfrac{1}{36^{3/2}}=\dfrac{1}{\left(\sqrt{36}\right)^3}=\dfrac{1}{6^3}=\dfrac{1}{216}$

32. $\left(-\dfrac{8}{27}\right)^{-2/3}=\dfrac{(-8)^{-2/3}}{27^{-2/3}}=\dfrac{27^{2/3}}{(-8)^{2/3}}=\dfrac{\left(\sqrt[3]{27}\right)^2}{\left(\sqrt[3]{-8}\right)^2}=\dfrac{3^2}{(-2)^2}=\dfrac{9}{4}$

33. $\dfrac{2^{5/3}2^{1/6}}{2^{1/2}}=2^{\frac{5}{3}+\frac{1}{6}-\frac{1}{2}}=2^{\frac{10}{6}+\frac{1}{6}-\frac{3}{6}}=2^{8/6}=2^{4/3}$

34. $\dfrac{\left(8x^3y\right)^{1/2}\left(8xy^5\right)^{1/2}}{\left(x^3y^6\right)^{1/3}}=\dfrac{\sqrt{8x^3y}\sqrt{8xy^5}}{\sqrt[3]{x^3y^6}}=\dfrac{\sqrt{64x^4y^6}}{xy^2}=\dfrac{8x^2y^3}{xy^2}=8x^{2-1}y^{3-2}=8xy$

35. 30°–60°–90° triangle.

Longer leg is shorter leg (y) times $\sqrt{3}$.

$$8 = y\sqrt{3} \quad \text{or} \quad y = \frac{8}{\sqrt{3}}$$

Hypotenuse is twice as long as shorter leg.

$$x = 2y$$

$$x = 2\frac{8}{\sqrt{3}} = \frac{16}{\sqrt{3}} \approx 9.24 \text{ cm}$$

36. 45°–45°–90° triangle.

Hypotenuse is length of leg times $\sqrt{2}$.

$$h = x\sqrt{2}$$

$$12.26 = x\sqrt{2}$$

$$8.67 = x$$

$$x = 8.67 \text{ cm}$$

37. $P(x_1, y_1) = P(-2, 5),\ Q(x_2, y_2) = (22, 12)$

$$d(PQ) = \sqrt{(x_2 - x_1)^2 + (y_2 - y_1)^2}$$

$$= \sqrt{[22-(-2)]^2 + (12-5)^2}$$

$$= \sqrt{24^2 + (7)^2}$$

$$= \sqrt{576 + 49}$$

$$= \sqrt{625}$$

$$= 25$$

38. Substitute $l = 4$ feet.

$$t = 2\pi\sqrt{\frac{l}{32}} = 2\pi\sqrt{\frac{4}{32}}$$

$$= 2\pi\sqrt{\frac{1}{8}} = 2\pi\left(\frac{1\sqrt{2}}{\sqrt{8}\sqrt{2}}\right)$$

$$= 2\pi\left(\frac{\sqrt{2}}{\sqrt{16}}\right) = 2\pi\left(\frac{\sqrt{2}}{4}\right)$$

$$= \frac{\pi\sqrt{2}}{2} \approx 2.2 \text{ sec}$$

39. Find the height of the box h when $a = 45$ in. and $c = 53$ in.

$$c^2 = a^2 + h^2$$

$$53^2 = 45^2 + h^2$$

$$53^2 - 45^2 = h^2$$

$$\sqrt{53^2 - 45^2} = \sqrt{h^2}$$

$$\sqrt{784} = h$$

$$h = 28 \text{ in.}$$

40. The principal square root cannot be negative.

Chapters 1–7 Cumulative Review Exercises

1. a. A rational number is any number that can be written as a fraction with an integer numerator and a nonzero integer denominator.
   b. An irrational number is a nonterminating, nonrepeating decimal.
   c. A real number is any number that is either a rational number or an irrational number.

2. $$\frac{|Ax_0 + By_0 + C|}{\sqrt{A^2 + B^2}} = \frac{|6(-2) + (-8)(-2) + (-5)|}{\sqrt{6^2 + (-8)^2}} = \frac{|-12 + 16 - 5|}{\sqrt{36 + 64}} = \frac{|-1|}{\sqrt{100}} = \frac{1}{10}$$

3. $$S = \frac{a - lr}{1 - r}$$

$$S(1 - r) = a - lr$$

$$S - Sr = a - lr$$

$$lr = a - S + Sr$$

$$l = \frac{a - S + Sr}{r}$$

4. $$\frac{x}{2} + (-8) + (-15) + x + (-10) = 0$$

$$x - 16 - 30 + 2x - 20 = 0$$

$$3x = 66$$

$$x = 22$$

5. **Analysis**: Find the number of cups of dressing required to make 10 cups of a 15% vinegar–oil dressing.

Equation: Let x represent the number of cups of 10% vinegar–oil dressing. Then $10 - x$ will represent the number of ounces of the 18% solution.

	percent vinegar–oil •	total quantity =	quantity of vinegar–oil
10% dressing	0.10	x	$0.10x$
18% dressing	0.18	$10-x$	$0.18(10-x)$
15% dressing	0.15	10	1.5

Quantity of vinegar–oil in 15% dressing	plus	quantity of vinegar–oil in 18% dressing	is	quantity of vinegar–oil in 15%dressing.
$0.10x$	+	$0.18(10-x)$	=	1.5

Solve:
$$\begin{aligned} 0.10x + 0.18(10-x) &= 1.5 \\ 0.10x + 1.8 - 0.18x &= 1.5 \\ -0.08x &= -0.3 \\ x &= 3.75 \end{aligned}$$

Conclusion: It would require $x = 3.75$ cups of 10% vinegar–oil dressing and $10 - x = 6.25$ cups of 18% vinegar–oil dressing.

6. a. The line goes through points (–5, 0) and (0, –3). $m = \dfrac{-3-0}{0-(-5)} = \dfrac{-3}{5} = -\dfrac{3}{5}$

b. The y–intercept is (0, –3).

c. Use $y = mx + b$. $y = -\dfrac{3}{5}x - 3$.

7. a. Using points $(x_1, y_1) = (0, -4)$ and $(x_2, y_2) = (1, -1)$ $\quad m = \dfrac{-1-(-4)}{1-0} = \dfrac{3}{1} = 3$

b. The slope is the negative reciprocal of 3, or $-\dfrac{1}{3}$. Use point A = (4, –2) and substitute in $y - y_1 = m(x - x_1)$.

$$\begin{aligned} y-(-2) &= -\frac{1}{3}(x-4) \\ y+2 &= -\frac{1}{3}x + \frac{4}{3} \\ y &= -\frac{1}{3}x - \frac{2}{3} \end{aligned}$$

8. $f(x) = 1.25 - x^5$. $f(2) = 1.25 - (2)^5 = 1.25 - 32 = -30.75$

9. D: all real numbers; $(-\infty, \infty)$ R: all real numbers less than or equal to –2; $(-\infty, -2]$

10.

x	$g(x) = 1 + x^3$	$g(x)$
–2	$g(-2) = 1 + (-2)^3$	–7
–1	$g(-1) = 1 + (-1)^3$	0
0	$g(0) = 1 + (0)^3$	1
1	$g(1) = 1 + (1)^3$	2
2	$g(2) = 1 + (2)^3$	9

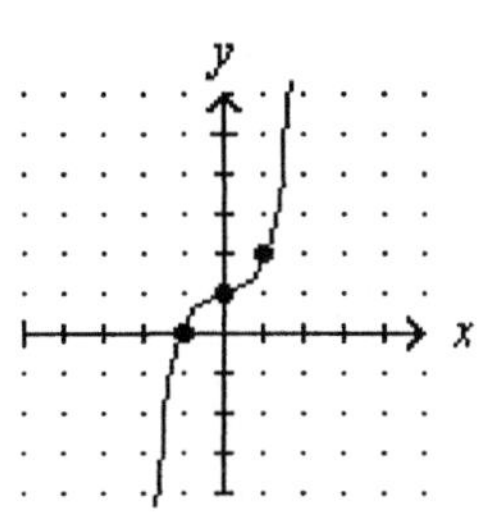

11. $\begin{cases} x = \dfrac{3}{2}y + 5 \\ 2x - 3y = 8 \end{cases}$

Substitute Equation 1 into Equation 2.

$$\begin{aligned} 2\left(\frac{3}{2}y + 5\right) - 3y &= 8 \\ 3y + 10 - 3y &= 8 \\ 0 &= -2 \end{aligned}$$

This is a contradiction. There is no solution.

12. **Analysis**: Find x, y, and z given the information in the illustration.

Equation: "x" represents savings, food, and utilities. "y" represent transportation. "z" represents clothing, health care, debt, and other. "$7z$" represents housing. The total of all categories is 100%.

3 "x" items	plus	1 "y" item	plus	4 "z" items plus housing	is	100%.
$3x$	$+$	y	$+$	$11z$	$=$	100

Utilities	plus	transportation	plus	housing	is	60%.
x	$+$	y	$+$	$7z$	$=$	60

food	plus	clothing	plus	savings	is	25%.
x	$+$	z	$+$	x	$=$	25

Solve:

$$\begin{cases} 3x + y + 11z = 100 \\ x + y + 7z = 60 \\ 2x \quad\;\; + z = 25 \end{cases}$$

Eliminate y.

Multiply Equation 2 by -1 and add Equations 1 and 2.

$$\begin{aligned} &1.\quad 3x + y + 11z = 100 \\ &2.\quad \underline{-x - y - 7z = -60} \\ &4.\quad 2x \quad\; + 4z = 40 \end{aligned}$$

Multiply Equation 3 by -1 and add Equations 3 and 4.

$$\begin{aligned} &3.\quad -2x - z = -25 \\ &4.\quad \underline{2x + 4z = 40} \\ &5.\quad 3z = 15 \\ &\qquad z = 5 \end{aligned}$$

Substitute $z = 5$ into Equation 3, solve for x.

$$\begin{aligned} 2x + (5) &= 25 \\ 2x &= 20 \\ x &= 10 \end{aligned}$$

Substitute $x = 10$ and $z = 5$ into Equation 2, solve for y.

$$\begin{aligned} (10) + y + 7(5) &= 60 \\ y &= 15 \end{aligned}$$

Conclusion: $x = 10$, $y = 15$, and $z = 5$

13. $5(x + 1) \le 4(x + 3$ and $x + 12 < -3$

$5x + 5 \le 4x + 12 \qquad x < -15$

$x \le 7$

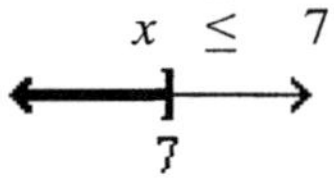

and

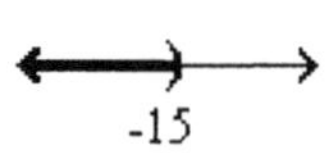

is

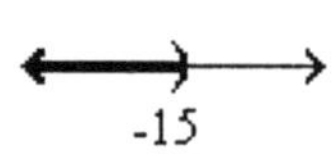

$(-\infty, -15)$

14. Rewrite $|-1 - 2x| > 5$ as

$-1 - 2x < -5$ or $-1 - 2x > 5$

$-2x < -4 \qquad -2x > 6$

$x > 2 \qquad x < -3$

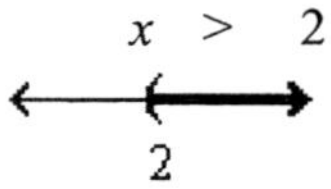

or

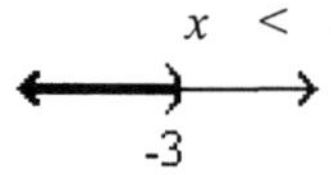

is

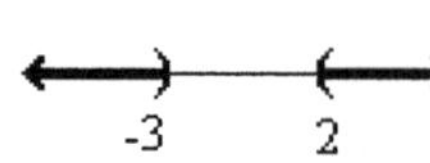

$(-\infty, -3) \cup (2, \infty)$

15. Inequality #1: $x + 2y < 3$

Graph: $x + 2y = 3$

x	y
-1	2
3	0

Broken line. Test point (0, 0).

$(0) + 2(0) < 3$

$0 < 3$ True

Inequality #2: $2x + 4y < 8$

Graph: $2x + 4y = 8$

x	y
0	2
4	0

Broken line. Test point (0, 0).

$2(0) + 4(0) < 8$

$0 < 8$ True

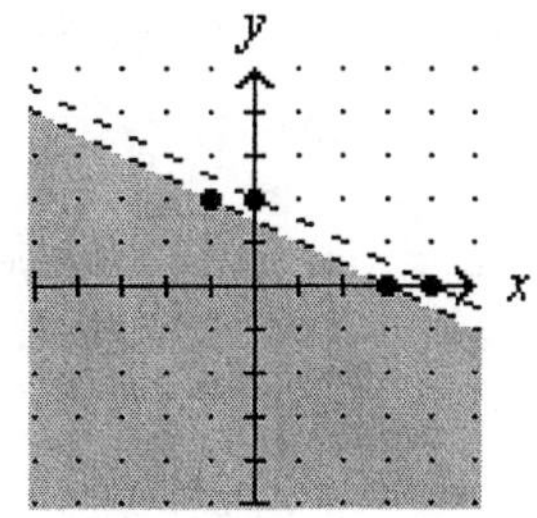

16. $\left(\dfrac{4a^{-2}b}{3ab^{-3}}\right)^3 = \left(\dfrac{4}{3}a^{-2-1}b^{1-(-3)}\right)^3 = \left(\dfrac{4}{3}a^{-3}b^4\right)^3 = \left(\dfrac{4}{3}\right)^3 a^{-3(3)}b^{4(3)} = \dfrac{64}{27}a^{-9}b^{12} = \dfrac{64b^{12}}{27a^9}$

17. $(6.1 \times 10^8)(3.9 \times 10^5) = (6.1 \bullet 3.9) \times 10^{8+5} = 23.79 \times 10^{13} = 2.379 \times 10^{13+1} = 2.379 \times 10^{14}$

18. $(-2x^2y^3 + 6xy + 5y^2) - (-4x^2y^3 - 7xy + 2y^2) = -2x^2y^3 + 6xy + 5y^2 + 4x^2y^3 + 7xy - 2y^2 = 2x^2y^3 + 13xy + 3y^2$

19. $(3y + 1)(2y^2 + 3y + 2) = 3y(2y^2 + 3y + 2) + 1(2y^2 + 3y + 2)$
$= 6y^3 + 9y^2 + 6y + 2y^2 + 3y + 2 = 6y^3 + 11y^2 + 9y + 2$

20. $(x + 3)(x - 3) + (2x - 1)(x + 2) = (x^2 - 9) + (2x^2 + 4x - x - 2) = 3x^2 + 3x - 11$

21. $3c - cd + 3d - c^2 = 3c + 3d - cd - c^2 = 3(c + d) - c(d + c) = (c + d)(3 - c)$

22. $x^3 - 8y^3 = x^3 - (2y)^3 = (x - 2y)(x^2 + 2xy + 4y^2)$

23. $(a + b)^2 - 2(a + b) + 1 = [(a + b) - 1][(a + b) - 1] = (a + b - 1)^2$ Perfect square.

24. $x^4 - 17x^2 + 16 = (x^2 - 16)(x^2 - 1) = (x + 4)(x - 4)(x + 1)(x - 1)$

25.
$$2z^3 - 200z = 0$$
$$2z(z^2 - 100) = 0$$
$$2z(z + 10)(z - 10) = 0$$

$2z = 0$ or $z + 10 = 0$ or $z - 10 = 0$

$z = 0$ $z = -10$ $z = 10$

Check:
$2(0)^3 - 200(0)$
$0 - 0 = 0$

Check:
$2(-10)^3 - 200(-10)$
$= -2000 + 2000 = 0$

Check:
$2(10)^3 - 200(10)$
$= 2000 - 2000 = 0$

26.
$$3m^2 + 10m = -3$$
$$3m^2 + 10m + 3 = 0$$
$$(3m + 1)(m + 3) = 0$$

$3m + 1 = 0$ or $m + 3 = 0$

$3m = -1$ $m = -3$

$m = -\frac{1}{3}$

Check:
$3(-\frac{1}{3})^2 + 10(-\frac{1}{3})$
$= \frac{1}{3} - \frac{10}{3} = -3$

Check:
$3(-3)^2 + 10(-3)$
$= 27 - 30 = -3$

27. $\dfrac{3x^2-10xy-8y^2}{4y^2-xy}=\dfrac{3x^2-12xy+2xy-8y^2}{y(4y-x)}=\dfrac{3x(x-4y)+2y(x-4y)}{-y(x-4y)}=\dfrac{(3x+2y)(x-4y)}{-y(x-4y)}=\dfrac{-3x-2y}{y}$

28. The fraction would be undefined when the denominator is zero.
$x^2-x-56 = (x-8)(x+7)$ would be zero when $x=8$ or $x=-7$.

29. $(2x^2-9x-5)\cdot\dfrac{x}{2x^2+x}=\dfrac{(2x+1)(x-5)}{1}\cdot\dfrac{x}{x(2x+1)}=x-5$

30.
$$\frac{2x}{x^2-4}-\frac{1}{x^2-3x+2}+\frac{x+1}{x^2+x-2}=\frac{2x((x-1)}{(x+2)(x-2)(x-1)}-\frac{1(x+2)}{(x-2)(x-1)(x+2)}+\frac{(x+1)(x-2)}{(x+2)(x-1)(x-2)}$$
$$=\frac{2x^2-2x-(x+2)+x^2-x-2}{(x+2)(x-2)(x-1)}=\frac{2x^2-2x-x-2+x^2-x-2}{(x+2)(x-2)(x-1)}$$
$$=\frac{3x^2-4x-4}{(x+2)(x-2)(x-1)}=\frac{(3x+2)(x-2)}{(x+2)(x-2)(x-1)}=\frac{3x+2}{(x+2)(x-1)}$$

31. **Analysis**: Find the amount of time it will take to fill the tank using both pipes.

Equation: Let x represent the amount of time it will take both pipes to finish the job.

Amount filled by one pipe in one hour	plus	amount filled by other pipe in one hour	is	Amount filled by both pipes in one hour.
$\frac{1}{4}$	$+$	$\frac{1}{6}$	$=$	$\frac{1}{x}$

Solve:
$$\frac{1}{4}+\frac{1}{6}=\frac{1}{x}$$
$$12x\left(\frac{1}{4}\right)+12x\left(\frac{1}{6}\right)=12x\left(\frac{1}{x}\right)$$
$$3x+2x=12$$
$$5x=12 \qquad x=\frac{12}{5}=2\frac{2}{5}$$

Conclusion: It will take $2\frac{2}{5}$ hours for both pipes to fill the tank.

32. Direct variation: As one quantity increases, the other increases, in a predictable way.
Inverse variation: As one quantity increases, the other decreases, in a predictable way.

33. $\left(-\dfrac{8x^3}{27}\right)^{-1/3}=\left(\dfrac{(-2x)^3}{3^3}\right)^{-1/3}=\dfrac{(-2x)^{-1}}{3^{-1}}=-\dfrac{3}{2x}$

34. $\sqrt{200x^4y^3z}=\sqrt{100x^4y^2\cdot 2yz}=10x^2y\sqrt{2yz}$

35. $\sqrt[3]{16}+\sqrt[3]{128}=\sqrt[3]{8\cdot 2}+\sqrt[3]{64\cdot 2}=2\sqrt[3]{2}+4\sqrt[3]{2}=6\sqrt[3]{2}$

36. $\left(\sqrt{5z}+\sqrt{3}\right)\left(\sqrt{5z}+\sqrt{3}\right)=\left(\sqrt{5z}\right)^2+2\left(\sqrt{5z}\sqrt{3}\right)+\left(\sqrt{3}\right)^2=5z+2\sqrt{15z}+3$

37. $\sqrt{-5x+24} = 6 - x$

$\left(\sqrt{-5x+24}\right)^2 = (6-x)^2$ Square both sides.

$-5x + 24 = 36 - 12x + x^2$ Do the squaring.

$0 = x^2 - 7x + 12$ Subtract 24 and add $5x$.

$0 = (x-4)(x-3)$ Factor.

$x - 4 = 0$ or $x - 3 = 0$

$x = 4$ $\quad$ $x = 3$

Check:

$\sqrt{-5(4)+24} \stackrel{?}{=} 6 - 4$

$\sqrt{4} = 2$

and

$\sqrt{-5(3)+24} \stackrel{?}{=} 6 - 3$

$\sqrt{9} = 3$

38. $\dfrac{\sqrt{3}}{\sqrt{50}} = \dfrac{\sqrt{3}\cdot\sqrt{2}}{\sqrt{50}\cdot\sqrt{2}} = \dfrac{\sqrt{6}}{\sqrt{100}} = \dfrac{\sqrt{6}}{10}$

Be actively involved in managing your learning. How much you learn depends on YOU- not a "terrific" instructor or a "terrible" instructor.

Chapter 8 Quadratic Equations, Functions, and Inequalities

Study Tip
Taking a Test

1. Look over the test- determine how many problems there are, and identify those you think you can finish quickly.
2. Keep an eye on the clock. If you have a 50 minute test with 25 problems, you know you have two minutes per problem. Do the quick problems first- this will put you ahead.
3. If you are working on a problem that takes more than double your allotted time per problem, and you aren't nearly done, stop working on the problem and go to the next one. Go back to that problem after you are done with the others.
4. Show all your work- make it easy for your instructor to give you some points for the part of the problem you complete correctly.
5. Read the directions carefully- this might prevent you from doing more work than is required, or from answering the wrong question.
6. Check your answers where possible- put solutions back into equations to verify that they make the equation true, check that your answers are reasonable in applied problems.
7. Don't leave a test early- redo as many problems as possible to look for errors.

Section 8.1 Completing the Square

VOCABULARY/CONCEPTS

1. quadratic

3. perfect

5. $x = \sqrt{c}$; $x = -\sqrt{c}$

7.
$$\begin{aligned} x^2 - 18 &= 0 \\ \left(-3\sqrt{2}\right)^2 - 18 &= 0 \\ 9\cdot 2 - 18 &\stackrel{?}{=} 0 \\ 18 - 18 &= 0 \end{aligned}$$

It is a solution.

9.

a. $\left[\frac{1}{2}(12)\right]^2 = 6^2 = 36$

b. $\left[\frac{1}{2}(-5)\right]^2 = \left(-\frac{5}{2}\right)^2 = \frac{25}{4}$

c. $\left[\frac{1}{2}\left(-\frac{1}{2}\right)\right]^2 = \left(-\frac{1}{4}\right)^2 = \frac{1}{16}$

11. a. Subtract 35 from both sides.

b. Add $\left[\frac{1}{2}(6)\right]^2 = 6^2 = 36$ to both sides.

13. $x - 1 = \pm\sqrt{\frac{5}{2}} = \pm\frac{\sqrt{5}}{\sqrt{2}} = \pm\frac{\sqrt{5}\sqrt{2}}{\sqrt{2}\sqrt{2}} = \pm\frac{\sqrt{10}}{2}$

15. Since 4 is not a factor of the numerator, it cannot be divided out.

NOTATION

17. A common denominator must be used as follows: $3 \pm \frac{\sqrt{10}}{5} = \frac{3 \cdot 5}{1 \cdot 5} \pm \frac{\sqrt{10}}{5} = \frac{15 \pm \sqrt{10}}{5}$. It does not equal $\frac{3 \pm \sqrt{10}}{5}$

19. a. 2 solutions; $2\sqrt{5}$; $-2\sqrt{5}$ b. ± 4.47

PRACTICE

21.
$$\begin{aligned} 6x^2 + 12x &= 0 \\ 6x(x+2) &= 0 \end{aligned}$$
$$6x = 0 \quad \text{or} \quad x + 2 = 0$$
$$x = 0 \quad\quad x = -2$$

23.
$$\begin{aligned} 2y^2 - 50 &= 0 \\ 2(y^2 - 25) &= 0 \\ 2(y+5)(y-5) &= 0 \end{aligned}$$
$$y + 5 = 0 \quad \text{or} \quad y - 5 = 0$$
$$y = -5 \quad\quad y = 5$$

25.
$$\begin{aligned} r^2 + 6r + 8 &= 0 \\ (r+4)(r+2) &= 0 \end{aligned}$$
$$r + 4 = 0 \quad \text{or} \quad r + 2 = 0$$
$$r = -4 \quad\quad r = -2$$

27.
$$\begin{aligned} 2z^2 &= -2 + 5z \\ 2z^2 - 5z + 2 &= 0 \\ (2z-1)(z-2) &= 0 \end{aligned}$$
$$2z - 1 = 0 \quad \text{or} \quad z - 2 = 0$$
$$2z = 1 \quad\quad z = 2$$
$$z =$$

29.
$$\begin{aligned} x^2 &= 36 \\ x &= \pm\sqrt{36} = \pm 6 \end{aligned}$$

31.
$$\begin{aligned} z^2 &= 5 \\ z &= \pm\sqrt{5} \end{aligned}$$

33.
$$\begin{aligned} 3x^2 - 16 &= 0 \\ 3x^2 &= 16 \\ x^2 &= \frac{16}{3} \\ x &= \pm\sqrt{\frac{16}{3}} = \pm\frac{4}{\sqrt{3}} = \pm\frac{4\sqrt{3}}{3} \end{aligned}$$

35.
$$\begin{aligned} (x+1)^2 &= 1 \\ x + 1 &= \pm 1 \end{aligned}$$
$$x + 1 = 1 \quad \text{or} \quad x + 1 = -1$$
$$x = 0 \quad\quad x = -2$$

37.
$$\begin{aligned} (s-7)^2 - 9 &= 0 \\ (s-7)^2 &= 9 \\ s - 7 &= \pm\sqrt{9} = \pm 3 \end{aligned}$$
$$s - 7 = 3 \quad \text{or} \quad s - 7 = -3$$
$$s = 3 + 7 \quad\quad s = -3 + 7$$
$$s = 10 \quad\quad s = 4$$

39.
$$\begin{aligned} (x+5)^2 - 3 &= 0 \\ (x+5)^2 &= 3 \\ x + 5 &= \pm\sqrt{3} \\ x &= -5 \pm \sqrt{3} \end{aligned}$$

41.
$$\begin{aligned} 2d^2 &= 3h \\ d^2 &= \frac{3h}{2} \\ d &= \sqrt{\frac{3h}{2}} = \frac{\sqrt{3h}\sqrt{2}}{\sqrt{2}\sqrt{2}} = \frac{\sqrt{6h}}{2} \end{aligned}$$

43.
$$\begin{aligned} E &= mc^2 \\ \frac{E}{m} &= c^2 \\ c &= \sqrt{\frac{E}{m}} = \frac{\sqrt{E}\sqrt{m}}{\sqrt{m}\sqrt{m}} = \frac{\sqrt{Em}}{m} \end{aligned}$$

45. $x^2 + 2x - 8 = 0$

$x^2 + 2x = 8$ — Add 8 to both sides.

$x^2 + 2x + 1 = 8 + 1$ — Square one–half of the coefficient of x and add it to both sides.

$(x+1)^2 = 9$ — Factor and combine like terms.

$x + 1 = \pm\sqrt{9} = \pm 3$ — Apply the square root property.

$x + 1 = -3$ or $x + 1 = 3$

$x = -3 - 1$ $\quad$ $x = 3 - 1$ — Subtract 1 from both sides.

$x = -4$ $\quad$ $x = 2$

47. $k^2 - 8k + 12 = 0$

$k^2 - 8k = -12$ — Move 12 to the other side.

$k^2 - 8k + 16 = -12 + 16$ — Square one–half of the coefficient of x and add it to both sides.

$(k-4)^2 = 4$ — Factor and combine like terms.

$k - 4 = \pm 2$ — Apply the square root property.

$k = 4 \pm 2$ — Add 4 to both sides

$k = 4 + 2 = 6 \; or \; k = 4 - 2 = 2$ — Write in concise form.

49. $g^2 + 5g - 6 = 0$

$g^2 + 5g = 6$ — Move 6 to the other side.

$g^2 + 5g + \frac{25}{4} = 6 + \frac{25}{4}$ — Square one–half of the coefficient of x and add it to both sides.

$\left(g + \frac{5}{2}\right)^2 = \frac{24+25}{4}$ — Factor and combine like terms.

$\left(g + \frac{5}{2}\right)^2 = \frac{49}{4}$ — Apply the square root property.

$\left(g + \frac{5}{2}\right) = \pm\frac{7}{2}$ — Add 5/2 to both sides

$g = -\frac{5}{2} \pm \frac{7}{2}$

$g = -\frac{5}{2} + \frac{7}{2} = \frac{2}{2} = 1 \; or \; g = -\frac{5}{2} - \frac{7}{2} = \frac{-12}{2} = -6$ — Write in concise form.

51. $x^2 - 3x - 4 = 0$

$x^2 - 3x = 4$ Move 4 to the other side.

$x^2 - 3x + \frac{9}{4} = 4 + \frac{9}{4} = \frac{16+9}{4}$ Square one–half of the coefficient of x and add it to both sides.

$\left(x - \frac{3}{2}\right)^2 = \frac{25}{4}$ Factor and combine like terms.

$x - \frac{3}{2} = \pm\frac{5}{2}$ Apply the square root property.

$x = \frac{3}{2} \pm \frac{5}{2}$ Add 3/2 to both sides

$x = \frac{3}{2} + \frac{5}{2} = \frac{8}{2} = 4 \quad or \quad \frac{3}{2} - \frac{5}{2} = \frac{-2}{2} = -1$ Write in concise form.

53. $2x^2 - x - 1 = 0$

$x^2 - \frac{1}{2}x = \frac{1}{2}$ Divide both sides by –2.

$x^2 - \frac{1}{2}x + \frac{1}{16} = \frac{1}{2} + \frac{1}{16}$ Square one–half of the coefficient of x and add it to both sides.

$\left(x - \frac{1}{4}\right)^2 = \frac{9}{16}$ Factor and combine like terms.

$x - \frac{1}{4} = \pm\sqrt{\frac{9}{16}} = \pm\frac{3}{4}$ Apply the square root property.

$x = \frac{1}{4} \pm \frac{3}{4}$ Add $\frac{1}{4}$ to both sides and write in concise form.

$x = \frac{1}{4} + \frac{3}{4} = \frac{4}{4} = 1$ or $x = \frac{1}{4} - \frac{3}{4} = -\frac{1}{2}$ Simplify.

55. $6x^2 + x - 2 = 0$

$x^2 + \frac{1}{6}x - \frac{1}{3} = 0$ Divide both sides by 6.

$x^2 + \frac{1}{6}x = \frac{1}{3}$ Add $\frac{1}{3}$ to both sides.

$x^2 + \frac{1}{6}x + \frac{1}{144} = \frac{1}{3} + \frac{1}{144}$ Square one–half of the coefficient of x and add it to both sides.

$\left(x + \frac{1}{12}\right)^2 = \frac{49}{144}$ Factor and combine like terms.

$x + \frac{1}{12} = \pm\sqrt{\frac{49}{144}} = \pm\frac{7}{12}$ Apply the square root property.

$x = -\frac{1}{12} \pm \frac{7}{12}$ Subtract $\frac{1}{12}$ from both sides and write in concise form.

$x = -\frac{1}{12} + \frac{7}{12} = \frac{6}{12} = \frac{1}{2}$ or $x = -\frac{1}{12} - \frac{7}{12} = -\frac{2}{3}$ Simplify.

57. $x^2+8x+6=0$

$x^2+8x=-6$ Subtract 6 from both sides.

$x^2+8x+16=-6+16$ Square one–half of the coefficient of x and add it to both sides.

$(x+4)^2=10$ Factor and combine like terms.

$x+4=\pm\sqrt{10}$ Apply the square root property.

$x=-4\pm\sqrt{10}$ Subtract 4 from both sides and write in concise form.

59. $x^2-2x-17=0$

$x^2-2x=17$ Add 17 to both sides.

$x^2-2x+1=17+1$ Square one–half of the coefficient of x and add it to both sides.

$(x-1)^2=18$ Factor and combine like terms.

$x-1=\pm\sqrt{18}=\pm3\sqrt{2}$ Apply the square root property.

$x=1\pm3\sqrt{2}$ Add 1 to both sides and write in concise form.

61. $b^2-3b=5$

$b^2-3b+\frac{9}{4}=5+\frac{9}{4}=\frac{20+9}{4}$ Square one–half of the coefficient of x and add it to both sides.

$\left(b-\frac{3}{2}\right)^2=\frac{29}{4}$ Factor and combine like terms.

$b-\frac{3}{2}=\pm\frac{\sqrt{29}}{2}$ Apply the square root property.

$b=\frac{3}{2}\pm\frac{\sqrt{29}}{2}=\frac{3\pm\sqrt{29}}{2}$ Add 1/2 to both sides and write in concise form.

63. $3x^2-6x=1$

$x^2-2x=\frac{1}{3}$ Divide both sides by 3.

$x^2-2x+1=\frac{1}{3}+1$ Square one–half of the coefficient of x and add it to both sides.

$(x-1)^2=\frac{4}{3}$ Factor and combine like terms.

$x-1=\pm\sqrt{\frac{4}{3}}=\pm\frac{\sqrt{4}}{\sqrt{3}}=\pm\frac{2\sqrt{3}}{3}$ Apply the square root property and simplify.

$x=1\pm\frac{2\sqrt{3}}{3}=\frac{3}{3}\pm\frac{2\sqrt{3}}{3}=\frac{3\pm2\sqrt{3}}{3}$ Add 1 to both sides and rewrite in concise form.

65. $4x^2 - 4x = 7$

$x^2 - 1x = \frac{7}{4}$ Divide both sides by 4.

$x^2 - x + \frac{1}{4} = \frac{7}{4} + \frac{1}{4}$ Square one–half of the coefficient of x and add it to both sides.

$\left(x - \frac{1}{2}\right)^2 = 2$ Factor and combine like terms.

$x - \frac{1}{2} = \pm\sqrt{2}$ Apply the square root property and simplify.

$x = \frac{1}{2} \pm \sqrt{2} = \frac{1}{2} \pm \frac{2\sqrt{2}}{2} = \frac{1 \pm 2\sqrt{2}}{2}$ Add $\frac{1}{2}$ to both sides and rewrite in concise form.

67. $2x^2 + 5x - 2 = 0$

$2x^2 + 5x = 2$ Add 2 to both sides.

$x^2 + \frac{5}{2}x = 1$ Divide both sides by 2.

$x^2 + \frac{5}{2}x + \frac{25}{16} = 1 + \frac{25}{16}$ Square one–half of the coefficient of x and add it to both sides.

$\left(x + \frac{5}{4}\right)^2 = \frac{41}{16}$ Factor and combine like terms.

$x + \frac{5}{4} = \pm\sqrt{\frac{41}{16}} = \pm\frac{\sqrt{41}}{4}$ Apply the square root property and simplify.

$x = -\frac{5}{4} \pm \frac{\sqrt{41}}{4} = \frac{-5 \pm \sqrt{41}}{4}$ Subtract $\frac{5}{4}$ from both sides and rewrite in concise form.

69. $\frac{7x+1}{5} = -x^2$

$7x + 1 = -5x^2$ Multiply both sides by 5.

$5x^2 + 7x = -1$ Add $5x^2$ to and subtract 1 from both sides.

$x^2 + \frac{7}{5}x = -\frac{1}{5}$ Divide both sides by 5.

$x^2 + \frac{7}{5}x + \frac{49}{100} = -\frac{1}{5} + \frac{49}{100} = -\frac{20}{100} + \frac{49}{100}$ Square one–half of the coefficient of x and add it to both sides.

$\left(x + \frac{7}{10}\right)^2 = \frac{29}{100}$ Factor and combine like terms.

$x + \frac{7}{10} = \pm\sqrt{\frac{29}{100}} = \pm\frac{\sqrt{29}}{10}$ Apply the square root property and simplify.

$x = -\frac{7}{10} \pm \frac{\sqrt{29}}{10} = \frac{-7 \pm \sqrt{29}}{10}$ Subtract $\frac{7}{10}$ from both sides and rewrite in concise form.

APPLICATIONS

71. Find width x and length $1.9x$ when area is 100 ft^2.

$$lw = A$$
$$1.9x(x) = 100$$
$$1.9x^2 = 100$$
$$x^2 = \frac{100}{1.9}$$
$$x = \sqrt{\frac{100}{1.9}} \approx 7.25 = 7\frac{1}{4} \text{ ft}$$
$$1.9x \approx 13.78 \approx 13\frac{3}{4} \text{ ft}$$

The dimensions are $7\frac{1}{4}$ ft by $13\frac{3}{4}$ ft.

73. Find t when $h = 5$ ft and $s = 4(12 \text{ ft}) = 48$ ft.

$$h = s - 16t^2$$
$$5 = 48 - 16t^2$$
$$-43 = -16t^2$$
$$t^2 = \frac{43}{16}$$
$$t = \sqrt{\frac{43}{16}} \approx 1.6 \text{ sec}$$

75. Find r when $h = 5.25$ in. and $V = 47.75$ in.^{3}

$$V = \pi r^2 h$$
$$47.75 = \pi r^2(5.25)$$
$$\frac{47.75}{5.25\pi} = r^2$$
$$r = \sqrt{\frac{47.75}{5.25\pi}} = \sqrt{2.895\ldots} \approx 1.70 \text{ in.}$$

77. **Analysis:** Write an equation representing the area of the matting in terms of the width of the matting. To do this, find the area of the entire piece and subtract the area of the picture. The area of the picture is (4 in.)(5 in.) = 20in.^{2} The area of the matting is equal to the area of the picture.

Equation: Let x represent the width of the matting. The length across the top of the entire piece is $5 + 2x$. The length down the side is $4 + 2x$. The area of the entire piece is: $(5 + 2x)(4 + 2x) = 20 + 18x + 4x^2$. Thus the area of the matting is $4x^2 + 18x - 20 - 20 = 4x^2 + 18x$. Set this equal to the area of the picture (20)

Solve:

$$4x^2 + 18x = 20$$
$$x^2 + \frac{9}{2}x = 5$$
$$x^2 + \frac{9}{2}x + \frac{81}{16} = 5 + \frac{81}{16} = \frac{80}{16} + \frac{81}{16}$$
$$\left(x + \frac{9}{4}\right)^2 = \frac{161}{16}$$
$$x + \frac{9}{4} = \sqrt{\frac{161}{16}} = \frac{\sqrt{161}}{4}$$
$$x = -\frac{9}{4} + \frac{\sqrt{161}}{4} \approx 0.92 \text{ in.}$$

Conclusion: The width of the matting is about 0.92 inches.

79. Let x represent the width, then $x + 4$ represents the length. The area is 20 ft^2.

$$lw = A$$
$$x(x+4) = 20$$
$$x^2 + 4x = 20$$
$$x^2 + 4x + 4 = 20 + 4$$
$$(x+2)^2 = 24$$
$$x + 2 = \sqrt{24}$$
$$x = -2 + \sqrt{24} \approx 2.9 \text{ ft}$$
$$x + 4 \approx 2.9 + 4 = 6.9 \text{ ft}$$

The dimensions are 2.9 ft by 6.9 ft.

WRITING

81. Write the equation in the form $ax^2 + bx = c$. Divide all terms by a. Find one–half of the coefficient of x. Add the square of this value to both sides of the equation. The left hand side of the equation is now a perfect square trinomial.

REVIEW

83. $\sqrt[3]{40a^3b^6} = \sqrt[3]{8a^3b^6}\sqrt[3]{5} = 2ab^2\sqrt[3]{5}$

85. $\sqrt[8]{x^{24}} = \sqrt[8]{(x^3)^8} = x^3$

87. $\sqrt{175a^2b^3} = \sqrt{25a^2b^2}\sqrt{7b} = 5ab\sqrt{7b}$

Section 8.2 The Quadratic Formula

VOCABULARY

1. quadratic

CONCEPTS

3. a. $x^2 + 2x + 5 = 0$ b. $3x^2 + 2x - 1 = 0$

5. a. true b. true

7. Let $x = c$, $a = r$, $b = s$, and $c = t$ in the normal quadratic formula, then $c = \dfrac{-s \pm \sqrt{s^2 - 4rt}}{2r}$

9. a. $x = \dfrac{3 \pm 6\sqrt{2}}{3} = \dfrac{3(1 \pm 2\sqrt{2})}{3} = 1 \pm 2\sqrt{2}$ b. $x = \dfrac{-12 \pm 4\sqrt{7}}{8} = \dfrac{4\left(-3 \pm \sqrt{7}\right)}{4 \cdot 2} = \dfrac{-3 \pm \sqrt{7}}{2}$

NOTATION

11. a. The fraction bar wasn't drawn under both parts of the numerator.
 b. A ± sign wasn't written between b and the radical.

PRACTICE

13. $x^2 + 3x + 2 = 0$

$a = 1;\ b = 3;\ c = 2$

$$x = \frac{-3 \pm \sqrt{3^2 - 4(1)(2)}}{2(1)} = \frac{-3 \pm \sqrt{9-8}}{2} = \frac{-3 \pm \sqrt{1}}{2} = \frac{-3 \pm 1}{2}$$

$x = \frac{-3+1}{2} = \frac{-2}{2} = -1$ or $x = \frac{-3-1}{2} = \frac{-4}{2} = -2$

15. $x^2 + 12x = -36$

$x^2 + 12x + 36 = 0$ Rearrange.

$a = 1;\ b = 12;\ c = 36$

$$x = \frac{-12 \pm \sqrt{12^2 - 4(1)(36)}}{2(1)} = \frac{-12 \pm \sqrt{144-144}}{2} = \frac{-12 \pm 0}{2}$$

$x = \frac{-12+0}{2} = -6$ or $x = \frac{-12-0}{2} = -6$

17. $2x^2 + 5x - 3 = 0$

$a = 2;\ b = 5;\ c = -3$

$$x = \frac{-5 \pm \sqrt{5^2 - 4(2)(-3)}}{2(2)} = \frac{-5 \pm \sqrt{25+24}}{4} = \frac{-5 \pm \sqrt{49}}{4} = \frac{-5 \pm 7}{4}$$

$x = \frac{-5+7}{4} = \frac{1}{2}$ or $x = \frac{-5-7}{4} = -3$

19. $5x^2 + 5x + 1 = 0$

$a = 5;\ b = 5;\ c = 1$

$$x = \frac{-5 \pm \sqrt{5^2 - 4(5)(1)}}{2(5)} = \frac{-5 \pm \sqrt{25-20}}{10}$$

$$x = \frac{-5 \pm \sqrt{5}}{10}$$

21. $8u = -4u^2 - 3$

$4u^2 + 8u + 3 = 0$ Rearrange.

$a = 4;\ b = 8;\ c = 3$

$$u = \frac{-8 \pm \sqrt{8^2 - 4(4)(3)}}{2(4)} = \frac{-8 \pm \sqrt{64-48}}{8} = \frac{-8 \pm \sqrt{16}}{8} = \frac{-8 \pm 4}{8}$$

$u = \frac{-8+4}{8} = -\frac{1}{2}$ or $u = \frac{-8-4}{8} = -\frac{3}{2}$

23. $-16y^2 - 8y + 3 = 0$

$16y^2 + 8y - 3 = 0$ Multiply by -1.

$a = 16;\ b = 8;\ c = -3$

$$y = \frac{-8 \pm \sqrt{8^2 - 4(16)(-3)}}{2(16)} = \frac{-8 \pm \sqrt{64+192}}{32} = \frac{-8 \pm \sqrt{256}}{32} = \frac{-8 \pm 16}{32}$$

$y = \frac{-8+16}{32} = \frac{1}{4}$ or $x = \frac{-8-16}{32} = -\frac{3}{4}$

25. $x^2 - \dfrac{14}{15}x = \dfrac{8}{15}$

$15x^2 - 14x = 8$ Multiply by 15.
$15x^2 - 14x - 8 = 0$ Rearrange.

$a = 15;\ b = -14;\ c = -8$

$$x = \frac{-(-14) \pm \sqrt{(-14)^2 - 4(15)(-8)}}{2(15)}$$
$$= \frac{14 \pm \sqrt{196 + 480}}{30}$$
$$= \frac{14 \pm \sqrt{676}}{30}$$
$$= \frac{14 \pm 26}{30}$$
$$x = \frac{14 + 26}{30} = \frac{4}{3} \quad \text{or} \quad x = \frac{14 - 26}{30} = -\frac{2}{5}$$

27. $\dfrac{x^2}{2} + \dfrac{5}{2}x = -1$

$x^2 + 5x = -2$ Multiply by 2.
$x^2 + 5x + 2 = 0$ Rearrange.

$a = 1;\ b = 5;\ c = 2$

$$x = \frac{-5 \pm \sqrt{5^2 - 4(1)(2)}}{2(1)}$$
$$= \frac{-5 \pm \sqrt{25 - 8}}{2}$$
$$x = \frac{-5 \pm \sqrt{17}}{2}$$

29. $2x^2 - 1 = 3x$
$2x^2 - 3x - 1 = 0$ Rearrange.

$a = 2;\ b = -3;\ c = -1$

$$x = \frac{-(-3) \pm \sqrt{(-3)^2 - 4(2)(-1)}}{2(2)}$$
$$= \frac{3 \pm \sqrt{9 + 8}}{4}$$
$$= \frac{3 \pm \sqrt{17}}{4}$$

31. $-x^2 + 10x = 18$
$x^2 - 10x + 18 = 0$ Multiply by –1 and rearrange.

$a = 1;\ b = -10;\ c = 18$

$$x = \frac{-(-10) \pm \sqrt{(-10)^2 - 4(1)(18)}}{2(1)}$$
$$= \frac{10 \pm \sqrt{100 - 72}}{2}$$
$$= \frac{10 \pm \sqrt{28}}{2}$$
$$x = \frac{10 \pm 2\sqrt{7}}{2} = \frac{2\left(5 \pm \sqrt{7}\right)}{2} = 5 \pm \sqrt{7}$$

33. $x^2 \quad 6x - 391$
$x^2 - 6x - 391 = 0$ Rearrange.

$a = 1;\ b = -6;\ c = -391$

$$x = \frac{-(-6) \pm \sqrt{(-6)^2 - 4(1)(-391)}}{2(1)}$$
$$= \frac{6 \pm \sqrt{36 + 1564}}{2}$$
$$= \frac{6 \pm \sqrt{1600}}{2}$$
$$= \frac{6 \pm 40}{2}$$

35. $x^2 - \dfrac{5}{3} = -\dfrac{11}{6}x$

$6x^2 - 10 = -11x$ Multiply by 6.
$6x^2 + 11x - 10 = 0$ Rearrange.

$a = 6;\ b = 11;\ c = -10$

$$x = \frac{-11 \pm \sqrt{11^2 - 4(6)(-10)}}{2(6)}$$
$$= \frac{-11 \pm \sqrt{121 + 240}}{12}$$
$$= \frac{-11 \pm \sqrt{361}}{12}$$
$$= \frac{-11 \pm 19}{12}$$

$$x = \frac{6 + 40}{2} = 23 \quad \text{or} \quad x = \frac{6 - 40}{2} = -17$$

37. $50x^2 + 30x - 10 = 0$
$5x^2 + 3x - 1 = 0$ Divide by 10.

$a = 5;\ b = 3;\ c = -1$

$$x = \frac{-3 \pm \sqrt{3^2 - 4(5)(-1)}}{2(5)} = \frac{-3 \pm \sqrt{9 + 20}}{10} = \frac{-3 \pm \sqrt{29}}{10}$$

39. $900x^2 - 8100x = 1800$
$x^2 - 9x - 2 = 0$ Divide by 900, rearrange.

$a = 1;\ b = -9;\ c = -2$

$$x = \frac{-(-9) \pm \sqrt{(-9)^2 - 4(1)(-2)}}{2(1)} = \frac{9 \pm \sqrt{81 + 8}}{2} = \frac{9 \pm \sqrt{89}}{2}$$

41. $-0.6x^2 - 0.03 = -0.4x$
$60x^2 + 3 = 40x$ Multiply by -100.
$60x^2 - 40x + 3 = 0$ Rearrange.

$a = 60;\ b = -40;\ c = 3$

$$x = \frac{-(-40) \pm \sqrt{(-40)^2 - 4(60)(3)}}{2(60)} = \frac{40 \pm \sqrt{1600 - 720}}{120} = \frac{40 \pm \sqrt{880}}{120}$$

$$x = \frac{40 \pm 4\sqrt{55}}{120} = \frac{4\left(10 \pm \sqrt{55}\right)}{4 \cdot 30} = \frac{10 \pm \sqrt{55}}{30}$$

43. $x^2 + 8x + 5 = 0$

$a = 1;\ b = 8;\ c = 5$

$$x = \frac{-8 \pm \sqrt{8^2 - 4(1)(5)}}{2(1)} = \frac{-8 \pm \sqrt{64 - 20}}{2} = \frac{-8 \pm \sqrt{44}}{2}$$

$$x = \frac{-8 + \sqrt{44}}{2} \approx -0.68 \quad \text{or} \quad x = \frac{-8 - \sqrt{44}}{2} \approx -7.32$$

45. $3x^2 - 2x - 2 = 0$

$a = 3;\ b = -2;\ c = -2$

$$x = \frac{-(-2) \pm \sqrt{(-2)^2 - 4(3)(-2)}}{2(3)} = \frac{2 \pm \sqrt{4 + 24}}{6} = \frac{2 \pm \sqrt{28}}{6}$$

$$x = \frac{2 + \sqrt{28}}{6} \approx 1.22 \quad \text{or} \quad x = \frac{2 - \sqrt{28}}{6} \approx -0.55$$

47. $0.7x^2 - 3.5x - 25 = 0$

$a = 0.7;\ b = -3.5;\ c = -25$

$$x = \frac{-(-3.5) \pm \sqrt{(-3.5)^2 - 4(0.7)(-25)}}{2(0.7)} = \frac{3.5 \pm \sqrt{12.25 + 70}}{1.4} = \frac{3.5 \pm \sqrt{82.25}}{1.4} \approx 8.98 \text{ or } -3.98$$

APPLICATIONS

49. **Analysis**: Find the dimensions of the screen with an area of 11,349 ft^2.

Equation: Let x represent the width of the screen and $x + 20$ represent the length of the screen.

Width	times	length	equals	area.
x	$\bullet$	$x + 20$	$=$	11,349

Solve:
$$x(x + 20) = 11{,}349$$
$$x^2 + 20x - 11{,}349 = 0$$

$a = 1$; $b = 20$; $c = -11{,}349$

$$x = \frac{-20 \pm \sqrt{20^2 - 4(1)(-11{,}349)}}{2(1)}$$
$$= \frac{-20 \pm \sqrt{400 + 45{,}396}}{2}$$
$$= \frac{-20 \pm \sqrt{45{,}796}}{2} = \frac{-20 \pm 214}{2}$$
$$x = \frac{-20 + 214}{2} = 97 \quad \text{or} \quad x = \frac{-20 - 214}{2} = -117$$

Discard negative solution.

Conclusion: The width of the screen is $x = 97$ ft and the length is $x + 20 = 117$ ft.

51. **Analysis**: Find the dimensions of Central Park that is a rectangle that is 5 times longer than it is wide.

Equation: Let x represent the width of the park and $5x$ represent the length of the park. The area is represented by $lw = 5x^2$. The perimeter is represented by $2w + 2l = 2(x) + 2(5x) = 12x$. Numerically, the perimeter exceeds the area by 4.75.

Area	plus	4.75	equals	perimeter.
$5x^2$	$+$	4.75	$=$	$12x$

Solve:
$$5x^2 + 4.75 = 12x$$
$$5x^2 - 12x + 4.75 = 0$$

$a = 5$; $b = -12$; $c = 4.75$

$$x = \frac{-(-12) \pm \sqrt{(-12)^2 - 4(5)(4.75)}}{2(5)}$$
$$= \frac{12 \pm \sqrt{144 - 95}}{10}$$
$$= \frac{12 \pm \sqrt{49}}{10} = \frac{12 \pm 7}{10}$$
$$x = \frac{12 - 7}{10} = 0.5 \quad \text{or} \quad x = \frac{12 + 7}{10} = 1.9$$

Discard $x = 1.9$ because the width is less than 1 mile.

Conclusion: The width of the park is $x = 0.5$ mi and the length is $5x = 2.5$ mi.

53. **Analysis**: Find the length of a string that forms the hypotenuse of a right triangle.

Equation: Let x represent the length of the string. Then $x - 4$ represents the distance on the pole and $\frac{1}{2}x - 1$ represents the distance on the ground.

a^2	plus	b^2	equals	c^2
$(x-4)^2$	$+$	$(\frac{1}{2}x-1)^2$	$=$	x^2

Solve:

$$\begin{aligned} (x-4)^2 + (\tfrac{1}{2}x-1)^2 &= x^2 \\ x^2 - 8x + 16 + \tfrac{1}{4}x^2 - x + 1 &= x^2 \\ \tfrac{1}{4}x^2 - 9x + 17 &= 0 \\ x^2 - 36x + 68 &= 0 \end{aligned}$$

$a = 1;\ b = -36;\ c = 68$

$$x = \frac{-(-36) \pm \sqrt{(-36)^2 - 4(1)(68)}}{2(1)}$$

$$= \frac{36 \pm \sqrt{1296 - 272}}{2}$$

$$= \frac{36 \pm \sqrt{1024}}{2} = \frac{36 \pm 32}{2}$$

$x = \frac{36+32}{2} = 34$ or $x = \frac{36-32}{2} = 2$ Discard $x = 2$ because $x - 4$ would be negative.

Conclusion: The length of the string would be $x = 34$ in.

55. **Analysis**: Find the ticket price that will make the receipts $1248 when increases are made in $0.10 increments and each increase decreases the attendance by 5.

Equation: Let x represent the number of $0.10 increments. Then the ticket price will be $\$(4 + 0.1x)$. The number attending will be $300 - 5x$.

Ticket price	times	attendees paying that price	equals	$1248.
$(4 + 0.1x)$	•	$(300 - 5x)$	$=$	1248

Solve:

$$\begin{aligned} (4 + 0.1x)(300 - 5x) &= 1248 \\ 1200 - 20x + 30x - 0.5x^2 &= 1248 \\ -0.5x^2 + 10x - 48 &= 0 \\ x^2 - 20x + 96 &= 0 \end{aligned}$$

$a = 1;\ b = -20;\ c = 96$

$$x = \frac{-(-20) \pm \sqrt{(-20)^2 - 4(1)(96)}}{2(1)}$$

$$= \frac{20 \pm \sqrt{400 - 384}}{2}$$

$$= \frac{20 \pm \sqrt{16}}{2} = \frac{20 \pm 4}{2}$$

$x = \frac{20+4}{2} = 12$ or $x = \frac{20-4}{2} = 8$

Conclusion: The number of increases could be either 8 or 12 making the amount of the increase either $1.20 or $0.80. The new ticket price to give an income of $1248 would be either $4.80 or $5.20.

57. **Analysis**: Find the number of subscribers needed to bring a profit of $120,000.

Equation: Let x represent the number of new subscribers. Then the profit per subscriber will be $(20 + 0.01$x$). The total number of subscribers will be $3000 + x$.

Profit per subscriber	times	total subscribers	equals	$120,000.
$(20 + 0.01x)$	•	$(3000 + x)$	=	120,000

Solve:

$$\begin{aligned} (20 + 0.01x)(3000 + x) &= 120{,}000 \\ 60{,}000 + 20x + 30x + 0.01x^2 &= 120{,}000 \\ 0.01x^2 + 50x - 60{,}000 &= 0 \\ x^2 + 5{,}000x + 6{,}000{,}000 &= 0 \end{aligned}$$

$a = 1;\ b = 5{,}000;\ c = 6{,}000{,}000$

$$x = \frac{-5{,}000 \pm \sqrt{(5{,}000)^2 - 4(1)(-6{,}000{,}000)}}{2(1)}$$

$$= \frac{-5{,}000 \pm \sqrt{25{,}000{,}000 + 24{,}000{,}000}}{2}$$

$$= \frac{-5{,}000 \pm \sqrt{49{,}000{,}000}}{2} = \frac{-5{,}000 \pm 7{,}000}{2}$$

$$x = \frac{-5{,}000 - 7{,}000}{2} = -6{,}000$$

or

$$x = \frac{-5{,}000 + 7{,}000}{2} = 1{,}000$$

Discard the negative solution.

Conclusion: The number of new subscribers would need to be $x = 1000$ making a total of 4000 subscribers.

59. Find r when the balance is $3368.10.

$$\begin{aligned} 1000(1 + r)^2 + 2000((1 + r) &= 3368.10 \\ 1000(1 + 2r + r^2) + 2000 + 2000r - 3368.10 &= 0 \\ 1000 + 2000r + 1000r^2 + 2000 + 2000r - 3368.10 &= 0 \\ 1000r^2 + 4000r - 368.10 &= 0 \end{aligned}$$

$a = 1000;\ b = 4000;\ c = -368.10$

$$r = \frac{-4000 \pm \sqrt{4000^2 - 4(1000)(-368.10)}}{2(1000)}$$

$$= \frac{-4000 \pm \sqrt{16{,}000{,}000 + 1{,}472{,}400}}{2000}$$

$$= \frac{-4000 \pm \sqrt{17{,}472{,}400}}{2000} = \frac{-4000 \pm 4180}{2000}$$

$x = \dfrac{-4000 - 4180}{2000} = -4.09$ or $x = \dfrac{-4000 + 4180}{2000} = 0.09 = 9\%$ Discard negative solution.

61. Find x when $P = 75\%$.

$$\begin{aligned} P &= 0.03x^2 - 1.37x + 82.51 \\ 75 &= 0.03x^2 - 1.37x + 82.51 \\ 0 &= 0.03x^2 - 1.37x + 7.51 \end{aligned} \qquad a = 0.03;\ b = -1.37;\ c = 7.51$$

$$x = \frac{-(-1.37) \pm \sqrt{(-1.37)^2 - 4(0.03)(7.51)}}{2(0.03)} = \frac{1.37 \pm \sqrt{1.8769 - 0.9012}}{0.06} = \frac{1.37 \pm \sqrt{0.9757}}{0.06} \approx \frac{1.37 \pm 0.988}{0.06}$$

$$x = \frac{1.37 - 0.988}{0.06} \approx 6.37 \quad \text{or} \quad x = \frac{1.37 + 0.988}{0.06} \approx 39.3$$

Since $0 < x < 30$, discard $x = 39.3$. In about 6.37 years after 1970, or early in 1976, the labor force participation rate will be 75%.

WRITING

63. When using the quadratic formula it is only necessary to substitute numbers into a formula.

REVIEW

65. $\sqrt{n} = n^{1/2}$

67. $\sqrt[4]{3b} = (3b)^{1/4}$

69. $t^{1/3} = \sqrt[3]{t}$

71. $(3t)^{1/4} = \sqrt[4]{3t}$

Section 8.3 Quadratic Functions and Their Graphs

VOCABULARY

1. quadratic

3. vertex

CONCEPTS

5. a. a parabola b. (1, 0), (3, 0) c. (0, –3) d. (2, 1) e. $x = 2$

7. $f(x) = x^2$

x	$f(x)$
–2	4
–1	1
0	0
1	1
2	4

$g(x) = 2x^2$

x	$g(x)$
–2	8
–1	2
0	0
1	2
2	8

$h(x) = \frac{1}{2}x^2$

x	$h(x)$
–4	8
–2	2
0	0
2	2
4	8

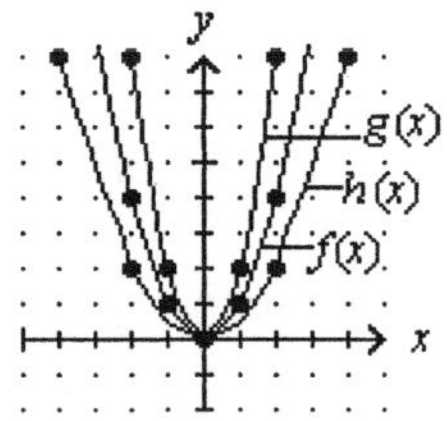

9. $f(x) = 4x^2$

x	$f(x)$
–2	16
–1	4
0	0
1	4
2	16

$g(x) = 4x^2 + 3$

Translate $f(x)$ upward 3 units.

$h(x) = 4x^2 - 2$

Translate $f(x)$ downward 2 units.

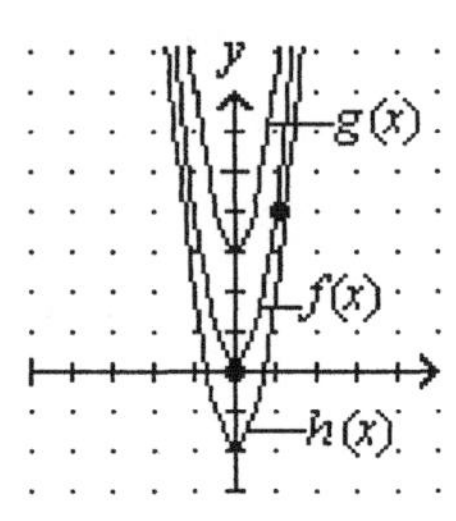

11. $f(x) = -3x^2$

x	$f(x)$
−2	−12
−1	−3
0	0
1	−3
2	−12

$g(x) = -3(x-2)^2 - 1$

Translate $f(x)$ 2 units to the right and then downward 1 unit.

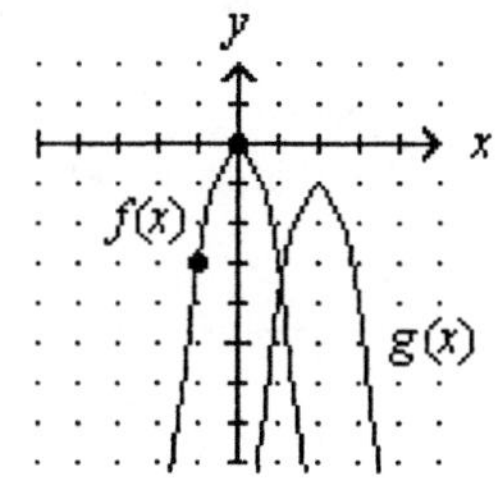

13. The values of x when $f(x) = \frac{x^2}{10} - \frac{x}{5} - \frac{3}{2}$ crosses the x–axis appear to be −3 and 5.
Substitute these values in $f(x)$ to verify that $f(x) = 0$.

$$f(-3) = \frac{(-3)^2}{10} - \frac{(-3)}{5} - \frac{3}{2} = \frac{9}{10} + \frac{6}{10} - \frac{15}{10} = 0 \qquad f(5) = \frac{(5)^2}{10} - \frac{(5)}{5} - \frac{3}{2} = \frac{25}{10} - \frac{10}{10} - \frac{15}{10} = 0$$

NOTATION

15. $h = -1$; The plus sign within the parentheses is a result of subtracting −1. $f(x) = 2[x - (-1)]^2 + 6$ in standard form.

PRACTICE

17.
$$\begin{aligned} f(x) &= (x-1)^2 + 2 \end{aligned}$$
Vertex: (1, 2)
Axis of symmetry: $x = 1$
Opens: Upward

19.
$$\begin{aligned} f(x) &= 2(x+3)^2 - 4 \\ f(x) &= 2[x-(-3)]^2 - 4 \end{aligned}$$
Vertex: (−3, −4)
Axis of symmetry: $x = -3$
Opens: Upward

21.
$$\begin{aligned} f(x) &= 2x^2 - 4x \\ f(x) &= 2(x^2 - 2x) \\ f(x) &= 2(x^2 - 2x + 1) - 2 \\ f(x) &= 2(x-1)^2 - 2 \end{aligned}$$
Vertex: (1, −2)
Axis of symmetry: $x = 1$
Opens: Upward

23.
$$\begin{aligned} f(x) &= -4x^2 + 16x + 5 \\ f(x) &= -4(x^2 - 4x) + 5 \\ f(x) &= -4(x^2 - 4x + 4) + 16 + 5 \\ f(x) &= -4(x-2)^2 + 21 \end{aligned}$$
Vertex: (2, 21)
Axis of symmetry: $x = 2$
Opens: Downward

25.
$$\begin{aligned} f(x) &= 3x^2 + 4x + 2 \\ f(x) &= 3(x^2 + \tfrac{4}{3}x) + 2 \\ f(x) &= 3(x^2 + \tfrac{4}{3}x + \tfrac{16}{36}) - \tfrac{4}{3} + 2 \\ f(x) &= 3(x + \tfrac{2}{3})^2 + \tfrac{2}{3} \\ f(x) &= 3[x - (-\tfrac{2}{3})] + \tfrac{2}{3} \end{aligned}$$
Vertex: $(-\frac{2}{3}, \frac{2}{3})$
Axis of symmetry: $x = -\frac{2}{3}$
Opens: Upward

27. $f(x) = (x-3)^2 + 2$

Vertex: (3, 2)
Axis of symmetry: $x = 3$

x	$f(x)$
1	6
2	3
3	**2**
4	3
5	6

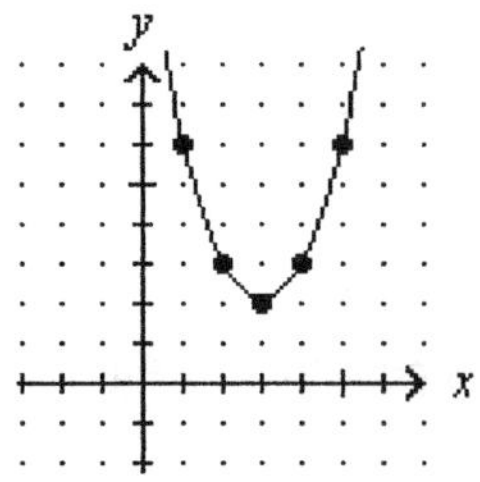

29. $f(x) = -(x-2)^2$

Vertex: (2, 0)
Axis of symmetry: $x = 2$

x	$f(x)$
0	−4
1	−1
2	**0**
3	−1
4	−4

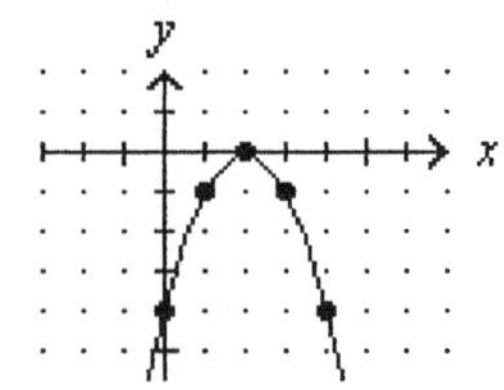

31. $f(x) = -2(x+3)^2 + 4$

Vertex: (−3, 4)
Axis of symmetry: $x = -3$

x	$f(x)$
−5	−4
−4	2
−3	**4**
−2	2
−1	−4

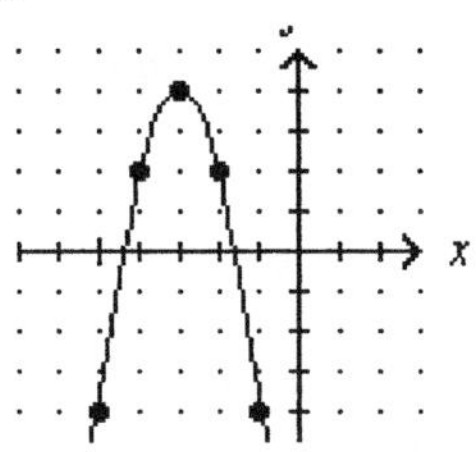

33. $f(x) = -(x+4)^2 - 1$

Vertex: (−4, −1)
Axis of symmetry: $x = -4$

x	$f(x)$
−6	−5
−5	−2
−4	**−1**
−3	−2
−2	−5

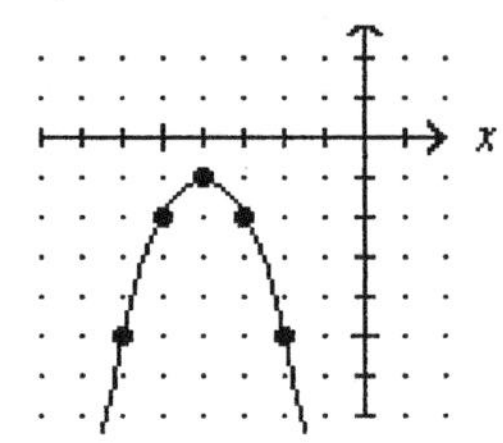

35. $f(x) = x^2 + x - 6$

$$x = -\frac{b}{2a} = -\frac{1}{2(1)} = -\frac{1}{2}$$

$$f\left(-\frac{b}{2a}\right) = f\left(-\frac{1}{2}\right) = \left(-\frac{1}{2}\right)^2 + \left(-\frac{1}{2}\right) - 6$$

$$= \frac{1}{4} - \frac{2}{4} - \frac{24}{4} = -\frac{25}{4}$$

Vertex: $(-\frac{1}{2}, -\frac{25}{4})$

Axis of symmetry: $x = -\frac{1}{2}$

x	$f(x)$
−3	0
−1	−6
$-\frac{1}{2}$	$-\frac{25}{4}$
0	−6
2	0

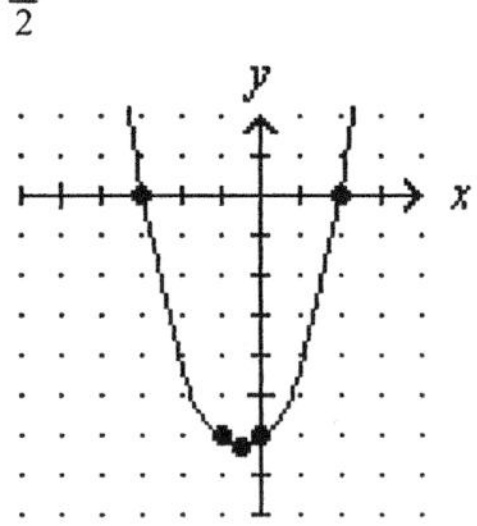

37. $f(x) = 3x^2 - 12x + 10$

$$x = -\frac{b}{2a} = -\frac{-12}{2(3)} = \frac{12}{6} = 2$$

$$f\left(-\frac{b}{2a}\right) = f(2) = 3(2)^2 - 12(2) + 10$$

$$= 12 - 24 + 10 = -2$$

Vertex: (2, −2)
Axis of symmetry: $x = 2$

x	$f(x)$
0	10
1	1
2	**−2**
3	1
4	10

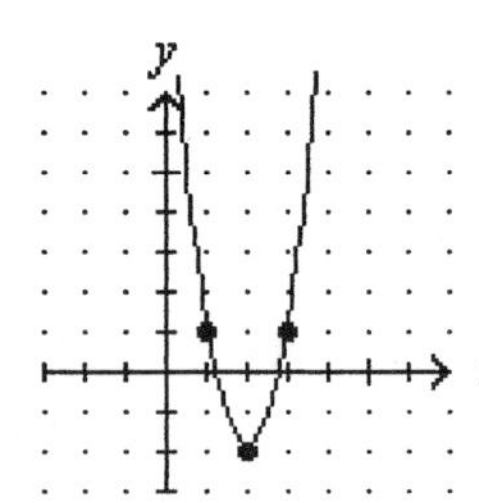

39. $f(x) = 4x^2 + 24x + 37$

$$x = -\frac{b}{2a} = -\frac{24}{2(4)} = -3$$

$$f\left(-\frac{b}{2a}\right) = f(-3) = 4(-3)^2 + 24(-3) + 37$$
$$= 36 - 72 + 37 = 1$$

Vertex: $(-3, 1)$
Axis of symmetry: $x = -3$

x	$f(x)$
−5	17
−4	5
−3	**1**
−2	5
−1	17

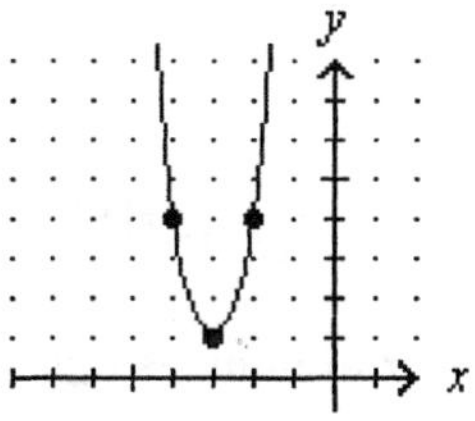

41. $f(x) = -x^2 + 2x + 3$

$$x = -\frac{b}{2a} = -\frac{2}{2(-1)} = 1$$

$$f\left(-\frac{b}{2a}\right) = f(1) = -(1)^2 + 2(1) + 3 = -1 + 2 + 3 = 4$$

Vertex: $(1, 4)$
Axis of symmetry: $x = 1$

x	$f(x)$
−1	0
0	3
1	**4**
2	3
3	0

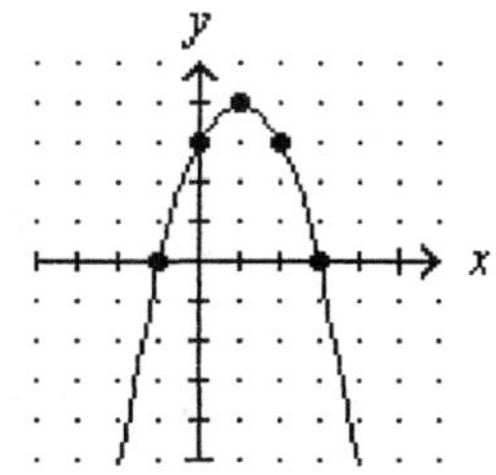

43. $f(x) = -3x^2 + 2x$

$$x = -\frac{b}{2a} = -\frac{2}{2(-3)} = \frac{1}{3}$$

$$f\left(-\frac{b}{2a}\right) = f\left(\frac{1}{3}\right) = -3\left(\frac{1}{3}\right)^2 + 2\left(\frac{1}{3}\right) = -\frac{1}{3} + \frac{2}{3} = \frac{1}{3}$$

Vertex: $(\frac{1}{3}, \frac{1}{3})$
Axis of symmetry: $x = \frac{1}{3}$

x	$f(x)$
−1	−5
0	0
$\frac{1}{3}$	$\frac{1}{3}$
1	−1
2	−8

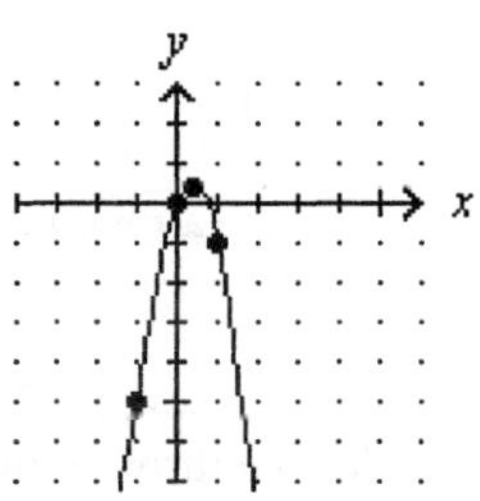

45. $f(x) = -12x^2 - 6x + 6$

$$x = -\frac{b}{2a} = -\frac{-6}{2(-12)} = -\frac{1}{4}$$

$$f\left(-\frac{b}{2a}\right) = f\left(-\frac{1}{4}\right) = -12\left(-\frac{1}{4}\right)^2 - 6\left(-\frac{1}{4}\right) + 6$$
$$= -\frac{3}{4} + \frac{3}{2} + 6 = -\frac{3}{4} + \frac{6}{4} + \frac{24}{4} = \frac{27}{4}$$

Vertex: $(-\frac{1}{4}, \frac{27}{4})$
Axis of symmetry: $x =$

x	$f(x)$
−2	−30
−1	0
$-\frac{1}{4}$	$\frac{27}{4}$
0	6
1	12

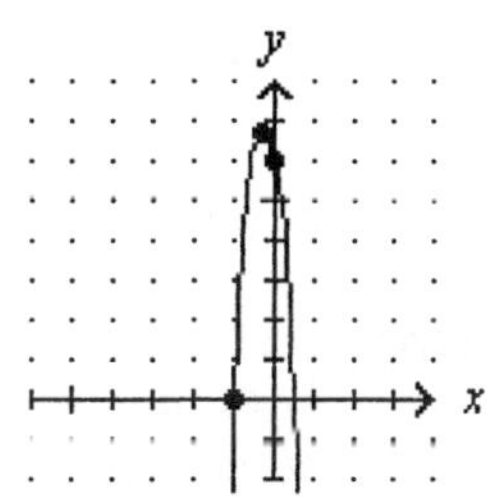

For exercises 47 – 54, follow the calculator instructions to determine the answers. The expected answer is given.

47. (0.25, 0.88) 49. (0.50, 7.25) 51. 2, −3 53. −1.85, 3.25

APPLICATIONS

55. The fireworks should explode when it reaches the vertex of the parabola given by the equation $s = 120t - 16t^2$.

$$t = -\frac{b}{2a} = -\frac{120}{2(-16)} = 3.75 \text{ sec}$$

$$s\left(-\frac{b}{2a}\right) = s(3.75) = -16(3.75)^2 + 120(3.75) = 225 \text{ ft}$$

The fireworks shell should explode 3.75 seconds after being fired at a height of 225 feet.

57. Let w represent the width of the field. Then the length is represented by $1000 - 2w$. (The total length of the fence minus the fence needed for the two ends.) The formula for the area of the field is $A = w(1000 - 2w)$. The w–value of the vertex of the equation of the area will be the width that gives the maximum area.

$$A = w(1000 - 2w) = 1000w - 2w^2$$

$$w = -\frac{b}{2a} = -\frac{1000}{2(-2)} = 250 \text{ ft}, \qquad \text{length} = 1000 - 2w = 1000 - 2(250) = 500 \text{ ft}$$

The dimensions of the fields would be 250 ft by 500 ft.

59. Find the vertex of $C(n) = 2.2n^2 - 66n + 655$.

$$n = -\frac{b}{2a} = -\frac{-66}{2(2.2)} = 15 \text{ min} \qquad C\left(-\frac{b}{2a}\right) = C(15) = 2.2(15)^2 - 66(15) + 655 = \$160$$

The cost of running the machine is a minimum when the number of minutes it is used are 15. This results in a cost of \$160.

61. Find the vertex of $N(x) = -0.0534x^2 + 0.337x + 0.969$.

$$x = -\frac{b}{2a} = -\frac{0.337}{2(-0.0534)} = 3.16 \text{ years}$$

$$N\left(-\frac{b}{2a}\right) = N(3.16) = -0.0534(3.16)^2 + 0.337(3.16) + 0.969 = 1.5 \text{ million}$$

The highest level of personnel was 3 years after 1965, or in 1968. The personnel level was 1.5 million. This time period was when the U.S. involvement in the war in Vietnam was at its peak.

63. Find the vertex of $R = -\frac{1}{5}x^2 + 80x - 1000$.

$$x = -\frac{b}{2a} = -\frac{80}{2(-\frac{1}{5})} = 200 \text{ stereos} \qquad R\left(-\frac{b}{2a}\right) = C(200) = -\frac{1}{5}(200)^2 + 80(200) - 1000 = \$7000$$

The number of stereos that needs to be sold is 200. This would produce a revenue of \$7000.

WRITING

65. A stream of water from a drinking fountain rises and falls in a parabolic shape. The maximum height of the water is at the vertex. The axis of symmetry is the vertical straight line through the vertex.

67. The vertex appears to be at (1, 1) because there are matching values to the right and left of (1, 1).

69. The mirror illustrates that a parabola is symmetric about its axis of symmetry.

REVIEW

71. $\sqrt{8a}\sqrt{2a^3b} = \sqrt{16a^4b} = 4a^2\sqrt{b}$

73. $\frac{\sqrt{3}}{\sqrt{50}} = \frac{\sqrt{3}\sqrt{2}}{\sqrt{50}\sqrt{2}} = \frac{\sqrt{6}}{\sqrt{100}} = \frac{\sqrt{6}}{10}$

75. $3\left(\sqrt{5b} - \sqrt{3}\right)^2 = 3\left(5b - 2\sqrt{15b} + 3\right) = 15b - 6\sqrt{15b} + 9$

Section 8.4 Complex Numbers

VOCABULARY

1. imaginary
3. real; imaginary
5. i or $\sqrt{-1}$

CONCEPTS

7. $\sqrt{-1}$
9. $-i$
11. FOIL

13.
$$x^2 + 50 = 0$$
$$\left(-5i\sqrt{2}\right)^2 + 50 \overset{?}{=} 0$$
$$25i^2(2) + 50 \overset{?}{=} 0$$
$$50(-1) + 50 \overset{?}{=} 0$$
$$-50 + 50 = 0$$

It is a solution.

15.

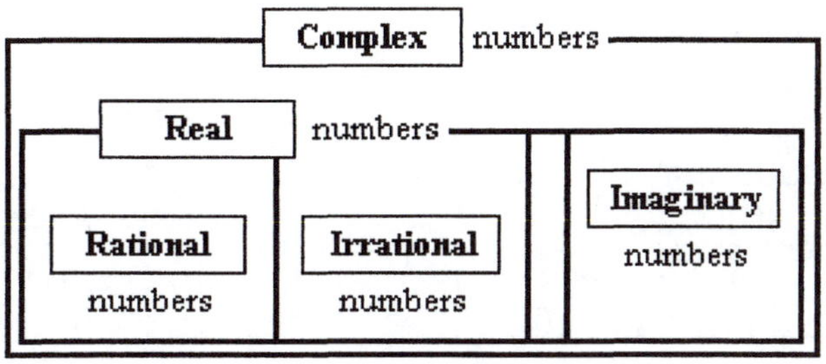

17. a.
$$x^2 - 1 = 0$$
$$x^2 = 1$$
$$\sqrt{x^2} = \sqrt{1}$$
$$x = \pm 1$$

b.
$$x^2 + 1 = 0$$
$$x^2 = -1$$
$$\sqrt{x^2} = \sqrt{-1}$$
$$x = \pm i$$

NOTATION

19.
$$(3 + 2i)(3 - i) = \mathbf{9} - 3i + \mathbf{6i} - 2i^2$$
$$= 9 + 3i + \mathbf{2}$$
$$= 11 + 3i$$

21. a. true b. false c. false d. false e. true f. false

PRACTICE

23. $\sqrt{-9} = \sqrt{-1}\sqrt{9} = i \cdot 3 = 3i$

25. $\sqrt{-7} = \sqrt{-1}\sqrt{7} = i \cdot \sqrt{7} = i\sqrt{7}$

27. $\sqrt{-24} = \sqrt{-1}\sqrt{4 \cdot 6} = i \cdot 2\sqrt{6} = 2i\sqrt{6}$

29. $-\sqrt{-24} = -\sqrt{-1}\sqrt{4 \cdot 6} = -i \cdot 2\sqrt{6} = -2i\sqrt{6}$

31. $5\sqrt{-81} = 5\sqrt{-1}\sqrt{81} = 5i \cdot 9 = 45i$

33. $\sqrt{-\frac{25}{9}} = \sqrt{-1 \cdot \frac{25}{9}} = \frac{\sqrt{-1}\sqrt{25}}{\sqrt{9}} = \frac{5}{3}i$

35. $\sqrt{-1}\sqrt{-36} = i\sqrt{-1}\sqrt{36} = i \cdot i \cdot 6 = i^2 \cdot 6 = -6$

37. $\sqrt{-2}\sqrt{-6} = i\sqrt{2}\left(i\sqrt{6}\right) = i^2 \cdot \sqrt{12} = -2\sqrt{3}$

39. $\dfrac{\sqrt{-25}}{\sqrt{-64}} = \dfrac{5i}{8i} = \dfrac{5}{8}$

41. $-\dfrac{\sqrt{-400}}{\sqrt{-1}} = -\dfrac{20i}{i} = -20$

43.
$$x^2 + 9 = 0$$
$$x^2 = -9$$
$$x = \pm\sqrt{-9} = \pm 3i$$

45.
$$3x^2 = -16$$
$$x^2 = -\frac{16}{3}$$
$$x = \pm\sqrt{-\frac{16}{3}} = \pm\frac{4\sqrt{3}}{\sqrt{3}\sqrt{3}}i = \pm\frac{4\sqrt{3}}{3}i$$

47. $x^2 + 2x + 2 = 0$

$$x = \frac{-2 \pm \sqrt{2^2 - 4(1)(2)}}{2(1)} = \frac{-2 \pm \sqrt{4-8}}{2}$$
$$= \frac{-2 \pm \sqrt{-4}}{2} = \frac{-2 \pm 2i}{2}$$
$$= \frac{2(-1 \pm i)}{2} = -1 \pm i$$

49. $2x^2 + x + 1 = 0$

$$x = \frac{-1 \pm \sqrt{1^2 - 4(2)(1)}}{2(2)} = \frac{-1 \pm \sqrt{1-8}}{4}$$
$$= \frac{-1 \pm \sqrt{-7}}{4} = \frac{-1 \pm i\sqrt{7}}{4}$$
$$= -\frac{1}{4} \pm \frac{\sqrt{7}}{4}i$$

51. $3x^2 - 4x = -2$
$3x^2 - 4x + 2 = 0$

$$x = \frac{-(-4) \pm \sqrt{(-4)^2 - 4(3)(2)}}{2(3)}$$
$$= \frac{4 \pm \sqrt{16-24}}{6} = \frac{4 \pm \sqrt{-8}}{6} = \frac{4 \pm 2i\sqrt{2}}{6}$$
$$= \frac{4}{6} \pm \frac{2\sqrt{2}}{6}i = \frac{2}{3} \pm \frac{\sqrt{2}}{3}i$$

53. $3x^2 - 2x = -3$
$3x^2 - 2x + 3 = 0$

$$x = \frac{-(-2) \pm \sqrt{(-2)^2 - 4(3)(3)}}{2(3)}$$
$$= \frac{2 \pm \sqrt{4-36}}{6} = \frac{2 \pm \sqrt{-32}}{6} = \frac{2 \pm 4i\sqrt{2}}{6}$$
$$= \frac{2}{6} \pm \frac{4\sqrt{2}}{6}i = \frac{1}{3} \pm \frac{2\sqrt{2}}{3}i$$

55. $(3 + 4i) + (5 - 6i) = (3 + 5) + (4 - 6)i = 8 - 2i$

57. $(7 - 3i) - (4 + 2i) = (7 - 4) + (-3 - 2)i = 3 - 5i$

59. $(6 - i) + (9 + 3i) = (6 + 9) + (-1 + 3)i = 15 + 2i$

61. $(8 + \sqrt{-25}) + (7 + \sqrt{-4}) = (8 + 5i) + (7 + 2i) = (8 + 7) + (5 + 2)i = 15 + 7i$

63. $3(2 - i) = 3(2) - 3(i) = 6 - 3i$

65. $-5i(5 - 5i) = -5i(5) - 5i(-5i) = -25i + 25i^2 = -25i + 25(-1) = -25i - 25 = -25 - 25i$

67. $(2 + i)(3 - i) = 6 - 2i + 3i - i^2 = 6 + i - (-1) = 6 + i + 1 = 7 + i$

69. $(3 - 2i)(2 + 3i) = 6 + 9i - 4i - 6i^2 = 6 + 5i - 6(-1) = 6 + 5i + 6 = 12 + 5i$

71. $(4 + i)(3 - i) = 12 - 4i + 3i - i^2 = 12 - i - (-1) = 12 - i + 1 = 13 - i$

73. $(2 - \sqrt{-16})(3 + \sqrt{-4}) = (2 - 4i)(3 + 2i) = 6 + 4i - 12i - 8i^2 = 6 - 8i + 8 = 14 - 8i$

75. $(2 + \sqrt{2}\,i)(3 - \sqrt{2}\,i) = 6 - 2\sqrt{2}\,i + 3\sqrt{2}\,i - 2i^2 = 6 + \sqrt{2}\,i + 2 = 8 + i\sqrt{2}$

77. $(2 + i)^2 = 4 + 2(2i) + i^2 = 4 + 4i - 1 = 3 + 4i$

79. $\frac{1}{i} = \frac{1}{i} \cdot \frac{i}{i} = \frac{i}{i^2} = \frac{i}{-1} = 0 - i$

81. $\frac{4}{5i^3} = \frac{4}{5i^3} \cdot \frac{i}{i} = \frac{4i}{5i^4} = \frac{4i}{5} = 0 + \frac{4}{5}i$

83. $\frac{3i}{8\sqrt{-9}} = \frac{3i}{8(3i)} = \frac{1}{8} = \frac{1}{8} + 0i$

85. $\frac{-3}{5i^5} = \frac{-3}{5i^4 i} = \frac{-3}{5i} \cdot \frac{i}{i} = \frac{-3i}{5i^2} = \frac{-3i}{-5} = 0 + \frac{3}{5}i$

87. $\frac{5}{2-i} = \frac{5}{2-i} \cdot \frac{2+i}{2+i} = \frac{5(2+i)}{4-i^2} = \frac{5(2+i)}{4+1} = \frac{5(2+i)}{5} = 2 + i$

89. $\frac{3}{5+i} = \frac{3}{5+i} \cdot \frac{5-i}{5-i} = \frac{3(5-i)}{25-i^2} = \frac{15-3i}{25+1} = \frac{15-3i}{26} = \frac{15}{26} - \frac{3}{26}i$

91. $\frac{-12}{7-\sqrt{-1}} = \frac{-12}{7-i} = \frac{-12}{7-i} \cdot \frac{7+i}{7+i} = \frac{-12(7+i)}{49-i^2} = \frac{-84-12i}{49+1} = \frac{2(-42-6i)}{50} = -\frac{42}{25} - \frac{6}{25}i$

93. $\frac{5i}{6+2i} = \frac{5i}{6+2i} \cdot \frac{6-2i}{6-2i} = \frac{5i(6-2i)}{36-4i^2} = \frac{30i-10i^2}{36+4} = \frac{30i+10}{40} = \frac{10(1+3i)}{40} = \frac{1}{4} + \frac{3}{4}i$

95. $\frac{-2i}{3+2i} = \frac{-2i}{3+2i} \cdot \frac{3-2i}{3-2i} = \frac{-2i(3-2i)}{9-4i^2} = \frac{-6i+4i^2}{9+4} = \frac{-6i-4}{13} = -\frac{4}{13} - \frac{6}{13}i$

97. $\frac{3-2i}{3+2i} = \frac{3-2i}{3+2i} \cdot \frac{3-2i}{3-2i} = \frac{(3-2i)(3-2i)}{9-4i^2} = \frac{9-12i+4i^2}{9+4} = \frac{9-12i-4}{13} = \frac{5}{13} - \frac{12}{13}i$

99. $\frac{3+2i}{3+i} = \frac{3+2i}{3+i} \cdot \frac{3-i}{3-i} = \frac{(3+2i)(3-i)}{9-i^2} = \frac{9+3i-2i^2}{9+1} = \frac{9+3i+2}{10} = \frac{11}{10} + \frac{3}{10}i$

101. $\frac{\sqrt{5}-\sqrt{3}i}{\sqrt{5}+\sqrt{3}i} = \frac{\sqrt{5}-\sqrt{3}i}{\sqrt{5}+\sqrt{3}i} \cdot \frac{\sqrt{5}-\sqrt{3}i}{\sqrt{5}-\sqrt{3}i} = \frac{5-2\sqrt{15}i+3i^2}{5-3i^2} = \frac{5-2\sqrt{15}i-3}{5+3} = \frac{2-2\sqrt{15}i}{8} = \frac{2(1-\sqrt{15}i)}{8} = \frac{1}{4} - \frac{\sqrt{15}}{4}i$

103. $i^{21} = i^{4 \bullet 5} \bullet i = 1^5 \bullet i = i$

105. $i^{27} = i^{4 \bullet 6} \bullet i^3 = 1^6(-i) = -i$

107. $i^{100} = i^{4 \bullet 25} = 1^{25} = 1$

109. $i^{97} = i^{4 \bullet 24} \bullet i^1 = 1^{24}(i) = i$

APPLICATIONS

111. Step 1: $i^2 + i = -1 + i$
Step 2: $(-1+i)^2 + i = 1 - 2i + i^2 + i = 1 - 2i - 1 + i = -i$
Step 3: $(-i)^2 + i = i^2 + i = -1 + i$

WRITING

113. It is a number containing the square root of –1.

REVIEW

115. 30°–60°–90° triangle.
Hypotenuse is twice as long as shorter leg.

$h = 2x$
$30 = 2x$
$15 = x$ The shorter leg is $x = 15$ units.

The long leg is the short leg times $\sqrt{3} = 15\sqrt{3}$ units

117. **Analysis**: Find the speed of the wind.

Equation: Let x represent the speed of the wind. Then $200 + x$ represents the speed of the plane going with a tail wind (with the wind) and $200 - x$ represents the speed of the plane going into the wind (against the wind).

If $r \cdot t = d$, then $t = \frac{d}{r}$

	r	• t	= d
With the wind	$200 + x$	$\frac{330}{200+x}$	330
Against the wind	$200 - x$	$\frac{330}{200-x}$	330

With the wind time plus against the wind time. equals total time

$$\frac{330}{200+x} + \frac{330}{200-x} = 3\frac{1}{3}$$

Solve:

$$\frac{330}{200+x} + \frac{330}{200-x} = \frac{10}{3}$$

$$3(200+x)(200-x)\left(\frac{330}{200+x} + \frac{330}{200-x}\right) = 3(200+x)(200-x)\left(\frac{10}{3}\right)$$

$$990(200-x) + 990(200+x) = 10(40{,}000 - x^2)$$

$$198{,}000 - 900x + 198{,}000 + 900x = 400{,}000 - 10x^2$$

$$396{,}000 = 400{,}000 - 10x^2$$

$$10x^2 = 4000$$

$$x^2 = 400$$

$$x = 20$$

Conclusion: The speed of the wind was $x = 20$ mph.

Section 8.5 The Discriminant and Equations That Can Be Written in Quadratic Form

VOCABULARY

1. $b^2 - 4ac$

CONCEPTS

3. conjugates

5. rational; unequal

7. a. $x^4 = (x^2)^2$ b. yes

NOTATION

9. $b^2 - 4ac = \mathbf{5}^2 - 4(1)(\mathbf{6})$
$= 25 - \mathbf{24}$
$= 1$

…the solutions are **rational** numbers…

PRACTICE

11. $4x^2 - 4x + 1 = 0$; $b^2 - 4ac = (-4)^2 - 4(4)(1) = 16 - 16 = 0$. Solutions are rational and equal.

13. $5x^2 + x + 2 = 0$; $b^2 - 4ac = 1^2 - 4(5)(2) = 1 - 40 = -39$. Solutions are complex conjugates.

15. $2x^2 - 4x + 1 = 0$ (rearrange); $b^2 - 4ac = (-4)^2 - 4(2)(1) = 16 - 8 = 8$. Solutions are irrational and unequal.

17. $x(2x - 3) = 20$ or $2x^2 - 3x = 20$ or $2x^2 - 3x - 20 = 0$ (rearranged);
$b^2 - 4ac = (-3)^2 - 4(2)(-20) = 9 + 160 = 169$. Solutions are rational and unequal.

19. $1492x^2 + 1776x - 2000 = 0$; $b^2 - 4ac = 1776^2 - 4(1492)(-2000) =$ positive number.
Solutions are real numbers.

21.
$$\begin{aligned} x^4 - 17x^2 + 16 &= 0 \\ (x^2)^2 - 17x^2 + 16 &= 0 && \text{Write } x^4 \text{ as } (x^2)^2. \\ y^2 - 17y + 16 &= 0 && \text{Let } y = x^2. \\ (y-16)(y-1) &= 0 && \text{Factor and solve.} \end{aligned}$$

$y - 16 = 0$ or $y - 1 = 0$
$y = 16$ $\quad$ $y = 1$

Undo the substitution.

$x^2 = 16$ or $x^2 = 1$
$x = \pm 4$ $\quad$ $x = \pm 1$

The solutions are $-4, 4, -1, 1$.

23.
$$\begin{aligned} x^4 &= 6x^2 - 5 \\ x^4 - 6x^2 + 5 &= 0 \\ (x^2)^2 - 6x^2 + 5 &= 0 && \text{Write } x^4 \text{ as } (x^2)^2. \\ y^2 - 6y + 5 &= 0 && \text{Let } y = x^2. \\ (y-5)(y-1) &= 0 && \text{Factor and solve.} \end{aligned}$$

$y - 5 = 0$ or $y - 1 = 0$
$y = 5$ $\quad$ $y = 1$

Undo the substitution.

$x^2 = 5$ or $x^2 = 1$
$x = \pm\sqrt{5}$ $\quad$ $x = \pm 1$

The solutions are $-\sqrt{5}, \sqrt{5}, -1, 1$

25.
$$\begin{aligned} t^4 + 3t^2 &= 28 \\ t^4 + 3t^2 - 28 &= 0 \\ (t^2)^2 + 3t^2 - 28 &= 0 && \text{Write } t^4 \text{ as } (t^2)^2. \\ y^2 + 3y - 28 &= 0 && \text{Let } y = t^2. \\ (y+7)(y-4) &= 0 && \text{Factor and solve.} \end{aligned}$$

$y + 7 = 0$ or $y - 4 = 0$
$y = -7$ $\quad$ $y = 4$

Undo the substitution.

$t^2 = -7$ or $t^2 = 4$
$t = \pm i\sqrt{7}$ $\quad$ $t = \pm 2$

The solutions are $-i\sqrt{7}, i\sqrt{7}, -2, 2$.

27.
$$\begin{aligned} 2x + \sqrt{x} - 3 &= 0 \\ 2(\sqrt{x})^2 + \sqrt{x} - 3 &= 0 && \text{Write } x \text{ as } (\sqrt{x})^2. \\ 2y^2 + y - 3 &= 0 && \text{Let } y = \sqrt{x}. \\ (2y+3)(y-1) &= 0 && \text{Factor and solve} \end{aligned}$$

$2y + 3 = 0$ or $y - 1 = 0$
$y = -\frac{3}{2}$ $\quad$ $y = 1$

Undo the substitution.

$\sqrt{x} = -\frac{3}{2}$ or $\sqrt{x} = 1$
$x = \frac{9}{4}$ $\quad$ $x = 1$

Check:

$2\left(\frac{9}{4}\right) + \sqrt{\frac{9}{4}} - 3$
$= \frac{9}{2} + \frac{3}{2} - \frac{6}{2} \neq 0$

Check:

$2(1) + \sqrt{1} - 3$
$= 2 + 1 - 3 = 0$

The solution is 1.

29.

$$\begin{aligned} 3x + 5\sqrt{x} + 2 &= 0 \\ 3(\sqrt{x})^2 + 5\sqrt{x} + 2 &= 0 && \text{Write } x \text{ as } (\sqrt{x})^2. \\ 3y^2 + 5y + 2 &= 0 && \text{Let } y = \sqrt{x}. \\ (3y+2)(y+1) &= 0 && \text{Factor and solve.} \end{aligned}$$

$3y + 2 = 0$, $y = -\frac{2}{3}$ or $y + 1 = 0$, $y = -1$

Undo the substitution.

$\sqrt{x} = -\frac{2}{3}$, $x = \frac{4}{9}$ or $\sqrt{x} = -1$, $x = 1$

Check:

$$3\left(\tfrac{4}{9}\right) + 5\sqrt{\tfrac{4}{9}} + 2 = \tfrac{4}{3} + \tfrac{10}{3} + \tfrac{6}{3} \neq 0$$

Check:

$$3(1) + 5\sqrt{1} + 2 = 3 + 5 + 2 \neq 0$$

No solution.

31.

$$\begin{aligned} x - 6\sqrt{x} &= -8 \\ x - 6\sqrt{x} + 8 &= 0 \\ (\sqrt{x})^2 - 6\sqrt{x} + 8 &= 0 && \text{Write } x \text{ as } (\sqrt{x})^2. \\ y^2 - 6y + 8 &= 0 && \text{Let } y = \sqrt{x}. \\ (y-4)(y-2) &= 0 && \text{Factor and solve.} \end{aligned}$$

$y - 4 = 0$, $y = 4$ or $y - 2 = 0$, $y = 2$

Undo the substitution.

$\sqrt{x} = 4$, $x = 16$ or $\sqrt{x} = 2$, $x = 4$

Check:

$$(16) - 6\sqrt{16} = 16 - 24 = -8$$

Check:

$$(4) - 6\sqrt{4} = 4 - 12 = -8$$

The solutions are 16 and 4.

33.

$$\begin{aligned} x^{2/3} + 5x^{1/3} + 6 &= 0 \\ (x^{1/3})^2 + 5x^{1/3} + 6 &= 0 && \text{Write } x^{2/3} \text{ as } (x^{1/3})^2. \\ y^2 + 5y + 6 &= 0 && \text{Let } y = x^{1/3}. \\ (y+3)(y+2) &= 0 && \text{Factor and solve.} \end{aligned}$$

$y + 3 = 0$, $y = -3$ or $y + 2 = 0$, $y = -2$

Undo the substitution.

$x^{1/3} = -3$, $x = -27$ or $x^{1/3} = -2$, $x = -8$

The solutions are –27 and –8.

35.

$$\begin{aligned} a^{2/3} - 2a^{1/3} - 3 &= 0 \\ (a^{1/3})^2 - 2a^{1/3} - 3 &= 0 && \text{Write } a^{2/3} \text{ as } (a^{1/3})^2. \\ y^2 - 2y - 3 &= 0 && \text{Let } y = a^{1/3}. \\ (y-3)(y+1) &= 0 && \text{Factor and solve.} \end{aligned}$$

$y - 3 = 0$, $y = 3$ or $y + 1 = 0$, $y = -1$

Undo the substitution.

$a^{1/3} = 3$, $a = 27$ or $a^{1/3} = -1$, $a = -1$

The solutions are 27 and –1.

37. $2(2x+1)^2 - 7(2x+1) + 6 = 0$

$$\begin{aligned} 2y^2 - 7y + 6 &= 0 && \text{Let } y = (2x+1). \\ (2y-3)(y-2) &= 0 && \text{Factor and solve.} \end{aligned}$$

$2y - 3 = 0$, $y = \frac{3}{2}$ or $y - 2 = 0$, $y = 2$

Undo the substitution.

$2x + 1 = \frac{3}{2}$, $2x = \frac{1}{2}$, $x = \frac{1}{4}$ or $2x + 1 = 2$, $2x = 1$, $x = \frac{1}{2}$

The solutions are $\frac{1}{4}$ and $\frac{1}{2}$.

39. $(c+1)^2 - 4(c+1) - 8 = 0$

$$y^2 - 4y - 8 = 0 \qquad \text{Let } y = (c+1).$$

Use quadratic equation.

$$y = \frac{-(-4) \pm \sqrt{(-4)^2 - 4(1)(-8)}}{2(1)} = \frac{4 \pm \sqrt{16+32}}{2}$$

$$= \frac{4 \pm \sqrt{48}}{2} = \frac{4 \pm 4\sqrt{3}}{2} = 2 \pm 2\sqrt{3}$$

Undo the substitution.

$$\begin{aligned} c + 1 &= 2 \pm 2\sqrt{3} \\ c &= 1 \pm 2\sqrt{3} \end{aligned}$$

The solutions are $1 + 2\sqrt{3}$ and $1 - 2\sqrt{3}$.

41.
$$x + 5 + \frac{4}{x} = 0$$

$$\begin{aligned} x^2 + 5x + 4 &= 0 && \text{Multiply by } x. \\ (x+4)(x+1) &= 0 && \text{Factor and solve.} \end{aligned}$$

$$\begin{aligned} x + 4 &= 0 \quad \text{or} \quad & x + 1 &= 0 \\ x &= -4 & x &= -1 \end{aligned}$$

The solutions are –4 and –1.

43.
$$\frac{1}{x+2} + \frac{24}{x+3} = 13 \qquad \text{Multiply by LCD.}$$

$$\begin{aligned} 1(x+3) + 24(x+2) &= 13(x^2 + 5x + 6) \\ x + 3 + 24x + 48 &= 13x^2 + 65x + 78 \\ 0 &= 13x^2 + 40x + 27 \end{aligned}$$

$$x = \frac{-40 \pm \sqrt{40^2 - 4(13)(27)}}{2(13)}$$

$$= \frac{-40 \pm \sqrt{1600 - 1404}}{26} = \frac{-40 \pm \sqrt{196}}{26} = \frac{-40 \pm 14}{26}$$

$$x = \frac{-40 + 14}{26} = -1 \quad \text{or} \quad x = \frac{-40 - 14}{26} = -\frac{27}{13}$$

The solutions are –1 and $-\frac{27}{13}$.

45.
$$\frac{2}{x-1} + \frac{1}{x+1} = 3 \qquad \text{Multiply by LCD.}$$

$$\begin{aligned} 2(x+1) + 1(x-1) &= 3(x^2 - 1) \\ 2x + 2 + x - 1 &= 3x^2 - 3 \\ 0 &= 3x^2 - 3x - 4 \end{aligned}$$

$$x = \frac{-(-3) \pm \sqrt{(-3)^2 - 4(3)(-4)}}{2(3)}$$

$$= \frac{3 \pm \sqrt{9 + 48}}{6} = \frac{3 \pm \sqrt{57}}{6}$$

The solutions are $\frac{3 + \sqrt{57}}{6}$ and $\frac{3 - \sqrt{57}}{6}$.

47.
$$\begin{aligned} x^{-4} - 2x^{-2} + 1 &= 0 \\ (x^{-2})^2 - 2x^{-2} + 1 &= 0 && \text{Write } x^{-4} \text{ as } (x^{-2})^2. \\ y^2 - 2y + 1 &= 0 && \text{Let } y = x^{-2}. \\ (y-1)(y-1) &= 0 && \text{Factor and solve.} \end{aligned}$$

$$\begin{aligned} y - 1 &= 0 \quad \text{or} \quad & y - 1 &= 0 \\ y &= 1 & y &= 1 \end{aligned}$$

Undo the substitution.

$$\begin{aligned} x^{-2} = \frac{1}{x^2} &= 1 \quad \text{or} \quad & x^{-2} = \frac{1}{x^2} &= 1 \\ 1 &= x^2 & 1 &= x^2 \\ \pm 1 &= x & \pm 1 &= x \end{aligned}$$

The solutions are 1, –1, 1, and –1.

49.
$$x + \frac{2}{x-2} = 0 \qquad \text{Multiply by LCD.}$$

$$\begin{aligned} x(x-2) + 2 &= 0 \\ x^2 - 2x + 2 &= 0 \end{aligned}$$

$$x = \frac{-(-2) \pm \sqrt{(-2)^2 - 4(1)(2)}}{2(1)} = \frac{2 \pm \sqrt{4 - 8}}{2}$$

$$= \frac{2 \pm \sqrt{-4}}{2} = \frac{2 \pm 2i}{2} = \frac{2(1 \pm i)}{2} = 1 \pm i$$

The solutions are $1 + i$ and $1 - i$.

APPLICATIONS

51. Find h when $w = 34$ in. and $r = 18$ in.

$$r = \frac{4h^2 + w^2}{8h}$$

$$18 = \frac{4h^2 + (34)^2}{8h}$$

$$144h = 4h^2 + 1156$$

$$0 = 4h^2 - 144h + 1156$$

$$0 = h^2 - 36h + 289$$

Use the quadratic formula:

$$h = \frac{-(-36) \pm \sqrt{(-36)^2 - 4(1)(289)}}{2(1)}$$

$$= \frac{36 \pm \sqrt{1296 - 1156}}{2} = \frac{36 \pm \sqrt{140}}{2}$$

$$h = \frac{36 + 11.83}{2} = 23.9 \text{ in.} \quad \text{or} \quad h = \frac{36 - 11.83}{2} = 12.1 \text{ in.}$$

Discard the 23.9 in. solution because it is more than the radius. The height of the flowers would be 12.1 in.

53. **Analysis**: Find the rate at which the woman originally drove her snowmobile

Equation: Let r represent the original rate of the snowmobile. Then $r + 20$ represents the increased rate of the snowmobile.

If $r \bullet t = d$, then $t = \frac{d}{r}$

	r	$\bullet\ t$	$= d$
Original speed	r	$\frac{150}{r}$	150
Faster speed	$r+20$	$\frac{150}{r+20}$	150

Original time less Faster time equals 2 hours.

$$\frac{150}{r} - \frac{150}{r+20} = 2$$

Solve:

$$\frac{150}{r} - \frac{150}{r+20} = 2$$

$$r(r+20)\left(\frac{150}{r} - \frac{150}{r+20}\right) = 2r(r+20) \quad \text{Multiply both sides by LCD} = r(r+20).$$

$$150(r+20) - 150r = 2r^2 + 40r$$

$$150r + 3000 - 150r = 2r^2 + 40r$$

$$0 = 2r^2 + 40r - 3000$$

Divide both sides by 2.

$$0 = r^2 + 20r - 1500$$

$$0 = (r+50)(r-30)$$

$$r - 30 = 0 \quad \text{or} \quad r + 50 = 0$$

$$r = 30 \qquad r = -50$$

Discard the negative solution.

Conclusion: The original rate was $r = 30$ mph.

55. **Analysis**: Find the amount of time it will take to clear the grandstand through the east exit only.

Equation: Let x represent the number of minutes necessary to clear the grandstand through the east exit. Then $x - 4$ will represent the minutes to clear the grandstand through the west exit. In 1 minute, $\frac{1}{x}$ of the grandstand will be cleared through the east exit and $\frac{1}{x-4}$ of the grandstand will be cleared through the west exit. Using both exits, $\frac{1}{6}$ of the grandstand will be cleared each minute.

East exit cleared in 1 minute plus west exit cleared in 1 minute equals both exits cleared together in 1minute.

$$\frac{1}{x} + \frac{1}{x-4} = \frac{1}{6}$$

Solve:

$$\frac{1}{x} + \frac{1}{x-4} = \frac{1}{6}$$

$$6x(x-4)\left(\frac{1}{x} + \frac{1}{x-4}\right) = 6x(x-4)\left(\frac{1}{6}\right)$$

$$6(x-4) + 6x = x^2 - 4x$$

$$6x - 24 + 6x = x^2 - 4x$$

$$0 = x^2 - 16x + 24$$

$$x = \frac{-(-16) \pm \sqrt{(-16)^2 - 4(1)(24)}}{2(1)}$$

$$= \frac{16 \pm \sqrt{256 - 96}}{2} = \frac{16 \pm \sqrt{160}}{2}$$

$$x = \frac{16 + 12.65}{2} = 14.3 \quad \text{or} \quad x = \frac{16 - 12.65}{2} = 1.8$$

Discard 1.8 because $x - 4$ would be negative.

Conclusion: It will take 14.3 minutes to clear the grandstand through the east exit only.

WRITING

57. Determine the value of the discriminant, $b^2 - 4ac$. If it is 0, the roots will be rational and equal. If it is a positive perfect square, the roots will be rational and unequal. If it is a positive real number that is not a perfect square, the roots will be irrational and unequal. If it is a negative number, the roots will be complex and conjugates.

REVIEW

59.
$$\frac{1}{4}+\frac{1}{t}=\frac{1}{2t}$$
$$4t\left(\frac{1}{4}+\frac{1}{t}\right)=4t\left(\frac{1}{2t}\right)$$
$$t+4=2$$
$$t=-2$$

61. Find slope through (–2, –4) and (3, 5).

$$m=\frac{y_2-y_1}{x_2-x_1}=\frac{5-(-4)}{3-(-2)}=\frac{9}{5}$$

Section 8.6 Quadratic and Other Nonlinear Inequalities

VOCABULARY

1. quadratic
3. interval

CONCEPTS

5. greater
7. undefined.
9. The solution of the inequality will be those numbers x for which the graph of $y = x^2 - x - 6$ lies above the x–axis. The graph appears to cross at $x = 3$ and $x = -2$. The solution of the inequality is $(-\infty, -2) \cup (3, \infty)$.

NOTATION

11. a. –2 b. –2 c. –2 d. 3 e. 3 f. 3

13. a. iii b. iv c. ii d. i

PRACTICE

15.
$$x^2 - 5x + 4 < 0$$
$$(x-4)(x-1) < 0$$

x – 1 ---0+++|+++

x – 4 ---|---0+++

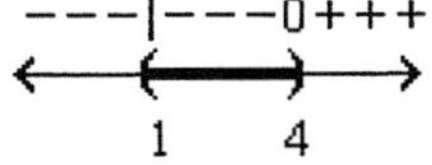

Solution: (1, 4)

17.
$$x^2 - 8x + 15 > 0$$
$$(x-5)(x-3) > 0$$

x – 3 ---0+++|+++

x – 5 ---|---0+++

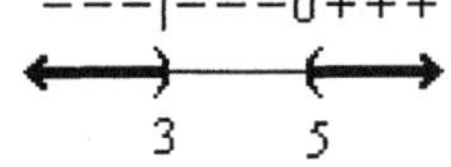

Solution: $(-\infty, 3) \cup (5, \infty)$

19.
$$\begin{aligned} x^2+x-12 &\le 0 \\ (x+4)(x-3) &\le 0 \end{aligned}$$

```
x + 4  ---0+++|+++
x - 3  ---|---0+++
```

Solution: $[-4, 3]$

21.
$$\begin{aligned} x^2+8x &< -16 \\ x^2+8x+16 &< 0 \\ (x+4)(x+4) &< 0 \end{aligned}$$

The square of a quantity is never negative, therefore there are no solutions.

23.
$$\begin{aligned} x^2 &\ge 9 \\ x^2-9 &\ge 0 \\ (x-3)(x+3) &\ge 0 \end{aligned}$$

```
x + 3  ---0+++|+++
x - 3  ---|---0+++
```

Solution: $(-\infty, -3] \cup [3, \infty)$

25.
$$\begin{aligned} 2x^2-50 &< 0 \\ x^2-25 &< 0 \\ (x-5)(x+5) &< 0 \end{aligned}$$

```
x + 5  ---0+++|+++
x - 5  ---|---0+++
```

Solution: $(-5, 5)$

27.
$$\frac{1}{x} < 2$$
$$\frac{1}{x} - 2 < 0$$
$$\frac{1}{x} - \frac{2x}{x} < 0$$
$$\frac{1-2x}{x} < 0$$

```
    x  ---0+++|+++
 1-2x  +++|+++0---
```

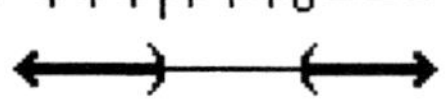

Solution: $(-\infty, 0) \cup (\frac{1}{2}, \infty)$

29.
$$-\frac{5}{x} < 3$$
$$-\frac{5}{x} - 3 < 0$$
$$-\frac{5}{x} - \frac{3x}{x} < 0$$
$$\frac{-5-3x}{x} < 0$$

```
-5-3x  +++0---|---
    x  ---|---0+++
```

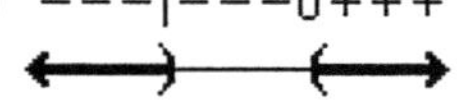

Solution: $(-\infty, -\frac{5}{3}) \cup (0, \infty)$

31.
$$\frac{x^2-x-12}{x-1} < 0$$
$$\frac{(x-4)(x+3)}{x-1} < 0$$

```
x+3  ---0+++|+++|+++
x-1  ---|---0+++|+++
x-4  ---|---|---0+++
```

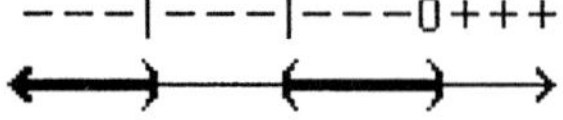

Solution: $(-\infty, -3) \cup (1, 4)$

33.
$$\frac{6x^2-5x+1}{2x+1} > 0$$
$$\frac{(3x-1)(2x-1)}{2x+1} > 0$$

```
2x+1  ---0+++|+++|+++
3x-1  ---|---0+++|+++
2x-1  ---|---|---0+++
```

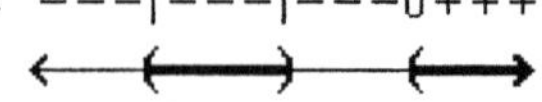

Solution: $(-\frac{1}{2}, \frac{1}{3}) \cup (\frac{1}{2}, \infty)$

35.

$$\frac{3}{x-2} < \frac{4}{x}$$

$$\frac{3}{x-2} - \frac{4}{x} < 0$$

$$\frac{3x}{x(x-2)} - \frac{4(x-2)}{x(x-2)} < 0$$

$$\frac{3x - 4x + 8}{x(x-2)} < 0$$

$$\frac{-x+8}{x(x-2)} < 0$$

```
x      ---0+++|+++|+++
x-2    ---|---0+++|+++
-x+8   +++|+++|+++0---
```

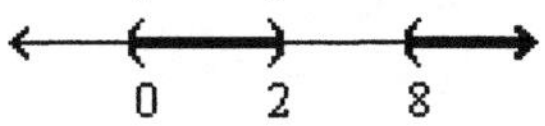

Solution: $(0, 2) \cup (8, \infty)$

37.

$$\frac{7}{x-3} \geq \frac{2}{x+4}$$

$$\frac{7}{x-3} - \frac{2}{x+4} \geq 0$$

$$\frac{7(x+4)}{(x-3)(x+4)} - \frac{2(x-3)}{(x+4)(x-3)} \geq 0$$

$$\frac{7x + 28 - 2x + 6}{(x-3)(x+4)} \geq 0$$

$$\frac{5x+34}{(x-3)(x+4)} \geq 0$$

```
5x+34  ---0+++|+++|+++
x+4    ---|---0+++|+++
x-3    ---|---|---0+++
```

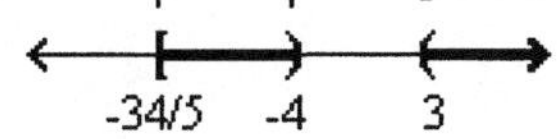

Solution: $[-\frac{34}{5}, -4) \cup (3, \infty)$

39.

$$\frac{x}{x+4} \leq \frac{1}{x+1}$$

$$\frac{x}{x+4} - \frac{1}{x+1} \leq 0$$

$$\frac{x(x+1)}{(x+4)(x+1)} - \frac{1(x+4)}{(x+4)(x+1)} \leq 0$$

$$\frac{x^2 + x - x - 4}{(x+4)(x+1)} \leq 0$$

$$\frac{x^2 - 4}{(x+4)(x+1)} \leq 0$$

$$\frac{(x-2)(x+2)}{(x+4)(x+1)} \leq 0$$

```
x+4    ---0+++|+++|+++|+++
x+2    ---|---0+++|+++|+++
x+1    ---|---|---0+++|+++
x-2    ---|---|---|---0+++
```

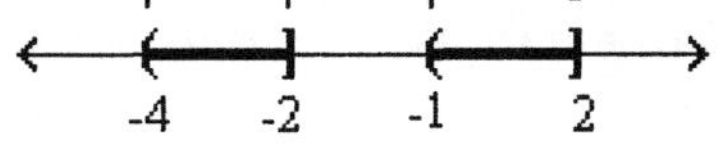

Solution: $(-4, -2] \cup (-1, 2]$

41. $(x+2)^2 > 0$

Any quantity squared is always positive.

The only value which does not make this statement true is -2.

Solution: $(-\infty, -2) \cup (-2, \infty)$

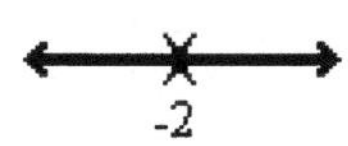

43. The solution of the inequality will be those numbers x for which the graph of $y = x^2 - 2x - 3$ lies below the x–axis.
The graph appears to cross at $x = -1$ and $x = 3$. The solution of the inequality is $(-1, 3)$.

45. The solution of the inequality will be those numbers x for which the graph of $y = \frac{x+3}{x-2}$ lies above the x–axis.
The graph appears to cross at $x = -3$ and $x = 2$. The solution of the inequality is $(-\infty, -3) \cup (2, \infty)$.

47. $y < x^2 + 1$ Vertex: (0, 1)

Broken line. Test point (0, 0): $0 < 1$ is true.

x	y
−2	5
−1	2
0	1
1	2
2	5

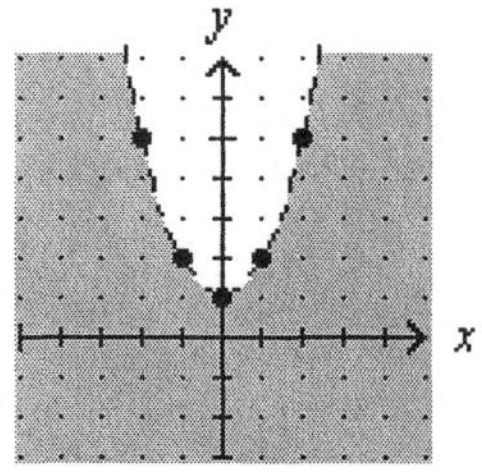

49. $y \le x^2 + 5x + 6$ Vertex: $(-\frac{5}{2}, -\frac{1}{4})$

Solid line. Test point (0, 0): $0 \le 6$ is true.

x	y
−4	2
−3	0
$-2\frac{1}{2}$	$-\frac{1}{4}$
−2	0
0	6

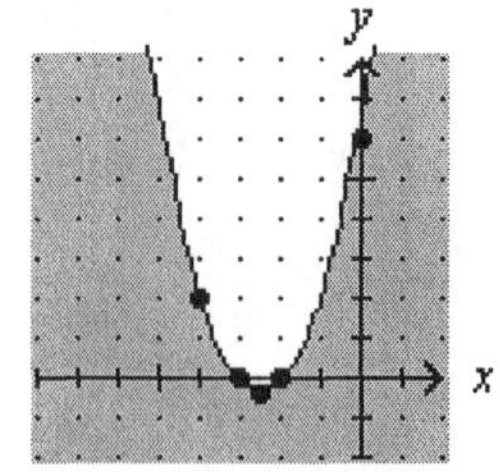

51. $y < |x + 4|$

Broken line. Test point (0, 0): $0 < 4$ is true.

x	y
−7	3
−5	1
−4	0
−2	2
0	4

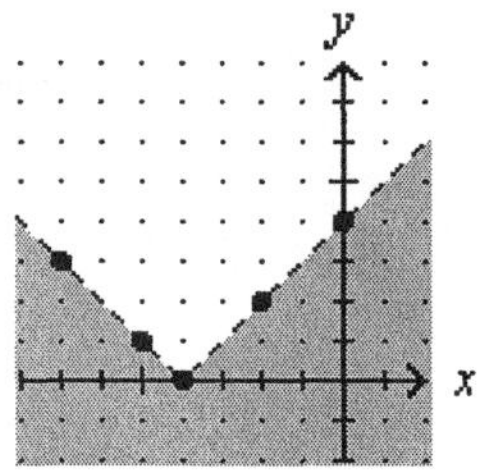

53. $y \le -|x| + 2$

Solid line. Test point (0, 0): $0 \le 2$ is true.

x	y
−4	−2
−2	0
0	2
2	0
4	−2

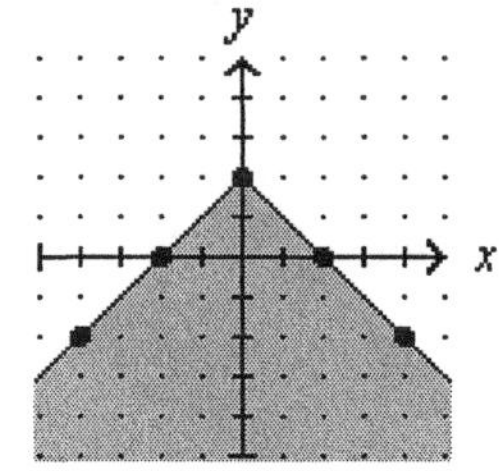

APPLICATIONS

55. Find the intervals for x when L is greater than 95 ft. Also $-2100 < x < 2100$.

$L = \frac{1}{9000}x^2 + 5$ and is > 95

$$\frac{1}{9000}x^2 + 5 > 95$$
$$x^2 + 45{,}000 > 855{,}000$$
$$x^2 - 810{,}000 > 0$$
$$(x + 900)(x - 900) > 0$$

Combine $-2100 < x < 2100$

with

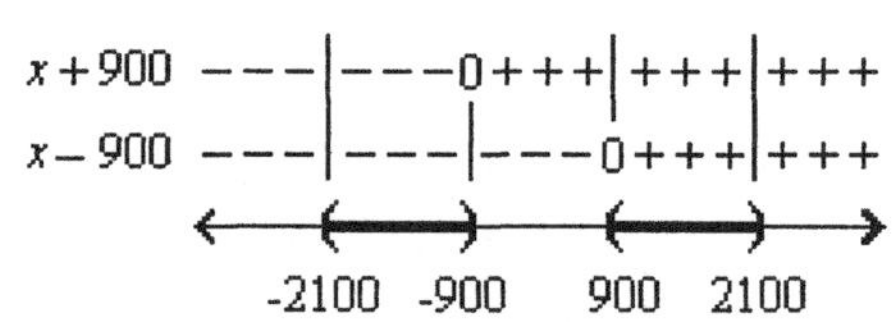

Solution: $(-2100, -900) \cup (900, 2100)$

WRITING

57. Two positive factors result in a positive product. Two negative factors result in a positive product. When factors have opposite signs the result is a negative product. Therefore the product will be positive only when the two factors have the same sign.

59. The graph shows that the function is never negative (less than 0).

REVIEW

61. $x = ky$ 63. $t = kxy$

65. Since the equation is written in slope–intercept form the slope is the coefficient of x. $m = 3$.

Factoring

1.
$$\begin{aligned} 4k^2 + 8k &= 0 \\ 4k(k+2) &= 0 \\ 4k = 0 \quad &\text{or} \quad k+2 = 0 \\ k = 0 \quad & \quad\quad k = -2 \end{aligned}$$

2.
$$\begin{aligned} z^2 + 8z + 15 &= 0 \\ (z+5)(z+3) &= 0 \\ z+5 = 0 \quad &\text{or} \quad z+3 = 0 \\ z = -5 \quad & \quad\quad z = -3 \end{aligned}$$

3.
$$\begin{aligned} 2r^2 + 5r &= -3 \\ 2r^2 + 5r + 3 &= 0 \\ (2r+3)(r+1) &= 0 \\ 2r+3 = 0 \quad &\text{or} \quad r+1 = 0 \\ 2r = -3 \quad & \quad\quad r = -1 \\ r = -\tfrac{3}{2} & \end{aligned}$$

The Square Root Method

4.
$$\begin{aligned} u^2 &= 24 \\ u &= \pm\sqrt{24} = \pm 2\sqrt{6} \end{aligned}$$

5.
$$\begin{aligned} (s-7)^2 - 9 &= 9 \\ (s-7)^2 &= 9 \\ s-7 &= \pm\sqrt{9} = \pm 3 \\ s-7 = -3 \quad &\text{or} \quad s-7 = 3 \\ s = 7-3 \quad & \quad\quad s = 7+3 \\ s = 4 \quad & \quad\quad s = 10 \end{aligned}$$

6.
$$\begin{aligned} 3x^2 - 16 &= 0 \\ 3x^2 &= 16 \\ x^2 &= \frac{16}{3} \\ x &= \pm\sqrt{\frac{16}{3}} = \pm\frac{4}{\sqrt{3}} = \pm\frac{4\sqrt{3}}{3} \end{aligned}$$

Completing the Square

7.
$x^2 + 10x - 7 = 0$

$x^2 + 10x = 7$ Add 8 to both sides.

$x^2 + 10x + 25 = 7 + 25$ Square one–half of the coefficient of x and add it to both sides.

$(x+5)^2 = 32$ Factor and combine like terms.

$x + 5 = \pm\sqrt{32} = \pm 4\sqrt{2}$ Apply the square root property.

$x = -5 \pm 4\sqrt{2}$ Subtract 5 from both sides and write in concise form.

8. $4x^2-4x-1=0$

$4x^2-4x=1$ Add 1 to both sides.

$x^2-x+\frac{1}{4}=\frac{1}{4}+\frac{1}{4}$ Square one–half of the coefficient of x and add it to both sides.

$\left(x-\frac{1}{2}\right)^2=\frac{1}{2}$ Factor and combine like terms.

$x-\frac{1}{2}=\pm\sqrt{\frac{1}{2}}=\pm\frac{\sqrt{2}}{2}$ Apply the square root property.

$x=\frac{1}{2}\pm\frac{\sqrt{2}}{2}=\frac{1\pm\sqrt{2}}{2}$ Add $\frac{1}{2}$ to both sides and write in concise form.

9. $x^2+2x+2=0$

$x^2+2x=-2$ Subtract 2 from both sides.

$x^2+2x+1=-2+1$ Square one–half of the coefficient of x and add it to both sides.

$(x+1)^2=-1$ Factor and combine like terms.

$x+1=\pm\sqrt{-1}=\pm i$ Apply the square root property.

$x=-1\pm i$ Subtract 1 from both sides and write in concise form.

The Quadratic Formula

10. $2x^2-1=3x$

$2x^2-3x-1=0$

$a=2;\ b=-3,\ c=-1$

$$x=\frac{-(-3)\pm\sqrt{(-3)^2-4(2)(-1)}}{2(2)}$$

$$=\frac{3\pm\sqrt{9+8}}{4}$$

$$=\frac{3\pm\sqrt{17}}{4}$$

11. $x^2-6x-391=0$

$a=1;\ b=-6;\ c=-391$

$$x=\frac{-(-6)\pm\sqrt{(-6)^2-4(1)(-391)}}{2(1)}$$

$$=\frac{6\pm\sqrt{36+1564}}{2}$$

$$=\frac{6\pm\sqrt{1600}}{2}$$

$$=\frac{6\pm 40}{2}$$

$x=\frac{6+40}{2}=23$ or $x=\frac{6-40}{2}=-17$

12. $3x^2+2x+1=0$

$a=3;\ b=2;\ c=1$

$$x=\frac{-(2)\pm\sqrt{(2)^2-4(3)(1)}}{2(3)}$$

$$=\frac{-2\pm\sqrt{4-12}}{6}$$

$$=\frac{-2\pm\sqrt{-8}}{6}$$

$$=\frac{-2\pm 2i\sqrt{2}}{6}=\frac{2\left(-1\pm i\sqrt{2}\right)}{2\cdot 3}=-\frac{1}{3}\pm\frac{\sqrt{2}}{3}i$$

The Graphing Method

13. Find the values of x where the graph crosses the x–axis. The values are $x=1$ and $x=-2$.

Section 8.1

1.
$$\begin{aligned} x^2+9x+20 &= 0 \\ (x+4)(x+5) &= 0 \end{aligned}$$
$$x+4 = 0 \quad \text{or} \quad x+5=0$$
$$x=-4 \qquad x=-5$$

2.
$$\begin{aligned} 6x^2+17x+5 &= 0 \\ (3x+1)(2x+5) &= 0 \end{aligned}$$
$$3x+1=0 \quad \text{or} \quad 2x+5=0$$
$$3x=-1 \qquad 2x=-5$$
$$x=-\tfrac{1}{3} \qquad x=-\tfrac{5}{2}$$

3.
$$\begin{aligned} x^2 &= 28 \\ \sqrt{x^2} &= \pm\sqrt{28} \\ x &= \pm 2\sqrt{7} \end{aligned}$$

4.
$$\begin{aligned} (t+2)^2 &= 36 \\ t+2 &= \pm\sqrt{36} = \pm 6 \end{aligned}$$
$$t+2=-6 \quad \text{or} \quad t+2=6$$
$$t=-8 \qquad t=4$$

5.
$$\begin{aligned} 5a^2+11a &= 0 \\ a(5a+11) &= 0 \end{aligned}$$
$$a=0 \quad \text{or} \quad 5a+11=0$$
$$a=-11$$
$$a=-\tfrac{11}{5}$$

6.
$$\begin{aligned} 5x^2-49 &= 0 \\ 5x^2 &= 49 \\ x^2 &= \frac{49}{5} \\ x &= \pm\sqrt{\frac{49}{5}} = \pm\frac{\sqrt{49}\sqrt{5}}{\sqrt{5}\sqrt{5}} = \pm\frac{7\sqrt{5}}{5} \end{aligned}$$

7. Add $\left[\frac{1}{2}(-1)\right]^2 = \left(-\frac{1}{2}\right)^2 = \frac{1}{4}$ to both sides.

8.
$$\begin{aligned} x^2+6x+8 &= 0 \\ x^2+6x &= -8 \\ x^2+6x+9 &= -8+9 \\ (x+3)^2 &= 1 \\ x+3 &= \pm 1 \end{aligned}$$
$$x+3=-1 \quad \text{or} \quad x+3=1$$
$$x=-4 \qquad x=-2$$

9.
$$\begin{aligned} 2x^2-6x+3 &= 0 \\ 2x^2-6x &= -3 \\ x^2-3x &= -\frac{3}{2} \\ x^2-3x+\frac{9}{4} &= -\frac{3}{2}+\frac{9}{4} \\ \left(x-\frac{3}{2}\right)^2 &= \frac{3}{4} \\ x-\frac{3}{2} &= \pm\sqrt{\frac{3}{4}} = \pm\frac{\sqrt{3}}{2} \\ x &= \frac{3}{2} \pm \frac{\sqrt{3}}{2} = \frac{3\pm\sqrt{3}}{2} \end{aligned}$$

10. Find t when $d = 605$.
$$\begin{aligned} d &= 16t^2 \\ 605 &= 16t^2 \\ t^2 &= \frac{605}{16} = 37.8125 \\ t &= \sqrt{37.8125} = 6.149\ldots \end{aligned}$$

The ball should be dropped about 6 seconds before midnight.

Section 8.2

11.
$$-x^2 + 10x - 18 = 0$$
$$x^2 - 10x + 18 = 0$$
$a = 1; b = -10; c = 18$
$$x = \frac{-(-10) \pm \sqrt{(-10)^2 - 4(1)(18)}}{2(1)} = \frac{10 \pm \sqrt{100-72}}{2} = \frac{10 \pm \sqrt{28}}{2} = \frac{10 \pm 2\sqrt{7}}{2} = \frac{2\left(5 \pm \sqrt{7}\right)}{2} = 5 \pm \sqrt{7}$$

12.
$$x^2 - 10x = 0$$
$a = 1; b = -10; c = 0$
$$x = \frac{-(-10) \pm \sqrt{(-10)^2 - 4(1)(0)}}{2(1)} = \frac{10 \pm \sqrt{100-0}}{2} = \frac{10 \pm 10}{2}$$
$$x = \frac{10-10}{2} = 0 \quad \text{or} \quad x = \frac{10+10}{2} = 10$$

13.
$$2x^2 + 13x = 7$$
$$2x^2 + 13x - 7 = 0 \qquad \text{Rearrange.}$$
$a = 2; b = 13; c = -7$
$$x = \frac{-13 \pm \sqrt{13^2 - 4(2)(-7)}}{2(2)} = \frac{-13 \pm \sqrt{169+56}}{4} = \frac{-13 \pm \sqrt{225}}{4} = \frac{-13 \pm 15}{4}$$
$$x = \frac{-13+15}{4} = \frac{1}{2} \quad \text{or} \quad x = \frac{-13-15}{4} = -7$$

14.
$$26y - 3y^2 = 2$$
$$0 = 3y^2 - 26y + 2 \qquad \text{Rearrange.}$$
$a = 3; b = -26; c = 2$
$$y = \frac{-(-26) \pm \sqrt{(-26)^2 - 4(3)(2)}}{2(3)} = \frac{26 \pm \sqrt{676-24}}{6} = \frac{26 \pm \sqrt{652}}{6} = \frac{26 \pm 2\sqrt{163}}{6} = \frac{2\left(13 \pm \sqrt{163}\right)}{6} = \frac{13 \pm \sqrt{163}}{3}$$

15. **Analysis**: Find the width of the border surrounding a 615 in.^{2} photograph. The borders on the sides are half as wide as those at the top and bottom. The width of the final poster is 23 in. and the length is 35 in.

Equation: Let x represent the width of the border on the sides and $2x$ is the width of the border at the top and bottom. The width of the photograph is $23 - 2x$ and the length of the photograph is $35 - 2(2x) = 35 - 4x$.

Width	times	length	equals	area.
$23 - 2x$	•	$35 - 4x$	=	615

Solve:
$$(23 - 2x)(35 - 4x) = 615$$
$$805 - 92x - 70x + 8x^2 = 615$$
$$8x^2 - 162x + 190 = 0$$
$a = 8; b = -162; c = 190$
$$x = \frac{-(-162) \pm \sqrt{(-162)^2 - 4(8)(190)}}{2(8)} = \frac{162 \pm \sqrt{26{,}244 - 6080}}{16} = \frac{162 \pm \sqrt{20{,}164}}{16} = \frac{162 \pm 142}{16}$$
$$x = \frac{162-142}{16} = 1.25$$
or
$$x = \frac{162+142}{16} = 19$$
Discard $x = 19$ because it is too large to be one of the side borders.

Conclusion: The width of the border on the sides will be $x = 1.25$ in. and the width of the borders on the top and bottom will be $2x = 2.5$ in.

16. Find t when d = 25 ft.

$$\begin{aligned} d &= -16t^2 + 40t + 5 \\ 25 &= -16t^2 + 40t + 5 \\ 16t^2 - 40t + 20 &= 0 \\ 4t^2 - 10t + 5 &= 0 \end{aligned}$$

$$t = \frac{-(-10) \pm \sqrt{(-10)^2 - 4(4)(5)}}{2(4)} = \frac{10 \pm \sqrt{100-80}}{8} = \frac{10 \pm \sqrt{20}}{8}$$

$$t = \frac{10+\sqrt{20}}{8} \approx 1.8 \text{ sec. or } t = \frac{10-\sqrt{20}}{8} \approx 0.7 \text{ sec,}$$

Section 8.3

17. Find the annual sales by substituting 4 for x in the given function.

$S(x) = 2.4375x^2 - 8.225x + 40.7$

$S(4) = 2.4375(4)^2 - 8.225(4) + 40.7 = \$46.8$ billion

18. $f(x) = 2x^2$

x	$f(x)$
−2	8
−1	2
0	0
1	2
2	8

$g(x) = 2x^2 - 3$

Translate $f(x)$ downward 3 units.

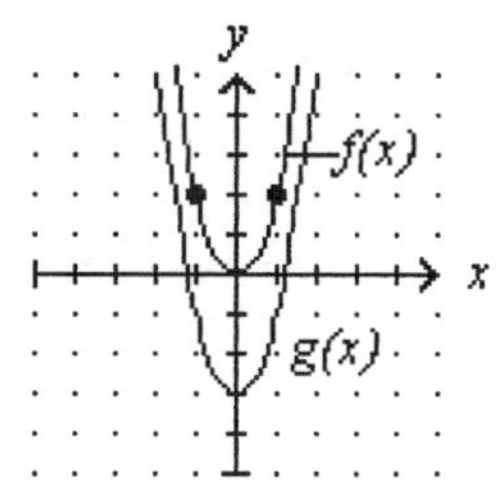

19. $f(x) = -4x^2$

x	$f(x)$
−2	−16
−1	−4
0	0
1	−4
2	−16

$g(x) = -4(x-2)^2 + 1$

Translate $f(x)$ 2 units to the right and then translate $f(x)$ upward 1 unit.

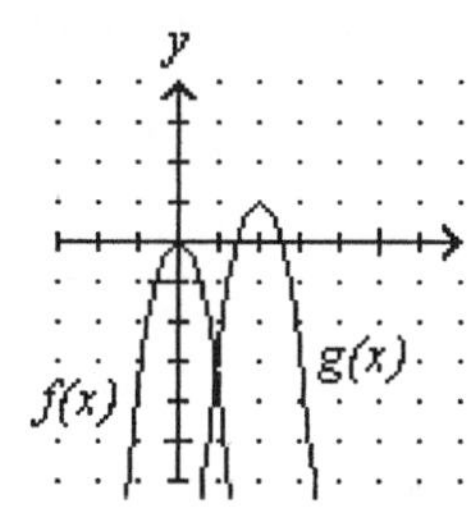

20. $$f(x) = -\left(x + \frac{3}{2}\right)^2 + \frac{5}{2}$$

Vertex: $\left(-\frac{3}{2}, \frac{5}{2}\right)$

Axis of symmetry: $x = -\frac{3}{2}$

x	$f(x)$
−4	$-3\frac{3}{4}$
−3	$\frac{1}{4}$
$-1\frac{1}{2}$	$2\frac{1}{2}$
0	$\frac{1}{4}$
1	$-3\frac{3}{4}$

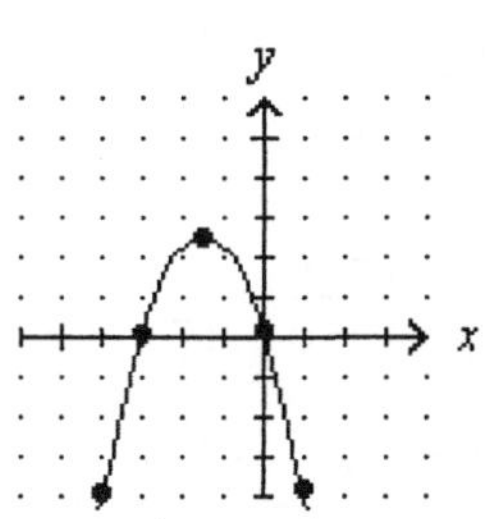

21. $f(x) = 5x^2 + 10x - 1$

$$x = -\frac{b}{2a} = -\frac{10}{2(5)} = -1$$

$$f\left(-\frac{b}{2a}\right) = f(-1) = 5(-1)^2 + 10(-1) - 1$$
$$= 5 - 10 - 1 = -6$$

Vertex: (−1, −6)

Axis of symmetry: $x = -1$

x	$f(x)$
−3	14
−2	−1
−1	−6
0	−1
1	14

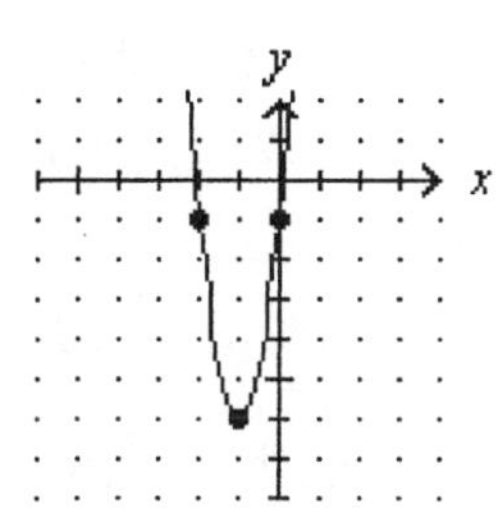

22. Find the vertex of $N(x) = -1.526x^2 + 155.652x + 2{,}500{,}200$.

$$x = -\frac{b}{2a} = -\frac{155.652}{2(-1.526)} = 51 \text{ years}$$

$$N\left(-\frac{b}{2a}\right) = N(51) = -1.526(51)^2 + 155.652(51) + 2{,}500{,}200 \approx 2{,}504{,}170$$

The highest number of family farms was 51 years after 1870, or in 1921. The number of farms was 2,504,170.

23. The solution would be the x value where the graph crosses the x–axis. These values are estimated to be $x = -2$ and $x = \frac{1}{3}$.

Section 8.4

24. $\sqrt{-4} = \sqrt{-1}\sqrt{4} = i \cdot 2 = 2i$

25. $\sqrt{-7} = \sqrt{-1}\sqrt{7} = i\sqrt{7}$

26. $-3\sqrt{-20} = -3\sqrt{-1}\sqrt{20} = -3i\left(2\sqrt{5}\right) = -6i\sqrt{5}$

27. $\sqrt{-\frac{36}{49}} = \frac{\sqrt{-1}\sqrt{36}}{\sqrt{49}} = \frac{6}{7}i$

28. $\sqrt{-3}\sqrt{-3} = i\sqrt{3}\left(i\sqrt{3}\right) = i^2\sqrt{9} = -1(3) = -3$

29. $\frac{\sqrt{-64}}{\sqrt{-9}} = \frac{8i}{3i} = \frac{8}{3}$

30.
$$a^2 = -25$$
$$a = \pm\sqrt{-25} = \pm 5i$$

31.
$$x^2 - 2x + 13 = 0$$
$$x = \frac{-(-2) \pm \sqrt{(-2)^2 - 4(1)(13)}}{2(1)}$$
$$= \frac{2 \pm \sqrt{4-52}}{2} = \frac{2 \pm \sqrt{-48}}{2} = \frac{2 \pm 4i\sqrt{3}}{2}$$
$$= \frac{2}{2} \pm \frac{4i\sqrt{3}}{2} = 1 \pm 2i\sqrt{3}$$

32. $(5 + 4i) + (7 - 12i) = (5 + 7) + (4 - 12)i = 12 - 8i$

33. $(-6 - 40i) - (-8 + 28i) = -6 - 40i + 8 - 28i = (-6 + 8) + (-40 - 28)i = 2 - 68i$

34. $(-8 + \sqrt{-8}) + (6 - \sqrt{-32}) = (-8 + 2i\sqrt{2}) + (6 - 4i\sqrt{2}) = (-8 + 6) + (2 - 4)i\sqrt{2} = -2 - 2i\sqrt{2}$

35. $2i(64 + 9i) = 2i(64) + 2i(9i) = 128i + 18i^2 = 128i + 18(-1) = -18 + 128i$

36. $(2 - 7i)(-3 + 4i) = -6 + 8i + 21i - 28i^2 = -6 + 29i + 28 = 22 + 29i$

37. $(5 - \sqrt{-27})(-6 + \sqrt{-12}) = (5 - 3i\sqrt{3})(-6 + 2i\sqrt{3})$
$= -30 + 10i\sqrt{3} + 18i\sqrt{3} - 6i^2(3) = -30 + 28i\sqrt{3} + 18 = -12 + 28i\sqrt{3}$

38. $\frac{6}{2+i} = \frac{6}{2+i} \cdot \frac{2-i}{2-i} = \frac{6(2-i)}{4-i^2} = \frac{12-6i}{4+1} = \frac{12-6i}{5} = \frac{12}{5} - \frac{6}{5}i$

39. $\frac{4+i}{4-i} = \frac{4+i}{4-i} \cdot \frac{4+i}{4+i} = \frac{16+2(4)(i)+i^2}{16-i^2} = \frac{16+8i-1}{16+1} = \frac{15+8i}{17} = \frac{15}{17} + \frac{8}{17}i$

40. $\frac{\sqrt{3}+\sqrt{-4}}{\sqrt{3}-\sqrt{-4}} = \frac{\sqrt{3}+2i}{\sqrt{3}-2i} = \frac{\sqrt{3}+2i}{\sqrt{3}-2i} \cdot \frac{\sqrt{3}+2i}{\sqrt{3}+2i} = \frac{3+2(\sqrt{3})(2i)+4i^2}{3-2i^2} = \frac{3+4i\sqrt{3}-4}{3+4} = \frac{-1+4i\sqrt{3}}{7} = -\frac{1}{7} + \frac{4\sqrt{3}}{7}i$

41. $\frac{-2}{5i^3} = \frac{-2}{5i^3}\cdot\frac{i}{i} = \frac{-2i}{5i^4} = \frac{-2i}{5} = 0 - \frac{2}{5}i$

42. $i^{65} = i^{4\bullet 16}\bullet i = 1^{16}\bullet i = i$

43. $i^{48} = i^{4\bullet 12} = 1^{12} = 1$

Section 8.5

44. $3x^2 + 4x - 3 = 0$; $b^2 - 4ac = (4)^2 - 4(3)(-3) = 16 + 36 = 52$. Solutions are irrational and unequal.

45. $4x^2 - 5x + 7 = 0$; $b^2 - 4ac = (-5)^2 - 4(1)(7) = 25 - 28 = -3$. Solutions are complex conjugates.

46. $9x^2 - 12x + 4 = 0$; $b^2 - 4\text{ac} = (-12)^2 - 4(9)(4) = 144 - 144 = 0$; Solutions are rational and equal.

47.

$x - 13\sqrt{x} + 12 = 0$

$(\sqrt{x})^2 - 13\sqrt{x} + 12 = 0$ Write x as $(\sqrt{x})^2$.

$y^2 - 13y + 12 = 0$ Let $y = \sqrt{x}$.

$(y-1)(y-12) = 0$ Factor and solve.

$y - 1 = 0$ or $y - 12 = 0$

$y = 1$ $\quad$ $y = 12$

Undo the substitution.

$\sqrt{x} = 1$ or $\sqrt{x} = 12$

$x = 1$ $\quad$ $x = 144$

Check: $(1) - 13\sqrt{1} + 12 = 1 - 13 + 12 = 0$

Check: $144 - 13\sqrt{144} + 12 = 144 - 156 + 12 = 0$

The solutions are 1 and 144.

48.

$a^{2/3} + a^{1/3} - 6 = 0$

$(a^{1/3})^2 + a^{1/3} - 6 = 0$ Write $a^{2/3}$ as $(a^{1/3})^2$.

$y^2 + y - 6 = 0$ Let $y = a^{1/3}$.

$(y-2)(y+3) = 0$ Factor and solve.

$y - 2 = 0$ or $y + 3 = 0$

$y = 2$ $\quad$ $y = -3$

Undo the substitution.

$a^{1/3} = 2$ or $a^{1/3} = -3$

$a = 8$ $\quad$ $a = -27$

The solutions are 8 and –27.

49.

$6x^4 - 19x^2 + 3 = 0$

$6(x^2)^2 - 19x^2 + 3 = 0$ Write x^4 as $(x^2)^2$.

$6y^2 - 19y + 3 = 0$ Let $y = x^2$.

$(6y-1)(y-3) = 0$ Factor and solve.

$6y - 1 = 0$ or $y - 3 = 0$

$y = 1$ $\quad$ $y = 3$

$y = \frac{1}{6}$

Undo the substitution.

$x^2 = \frac{1}{6}$ or $x^2 = 3$

$x = \pm\sqrt{\frac{1}{6}} = \pm\frac{\sqrt{6}}{6}$ $\quad$ $x = \pm\sqrt{3}$

The solutions are $-\frac{\sqrt{6}}{6}, \frac{\sqrt{6}}{6}, -\sqrt{3}, \sqrt{3}$.

50.

$\frac{6}{x+2} + \frac{6}{x+1} = 5$ Multiply by LCD.

$6(x+1) + 6(x+2) = 5(x^2 + 3x + 2)$

$6x + 6 + 6x + 12 = 5x^2 + 15x + 10$

$0 = 5x^2 + 3x - 8$

$0 = (5x+8)(x-1)$

$5x + 8 = 0$ or $x - 1 = 0$

$5x = -8$ $\quad$ $x = 1$

$x = -\frac{8}{5}$

The solutions are $-\frac{8}{5}$ and 1.

51.

$(x-3)^2 - 8(x-3) + 7 = 0$

$y^2 - 8y + 7 = 0$ Let $y = (x-3)$.

$(y-1)(y-7) = 0$ Factor and solve.

$y - 1 = 0$ or $y - 7 = 0$

$y = 1$ $\quad$ $y = 7$

Undo the substitution.

$x - 3 = 1$ or $x - 3 = 7$

$x = 4$ $\quad$ $x = 10$

The solutions are 4 and 10.

52. **Analysis**: Find the amount of time it takes the older girl to do the yard work when she works alone.

Equation: Let x represent the number of minutes the older girl takes to do the yard work. Then $x + 20$ will represent the minutes for the younger girl to do the yard work. Working together it would take them 45 minutes.

Amount the older girl does in 1 minute	plus	amount the younger girl does in 1 minute	equals	amount done together in 1minute.
$\frac{1}{x}$	$+$	$\frac{1}{x+20}$	$=$	$\frac{1}{45}$

Solve:

$$\frac{1}{x} + \frac{1}{x+20} = \frac{1}{45}$$

$$45x(x+20)\left(\frac{1}{x} + \frac{1}{x+20}\right) = 45x(x+20)\left(\frac{1}{45}\right)$$

$$45(x+20) + 45x = x(x+20)$$

$$45x + 900 + 45x = x^2 + 20x$$

$$90x + 900 = x^2 + 20x$$

$$0 = x^2 - 70x - 900$$

$$x = \frac{-(-70) \pm \sqrt{(-70)^2 - 4(1)(-900)}}{2(1)}$$

$$= \frac{70 \pm \sqrt{4900 + 3600}}{2} = \frac{70 \pm \sqrt{8500}}{2}$$

$$x = \frac{70 + 92.195}{2} \approx 81 \quad \text{or} \quad x = \frac{70 - 92.195}{2} \approx -11$$

Discard the negative solution.

Conclusion: It takes the older girl about 81 minutes to do the yard work.

Section 8.6

53.

$$x^2 + 2x - 35 > 0$$
$$(x+7)(x-5) > 0$$

```
x + 7  ---0+++|+++
x - 5  ---|---0+++
```

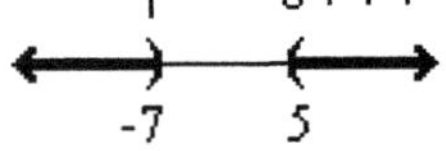

Solution: $(-\infty, -7) \cup (5, \infty)$

54.

$$x^2 - 81 \leq 0$$
$$(x+9)(x-9) \leq 0$$

```
x + 9  ---0+++|+++
x - 9  ---|---0+++
```

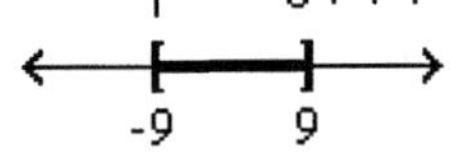

Solution: $[-9, 9]$

55.

$$\frac{3}{x} \leq 5$$

$$\frac{3}{x} - 5 \leq 0$$

$$\frac{3}{x} - \frac{5x}{x} \leq 0$$

$$\frac{3-5x}{x} \leq 0$$

```
x       ---0+++|+++
3 - 5x  +++|+++0---
```

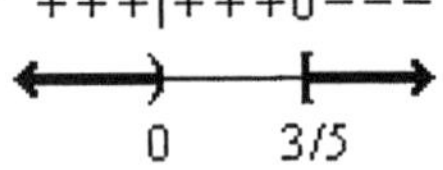

Solution: $(-\infty, 0) \cup [\frac{3}{5}, \infty)$

56.

$$\frac{2x^2 - x - 28}{x-1} > 0$$

$$\frac{(2x+7)(x-4)}{x-1} > 0$$

```
2x+7  ---0+++|+++|+++
x-1   ---|---0+++|+++
x-4   ---|---|---0+++
```

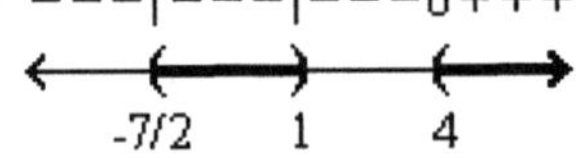

Solution: $(-\frac{7}{2}, 1) \cup (4, \infty)$

57. $x^2 + 2x - 35 > 0$

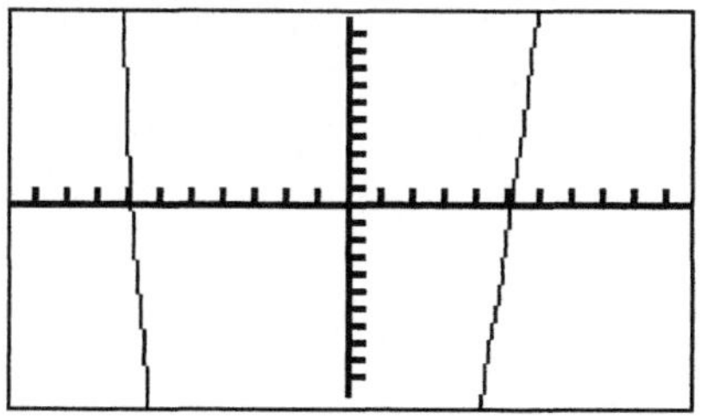

The solution will be those numbers x for which the graph lies above the x–axis. The graph appears to cross at $x = -7$ and $x = 5$ and is above on the outside of those values.

Solution: $(-\infty, -7) \cup (5, \infty)$

58. $\dfrac{2x^2 - x - 28}{x - 1} > 0$

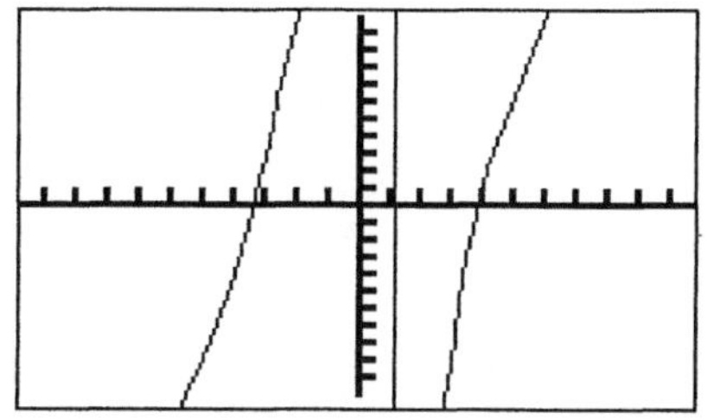

The solution will be those numbers x for which the graph lies above the x–axis. The graph appears to cross at $x = -3.5$, $x = 1$ (an asymptote), and $x = 4$. The solution is above between –3 and 1 and to the right of 4. Compared to Exercise 21 it is not possible to read as accurately as it was calculated.

Solution: $(-3.5, 1) \cup (4, \infty)$

59. $y < \frac{1}{2}x^2 - 1$ Vertex: $(0, -1)$

Broken line. Test point (0, 0): $0 < -1$ is false.

x	y
–4	7
–2	1
0	–1
2	1
4	7

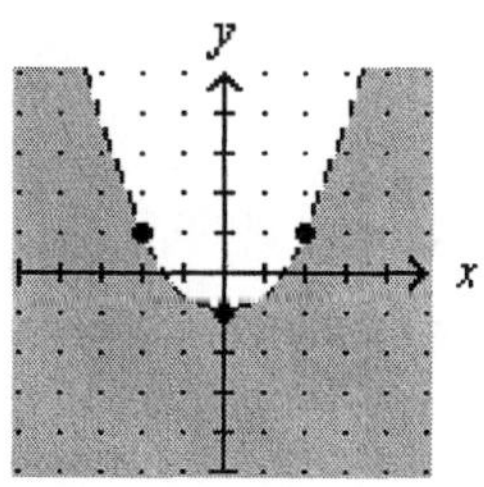

60. $y \geq -|x|$

Solid line. Test point (2, 1): $1 \geq -2$ is true.

x	y
–4	–4
–2	–2
0	0
2	–2
4	–4

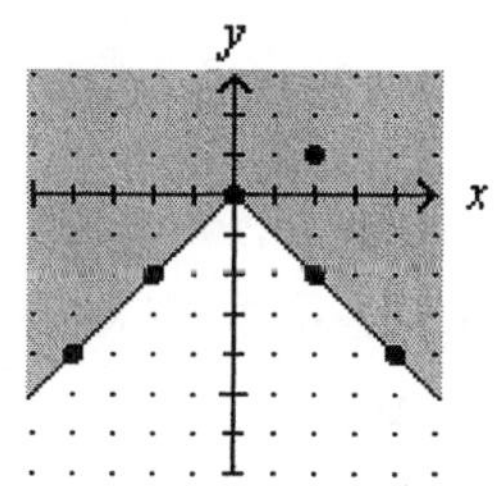

1.
$$\begin{aligned} 3x^2 + 18x &= 0 \\ 3x(x+6) &= 0 \end{aligned}$$

$3x = 0$ or $x + 6 = 0$

$x = 0$ $\quad x = -6$

2.
$$\begin{aligned} x(6x+19) &= -15 \\ 6x^2 + 19x + 15 &= 0 \\ (2x+3)(3x+5) &= 0 \end{aligned}$$

$2x + 3 = 0$ or $3x + 5 = 0$

$2x = -3$ $\quad 3x = -5$

$x = -\frac{3}{2}$ $\quad x = -\frac{5}{3}$

3. $\left[\frac{1}{2}(24)\right]^2 = 12^2 = 144$

4.
$$\begin{aligned} x^2 - x - 1 &= 0 \\ x^2 - x &= 1 \\ x^2 - x + \frac{1}{4} &= 1 + \frac{1}{4} \\ \left(x - \frac{1}{2}\right)^2 &= \frac{5}{4} \\ x - \frac{1}{2} &= \pm\sqrt{\frac{5}{4}} = \pm\frac{\sqrt{5}}{2} \\ x &= \frac{1}{2} \pm \frac{\sqrt{5}}{2} = \frac{1 \pm \sqrt{5}}{2} \end{aligned}$$

5.
$$\begin{aligned} 2x^2 - 8x &= -5 \\ 2x^2 - 8x + 5 &= 0 \end{aligned}$$
Rearrange.

$a = 2;\ b = -8;\ c = 5$

$$\begin{aligned} x &= \frac{-(-8) \pm \sqrt{(-8)^2 - 4(2)(5)}}{2(2)} \\ &= \frac{8 \pm \sqrt{64 - 40}}{4} \\ &= \frac{8 \pm \sqrt{24}}{4} = \frac{8 \pm 2\sqrt{6}}{4} \\ &= \frac{2\left(4 \pm \sqrt{6}\right)}{4} = \frac{4 \pm \sqrt{6}}{2} \end{aligned}$$

6. **Analysis**: Find the dimensions of a tablecloth that covers an area of 6759 ft^2.

Equation: Let x represent the width of the screen and $333x + 3.5$ represent the length of the tablecloth.

Width	times	length	equals	area.
x	•	$333x + 3.5$	=	6759

Solve:
$$\begin{aligned} x(332x + 8) &= 6759 \\ 332x^2 + 8x - 6759 &= 0 \end{aligned}$$

$a = 332;\ b = 8;\ c = -6759$

$$\begin{aligned} x &= \frac{-8 \pm \sqrt{8^2 - 4(332)(-6759)}}{2(332)} \\ &= \frac{-8 \pm \sqrt{64 + 8{,}975{,}952}}{664} \\ &= \frac{-8 \pm \sqrt{8{,}976{,}016}}{664} = \frac{-8 \pm 2996}{664} \end{aligned}$$

$x = \frac{-8 - 2996}{664} \approx -4.52$ *or* $x = \frac{-8 + 2996}{664} = 4.5$ Discard negative solution

Conclusion: The width of the tablecloth was $x = 4.5$ ft and the length was $332x + 8 = 1{,}502$ ft.

7. $\sqrt{-48} = \sqrt{-1}\sqrt{48} = i\sqrt{16 \bullet 3} = 4i\sqrt{3}$

8. $i^{54} = i^{4 \bullet 13} \bullet i^2 = 1 \bullet (-1) = -1$

9. $(2 + 4i) + (-3 + 7i) = (2 - 3) + (4 + 7)i = -1 + 11i$

10. $(3 - \sqrt{-9}) - (-1 + \sqrt{-16}) = (3 - 3i) - (-1 + 4i) = 3 - 3i + 1 - 4i = 4 - 7i$

11. $2i(3 - 4i) = 6i - 8i^2 = 6i + 8 = 8 + 6i$

12. $(3 + 2i)(-4 - i) = -12 - 3i - 8i - 2i^2 = -12 - 11i + 2 = -10 - 11i$

13. $\frac{1}{i^3} = \frac{1 \cdot i}{i^3 \cdot i} = \frac{i}{i^4} = \frac{i}{1} = i = 0 + i$

14. $\frac{2+i}{3-i} = \frac{2+i}{3-i} \cdot \frac{3+i}{3+i} = \frac{6+2i+3i+i^2}{9-i^2} = \frac{6+5i-1}{9+1} = \frac{5+5i}{10} = \frac{5(1+i)}{5 \cdot 2} = \frac{1+i}{2} = \frac{1}{2} + \frac{1}{2}i$

15. $3x^2 + 5x + 17 = 0$; $b^2 - 4ac = 5^2 - 4(3)(17) = 25 - 204 = -179$. Solutions are nonreal.

16. $x^2 = -12$

$x = \pm\sqrt{-12} = \pm 2i\sqrt{3}$

17.

$13 = 4t - t^2$

$t^2 - 4t + 13 = 0$ Rearrange.

$a = 1$; $b = -4$; $c = 13$

$$x = \frac{-(-4) \pm \sqrt{(-4)^2 - 4(1)(13)}}{2(1)}$$

$$= \frac{4 \pm \sqrt{16-52}}{2} = \frac{4 \pm \sqrt{-36}}{2} = \frac{4 \pm 6i}{2}$$

$$= \frac{2(2 \pm 3i)}{2} = 2 \pm 3i$$

18.

$2y - 3\sqrt{y} + 1 = 0$

$2(\sqrt{y})^2 - 3\sqrt{y} + 1 = 0$ Write y as $(\sqrt{y})^2$.

$2a^2 - 3a + 1 = 0$ Let $a = \sqrt{y}$.

$(2a - 1)(a - 1) = 0$ Factor and solve.

$2a - 1 = 0$ or $a - 1 = 0$

$a = \frac{1}{2}$ $a = 1$

Undo the substitution.

$\sqrt{y} = \frac{1}{2}$ or $\sqrt{y} = 1$

$y = \frac{1}{4}$ $y = 1$

Check:

$2(\frac{1}{4}) - 3\sqrt{\frac{1}{4}} + 1$

$= \frac{1}{2} - \frac{3}{2} + 1 = 0$

Check:

$2(1) - \sqrt{1} - 1$

$= 2 - 3 + 1 = 0$

The solutions are 1 and $\frac{1}{4}$.

19.

$x^4 - x^2 - 12 = 0$ let $u = x^2$

$u^2 - u - 12 = 0$

$(u - 4)(u + 3) = 0$

$u = 4$ or $u = -3$

$x^2 = 4$ or $x^2 = -3$

$x = \pm 2$ or $x = \pm\sqrt{-3} = \pm\sqrt{3}\,i$

20.

$4\left(\frac{x+2}{3x}\right)^2 - 4\left(\frac{x+2}{3x}\right) - 3 = 0$ let $u = \frac{x+2}{3x}$

$4u^2 - 4u - 3 = 0$

$(2u - 3)(2u + 1) = 0$

$u = \frac{3}{2}$ or $u = -\frac{1}{2}$

$\frac{x+2}{3x} = \frac{3}{2}$ or $\frac{x+2}{3x} = -\frac{1}{2}$

$9x = 2(x + 2)$

$9x = 2x + 4$

$7x = 4$

$x = \frac{4}{7}$

$-3x = 2(x + 2)$

$-3x = 2x + 4$

$-5x = 4$

$x = -\frac{4}{5}$

21. $f(x) = 2x^2 + x - 1$

$$x = -\frac{b}{2a} = -\frac{1}{2(2)} = -\frac{1}{4}$$

$$f\left(-\frac{b}{2a}\right) = f\left(-\frac{1}{4}\right) = 2\left(-\frac{1}{4}\right)^2 + \left(-\frac{1}{4}\right) - 1$$

$$= 2\left(\frac{1}{16}\right) - \frac{1}{4} - 1 = -\frac{9}{8}$$

Vertex: $(-\frac{1}{4}, -\frac{9}{8})$

Axis of symmetry: $x = -\frac{1}{4}$

x	$f(x)$
-2	5
-1	0
$-\frac{1}{4}$	$-\frac{9}{8}$
0	-1
1	2

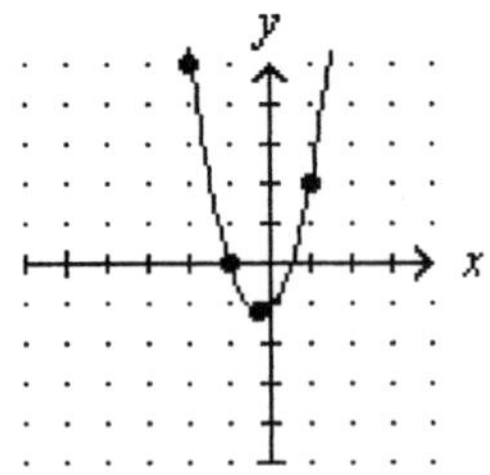

22. $f(x) = -3(x-1)^2 - 2$

Vertex: $(1, -2)$

Axis of symmetry: $x = 1$

x	$f(x)$
-1	-10
0	-5
1	-2
2	-5
3	-10

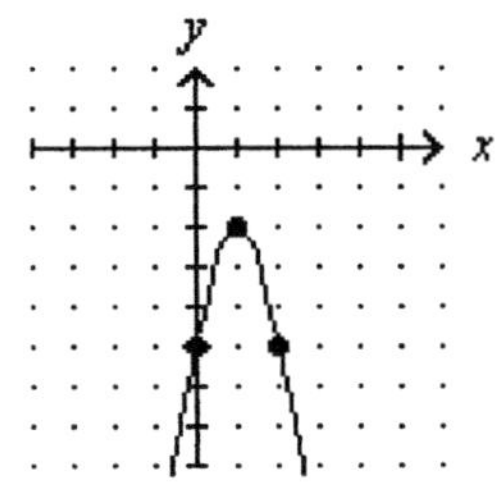

23.

$$x^2 - 2x - 8 > 0$$
$$(x-4)(x+2) > 0$$

$x + 2$ ---0+++|+++

$x - 4$ ---|---0+++

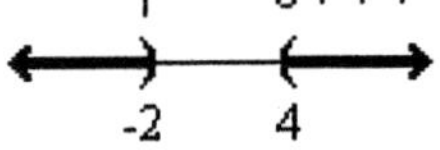

Solution: $(-\infty, -2) \cup (4, \infty)$

24. $\frac{x-2}{x+3} \le 0$

$x + 3$ ---0+++|+++

$x - 2$ ---|---0+++

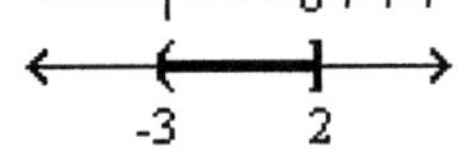

Solution: $(-3, 2]$

25. $y \le -x^2 + 3$ Vertex: $(0, 3)$

Solid line. Test point $(0, 0)$: $0 \le 3$ is true.

x	y
-2	-1
-1	2
0	3
1	2
2	-1

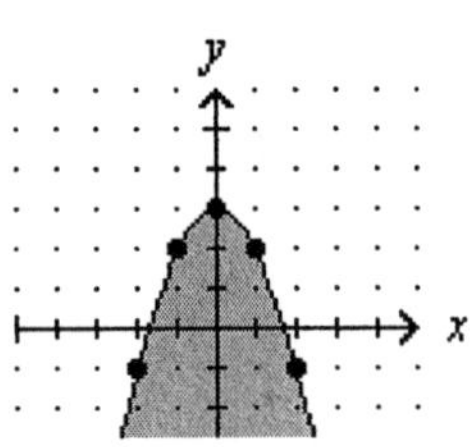

26. **Analysis**: Find the length of the centerline of the drawing which is two times the length of the shorter leg of the triangle.

Equation: Let x represent the length of the shorter leg, then $x + 14$ represents the longer leg. The length of the hypotenuse is 26 in. Use the Pythagorean theorem.

a^2	plus	b^2	equals	c^2.
x^2	+	$(x+14)^2$	=	26^2

Solve:

$$\begin{aligned} x^2 + (x+14)^2 &= 26^2 \\ x^2 + x^2 + 28x + 196 &= 676 \\ 2x^2 + 28x - 480 &= 0 \end{aligned}$$

$a = 2;\ b = 28;\ c = -480$

$$x = \frac{-28 \pm \sqrt{28^2 - 4(2)(-480)}}{2(2)}$$

$$= \frac{-28 \pm \sqrt{784 + 3840}}{4}$$

$$= \frac{-28 \pm \sqrt{4624}}{4} = \frac{-28 \pm 68}{4}$$

$$x = \frac{-28 - 68}{4} = -24$$

or

$$x = \frac{-28 + 68}{4} = 10$$

Discard negative solution.

Conclusion: The length of the shorter leg is $x = 10$ in. The length of the segment extending from the ground to the top of the building is $2x = 20$ in.

27. Number iii. is the best model. The negative coefficient for x^2 indicates parabola curved downward and the vertex at (0,6) indicates a standard parabola, shifted up 6 units.

28. **Analysis**: Determine how long it takes the chef alone to make a dessert if together he and his assistant can make the dessert in 25 minutes.

Equation: Let x represent the length of time it takes the chef alone. Then $x + 8$ represents the time it takes his assistant alone.

Chef rate: 1 dessert/ x min.		Asst rate: 1 dessert/x+8 min.		Combined rate: 1 dessert/ 25 min.
$\frac{1}{x}$	+	$\frac{1}{x+8}$	=	$\frac{1}{25}$

Solve:

$$\frac{1}{x} + \frac{1}{x+8} = \frac{1}{25}$$

$$25x(x+8)\left(\frac{1}{x} + \frac{1}{x+8}\right) = \left(\frac{1}{25}\right)25x(x+8)$$

$$25(x+8) + 25x = x(x+8)$$

$$25x + 200 + 25x = x^2 + 8x$$

$$x^2 - 42x - 200 = 0$$

$a = 1 \quad b = -42 \quad c = -200$

$$x = \frac{42 \pm \sqrt{(-42)^2 - 4(1)(-200)}}{2(1)} = \frac{42 \pm \sqrt{2564}}{2} \approx \frac{42 \pm 50.63}{2}$$

$$x \approx \frac{42 - 50.63}{2} = -4.3 \quad \text{or} \quad x \approx \frac{42 + 50.63}{2} = 46.315$$

Discard the negative answer

Conclusion: The chef can make one dessert in about 46 minutes.

29. The flare should explode when it reaches the vertex of the parabola given by the equation $h = -16t^2 + 112t + 15$.

$$t = -\frac{b}{2a} = -\frac{112}{2(-16)} = 3.5 \text{ sec} \qquad h\left(-\frac{b}{2a}\right) = h(3.5) = -16(3.5)^2 + 112(3.5) = 211 \text{ ft}$$

The flare should explode 3.5 seconds after being fired at a height of 211 feet.

30. An imaginary number is the square root of a negative number. The imaginary number $\sqrt{-1}$ is denoted by i. Examples will vary.

31. Find the values of x where the graph crosses the x–axis. The values are $x = -3$ and $x = 2$.

32. The solution of the inequality $ax^2 + bx + c \leq 0$ would be those values of x where the graph is below the x–axis. The interval will include the points where the graph crosses because the inequality is less than or equal. The interval solution is $[-3, 2]$.

Chapters 1–8 Cumulative Review Exercises

1. $m = 3$ and $(x_1, y_1) = (-2, -4)$.

$$\begin{aligned} y - y_1 &= m(x - x_1) \\ y - (-4) &= 3[x - (-2)] \\ y + 4 &= 3(x + 2) \\ y + 4 &= 3x + 6 \\ y &= 3x + 2 \end{aligned}$$

2. Find the slope by solving $2x + 3y = 6$ for y.

$$\begin{aligned} 2x + 3y &= 6 \\ 3y &= -2x + 6 \\ y &= -\tfrac{2}{3}x + 2 \qquad m = -\tfrac{2}{3} \end{aligned}$$

Use the point $(0, -2)$.

$$\begin{aligned} y - y_1 &= m(x - x_1) \\ y - (-2) &= -\tfrac{2}{3}[x - (0)] \\ y + 2 &= -\tfrac{2}{3}x \\ y &= -\tfrac{2}{3}x - 2 \end{aligned}$$

3. Data points (y in thousands): $(2000, 800)$ and $(2015, 1000)$.

$$m = \frac{1000 - 800}{2015 - 2000} = \frac{200}{15} = 13.333 \text{ thousands}$$

The increase is about 13,333 a year.

4. Equation #1: $y = -\frac{5}{2}x + \frac{1}{2}$

Graph: $2y = -5x + 1$

x	y
1	−2
−1	3

Equation #2: $2x - \frac{3}{2}y = 5$

Graph: $4x - 3y = 10$

x	y
1	−2
4	2

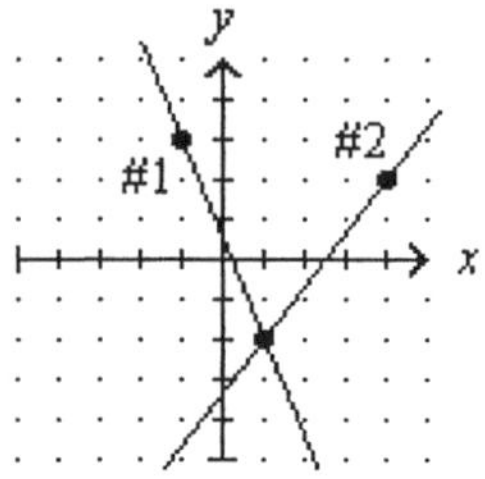

Solution: $(1, -2)$

5. $\begin{cases} x - y + z = 4 \\ x + 2y - z = -1 \\ x + y - 3z = -2 \end{cases}$

$$x = \frac{D_x}{D} = \frac{\begin{vmatrix} 4 & -1 & 1 \\ -1 & 2 & -1 \\ -2 & 1 & -3 \end{vmatrix}}{\begin{vmatrix} 1 & -1 & 1 \\ 1 & 2 & -1 \\ 1 & 1 & -3 \end{vmatrix}} = \frac{4\begin{vmatrix} 2 & -1 \\ 1 & -3 \end{vmatrix} - (-1)\begin{vmatrix} -1 & 1 \\ 1 & -3 \end{vmatrix} + (-2)\begin{vmatrix} -1 & 1 \\ 2 & -1 \end{vmatrix}}{1\begin{vmatrix} 2 & -1 \\ 1 & -3 \end{vmatrix} - 1\begin{vmatrix} -1 & 1 \\ 1 & -3 \end{vmatrix} + 1\begin{vmatrix} -1 & 1 \\ 2 & -1 \end{vmatrix}} = \frac{4(-6+1) + 1(3-1) - 2(1-2)}{1(-6+1) - 1(3-1) + 1(1-2)} = \frac{-20+2+2}{-5-2-1} = \frac{-16}{-8} = 2$$

$$y = \frac{D_y}{D} = \frac{\begin{vmatrix} 1 & 4 & 1 \\ 1 & -1 & -1 \\ 1 & -2 & -3 \end{vmatrix}}{-8} = \frac{1\begin{vmatrix} -1 & -1 \\ -2 & -3 \end{vmatrix} - 1\begin{vmatrix} 4 & 1 \\ -2 & -3 \end{vmatrix} + 1\begin{vmatrix} 4 & 1 \\ -1 & -1 \end{vmatrix}}{-8} = \frac{1(3-2) - 1(-12+2) + 1(-4+1)}{-8} = \frac{1+10-3}{-8} = \frac{8}{-8} = -1$$

$$z = \frac{D_z}{D} = \frac{\begin{vmatrix} 1 & -1 & 4 \\ & 2 & -1 \\ 1 & 1 & -2 \end{vmatrix}}{-8} = \frac{1\begin{vmatrix} 2 & -1 \\ 1 & -2 \end{vmatrix} - 1\begin{vmatrix} -1 & 4 \\ 1 & -2 \end{vmatrix} + 1\begin{vmatrix} -1 & 4 \\ 2 & -1 \end{vmatrix}}{-8} = \frac{1(-4+1) - 1(2-4) + 1(1-8)}{-8} = \frac{-3+2-7}{-8} = \frac{-8}{-8} = 1$$

The solution is $(2, -1, 1)$.

6. Inequality #1: $3x + 2y > 6$
Graph: $3x + 2y = 6$

x	y
0	3
2	0

Broken line. Test point (0, 0).
$3(0) + 2(0) > 6$
$0 > 6$ False

Inequality #2: $x + 3y \le 2$
Graph: $x + 3y = 2$

x	y
−1	1
2	0

Solid line. Test point (0, 0).
$(0) + 3(0) < 2$
$0 < 2$ True

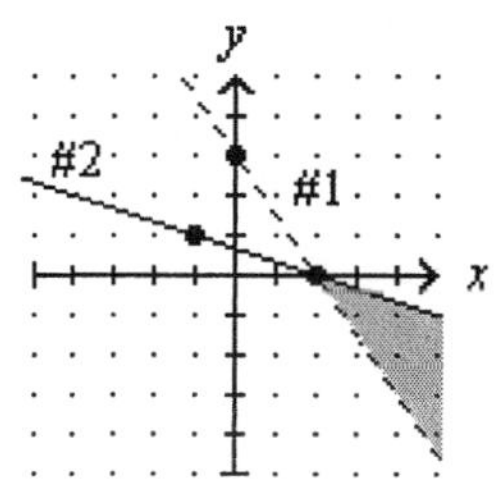

7. $-9x + 6 > 16$
$-9x > 10$ Subtract 6 from both sides.
$x < -\frac{10}{9}$ Divide by −9 and reverse the symbol.

$(-\infty, -\frac{10}{9})$

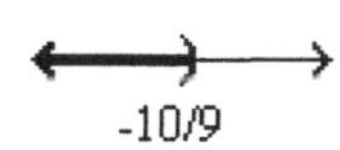

8. Rewrite $|2x - 5| \ge 25$ as
$2x - 5 \le -25$ or $2x - 5 \ge 25$
$2x \le -20$ $\quad$ $2x \ge 30$
$x \le -10$ $\quad$ $x \ge 15$

$(-\infty, -10] \cup [15. \infty)$

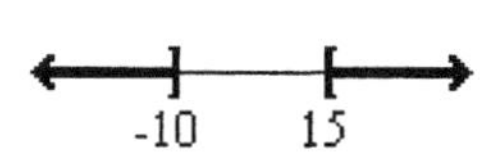

9.

x	$f(x) = 2x^2 - 3$	$f(x)$
−2	$f(-2) = 2(-2)^2 - 3$	5
−1	$f(-1) = 2(-1)^2 - 3$	−1
0	$f(0) = 2(0)^2 - 3$	−3
1	$f(1) = 2(1)^2 - 3$	−1
2	$f(2) = 2(2)^2 - 3$	5

D = $(-\infty, \infty)$ R = $[-3, \infty)$

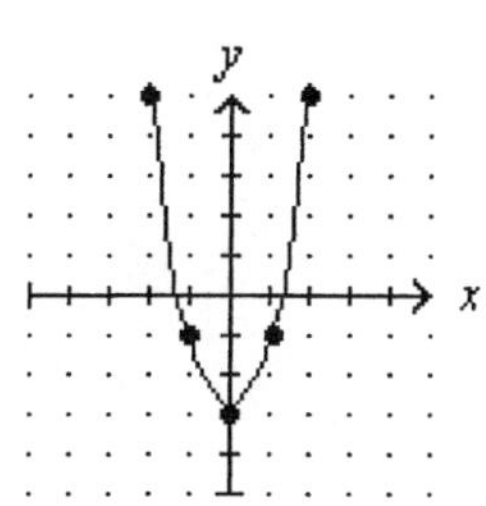

10. There are no values of x which need to be excluded, so D $= (-\infty, \infty)$.
The largest value of y that is possible is 0 when $x = 4$ because the result of taking the absolute value will become negative in the final calculation. So, $R = (-\infty, 0]$.

11. $(2a^2 + 4a - 7) - 2(3a^2 - 4a) = 2a^2 + 4a - 7 - 6a^2 + 8a = -4a^2 + 12a - 7$

12. $8(3x + 2)(2x - 3) = 8(6x^2 - 9x + 4x - 6) = 8(6x^2 - 5x - 6) = 48x^2 - 40x - 48$

13. $x^4 - 16y^4 = (x^2)^2 - (4y^2)^2 = (x^2 + 4y^2)(x^2 - 4y^2) = (x^2 + 4y^2)(x + 2y)(x - 2y)$ Difference of 2 squares, twice.

14. $15x^2 - 2x - 8 = 15x^2 - 12x + 10x - 8 = 3x(5x - 4) + 2(5x - 4) = (5x - 4)(3x + 2)$

15. $x^2 z + 4yz - xyz - 4x = x^2 z - xyz - 4xz + 4yz = xz(x - y) - 4z(x - y) = z(x - y)(x - 4)$ Rearrange, then group.

16. $8x^6 + 125y^3 = (2x^2)^3 + (5y)^3 = (2x^2 + 5y)(4x^4 - 10x^2y + 25y^2)$

17.
$$x^2 - 5x - 6 = 0$$
$$(x - 6)(x + 1) = 0$$
$$x - 6 = 0 \quad \text{or} \quad x + 1 = 0$$
$$x = 6 \qquad\qquad x = -1$$

18.
$$6a^3 - 2a = a^2$$
$$6a^3 - a^2 - 2a = 0$$
$$a(6a^2 - a - 2) = 0$$
$$a(2a + 1)(3a - 2) = 0$$
$$a = 0 \quad \text{or} \quad 2a + 1 = 0 \quad \text{or} \quad 3a - 2 = 0$$
$$2a = -1 \qquad 3a = 2$$
$$a = -\tfrac{1}{2} \qquad a = \tfrac{2}{3}$$

19.
$$\frac{x-4}{x-3} + \frac{x-2}{x-3} = x-3$$
$$(x-3)\left(\frac{x-4}{x-3} + \frac{x-2}{x-3}\right) = (x-3)(x-3)$$
$$(x - 4) + (x - 2) = x^2 - 6x + 9$$
$$2x - 6 = x^2 - 6x + 9$$
$$0 = x^2 - 8x + 15$$
$$0 = (x - 3)(x - 5)$$
$$x - 3 = 0 \quad \text{or} \quad x - 5 = 0$$
$$x = 3 \qquad\qquad x = 5$$

Extraneous. Makes denominator zero.

20.
$$P + \frac{a}{V^2} = \frac{RT}{V - b}$$
$$V^2(V-b)\left(P + \frac{a}{V^2}\right) = V^2(V-b)\left(\frac{RT}{V-b}\right)$$
$$PV^2(V - b) + a(V - b) = RTV^2$$
$$PV^3 - bPV^2 + aV - ab = RTV^2$$
$$-bPV^2 - ab = RTV^2 - aV - PV^3$$
$$-b(PV^2 + a) = RTV^2 - aV - PV^3$$
$$b = \frac{RTV^2 - aV - PV^3}{-(PV^2 + a)}$$
$$b = \frac{-RTV^2 + aV + PV^3}{PV^2 + a}$$

21. $$\frac{x^3 + y^3}{x^3 - y^3} \div \frac{x^2 - xy + y^2}{x^2 + xy + y^2} = \frac{(x+y)(x^2 - xy + y^2)}{(x-y)(x^2 + xy + y^2)} \bullet \frac{x^2 + xy + y^2}{x^2 - xy + y^2} = \frac{x+y}{x-y}$$

22. $$\frac{1}{x+y} - \frac{1}{x-y} - \frac{2y}{y^2 - x^2} = \frac{1(x-y)}{(x+y)(x-y)} - \frac{1(x+y)}{(x-y)(x+y} - \frac{(-1)2y}{(-1)(y^2 - x^2)}$$
$$= \frac{(x-y) - (x+y) - (-2y)}{(x+y)(x-y)} = \frac{x - y - x - y + 2y}{(x+y)(x-y)} = 0$$

23.

x	$f(x) = x^3 + x^2 - 6x$	$f(x)$
-3	$f(-3) = (-3)^3 + (-3)^2 - 6(-3)$	0
-2	$f(-2) = (-2)^3 + (-2)^2 - 6(-2)$	8
-1	$f(-1) = (-1)^3 + (-1)^2 - 6(-1)$	6
0	$f(0) = (0)^3 + (0)^2 - 6(0)$	0
1	$f(1) = (1)^3 + (1)^2 - 6(1)$	-4
2	$f(2) = (2)^3 + (2)^2 - 6(2)$	0
3	$f(3) = (3)^3 + (3)^2 - 6(3)$	18

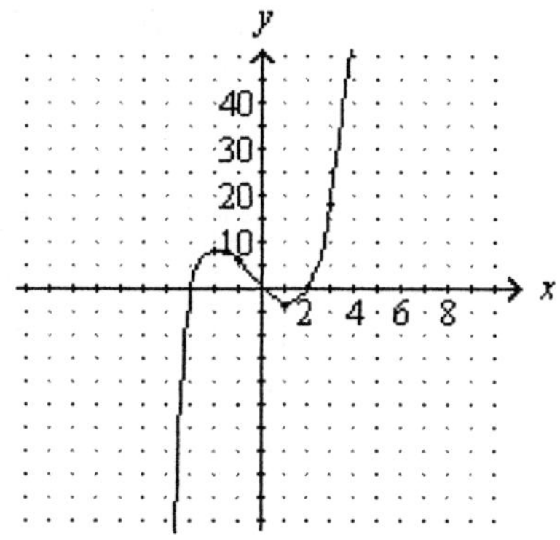

D = $(-\infty, \infty)$
R = $(-\infty, \infty)$

24.

x	$f(x) = \frac{4}{x}$	$f(x)$
$\frac{1}{4}$	$f(\frac{1}{4}) = \frac{4}{1/4}$	16
$\frac{1}{2}$	$f(\frac{1}{2}) = \frac{4}{1/2}$	8
1	$f(1) = \frac{4}{1}$	4
2	$f(2) = \frac{4}{2}$	2
4	$f(4) = \frac{4}{4}$	1
8	$f(8) = \frac{4}{8}$	$\frac{1}{2}$
16	$f(16) = \frac{4}{16}$	$\frac{1}{4}$

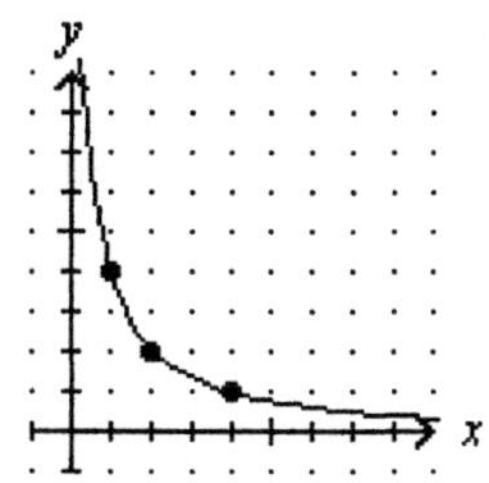

D = $(0, \infty)$
R = $(0, \infty)$

25. Let A represent the area of the spread of the intensity of the light energy and d the distance from the source. The area varies directly with the distance from the source.

Direct variation model

$$A = kd^2$$

Determine k.

$$4 = k(2)^2$$
$$4 = k(4)$$
$$1 = k$$

Solve for A when $d = 3$ ft.

$$A = 1(3)^2$$
$$= 9$$

26.

x	$f(x) = \sqrt{x-2}$	$f(x)$
2	$f(2) = \sqrt{2-2} = \sqrt{0}$	0
3	$f(3) = \sqrt{3-2} = \sqrt{1}$	1
6	$f(6) = \sqrt{6-2} = \sqrt{4}$	2
11	$f(11) = \sqrt{11-2} = \sqrt{9}$	3

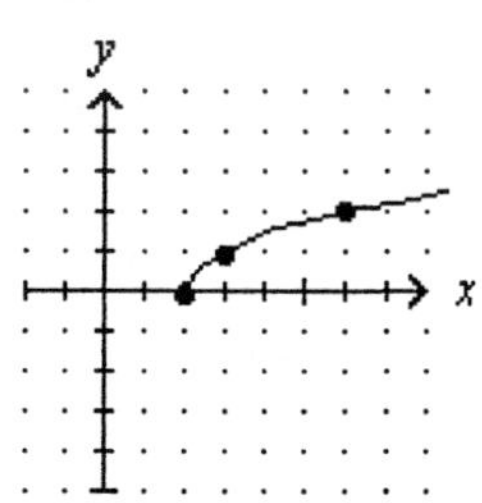

D = $[2, \infty)$
R = $[0, \infty)$

27. $\sqrt[3]{-27x^3} = \sqrt[3]{(-3x)^3} = -3x$

28. $\sqrt{48t^3} = \sqrt{16t^2 \bullet 3t} = 4t\sqrt{3t}$

29. $64^{-2/3} = \frac{1}{64^{2/3}} = \frac{1}{\left(\sqrt[3]{64}\right)^2} = \frac{1}{4^2} = \frac{1}{16}$

30. $\frac{x^{5/3}x^{1/2}}{x^{3/4}} = x^{\frac{5}{3}+\frac{1}{2}-\frac{3}{4}} = x^{\frac{20}{12}+\frac{6}{12}-\frac{9}{12}} = x^{17/12}$

31. $-3\sqrt[4]{32}-2\sqrt[4]{162}+5\sqrt[4]{48}=-3\sqrt[4]{16\cdot 2}-2\sqrt[4]{81\cdot 2}+5\sqrt[4]{16\cdot 3}$

$$=-3\cdot 2\sqrt[4]{2}-2\cdot 3\sqrt[4]{2}+5\cdot 2\sqrt[4]{3}=-6\sqrt[4]{2}-6\sqrt[4]{2}+10\sqrt[4]{3}=-12\sqrt[4]{2}+10\sqrt[4]{3}$$

32. $3\sqrt{2}\left(2\sqrt{3}-4\sqrt{12}\right)=3\cdot 2\sqrt{2\cdot 3}-3\cdot 4\sqrt{2\cdot 12}=6\sqrt{6}-12\left(2\sqrt{6}\right)=6\sqrt{6}-24\sqrt{6}=-18\sqrt{6}$

33. $\dfrac{\sqrt{x}+2}{\sqrt{x}-1}=\dfrac{\left(\sqrt{x}+2\right)\left(\sqrt{x}+1\right)}{\left(\sqrt{x}-1\right)\left(\sqrt{x}+1\right)}=\dfrac{x+\sqrt{x}+2\sqrt{x}+1}{x-1}=\dfrac{x+3\sqrt{x}+1}{x-1}$

34. $\dfrac{5}{\sqrt[3]{x}}=\dfrac{5\sqrt[3]{x^2}}{\sqrt[3]{x}\sqrt[3]{x^2}}=\dfrac{5\sqrt[3]{x^2}}{\sqrt[3]{x^3}}=\dfrac{5\sqrt[3]{x^2}}{x}$

35.

$5\sqrt{x+2}=x+8$	
$\left(5\sqrt{x+2}\right)^2=(x+8)^2$	Square both sides.
$25(x+2)=x^2+16x+64$	Do the squaring.
$25x+50=x^2+16x+64$	Simplify.
$0=x^2-9x+14$	Subtract 50 and $25x$.
$0=(x-7)(x-2)$	Factor.

$x-7=0$ or $x-2=0$

$x=7$ $\quad$ $x=2$

Check:

$5\sqrt{(7)+2}\overset{?}{=}7+8$

$5\sqrt{9}\overset{?}{=}15$

$5(3)=15$

and

$5\sqrt{(2)+2}\overset{?}{=}2+8$

$5\sqrt{4}=5(2)=10$

36.

$\sqrt{x}+\sqrt{x+2}=2$	
$\sqrt{x+2}=2-\sqrt{x}$	Subtract $\sqrt{x}$.
$\left(\sqrt{x+2}\right)^2=\left(2-\sqrt{x}\right)^2$	Square both sides.
$x+2=4-4\sqrt{x}+x$	Do the squaring.
$-2=-4\sqrt{x}$	Simplify.
$1=2\sqrt{x}$	Divide by −2.
$1^2=\left(2\sqrt{x}\right)^2$	Square both sides.
$1=4x$	Do the squaring.
$\dfrac{1}{4}=x$	Divide by 4.

Check:

$\sqrt{\tfrac{1}{4}}+\sqrt{\left(\tfrac{1}{4}\right)+2}\overset{?}{=}2$

$\tfrac{1}{2}+\sqrt{\tfrac{9}{4}}\overset{?}{=}2$

$\tfrac{1}{2}+\tfrac{3}{2}=2$

37. 45°–45°–90° triangle.

Hypotenuse is length of leg times $\sqrt{2}$.

$h=3\sqrt{2}$ in.

38. 30°–60°–90° triangle.

Longer leg is shorter leg times $\sqrt{3}$.

$3=x\sqrt{3}$

$x=\dfrac{3}{\sqrt{3}}=\dfrac{3\sqrt{3}}{3}=\sqrt{3}$

Hypotenuse is twice as long as shorter leg.

$h=2\left(\sqrt{3}\right)=2\sqrt{3}$

39. $P(x_1, y_1) = P(-2, 6),\ Q(x_2, y_2) = (4, 14)$

$$\begin{aligned} d(PQ) &= \sqrt{(x_2 - x_1)^2 + (y_2 - y_1)^2} \\ &= \sqrt{[4-(-2)]^2 + (14-6)^2} \\ &= \sqrt{6^2 + 8^2} \\ &= \sqrt{36+64} \\ &= \sqrt{100} \\ &= 10 \end{aligned}$$

40. $\left[\frac{1}{2}(6)\right]^2 = 3^2 = 9$

41.

$2x^2 + x - 3 = 0$

$x^2 + \frac{1}{2}x - \frac{3}{2} = 0$ Divide both sides by 2.

$x^2 + \frac{1}{2}x = \frac{3}{2}$ Add $\frac{3}{2}$ to both sides.

$x^2 + \frac{1}{2}x + \frac{1}{16} = \frac{3}{2} + \frac{1}{16}$ Square one–half of the coefficient of x and add it to both sides.

$\left(x + \frac{1}{4}\right)^2 = \frac{25}{16}$ Factor and combine like terms.

$x + \frac{1}{4} = \pm\sqrt{\frac{25}{16}} = \pm\frac{5}{4}$ Apply the square root property.

$x = -\frac{1}{4} \pm \frac{5}{4}$ Subtract $\frac{1}{4}$ from both sides.

$x = -\frac{1}{4} + \frac{5}{4} = \frac{4}{4} = 1$ or $x = -\frac{1}{4} - \frac{5}{4} = -\frac{6}{4} = -\frac{3}{2}$ Simplify.

42. $3x^2 + 4x - 1 = 0$

$a = 3;\ b = 4;\ c = -1$

$$\begin{aligned} x &= \frac{-4 \pm \sqrt{4^2 - 4(3)(-1)}}{2(3)} \\ &= \frac{-4 \pm \sqrt{16+12}}{6} \\ &= \frac{-4 \pm \sqrt{28}}{6} = \frac{-4 \pm 2\sqrt{7}}{6} \\ &= \frac{2\left(-2 \pm \sqrt{7}\right)}{2 \bullet 3} = \frac{-2 \pm \sqrt{7}}{3} \end{aligned}$$

43. $f(x) = \frac{1}{2}x^2 - x + 1$

$$x = -\frac{b}{2a} = -\frac{-1}{2(\frac{1}{2})} = -\frac{1}{1} = 1$$

$$f(1) = \frac{1}{2}(1)^2 - (1) + 1 = \frac{1}{2}$$

Vertex: $(1, \frac{1}{2})$

x	$f(x)$
−2	5
0	1
1	$\frac{1}{2}$
2	1
4	5

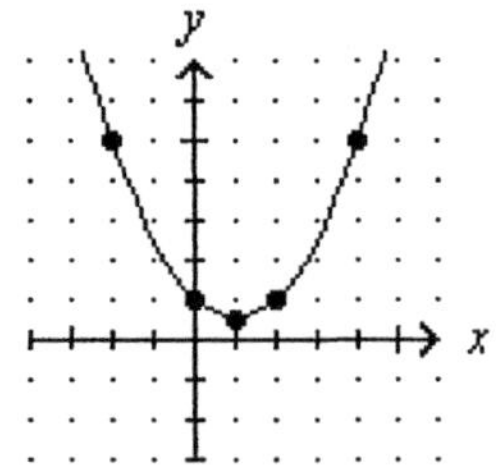

44. $f(x) = -x^2 - 4x$

$$x = -\frac{b}{2a} = -\frac{-4}{2(-1)} = -\frac{4}{2} = -2$$

$$f(-2) = -(-2)^2 - 4(-2) = 4$$

Vertex: $(-2, 4)$

x	$f(x)$
−4	0
−3	3
−2	**4**
−1	3
0	0

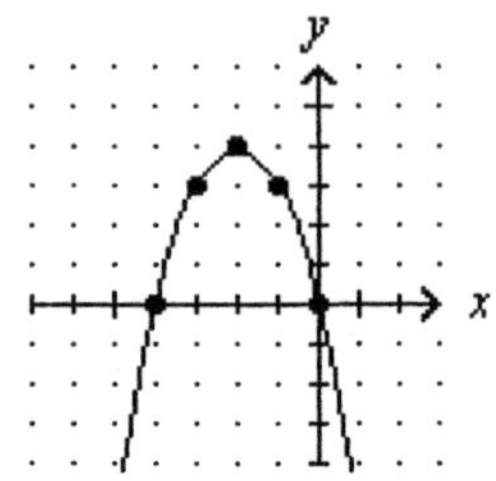

45. $(3 + 5i) + (4 - 3i) = (3 + 4) + (5 - 3)i = 7 + 2i$

46. $\frac{5}{3-i} = \frac{5(3+i)}{(3-i)(3+i)} = \frac{15+5i}{9-i^2} = \frac{15+5i}{9+1} = \frac{5(3+i)}{10} = \frac{3+i}{2} = \frac{3}{2} + \frac{1}{2}i$

47. $\sqrt{-64} = \sqrt{-1}\sqrt{64} = 8i$

48.

$$\begin{aligned} a - 7a^{1/2} + 12 &= 0 \\ (a^{1/2})^2 - 7a^{1/2} + 12 &= 0 && \text{Write } a \text{ as } (a^{1/2})^2. \\ y^2 - 7y + 12 &= 0 && \text{Let } y = a^{1/2}. \\ (y - 4)(y - 3) &= 0 && \text{Factor and solve.} \end{aligned}$$

$y - 4 = 0$ or $y - 3 = 0$

$y = 4$ $\quad$ $y = 3$

Undo the substitution.

$a^{1/2} = 4$ or $a^{1/2} = 3$

$a = 16$ $\quad$ $a = 9$

Check: $\quad$ Check:

$(16) - 7(16)^{1/2} + 4$ $\quad$ $(9) - 7(9)^{1/2} + 12$

$= 16 - 20 + 4 = 0$ $\quad$ $= 9 - 21 + 12 = 0$

The solutions are 16 and 9.

49.

$$\begin{aligned} 2x^2 + 25 &= 0 \\ 2x^2 &= -25 && \text{Rearrange.} \\ x^2 &= -\frac{25}{2} \\ x &= \end{aligned}$$

$$= \pm\frac{5\sqrt{2}}{\sqrt{2}\sqrt{2}}i = \pm\frac{5\sqrt{2}}{2}i$$

50. $i^{42} = i^{4 \bullet 10} \bullet i^2 = 1 \bullet (-1) = -1$

51. Find the value of x where the graph touches the x–axis. The value is between 0 and −1. Estimated at $x = -\frac{3}{4}$.

52. The solution of the inequality would be those values x below the x–axis. Since the graph does not extend below the x–axis there is no solution.

Homework is not all about the right answers.
Homework is about finding out what works and what doesn't.

Chapter 9 Exponential and Logarithmic Functions

Study Tip
Preparing for a Cumulative Final Exam

You should start preparing for the final exam at least two weeks prior to the Exam Day. Here are some suggestions:

1. Redo all the regular tests you took throughout the semester.
2. Read the directions for all the review exercises for chapters you covered in your class. Try to decide how to handle each type of problem. Do one problem from each category of problems. If you don't remember how to do a type of problem, go back and look in the section covering that type and review the material.
3. Do all the Chapter Tests for the chapters you covered. (And check your answers)
4. Do all the Cumulative Review problem sets for the chapters you covered. (And check your answers)
5. Ask your instructor for tips on how to prepare.

Section 9.1 Algebra and Composition of Functions

VOCABULARY

1. sum 3. product 5. domain 7. identity

CONCEPTS

9. a. $g(x)$ b. $g(x)$

11.

x	$I(x) = x$
−3	**−3**
−2	**−2**
−1	**−1**
0	**0**
1	**1**
2	**2**
3	**3**

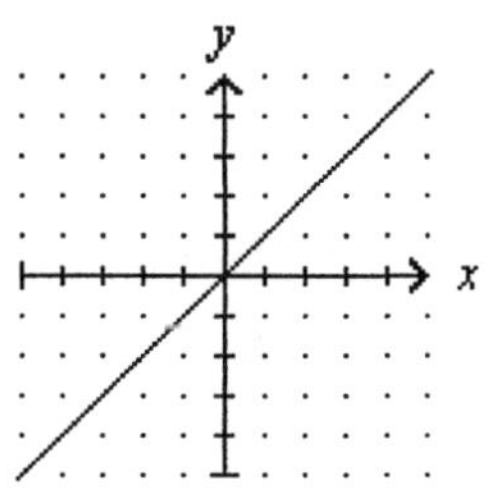

NOTATION

13. $(f \bullet g)(x) = f(x) \bullet g(x)$
$= \mathbf{(3x - 1)}\ (2x + 3)$
$= 6x^2 + \mathbf{9x} - \mathbf{2x} - 3$
$= 6x^2 + 7x - 3$

PRACTICE

15. $f + g = f(x) + g(x) = 3x + 4x = 7x$ D $= (-\infty, \infty)$

17. $f \bullet g = f(x) \bullet g(x) = 3x(4x) = 12x^2$ D $= (-\infty, \infty)$

19. $g - f = g(x) - f(x) = 4x - 3x = x$ D $= (-\infty, \infty)$

21. $g/f = \dfrac{g(x)}{f(x)} = \dfrac{4x}{3x} = \dfrac{4}{3}$ D $= (-\infty, 0) \cup (0, \infty)$

23. $f + g = f(x) + g(x) = (2x + 1) + (x - 3) = 3x - 2$ D $= (-\infty, \infty)$

25. $f \bullet g = f(x) \bullet g(x) = (2x + 1)(x - 3) = 2x^2 - 6x + x - 3 = 2x^2 - 5x - 3$ D $= (-\infty, \infty)$

27. $g - f = g(x) - f(x) = (x - 3) - (2x + 1) = x - 3 - 2x - 1 = -x - 4$ D $= (-\infty, \infty)$

29. $g/f = \dfrac{g(x)}{f(x)} = \dfrac{x-3}{2x+1}$ D $= (-\infty, -\frac{1}{2}) \cup (-\frac{1}{2}, \infty)$

31. $f - g = f(x) - g(x) = (3x - 2) - (2x^2 + 1) = 3x - 2 - 2x^2 - 1 = -2x^2 + 3x - 3$ D = $(-\infty, \infty)$

33. $f/g = \dfrac{f(x)}{g(x)} = \dfrac{3x-2}{2x^2+1}$ D = $(-\infty, \infty)$

35. $f - g = f(x) - g(x) = (x^2 - 1) - (x^2 - 4) = x^2 - 1 - x^2 + 4 = 3$ D = $-\infty, \infty)$

37. $g/f = \dfrac{g(x)}{f(x)} = \dfrac{x^2-4}{x^2-1}$ D = $(-\infty, -1) \cup (-1, 1) \cup (1, \infty)$

39. $(f \circ g)(2) = f(g(2)) = f(2^2 - 1) = f(3) = 2(3) + 1 = 7$

41. $(g \circ f)(-3) = g(f(-3)) = g(2(-3) + 1) = g(-5) = (-5)^2 - 1 = 24$

43. $(f \circ g)(0) = f(g(0)) = f(-1) = 2(-1) + 1 = -1$

45. $(f \circ g)\left(\frac{1}{2}\right) = f\left(g\left(\frac{1}{2}\right)\right) = f\left(\left(\frac{1}{2}\right)^2 - 1\right) = f\left(-\frac{3}{4}\right) = 2\left(-\frac{3}{4}\right) + 1 = -\frac{1}{2}$

47. $(f \circ g)(x) = f(g(x)) = f(x^2 - 1) = 2(x^2 - 1) + 1 = 2x^2 - 2 + 1 = 2x^2 - 1$

49. $(g \circ f)(2x) = g(f(2x)) = g(2(2x) + 1) = g(4x + 1) = (4x + 1)^2 - 1 = 16x^2 + 8x + 1 - 1 = 16x^2 + 8x$

51. $(f \circ g)(4) = f(g(4)) = f(4^2 + 4) = f(20) = 3(20) - 2 = 58$

53. $(g \circ f)(-3) = g(f(-3)) = g(3(-3) - 2) = g(-11) = (-11)^2 + (-11) = 110$

55. $(g \circ f)(0) = g(f(0)) = g(3(0)–2) = g(-2) = (-2)^2 + (-2) = 2$

57. $(g \circ f)(x) = g(f(x)) = g(3x - 2) = (3x - 2)^2 + (3x - 2) = 9x^2 -12x + 4 + 3x - 2 = 9x^2 - 9x + 2$

59. $(f \circ g)(x) = f(g(x) = f(2x - 5) = (2x - 5) + 1 = 2x - 4$

$(g \circ f)(x) = g(f(x) = g(x + 1) = 2(x + 1) - 5 = 2x + 2 - 5 = 2x - 3$

APPLICATIONS

61. The temperature in Fahrenheit at any time t can be represented by the function: $F(t) = 2700 - 200t$.
The formula for converting Fahrenheit to Celsius is $C(F) = \frac{5}{9}(F - 32)$. Find the composite function $(C \circ F)(t)$.

$$C(t) = (C \circ F)(t) = C(F(t)) = C(2700 - 200t) = \frac{5}{9}[(2700 - 200t) - 32] = \frac{5}{9}(2668 - 200t)$$

63. a. $(C \circ G)(500) = C(G(500)) = C\left(\frac{500}{20}\right) = C(25) = 1.50(25) = \$37.50$

b. $C(m) = (C \circ G)(m) = C(G(m)) = C\left(\frac{m}{20}\right) = 1.50\left(\frac{m}{20}\right) = \frac{1.50m}{20}$

WRITING

65. Answers will vary.

67. $(f \circ g)(2)$ would be "f of g of 2." $g(f(-8))$ would be "g of f of –8."

REVIEW

69. $\dfrac{3x^2 + x - 14}{4 - x^2} = \dfrac{(3x+7)(x-2)}{(2+x)(2-x)} = \dfrac{x-2}{2-x} \cdot \dfrac{3x+7}{x+2} = -\dfrac{3x+7}{x+2}$

71. $\frac{x^2-2x-8}{3x^2-x-12} \div \frac{3x^2+5x-2}{3x-1} = \frac{x^2-2x-8}{3x^2-x-12} \cdot \frac{3x-1}{3x^2+5x-2} = \frac{(x-4)(x+2)}{3x^2-x-12} \cdot \frac{3x-1}{(3x-1)(x+2)} = \frac{x-4}{3x^2-x-12}$

Section 9.2 Inverses of Functions

VOCABULARY

1. one–to–one
3. inverses

CONCEPTS

5. once
7. 2
9. x

11. Interchange values of x and y.

x	$f(x)$
−6	−3
−4	−2
0	0
2	1
8	4

x	$f^{-1}(x)$
−3	**−6**
−2	**−4**
0	**0**
1	**2**
4	**8**

13. no
15. 2
17. The graphs are not symmetric about the line $y = x$.

NOTATION

19.
$$y = \mathbf{2x} - 3$$
$$x = \mathbf{2y} - 3$$
$$x + \mathbf{3} = 2y$$
$$\frac{x+3}{2} = \mathbf{y}$$

The inverse of $f(x)$ is $\boldsymbol{f^{-1}(x)} = \frac{x+3}{2}$

21. the inverse of; inverse

PRACTICE

23. Yes, it is one–to–one because different input values x determine different output values y.
25. No, because different input values x can produce the same output value y. For example, inputs of 3 and −3 produce the same output value of 81.
27. Yes, because there are no horizontal lines that will intersect the graph more than once.
29. No, because there are many horizontal lines that will intersect the graph more than once.
31. No, because one horizontal line would be the same as the given line.
33. It is a function. The inverse is: $\{(2, 3), (1, 2), (0, 1)\}$. The inverse is a function.
35. It is a function. The inverse is: $\{(1, 1), (1, 2), (1, 3)\ (1, 4)\}$. The inverse is not a function because $x = 1$ has many different y values.

37.

$f(x) = \frac{4}{5}x - 4$

$y = \frac{4}{5}x - 4$ Replace $f(x)$ with y.

$5y = 4x - 20$ Multiply by LCD 5.

$5x = 4y - 20$ Interchange x and y.

$5x + 20 = 4y$ Add 20.

$\frac{5}{4}x + 5 = y$ Divide by 4.

$f^{-1}(x) = \frac{5}{4}x + 5$ Rewrite.

39.

$f(x) = \frac{x}{5} + \frac{4}{5}$

$y = \frac{x}{5} + \frac{4}{5}$ Replace $f(x)$ with y.

$5y = x + 4$ Multiply by LCD 5.

$5x = y + 4$ Interchange x and y.

$5x - 4 = y$ Subtract 4.

$f^{-1}(x) = 5x - 4$ Rewrite.

41.

$f(x) = \frac{x-4}{5}$

$y = \frac{x-4}{5}$ Replace $f(x)$ with y.

$x = \frac{y-4}{5}$ Interchange x and y.

$5x = y - 4$ Multiply by 5.

$5x + 4 = y$ Add 4.

$f^{-1}(x) = 5x + 4$ Rewrite.

43.

$f(x) = 4x + 3$

$y = 4x + 3$

$x = 4y + 3$

$x - 3 = 4y$

$\frac{x-3}{4} = y = f^{-1}(x)$

Function (#1)		Inverse (#2)	
x	y	x	y
−2	−5	−5	−2
−1	−1	−1	−1
0	3	3	0

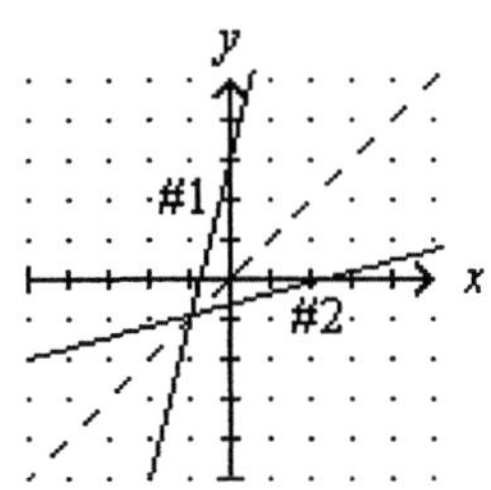

45.

$f(x) = -\frac{2}{3}x + 3$

$y = -\frac{2}{3}x + 3$

$3y = -2x + 9$

$3x = -2y + 9$

$2y = -3x + 9$

$y = -\frac{3}{2}x + \frac{9}{2} = f^{-1}(x)$

Function (#1)		Inverse (#2)	
x	y	x	y
−3	5	5	−3
0	3	3	0
3	1	1	3

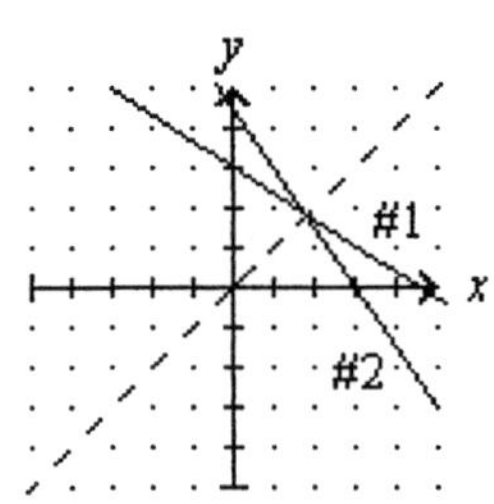

47.
$$\begin{aligned} f(x) &= x^3 \\ y &= x^3 \\ x &= y^3 \\ \sqrt[3]{x} &= y = f^{-1}(x) \end{aligned}$$

Function (#1)		Inverse (#2)	
x	y	x	y
−2	−8	−8	−2
−1	−1	−1	−1
0	0	0	0
1	1	1	1
2	8	8	2

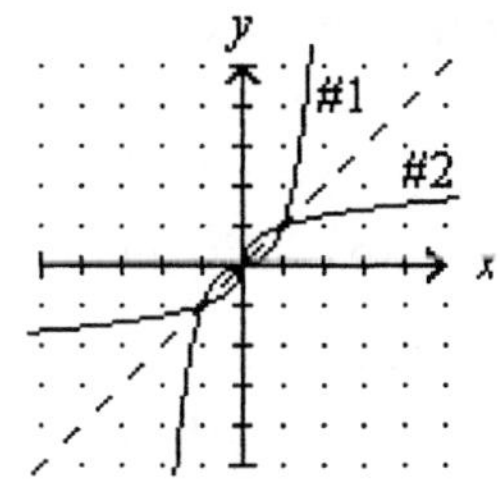

49.
$$\begin{aligned} f(x) &= x^2 - 1,\ (x \geq 0) \\ y &= x^2 - 1 \\ x &= y^2 - 1 \\ x + 1 &= y^2 \\ \sqrt{x+1} &= y = f^{-1}(x) \end{aligned}$$

Function (#1)		Inverse (#2)	
x	y	x	y
0	−1	−1	0
1	0	0	1
2	3	3	2

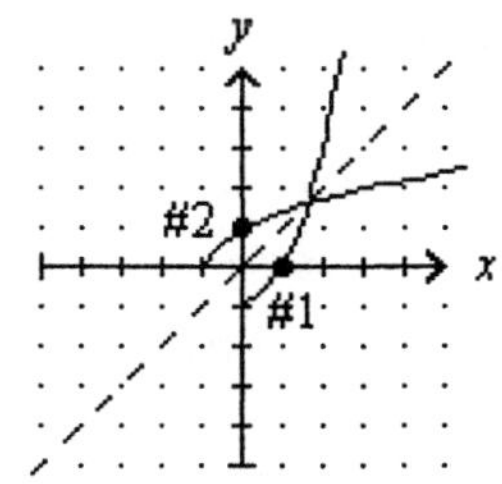

51.
$$\begin{aligned} f(x) &= x^2 + 4 & \\ y &= x^2 + 4 & \\ x &= y^2 + 4 & \text{Interchange } x \text{ and y.} \\ x - 4 &= y^2 & \text{Subtract 4.} \\ \pm\sqrt{x-4} &= y & \text{Square root} \\ y &= \pm\sqrt{x-4} & \text{Rewrite.} \end{aligned}$$

Inverse is not a function.

53.
$$\begin{aligned} f(x) &= x^3 & \\ y &= x^3 & \\ x &= y^3 & \text{Interchange } x \text{ and } y. \\ \sqrt[3]{x} &= y & \text{Take cube roots.} \\ f^{-1}(x) &= \sqrt[3]{x} & \text{Rewrite.} \end{aligned}$$

Inverse is a function.

55.
$$\begin{aligned} f(x) &= |x| & \\ y &= |x| & \\ x &= |y| & \text{Interchange } x \text{ and y.} \end{aligned}$$

Inverse is not a function.

57.
$$\begin{aligned} f(x) &= 2x^3 - 3 & \\ y &= 2x^3 - 3 & \\ x &= 2y^3 - 3 & \text{Interchange } x \text{ and } y. \\ x + 3 &= 2y^3 & \text{Add 3.} \\ \frac{x+3}{2} &= y^3 & \text{Divide by 2.} \\ \sqrt[3]{\frac{x+3}{2}} &= y & \text{Take cube roots.} \\ f^{-1}(x) &= \sqrt[3]{\frac{x+3}{2}} & \text{Rewrite.} \end{aligned}$$

Inverse is a function

APPLICATIONS

59. a. Yes, it is a function. No, the inverse would not be a function because there are many horizontal lines which would intersect the graph more than once.

    b. No. Twice during this period, the person's anxiety level was at the maximum threshold.

WRITING

61. It means that if the graph were folded along this line of symmetry, the "two halves" of the graph would lie on top of each other.

63. If the two functions are symmetric about the line $y = x$, the two functions are inverses of each other.

REVIEW

65. $3-\sqrt{-64}=3-\sqrt{-1}\sqrt{64}=3-8i$

67. $(3+4i)(2-3i) = 6-9i+8i-12i^2 = 6-i+12 = 18-i$

69. $(6-8i)^2 = 36-2(48i)+64i^2 = 36-96i-64 = -28-96i$

Section 9.3 Exponential Functions

VOCABULARY

1. exponential
3. $(0, \infty)$
5. none
7. increasing

CONCEPTS

9.

$f(x) = x^2$

x	$f(x)$
−2	4
−1	1
0	0
1	1
2	4
9	3

$g(x) = 2^x$

x	$g(x)$
−1	$\frac{1}{2}$
0	1
1	2
2	4
3	8
1	$\frac{1}{2}$
2	$\frac{1}{4}$

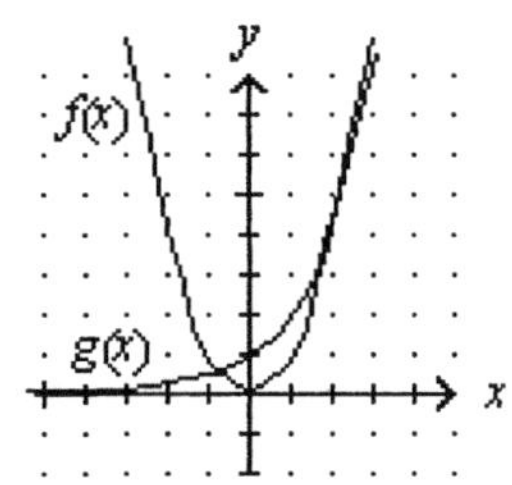

11. $A=P\left(1+\frac{r}{k}\right)^{kt}$; $FV = PV(1+i)^n$

13. $g(x) = 2^x + 3$; $h(x) = 2^x - 2$

NOTATION

15. The base is $\left(1+\frac{r}{k}\right)$ and the exponent is kt.

PRACTICE

17. $2^{\sqrt{2}}=2.6651$

19. $5^{\sqrt{5}}=36.5548$

21. $\left(2^{\sqrt{3}}\right)^{\sqrt{3}}=2^{\sqrt{3}\bullet\sqrt{3}}=2^{\sqrt{9}}=2^3=8$

23. $7^{\sqrt{3}}7^{\sqrt{12}}=7^{\sqrt{3}+\sqrt{12}}=7^{\sqrt{3}+2\sqrt{3}}=7^{3\sqrt{3}}$

25. $f(x) = 5^x$

x	$f(x)$
−1	$\frac{1}{5}$
0	1
1	5
2	25

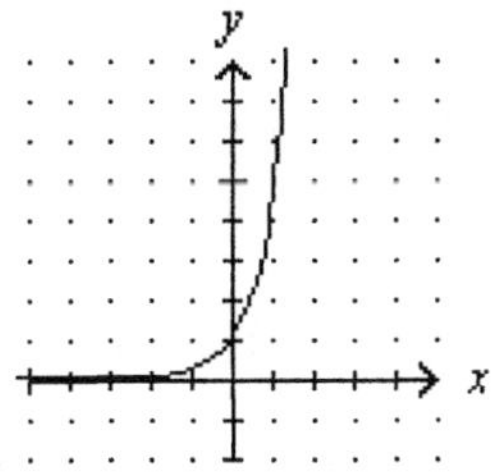

27. $f(x) = \left(\frac{1}{4}\right)^x$

x	$f(x)$
−2	16
−1	4
0	1
1	$\frac{1}{4}$

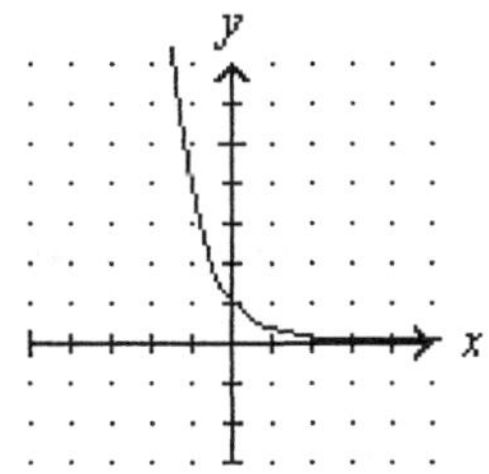

29. $f(x) = 3^x - 2$

x	$f(x)$
−1	$-1\frac{2}{3}$
0	−1
1	1
2	7

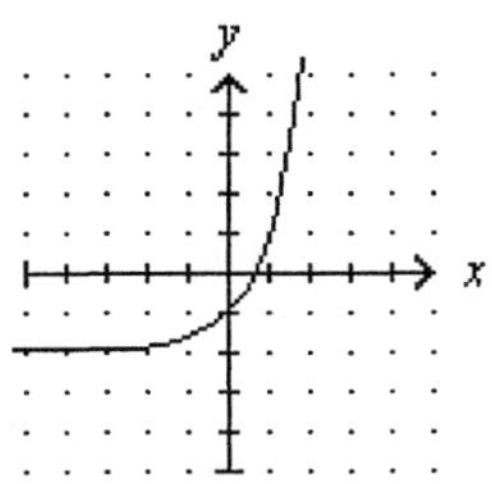

31. $f(x) = 3^{x-1}$

x	$f(x)$
0	$\frac{1}{3}$
1	1
2	3
3	9

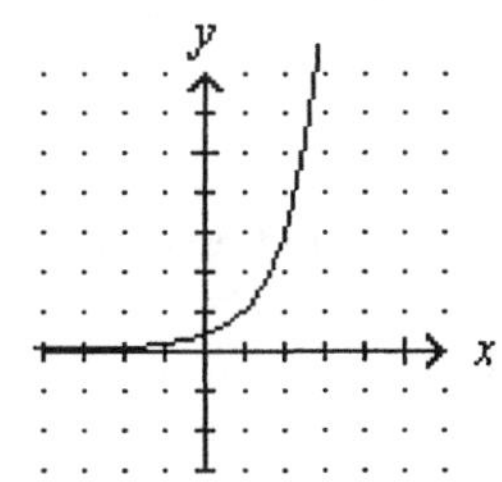

33. $f(x) = \frac{1}{2}\left(3^{x/2}\right)$; increasing function

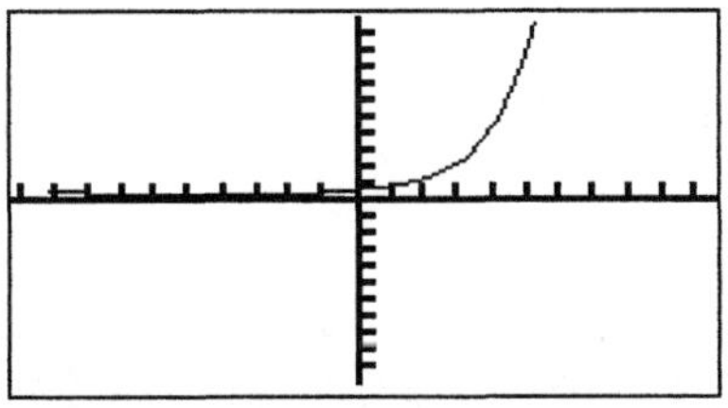

35. $f(x) = 2\left(3^{-x/2}\right)$; decreasing function

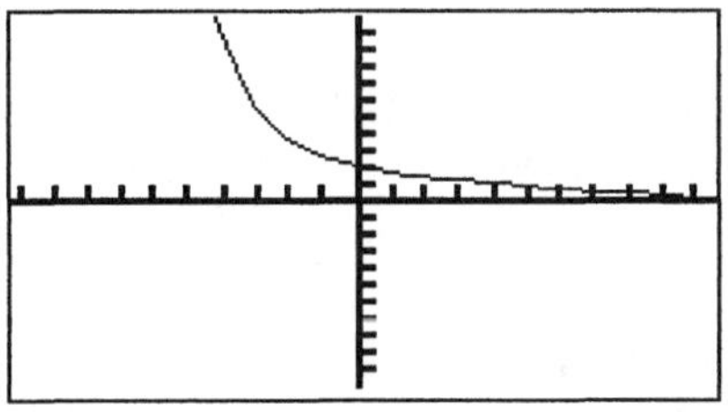

APPLICATIONS

37. Find A when $P = \$10{,}000$, $r = 8\% = 0.08$, $k = 4$, and $t = 10$.

$$A = P\left(1+\frac{r}{k}\right)^{kt} = 10{,}000\left(1+\frac{0.08}{4}\right)^{4(10)} = 10{,}000(1.02)^{40} = \$22{,}080.40$$

39. Find A when $P = \$1000$, , $k = 4$, $t = 5$, and $r = 5\% = 0.05$.

$$A = P\left(1+\frac{r}{k}\right)^{kt} = 1000\left(1+\frac{0.05}{4}\right)^{4(5)} = 1000(1.0125)^{20} = \$1282.04$$

Also, find A when $P = \$1000$, , $k = 4$, $t = 5$, and $r = 5.5\% = 0.055$.

$$A = P\left(1+\frac{r}{k}\right)^{kt} = 1000\left(1+\frac{0.055}{4}\right)^{4(5)} = 1000(1.01375)^{20} = \$1314.07$$

Find the increase in interest: Subtract the amount at 5% from the amount at 5.5%.
$1314.07 – $1282.04 = $32.03

41. Find A when $P = \$1$, $r = 5\% = 0.05$, $k = 1$, and $t = 2076 - 1776 = 300$.

$$A = P\left(1+\frac{r}{k}\right)^{kt} = 1\left(1+\frac{0.05}{1}\right)^{1(300)} = 1(1.05)^{300} = \$2{,}273{,}996.13$$

43. a. about 1500; about 1825 b. 6.5 billion c. exponential

45. a. at the end of the 2nd year. b. at the end of the 4th year c. during the 7th year

47. Find P when $t = 4$ hours.

$$P = (6 \times 10^6)(2.3)^t = (6 \times 10^6)(2.3)^4 = (6 \times 10^6)(27.9841) = 167.9046 \times 10^6 = 1.679046 \times 10^8$$

49. Find C when $t = 5$ days.

$$C = (3 \times 10^{-4})(0.7)^t = (3 \times 10^{-4})(0.7)^5 = (3 \times 10^{-4})(0.16807) = 0.50421 \times 10^{-4} = 5.0421 \times 10^{-5} \text{ coulombs}$$

51. Find the salvage value after 5 years when each year the value is 75% less than the previous year. The original value was \$4700. Multiply the original value by 0.75 five times (the same as raising 0.75 to the 5th power).

$$\text{Salvage value} = 4700(0.75)^5 = 4700(0.2373046) = \$1115.33$$

WRITING

53. The population could grow so large that there would not be enough resources to support it.

REVIEW

55. Since lines r and s are parallel, the measure of the two angles containing the variable x are supplementary (the sum is 180°).

$$\begin{aligned} 3x + (2x - 20) &= 180 \\ 5x - 20 &= 180 \\ 5x &= 200 \\ x &= 40 \end{aligned}$$

57. The measure of angle 2 is the same as the angle having a measure of $3x°$. Since $x = 40$ from problem 55, the measure of angle 2 is $3(40) = 120°$.

Section 9.4 Base-*e* Exponential Functions

VOCABULARY

1. the natural exponential function 3. $(0, \infty)$ 5. none 7. increasing

CONCEPTS

9. continuous 11. 2.72

13.

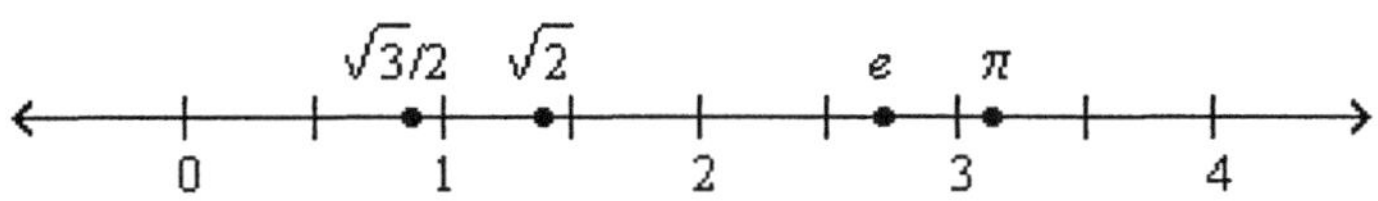

15.

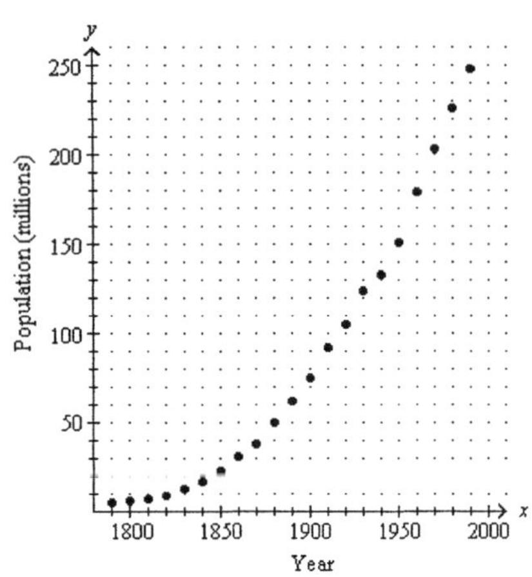

This appears to be an exponential function.

17. $y = 2.7182818$; e

NOTATION

19. $A = \mathbf{1{,}000}\ e^{(0.09)(\mathbf{10})}$
$= 1{,}000e^{0.9}$
$\approx \mathbf{1{,}000}\ (2.459603111)$
$\approx 2{,}459.603111$

PRACTICE

21. $f(x) = e^x + 1$

x	$f(x)$
−2	1.1
−1	1.4
0	2
1	3.7
2	8.4

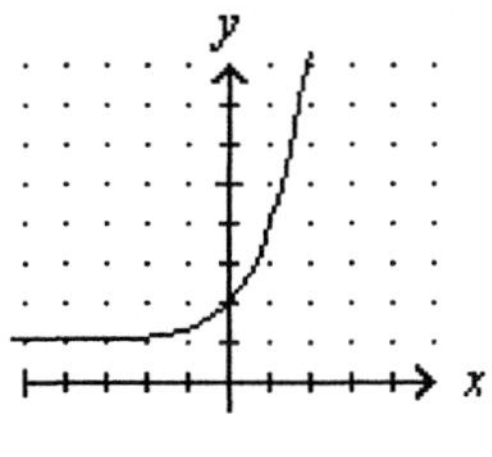

$f(x)$ is translated 1 unit up.

23. $f(x) = e^{x+3}$

x	$f(x)$
−4	0.4
−3	1
−2	2.7
−1	7.4
0	20.1

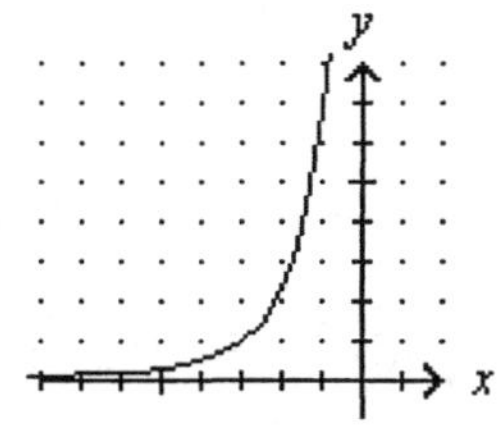

$f(x)$ is translated 3 units left.

25. $f(x) = -e^x$

x	$f(x)$
−2	−0.1
−1	−0.4
0	1
1	−2.7
2	−7.4

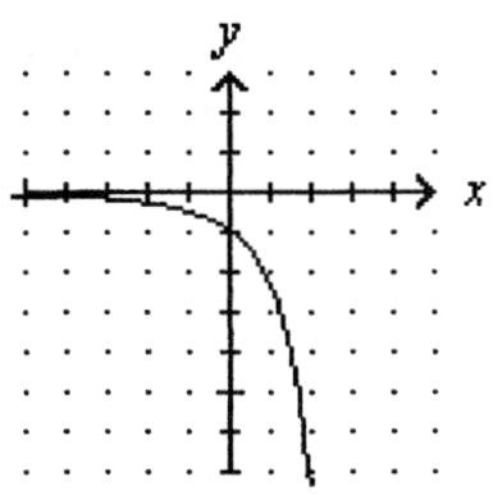

$f(x)$ is reflected around the x– axis.

27. $f(x) = 2e^x$

x	$f(x)$
−2	0.3
−1	0.7
0	2
1	5.4
2	14.8

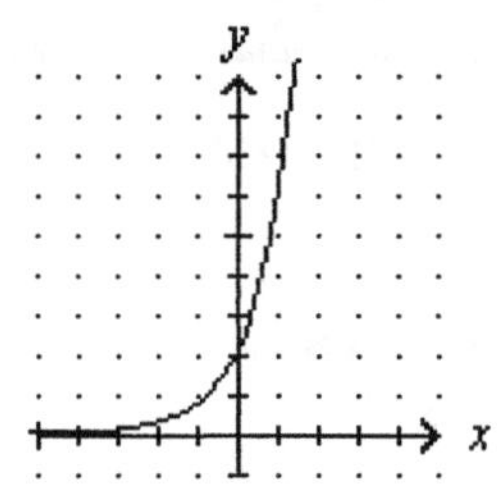

$f(x)$ crosses at $y = 2$ and increases more rapidly.

APPLICATIONS

29. Find A when $P = \$5000$, $t = 12$ years, and $r = 8.2\% = 0.082$ compounded continuously.

$$A = Pe^{rt} = 5000e^{0.082(12)} = 5000(2.6751354) = \$13{,}375.68$$

31. Find A when $P = \$5000$, $t = 5$ years, and $r = 8.5\% = 0.085$ compounded annually.

$$A = P\left(1 + \frac{r}{k}\right)^{kt} = 5000\left(1 + \frac{0.085}{1}\right)^{1(5)} = 5000(1.085)^5 = \$7{,}518.28$$

Find A when $P = \$5000$, $t = 5$ years, and $r = 8.5\% = 0.085$ compounded continuously.

$$A = Pe^{rt} = 5000e^{0.085(5)} = 5000(1.5295904) = \$7{,}647.95$$

33. Find P when $A = \$11{,}180$, $t = 7$ years, and $r = 7\% = 0.07$ compounded continuously.

$$11{,}180 = Pe^{0.07(7)} = P(1.6323162)$$

$$P = \frac{11{,}180}{1.6323162} = \$6849.16$$

35. Find A when $P = 6.1$ billion, $t = 30$ years, and growth rate $r = 1.4\% = 0.019$.

$$A = Pe^{rt} = 6.1e^{0.014(30)} < 6.1(1.521961556) < 9.3 \text{ billion}$$

37. Find P when $t = 20$. $\quad P = 173e^{0.03t} = 173e^{0.03(20)} = 173(1.8221188) = 315$

39. Find w when $n = 6$.

$$w = 1.54e^{0.503n} = 1.54e^{0.503(6)} < 1.54(20.45035) < 31.5 \text{ mm}$$

41. Read the graph to find x when $y = 50$, so $x = 12$ hours.

43. Find v when $t = 20$ seconds.

$$v = 50(1 - e^{-0.2t}) = 50(1 - e^{-0.2(20)}) = 50(1 - 0.0183156) = 49 \text{ mps}$$

45. The change in the linear equation does not significantly change the graph from that shown in Example 3. The food supply would be adequate for about 72 years.

WRITING

47. The –5 changes the vertical placement of any graph. It does not change the shape of the graph.

REVIEW

49. $\sqrt{240x^5} = \sqrt{16x^4 \cdot 15x} = 4x^2\sqrt{15x}$

51. $4\sqrt{48y^3} - 3y\sqrt{12y} = 4\sqrt{16y^2 \cdot 3y} - 3y\sqrt{4 \cdot 3y} = 4 \cdot 4y\sqrt{3y} - 3y \cdot 2\sqrt{3y} = 16y\sqrt{3y} - 6y\sqrt{3y} = 10y\sqrt{3y}$

Section 9.5 Logarithmic Functions

VOCABULARY

1. logarithmic

3. $(-\infty, \infty)$

5. $(1, 0)$

7. increasing

CONCEPTS

9. $x = b^y$

11. inverse

13. $y = \log x$

x	y
$\frac{1}{100}$	**–2**
$\frac{1}{10}$	**–1**
1	**0**
10	**1**
100	**2**

15. $y = \log_6 x$

x	$f(x)$
–6	**none**
0	**none**
$\frac{1}{216}$	**–3**
$\sqrt{6}$	$\frac{1}{2}$
6^8	**8**

17. $y = \log x$

x	y
0.5	**–0.30**
1	**0**
2	**0.30**
3	**0.48**
4	**0.60**
5	**0.70**
6	**0.78**
7	**0.85**
8	**0.90**
9	**0.95**
10	**1**

19. The values of $f(x)$ decrease.

NOTATION

21. The notation $\log x$ means $\log_{10} x$.

PRACTICE

23. If $\log_3 81 = 4$, then $3^4 = 81$.

25. If $\log_{12} 12 = 1$, then $12^1 = 12$.

27. If $\log_4 \frac{1}{64} = -3$, then $4^{-3} = \frac{1}{64}$.

29. If $\log 0.001 = -3$, then $10^{-3} = 0.001$ ($\log x = \log_{10} x$).

31. If $8^2 = 64$, then $\log_8 64 = 2$.

33. If $4^{-2} = \frac{1}{16}$, then $\log_4 \frac{1}{16} = -2$.

35. If $\left(\frac{1}{2}\right)^{-5} = 32$, then $\log_{1/2} 32 = -5$.

37. If $x^y = z$, then $\log_x z = y$.

39. Rewrite $\log_2 8 = x$ as $2^x = 8$. Since $8 = 2^3$, then $2^x = 2^3$ and $x = 3$.

41. Rewrite $\log_4 64 = x$ as $4^x = 64$. Since $64 = 4^3$, then $4^x = 4^3$ and $x = 3$.

43. Rewrite $\log_{1/2} \frac{1}{8} = x$ as $\left(\frac{1}{2}\right)^x = \frac{1}{8}$. Since $\frac{1}{8} = \left(\frac{1}{2}\right)^3$, then $\left(\frac{1}{2}\right)^x = \left(\frac{1}{2}\right)^3$ and $x = 3$.

45. Rewrite $\log_9 3 = x$ as $9^x = 3$. Since $3 = 9^{1/2}$, then $9^x = 9^{1/2}$ and $x = \frac{1}{2}$.

47. Rewrite $\log_8 x = 2$ as $8^2 = x$. So $x = 64$.

49. Rewrite $\log_{25} x = \frac{1}{2}$ as $25^{1/2} = x$. So $x = \sqrt{25} = 5$.

51. Rewrite $\log_5 x = -2$ as $5^{-2} = x$. So $x = \frac{1}{25}$.

53. Rewrite $\log_{36} x = -\frac{1}{2}$ as $36^{-1/2} = x$. So $x = \frac{1}{\sqrt{36}} = \frac{1}{6}$.

55.

$\log_{100} \frac{1}{1000} = x$		
$100^x = \frac{1}{1000}$	Rewrite.
$(10^2)^x = 10^{-3}$	Change to base 10.
$2x = -3$	Exponents are equal.
$x = -\frac{3}{2}$	Solve for x.

57.

$\log_{27} 9 = x$	
$27^x = 9$	Rewrite.
$(3^3)^x = 3^2$	Change to base 3.
$3x = 2$	Exponents are equal.
$x = \frac{2}{3}$	Solve for x.

59.

$\log_x 5^3 = 3$	
$x^3 = 5^3$	Rewrite.
$x = 5$	

Exponents are equal, so bases are equal.

61.

$\log_x \frac{9}{4} = 2$	
$x^2 = \frac{9}{4}$	Rewrite.
$x = \frac{3}{2}$	Solve for x.

63.

$\log_x \frac{1}{64} = -3$	
$x^{-3} = \frac{1}{64}$	Rewrite.
$x^{-3} = \frac{1}{4^3} = 4^{-3}$	Write in exponents.
$x = 4$	

Exponents are equal, so bases are equal.

65.

$\log_8 x = 0$	
$8^0 = x$	Rewrite.
$1 = x$	Solve for x.

67. $\log 3.25 = 0.5119$

69. $\log 0.00467 = -2.3307$

71. If $\log x = 1.4023$, then $10^{1.4023} = 25.25$

73. If $\log x = -1.71$, then $10^{-1.71} = 0.02$.

75. $f(x) = \log_3 x$

x	$f(x)$
$\frac{1}{9}$	-2
$\frac{1}{3}$	-1
1	0
3	1
9	2

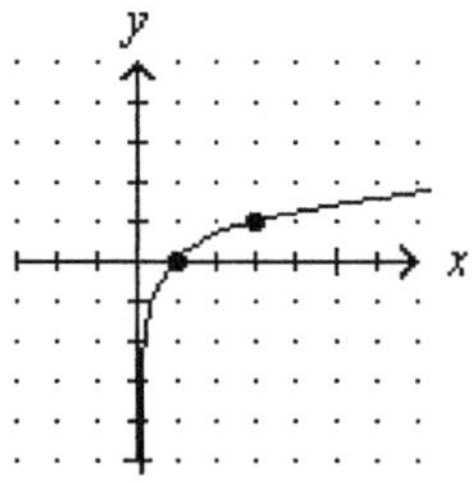

The function is increasing.

77. $y = \log_{1/2} x$

x	y
4	-2
2	-1
1	0
$\frac{1}{2}$	1
$\frac{1}{4}$	2

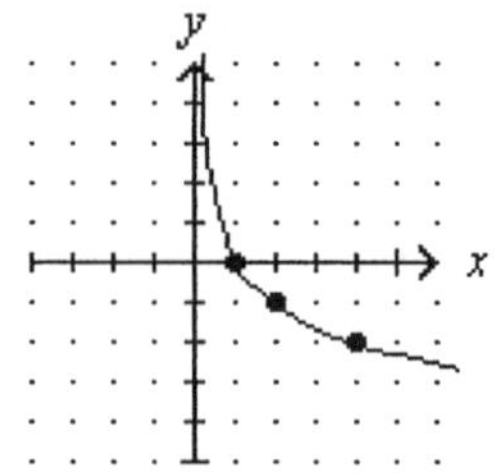

The function is decreasing.

79. $f(x) = 3 + \log_3 x$

x	$f(x)$
$\frac{1}{9}$	1
$\frac{1}{3}$	2
1	3
3	4
9	5

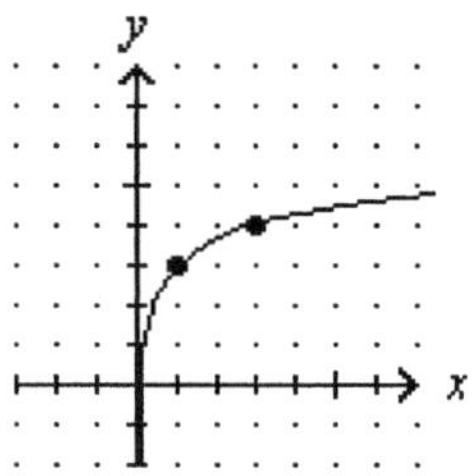

81. $y = \log_{1/2} (x - 2)$

x	y
6	-2
4	-1
3	0
$2\frac{1}{2}$	1
$2\frac{1}{4}$	2

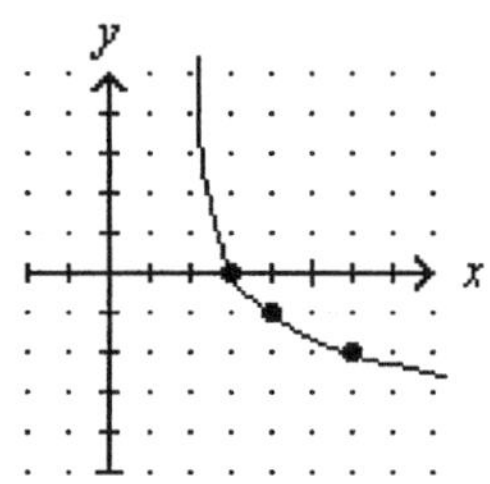

83.

#1: $f(x) = 6^x$

x	$f(x)$
-2	$\frac{1}{36}$
-1	$\frac{1}{6}$
0	1
1	6
2	36

#2: $f^{-1}(x) = \log_6 x$

x	$f^{-1}(x)$
$\frac{1}{36}$	-2
$\frac{1}{6}$	-1
1	0
6	1
36	2

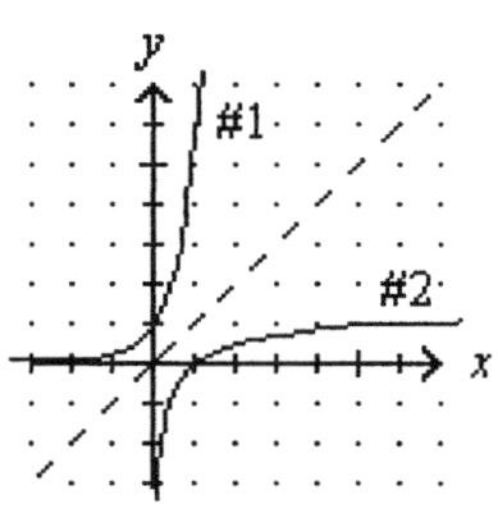

85.

#1: $f(x) = 5^x$

x	$f(x)$
-2	$\frac{1}{25}$
-1	$\frac{1}{5}$
0	1
1	5
2	25

#2: $f^{-1}(x) = \log_5 x$

x	$f^{-1}(x)$
$\frac{1}{25}$	-2
$\frac{1}{5}$	-1
1	0
5	1
25	2

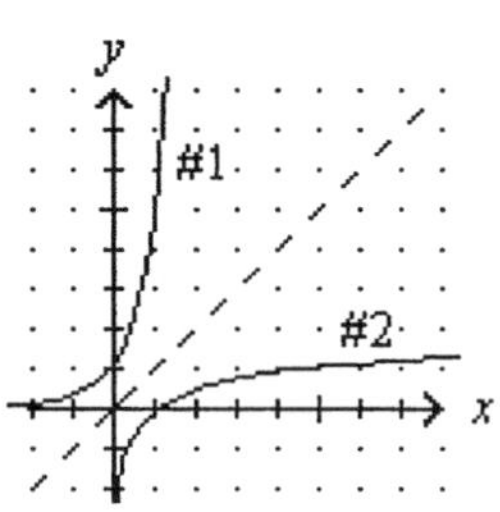

APPLICATIONS

87. Find the db gain when $E_I = 0.71$ volts and $E_O = 20$ volts.

$$\text{db gain} = 20 \log \frac{E_O}{E_I} = 20 \log \frac{20}{0.71} = 20 \log 28.17 = 29 \text{ db}$$

89. Find the db gain when $E_I = 0.1$ volts and $E_O = 30$ volts.

$$\text{db gain} = 20 \log \frac{E_O}{E_I} = 20 \log \frac{30}{0.1} = 20 \log 300 = 49.542 \approx 49.5 \text{ db}$$

91. Find R when $A = 5000$ micrometers and $P = 0.2$ second.

$$R = \log\frac{A}{P} = \log\frac{5000}{0.2} = \log 25{,}000 \approx 4.39794 \approx 4.4 \text{ on the Richter scale.}$$

93. Find A when $P = 0.25$ second and $R = 4$.

$$R = \log\frac{A}{P}$$
$$4 = \log\frac{A}{0.25}$$
$$10^4 = \frac{A}{0.25}$$
$$A = 0.25 \times 10^4 = 2500 \text{ micrometers}$$

95. Find n when $V = \$8000$, $C = \$37{,}000$, and $N = 5$ years.

$$n = \frac{\log V - \log C}{\log\left(1 - \frac{2}{N}\right)} = \frac{\log 8000 - \log 37{,}000}{\log(1 - \frac{2}{5})} = \frac{3.90309 - 4.5682017}{-0.2218487} \approx 2.8726\ldots \text{ or 3 years old.}$$

97. Find n when $A = \$20{,}000$, $r = 12\% = 0.12$, and $P = \$1000$.

$$n = \frac{\log\left[\frac{Ar}{P} + 1\right]}{\log(1 + r)} = \frac{\log\left[\frac{20{,}000(0.12)}{1000} + 1\right]}{\log(1 + 0.12)} = \frac{\log[3.4]}{\log(1.12)} = 10.8 \text{ yr.}$$

WRITING

99. These equations are inverses of each other.

REVIEW

101.
$$\sqrt[3]{6x+4} = 4$$
$$\left(\sqrt[3]{6x+4}\right) = 4^3$$
$$6x + 4 = 64$$
$$6x = 60$$
$$x = 10$$

103.
$$\sqrt{a+1} - 1 = 3a$$
$$\sqrt{a+1} = 3a + 1$$
$$\left(\sqrt{a+1}\right)^2 = (3a+1)^2$$
$$a + 1 = 9a^2 + 6a + 1$$
$$0 = 9a^2 + 5a$$
$$0 = a(9a + 5)$$
$$a = 0 \quad \text{or} \quad 9a + 5 = 0$$
$a = -\frac{5}{9}$ is extraneous.

Check:
$$\sqrt{0+1} - 1 = 3(0)$$
$$\sqrt{1} - 1 = 0$$

Check:
$$\sqrt{-\frac{5}{9} + 1} - 1 = 3\left(-\frac{5}{9}\right)$$
$$\sqrt{\frac{4}{9}} - 1 = -\frac{5}{3}$$
$$\frac{2}{3} - 1 \neq -\frac{5}{3}$$

VOCABULARY

1. natural

CONCEPTS

3. $f(x) = \ln x$

x	$f(x)$
0.5	**−0.69**
1	**0**
2	**0.69**
3	**1.10**
4	**1.39**
5	**1.61**
6	**1.79**
7	**1.95**
8	**2.08**
9	**2.20**
10	**2.30**

5. y–axis

7. $(-\infty, \infty)$

9. $e^y = x$

11. The logarithm of a negative number or zero is not defined.

13. $x = 2.7182818$, e

NOTATION

15. ln 2 means $\log_e 2$

17. ...given by the formula $t = \dfrac{\ln 2}{r}$.

PRACTICE

19. $\ln 35.15 = 3.5596$

21. $\ln 0.00465 = -5.3709$

23. $\ln 1.72 = 0.5423$

25. $\ln(-0.1)$ is undefined. (ln of negative number)

27. If $\ln x = 1.4023$, then $e^{1.4023} = 4.0645$.

29. If $\ln x = 4.24$, then $e^{4.24} = 69.4079$.

31. If $\ln x = -3.71$, then $e^{-3.71} = 0.0245$.

33. If $1.001 = \ln x$, then $e^{1.001} = 2.7210$.

35. $y = \ln\left(\dfrac{1}{2}x\right)$

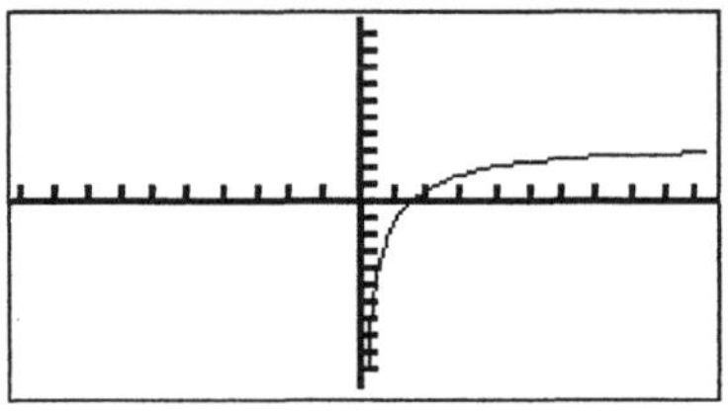

37. $f(x) = \ln(-x)$

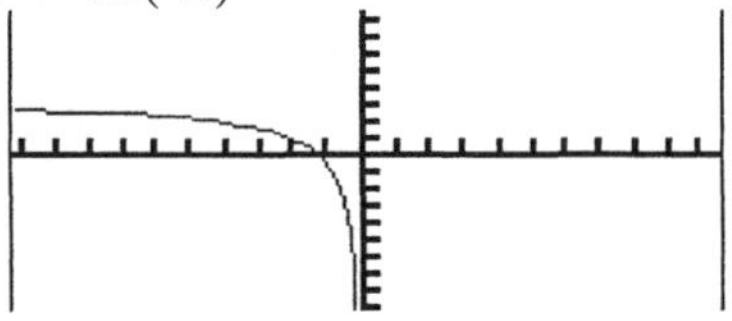

APPLICATIONS

39. Find t when $r = 12\% = 0.12$. $t = \dfrac{\ln 2}{r} = \dfrac{\ln 2}{0.12} \approx 5.8$ years.

41. Find t when $r = 12\% = 0.12$. $t = \dfrac{\ln 3}{r} = \dfrac{\ln 3}{0.12} \approx 9.2$ years.

43. Find t when $T_s = 70°$ F.

$$t = \frac{1}{0.25}\ln\frac{98.6 - T_s}{82 - T_s} = \frac{1}{0.25}\ln\frac{98.6 - 70}{82 - 70} = \frac{1}{0.25}\ln\frac{28.6}{12} = \frac{1}{0.25}\ln 2.3833 \approx 3.5 \text{ hours.}$$

WRITING

45. $\log x$ indicates a base of 10 while $\ln x$ indicates a base of e.

REVIEW

47.
$$\begin{aligned} y &= 5x - 8 \\ y - 0 &= 5(x - 0) \\ y &= 5x \end{aligned}$$
The slope of this line is 5. The slope of the parallel line is 5. The point is (0, 0). Use the point–slope form to find the equation.

49.
$$\begin{aligned} y &= \tfrac{2}{3}x - 12 \\ y - 2 &= -\tfrac{3}{2}(x - 3) \\ y - 2 &= -\tfrac{3}{2}x + \tfrac{9}{2} \\ y &= -\tfrac{3}{2}x + \tfrac{13}{2} \end{aligned}$$
The slope of this line is $\frac{2}{3}$. The slope of the perpendicular line is $-\frac{3}{2}$. The point is (3, 2). Use the point–slope form to find the equation.

51. The format of a vertical line has the format $x = k$ where k is a constant value for x. In this case that value is 2. The equation is $x = 2$.

Section 9.7 Properties of Logarithms

VOCABULARY

1. product
3. power

CONCEPTS

5. $\log_b 1 = \mathbf{0}$, because $b^0 = 1$.
7. $\log_b MN = \log_b \boldsymbol{M} + \log_b \boldsymbol{N}$
9. If $\log_b x = \log_b y$, then $\boldsymbol{x} = \boldsymbol{y}$.
11. $\log_b x^p = \boldsymbol{p} \bullet \log_b \boldsymbol{x}$
13. $\log_b (A + B) \neq \log_b A + \log_b B$
15. $\log_b x = \dfrac{\log_a x}{\log_a b}$

17. $\log(1) = 0$ is $10^0 = 1$; $\log(10) = 1$ is $10^1 = 10$; $\log(10^2) = 2$ is $10^2 = 10^2$

19. $\log_4 1 = 0$, because $4^0 = 1$.
21. $\log_4 4^7 = 7$, because $4^7 = 4^7$.

23. $5^{\log_5 10} = 10$, because $\log_b x$ is the exponent to which b is raised to get x.

25. $\log_5 5^2 = 2$, because $5^2 = 5^2$.
27. $\ln e = 1$, because $e^1 = e$.

29. $\log_3 3^7 = 7$, because $3^7 = 3^7$.

NOTATION

31.
$$\begin{aligned} \log_b rst &= \log_b (\ rs\ )t \\ &= \log_b (rs) + \log_b t \\ &= \log_b r + \log_b s + \log_b t \end{aligned}$$

PRACTICE

33.
$$\begin{aligned} \log [(2.5)(3.7)] &= \log 2.5 + \log 3.7 \\ \log [9.25] &= 0.39794 + 0.5682017 \\ 0.9661417 &= 0.9661417 \end{aligned}$$

35. $\ln(2.25)^4 = 4\ln 2.25$
$\ln 25.628906 = 4(0.8109302)$
$3.2437209 = 3.2437209$

37. $\log\sqrt{24.3} = \frac{1}{2}\log 24.3$
$\log 4.929503 = \frac{1}{2}(1.3856063)$
$0.6928031 = 0.6928031$

39. $\log_2(4 \bullet 5) = \log_2 4 + \log_2 5$ — The log of a product is the sum of the logs.
$= 2 + \log_2 5$ — $\log_2 4 = 2$.

41. $\log_6 \frac{x}{36} = \log_6 x - \log_6 36$ — The log of a quotient is the difference of the logs.
$= \log_6 x - 2$ — $\log_6 36 = 2$

43. $\ln y^4 = 4\ln y$ — The log of a power is the power times the log.

45. $\log\sqrt{5} = \log 5^{1/2}$ — Write $\sqrt{5}$ as $5^{1/2}$.
$= \frac{1}{2}\log 5$ — The log of a power is the power times the log.

47. $\log xyz = \log x + \log y + \log z$ — The log of a product is the sum of the logs.

49. $\log_2 \frac{2x}{y} = \log_2 2x - \log_2 y$ — The log of a quotient is the difference of the logs.
$= \log_2 2 + \log_2 x - \log_2 y$ — The log of a product is the sum of the logs.
$= 1 + \log_2 x - \log_2 y$ — $\log_2 2 = 1$.

51. $\log x^3y^2 = \log x^3 + \log y^2$ — The log of a product is the sum of the logs.
$= 3\log x + 2\log y$ — The log of a power is the power times the log.

53. $\log_b(xy)^{1/2} = \frac{1}{2}\log_b(xy)$ — The log of a power is the power times the log.
$= \frac{1}{2}(\log_b x + \log_b y)$ — The log of a product is the sum of the logs.

55. $\log_a \frac{\sqrt[3]{x}}{\sqrt[4]{yz}} = \log_a \sqrt[3]{x} - \log_a \sqrt[4]{yz}$ — The log of a quotient is the difference of the logs.
$= \log_a x^{1/3} - \log_a (yz)^{1/4}$ — Write $\sqrt[3]{x}$ as $x^{1/3}$ and $\sqrt[4]{yz}$ as $(yz)^{1/4}$.
$= \frac{1}{3}\log_a x - \frac{1}{4}\log_a(yz)$ — The log of a power is the power times the log.
$= \frac{1}{3}\log_a x - \frac{1}{4}(\log_a y + \log_a z)$ — The log of a product is the sum of the logs.
$= \frac{1}{3}\log_a x - \frac{1}{4}\log_a y - \frac{1}{4}\log_a z$ — Distribute the $\frac{1}{4}$.

57. $\ln x\sqrt{z} = \ln x + \ln\sqrt{z}$ — The log of a product is the sum of the logs.
$= \ln x + \ln z^{1/2}$ — Write $\sqrt{z}$ as $z^{1/2}$.
$= \ln x + \frac{1}{2}\ln z$ — The log of a power is the power times the log.

59. $\log_2(x+1) - \log_2 x = \log_2 \frac{x+1}{x}$ — The difference of two logs is the log of the quotient.

61. $2 \log x + \frac{1}{2} \log y = \log x^2 + \log y^{1/2}$ A power times a log is the log of the power.

$= \log x^2 y^{1/2}$ The sum of two logs is the log of the product.

63. $-3 \log_b x - 2 \log_b y + \frac{1}{2} \log_b z = \log_b x^{-3} + \log_b y^{-2} + \log_b z^{1/2}$ A power times a log is the log of the power.

$= \log_b x^{-3} y^{-2} z^{1/2}$ The sum of logs is the log of the products

$= \log_b \frac{z^{1/2}}{x^3 y^2}$ Write without negative exponents.

65. $\ln\left(\frac{x}{z} + x\right) - \ln\left(\frac{y}{z} + y\right) = \ln \frac{\frac{x}{z} + x}{\frac{y}{z} + y}$ The difference of two logs is the log of the quotient.

$= \ln \frac{\frac{x + xz}{z}}{\frac{y + yz}{z}} = \ln \frac{x(1+z)}{y(1+z)} = \ln \frac{x}{y}$ Simplify.

67. false; The log of a product equals the sum of the logs of the factors, not the product of the logs.

69. false; the log of a quotient equals the difference of the logs of the dividend and the divisor, not the log of the difference.

71. true.

73. $\log_b 28 = \log_b (4 \bullet 7) = \log_b 4 + \log_b 7 = 0.6021 + 0.8451 = 1.4472$

75. $\log_b \frac{4}{63} = \log_b \frac{4}{7 \cdot 9} = \log_b 4 - \log_b 7 - \log_b 9 = 0.6021 - 0.8451 - 0.9542 = -1.1972$

77. $\log_b \frac{63}{4} = \log_b \frac{7 \cdot 9}{4} = \log_b 7 + \log_b 9 - \log_b 4 = 0.8451 + 0.9542 - 0.6021 = 1.1972$

79. $\log_b 64 = \log_b 4^3 = 3 \log_b 4 = 3(0.6021) = 1.8063$

81. $\log_3 7 = \frac{\log_{10} 7}{\log_{10} 3} = 1.7712$

83. $\log_{1/3} 3 = \frac{\log_{10} 3}{\log_{10} \left(\frac{1}{3}\right)} = -1.0000$

85. $\log_3 8 = \frac{\log_{10} 8}{\log_{10} 3} = 1.8928$

87. $\log_{\sqrt{2}} \sqrt{5} = \frac{\log_{10} \sqrt{5}}{\log_{10} \sqrt{2}} = 2.3219$

APPLICATIONS

89. Find the pH of a solution when $[H^+] = 1.7 \times 10^{-5}$ gram–ions per liter.

$\text{pH} = -\log [H^+] = -\log (1.7 \times 10^{-5}) = -(-4.7695) = 4.8$

91. Find $[H^+]$ when pH $= 6.8$.

$\text{pH} = -\log [H^+]$
$6.8 = -\log [H^+]$
$-6.8 = \log [H^+]$
$[H^+] = 10^{-6.8} = 1.6 \times 10^{-7}$

Find $[H^+]$ when pH $= 7.6$.

$\text{pH} = -\log [H^+]$
$7.6 = -\log [H^+]$
$-7.6 = \log [H^+]$
$[H^+] = 10^{-7.6} = 2.5 \times 10^{-8}$

The range of the hydrogen ion concentrations is 2.5×10^{-8} to 1.6×10^{-7}.

WRITING

93. When determining the *logarithm of a product*, find the product then take the logarithm. When determining the *product of logarithms*, find the individual logarithms then multiply.

REVIEW

95. $P(x_1, y_1) = (-2, 3)$ and $Q(x_2, y_2) = (4, -4)$. $m = \dfrac{y_2 - y_1}{x_2 - x_1} = \dfrac{-4-3}{4-(-2)} = \dfrac{-7}{6} = -\dfrac{7}{6}$

97. $P(x_1, y_1) = (-2, 3)$ and $Q(x_2, y_2) = (4, -4)$.

$$\text{midpoint} = \left(\frac{x_2 + x_1}{2}, \frac{y_2 + y_1}{2}\right) = \left(\frac{4-2}{2}, \frac{-4+3}{2}\right) = \left(1, -\frac{1}{2}\right)$$

Section 9.8 Exponential and Logarithmic Equations

VOCABULARY

1. exponential

CONCEPTS

3. $A_0 2^{-t/h}$
5. about 1.8
7. logarithm; exponent
9. 1.2920

11. $\ln e = 1$, because $\ln e = \log_e e = 1$ and $e^1 = e$.

13. $\log_{10}(x+1) = 2$, so $10^2 = x + 1$.

15. a.
$$\begin{aligned} x^2 &= 12 \\ \sqrt{x^2} &= \pm\sqrt{12} \\ x &\approx \pm 3.46 \end{aligned}$$

b.
$$\begin{aligned} 2^x &= 12 \\ \log 2^x &= \log 12 \\ x \log 2 &= \log 12 \\ x &= \frac{\log 12}{\log 2} \approx 3.58 \end{aligned}$$

17. Substitute $x = -4$: $\log_5(x+3) = \log_5(-4+3) = \log_5(-1)$ does not exist so $x = -4$ is not a solution.

NOTATION

19. Solve $2^x = 7$.
$$\begin{aligned} 2^x &= 7 \\ \mathbf{log}\ 2^x &= \log 7 \\ x\ \mathbf{log\ 2} &= \log 7 \\ x &= \frac{\log 7}{\log 2} \end{aligned}$$

PRACTICE

21.
$$\begin{aligned} 4^x &= 5 \\ \log 4^x &= \log 5 && \text{Take log of both sides.} \\ x \log 4 &= \log 5 && \text{Use power rule.} \\ x &= \frac{\log 5}{\log 4} && \text{Divide by log 4.} \\ x &\approx 1.1610 && \text{Use a calculator.} \end{aligned}$$

23.

$$\begin{aligned} 13^{x-1} &= 2 \\ \log 13^{x-1} &= \log 2 && \text{Take log of both sides.} \\ (x-1)\log 13 &= \log 2 && \text{Use power rule.} \\ x-1 &= \frac{\log 2}{\log 13} && \text{Divide by log 13.} \\ x-1 &= 0.2702 && \text{Use calculator.} \\ x &\approx 1.2702 && \text{Add 1.} \end{aligned}$$

25.

$$\begin{aligned} 2^{x+1} &= 3^x \\ \log 2^{x+1} &= \log 3^x && \text{Take log of both sides.} \\ (x+1)\log 2 &= x\log 3 && \text{Use power rule.} \\ x\log 2 + \log 2 &= x\log 3 && \text{Use the distributive property.} \\ x\log 2 - x\log 3 &= -\log 2 && \text{Subtract } x \log 3 \text{ and } \log 2. \\ x(\log 2 - \log 3) &= -\log 2 && \text{Factor.} \\ x &= \frac{-\log 2}{\log 2 - \log 3} && \text{Divide by } (\log 2 - \log 3). \\ x &\approx 1.7095 && \text{Use a calculator.} \end{aligned}$$

27.

$$\begin{aligned} 2^x &= 3^x \\ \log 2^x &= \log 3^x && \text{Take log of both sides.} \\ x\log 2 &= x\log 3 && \text{Use power rule.} \\ x\log 2 - x\log 3 &= 0 && \text{Subtract } x\log 3. \\ x(\log 2 - \log 3) &= 0 && \text{Factor.} \\ x &= \frac{0}{\log 2 - \log 3} && \text{Divide by } (\log 2 - \log 3). \\ x &= 0 \end{aligned}$$

29.

$$\begin{aligned} 7^{x^2} &= 10 \\ \log 7^{x^2} &= \log 10 && \text{Take log of both sides.} \\ x^2\log 7 &= 1 && \text{Use power rule and } \log 10 = 1. \\ x^2 &= \frac{1}{\log 7} && \text{Divide by log 7.} \\ x &= \pm\sqrt{\frac{1}{\log 7}} && \text{Take the square root.} \\ x &\approx \pm 1.0878 && \text{Use calculator.} \end{aligned}$$

31.

$$\begin{aligned} 8^{x^2} &= 9^x \\ \log 8^{x^2} &= \log 9^x && \text{Take log of both sides.} \\ x^2\log 8 &= x\log 9 && \text{Use power rule.} \\ x^2\log 8 - x\log 9 &= 0 && \text{Subtract } x\log 9. \\ x(x\log 8 - \log 9) &= 0 && \text{Factor.} \end{aligned}$$

$x = 0$ or

$$\begin{aligned} x\log 8 - \log 9 &= 0 \\ x\log 8 &= \log 9 \\ x &= \frac{\log 9}{\log 8} \approx 1.0566 \end{aligned}$$

33.
$$\begin{aligned} e^{3x} &= 9 \\ \ln e^{3x} &= \ln 9 && \text{Take ln of both sides.} \\ 3x \ln e &= \ln 9 && \text{Use power rule.} \\ 3x &= \ln 9 && \ln e = 1 \\ x &= \frac{\ln 9}{3} && \text{Divide by 3.} \\ x &\approx 0.7324 && \text{Use calculator.} \end{aligned}$$

35.
$$\begin{aligned} e^{-0.2t} &= 14.2 \\ \ln e^{-0.2t} &= \ln 14.2 && \text{Take ln of both sides.} \\ -0.2t \ln e &= \ln 14.2 && \text{Use power rule.} \\ -0.2t &= \ln 14.2 && \ln e = 1 \\ t &= \frac{\ln 14.2}{-0.2} && \text{Divide by } -0.2. \\ t &\approx -13.2662 && \text{Use calculator.} \end{aligned}$$

37.
$$\begin{aligned} 2^{x-2} &= 64 \\ 2^{x-2} &= 2^6 && \text{Write each side with base of 2.} \\ x-2 &= 6 && \text{Equate the exponents.} \\ x &= 8 && \text{Add 2 to both sides.} \end{aligned}$$

39.
$$\begin{aligned} 5^{4x} &= \frac{1}{125} \\ 5^{4x} &= 5^{-3} && \text{Write each side with base of 5.} \\ 4x &= -3 && \text{Equate the exponents.} \\ x &= -\frac{3}{4} && \text{Divide by 4.} \end{aligned}$$

41.
$$\begin{aligned} 2^{x^2-2x} &= 8 \\ 2^{x^2-2x} &= 2^3 && \text{Write each side with base of 2.} \\ x^2-2x &= 3 && \text{Equate the exponents.} \\ x^2-2x-3 &= 0 && \text{Subtract 3.} \\ (x-3)(x+1) &= 0 && \text{Factor.} \end{aligned}$$

$$x-3 = 0 \quad \text{or} \quad x+1 = 0$$
$$x = 3 \qquad\qquad x = -1$$

43.
$$\begin{aligned} 3^{x^2+4x} &= \frac{1}{81} \\ 3^{x^2+4x} &= 3^{-4} && \text{Write each side with base of 3.} \\ x^2+4x &= -4 && \text{Equate the exponents.} \\ x^2+4x+4 &= 0 && \text{Add 4.} \\ (x+2)(x+2) &= 0 && \text{Factor.} \end{aligned}$$

$$x+2 = 0 \quad \text{or} \quad x+2 = 0$$
$$x = -2 \qquad\qquad x = -2$$

45. Use an appropriate method for your calculator. The answer should be 1.8.

46. Use an appropriate method for your calculator. The answer should be 2.7.

47. Use an appropriate method for your calculator. The answers should be 8.8 and 0.2.

49.
$$\begin{aligned} \log(x+2) &= 4 \\ x+2 &= 10^4 && \text{Change to exponential form.} \\ x+2 &= 10{,}000 && \text{Simplify.} \\ x &= 9998 && \text{Subtract 2.} \end{aligned}$$

51.
$$\begin{aligned} \log(7-x) &= 2 \\ 7-x &= 10^2 && \text{Change to exponential form.} \\ 7-x &= 100 && \text{Simplify.} \\ -x &= 93 && \text{Subtract 7.} \\ x &= -93 && \text{Multiply by } -1. \end{aligned}$$

53.
$$\begin{aligned} \ln x &= 1 \\ x &= e^1 && \text{Change to exponential form.} \\ x &\approx 2.7183 && \text{Use a calculator.} \end{aligned}$$

55.
$$\begin{aligned} \ln(x+1) &= 3 \\ x+1 &= e^3 && \text{Change to exponential form.} \\ x &= e^3-1 && \text{Subtract 1.} \\ x &\approx 19.0855 && \text{Use a calculator.} \end{aligned}$$

57.
$$\begin{aligned} \log 2x &= \log 4 \\ 2x &= 4 && \text{The expressions in the logarithm are equal.} \\ x &= 2 && \text{Divide by 2.} \end{aligned}$$

59.
$$\begin{aligned} \ln(3x+1) &= \ln(x+7) \\ 3x+1 &= x+7 && \text{The expressions in the logarithm are equal.} \\ 2x &= 6 && \text{Subtract } x \text{ and 1.} \\ x &= 3 && \text{Divide by 2.} \end{aligned}$$

61.
$$\begin{aligned} \log(3-2x)-\log(x+24) &= 0 \\ \log(3-2x) &= \log(x+24) && \text{Add } \log(x+24). \\ 3-2x &= x+24 && \text{The expressions in the logarithm are equal.} \\ -3x &= 21 && \text{Subtract } x \text{ and 3.} \\ x &= -7 && \text{Divide by } -3. \end{aligned}$$

63.
$$\begin{aligned} \log\frac{4x+1}{2x+9} &= 0 \\ \log(4x+1)-\log(2x+9) &= 0 && \text{Use the quotient rule of logarithms.} \\ \log(4x+1) &= \log(2x+9) && \text{Add } \log(2x+9). \\ 4x+1 &= 2x+9 && \text{The expressions in the logarithm are equal.} \\ 2x &= 8 && \text{Subtract } 2x \text{ and 1.} \\ x &= 4 && \text{Divide by 2.} \end{aligned}$$

65.
$$\begin{aligned} \log x^2 &= 2 \\ x^2 &= 10^2 && \text{Change to exponential form.} \\ x &= \pm 10 && \text{Take the square root.} \end{aligned}$$

67.
$$\begin{aligned} \log x + \log(x-48) &= 2 \\ \log x(x-48) &= 2 && \text{Use the product rule of logarithms.} \\ x(x-48) &= 10^2 && \text{Change to exponential form.} \\ x^2 - 48x - 100 &= 0 && \text{Remove parentheses and subtract 100.} \\ (x-50)(x+2) &= 0 && \text{Factor.} \end{aligned}$$
$x - 50 = 0$ or $x + 2 = 0$
$x = 50$ $\quad x = -2$ Not possible. $\log(-2)$ is undefined.

69.
$$\begin{aligned} \log x + \log(x-15) &= 2 \\ \log x(x-15) &= 2 && \text{Use the product rule of logarithms.} \\ x(x-15) &= 10^2 && \text{Change to exponential form.} \\ x^2 - 15x - 100 &= 0 && \text{Remove parentheses and subtract 100.} \\ (x-20)(x+5) &= 0 && \text{Factor.} \end{aligned}$$
$x - 20 = 0$ or $x + 5 = 0$
$x = 20$ $\quad x = -5$ Not possible. $\log(-5)$ is undefined.

71.
$$\begin{aligned} \log(x+90) &= 3 - \log x \\ \log x + \log(x+90) &= 3 && \text{Add } \log x \text{ to both sides.} \\ \log x(x+90) &= 3 && \text{Use the product rule of logarithms.} \\ x(x+90) &= 10^3 && \text{Change to exponential form.} \\ x^2 + 90x - 1000 &= 0 && \text{Remove parentheses and subtract 1000.} \\ (x-10)(x+100) &= 0 && \text{Factor.} \end{aligned}$$
$x - 10 = 0$ or $x + 100 = 0$
$x = 10$ $\quad x = -100$ Not possible. $\log(-100)$ is undefined.

73.
$$\begin{aligned} \log(x-6) - \log(x-2) &= \log\frac{5}{x} \\ \log\frac{x-6}{x-2} &= \log\frac{5}{x} && \text{Use the quotient rule of logarithms.} \\ \frac{x-6}{x-2} &= \frac{5}{x} && \text{The expressions in the logarithm are equal.} \\ x(x-6) &= 5(x-2) && \text{Multiply the means and extremes.} \\ x^2 - 6x &= 5x - 10 && \text{Remove parentheses.} \\ x^2 - 11x + 10 &= 0 && \text{Subtract } 5x \text{ and add 10 to both sides.} \\ (x-10)(x-1) &= 0 && \text{Factor.} \end{aligned}$$
$x - 10 = 0$ or $x - 1 = 0$
$x = 10$ $\quad x = 1$ Not possible. $\log(1-6)$ is undefined.

75.
$$\begin{aligned} \frac{\log(3x-4)}{\log x} &= 2 \\ \log(3x-4) &= 2\log x && \text{Multiply by } \log x. \\ \log(3x-4) &= \log x^2 && \text{Use the power rule of logarithms.} \\ 3x - 4 &= x^2 && \text{The expressions in the logarithm are equal.} \\ 0 &= x^2 - 3x + 4 && \text{Subtract } 3x \text{ and add 4.} \end{aligned}$$
$$x = \frac{-(-3) \pm \sqrt{(-3)^2 - 4(1)(4)}}{2(1)} = \frac{3 \pm \sqrt{-7}}{2}$$
Since this is a complex value there is no solution.

77.

$$\begin{aligned}
\frac{\log(5x+6)}{2} &= \log x \\
\log(5x+6) &= 2\log x && \text{Multiply by 2.} \\
\log(5x+6) &= \log x^2 && \text{Use the power rule of logarithms.} \\
5x+6 &= x^2 && \text{The expressions in the logarithm are equal.} \\
0 &= x^2-5x-6 && \text{Subtract } 5x \text{ and } 6. \\
0 &= (x-6)(x+1) && \text{Factor.}
\end{aligned}$$

$x - 6 = 0$ or $x + 1 = 0$

$x = 6$ $\quad$ $x = -1$ Not possible. $\log(-1)$ is undefined.

79.

$$\begin{aligned}
\log_3 x &= \log_3\left(\frac{1}{x}\right)+4 \\
\log_3 x - \log_3\left(\frac{1}{x}\right) &= 4 && \text{Subtract } \log_3\left(\frac{1}{x}\right). \\
\log_3\left(\frac{x}{\frac{1}{x}}\right) &= 4 && \text{Use the quotient rule of logarithms.} \\
\log_3 x^2 &= 4 && \text{Simplify.} \\
x^2 &= 3^4 && \text{Change to exponential form.} \\
x^2 &= 81 && \text{Simplify.} \\
x &= \pm 9 && \text{Take the square root.}
\end{aligned}$$

Discard -9 because $\log_3(-9)$ is undefined.

81.

$$\begin{aligned}
2\log_2 x &= 3+\log_2(x-2) \\
\log_2 x^2 &= 3+\log_2(x-2) && \text{Use the power rule of logarithms.} \\
\log_2 x^2 - \log_2(x-2) &= 3 && \text{Subtract } \log_2(x-2). \\
\log_2\left(\frac{x^2}{x-2}\right) &= 3 && \text{Use the quotient rule of logarithms.} \\
\frac{x^2}{x-2} &= 2^3 = 8 && \text{Change to exponential form.} \\
x^2 &= 8(x-2) && \text{Multiply by } x-2. \\
x^2 &= 8x-16 && \text{Remove the parentheses.} \\
x^2-8x+16 &= 0 && \text{Subtract } 8x \text{ and add } 16. \\
(x-4)(x-4) &= 0 && \text{Factor.}
\end{aligned}$$

$x - 4 = 0$ or $x - 4 = 0$

$x = 4$ $\quad$ $x = 4$

83.

$$\begin{aligned}
\log(7y+1) &= 2\log(y+3)-\log 2 \\
\log(7y+1) &= \log(y+3)^2-\log 2 && \text{Use the power rule of logarithms.} \\
\log(7y+1) &= \log\left(\frac{(y+3)^2}{2}\right) && \text{Use the quotient rule of logarithms.} \\
7y+1 &= \frac{(y+3)^2}{2} && \text{The expressions in the logarithm are equal.} \\
14y+2 &= y^2+6y+9 && \text{Multiply by 2 and remove the parentheses.} \\
0 &= y^2-8y+7 && \text{Subtract 14y and 2.} \\
0 &= (y-7)(y-1) && \text{Factor.}
\end{aligned}$$

$y - 7 = 0$ or $y - 1 = 0$

$y = 7$ $\quad$ $y = 1$

85. Use an appropriate method for your calculator. The answer should be 20.

87. Use an appropriate method for your calculator. The answer should be 8.

APPLICATIONS

89. Find t when 25% of the tritium has decomposed. This leaves 75%. So, $A = 0.75A_0$ and $h = 12.4$ years.

$$\begin{aligned} A &= A_0 2^{-t/h} \\ 0.75A_0 &= A_0 2^{-t/12.4} && \text{Substitute given values.} \\ 0.75 &= 2^{-t/12.4} && \text{Divide by } A_0. \\ \log 0.75 &= -\frac{t}{12.4}\log 2 && \text{Take the logarithm of both sides.} \\ -12.4\log 0.75 &= t\log 2 && \text{Multiply by } -12.4. \\ t &= \frac{-12.4\log 0.75}{\log 2} \approx 5.1 \text{ years} \end{aligned}$$

91. Find t when 80% of the thorium has decomposed. This leaves 20%. So, $A = 0.20A_0$ and $h = 18.4$ days.

$$\begin{aligned} A &= A_0 2^{-t/h} \\ 0.20A_0 &= A_0 2^{-t/18.4} && \text{Substitute given values.} \\ 0.20 &= 2^{-t/18.4} && \text{Divide by } A_0. \\ \log 0.20 &= -\frac{t}{18.4}\log 2 && \text{Take the logarithm of both sides.} \\ -18.4\log 0.20 &= t\log 2 && \text{Multiply by } -18.4. \\ t &= \frac{-18.4\log 0.20}{\log 2} \approx 42.7 \text{ days} \end{aligned}$$

93. Find t when 60% of the carbon–14 is left. So, $A = 0.60A_0$ and $h = 5700$ years.

$$\begin{aligned} A &= A_0 2^{-t/h} \\ 0.60A_0 &= A_0 2^{-t/5700} && \text{Substitute given values.} \\ 0.60 &= 2^{-t/5700} && \text{Divide by } A_0. \\ \log 0.60 &= -\frac{t}{5700}\log 2 && \text{Take the logarithm of both sides.} \\ -5700\log 0.60 &= t\log 2 && \text{Multiply by } -5700. \\ t &= \frac{-5700\log 0.60}{\log 2} \approx 4200 \text{ years} \end{aligned}$$

95. Find t when $P = \$500$, $A = \$800$, and $r = 8.5\% = 0.085$ compounded semiannually ($k = 2$). Use formula from Section 9.4.

$$\begin{aligned} A &= P\left(1+\frac{r}{k}\right)^{kt} \\ 800 &= 500\left(1+\frac{0.085}{2}\right)^{2t} && \text{Substitute given values.} \\ 1.6 &= (1.0425)^{2t} && \text{Divide by 500 and simplify.} \\ \log 1.6 &= \log (1.0425)^{2t} && \text{Take the logarithm of both sides.} \\ \log 1.6 &= 2t\log 1.0425 && \text{Use the power rule of logarithms.} \\ t &= \frac{\log 1.6}{2\log 1.0425} \approx 5.6 \text{ years.} \end{aligned}$$

97. Find t when $P = \$1300$, $A = \$2100$, and $r = 9\% = 0.09$ compounded quarterly ($k = 4$).

$$A = P\left(1+\frac{r}{k}\right)^{kt}$$

$$2100 = 1300\left(1+\frac{0.09}{4}\right)^{4t} \quad \text{Substitute given values.}$$

$$\frac{21}{13} = (1.0225)^{4t} \quad \text{Divide by 1300 and simplify.}$$

$$\log\left(\frac{21}{13}\right) = \log(1.0225)^{4t} \quad \text{Take the logarithm of both sides.}$$

$$\log\left(\frac{21}{13}\right) = 4t\log 1.0225 \quad \text{Use the power rule of logarithms.}$$

$$t = \frac{\log\left(\frac{21}{13}\right)}{4\log 1.0225} \approx 5.4 \text{ years.}$$

99. Find t when $A = 2P$, and r varies but is compounded continuously. Use formula from Section 9.4.

$$A = Pe^{rt}$$

$$2P = Pe^{rt} \quad \text{Substitute given values.}$$

$$2 = e^{rt} \quad \text{Divide by } P.$$

$$\ln 2 = \ln e^{rt} \quad \text{Take the natural logarithm of both sides.}$$

$$\ln 2 = rt\ln e \quad \text{Use the power rule of logarithms.}$$

$$\ln 2 = rt(1) \quad \ln e = 1.$$

$$t = \frac{\ln 2}{r} \approx \frac{0.70}{r} \quad \text{So, the rule of seventy works because the } \ln 2 \approx 0.70.$$

101. Find t when $P = 1$ million, $P_0 = 30{,}000$ and doubles every 5 years.

Find k when $P_0 = 30{,}000$, $P = 60{,}000$, and $t = 5$.

$$P = P_0e^{kt}$$

$$60{,}000 = 30{,}000\, e^{k(5)}$$

$$2 = e^{5k}$$

$$\ln 2 = 5k$$

$$k = \frac{\ln 2}{5} \approx 0.1386$$

Find t when $P_0 = 30{,}000$, $P = 1$ million, and $k = 0.1386$

$$P = P_0e^{kt}$$

$$1{,}000{,}000 = 30{,}000\, e^{0.1386t}$$

$$\frac{100}{3} = e^{0.1386t}$$

$$\ln\left(\frac{100}{3}\right) = 0.1386t$$

$$t = \frac{\ln\left(\frac{100}{3}\right)}{0.1386} \approx 25.3 \text{ years}$$

103. Find how much P increases in 36 hours when P_0 doubles every 24 hours.

Find k when $P = 2P_0$ and $t = 24$.

$$P = P_0e^{kt}$$

$$2P_0 = P\, e^{k(24)}$$

$$2 = e^{24k}$$

$$\ln 2 = 24k$$

$$k = \frac{\ln 2}{24} \approx 0.0289$$

Find P in comparison to P_0 when $t = 36$ and $k = 0.0289$

$$P = P_0e^{kt}$$

$$P = P_0\, e^{0.0289(36)}$$

$$P = P_0\, e^{1.0397}$$

$$P \approx 2.828P_0$$

P would be about 2.828 times larger.

105. Find the number of generations n when $b = 500$ cells per milliliter and $B = 5 \times 10^6$ cells per milliliter.

$$\begin{aligned} n &= \frac{1}{\ln 2}\left(\ln \frac{B}{b}\right) \\ &= \frac{1}{\ln 2}\left(\ln \frac{5 \times 10^6}{500}\right) \\ &\approx 13.3 \text{ generations} \end{aligned}$$

107. Find k when $T = 90°C$ and $t = 3$ minutes.

$$\begin{aligned} T &= 60 + 40e^{kt} \\ 90 &= 60 + 40e^{k(3)} \\ 30 &= 40e^{3k} \\ 0.75 &= e^{3k} \\ \ln 0.75 &= 3k \ln e = 3k \\ k &= \frac{\ln 0.75}{3} = \frac{1}{3}\ln 0.75 \end{aligned}$$

WRITING

109. First, take the logarithm of both sides of the equation. Second, apply the power rule of logarithms. Then, divide both sides by log 2.

REVIEW

111.
$$\begin{aligned} 5x^2 - 25x &= 0 \\ 5x(x-5) &= 0 \end{aligned}$$
$$5x = 0 \quad \text{or} \quad x - 5 = 0$$
$$x = 0 \qquad\qquad x = 5$$

113.
$$\begin{aligned} 3p^2 + 10p &= 8 \\ 3p^2 + 10p - 8 &= 0 \\ (3p-2)(p+4) &= 0 \end{aligned}$$
$$3p - 2 = 0 \quad \text{or} \quad p + 4 = 0$$
$$3p = 2 \qquad\qquad p = -4$$
$$p = \tfrac{2}{3}$$

115. Use the Pythagorean Theorem where $a = \sqrt{7}$ in. and $c = 12$ in. Find b.

$$\begin{aligned} a^2 + b^2 &= c^2 \\ \left(\sqrt{7}\right)^2 + b^2 &= 12^2 \\ 7 + b^2 &= 144 \\ b^2 &= 137 \\ b &= \sqrt{137} \text{ in.} \end{aligned}$$

One–to–One Functions

1. $f(x) = x^2$ is not a one–to–one function because both $x = 3$ and $x = -3$ would make $f(x) = 9$.

2. $f(x) = |x|$ is not a one–to–one function because both $x = 3$ and $x = -3$ would make $f(x) = 3$.

3. $x = 4 - 2y$ is one–to–one.

The Inverse of a Function

4.
$$\begin{aligned} f(x) &= -2x - 1 \\ y &= -2x - 1 \\ x &= -2y - 1 \quad \text{Interchange } x \text{ and } y. \\ x + 1 &= -2y \\ \frac{x+1}{-2} &= y \quad \text{so, } f^{-1}(x) = -\frac{x+1}{2} \text{ or } \frac{-x-1}{2} \end{aligned}$$

5. Interchange the values of x and y.

x	4	−2	-6
$f^{-1}(x)$	**−2**	**1**	**3**

Exponential and Logarithmic Functions

6. $10^3 = 1000$ would be rewritten $\log 1000 = 3$. (The base equal to 10 is understood.)

7. $\log_2 \frac{1}{8} = -3$ would be rewritten $2^{-3} = \frac{1}{8}$.

8. $\log_4 x = \frac{1}{2}$, so $4^{1/2} = x$ and $x = \sqrt{4} = 2$.

9. $\log_b \frac{9}{4} = 2$, so $b^2 = \frac{9}{4}$ and $b = \frac{3}{2}$.

The Natural Exponential and Natural Logarithmic Functions

10. $e = 2.71828$

11. The base of $f(x) = \ln x$ is e.

12. If $\ln x = -0.28$, then $e^{-0.28} = 0.7558$

13.

#1: $f(x) = e^x$

x	y
−2	0.1
−1	0.4
0	1
1	2.7
2	7.4

#2: $g(x) = \ln x$

x	y
0.1	−2
0.4	−1
1	0
2.7	1
7.4	2

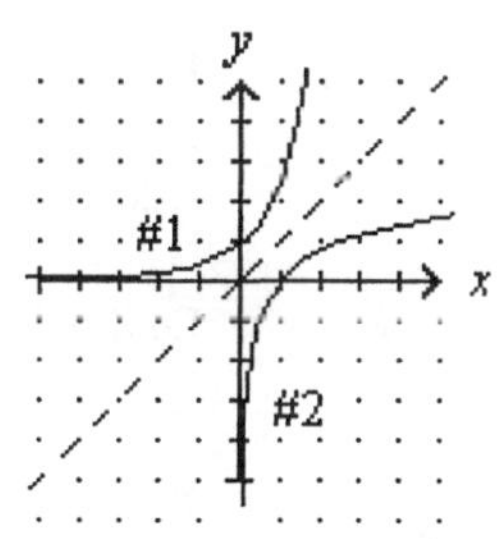

Chapter 9 Chapter Review

Section 9.1

1. $f+g = f(x)+g(x) = (2x)+(x+1) = 3x+1$

2. $f-g = f(x)-g(x) = (2x)-(x+1) = 2x-x-1 = x-1$

3. $f \bullet g = f(x) \bullet g(x) = 2x(x+1) = 2x^2+2x$

4. $f/g = \dfrac{f(x)}{g(x)} = \dfrac{2x}{x+1}$

5. $(f \circ g)(1) = f(g(1)) = f(1+1) = f(2) = 2(2) = 4$

6. $g(f(1)) = g(2(1)) = g(2) = 2+1 = 3$

7. $(f \circ g)(x) = f(g(x) = f(x+1) = 2(x+1) = 2x+2$

8. $(g \circ f)(x) = g(f(x) = g(2x) = 2x+1$

Section 9.2

9. No, it is not a function because $x = 2$ and $x = -2$ would give the same value of $f(x)$.
10. Yes.
11. Yes, it is one–to–one because there are no horizontal lines that will intersect the graph more than once.
12. No, it is not one–to–one because there are many horizontal lines that will intersect the graph more than once.

13.

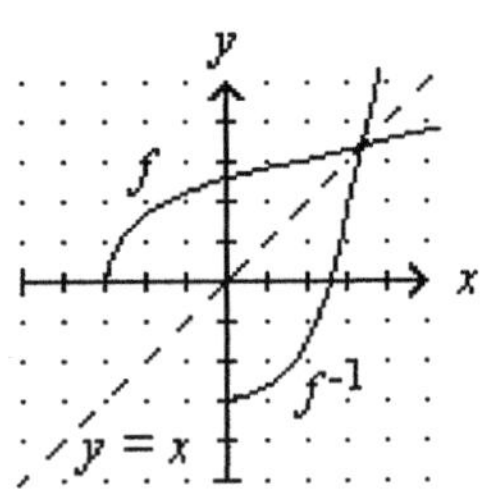

14.

$$\begin{aligned}
f(x) &= 6x-3 \\
y &= 6x-3 && \text{Replace } f(x) \text{ with y.} \\
x &= 6y-3 && \text{Interchange } x \text{ and y.} \\
x+3 &= 6y && \text{Add 3.} \\
\frac{x+3}{6} &= y && \text{Divide by 6.} \\
f^{-1}(x) &= \frac{x+3}{6} && \text{Rewrite.}
\end{aligned}$$

15.

$$\begin{aligned}
f(x) &= 4x+5 \\
y &= 4x+5 && \text{Replace } f(x) \text{ with } y. \\
x &= 4y+5 && \text{Interchange } x \text{ and } y. \\
x-5 &= 4y && \text{Subtract 5.} \\
\frac{x-5}{4} &= y && \text{Divide by 4.} \\
f^{-1}(x) &= \frac{x-5}{4} && \text{Rewrite.}
\end{aligned}$$

16.

$$\begin{aligned}
f(x) &= x^3 \\
y &= x^3 && \text{Replace } f(x) \text{ with y.} \\
x &= y^3 && \text{Interchange } x \text{ and y.} \\
\sqrt[3]{x} &= y && \text{Take cube roots.} \\
f^{-1}(x) &= \sqrt[3]{x} && \text{Rewrite.}
\end{aligned}$$

17.

$$\begin{aligned}
f(x) &= 2x^2-1 && (x \geq 0) \\
y &= 2x^2-1 && \text{Replace } f(x) \text{ with } y. \\
x &= 2y^2-1 && \text{Interchange } x \text{ and } y. \\
x+1 &= 2y^2 && \text{Add 1.} \\
\frac{x+1}{2} &= y^2 && \text{Divide by 2.} \\
\sqrt{\frac{x+1}{2}} &= y && \text{Take square roots.} \\
f^{-1}(x) &= \sqrt{\frac{x+1}{2}} && \text{Rewrite.}
\end{aligned}$$

Section 9.3

18. $5^{\sqrt{2}} \cdot 5^{\sqrt{2}} = 5^{\sqrt{2}+\sqrt{2}} = 5^{2\sqrt{2}}$

19. $\left(2^{\sqrt{5}}\right)^{\sqrt{2}} = 2^{\sqrt{5}\bullet\sqrt{2}} = 2^{\sqrt{10}}$

20. $f(x) = 3^x$

x	$f(x)$
-1	$\frac{1}{3}$
0	1
1	3
2	9

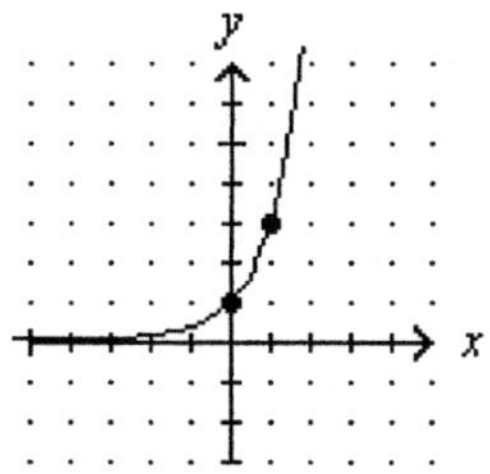

D: $(-\infty, \infty)$, R: $(0, \infty)$

21. $f(x) = \left(\frac{1}{3}\right)^x$

x	$f(x)$
-2	9
-1	3
0	1
1	$\frac{1}{3}$

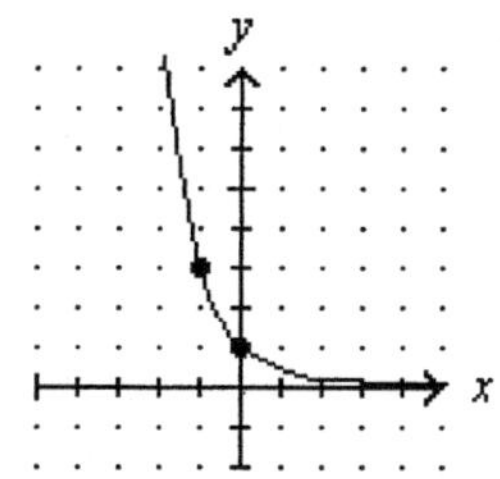

D: $(-\infty, \infty)$, R: $(0, \infty)$

22. $f(x) = \left(\frac{1}{2}\right)^x$

x	$f(x)$
-2	4
-1	2
0	1
1	$\frac{1}{2}$
2	$\frac{1}{4}$

$h(x) = \left(\frac{1}{2}\right)^x - 2$

Translate $f(x)$ 2 units downward.

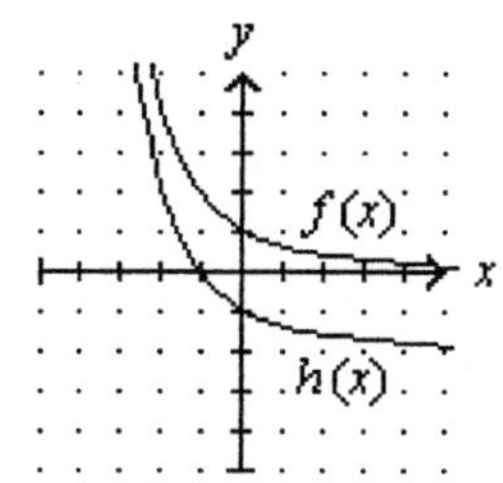

23. $f(x) = \left(\frac{1}{2}\right)^x$

x	$f(x)$
-2	4
-1	2
0	1
1	$\frac{1}{2}$
2	$\frac{1}{4}$

$g(x) = \left(\frac{1}{2}\right)^{x+2}$

Translate $f(x)$ 2 units to the left.

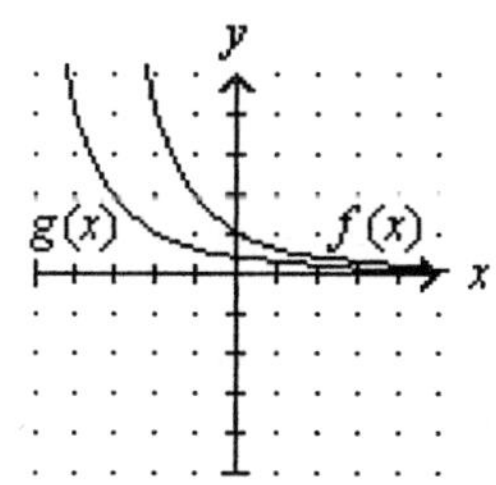

24.

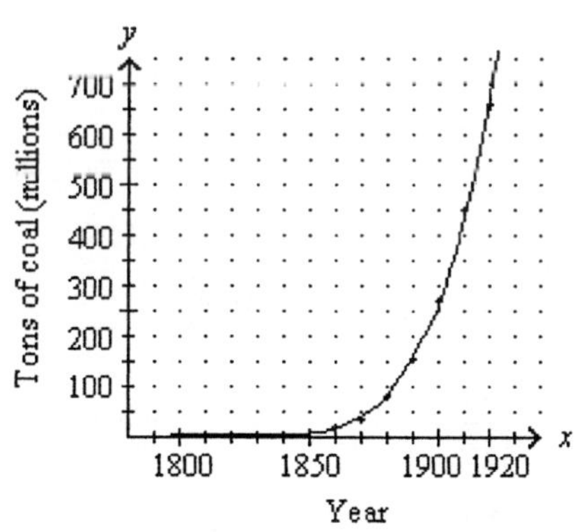

This appears to be an exponential function.

25. Find A when $P = \$10,500$, $r = 9\% = 0.09$, $k = 4$, and $t = 60$.

$$A = P\left(1+\frac{r}{k}\right)^{kt} = 10,500\left(1+\frac{0.09}{4}\right)^{4(60)} = 10,500(1.0225)^{240} = \$2,189,703.45$$

26. Find V when t=5

$V = 12,000\left(10^{-0.155(5)}\right) < \$2,015$

Section 9.4

27. $t = 1995 - 1980 = 15$ years. $r(t) = 13.9e^{-0.035t}$
$r(15) = 13.9e^{-0.035(15)} = 13.9e^{-0.525} = 8.22\%$

28. Find A when $P = \$10{,}500$, $t = 60$ years, and $r = 9\% = 0.09$ compounded continuously.
$A = Pe^{rt} = 10{,}500e^{0.09(60)} = 10{,}500(221.40642) = \$2{,}324{,}767.37$

29. $f(x) = e^x + 1$

x	$f(x)$
−2	1.1
−1	1.4
0	2
1	3.7
2	8.4

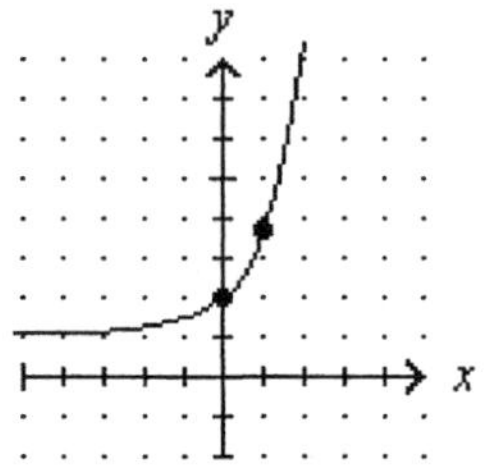

D: $(-\infty, \infty)$, R: $(1, \infty)$

30. $f(x) = e^{x-3}$

x	$f(x)$
1	0.1
2	0.4
3	1
4	2.7
5	7.4

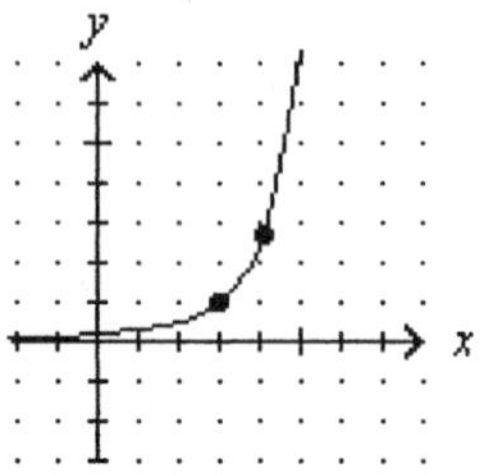

D: $(-\infty, \infty)$, R: $(0, \infty)$

31. Find A when $P = 290{,}340{,}000$, $t = 50$ years, and growth rate $k = 0.0092$.
$A = Pe^{kt} = 290{,}340{,}000\, e^{0.0092(50)} = 290{,}340{,}000(1.584) < 459{,}920{,}040$

32. When the exponent on the base e is negative, the function is decreasing.

Section 9.5

33. Domain: $(0, \infty)$; Range: $(-\infty, \infty)$

34. $\log_4 64 = 3$ becomes $4^3 = 64$.

35. $7^{-1} = \frac{1}{7}$ becomes $\log_7 \frac{1}{7} = -1$.

36. $\log_3 9 = 2$, because $3^2 = 9$.

37. $\log_9 \dfrac{1}{81} = -2$, because $9^{-2} = \dfrac{1}{81}$.

38. $\log_8 1 = 0$, because $8^0 = 1$.

39. $\log_5(-25)$ is not possible because -25 is not in the domain.

40. $\log_6 \sqrt{6} = \dfrac{1}{2}$, because $6^{1/2} = \sqrt{6}$.

41. $\log 1{,}000 = 3$, because $10^3 = 1{,}000$.

42. Rewrite $\log_2 x = 5$ as $2^5 = x$. So $x = 32$.

43. Rewrite $\log_3 x = -4$ as $3^{-4} = x$. So $x = \dfrac{1}{3^4} = \dfrac{1}{81}$

44.
$\log_x 16 = 2$
$x^2 = 16$ Rewrite.
$x^2 = 4^2$ Write in exponents.
$x = 4$
Exponents are equal, so bases are equal.

45.
$\log_x \dfrac{1}{100} = -2$
$x^{-2} = \dfrac{1}{100}$ Rewrite.
$x^{-2} = \dfrac{1}{10^2} = 10^{-2}$ Write in exponents.
$x = 10$
Exponents are equal, so bases are equal.

46.

$$\begin{aligned} \log_9 3 &= x & \\ 9^x &= 3 & \text{Rewrite.} \\ (3^2)^x &= 3^1 & \text{Change to base 3.} \\ 2x &= 1 & \text{Exponents are equal.} \\ x &= \frac{1}{2} & \text{Solve for } x. \end{aligned}$$

47.

$$\begin{aligned} \log_{27} x &= \frac{2}{3} & \\ 27^{2/3} &= x & \text{Rewrite.} \\ \left(\sqrt[3]{27}\right)^2 &= x & \text{Solve for } x. \\ c &= 9 & \end{aligned}$$

48. $\log 4.51 = x, \ x = 0.6542$

49. If $\log x = 1.43$, then $10^{1.43} = 26.9153$

50.

#1: $y = 4^x$

x	y
−2	$\frac{1}{16}$
−1	$\frac{1}{4}$
0	1
1	4
2	16

#2: $y = \log_4 x$

x	y
$\frac{1}{16}$	−2
$\frac{1}{4}$	−1
1	0
4	1
16	2

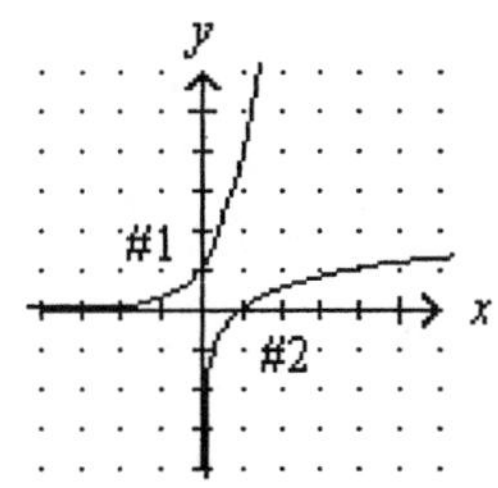

51.

#1: $f(x) = \left(\frac{1}{3}\right)^x$

x	$f(x)$
−2	9
−1	3
0	1
1	$\frac{1}{3}$
2	$\frac{1}{9}$

#2: $g(x) = \log_{1/3} x$

x	$g(x)$
9	−2
3	−1
1	0
$\frac{1}{3}$	1
$\frac{1}{9}$	2

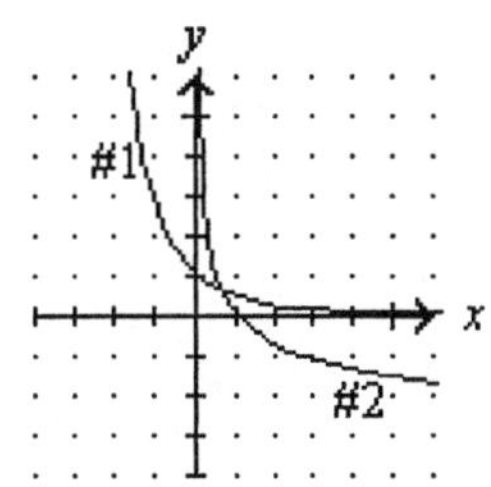

52. $f(x) = \log(x - 2)$

x	$f(x)$
2.1	−1
3	0
4	0.3
5	0.5
6	0.6

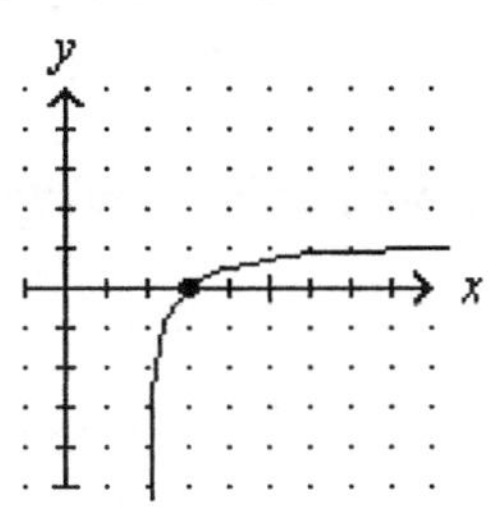

53. $f(x) = 3 + \log x$

x	y
0.01	1
0.1	2
1	3
10	4
100	5

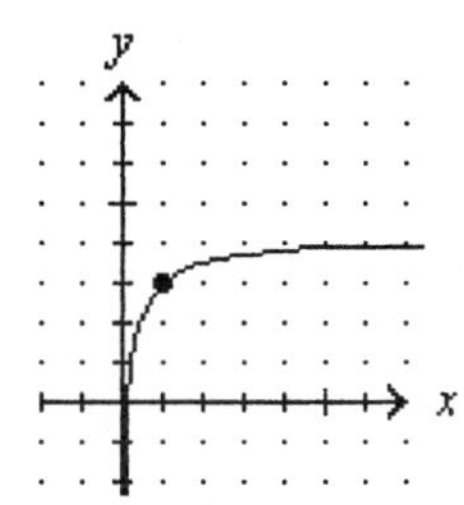

54. Find the db gain when $E_I = 0.04$ volts and $E_O = 18$ volts.

$$\text{db gain} = 20\log\frac{E_O}{E_I} = 20\log\frac{18}{0.04} = 20\log 450 = 53 \text{ db}$$

55. Find R when $A = 7500$ micrometers and $P = 0.3$ second.

$$R = \log\frac{A}{P} = \log\frac{7500}{0.3} = \log 25{,}000 \approx 4.39794 \approx 4.4 \text{ on the Richter scale.}$$

Section 9.6

56. $\ln 452 = 6.1137$

57. $\ln 0.85 = -0.1625$

58. If $\ln x = 2.336$, then $e^{2.336} = 10.3398$

59. If $\ln x = -8.8$, then $e^{-8.8} = 0.0002$

60. The inverse of $f(x) = \ln x$ is $f^{-1}(x) = e^x$.

61. $f(x) = 1 + \ln x$

x	$f(x)$
2.1	–1
3	0
4	0.3
5	0.5
6	0.6

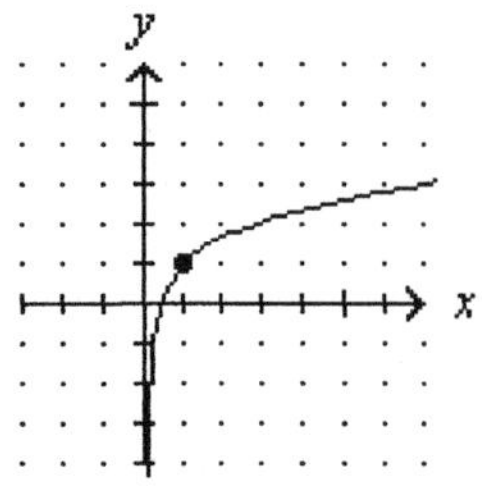

62. $f(x) = \ln(x + 1)$

x	$f(x)$
0.01	1
0.1	2
1	3
10	4
100	5

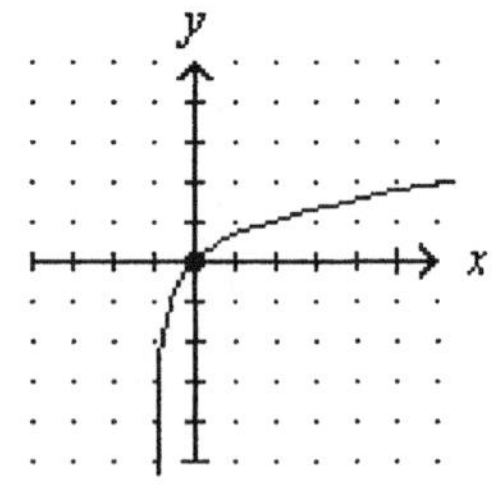

63. Find t when $r = 1.43\% = 0.0143$. $t = \dfrac{\ln 2}{r} = \dfrac{\ln 2}{0.0143} \approx 48.47 \approx 48\frac{1}{2}$ years.

64. Find H when $a = 19$. $H(19) = 13 + 20.03 \ln(19) < 71.98 < 72$ in. or 6 feet.

Section 9.7

65. $\log_7 1 = 0$, because $7^0 = 1$.

66. $\log_7 7 = 1$, because $7^1 = 7$.

67. $\log_7 7^3 = 3$, because $7^3 = 7^3$.

68. $7^{\log_7 4} = 4$, because $\log_b x$ is the exponent to which b is raised to get x.

69. $\ln e^4 = 4$, because $e^4 = e^4$.

70. $\ln e = 1$, because $e^1 = e$.

71. $\log_3 27x = \log_3 27 + \log_3 x$ — The log of a product is the sum of the logs.
$= 3 + \log_3 x$ — $\log_3 27 = 3$.

72. $\log \dfrac{100}{x} = \log 100 - \log x$ — The log of a quotient is the difference of the logs.
$= 2 - \log x$ — $\log_{10} 100 = 2$.

73. $\log_5 \sqrt{27} = \log_5 27^{1/2}$ — Write $\sqrt{27}$ as $27^{1/2}$.
$= \frac{1}{2} \log_5 27$ — The log of a power is the power times the log.

74. $\log 10ab = \log 10 + \log a + \log b$ — The log of a product is the sum of the logs.
$= 1 + \log a + \log b$ — $\log_{10} 10 = 1$.

75. $\log_b \dfrac{x^2y^3}{z} = \log_b x^2y^3 - \log_b z = \log_b x^2 + \log_b y^3 - \log_b z = 2\log_b x + 3\log_b y - \log_b z$

76. $\ln\sqrt{\dfrac{x}{yz^2}} = \ln\left(\dfrac{x}{yz^2}\right)^{1/2} = \dfrac{1}{2}\ln\dfrac{x}{yz^2} = \dfrac{1}{2}[\ln x - (\ln y + \ln z^2)] = \dfrac{1}{2}(\ln x - \ln y - 2\ln z)$

77. $3\log_2 x - 5\log_2 y + 7\log_2 z = \log_2 x^3 + \log_2 y^{-5} + \log_2 z^7$ — A power times a log is log of power.
$= \log_2 x^3y^{-5}z^7$ — The sum of logs is log of the products.
$= \log_2 \dfrac{x^3z^7}{y^5}$ — Write with positive exponents.

78. $-3\log_b y - 7\log_b z + \dfrac{1}{2}\log_b(x + 2) = \log_b y^{-3} + \log_b z^{-7} + \log_b(x + 2)^{1/2}$ — The power rule.
$= \log_b y^{-3}z^{-7}(x + 2)^{1/2}$ — The product rule.
$= \log_b \dfrac{\sqrt{x + 2}}{y^3z^7}$ — Write with positive exponents.

79. $\log_b 40 = \log_b 5 \cdot 8 = \log_b 5 + \log_b 8 = 1.1609 + 1.5000 = 2.6609$

80. $\log_b 64 = \log_b 8^2 = 2\log_b 8 = 2(1.5000) = 3.0000$

81. $\log_5 17 = \dfrac{\log_{10} 17}{\log_{10} 5} = 1.7604$

82. Find $[H^+]$ when pH = 3.1.

$$\begin{aligned} pH &= -\log [H^+] \\ 3.1 &= -\log [H^+] \\ -3.1 &= \log [H^+] \\ [H^+] &= 10^{-3.1} = 7.9 \times 10^{-4} \text{ gram–ions per liter.} \end{aligned}$$

Section 9.8

83.

3^x	=	7	
$\log 3^x$	=	$\log 7$	Take log of both sides.
$x \log 3$	=	$\log 7$	Use power rule.
x	=	$\dfrac{\log 7}{\log 3}$	Divide by log 3.
x	≈	1.7712	Use a calculator.

84.

5^{x+2}	=	625	
5^{x+2}	=	5^4	Write each side with base of 5.
$x+2$	=	4	Equate the exponents.
x	=	2	Subtract 2 from both sides.

85.

$2^x = 3^{x-1}$	
$\log 2^x = \log 3^{x-1}$	Take log of both sides
$x\log 2 = (x-1)\log 3$	Use power rule
$x\log 2 = x\log 3 - \log 3$	Move x terms to one side
$x(\log 2 - \log 3) = -\log 3$	Combine like terms
$x = \dfrac{-\log 3}{\log 2 - \log 3} \approx 2.7095$	Divide both sides by log 2-log 3; use calculator

86.

2^{x^2+4x}	=	$\dfrac{1}{8}$	
2^{x^2+4x}	=	2^{-3}	Write each side with base of 2.
x^2+4x	=	-3	Equate the exponents.
x^2+4x+3	=	0	Add 3.
$(x+3)(x+1)$	=	0	Factor.

$x+3 = 0$ or $x+1 = 0$

$x = -3$ $x = -1$

87.

e^x	=	7	
$\ln e^x$	=	$\ln 7$	Take ln of both sides.
$x \ln e$	=	$\ln 7$	Use power rule.
x	=	$\ln 7$	$\ln e = 1$
x	≈	1.9459	Use calculator.

88.

$e^{-0.4t}$	=	25	
$\ln e^{-0.4t}$	=	$\ln 25$	Take ln of both sides.
$-0.4t \ln e$	=	$\ln 25$	Use power rule.
$-0.4t$	=	$\ln 25$	$\ln e = 1$
t	=	$\dfrac{\ln 25}{-0.4}$	Divide by −0.4.
t	≈	−8.0472	Use calculator.

89.

$\log(x-4)$	=	2	
$x-4$	=	10^2	Change to exponential form.
$x-4$	=	100	Simplify.
x	=	104	Add 4.

90.

$$\begin{aligned} \ln(2x-3) &= \ln 15 \\ 2x-3 &= 15 && \text{The expressions in the logarithm are equal.} \\ 2x &= 18 && \text{Add 3.} \\ x &= 9 && \text{Divide by 2.} \end{aligned}$$

91.

$$\begin{aligned} \log x + \log(29-x) &= 2 \\ \log x(29-x) &= 2 && \text{Use the product rule of logarithms.} \\ x(29-x) &= 10^2 && \text{Change to exponential form.} \\ 29x - x^2 - 100 &= 0 && \text{Remove parentheses and subtract 100.} \\ x^2 - 29x + 100 &= 0 && \text{Multiply by } -1 \text{ and rearrange.} \\ (x-25)(x-4) &= 0 && \text{Factor.} \end{aligned}$$

$x - 25 = 0$ or $x - 4 = 0$

$x = 25$ $\quad$ $x = 4$

92.

$$\begin{aligned} \log_2 x + \log_2(x-2) &= 3 \\ \log_2 x(x-2) &= 3 && \text{Use the product rule of logarithms.} \\ x(x-2) &= 2^3 && \text{Change to exponential form.} \\ x^2 - 2x - 8 &= 0 && \text{Remove parentheses and subtract 8.} \\ (x-4)(x+2) &= 0 && \text{Factor.} \end{aligned}$$

$x - 4 = 0$ or $x + 2 = 0$

$x = 4$ $\quad$ $x = -2$ Not possible. $\log(-2)$ is undefined.

93.

$$\frac{\log(7x-12)}{\log x} = 2$$

$$\begin{aligned} \log(7x-12) &= 2\log x && \text{Multiply by } \log x. \\ \log(7x-12) &= \log x^2 && \text{Use the power rule of logarithms.} \\ 7x - 12 &= x^2 && \text{The expressions in the logarithm are equal.} \\ 0 &= x^2 - 7x + 12 && \text{Subtract } 7x \text{ and add 12.} \\ 0 &= (x-3)(x-4) && \text{Factor.} \end{aligned}$$

$x - 3 = 0$ or $x - 4 = 0$

$x = 3$ $\quad$ $x = 4$

94.

$$\begin{aligned} \log_2(x+2) + \log_2(x-1) &= 2 \\ \log_2(x+2)(x-1) &= 2 && \text{Use the product rule of logarithms.} \\ (x+2)(x-1) &= 2^2 && \text{Change to exponential form.} \\ x^2 + x - 2 &= 4 && \text{Remove parentheses.} \\ x^2 + x - 6 &= 0 && \text{Subtract 4.} \\ (x-2)(x+3) &= 0 && \text{Factor.} \end{aligned}$$

$x - 2 = 0$ or $x + 3 = 0$

$x = 2$ $\quad$ $x = -3$ Not possible. $\log_2(-3+2)$ is undefined.

95.

$$\begin{aligned} \log x + \log(x-5) &= \log 6 \\ \log x(x-5) &= \log 6 && \text{Use the product rule of logarithms.} \\ x(x-5) &= 6 && \text{The expressions in the logarithm are equal.} \\ x^2 - 5x &= 6 && \text{Remove parentheses.} \\ x^2 - 5x - 6 &= 0 && \text{Subtract 6 from both sides.} \\ (x-6)(x+1) &= 0 && \text{Factor.} \end{aligned}$$

$x - 6 = 0$ or $x + 1 = 0$

$x = 6$ $\quad$ $x = -1$ Not possible. $\log(-1)$ is undefined.

96.

$$\begin{aligned}
\log 3 - \log(x-1) &= -1 \\
\log\left(\frac{3}{x-1}\right) &= -1 && \text{Use the quotient rule of logarithms.} \\
\frac{3}{x-1} &= 10^{-1} = \frac{1}{10} && \text{Change to exponential form.} \\
x - 1 &= 30 && \text{Multiply the means and extremes.} \\
x &= 31 && \text{Add 1.}
\end{aligned}$$

97. Find t when $\frac{2}{3}$ of the carbon–14 is left. So, $A = \frac{2}{3}A_0$ and $h = 5700$ years.

$$\begin{aligned}
A &= A_0 2^{-t/h} \\
\tfrac{2}{3}A_0 &= A_0 2^{-t/5700} && \text{Substitute given values.} \\
\tfrac{2}{3} &= 2^{-t/5700} && \text{Divide by } A_0. \\
\log \tfrac{2}{3} &= -\frac{t}{5700}\log 2 && \text{Take the logarithm of both sides.} \\
-5700 \log \tfrac{2}{3} &= t \log 2 && \text{Multiply by } -5700. \\
t &= \frac{-5700 \log \frac{2}{3}}{\log 2} = 3{,}334 \text{ years or about } 3{,}300 \text{ years}
\end{aligned}$$

98. The initial population is 800, and after $\frac{n}{14}$ days, the population is $800(3^{n/14})$. When is the population 1,000,000?

$$\begin{aligned}
1{,}000{,}000 &= (800)3^{n/14} \\
\frac{1{,}000{,}000}{800} &= 3^{n/14} \\
\ln\frac{1{,}000{,}000}{800} &= \ln 3^{n/14} \\
7.131 &\approx \frac{n}{14}(1.099) \\
n &\approx 91.84 \text{ or } 91 \text{ days}
\end{aligned}$$

99. The solution is $x = 2$.

Check:

$$\begin{aligned}
\log x &= 1 - \log(7 - x) \\
\log(2) &= 1 - \log(7-2) \\
\log 2 &= 1 - \log 5 \\
\log 2 + \log 5 &= 1 \\
\log 2(5) &= 1 \\
\log 10 &= 1 \\
1 &= 1
\end{aligned}$$

The solution is $x = 5$.

Check:

$$\begin{aligned}
\log x &= 1 - \log(7 - x) \\
\log(5) &= 1 - \log(7-5) \\
\log 5 &= 1 - \log 2 \\
\log 5 + \log 2 &= 1 \\
\log 5(2) &= 1 \\
\log 10 &= 1 \\
1 &= 1
\end{aligned}$$

Chapter 9 Chapter Test

1. $g + f = g(x) + f(x) = (x - 1) + (4x) = 5x - 1$

2. $g \bullet f = g(x) \bullet f(x) = (x - 1)(4x) = 4x^2 - 4x$

3. $(g \circ f)(1) = g(f(1)) = g(4 \bullet 1) = g(4) = 4 - 1 = 3$

4. $(f \circ g)(x) = f(g(x)) = f(x - 1) = 4(x - 1) = 4x - 4$

5.
$$\begin{aligned} f(x) &= -\frac{3}{2}x + 6 \\ y &= -\frac{3}{2}x + 6 \\ 2y &= -3x + 12 \\ 3x + 2y &= 12 \\ 3y + 2x &= 12 && \text{Interchange } x \text{ and y.} \\ 3y &= 12 - 2x && \text{Subtract } 2x. \\ y &= \frac{12 - 2x}{3} && \text{Divide by 3.} \\ f^{-1}(x) &= \frac{12 - 2x}{3} && \text{Rewrite.} \end{aligned}$$

6.
$$\begin{aligned} f(x) &= 3x^2 + 4 \\ y &= 3x^2 + 4 \\ x &= 3y^2 + 4 && \text{Interchange } x \text{ and y.} \\ x - 4 &= 3y^2 && \text{Subtract 4.} \\ \frac{x - 4}{3} &= y^2 && \text{Divide by 3.} \\ y &= \sqrt{\frac{x - 4}{3}} && \text{Solve for } y. \\ f^{-1}(x) &= \sqrt{\frac{x - 4}{3}} && \text{Rewrite.} \end{aligned}$$

7. $f(x) = 2^x + 1$

x	$f(x)$
−2	1.25
−1	1.5
0	2
1	3
2	5

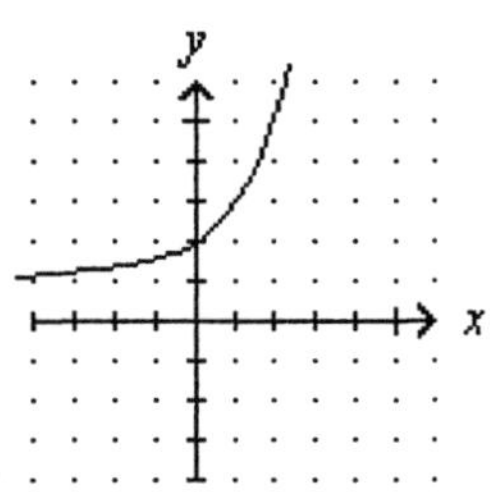

8. $y = 3^{-x}$

x	y
−2	9
−1	3
0	1
1	$\frac{1}{3}$
2	$\frac{1}{9}$

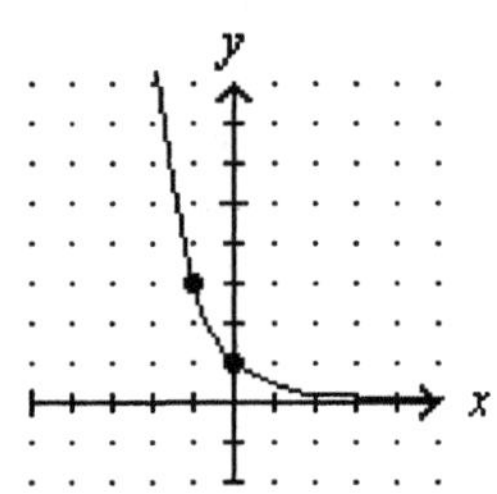

9. Find A when $A_0 = 3$ grams and $t = 6$ years.

$$A = A_0(2)^{-t} = 3(2)^{-6} = 3\left(\frac{1}{2^6}\right) = \frac{3}{64}\ g = 0.046875\ g$$

10. Find A when $P = \$1000$, , $k = 2$, $t = 1$, and $r = 6\% = 0.06$.

$$A = P\left(1 + \frac{r}{k}\right)^{kt} = 1000\left(1 + \frac{0.06}{2}\right)^{2(1)} = 1000(1.03)^2 = \$1060.90$$

11. $f(x) = e^x$

x	$f(x)$
−1	0.4
0	1
1	2.7
2	7.4

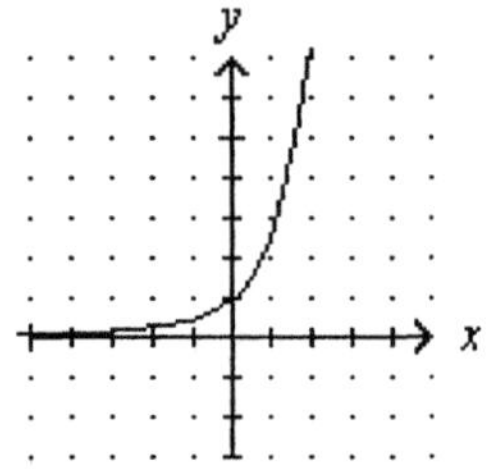

12. Find A when $P = \$2000$, $t = 10$ years, and $r = 8\% = 0.08$ compounded continuously.

$$A = Pe^{rt} = 2000e^{0.08(10)} = 2000(2.2255409) = \$4451.08$$

13. Rewrite $\log_4 16 = x$ as $4^x = 16$. Since $16 = 4^2$, then $4^x = 4^2$ or $x = 2$.

14. Rewrite $\log_x 81 = 4$ as $x^4 = 81$. Since $81 = 3^4$, then $x^4 = 3^4$ or $x = 3$.

15. Rewrite $\log_3 x = -3$ as $3^{-3} = x$. So $x = \dfrac{1}{3^3} = \dfrac{1}{27}$.

16. Rewrite $\ln x = 1$ as $e^1 = x$. So $x = e$.

17. $\log_6 \frac{1}{36} = -2$ would be rewritten $6^{-2} = \dfrac{1}{36}$.

18. D: $(0, \infty)$; R: $(-\infty, \infty)$

19. $f(x) = -\log_3 x = \log_3 x^{-1}$

x	$f(x)$
9	−2
3	−1
1	0
0.3	1
0.1	2

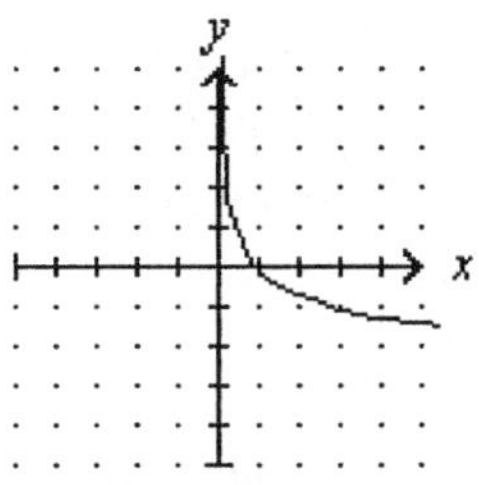

20. $f(x) = \ln x$

x	$f(x)$
0.1	−2
0.4	−1
1	0
2.7	1
7.4	2

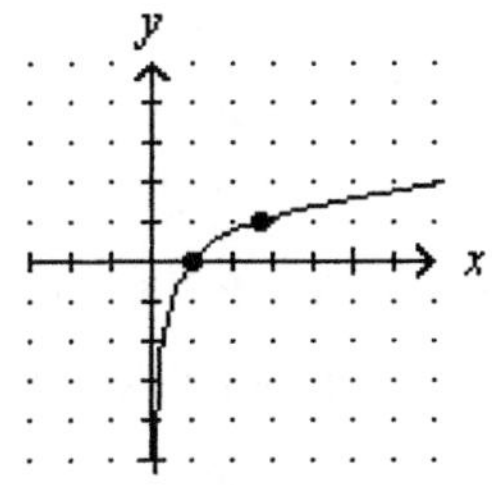

21. $\log a^2bc^3 = \log a^2 + \log b + \log c^3 = 2\log a + \log b + 3\log c$

22. $\frac{1}{2}\ln(a+2) + \ln b - 3\ln c = \ln(a+2)^{1/2} + \ln b + \ln c^{-3} = \ln(a+2)^{1/2}bc^{-3} = \ln\dfrac{b\sqrt{a+2}}{c^3}$

23.Use $\log_b x = \dfrac{\log_a x}{\log_a b}$. $\log_7 3 = \dfrac{\log_{10} 3}{\log_{10} 7} = 0.5646$

24. The inverse of $y = 10^x$ is $y = \log x$.

25. Find the pH of a solution when $[H^+] = 3.7 \times 10^{-7}$ gram–ions per liter.
$\text{pH} = -\log[H^+] = -\log(3.7 \times 10^{-7}) = -(-6.4318) \approx 6.4$

26. Find the db gain when $E_I = 0.3$ volts and $E_O = 60$ volts.

$$\text{db gain} = 20\log\frac{E_O}{E_I} = 20\log\frac{60}{0.3} = 20\log 200 = 46.021 \approx 46 \text{ db}$$

27.
$5^x = 3$
$\log 5^x = \log 3$ Take log of both sides.
$x\log 5 = \log 3$ Use power rule.
$x = \dfrac{\log 3}{\log 5}$ Divide by log 5.
$x \approx 0.6826$ Use a calculator.

28.
$3^{x-1} = 27$
$3^{x-1} = 3^3$ Write each side with base of 3.
$x - 1 = 3$ Equate the exponents.
$x = 4$ Add 1 to both sides.

29.
$\ln(5x+2) = \ln(2x+5)$
$5x + 2 = 2x + 5$ The expressions in the logarithm are equal.
$3x = 3$ Subtract $2x$ and 2.
$x = 1$ Divide by 3.

30.
$$\begin{aligned} \log x + \log(x-9) &= 1 \\ \log x(x-9) &= 1 && \text{Use the product rule of logarithms.} \\ x(x-9) &= 10^1 && \text{Change to exponential form.} \\ x^2 - 9x - 10 &= 0 && \text{Remove parentheses and subtract 10.} \\ (x-10)(x+1) &= 0 && \text{Factor.} \end{aligned}$$

$x - 10 = 0$ or $x + 1 = 0$

$x = 10$ $\quad x = -1$ Not possible. $\log(-1)$ is undefined.

31. The solution would be the value of x which is $x = 5$.

32. Substitute $x = 5$. $\frac{1}{2}\ln(5-1) = \frac{1}{2}\ln 4 = 0.693147\ldots = \ln 2$

33. The exponential growth of money invested at a compounded rate would be an example of this type of graph. The x–axis would be labeled for time, usually in years, and the y–axis would be labeled in dollars.

34. a. yes b. yes c. 80; When the temperature of the tire tread is 260°, the vehicle is traveling 80 mph.

35. Use $A = Pe^{rt}$ where $P = 1{,}050{,}000{,}000$, $r = 1.47\% = 0.0147$, and $t = 30$

$A = 1{,}050{,}000{,}000\, e^{30\,(0.0147)} < 1{,}631{,}973{,}737$

36. The number of insects after t seconds is $(5)4^{t/6}$. Determine at what time t, this equals 500.

$$500 = (5)4^{t/6}$$
$$100 = 4^{t/6}$$
$$\ln 100 = \ln 4^{t/6}$$
$$\ln 100 = \frac{t}{6}\ln 4$$
$$\frac{\ln 100}{\ln 4} = \frac{t}{6}$$
$19.93 \approx t$ or about 20 min

Chapters 1–9 Cumulative Review Exercises

1. a. $P = 2l + 2w$ b. $A = \pi r^2$ c. $A = \frac{1}{2}bh$ d. $V = s^3$

e. $I = Prt$ f. $d = rt$ g. $\left(\frac{x_1 + x_2}{2}, \frac{y_1 + y_2}{2}\right)$ h. $y = mx + b$

i. $y - y_1 = m(x - x_1)$ j. $m = \frac{y_2 - y_1}{x_2 - x_1}$ k. $a^2 + b^2 = c^2$

l. $d = \sqrt{(x_2 - x_1)^2 + (y_2 - y_1)^2}$ m. $y = kx$ n. $y = \frac{k}{x}$

o. $x = \frac{-b \pm \sqrt{b^2 - 4ac}}{2a}$ p. $A = Pe^{rt}$ q. $\log_b x = \frac{\log_a x}{\log_a b}$

2. a. $x^1 = x$ b. $x^m x^n = x^{m+n}$ c. $(x^m)^n = x^{mn}$ d. $(xy)^n = x^n y^n$

e. $x^0 = 1$ f. $\left(\frac{x}{y}\right)^n = \frac{x^n}{y^n}$ g. $\frac{x^m}{x^n} = x^{m-n}$ h. $x^{-n} = \frac{1}{x^n}$

i. $\frac{1}{x^{-n}} = x^n$ j. $x^{m/n} = \left(\sqrt[n]{x}\right)^m$

3. a. $x^2 - y^2 = (x + y)(x - y)$ b. $x^3 - y^3 = (x - y)(x^2 + xy + y^2)$ c. $x^3 + y^3 = (x + y)(x^2 - xy + y^2)$

4. a. $(x + y)^2 = x^2 + 2xy + y^2$ b. $(x - y)^2 = x^2 - 2xy + y^2$ c. $(x + y)(x - y) = x^2 - y^2$

5. a. $\sqrt[n]{ab} = \sqrt[n]{a}\sqrt[n]{b}$ b. $\sqrt[n]{\frac{a}{b}} = \frac{\sqrt[n]{a}}{\sqrt[n]{b}}$

6. a. $\pi = 3.14$ b. $\sqrt{2} = 1.41$ c. $e = 2.72$

7. Pick a test point on one side of the boundary line. Replace x and y with the coordinates of that point. If the inequality is satisfied, shade the side that contains that point. If the inequality is not satisfied, shade the other side.

8. a. $\log_b 1 = 0$ b. $\log_b b = 1$ c. $\log_b b^x = x$

d. $b^{\log_b x} = x$ e. $\log_b MN = \log_b M + \log_b N$

f. $\log_b \frac{M}{N} = \log_b \mathrm{M} - \log_b N$ g. $\log_b M^P = P \log_b M$

9. If a, b, and k represent real numbers, and $b \neq 0$ and $k \neq 0$ then $\frac{a}{b} = \frac{a \cdot k}{b \cdot k}$

10. a. Multiplying each fraction by a lowest common denominator is used to build up the fractions so they have the same denominator.

b. Multiplying by the same radical is used to rationalize the denominator.

c. Multiplying by a common denominator is used to simplify a complex fraction.

11. It was used to simplify a rational expression. The slashes and 1's show a common factor of 3a – 2 being divided out.

12. If $\frac{a}{b} = \frac{c}{d}$, then $ad = bc$.

13. A quadratic equation is an equation that is in the form $ax^2 + bx + c = 0$, where $a \neq 0$.

14. The zero–factor property is used to set each factor equal to 0. In this case, $x + 1 = 0$ or $x - 7 = 0$.

15. Multiply both sides of the equation by the LCD, which is $x(x - 2)$.

16. The addition method would be the most efficient method because it is easier to multiply to eliminate the y variable than to solve one of the equation for a variable and substitute a fractional equation.

17. For any real number x, $\begin{cases} \text{if } x \geq 0, \text{ then } |x| = x \\ \text{if } x \leq 0, \text{ then } |x| = -x \end{cases}$

18. The number b is a square root of a if $b^2 = a$.

19. If x can be any real number, then $\sqrt{x} = |x|$.

20. If x is any real number, then $\sqrt[3]{x} = x$.

21. If a and b are real numbers, $a - b = a + (-b)$.

22. $i = \sqrt{-1}$

23. domain; range; one

24. a. both b. one; other; both

Chapter 10 Conic Sections; More Graphing

Study Tip
Getting ready for next semester

You've made it through the semester, and now you need to make plans for your next math class. Here are some suggestions:

1. Look for an instructor from whom you believe you can learn- but remember- how much you learn depends much more on YOU than on your instructor.
3. Choose a time for your class in which you work well. If you are not an early bird, avoid the 8:00 am class. If you get sleepy after lunch, schedule a morning class.
2. If you plan to take a summer course, remember that in a summer course, you will be responsible for the same amount of material as in a regular course- it will just go FASTER!
3. You will be held responsible for everything you learned in this course- no excuses! Keep your notes from the class so you can go back and refer to them if you need to.
4. Be ready to begin your class on the very first day. Read the first section of your new book before you come to class on the first day so you know what your instructor will be talking about.

Section 10.1 The Circle and the Parabola

VOCABULARY

1. conic sections
3. radius; center

CONCEPTS

5. vertex; left
7. parabola; (3, 2); right
9. $(x-h)^2+(y-k)^2 = r^2$
11. $y=-ax^2$

NOTATION

13.
$$\begin{aligned}(x-h)^2+(y-k)^2 &= r^2\\ (x-\mathbf{2})^2+(y-3)^2 &= \mathbf{6}^2\\ x^2-\mathbf{4}x+4+y^2-6y+9 &= 36\\ x^2+y^2-4x-6y-23 &= 0\end{aligned}$$

PRACTICE

15. $x^2+y^2 = 9$

Center: (0, 0); Radius = 3

x	y
0	−3
0	3
−3	0
3	0

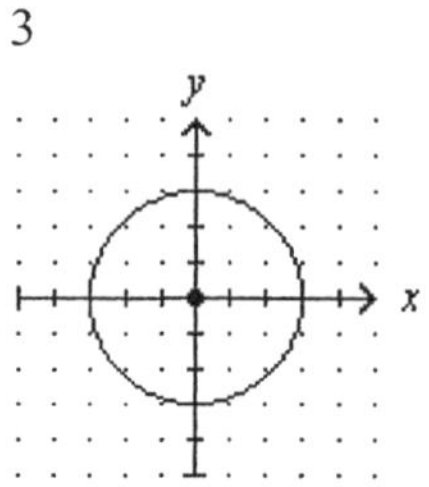

17. $(x-2)^2+y^2 = 9$

Center: (2, 0); Radius = 3

x	y
2	−3
2	3
−1	0
5	0

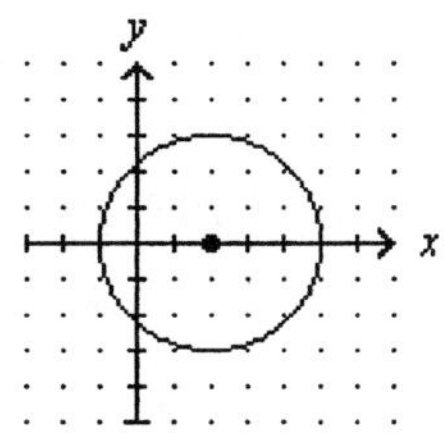

19. $(x-2)^2+(y-4)^2 = 4$

Center: (2, 4); Radius = 2

x	y
2	2
2	6
0	4
4	4

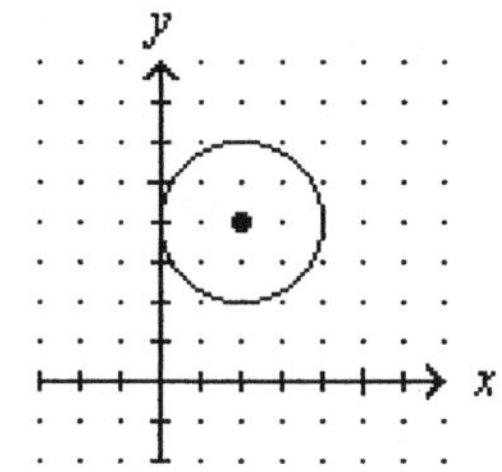

21. $(x+3)^2+(y-1)^2 = 16$

Center: (−3, 1); Radius = 4

x	y
−3	5
−3	−3
1	1
−7	1

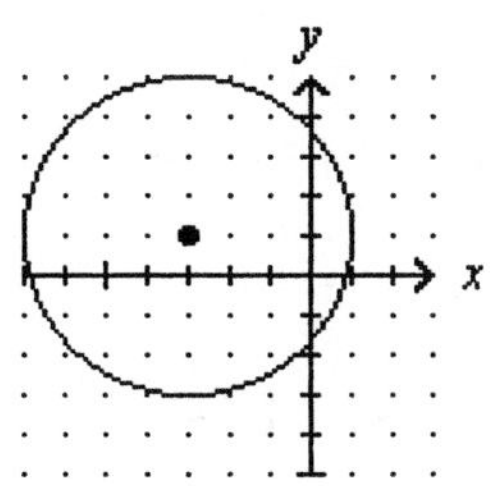

23. $x^2+(y+3)^2 = 1$

Center: (0, −3); Radius = 1

x	y
0	−4
0	−2
−1	−3
1	−3

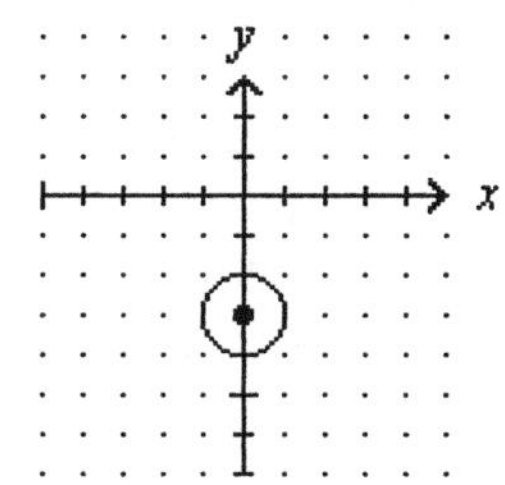

25. Solve $3x^2+3y^2 = 16$ for y.

$$3y^2 = 16-3x^2$$

$$y^2 = \frac{16-3x^2}{3}$$

$$y = \pm\sqrt{\frac{16-3x^2}{3}}$$

Graph $y=\sqrt{\frac{16-3x^2}{3}}$ and $y=-\sqrt{\frac{16-3x^2}{3}}$.

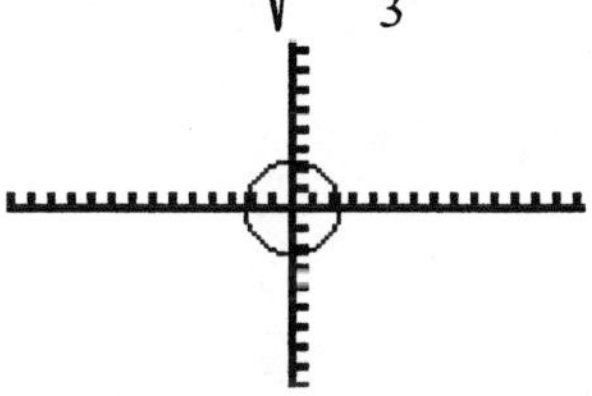

27. Solve $(x+1)^2+y^2 = 16$ for y.

$$y^2 = 16-(x+1)^2$$

$$y = \pm\sqrt{16-(x+1)^2}$$

Graph $y=\sqrt{16-(x+1)^2}$ and $y=-\sqrt{16-(x+1)^2}$.

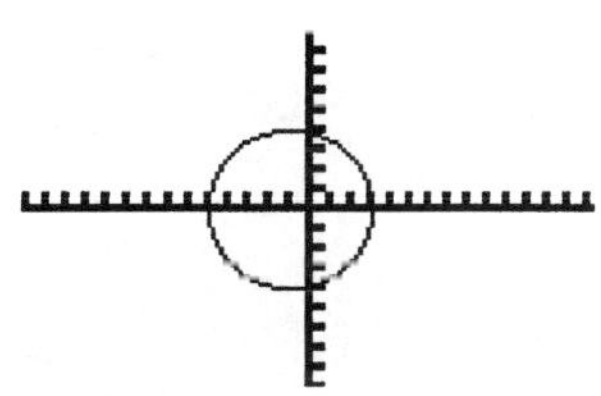

29. Center at the origin (0, 0) and radius 1.
Substitute 1 for r, 0 for h, and 0 for k.

$$(x-h)^2+(y-k)^2 = r^2$$
$$(x-0)^2+(y-0)^2 = 1^2$$
$$x^2+y^2 = 1$$

31. Center at (6, 8) and radius 5.
Substitute 5 for r, 6 for h, and 8 for k.

$$(x-h)^2+(y-k)^2 = r^2$$
$$(x-6)^2+(y-8)^2 = 5^2$$
$$(x-6)^2+(y-8)^2 = 25$$

33. Center at (−2, 6) and radius 12.
Substitute 12 for r, −2 for h, and 6 for k.

$$(x-h)^2+(y-k)^2 = r^2$$
$$[x-(-2)]^2+(y-6)^2 = 12^2$$
$$(x+2)^2+(y-6)^2 = 144$$

35. Center at the origin (0, 0) and diameter $4\sqrt{2}$.
Substitute $2\sqrt{2}$ for r, 0 for h, and 0 for k.

$$(x-h)^2+(y-k)^2 = r^2$$
$$(x-0)^2+(y-0)^2 = \left(2\sqrt{2}\right)^2$$
$$x^2+y^2 = 8$$

37.

$$\begin{aligned} x^2 + y^2 + 2x - 8 &= 0 \\ x^2 + 2x + y^2 &= 8 \\ x^2 + 2x + 1 + y^2 &= 8 + 1 \\ (x + 1)^2 + (y - 0)^2 &= 9 \\ [x - (-1)]^2 + (y - 0)^2 &= 3^2 \end{aligned}$$

Center: $(-1, 0)$; Radius $= 3$

x	y
−1	−3
−1	3
−4	0
2	0

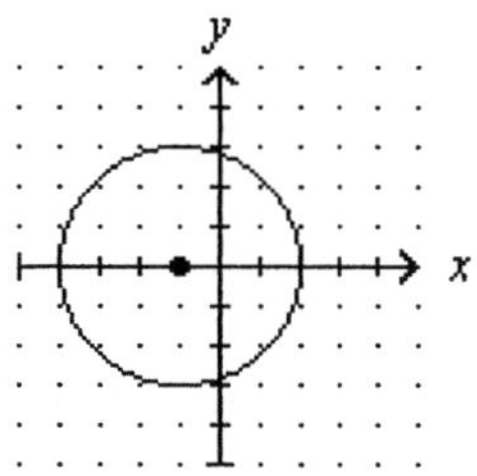

39.

$$\begin{aligned} x^2 + y^2 - 2x + 4y &= -1 \\ x^2 - 2x + y^2 + 4y &= -1 \\ x^2 - 2x + 1 + y^2 + 4y + 4 &= -1 + 1 + 4 \\ (x - 1)^2 + (y + 2)^2 &= 4 \\ (x - 1)^2 + [y - (-2)]^2 &= 22 \end{aligned}$$

Center: $(1, -2)$; Radius $= 2$

x	y
1	0
1	−4
−1	−2
3	−2

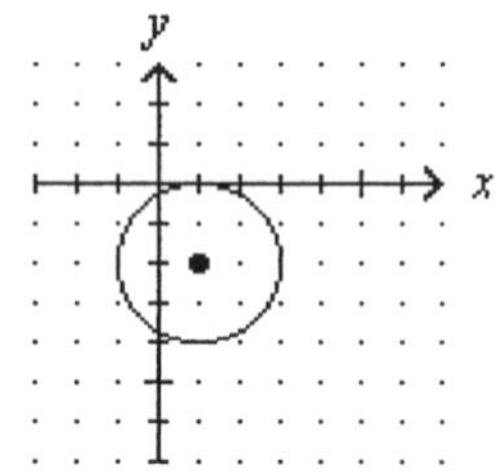

41.

$$\begin{aligned} x^2 + y^2 + 6x - 4y &= -12 \\ x^2 + 6x + y^2 - 4y &= -12 \\ x^2 + 6x + 9 + y^2 - 4y + 4 &= -12 + 9 + 4 \\ (x + 3)^2 + (y - 2)^2 &= 1 \\ [x - (-3)]^2 + (y - 2)^2 &= 1^2 \end{aligned}$$

Center: $(-3, 2)$; Radius $= 1$

x	y
−3	3
−3	1
−4	2
−2	2

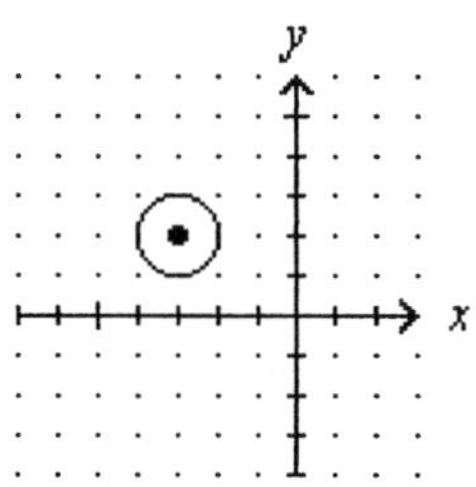

43.

$$\begin{aligned} x^2 + y^2 - 2x + 4y &= -1 \\ x^2 - 2x + y^2 + 4y &= -1 \\ x^2 - 2x + 1 + y^2 + 4y + 4 &= -1 + 1 + 4 \\ (x - 1)^2 + (y + 2)^2 &= 4 \\ (x - 1)^2 + [y - (-2)]^2 &= 2^2 \end{aligned}$$

Center: $(1, -2)$; Radius $= 2$

x	y
1	0
1	−4
−1	−2
3	−2

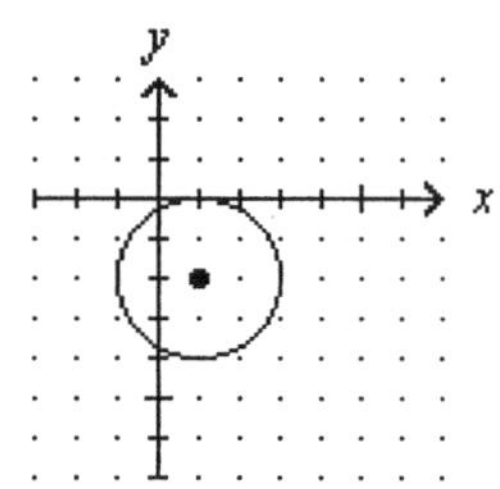

45. $x = y^2$

Vertex: $(0, 0)$ Opens to the right.

x	y
4	−2
1	−1
0	**0**
1	1
4	2

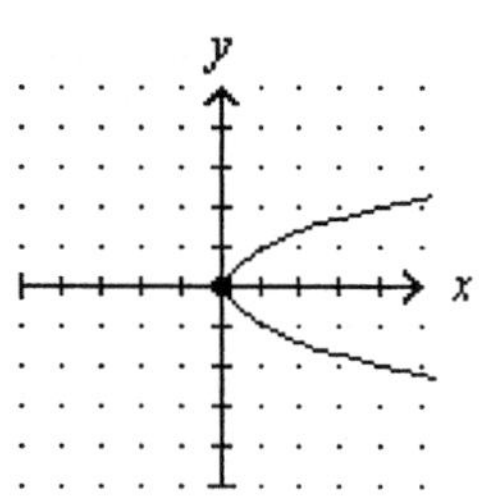

47. $x = -\frac{1}{4}y^2$

Vertex: $(0, 0)$ Opens to the left.

x	y
−4	−4
−1	−2
0	**0**
−1	2
−4	4

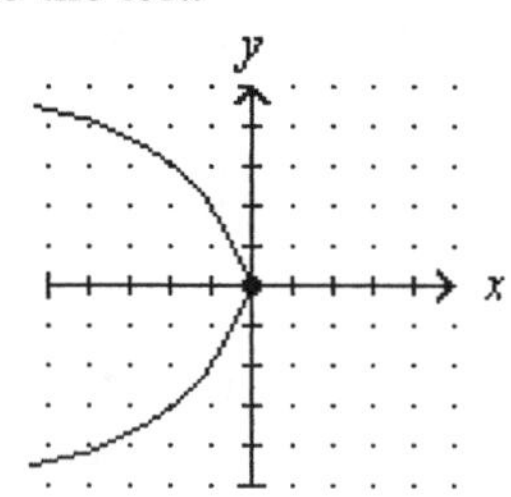

49.
$$\begin{aligned} y &= x^2 + 4x + 5 \\ y &= x^2 + 4x + 4 + 5 - 4 \\ y &= (x + 2)^2 + 1 \\ y &= [x - (-2)]^2 + 1 \end{aligned}$$

Vertex: $(-2, 1)$; Opens upward.

x	y
-4	5
-3	2
-2	**1**
-1	2
0	5

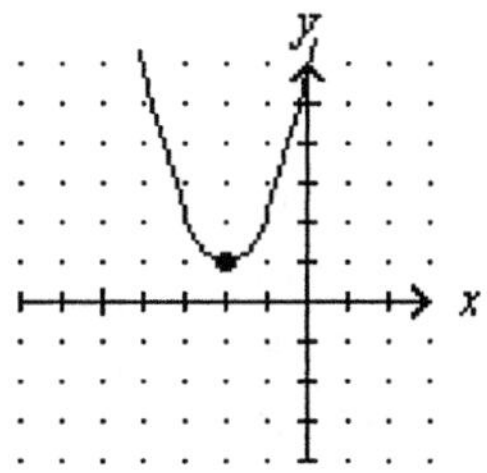

51.
$$\begin{aligned} y &= -x^2 - x + 1 \\ y &= -(x^2 + x) + 1 \\ y &= -(x^2 + x + \tfrac{1}{4}) + 1 + \tfrac{1}{4} \\ y &= -(x + \tfrac{1}{2})^2 + \tfrac{5}{4} \\ y &= -[x - (-\tfrac{1}{2})]^2 + \tfrac{5}{4} \end{aligned}$$

Vertex: $(-\frac{1}{2}, \frac{5}{4})$; Opens downward.

x	y
-2	-1
-1	1
$-\frac{1}{2}$	$\frac{5}{4}$
0	1
1	-1

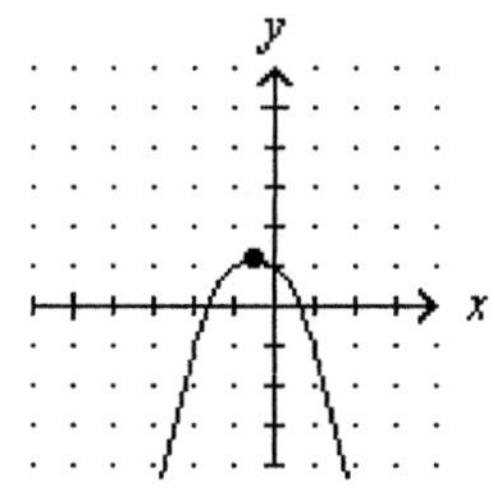

53.
$$\begin{aligned} y^2 + 4x - 6y &= -1 \\ 4x &= -y^2 + 6y - 1 \\ 4x &= -(y^2 - 6y + 9) - 1 + 9 \\ 4x &= -(y - 3)^2 + 8 \\ x &= -\tfrac{1}{4}(y - 3)^2 + 2 \end{aligned}$$

Vertex: $(2, 3)$; Opens to the left.

x	y
$-\frac{1}{4}$	0
1	1
2	**3**
1	5
$-\frac{1}{4}$	6

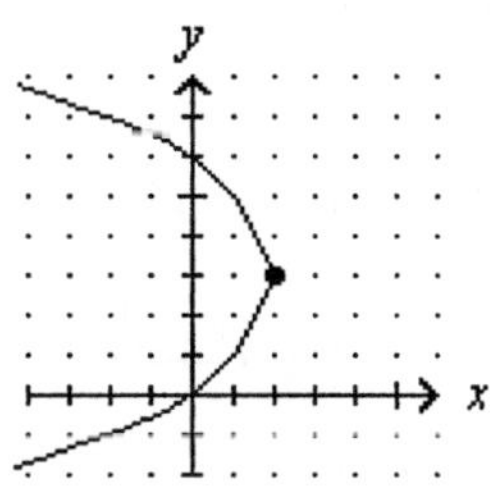

55.
$$y = 2(x - 1)^2 + 3$$

Vertex: $(1, 3)$; Opens upward.

x	y
-1	11
0	5
1	**3**
2	5
3	11

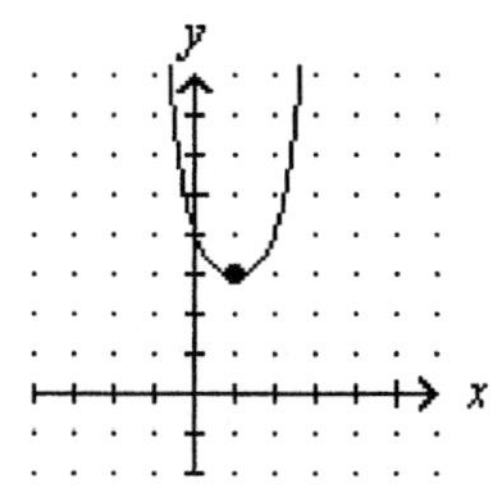

57. Solve $x = 2y^2$ for y.
$$\begin{aligned} 2y^2 &= x \\ y^2 &= \frac{x}{2} \\ y &= \pm\sqrt{\frac{x}{2}} \end{aligned}$$

Graph $y = \sqrt{\frac{x}{2}}$ and $y = -\sqrt{\frac{x}{2}}$.

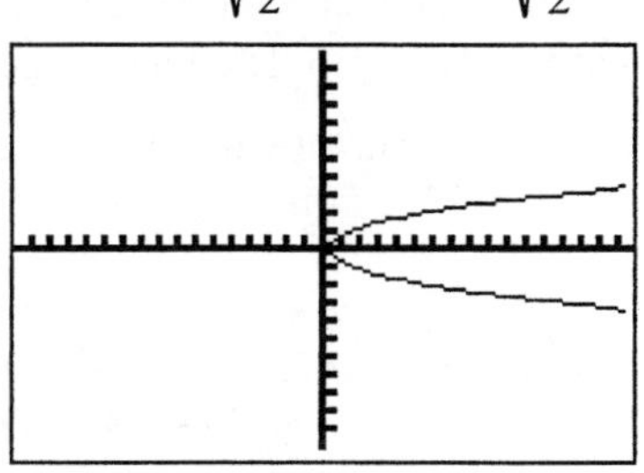

59. Solve $x^2 - 2x + y = 6$ for y.
$$y = -x^2 + 2x + 6$$

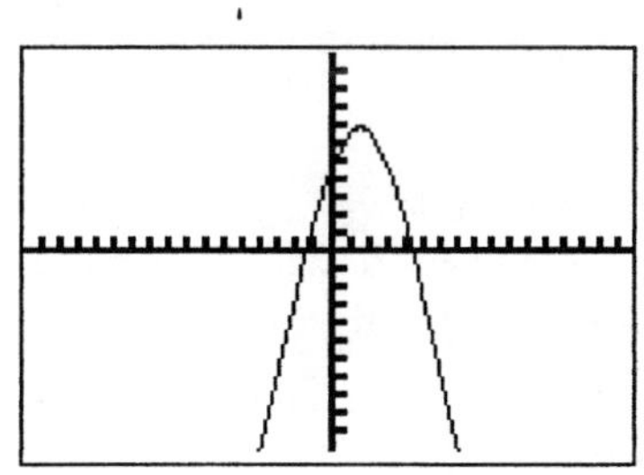

APPLICATIONS

61. The large gear has a radius of 4, therefore the point where the two circles are tangent is at (4, 0). Since the center of the small gear is at (7, 0) the radius of the small gear is 3. Substitute $r = 3$, $h = 7$, and $k = 0$ in the standard form of the circle.

$$(x-7)^2 + (y-0)^2 = 3^2$$
$$(x-7)^2 + y^2 = 9$$

63. Find the distance between the centers of the two broadcast areas. If this distance is less than the sum of the radius of the two broadcast areas they may not be licensed for the same frequency.

Station 1:

$$x^2 + y^2 - 8x - 20y + 16 = 0$$
$$x^2 - 8x + y^2 - 20y = -16$$
$$x^2 - 8x + 16 + y^2 - 20y + 100 = -16 + 16 + 100$$
$$(x-4)^2 + (y-10)^2 = 100$$
$$(x-4)^2 + (y-10)^2 = 10^2$$

Center: (4, 10); Radius = 10

Station 2:

$$x^2 + y^2 + 2x + 4y - 11 = 0$$
$$x^2 + 2x + y^2 + 4y = 11$$
$$x^2 + 2x + 1 + y^2 + 4y + 4 = 11 + 1 + 4$$
$$(x+1)^2 + (y+2)^2 = 16$$
$$[x-(-1)]^2 + [y-(-2)]^2 = 4^2$$

Center: (–1, –2); Radius = 4

Distance between (4, 10) and (–1, –2) is

$$d = \sqrt{[4-(-1)]^2 + [10-(-2)]^2} = \sqrt{5^2 + 12^2} = \sqrt{25 + 144} = \sqrt{169} = 13$$

The sum of the radius of the two broadcast areas is

Sum = 10 + 4 = 14.

Therefore, the two broadcast areas would overlap and they may **not** be licensed for the same frequency.

65. The cannonball will land when $y = 0$. Solve $y = 30x - x^2$ for x when $y = 0$.

$$0 = 30x - x^2$$
$$0 = x(30 - x)$$
$$30 - x = 0 \quad \text{or} \quad x = 0$$
$$x = 30$$

Discard $x = 0$ because that is the beginning of the flight.

The cannonball will land 30 feet away. Therefore, it will fall 5 feet short of the castle.

67. Find the vertex of $2y^2 - 9x = 18$. Assume the sun is at the origin (0, 0).

$$2y^2 - 9x = 18$$
$$-9x = -2y^2 + 18$$
$$-9x = -2(y-0)^2 + 18$$
$$x = \tfrac{2}{9}(y-0)^2 - 2$$

The vertex is at (–2, 0). Therefore the vertex is 2 AU's away from the sun..

WRITING

69. A parabola opening up or down has the form $y = a(x-h)^2 + k$. If a is positive, the parabola opens up. If a is negative, the parabola opens down. For a parabola to open to the right or the left, the equation has the form $x = a(y-k)^2 + h$. If a is positive, the parabola opens to the right. If a is negative, the parabola opens to the left.

71. No one is allowed to campaign within a circle with a radius of 1000 feet that could be drawn with the polling place at the center.

REVIEW

73. Rewrite $|3x - 4| = 11$ as

$$3x - 4 = -11 \quad \text{or} \quad 3x - 4 = 11$$
$$3x = -7 \qquad\qquad 3x = 15$$
$$x = -\tfrac{7}{3} \qquad\qquad x = 5$$

75. Rewrite $|3x + 4| = |5x - 2|$ as

$$3x + 4 = -(5x - 2) \quad \text{or} \quad 3x + 4 = 5x - 2$$
$$3x + 4 = -5x + 2 \qquad\qquad -2x = -6$$
$$8x = -2 \qquad\qquad x = 3$$
$$x = -\tfrac{1}{4}$$

VOCABULARY

1. ellipse; sum
3. center

CONCEPTS

5. $(\pm a, 0)$
7. $(0, 0)$

NOTATION

9.
$$16x^2 + 9y^2 - 144 = 0$$
$$16x^2 + 9y^2 = \mathbf{144}$$
$$\frac{x^2}{9} + \frac{y^2}{16} = 1$$

PRACTICE

11. $\frac{x^2}{4} + \frac{y^2}{9} = 1$

$\frac{x^2}{2^2} + \frac{y^2}{3^2} = 1$

Center: (0, 0)

x	y
0	−3
0	3
−2	0
2	0

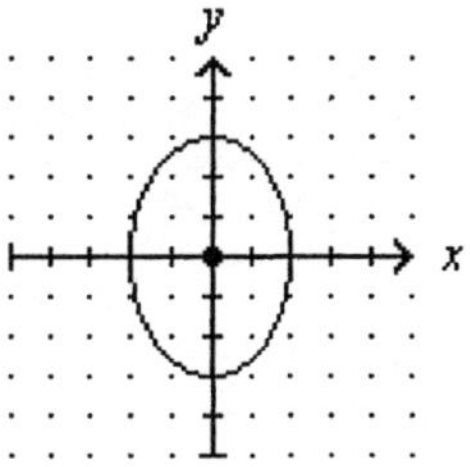

13. $\frac{x^2}{9} + \frac{y^2}{1} = 1$

$\frac{x^2}{3^2} + \frac{y^2}{1^2} = 1$

Center: (0, 0)

x	y
0	−1
0	1
−3	0
3	0

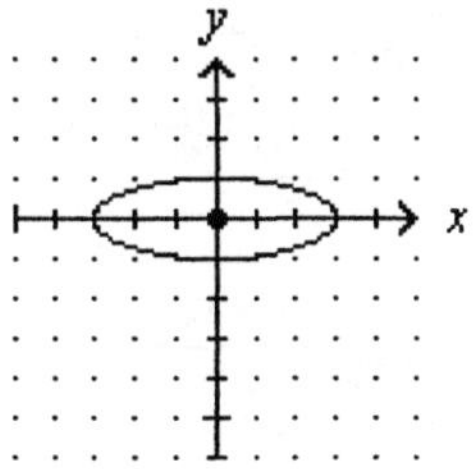

15. $16x^2 + 4y^2 = 64$

$\frac{x^2}{4} + \frac{y^2}{16} = 1$

$\frac{x^2}{2^2} + \frac{y^2}{4^2} = 1$

Center: (0, 0)

x	y
0	−4
0	4
−2	0
2	0

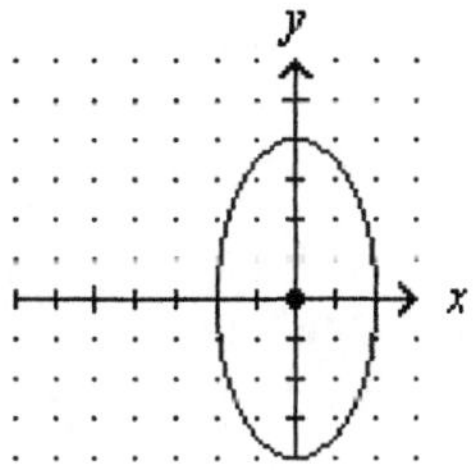

17. $\frac{(x-2)^2}{9} + \frac{(y-1)^2}{4} = 1$

$\frac{(x-2)^2}{3^2} + \frac{(y-1)^2}{2^2} = 1$

Center: (2, 1)

x	y
2	−1
2	3
−1	1
5	1

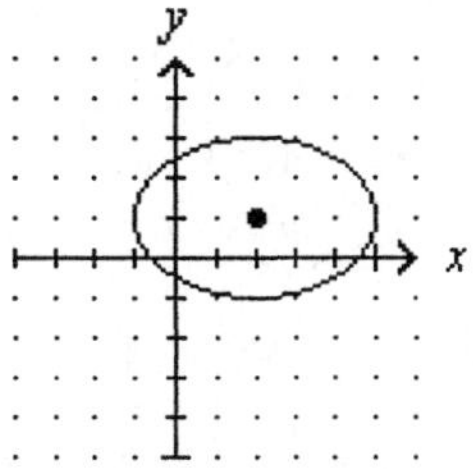

19.

$$\frac{(x+1)^2}{4}+\frac{(y+2)^2}{1}=1$$

$$\frac{[x-(-1)]^2}{2^2}+\frac{[y-(-2)]^2}{1^2}=1$$

Center: (−1, −2)

x	y
−1	−3
−1	−1
−3	−2
1	−2

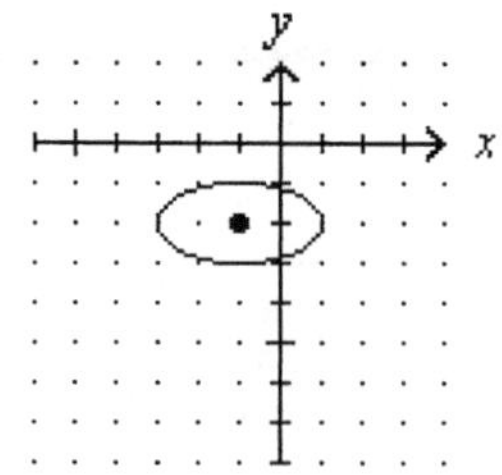

21.

$$\frac{x^2}{9}+\frac{y^2}{4}=1$$

$$4x^2+9y^2=36$$

$$9y^2=36-4x^2$$

$$y^2=\frac{36-4x^2}{9}$$

$$y=\pm\sqrt{\frac{36-4x^2}{9}}=\pm\frac{\sqrt{36-4x^2}}{3}$$

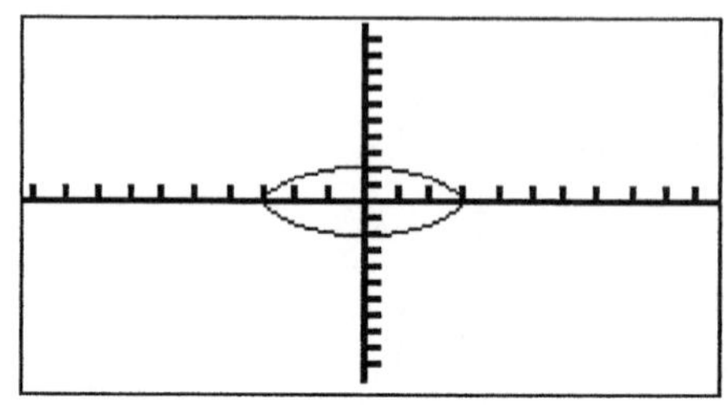

23.

$$\frac{x^2}{4}+\frac{(y-1)^2}{9}=1$$

$$9x^2+4(y-1)^2=36$$

$$4(y-1)^2=36-9x^2$$

$$(y-1)^2=\frac{36-9x^2}{4}$$

$$y-1=\pm\sqrt{\frac{36-9x^2}{4}}$$

$$y=1\pm\sqrt{\frac{36-9x^2}{4}}=1\pm\frac{\sqrt{36-9x^2}}{2}$$

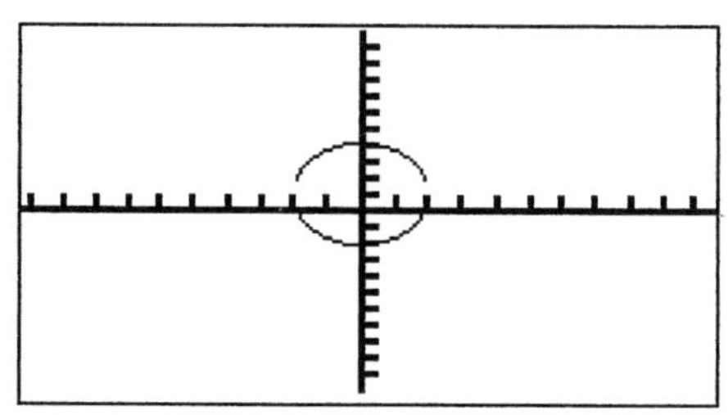

25.

$$\begin{aligned} x^2+4y^2-4x+8y+4 &= 0\\ x^2-4x+4y^2+8y &= -4\\ x^2-4x+4(y^2+2y) &= -4\\ x^2-4x+4+4(y^2+2y+1) &= -4+4+4\\ (x-2)^2+4(y+1)^2 &= 4\\ \frac{(x-2)^2}{4}+\frac{(y+1)^2}{1} &= 1 \end{aligned}$$

Center: (2, −1)

x	y
2	0
2	−2
0	−1
4	−1

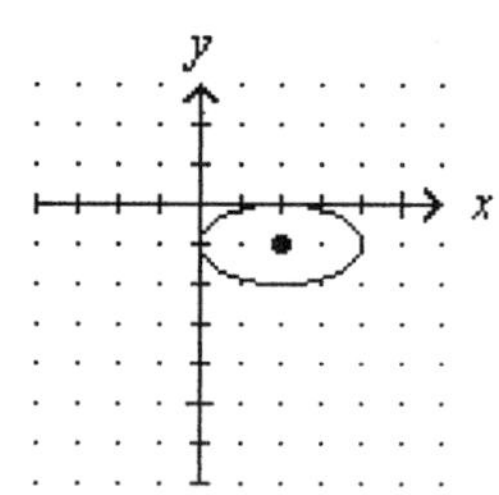

27.

$$\begin{aligned} 9x^2+4y^2-18x+16y &= 11\\ 9x^2-18x+4y^2+16y &= 11\\ 9(x^2-2x)+4(y^2+4y) &= 11\\ 9(x^2-2x+1)+4(y^2+4y+4) &= 11+9+16\\ 9(x-1)^2+4(y+2)^2 &= 36\\ \frac{(x-1)^2}{4}+\frac{(y+2)^2}{9} &= 1 \end{aligned}$$

Center: (1, −2)

x	y
1	−5
1	1
−1	−2
3	−2

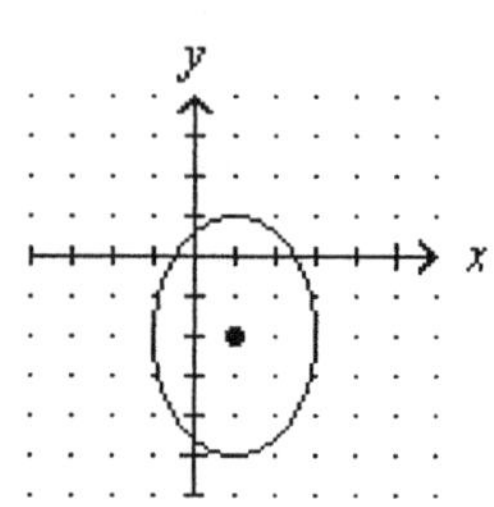

APPLICATIONS

29. Place the underpass in a coordinate system with (0, 0) at the center of the underpass. The vertices are the points V_1 (−20, 0) and V_2 (20, 0). The y–intercept is (0, 10). Since only the top part of the ellipse is shown only the positive root is needed when solved for y.

Substitute 20 for a and 10 for b in the equation for an ellipse and solve for positive y.

$$\frac{x^2}{20^2}+\frac{y^2}{10^2}=1$$
$$\frac{x^2}{400}+\frac{y^2}{100}=1$$
$$x^2+4y^2=400$$
$$4y^2=400-x^2$$
$$y^2=\frac{400-x^2}{4}$$
$$y=\sqrt{\frac{400-x^2}{4}}=\frac{1}{2}\sqrt{400-x^2}$$

31. Write the equation in standard form to find a and b and then substitute in the area formula.

$$\frac{x^2}{16}+\frac{y^2}{9}=1$$
$$\frac{x^2}{4^2}+\frac{y^2}{3^2}=1$$

$A = \pi ab = \pi(4)(3) = 12\pi$ square units

WRITING

33. Let $y = 0$ and solve for x. $x = \pm a$, so the x–intercepts are $(a, 0)$ and $(-a, 0)$.
Let $x = 0$ and solve for y. $y = \pm b$, so the y–intercepts are $(0, b)$ and $(0, -b)$.\

REVIEW

35. $3x^{-2}y^2(4x^2+3y^{-2}) = 3\bullet 4\bullet x^{-2+2}y^2 + 3\bullet 3\bullet x^{-2}y^{2-2} = 12y^2+9x^{-2} = 12y^2+\dfrac{9}{x^2}$

37. $$\frac{x^{-2}+y^{-2}}{x^{-2}-y^{-2}}=\frac{\frac{1}{x^2}+\frac{1}{y^2}}{\frac{1}{x^2}-\frac{1}{y^2}}=\frac{x^2y^2\left(\frac{1}{x^2}+\frac{1}{y^2}\right)}{x^2y^2\left(\frac{1}{x^2}-\frac{1}{y^2}\right)}=\frac{y^2+x^2}{y^2-x^2}$$

VOCABULARY

1. hyperbola; difference

3. center

CONCEPTS

5 $(\pm a, 0)$

7. $(0, 0)$

NOTATION

9.
$$\begin{aligned} 4x^2 - 9y^2 - 36 &= 0 \\ 4x^2 - 9y^2 &= \mathbf{36} \\ \frac{4x^2}{36} - \frac{9y^2}{36} &= 1 \\ \frac{x^2}{9} - \frac{y^2}{4} &= 1 \end{aligned}$$

PRACTICE

11.
$$\frac{x^2}{9} - \frac{y^2}{4} = 1$$
$$\frac{x^2}{3^2} - \frac{y^2}{2^2} = 1$$

Center: $(0, 0)$

Vertices ($y = 0$):

$x^2 = 9$

$x = \pm 3$

$V_1(-3, 0)$ and $V_2(3, 0)$

Fundamental Rectangle:

Use vertices and $(0, -2)$ and $(0, 2)$ on y–axis.

Opens: left and right.

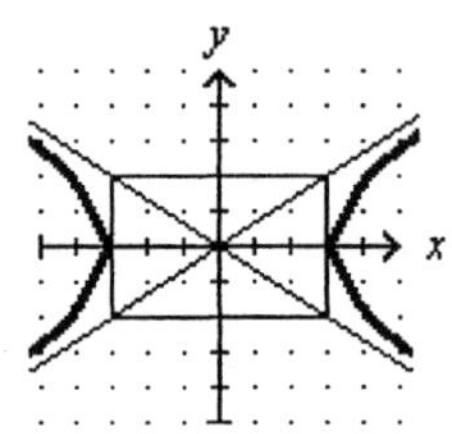

13.
$$\frac{y^2}{4} - \frac{x^2}{9} = 1$$
$$\frac{y^2}{2^2} - \frac{x^2}{3^2} = 1$$

Center: $(0, 0)$

Vertices ($x = 0$):

$y^2 = 4$

$y = \pm 2$

$V_1(0, -2)$ and $V_2(0, 2)$

Fundamental Rectangle:

Use vertices and $(-3, 0)$ and $(3, 0)$ on x–axis.

Opens: up and down.

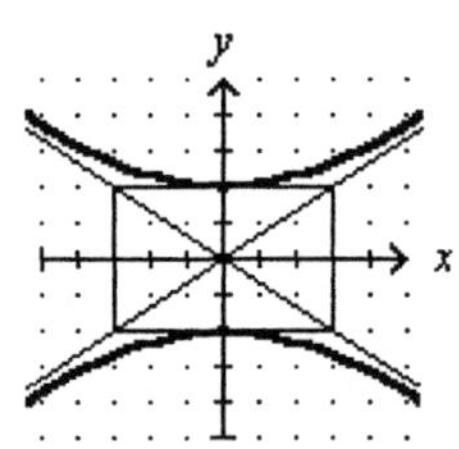

15.
$$25x^2 - y^2 = 25$$
$$\frac{25x^2}{25} - \frac{y^2}{25} = 1$$
$$\frac{x^2}{1^2} - \frac{y^2}{5^2} = 1$$

Center: $(0, 0)$

Vertices ($y = 0$):

$x^2 = 1$

$x = \pm 1$

$V_1(-1, 0)$ and $V_2(1, 0)$

Fundamental Rectangle:

Use vertices and $(0, -5)$ and $(0, 5)$ on y–axis.

Opens: left and right.

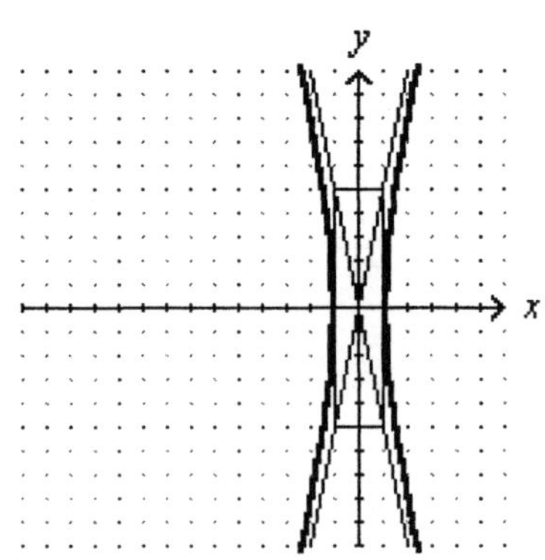

17. $\dfrac{(x-2)^2}{9}-\dfrac{y^2}{16}=1$

$\dfrac{(x-2)^2}{3^2}-\dfrac{y^2}{4^2}=1$

Center: (2, 0)

Vertices ($y = 0$):

$(x-2)^2 = 3^2$

$x = 2 \pm 3$

$V_1(-1, 0)$ and $V_2(5, 0)$

Fundamental Rectangle:

Use vertices and (2, –4) and (2, 4).

Opens: left and right.

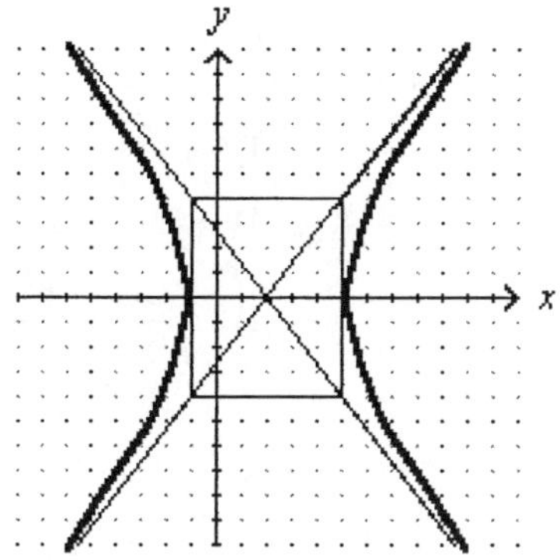

19. $4(x+3)^2-(y-1)^2=4$

$\dfrac{4(x+3)^2}{4}-\dfrac{(y-1)^2}{4}=1$

$\dfrac{(x+3)^2}{1^2}-\dfrac{(y-1)^2}{2^2}=1$

Center: (–3, 1)

Vertices ($y = 1$):

$(x+3)^2 = 1$

$x = -3 \pm 1$

$V_1(-4, 1)$ and $V_2(-2, 1)$

Fundamental Rectangle:

Use vertices and (–3, –1) and (–3, 3).

Opens: left and right.

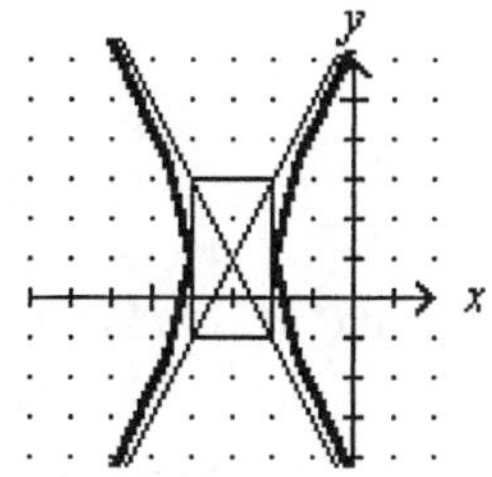

21. $xy = 8$

x	y
–8	–1
–4	–2
–2	–4
–1	–8
1	8
2	4
4	2
8	1

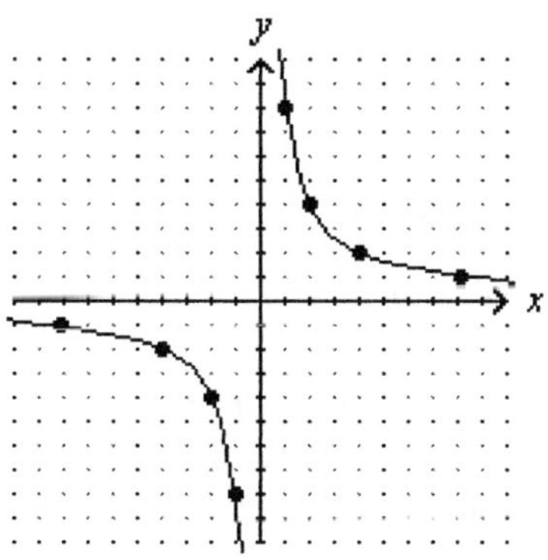

23. $\dfrac{x^2}{9}-\dfrac{y^2}{4}=1$

$4x^2 \quad 9y^2 = 36$

$-9y^2 = 36-4x^2$

$y^2=\dfrac{36-4x^2}{-9}=\dfrac{4x^2-36}{9}$

$y=\pm\sqrt{\dfrac{4x^2-36}{9}}=\pm\dfrac{\sqrt{4x^2-36}}{3}$

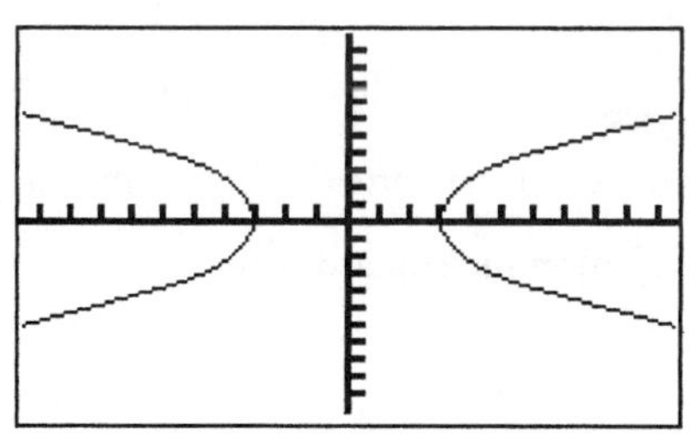

25.

$$\frac{x^2}{4}-\frac{(y-1)^2}{9}=1$$

$$9x^2-4(y-1)^2=36$$

$$-4(y-1)^2=36-9x^2$$

$$(y-1)^2=\frac{36-9x^2}{-4}=\frac{9x^2-36}{4}$$

$$y-1=\pm\sqrt{\frac{9x^2-36}{4}}$$

$$y=1\pm\sqrt{\frac{9x^2-36}{4}}=1\pm\frac{\sqrt{9x^2-36}}{2}$$

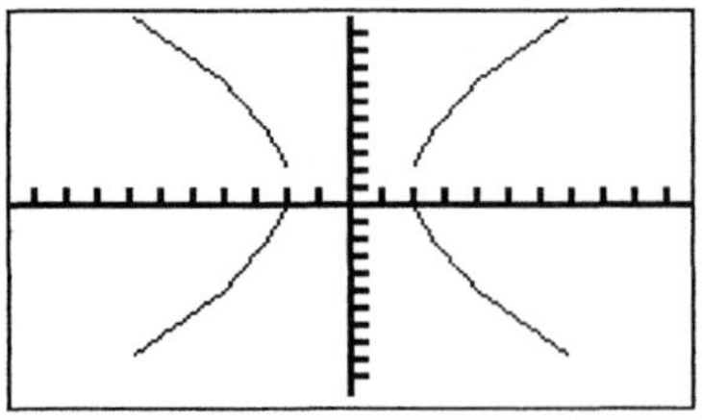

27.

$$\begin{aligned}4x^2-y^2+8x-4y &= 4\\ 4x^2+8x-y^2-4y &= 4\\ 4(x^2+2x)-(y^2+4y) &= 4\\ 4(x^2+2x+1)-(y^2+4y+4) &= 4+4-4\\ 4(x+1)^2-(y+2)^2 &= 4\\ \frac{4(x+1)^2}{4}-\frac{(y+2)^2}{4} &= 1\\ \frac{(x+1)^2}{1^2}-\frac{(y+2)^2}{2^2} &= 1\end{aligned}$$

Center: $(-1, -2)$

Vertices $(y=-2)$:

$$(x+1)^2 = 1$$
$$x = -1\pm 1$$

$V_1(-2, -2)$ and $V_2(0, -2)$

Fundamental Rectangle:

Use vertices and $(-1, -4)$ and $(-1, 0)$.

Opens: left and right.

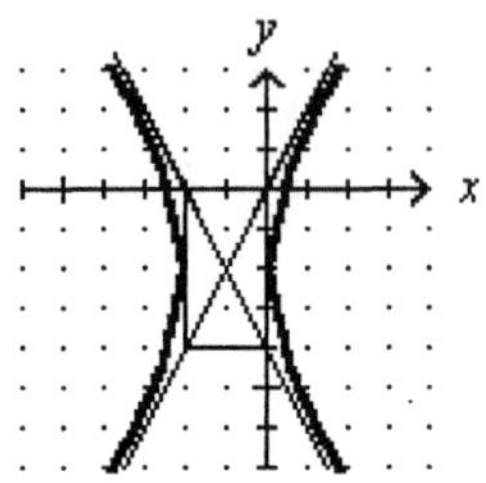

29.

$$\begin{aligned}4y^2-x^2+8y+4x &= 4\\ 4y^2+8y-x^2+4x &= 4\\ 4(y^2+2y)-(x^2-4x) &= 4\\ 4(y^2+2y+1)-(x^2-4x+4) &= 4+4-4\\ 4(y+1)^2-(x-2)^2 &= 4\\ \frac{4(y+1)^2}{4}-\frac{(x-2)^2}{4} &= 1\\ \frac{(y+1)^2}{1^2}-\frac{(x-2)^2}{2^2} &= 1\end{aligned}$$

Center: $(2, -1)$

Vertices $(x=2)$:

$$(y+1)^2 = 1$$
$$y = -1\pm 1$$

$V_1(2, -2)$ and $V_2(2, 0)$

Fundamental Rectangle:

Use vertices and $(0, -1)$ and $(4, -1)$.

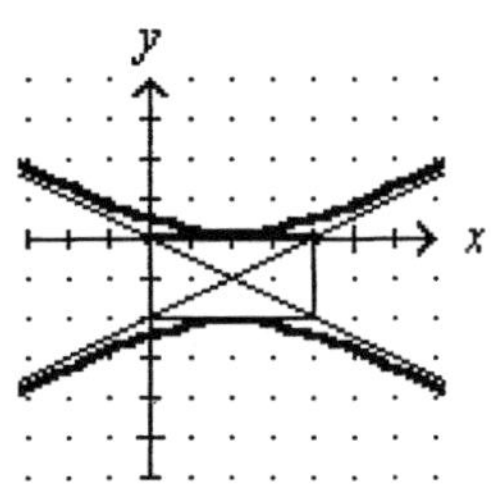

Opens: up and down.

APPLICATIONS

31. The coordinates of the vertex will give the closest distance between the nucleus at the origin and the particle.

With the equation in standard form:

$$\frac{y^2}{9}-\frac{x^2}{81}=1$$

Find the y–value of the vertex $(x=0)$.

$$y^2=3^2$$
$$y=\pm 3$$

Discard the negative solution. The distance would be 3 units.

33. In standard form the equation of the hyperbola is $\frac{y^2}{25} - \frac{x^2}{25} = 1$. This indicates that the vertex of the portion of the hyperbola shown in the illustration is at (0, 5). Five miles from this vertex means the y–value will be 10. By substituting 10 for y into the equation, the x–value of the points on the hyperbola can be determined.

$$\begin{aligned} y^2 - x^2 &= 25 \\ 10^2 - x^2 &= 25 \\ -x^2 &= 25 - 100 = -75 \\ x^2 &= 75 \\ x &= \pm\sqrt{75} = \pm 5\sqrt{3} \end{aligned}$$

The distance between the points ($5\sqrt{3}$, 5) and ($-5\sqrt{3}$, 5) is $10\sqrt{3}$ units.

WRITING

35. A hyperbola is the set of all points P in a plane for which the difference of the distances of each point from two fixed points (called foci) is a constant.

REVIEW

37. $-6x^4 + 9x^3 - 6x^2 = -3x^2(2x^2 - 3x + 2)$

39. $15a^2 - 4ab - 4b^2 = 15a^2 - 10ab + 6ab - 4b^2 = 5a(3a - 2b) + 2b(3a - 2b) = (3a - 2b)(5a + 2b)$

Section 10.4 Solving Simultaneous Second Degree Equations

VOCABULARY / CONCEPTS

1. graphing; substitution

3. tangent

5. two; four

7. Substitute $x = 1$ and $y = -1$ into each equation:

$2(1) + (-1) - 1 = 2 - 1 - 1 = 0$

$(1)^2 + (-1)^2 = 2 \neq 3$

No, (1,–1) is not a solution of the system.

NOTATION

9.
$$\begin{aligned} x^2 + y^2 &= 5 \\ x^2 + (\mathbf{2x})^2 &= 5 \\ x^2 + 4x^2 &= \mathbf{5} \\ \mathbf{5}\,x^2 &= 5 \\ x^2 &= \mathbf{1} \end{aligned}$$

$x = 1$ or $x = -1$

If $x = 1$, then
$y = 2(\ \mathbf{1}\ ) = 2$

If $x = -1$, then
$y = 2(\ \mathbf{-1}\ ) = -2$

The solutions are (1, 2) and (–1, –2).

PRACTICE

11. Equation #1: $8x^2 + 32y^2 = 256$

Ellipse: $\frac{x^2}{32} + \frac{y^2}{8} = 1$

x	y
0	$\pm\sqrt{8} = \pm 2.8$
$\pm\sqrt{32} = \pm 5.7$	0

Equation #2: $x = 2y$

Line

x	y
0	0
6	3

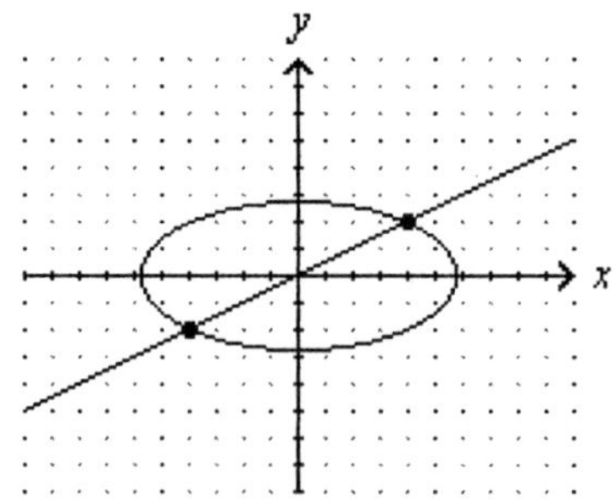

The solutions are $(-4, -2)$ and $(4, 2)$.

13. Equation #1: $x^2 + y^2 = 10$

Circle: Center is (0, 0)

x	y
0	$\pm\sqrt{10} = \pm 3.2$
$\pm\sqrt{10} = \pm 3.2$	0

Equation #2: $y = 3x^2$

Parabola

x	y
0	0
± 1	3
± 2	12

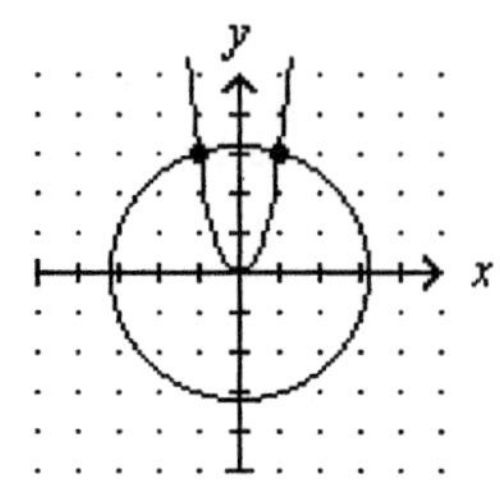

The solutions are $(-1, 3)$ and $(1, 3)$.

15. Equation #1: $x^2 + y^2 = 25$

Circle: Center is (0, 0)

x	y
0	$\pm\sqrt{25} = \pm 5$
$\pm\sqrt{25} = \pm 5$	0

Equation #2: $12x^2 + 64y^2 = 768$

Ellipse: $\frac{x^2}{64} + \frac{y^2}{12} = 1$

x	y
0	$\pm\sqrt{12} = \pm 3.5$
$\pm\sqrt{64} = \pm 8$	0

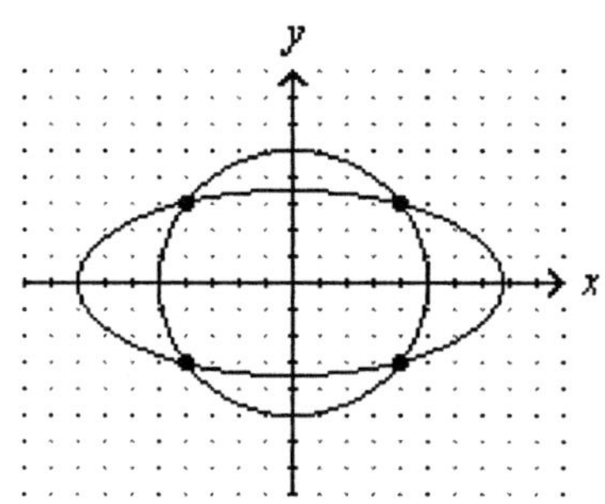

The solutions are $(-4, -3)$, $(-4, 3)$, $(4, -3)$ and $(4, 3)$.

17. Equation #1: $x^2 - 13 = -y^2$

$x^2 + y^2 = 13$

Circle: Center is (0, 0)

x	y
0	$\pm\sqrt{13} = \pm 3.6$
$\pm\sqrt{13} = \pm 3.6$	0

Equation #2: $y = \frac{2}{3}x$

Line

x	y
0	0
3	2

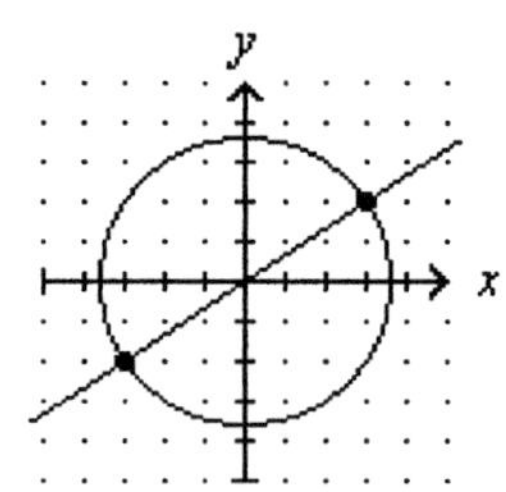

The solutions are $(-3, -2)$ and $(3, 2)$.

19. Equation #1: $x^2 - 6x - y = -5$

$y = x^2 - 6x + 5$

Parabola: Opens up.

Equation #2: $x^2 - 6x + y = -5$

$y = -x^2 + 6x - 5$

Parabola: Opens down.

The solutions are $(1, 0)$ and $(5, 0)$.

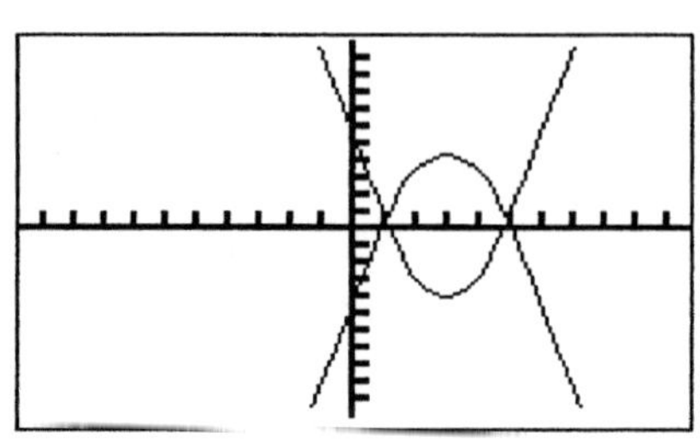

21. $\begin{cases} 25x^2 + 9y^2 = 225 \\ 5x + 3y = 15 \end{cases}$ or $\begin{cases} 25x^2 + (3y)^2 = 225 \\ 5x + 3y = 15 \end{cases}$

Solve the linear equation for $3y$.

$3y = -5x + 15$

Substitute into the quadratic equation.

$$\begin{aligned} 25x^2 + (-5x + 15)^2 &= 225 \\ 25x^2 + (25x^2 - 150x + 225) &= 225 \\ 50x^2 - 150x &= 0 \\ x^2 - 3x &= 0 \\ x(x - 3) &= 0 \end{aligned}$$

$x = 0$ or $x - 3 = 0$
$x = 3$

Substitute 0 and 3 for x into $5x + 3y = 15$ and solve for y.

$$\begin{aligned} 5(0) + 3y &= 15 \\ 3y &= 15 \\ y &= 5 \end{aligned} \qquad \begin{aligned} 5(3) + 3y &= 15 \\ 15 + 3y &= 15 \\ 3y &= 0 \\ y &= 0 \end{aligned}$$

The solutions are $(0, 5)$ and $(3, 0)$.

23. $\begin{cases} x^2 + y^2 = 2 \\ x + y = 2 \end{cases}$

Solve the linear equation for y.

$y = -x + 2$

Substitute into the quadratic equation.

$$\begin{aligned} x^2 + (-x + 2)^2 &= 2 \\ x^2 + (x^2 - 4x + 4) &= 2 \\ 2x^2 - 4x + 2 &= 0 \\ x^2 - 2x + 1 &= 0 \\ (x - 1)(x - 1) &= 0 \end{aligned}$$

$x - 1 = 0$ or $x - 1 = 0$
$x = 1$ $\quad x = 1$

Substitute 1 for x into $x + y = 2$ and solve for y.

$$\begin{aligned} (1) + y &= 2 \\ y &= 1 \end{aligned}$$

The solution is $(1, 1)$.

25. $\begin{cases} x^2 + y^2 = 5 \\ x + y = 3 \end{cases}$

Solve the linear equation for y.

$y = -x + 3$

Substitute into the quadratic equation.

$$\begin{aligned} x^2 + (-x + 3)^2 &= 5 \\ x^2 + (x^2 - 6x + 9) &= 5 \\ 2x^2 - 6x + 4 &= 0 \\ x^2 - 3x + 2 &= 0 \\ (x - 1)(x - 2) &= 0 \end{aligned}$$

$x - 1 = 0$ or $x - 2 = 0$
$x = 1$ $\quad x = 2$

Substitute 1 and 2 for x into $x + y = 3$ and solve for y.

$$\begin{aligned} (1) + y &= 3 \\ y &= 2 \end{aligned} \qquad \begin{aligned} (2) + y &= 3 \\ y &= 1 \end{aligned}$$

The solutions are $(1, 2)$ $(2, 1)$.

27. $\begin{cases} x^2 + y^2 = 13 \\ y = x^2 - 1 \end{cases}$

Solve the second equation for x^2.

$x^2 = y + 1$

Substitute into the first equation.

$$\begin{aligned} (y + 1) + y^2 &= 13 \\ y^2 + y + 1 &= 13 \\ y^2 + y - 12 &= 0 \\ (y + 4)(y - 3) &= 0 \end{aligned}$$

$y + 4 = 0$ or $y - 3 = 0$
$y = -4$ $\quad y = 3$

Substitute -4 and 3 for y into $x^2 = y + 1$ and solve for x.

$$\begin{aligned} x^2 &= -4 + 1 \\ x^2 &= -3 \end{aligned} \qquad \begin{aligned} x^2 &= 3 + 1 \\ x^2 &= 4 \\ x &= \pm 2 \end{aligned}$$

Not possible

The solutions are $(-2, 3)$ and $(2, 3)$.

29. $\begin{cases} x^2 + y^2 = 30 \\ y = x^2 \end{cases}$

Substitute y for x^2 into the first equation.

$$\begin{aligned} (y) + y^2 &= 30 \\ y^2 + y - 30 &= 0 \\ (y+6)(y-5) &= 0 \end{aligned}$$

$y + 6 = 0$ or $y - 5 = 0$

$y = -6$ $\quad$ $y = 5$

Substitute -6 and 5 for y into $y = x^2$ and solve for x.

$x^2 = -6$ $\quad$ $x^2 = 5$

Not possible $\quad$ $x = \pm\sqrt{5}$

The solutions are $(-\sqrt{5}, 5)$ $(\sqrt{5}, 5)$.

31. $\begin{cases} x^2 + y^2 = 13 \\ x^2 - y^2 = 5 \end{cases}$

Add the two equations and solve for x.

$$\begin{aligned} 2x^2 &= 18 \\ x^2 &= 9 \\ x &= \pm 3 \end{aligned}$$

Substitute -3 and 3 for x into $x^2 - y^2 = 5$ and solve for y.

$$\begin{aligned} (-3)^2 - y^2 &= 5 \\ -y^2 &= -4 \\ y^2 &= 4 \\ y &= \pm 2 \end{aligned} \qquad \begin{aligned} (3)^2 - y^2 &= 5 \\ -y^2 &= -4 \\ y^2 &= 4 \\ y &= \pm 2 \end{aligned}$$

The solutions are $(-3, -2)$, $(-3, 2)$, $(3, -2)$ and $(3, 2)$.

33. $\begin{cases} x^2 + y^2 = 20 \\ x^2 - y^2 = -12 \end{cases}$

Add the two equations and solve for x.

$$\begin{aligned} 2x^2 &= 8 \\ x^2 &= 4 \\ x &= \pm 2 \end{aligned}$$

Substitute -2 and 2 for x into $x^2 - y^2 = -12$ and solve for y.

$$\begin{aligned} (-2)^2 - y^2 &= -12 \\ -y^2 &= -16 \\ y^2 &= 16 \\ y &= \pm 4 \end{aligned} \qquad \begin{aligned} (2)^2 - y^2 &= -12 \\ -y^2 &= -16 \\ y^2 &= 16 \\ y &= \pm 4 \end{aligned}$$

The solutions are $(-2, -4)$, $(-2, 4)$, $(2, -4)$ and $(2, 4)$.

35. $\begin{cases} y^2 = 40 - x^2 \\ y = x^2 - 10 \end{cases}$

Add the two equations and solve for y.

$$\begin{aligned} y^2 + y &= 30 \\ y^2 + y - 30 &= 0 \\ (y+6)(y-5) &= 0 \end{aligned}$$

$y + 6 = 0$ or $y - 5 = 0$

$y = -6$ $\quad$ $y = 5$

Substitute -6 and 5 for y into $x^2 = 40 - y^2$ and solve for x.

$$\begin{aligned} x^2 &= 40 - (-6)^2 \\ x^2 &= 4 \\ x &= \pm 2 \end{aligned} \qquad \begin{aligned} x^2 &= 40 - (5)^2 \\ x^2 &= 15 \\ x &= \pm\sqrt{15} \end{aligned}$$

The solutions are $(-2, -6)$, $(2, -6)$, $(-\sqrt{15}, 5)$ and $(\sqrt{15}, 5)$.

37. $\begin{cases} y = x^2 - 4 \\ x^2 - y^2 = -16 \end{cases}$

Solve the first equation for x^2.

$$x^2 = y + 4$$

Substitute into the second equation.

$$\begin{aligned} (y+4) - y^2 &= -16 \\ -y^2 + y + 4 &= -16 \\ y^2 - y - 20 &= 0 \\ (y+4)(y-5) &= 0 \end{aligned}$$

$y + 4 = 0$ or $y - 5 = 0$

$y = -4$ $\quad y = 5$

Substitute –4 and 5 for y into $x^2 = y + 4$ and solve for x.

$$\begin{aligned} x^2 &= (-4) + 4 & x^2 &= 5 + 4 \\ x^2 &= 0 & x^2 &= 9 \\ x &= 0 & x &= \pm 3 \end{aligned}$$

The solutions are $(0, -4)$, $(-3, 5)$, and $(3, 5)$.

39. $\begin{cases} x^2 - y^2 = -5 \\ 3x^2 + 2y^2 = 30 \end{cases}$

Solve the first equation for x^2.

$$x^2 = y^2 - 5$$

Substitute into the second equation.

$$\begin{aligned} 3(y^2 - 5) + 2y^2 &= 30 \\ 3y^2 - 15 + 2y^2 &= 30 \\ 5y^2 &= 45 \\ y^2 &= 9 \\ y &= \pm 3 \end{aligned}$$

Substitute –3 and 3 for y into $x^2 = y^2 - 5$ and solve for x.

$$\begin{aligned} x^2 &= (-3)^2 - 5 & x^2 &= (3)^2 - 5 \\ x^2 &= 4 & x^2 &= 4 \\ x &= \pm 2 & x &= \pm 2 \end{aligned}$$

The solutions are $(-2, -3)$, $(2, -3)$, $(-2, 3)$, and $(2, 3)$.

41. $\begin{cases} \dfrac{1}{x} + \dfrac{2}{y} = 1 \\ \dfrac{2}{x} - \dfrac{1}{y} = \dfrac{1}{3} \end{cases}$

Multiply Equation 2 by 2.

$$\begin{cases} \dfrac{1}{x} + \dfrac{2}{y} = 1 \\ \dfrac{4}{x} - \dfrac{2}{y} = \dfrac{2}{3} \end{cases}$$

Add the two equations and solve for x.

$$\frac{5}{x} = \frac{5}{3}$$

$$5x = 15$$

$$x = 3$$

Substitute 3 for x into Equation 1 and solve for y.

$$\frac{1}{3} + \frac{2}{y} = 1$$

$$\frac{2}{y} = \frac{2}{3}$$

$$2y = 6$$

$$y = 3$$

The solution is $(3, 3)$.

43. $\begin{cases} 3y^2 = xy \\ 2x^2 + xy - 84 = 0 \end{cases}$

Factor the first equation.

$$\begin{aligned} 3y^2 - xy &= 0 \\ y(3y - x) &= 0 \end{aligned}$$

$y = 0$ or $3y - x = 0$

$3y = x$

Substitute 0 for y in the second equation.

$$\begin{aligned} 2x^2 + x(0) - 84 &= 0 \\ 2x^2 - 84 &= 0 \\ 2x^2 &= 84 \\ x^2 &= 42 \\ x &= \pm\sqrt{42} \end{aligned}$$

Substitute $3y$ for x into the second equation.

$$\begin{aligned} 2(3y)^2 + (3y)y - 84 &= 0 \\ 18y^2 + 3y^2 - 84 &= 0 \\ 21y^2 &= 84 \\ y^2 &= 4 \\ y &= \pm 2 \end{aligned}$$

Substitute –2 and 2 for y in $x = 3y$.

$$\begin{aligned} x &= 3(-2) & x &= 3(2) \\ x &= -6 & x &= 6 \end{aligned}$$

The solutions are $(-\sqrt{42}, 0)$, $(\sqrt{42}, 0)$, $(-6, -2)$, and $(6, 2)$.

45. $\begin{cases} xy = \dfrac{1}{6} \\ y + x = 5xy \end{cases}$

Solve the first equation for y.

$$y = \frac{1}{6x}$$

Substitute $\frac{1}{6x}$ for y in the second equation and solve for x.

$$\frac{1}{6x} + x = 5x\left(\frac{1}{6x}\right)$$
$$\frac{1}{6x} + x = \frac{5}{6}$$
$$6x\left(\frac{1}{6x} + x\right) = 6x\left(\frac{5}{6}\right)$$
$$1 + 6x^2 = 5x$$
$$6x^2 - 5x + 1 = 0$$
$$(3x - 1)(2x - 1) = 0$$

$3x - 1 = 0$ or $2x - 1 = 0$

$3x = 1$ $\quad$ $2x = 1$

$x = \frac{1}{3}$ $\quad$ $x = \frac{1}{2}$

Substitute $\frac{1}{3}$ and $\frac{1}{2}$ for x in the first equation.

$\left(\frac{1}{3}\right)y = \frac{1}{6}$ $\quad$ $\left(\frac{1}{2}\right)y = \frac{1}{6}$

$2y = 1$ $\quad$ $3y = 1$

$y = \frac{1}{2}$ $\quad$ $y = \frac{1}{3}$

The solutions are $\left(\frac{1}{3}, \frac{1}{2}\right)$ and $\left(\frac{1}{2}, \frac{1}{3}\right)$.

47. Let x represent the first integer and y represent the second integer. The system of equations is:

$\begin{cases} xy = 32 \\ x + y = 12 \end{cases}$

Solve the first equation for y.

$$y = \frac{32}{x}$$

Substitute $\frac{32}{x}$ for y in the second equation and solve for x.

$$x + \frac{32}{x} = 12$$
$$x^2 + 32 = 12x$$
$$x^2 - 12x + 32 = 0$$
$$(x - 8)(x - 4) = 0$$

$x - 8 = 0$ or $x - 4 = 0$

$x = 8$ $\quad$ $x = 4$

Substitute 8 and 4 for x in the first equation and solve for y.

$y = \frac{32}{8} = 4$ $\quad$ $y = \frac{32}{4} = 8$

The integers are 4 and 8.

APPLICATIONS

49. **Analysis**: Find the dimensions of a rectangle.

Equation: Let x represent the width of the rectangle and y represent the length. Use the formulas for perimeter and area.

Width	times	length	equals	area
x	$\bullet$	y	$=$	63
2 widths	plus	two lengths	equals	perimeter
$2x$	$+$	$2y$	$=$	32

Solve: $\begin{cases} xy = 63 \\ 2x + 2y = 32 \end{cases}$ $\quad$ Divide second equation by 2 and solve for y. $\quad$ $x + y = 16$

$y = 16 - x$

49. cont.

Substitute into the first equation and solve for x.

$$\begin{aligned} x(16-x) &= 63 \\ 16x - x^2 &= 63 \\ 0 &= x^2 - 16x + 63 \\ 0 &= (x-9)(x-7) \end{aligned}$$

$x - 9 = 0$ or $x - 7 = 0$

$x = 9$ $\quad$ $x = 7$

$y = 16 - x$ $\quad$ $y = 16 - x$ $\quad$ Substitute 9 and 7 for x in the second equation.

$y = 16 - 9$ $\quad$ $y = 16 - 7$

$y = 7$ $\quad$ $y = 9$

Conclusion: The rectangle would measure 7 cm by 9 cm.

51. **Analysis**: Find the amount and rate of a given investment.

Equation: Let x represent the number of dollars invested by Carol and y the annual rate of that investment. Then $\$(x + 150)$ represents the amount invested by John at a rate of $(y + 0.015)\%$.

	P •	r •	t =	I
Carol	x	y	1	67.50
John	$x + 150$	$y + 0.015$	1	94.50

Solve: $\begin{cases} xy = 67.50 \\ (x+150)(y+0.015) = 94.50 \end{cases}$ Remove parentheses. $\begin{cases} xy = 67.50 \\ xy + 150y + 0.015x + 2.25 = 94.50 \end{cases}$

Solve first equation for y. $y = \dfrac{67.50}{x}$ Substitute into the second equation and solve for x.

$$\begin{aligned} x\left(\frac{67.50}{x}\right) + 150\left(\frac{67.50}{x}\right) + 0.015x &= 92.25 \\ 67.50 + \frac{10{,}125}{x} + 0.015x &= 92.25 \\ \frac{10{,}125}{x} + 0.015x - 24.75 &= 0 \\ 10{,}125 + 0.015x^2 - 24.75x &= 0 \\ 0.015x^2 - 24.75x + 10{,}125 &= 0 \end{aligned}$$

$$x = \frac{-(-24.75) \pm \sqrt{(-24.75)^2 - 4(0.015)(10{,}125)}}{2(0.015)}$$

$$= \frac{24.75 \pm \sqrt{612.5625 - 607.5}}{0.03} = \frac{24.75 \pm 2.25}{0.03}$$

$x = \dfrac{24.75 - 2.25}{0.03} = 750$ or $x = \dfrac{24.75 + 2.25}{0.03} = 900$

$y = \dfrac{67.50}{750} = 0.09$ $\quad$ $y = \dfrac{67.50}{900} = 0.075$ $\quad$ Substitute to find y.

Conclusion: Carol either had \$750 invested at 9% or \$900 invested at 7.5%.

53. **Analysis**: Find Jim's rate and time required to drive 306 miles.

Equation: Let x represent Jim's rate and y represent Jim's time. Then $x - 17$ represents the brother's rate and $y + 1.5$ hours represents the brother's time.

	r •	t =	d
Jim	x	y	306
Brother	$x - 17$	$y + 1.5$	306

Solve: $\begin{cases} xy = 306 \\ (x-17)(y+1.5) = 306 \end{cases}$ Remove parentheses. $\begin{cases} xy = 67.50 \\ xy - 17y + 1.5x - 25.5 = 306 \end{cases}$

Solve first equation for y. $y = \dfrac{306}{x}$

Substitute into the second equation and solve for x.

$$x\left(\frac{306}{x}\right) - 17\left(\frac{306}{x}\right) + 1.5x - 25.5 = 306$$

$$306 - \frac{5202}{x} + 1.5x = 331.5$$

$$306x - 5202 + 1.5x^2 - 331.5x = 0$$

$$1.5x^2 - 25.5x - 5202 = 0$$

$$x = \frac{-(-25.5) \pm \sqrt{(-25.5)^2 - 4(1.5)(-5202)}}{2(1.5)}$$

$$= \frac{25.5 \pm \sqrt{650.25 + 31212}}{3}$$

$$= \frac{25.5 \pm 178.5}{3}$$

$x = \dfrac{25.5 - 178.5}{3} = -51$ or $x = \dfrac{25.5 + 178.5}{3} = 68$

Not possible $y = \dfrac{306}{68} = 4.5$ Substitute to find y.

Conclusion: Jim's rate was 68 mph and his time was 4.5 hours.

WRITING

55. The benefit is that you can see how many solutions the system has by counting the number of intersections of the graphs.

REVIEW

57. $\sqrt{200x^2} - 3\sqrt{98x^2} = \sqrt{2 \cdot 100x^2} - 3\sqrt{2 \cdot 49x^2} = 10x\sqrt{2} - 3\left(7x\sqrt{2}\right) = 10x\sqrt{2} - 21x\sqrt{2} = -11x\sqrt{2}$

59. $\dfrac{3t\sqrt{2t} - 2\sqrt{2t^3}}{\sqrt{18t} - \sqrt{2t}} = \dfrac{3t\sqrt{2t} - 2t\sqrt{2t}}{3\sqrt{2t} - \sqrt{2t}} = \dfrac{t\sqrt{2t}}{2\sqrt{2t}} = \dfrac{t}{2}$

Classify equations

1. ellipse; Equation is in standard form.
2. circle; Equation is in standard form.
3. parabola; Complete square on x to obtain standard form.
   $y = 4(x^2 - \frac{1}{2}x + \frac{1}{4}) - 1 + 3 = 4(x - \frac{1}{2})^2 + 2$
4. hyperbola; Equation is in standard form.
5. hyperbola; Complete squares on x and y to obtain standard form.
$$\begin{aligned} 4x^2 + 8x - (y^2 + 4y) &= 4 \\ 4(x^2 + 2x + 1) - (y^2 + 4y + 4) &= 4 + 4 - 4 \\ 4(x + 1)^2 - (y + 2)^2 &= 4 \\ \frac{(x+1)^2}{1^2} - \frac{(y+2)^2}{2^2} &= 1 \end{aligned}$$
6. ellipse; Equation is in standard form.
7. ellipse; Complete squares on x and y to obtain standard form.
$$\begin{aligned} 9x^2 - 18x + 4y^2 + 16y &= 11 \\ 9(x^2 - 2x + 1) + 4(y^2 + 4y + 4) &= 11 + 9 + 16 \\ 9(x - 1)^2 + 4(y + 2)^2 &= 36 \\ \frac{(x-1)^2}{2^2} + \frac{(y+2)^2}{3^2} &= 1 \end{aligned}$$
8. parabola; Solve for y and complete square on x to obtain standard form.
$$\begin{aligned} x^2 - 2y - 2x &= -7 \\ x^2 - 2x + 7 &= 2y \\ (x^2 - 2x + 1) + 7 - 1 &= 2y \\ (x - 1)^2 + 6 &= 2y \\ \tfrac{1}{2}(x - 1)^2 + 3 &= y \end{aligned}$$
9. circle; Complete squares on x and y to obtain standard form.
$$\begin{aligned} x^2 + 8x + y^2 + 2y &= -13 \\ (x^2 + 8x + 16) + (y^2 + 2y + 1) &= -13 + 16 + 1 \\ (x + 4)^2 + (y + 1)^2 &= 4 = 2^2 \end{aligned}$$

Equations of Conic Sections

10. Compare $(x + 1)^2 + (y - 2)^2 = 16$ to $(x - h)^2 + (y - k)^2 = r^2$
    a. The center is at $(h, k) = (-1, 2)$.
    b. The radius is $r = 4$.
11. Compare $x = \frac{1}{2}(y - 1)^2 - 2$ to $x = a(y - k)^2 + h$.
    a. The vertex is at $(h, k) = (-2, 1)$.
    b. The parabola will open to the right.
12. Compare $\frac{x^2}{4} + \frac{y^2}{16} = 1$ to $\frac{(x-h)^2}{a^2} + \frac{(y-k)^2}{b^2} = 1$.
    a. The center is at $(h, k) = (0, 0)$.
    b. The major axis would be parallel to the y–axis, because $b > a$.
    c. The length of the major axis is $2b = 2(4) = 8$ units.
    d. The length of the minor axis is $2a = 2(2) = 4$ units.

13. Compare $\dfrac{(x+2)^2}{9}-\dfrac{(y-1)^2}{4}=1$ to $\dfrac{(x-h)^2}{a^2}-\dfrac{(y-k)^2}{b^2}=1$.

a. The center is at $(h, k) = (-2, 1)$.
b. The branches will open to the left and the right.
c. The dimensions of the fundamental rectangle will be $2a$ by $2b$ which is 6 units horizontally and 4 units vertically.

Graphing Conic Sections

14. $(x+1)^2+(y-2)^2 = 16$

Center: $(-1, 2)$; Radius $= 4$

x	y
−1	−2
−1	6
−5	2
3	2

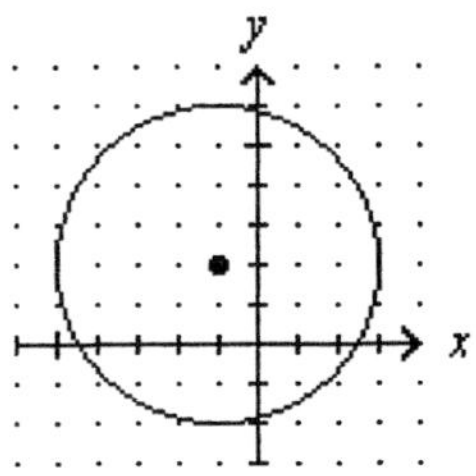

15. $x = \frac{1}{2}(y-1)^2-2$

Vertex: $(-2, 1)$; Opens to the right.

x	y
6	−3
0	−1
−2	**1**
0	3
6	5

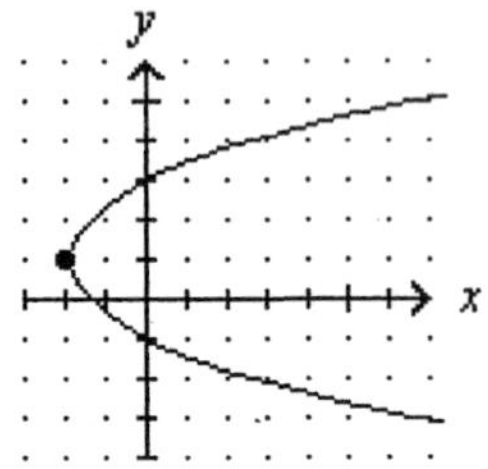

16. $\dfrac{x^2}{4}+\dfrac{y^2}{16}=1$

Center: $(0, 0)$

x	y
0	−4
0	4
−2	0
2	0

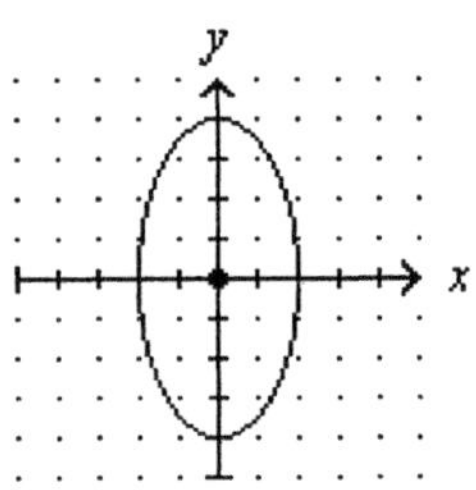

17. $\dfrac{(x+2)^2}{9}-\dfrac{(y-1)^2}{4}=1$ Center at $(-2, 1)$.

Vertices: $(x+2)^2 = 9$, or $x = -2 \pm 3$
$V_1(-5, 1)$ and $V_2(1, 1)$.

Fundamental Rectangle:
Vertices and $(-2, -1)$ and $(-2, 3)$.

Opens left and right.

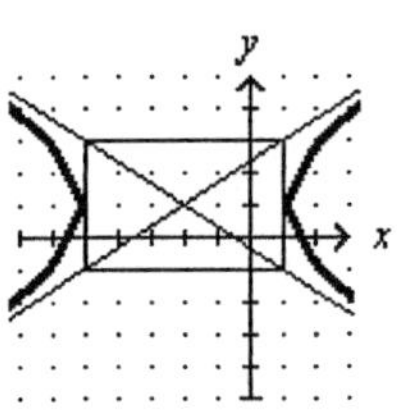

Section 10.1

1. $(x-1)^2+(y+2)^2 = 9$

Center: $(1, -2)$; Radius $= 3$

x	y
1	−5
1	1
−2	−2
4	−2

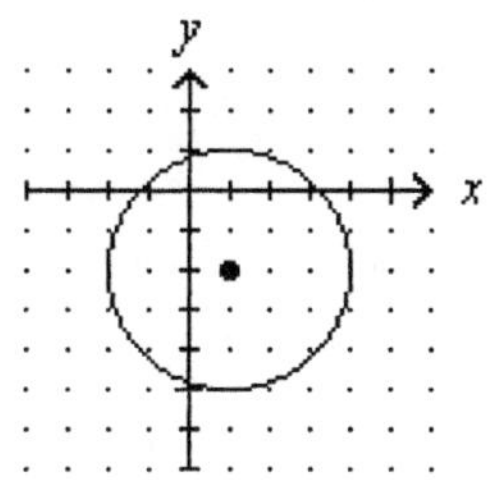

2. $x^2+y^2 = 16$

Center: $(0, 0)$; Radius $= 4$

x	y
0	−4
0	4
−4	0
4	0

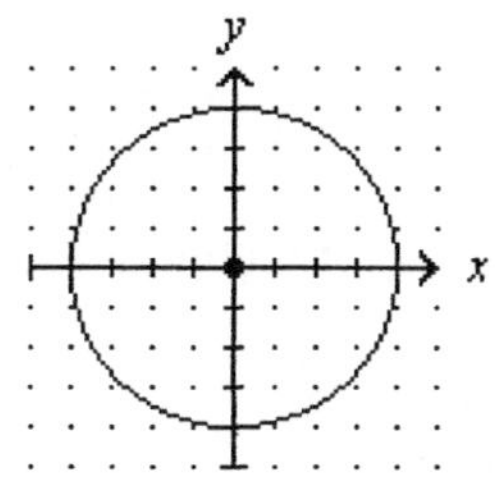

3.
$$\begin{aligned} x^2+y^2+4x-2y &= 4 \\ x^2+4x+y^2-2y &= 4 \\ (x^2+4x+4)+(y^2-2y+1) &= 4+4+1 \\ (x+2)^2+(y-1)^2 &= 9 \end{aligned}$$

Center $(-2, 1)$; Radius $= 3$

x	y
−2	−2
−2	4
−5	1
1	1

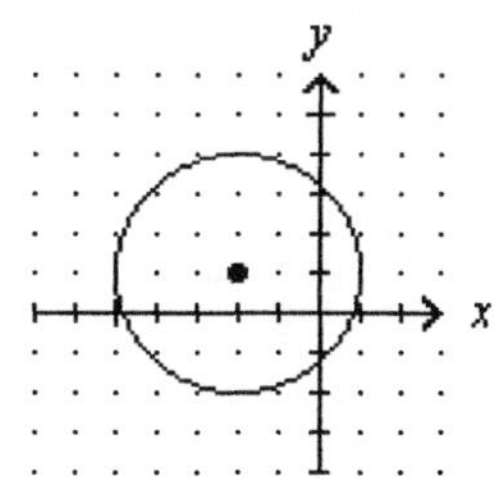

4. $x = -3(y-2)^2+5$

Vertex: $(5, 2)$; Opens to the left.

x	y
−7	0
2	1
5	**2**
2	3
−7	4

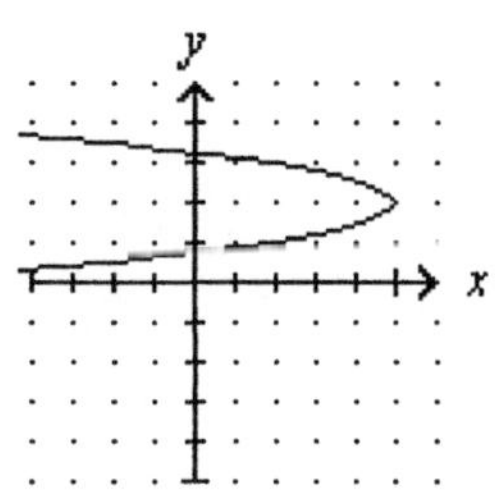

5. $x = 2(y+1)^2-2$

Vertex: $(-2, -1)$; Opens to the right.

x	y
6	−3
0	−2
−2	**−1**
0	0
6	1

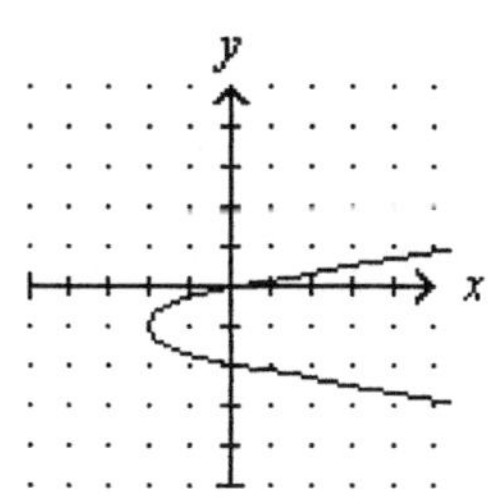

6. The jumper will land when $y = 0$. We can verify this by substituting $x = 22$ into the equation:

$$-\frac{5}{121}(22-11)^2+5 = -\frac{5}{121}(121)+5 = -5+5 = 0 = y$$

Section 10.2

7. $9x^2+16y^2 = 144$

$$\frac{x^2}{16}+\frac{y^2}{9} = 1$$

Center: $(0, 0)$

x	y
0	−3
0	3
−4	0
4	0

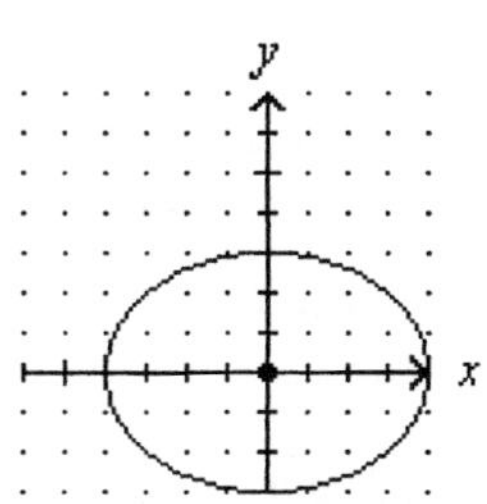

8. $$\frac{(x-2)^2}{4}+\frac{(y-1)^2}{9} = 1$$

Center: $(2, 1)$

x	y
2	−2
2	4
0	1
4	1

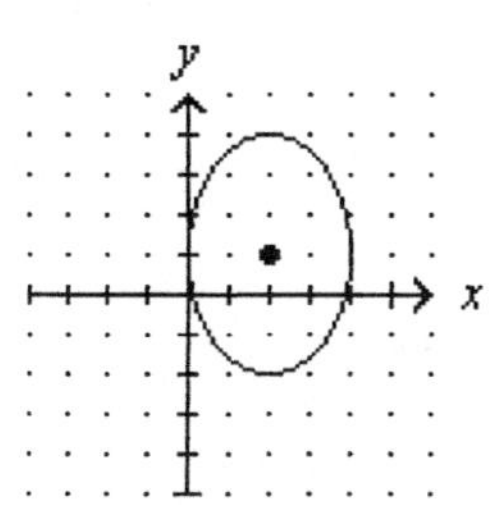

9.

$$\begin{aligned} 4x^2 + 9y^2 + 8x - 18y &= 23 \\ 4x^2 + 8x + 9y^2 - 18y &= 23 \\ 4(x^2 + 2x) + 9(y^2 - 2y) &= 23 \\ 4(x^2 + 2x + 1) + 9(y^2 - 2y + 1) &= 23 + 4 + 9 \\ 4(x + 1)^2 + 9(y - 1)^2 &= 36 \\ \frac{(x+1)^2}{9} + \frac{(y-1)^2}{4} &= 1 \end{aligned}$$

Center: (–1, 1)

x	y
–1	–1
–1	3
–4	1
2	1

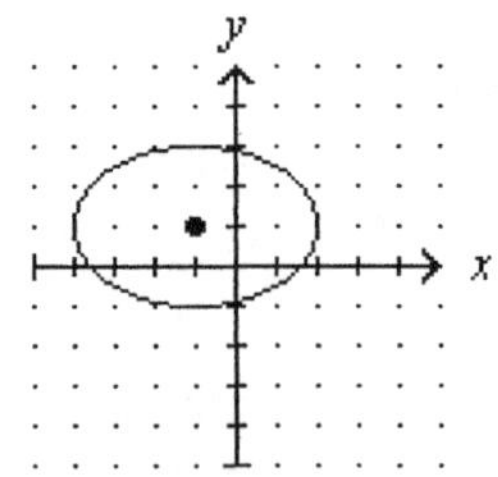

10. The equation for an ellipse centered at (0,0) and extending a distance of 5 along the x- axis and 3 along the y-axis is $\frac{x^2}{25} + \frac{y^2}{9} = 1$

Section 10.3

11.

$$\begin{aligned} 9x^2 - y^2 &= -9 \\ \frac{x^2}{-1} - \frac{y^2}{-9} &= 1 \\ \frac{y^2}{3^2} - \frac{x^2}{1^2} &= 1 \end{aligned}$$

Center: (0, 0)

Vertices ($x = 0$):

$y^2 = 9$

$y = \pm 3$

$V_1\,(0, -3)$ and $V_2\,(0, 3)$

Fundamental Rectangle:

Use vertices and (–1, 0) and (1, 0)

Opens: up and down.

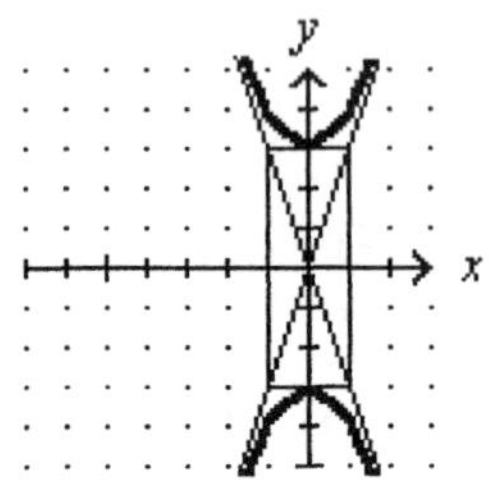

12. $xy = 9$

x	y
–9	–1
–3	–3
–1	–9
1	9
3	3
9	1

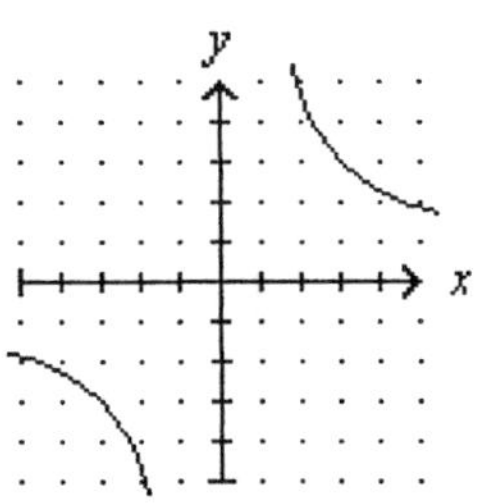

13.

$$\begin{aligned} 4x^2 - 2y^2 + 8x - 8y &= 8 \\ 4(x^2 + 2x) - 2(y^2 - 4y) &= 8 \\ 4(x^2 + 2x + 1) - 2(y^2 + 4y + 4) &= 8 + 4 - 8 \\ 4(x + 1)^2 - 2(y + 2)^2 &= 4 \\ \frac{(x+1)^2}{1} - \frac{(y+2)^2}{2} &= 1 \end{aligned}$$

hyperbola, because of the negative sign

14.

$$\begin{aligned} 9x^2 - 4y^2 - 18x - 8y &= 31 \\ 9x^2 - 18x - 4y^2 - 8y &= 31 \\ 9(x^2 - 2x) - 4(y^2 + 2y) &= 31 \\ 9(x^2 - 2x + 1) - 4(y^2 + 2y + 1) &= 31 + 9 - 4 \\ 9(x - 1)^2 - 4(y + 1)^2 &= 36 \\ \frac{(x-1)^2}{4} - \frac{(y+1)^2}{9} &= 1 \end{aligned}$$

Center: (1, –1)

Vertices ($y = -1$):

$(x - 1)^2 = 4$

$x = 1 \pm 2$

$V_1\,(-1, -1)$ and $V_2\,(3, -1)$

Fundamental Rectangle:

Use vertices and (1, –4) and (1, 2).

Opens: left and right.

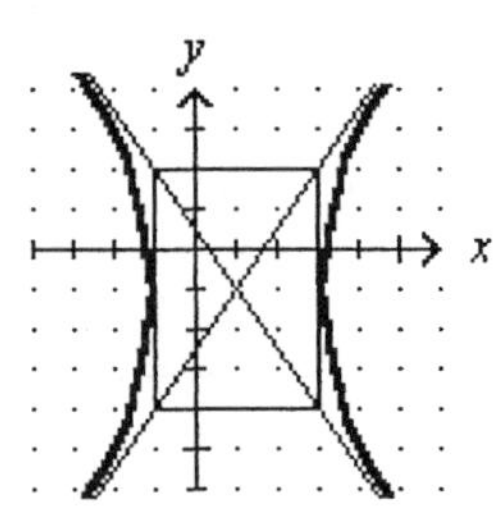

15. Write the equation in standard form:

$$x^2 - 4y^2 = 4$$

$$\frac{x^2}{4} - \frac{y^2}{1} = 1$$

The vertices of the hyperbola are (0,2) and (0,–2) which are a distance of 4 units apart. Thus the particles were at most 4 units apart.

16. a. Since the coefficients of x^2 and y^2 are both positive and the denominators are not equal, this is an ellipse.
b. Since one of the coefficients of x^2 and y^2 is positive and the other is negative, this is a hyperbola.
c. Since only one of the variables is squared, this is a parabola.
d. Since the coefficients of x^2 and y^2 are both positive and the same, this is a circle.

Section 10.4

17. The intersection of the two graphs and therefore the solutions appear to be (0, 3) and (0, –3).

18. $$\begin{cases} 3x^2 + y^2 = 52 \\ x^2 - y^2 = 12 \end{cases}$$

Add the two equations and solve for x.

$$\begin{aligned} 4x^2 &= 64 \\ x^2 &= 16 \\ x &= \pm 4 \end{aligned}$$

Substitute –4 and 4 for x into $x^2 - y^2 = 12$ and solve for y.

$$\begin{aligned} (-4)^2 - y^2 &= 12 \\ -y^2 &= -4 \\ y^2 &= 4 \\ y &= \pm 2 \end{aligned} \qquad \begin{aligned} (4)^2 - y^2 &= 12 \\ -y^2 &= -4 \\ y^2 &= 4 \\ y &= \pm 2 \end{aligned}$$

The solutions are (–4, –2), (–4, 2), (4, –2) and (4, 2).

19. $$\begin{cases} \dfrac{x^2}{16} + \dfrac{y^2}{12} = 1 \\ x^2 - \dfrac{y^2}{3} = 1 \end{cases} \quad \text{Becomes} \quad \begin{cases} 3x^2 + 4y^2 = 48 \\ 3x^2 - y^2 = 3 \end{cases}$$

Subtract the two equations and solve for y.

$$\begin{aligned} 5y^2 &= 45 \\ y^2 &= 9 \\ y &= \pm 3 \end{aligned}$$

Substitute –3 and 3 for y into $3x^2 - y^2 = 3$ and solve for x.

$$\begin{aligned} 3x^2 - (-3)^2 &= 3 \\ 3x^2 &= 12 \\ x^2 &= 4 \\ x &= \pm 2 \end{aligned} \qquad \begin{aligned} 3x^2 - (3)^2 &= 3 \\ 3x^2 &= 12 \\ x^2 &= 4 \\ x &= \pm 2 \end{aligned}$$

The solutions are (–2, –3), (–2, 3), (2, –3), and (2, 3).

20. $$\begin{cases} y = -x^2 + 2 \\ x^2 - y - 2 = 0 \end{cases}$$ Solve the second equation for y. $y = x^2 - 2$

Substitute this expression for y in the first equation and solve for x.

$$x^2 - 2 = -x^2 + 2$$

$$2x^2 - 4 = 0$$

$$(\sqrt{2}x - 2)(\sqrt{2}x + 2) = 0$$

$$\sqrt{2}x - 2 = 0 \quad or \quad \sqrt{2}x + 2 = 0$$

$$x = \pm\frac{2}{\sqrt{2}} \cdot \frac{\sqrt{2}}{\sqrt{2}} = \pm\sqrt{2}$$

Now substitute back in to find y:

if $x = \sqrt{2}$, then $y = \left(\sqrt{2}\right)^2 - 2 = 0$

if $x = -\sqrt{2}$, then $y = \left(-\sqrt{2}\right)^2 - 2 = 0$

solutions: $\left(\sqrt{2}, 0\right), \left(-\sqrt{2}, 0\right)$

21. $\begin{cases} x^2+2y^2=12 \\ 2x-y=2 \end{cases}$ Solve the second equation for y: $y=2x-2$

Substitute this expression for y in the first equation and solve for x.

$$\begin{aligned} x^2+2(2x-2)^2&=12 \\ x^2+2(4x^2-8x+4)&=12 \\ x^2+8x^2-16x+8&=12 \\ 9x^2-16x-4&=0 \\ (9x+2)(x-2)&=0 \\ x=-\frac{2}{9} \text{ or } x&=2 \end{aligned}$$

Now substitute back in to find y.

if $x=-\frac{2}{9}$ then $y=2\left(-\frac{2}{9}\right)-2=\frac{-4}{9}-\frac{18}{9}=-\frac{22}{9}$

if $x=2$ then $y=2(2)-2=2$

solutions: $\left(-\frac{2}{9},-\frac{22}{9}\right)$, $(2,2)$

Chapter 10 Chapter Test

1. Compare $(x-2)^2+(y+3)^2 = 4$ to $(x-h)^2+(y-k)^2 = r^2$. The center is $(h, k) = (2, -3)$. The radius is $r = 2$.

2. Complete the squares on x and y.

$$\begin{aligned} x^2+y^2+4x-6y &= 3 \\ x^2+4x+y^2-6y &= 3 \\ (x^2+4x+4)+(y^2-6y+9) &= 3+4+9 \\ (x+2)^2+(y-3)^2 &= 16 \end{aligned}$$

The center is at $(-2, 3)$ and the radius is 4.

3. $(x+1)^2+(y-2)^2 = 9$

Center: $(-1, 2)$; Radius $= 3$

x	y
−1	−1
−1	5
−4	2
2	2

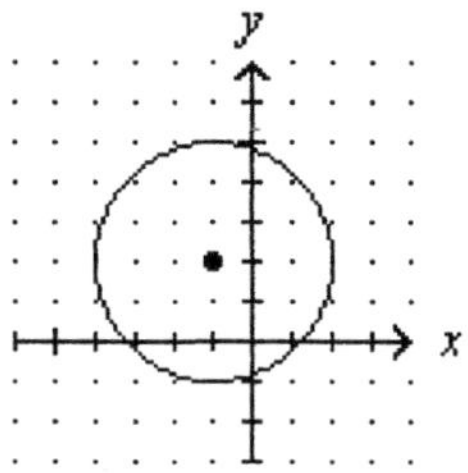

4. $x = (y-2)^2-1$

Vertex: $(-1, 2)$; Opens to the right.

x	y
3	0
0	1
−1	**2**
0	3
3	4

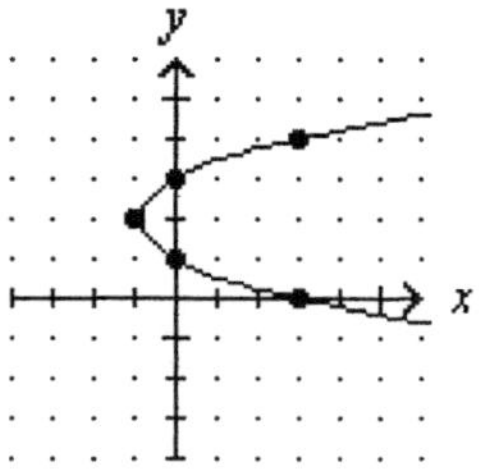

5. $9x^2+4y^2=36$

$$\frac{x^2}{4}+\frac{y^2}{9}=1$$

$$\frac{x^2}{2^2}+\frac{y^2}{3^2}=1$$

Center: $(0, 0)$

x	y
0	−3
0	3
−2	0
2	0

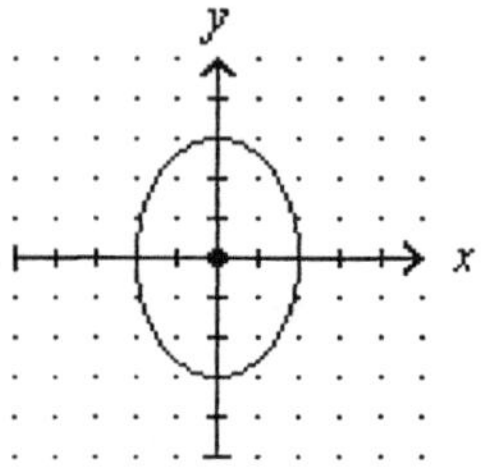

6. $\dfrac{(x-2)^2}{9} - y^2 = 1$

Center: (2, 0)

Vertices ($y = 0$):

$(x-2)^2 = 9$
$x = 2 \pm 3$

$V_1(-1, 0)$ and
$V_2(5, 0)$

Fundamental Rectangle:

Use vertices and
(2, –1) and (2, 1).

Opens: left and right.

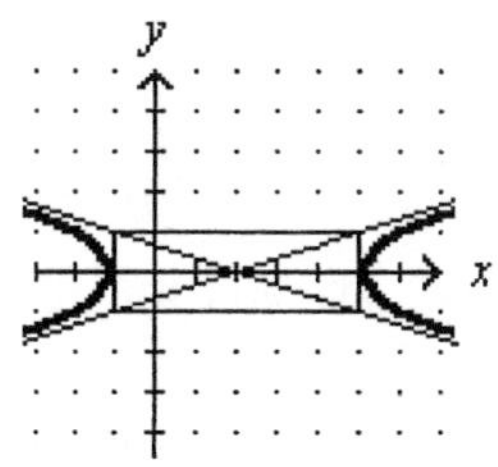

7. $xy = -4$
Special hyperbola- graph by plotting points

x	y
1	–4
–1	4
2	–2
–2	2

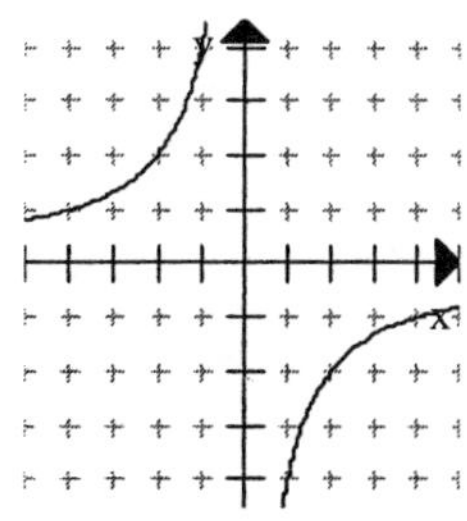

8.
$$
\begin{aligned}
x^2 + 4y^2 + 6x + 16y &= -9 \\
x^2 + 6x + 4y^2 + 16y &= -9 \\
(x^2 + 6x) + 4(y^2 + 4y) &= -9 \\
(x^2 + 6x + 9) + 4(y^2 + 4y + 4) &= -9 + 9 + 16 \\
(x+3)^2 + 4(y+2)^2 &= 16 \\
\frac{(x+3)^2}{16} + \frac{(y+2)^2}{4} &= 1
\end{aligned}
$$

Center: (–3, –2)

x	y
–3	–4
–3	0
1	–2
–7	–2

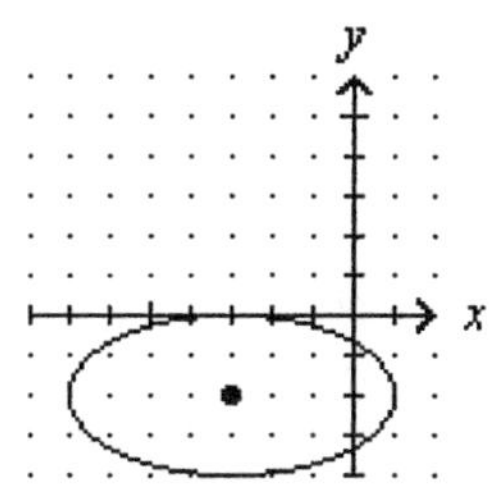

9.
$$
\begin{aligned}
4y^2 - 9x^2 - 8y &= 32 \\
4y^2 - 8y - 9x^2 &= 32 \\
4(y^2 - 2y) - 9x^2 &= 32 \\
4(y^2 - 2y + 1) - 9x^2 &= 32 + 4 \\
4(y-1)^2 - 9x^2 &= 36 \\
\frac{(y-1)^2}{9} - \frac{x^2}{4} &= 1
\end{aligned}
$$

Center: (0, 1)

Vertices ($x = 0$):

$(y-1)^2 = 9$
$y = 1 \pm 3$

$V_1(0, -2)$ and $V_2(0, 4)$

Fundamental Rectangle:

Use vertices and
(–2, 1) and (2, 1).

Opens: up and down.

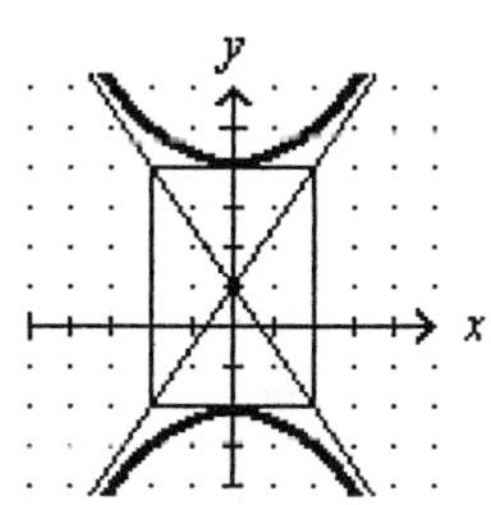

10. $\begin{cases} x^2 + y^2 = 25 \\ y - x = 1 \end{cases}$

The line intersects the circle at the points (3,4) and (–4,–3).

The first equation is a circle centered at the origin, with radius of 5

The second equation is a line

x	y
0	1
–1	0

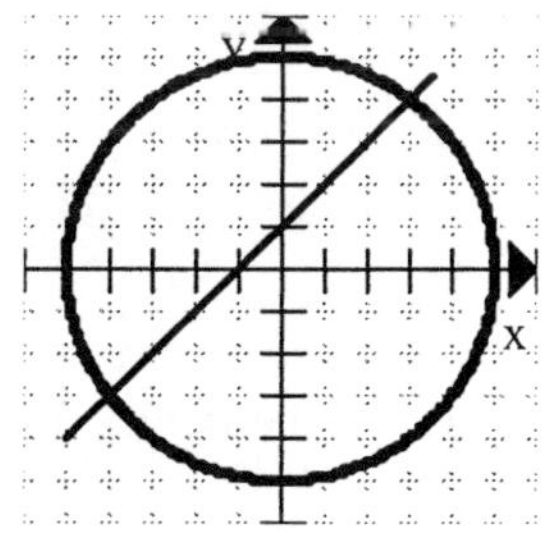

11. $\begin{cases} 2x - y = -2 \\ x^2 + y^2 = 16 + 4y \end{cases}$

Solve the linear equation for y.

$$y = 2x + 2$$

Substitute into the quadratic equation.

$$\begin{aligned} x^2 + (2x+2)^2 &= 16 + 4(2x+2) \\ x^2 + (4x^2 + 8x + 4) &= 16 + 8x + 8 \\ 5x^2 &= 20 \\ x^2 &= 4 \\ x &= \pm 2 \end{aligned}$$

Substitute –2 and 2 for x into $y = 2x + 2$ and solve for y.

$$\begin{aligned} y &= 2(-2) + 2 \\ y &= -2 \end{aligned} \qquad \begin{aligned} y &= 2(2) + 2 \\ y &= 6 \end{aligned}$$

The solutions are $(-2, -2)$ $(2, 6)$.

12. $\begin{cases} x^2 + y^2 = 25 \\ 4x^2 - 9y = 0 \end{cases}$

Solve Equation 1 for x^2.

$$x^2 = 25 - y^2$$

Substitute into Equation 2.

$$\begin{aligned} 4(25 - y^2) - 9y &= 0 \\ 100 - 4y^2 - 9y &= 0 \\ -4y^2 - 9y + 100 &= 0 \\ 4y^2 + 9y - 100 &= 0 \\ (4y + 25)(y - 4) &= 0 \end{aligned}$$

$$\begin{aligned} 4y + 25 &= 0 \\ y &= -\tfrac{25}{4} \end{aligned} \qquad \begin{aligned} y - 4 &= 0 \\ y &= 4 \end{aligned}$$

Substitute $-\frac{25}{4}$ and 4 into $x^2 = 25 - y^2$.

$$\begin{aligned} x^2 &= 25 - \left(-\tfrac{25}{4}\right)^2 \\ x^2 &= 25 - \tfrac{625}{16} \\ x^2 &= -\tfrac{225}{16} \end{aligned} \qquad \begin{aligned} x^2 &= 25 - (4)^2 \\ x^2 &= 9 \\ x &= \pm 3 \end{aligned}$$

No real solution.

The solutions are $(-3, 4)$ and $(3, 4)$.

13. $\begin{cases} 5x^2 - y^2 - 3 = 0 \\ x^2 + 2y^2 = 5 \end{cases}$ Multiply equation 1 by 2 and add the two equations together.

$$\begin{cases} 10x^2 - 2y^2 = 6 \\ x^2 + 2y^2 = 5 \end{cases}$$

$$11x^2 = 11$$

$$x = \pm 1$$

Substitute $x = \pm 1$ into one of the equations (we'll use equation 2) and solve for y

$$\begin{aligned} &(1)^2 + 2y^2 = 5 \\ &2y^2 = 4 \\ &y^2 = 2 \\ &y = \pm\sqrt{2} \\ &(1, \sqrt{2}),\ \left(1, -\sqrt{2}\right) \end{aligned} \qquad \begin{aligned} &(-1)^2 + 2y^2 = 5 \\ &2y^2 = 4 \\ &y^2 = 2 \\ &y = \pm\sqrt{2} \\ &(-1, \sqrt{2}),\ \left(-1, -\sqrt{2}\right) \end{aligned}$$

14. From the picture, we can find the center of the circle is (4,3). We also can observe that the radius is 3. Using the standard equation for a circle: $(x-h)^2 + (y-k)^2 = r^2$, obtain the equation:

$$(x-4)^2 + (y-3)^2 = 9$$

15. $x = -\frac{1}{10}y^2$

x	y
0	0
-1	$\pm\sqrt{10}$
$-\frac{1}{10}$	± 1

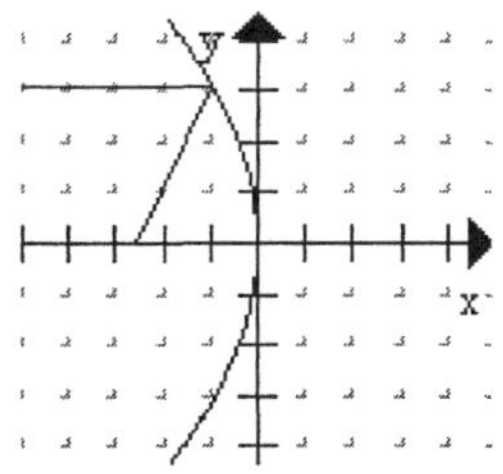

16. The major axis is parallel to the x-axis and is of length 8 ($a = 4$), and its center is at $x = 1$
The minor axis is parallel to the y-axis and is of length 6 ($b = 3$), and its center is at $y = -2$
Using the standard equation: $\frac{(x-h)^2}{a^2} + \frac{(y-k)^2}{b^2} = 1$ we obtain: $\frac{(x-1)^2}{16} + \frac{(y+2)^2}{9} = 1$

17. Since the vertices are on the y-axis and the center is the origin, we will use the standard equation: $\frac{y^2}{a^2} - \frac{x^2}{b^2} = 1$
The values for a and b come from the fundamental rectangle. On the y-axis, $a = 4$; on the x-axis, $b = 6$.

Therefore, the equation desired is: $\frac{y^2}{16} - \frac{x^2}{36} = 1$

18. a. ellipse because: addition between the squared terms and the coefficients of the squared terms are not equal.
b. hyperbola because: subtraction between the squared terms.
c. circle because: addition between the squared terms and the coefficients of the squared terms are equal.
d. ellipse because: only one squared term.

Chapters 1–10 Cumulative Review Exercises

1. 0
2. $-\frac{4}{3}$, 5.6, 0, -23
3. π, $\sqrt{2}$, e
4. $-\frac{4}{3}$, π, 5.6, $\sqrt{2}$, 0, -23, e

5. **Analysis**: Find the amount that Ana has for her original investment. If she invested \$3000 more, she could double the annual income.

Equation: Let x represent the number of dollars that could be originally invested in at 7.5%. Then $\$(x + 3000)$ would be the number of dollars invested at 11.0%.

	P •	r •	t =	I
Original investment	x	0.075	1	$0.075x$
Larger investment	$x + 3000$	0.11	1	$0.11(x + 3000)$

Two	times	interest earned at 7.5%	is	interest earned 11%.
2	•	$0.075x$	=	$0.11(x + 3000)$

Solve:

$$\begin{aligned} 2(0.075x) &= 0.11(x + 3000) \\ 0.15x &= 0.11x + 330 \\ 0.04x &= 330 \\ x &= 8250 \end{aligned}$$

Conclusion: The original investment would have been \$8250.

6. Two points on the line are (1600, 68) and (2800, 78). The average rate of change is the slope of this line.

$$\text{Average rate of change} = \frac{\text{Change in sound level}}{\text{Change in RPM}} = \frac{78-68}{2800-1600} = \frac{10}{1200} = \frac{1}{120} \text{ db/rpm}$$

7. Compare both equations to $y = mx + b$.

$$\begin{aligned} 3x - 4y &= 12 \\ -4y &= -3x + 12 \\ y &= \tfrac{3}{4}x + 12 \end{aligned} \qquad \text{and} \qquad y = \tfrac{3}{4}x - 5$$

Since the slope is $m = \frac{3}{4}$ in both equations, the graphs of the equations are parallel.

8. Compare both equations to $y = mx + b$.

$$y = 3x + 4 \qquad \text{and} \qquad \begin{aligned} x &= -3y + 4 \\ 3y &= -x + 4 \\ y &= -\tfrac{1}{3}x + \tfrac{4}{3} \end{aligned}$$

Since the slopes are negative reciprocals of each other, the graphs of the equations are perpendicular.

9. Use the point–slope formula:

$$\begin{aligned} y - y_1 &= m(x - x_1) \\ y - (5) &= -2(x - 0) \\ y &= -2x + 5 \end{aligned}$$

where $m = -2$ and $(x_1, y_1) = (0, 5)$

10. Find the slope and then use the point–slope formula with points (8, –5) and (–5, 4).

$$m = \frac{y_2 - y_1}{x_2 - x_1} = \frac{4-(-5)}{-5-8} = -\frac{9}{13}$$

$$\begin{aligned} y - y_1 &= m(x - x_1) \\ y - (-5) &= -\tfrac{9}{13}(x - 8) \\ y + 5 &= -\tfrac{9}{13}x + \tfrac{72}{13} \\ y &= -\tfrac{9}{13}x + \tfrac{7}{13} \end{aligned}$$

11. $\begin{cases} 3x + y = 4 \\ 2x - 3y = -1 \end{cases}$

Solve Equation 1 for y.

$$\begin{aligned} 3x + y &= 4 \\ y &= -3x + 4 \end{aligned}$$

Substitute $-3x + 4$ for y in Equation 2.

$$\begin{aligned} 2x - 3(-3x + 4) &= -1 \\ 2x + 9x - 12 &= -1 \\ 11x &= 11 \\ x &= 1 \end{aligned}$$

Substitute 1 for x in Equation 1.

$$\begin{aligned} 3(1) + y &= 4 \\ 3 + y &= 4 \\ y &= 1 \end{aligned}$$

The solution is (1, 1).

12. $\begin{cases} x + 2y = -2 \\ 2x - y = 6 \end{cases}$

Multiply Equation 2 by 2.

$$\begin{cases} x + 2y = -2 \\ 4x - 2y = 12 \end{cases}$$

Add the equations and solve for x.

$$\begin{aligned} 5x &= 10 \\ x &= 2 \end{aligned}$$

Substitute 2 for x in Equation 2.

$$\begin{aligned} 2(2) - y &= 6 \\ -y &= 2 \\ y &= -2 \end{aligned}$$

The solution is (2, –2).

13. $\begin{cases} 4x - 3y = -1 \\ 3x + 4y = -7 \end{cases}$

$$x = \frac{D_x}{D} = \frac{\begin{vmatrix} -1 & -3 \\ -7 & 4 \end{vmatrix}}{\begin{vmatrix} 4 & -3 \\ 3 & 4 \end{vmatrix}} = \frac{-4-(21)}{16-(-9)} = \frac{-25}{25} = -1 \qquad y = \frac{D_y}{D} = \frac{\begin{vmatrix} 4 & -1 \\ 3 & -7 \end{vmatrix}}{\begin{vmatrix} 4 & -3 \\ 3 & 4 \end{vmatrix}} = \frac{-28-(-3)}{16-(-9)} = \frac{-25}{25} = -1$$

The solution is $(-1, -1)$.

14. Answers vary

15. The x–value of the intersection of the two graphs is the solution $x = 3$.

16. $\begin{vmatrix} 3 & -2 \\ 1 & -1 \end{vmatrix} = 3(-1) - (-2)(1) = -3 + 2 = -1$

17. Let c represent the measure of angle C. Then $5c + 5$ represents the measure of angle B and $5c + 5 + 5$ represents the measure of angle A respectively. The sum of the measures of the angles of a triangle is 180°.

$$\begin{aligned} \text{Angle A} + \text{Angle } B + \text{Angle } C &= 180 \\ (5c + 10) + (5c + 5) + c &= 180 \\ 11c + 15 &= 180 \\ 11c &= 165 \\ c &= 15 \end{aligned}$$

The measure of angle A $= 5(15) + 10 = 75 + 10 = 85°$, the measure of angle B $= 5(15) + 5 = 80°$, and the measure of angle C $= 15°$.

18. Inequality #1: $3x - 2y \le 6$

Graph: $3x - 2y = 6$

x	y
0	−3
2	0

Solid line. Test point (0, 0).

$3(0) - 2(0) \le 6$

$0 \le 6$ True

Inequality #2: $y < -x + 2$

Graph: $y = -x + 2$

x	y
0	2
4	−2

Broken line. Test point (0, 0).

$(0) < -(0) + 2$

$0 < 2$ True

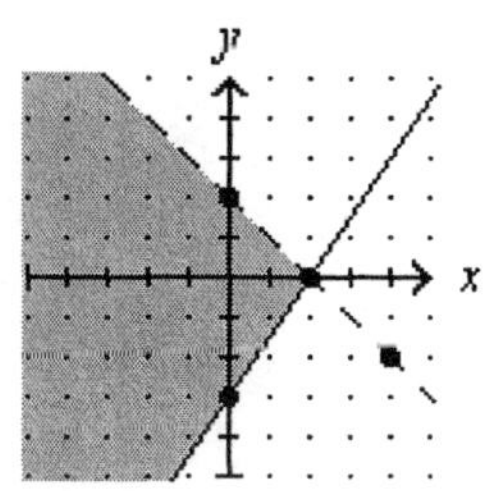

19. Rewrite $|5 - 3x| \le 14$ as

$$\begin{aligned} -14 \le\ & 5 - 3x \le 14 \\ -19 \le\ & -3x \le 9 \\ \frac{19}{3} \ge\ & x \ge -3 \end{aligned}$$

$[-3, \frac{19}{3}]$

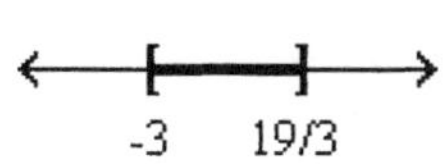

20. $4.5x - 1 < -10$ or $6 - 2x \ge 12$

$4.5x < -9$ $\quad -2x \ge 6$

$x < -2$ $\quad x \le -3$

$(-\infty, -2)$

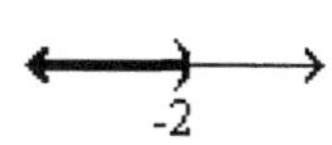

21. $(4x - 3y)(3x + y) = 4x(3x) + 4x(y) - 3y(3x) - 3y(y) = 12x^2 + 4xy - 9xy - 3y^2 = 12x^2 - 5xy - 3y^2$

22. $(-2x^2y^3 + 6xy + 5y^2) - (-4x^2y^3 - 7xy + 2y^2) = -2x^2y^3 + 6xy + 5y^2 + 4x^2y^3 + 7xy - 2y^2 = 2x^2y^3 + 13xy + 3y^2$

23. $(a - 2b)^2 = a^2 - 2(a)(2b) + (2b)^2 = a^2 - 4ab + 4b^2$

24. $(a + 2)(3a^2 + 4a - 2) = 3a^3 + 4a^2 - 2a + 6a^2 + 8a - 4 = 3a^3 + 10a^2 + 6a - 4$

25. $3x^3y - 4x^2y^2 - 6x^2y + 8xy^2 = xy(3x^2 - 4xy - 6x + 8y) = xy[x(3x - 4y) - 2(3x - 4y) = xy(3x - 4y)(x - 2)$

26. $256x^4y^4 - z^8 = (16x^2y^2)^2 - (z^4)^2 = (16x^2y^2 + z^4)(16x^2y^2 - z^4) = (16x^2y^2 + z^4)(4xy + z^2)(4xy - z^2)$

27. Solve for λ.

$$\begin{aligned}
\frac{A\lambda}{2} + 1 &= 2d + 3\lambda \\
A\lambda + 2 &= 4d + 6\lambda \\
A\lambda - 6\lambda &= 4d - 2 \\
\lambda(A - 6) &= 4d - 2 \\
\lambda &= \frac{4d - 2}{A - 6}
\end{aligned}$$

28.

x	$f(x) = -x^3 - x^2 + 6x$	$f(x)$
-4	$f(-4) = -(-4)^3 - (-4)^2 + 6(-4)$	24
-3	$f(-3) = -(-3)^3 - (-3)^2 + 6(-3)$	0
-2	$f(-2) = -(-2)^3 - (-2)^2 + 6(-2)$	-8
-1	$f(-1) = -(-1)^3 - (-1)^2 + 6(-1)$	-6
0	$f(0) = -(0)^3 - (0)^2 + 6(0)$	0
1	$f(1) = -(1)^3 - (1)^2 + 6(1)$	4
2	$f(2) = -(2)^3 - (2)^2 + 6(2)$	0
3	$f(3) = -(3)^3 - (3)^2 + 6(3)$	-18

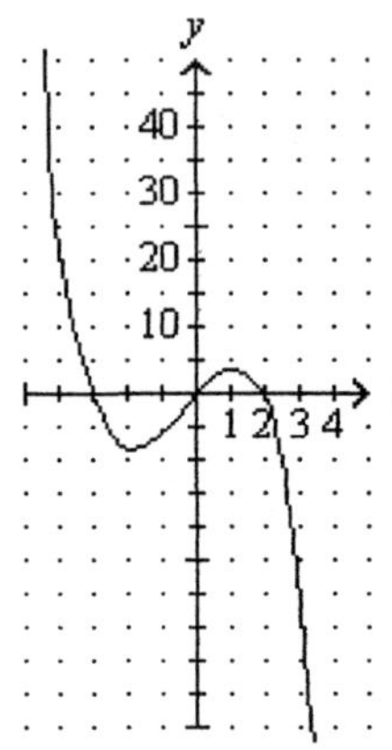

x–intercepts: $(-3, 0)$, $(0, 0)$, $(2, 0)$
y–intercept: $(0, 0)$

29. $\left(\dfrac{4a^{-2}b}{3ab^{-3}}\right)^3 = \left(\dfrac{4a^{-2-1}b^{1-(-3)}}{3}\right)^3 = \left(\dfrac{4a^{-3}b^4}{3}\right)^3 = \left(\dfrac{4b^4}{3a^3}\right)^3 = \dfrac{4^3 b^{4\bullet 3}}{3^3 a^{3\bullet 3}} = \dfrac{64b^{12}}{27a^9}$

30. $\dfrac{6x^2 + 13x + 6}{6 - 5x - 6x^2} = \dfrac{(2x+3)(3x+2)}{-(6x^2 + 5x - 6)} = \dfrac{(2x+3)(3x+2)}{-(2x+3)(3x-2)} = -\dfrac{3x+2}{3x-2}$

31. $\dfrac{p^3 - q^3}{q^2 - p^2} \cdot \dfrac{q^2 + pq}{p^3 + p^2q + pq^2} = \dfrac{(p-q)(p^2 + pq + q^2)}{(q-p)(q+p)} \cdot \dfrac{q(q+p)}{p(p^2 + pq + q^2)} = \dfrac{(p-q)}{(q-p)} \cdot \dfrac{q}{p} = -\dfrac{q}{p}$

32. $\dfrac{2}{a-2} + \dfrac{3}{a+2} - \dfrac{a-1}{a^2-4} = \dfrac{2(a+2)}{(a-2)(a+2)} + \dfrac{3(a-2)}{(a+2)(a-2)} - \dfrac{a-1}{(a+2)(a-2)} = \dfrac{2a + 4 + 3a - 6 - a + 1}{(a+2)(a-2)} = \dfrac{4a - 1}{(a+2)(a-2)}$

33.

$$\begin{aligned}
\frac{x-4}{x-3} + \frac{x-2}{x-3} &= x - 3 \\
(x-3)\left(\frac{x-4}{x-3} + \frac{x-2}{x-3}\right) &= (x-3)(x-3) \qquad \text{Multiply by LCD } x - 3. \\
(x-4) + (x-2) &= x^2 - 6x + 9 \\
0 &= x^2 - 8x + 15 \\
0 &= (x-3)(x-5)
\end{aligned}$$

$x - 3 = 0$ or $x - 5 = 0$

$x = 3$ $\quad$ $x = 5$

$x = 3$ would make the denominator 0 so it is extraneous and the solution is $x = 5$.

34.
$$\frac{1}{R}=\frac{1}{R_1}+\frac{1}{R_2}+\frac{1}{R_3}$$ Solve for R.

$$RR_1R_2R_3\left(\frac{1}{R}\right)=RR_1R_2R_3\left(\frac{1}{R_1}+\frac{1}{R_2}+\frac{1}{R_3}\right)$$ Multiply by the LCD.

$$R_1R_2R_3=RR_2R_3+RR_1R_3+RR_1R_2$$ Do the multiplication.

$$R_1R_2R_3=R(R_2R_3+R_1R_3+R_1R_2)$$ Factor.

$$R=\frac{R_1R_2R_3}{R_2R_3+R_1R_3+R_1R_2}$$ Divide by $R_2R_3+R_1R_3+R_1R_2$.

35. a. a quadratic function

b. The tire will offer only 90% of its service at about 85% and 120% of the recommended inflation.

36. Find the hypotenuse of a right triangle where the legs are both 16 in. Then, subtract 1 inch to allow for the overlap.

$$\begin{aligned} c^2 &= a^2+b^2 \\ c^2 &= (16)^2+(16)^2 \\ c^2 &= 256+256 \\ c^2 &= 512 \\ c &= 22.6 \text{ in.} \end{aligned}$$

The largest waist size would be about 22.6 – 1 = 21.6 inches.

37. Rearrange the terms before dividing.

$$\begin{array}{r|l} & -x^2+x+5 \\ \hline x-1 & -x^3+2x^2+4x+3 \\ & -x^3+x^2 \\ \hline & x^2+4x \\ & x^2-x \\ \hline & 5x+3 \\ & 5x-5 \\ \hline & 8 \end{array}$$

Solution: $-x^2+x+5+\dfrac{8}{x-1}$

38. When $f(x)=\sqrt{2x+5}+\sqrt{x+2}-5=0$ is equivalent to $\sqrt{2x+5}=-\sqrt{x+2}+5$. From the illustration this would be the value of x where the graph of f(x) crosses the x–axis. The x–value would be 2.

39. $\sqrt{98}+\sqrt{8}-\sqrt{32}=\sqrt{49\cdot2}+\sqrt{4\cdot2}-\sqrt{16\cdot2}=7\sqrt{2}+2\sqrt{2}-4\sqrt{2}=5\sqrt{2}$

40. $12\sqrt[3]{648x^4}+3\sqrt[3]{81x^4}=12\sqrt[3]{216x^3\cdot3x}+3\sqrt[3]{27x^3\cdot3x}=12\cdot6x\sqrt[3]{3x}+3\cdot3x\sqrt[3]{3x}=72x\sqrt[3]{3x}+9x\sqrt[3]{3x}=81x\sqrt[3]{3x}$

41. $\left(\dfrac{25}{49}\right)^{-3/2}=\dfrac{25^{-3/2}}{49^{-3/2}}=\dfrac{49^{3/2}}{25^{3/2}}=\dfrac{\left(\sqrt{49}\right)^3}{\left(\sqrt{25}\right)^3}=\dfrac{7^3}{5^3}=\dfrac{343}{125}$

42. $\dfrac{3t-1}{\sqrt{3t}+1}=\dfrac{(3t-1)\left(\sqrt{3t}-1\right)}{\left(\sqrt{3t}+1\right)\left(\sqrt{3t}-1\right)}=\dfrac{(3t-1)\left(\sqrt{3t}-1\right)}{3t-1}=\sqrt{3t}-1$

43. $(-7+\sqrt{-81})-(-2-\sqrt{-64})=(-7+9i)-(-2-8i)=-7+9i+2+8i=-5+17i$

44. $\frac{2-5i}{2+5i}=\frac{(2-5i)(2-5i)}{(2+5i)(2-5i)}=\frac{4-20i+25i^2}{4-25i^2}=\frac{4-20i-25}{4+25}=\frac{-21-20i}{29}=-\frac{21}{29}-\frac{20}{29}i$

45.

$\sqrt{3a+1}=a-1$

$\left(\sqrt{3a+1}\right)^2=(a-1)^2$ Square both sides.

$3a+1=a^2-2a+1$ Do the squaring.

$0=a^2-5a$ Subtract $3a$ and 1.

$0=a(a-5)$ Factor.

$a-5=0$ or $a=0$ is an extraneous solution.

$a=5$

Check:

$\sqrt{3(5)+1}\overset{?}{=}(5)-1$

$\sqrt{16}\overset{?}{=}4$

$4=4$

and

$\sqrt{3(0)+1}\overset{?}{=}(0)-1$

$\sqrt{1}\overset{?}{=}-1$

$1\neq-1$

46.

$\sqrt{x+3}-\sqrt{3}=\sqrt{x}$

$\sqrt{x+3}=\sqrt{x}+\sqrt{3}$ Add $\sqrt{3}$ to both sides.

$\left(\sqrt{x+3}\right)^2=\left(\sqrt{x}+\sqrt{3}\right)^2$ Square both sides.

$x+3=x+2\sqrt{3x}+3$ Do the squaring.

$0=2\sqrt{3x}$ Subtract x and 3.

$0=\sqrt{3x}$ Divide by 2.

$0=\left(\sqrt{3x}\right)^2$ Square both sides.

$0=3x$ Do the squaring

$0=x$ Divide by 3.

Check:

$\sqrt{(0)+3}-\sqrt{3}\overset{?}{=}\sqrt{0}$

$\sqrt{3}-\sqrt{3}\overset{?}{=}0$

$0=0$

47.

$$\begin{aligned}6a^2+5a-6&=0\\6a^2-4a+9a-6&=0\\2a(3a-2)+3(3a-2)&=0\\(3a-2)(2a+3)&=0\end{aligned}$$

$3a-2=0$ or $2a+3=0$

$3a=2$ $\qquad 2a=-3$

$a=\frac{2}{3}$ $\qquad a=-\frac{3}{2}$

The solutions are $\frac{2}{3}$ and $-\frac{3}{2}$.

48. $4w^2+6w+1=0$

$a=4,\ b=6,\ c=1$

$$w=\frac{-(6)\pm\sqrt{6^2-4(4)(1)}}{2(4)}$$

$$=\frac{-6\pm\sqrt{36-16}}{8}=\frac{-6\pm\sqrt{20}}{8}$$

$$=\frac{-6\pm2\sqrt{5}}{8}=\frac{2\left(-3\pm\sqrt{5}\right)}{2\cdot4}$$

$$=\frac{-3\pm\sqrt{5}}{4}$$

49. $2(2x+1)^2 - 7(2x+1) + 6 = 0$

$$2y^2 - 7y + 6 = 0 \quad \text{Let } y = (2x+1).$$
$$(2y-3)(y-2) = 0$$

$2y - 3 = 0$ or $y - 2 = 0$

$y = \frac{3}{2}$ $\qquad y = 2$

Undo the substitution.

$2x + 1 = \frac{3}{2}$ or $2x + 1 = 2$

$2x = \frac{1}{2}$ $\qquad 2x = 1$

$x = \frac{1}{4}$ $\qquad x = \frac{1}{2}$

The solutions are $\frac{1}{4}$ and $\frac{1}{2}$.

50. $3x^2 - 4x = -2$

$$3x^2 - 4x + 2 = 0$$

$a = 3, b = -4, c = -2$

$$x = \frac{-(-4) \pm \sqrt{(-4)^2 - 4(3)(2)}}{2(3)}$$
$$= \frac{4 \pm \sqrt{16-24}}{6} = \frac{4 \pm \sqrt{-8}}{6}$$
$$= \frac{4 \pm 2i\sqrt{2}}{6} = \frac{2(2 \pm i\sqrt{2})}{2 \cdot 3}$$
$$= \frac{2 \pm i\sqrt{2}}{3}$$

51. $(f \circ g)(x) = f(g(x)) = f(2x+1) = (2x+1)^2 - 2 = 4x^2 + 4x + 1 - 2 = 4x^2 + 4x - 1$

52.
$$f(x) = 2x^3 - 1$$
$$y = 2x^3 - 1 \quad \text{Rewrite.}$$
$$x = 2y^3 - 1 \quad \text{Interchange } x \text{ and } y.$$
$$x + 1 = 2y^3$$
$$\frac{x+1}{2} = y^3$$
$$\sqrt[3]{\frac{x+1}{2}} = y \quad \text{or} \quad f^{-1}(x) = \sqrt[3]{\frac{x+1}{2}}$$

53. $f(x) - \left(\frac{1}{2}\right)^x$

x	$f(x)$
−2	4
−1	2
0	1
1	$\frac{1}{2}$

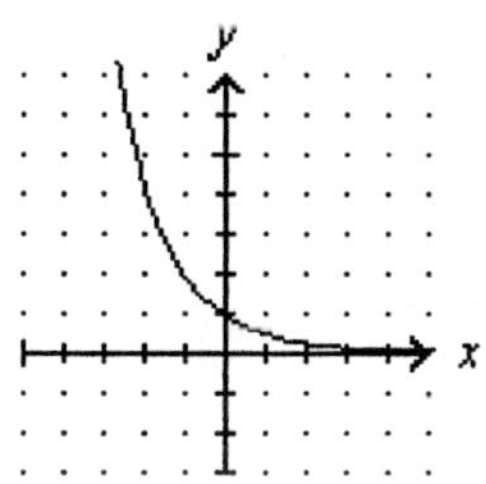

54. $y = e^x$

x	y
−1	0.4
0	1
1	2.7
2	7.4

$y = \ln x$

x	y
0.4	−1
1	0
2.7	1
7.4	2

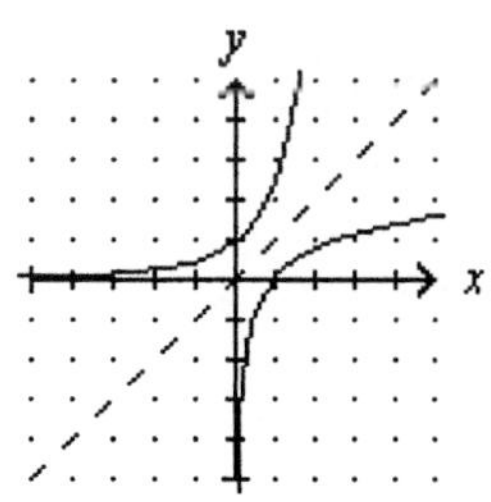

55. If $y = \log_2 x$, then $2^y = x$.

56. $\log_6 \frac{x}{36} = \log_6 x - \log_6 36 = \log_6 x - \log_6 6^2 = \log_6 x - 2\log_6 6 = \log_6 x - 2$

57. Rewrite $\log_x 25 = 2$ as $x^2 = 25$. So $x = 5$.

58. Rewrite $\log_5 125 = x$ as $5^x = 125$. Since $125 = 5^3$, $5^x = 5^3$ and $x = 3$.

59. Rewrite $\log_3 x = -3$ as $3^{-3} = x$. So $x = \dfrac{1}{3^3} = \dfrac{1}{27}$.

60. Rewrite $\ln e = x$ as $\log_e e = x$ or $e^x = e$. Thus, $x = 1$.

61.
$$\begin{aligned} y &= \log_2 x \\ x &= \log_2 y && \text{Interchange } x \text{ and } y. \\ 2^x &= y && \text{Rewrite in exponential form.} \end{aligned}$$

62. Rewrite $\log_{10} 10^x = y$ as $10^y = 10^x$. So $y = x$.

63. $\log 98 = \log(7 \bullet 14) = \log 7 + \log 14 = 0.8451 + 1.1461 = 1.9912$

64. $\log 2 = \log \dfrac{14}{7} = \log 14 - \log 7 = 1.1461 - 0.8451 = 0.301$

65.
$$\begin{aligned} 2^{x+2} &= 3^x \\ \log 2^{x+2} &= \log 3^x && \text{Take log of both sides.} \\ (x+2)\log 2 &= x \log 3 && \text{Use power rule.} \\ x\log 2 + 2\log 2 &= x\log 3 && \text{Use the distributive property.} \\ x\log 2 - x\log 3 &= -2\log 2 && \text{Subtract } x\log 3 \text{ and } 2\log 2. \\ x(\log 2 - \log 3) &= -2\log 2 && \text{Factor.} \\ x &= \frac{-2\log 2}{\log 2 - \log 3} && \text{Divide by } (\log 2 - \log 3). \\ x &\approx 3.4190 && \text{Use a calculator.} \end{aligned}$$

66.
$$\begin{aligned} \log x + \log(x+9) &= 1 \\ \log x(x+9) &= 1 && \text{Use the product rule of logarithms.} \\ x(x+9) &= 10^1 && \text{Change to exponential form.} \\ x^2 + 9x - 10 &= 0 && \text{Remove parentheses and subtract 10.} \\ (x-1)(x+10) &= 0 && \text{Factor.} \end{aligned}$$

$x - 1 = 0$ or $x + 10 = 0$

$x = 1$ $\quad$ $x = -10$ Not possible. $\log(-10)$ is undefined.

67.
$$\begin{aligned} 5^{4x} &= \frac{1}{125} \\ 5^{4x} &= 5^{-3} && \text{Write each side with base of 5.} \\ 4x &= -3 && \text{Equate the exponents.} \\ x &= -\frac{3}{4} && \text{Divide by 4.} \end{aligned}$$

68.
$$\begin{aligned} \log_3 x &= \log_3\left(\frac{1}{x}\right) + 4 \\ \log_3 x - \log_3\left(\frac{1}{x}\right) &= 4 && \text{Subtract } \log_3\left(\frac{1}{x}\right). \\ \log_3\left(\frac{x}{\frac{1}{x}}\right) &= 4 && \text{Use the quotient rule of logarithms.} \\ \log_3 x^2 &= 4 && \text{Simplify.} \\ x^2 &= 3^4 && \text{Change to exponential form.} \\ x^2 &= 81 && \text{Simplify.} \\ x &= \pm 9 && \text{Take the square root.} \end{aligned}$$

Discard -9 because $\log_3(-9)$ is undefined.

69. Find V using the formula for compound interest with a negative rate $r = -12\% = -0.12$, A $= \$9000$, and $t =$ 9 years.

$$V = A(1+r)^t$$
$$V = 9000[1+(-0.12)]^9 = \$2848.31$$

70. $\log_6 8 = \dfrac{\log_{10} 8}{\log_{10} 6} = 1.16056$

71. Center at (1, 3) and passes through (−2, −1).
Find r^2 by substituting −2 for x, −1 for y, 1 for h, and 3 for k.

$$\begin{aligned}(x-h)^2+(y-k)^2 &= r^2\\ [(-2)-(1)]^2+[(-1)-(3)]^2 &= r^2\\ (-3)^2+(-4)^2 &= r^2\\ 9+16 &= r^2\\ 25 &\quad r^2\end{aligned}$$

Then substitute 25 for r^2, 1 for h and 3 for k and simplify.

$$\begin{aligned}(x-h)^2+(y-k)^2 &= r^2\\ (x-1)^2+(y-3)^2 &= 25\\ x^2-2x+1+y^2-6y+9 &= 25\\ x^2+y^2-2x-6y-15 &= 0\end{aligned}$$

72.
$$\begin{aligned}x^2-9y^2+2x+36y &= 44\\ x^2+2x-9y^2+36y &= 44\\ (x^2+2x)-9(y^2-4y) &= 44\\ (x^2+2x+1)-9(y^2-4y+4) &= 44+1-36\\ (x+1)^2-9(y-2)^2 &= 9\\ \frac{(x+1)^2}{9}-\frac{(y-2)^2}{1} &= 1\end{aligned}$$

Center: (−1, 2)

Vertices ($y = 2$):

$(x+1)^2 = 9$

$x = -1 \pm 3$

$V_1(-4, 2)$ and $V_2(2, 2)$

Fundamental Rectangle:

Use vertices and (−1, 1) and (−1, 3).

Opens: left and right.

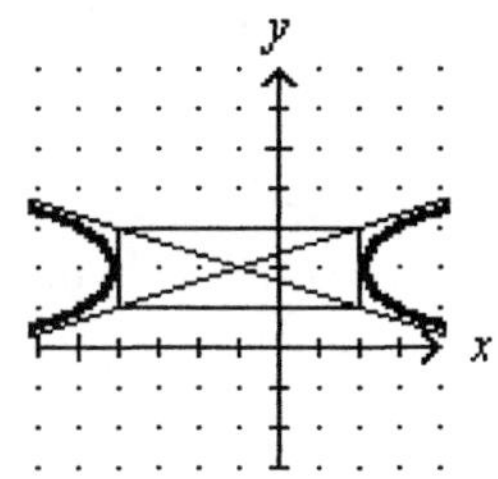

73.
$$\frac{(x-1)^2}{9}+\frac{(y-3)^2}{4}=1$$
$$\frac{(x-1)^2}{3^2}+\frac{(y-3)^2}{2^2}=1$$

Center: (1, 3)

x	y
1	1
1	5
−2	3
4	3

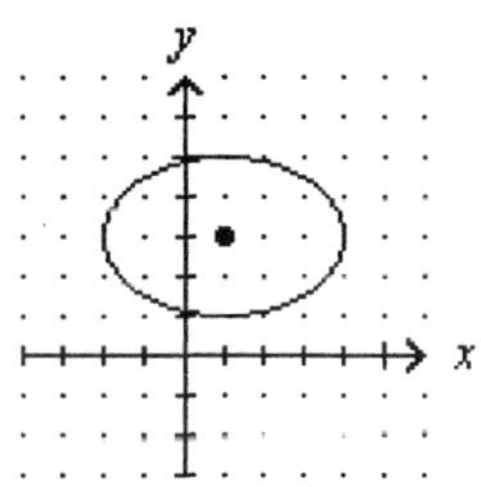

74.
$$\begin{aligned}y^2+4x-6y &= -1\\ 4x &= -y^2+6y-1\\ 4x &= -(y^2-6y+9)-1+9\\ 4x &= -(y-3)^2+8\\ x &= -\tfrac{1}{4}(y-3)^2+2\end{aligned}$$

Vertex: (2, 3); Opens to the left.

x	y
$-\frac{1}{4}$	0
1	1
2	**3**
1	5
$-\frac{1}{4}$	6

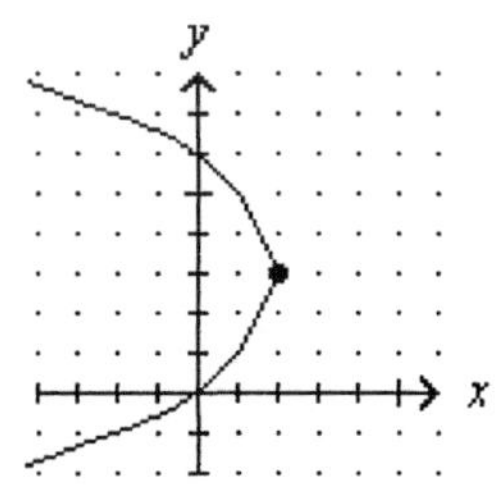

Now, get going and make it happen!

Appendix I Fractions and Decimals

Appendix I Fractions and Decimals

VOCABULARY

1. numerator; denominator
2. simplified or reduced; lowest
3. equivalent
4. prime
5. factored; prime–factored
6. reciprocal
7. terminating; repeating
8. least or lowest

CONCEPTS

9. $\frac{4}{12}=\frac{1}{3}$
10. $2 \cdot 2 \cdot 3 \cdot 5 = 60$
11. 5

12. a. 12, 24, 36 (answers will vary) b. 12

13. a. 3 times b. 2 times

14. $2\frac{15}{16}=\frac{2(16)+15}{16}=\frac{32+15}{16}=\frac{47}{16}$ b. $\frac{49}{12}=\frac{48+1}{12}=(48\div 12)+\frac{1}{12}=4\frac{1}{12}$

NOTATION

15. Common factors of 7 and 5 in the numerator and the denominator have been divided out.

16. a. 2 b. 8

PRACTICE

17. Factors of 20 are 1, 2, 4, 5, 10, 20.
18. Factors of 50 are 1, 2, 5, 10, 25, 50
19. Factors of 28 are 1, 2, 4, 7, 14, 28.
20. Factors of 36 are 1, 2, 3, 4, 6, 9, 12, 18, 36
21. $75 = 3 \cdot 25 = 3 \cdot 5 \cdot 5$
22. $20 = 2 \cdot 10 = 2 \cdot 2 \cdot 5$
23. $28 = 4 \cdot 7 = 2 \cdot 2 \cdot 7$
24. $54 = 2 \cdot 27 = 2 \cdot 3 \cdot 9 = 2 \cdot 3 \cdot 3 \cdot 3$
25. $117 = 3 \cdot 39 = 3 \cdot 3 \cdot 13$
26. $147 = 3 \cdot 49 = 3 \cdot 7 \cdot 7$
27. $220 = 22 \cdot 10 = 2 \cdot 11 \cdot 2 \cdot 5 = 2 \cdot 2 \cdot 5 \cdot 11$
28. $270 = 10 \cdot 27 = 2 \cdot 5 \cdot 3 \cdot 3 \cdot 3$
29. $\frac{1}{3}=\frac{1\cdot 3}{3\cdot 3}=\frac{3}{9}$
30. $\frac{3}{8}=\frac{3\cdot 3}{8\cdot 3}=\frac{9}{24}$
31. $\frac{4}{9}=\frac{4\cdot 6}{9\cdot 6}=\frac{24}{54}$
32. $\frac{9}{16}=\frac{9\cdot 4}{16\cdot 4}=\frac{36}{64}$
33. $7=\frac{7\cdot 5}{1\cdot 5}=\frac{35}{5}$
34. $12=\frac{12}{1}=\frac{12\cdot 3}{1\cdot 3}=\frac{36}{3}$
35. $\frac{6}{12}=\frac{\overset{1}{\cancel{2}}\cdot\overset{1}{\cancel{3}}}{\underset{1}{\cancel{2}}\cdot 2\cdot\underset{1}{\cancel{3}}}=\frac{1}{2}$
36. $\frac{3}{9}=\frac{1\cdot\overset{1}{\cancel{3}}}{3\cdot\underset{1}{\cancel{3}}}=\frac{1}{3}$
37. $\frac{24}{18}=\frac{4\cdot\overset{1}{\cancel{6}}}{3\cdot\underset{1}{\cancel{6}}}=\frac{4}{3}$
38. $\frac{35}{14}=\frac{5\cdot\overset{1}{\cancel{7}}}{2\cdot\underset{1}{\cancel{7}}}=\frac{5}{2}$
39. $\frac{15}{20}=\frac{3\cdot\overset{1}{\cancel{5}}}{4\cdot\underset{1}{\cancel{5}}}=\frac{3}{4}$
40. $\frac{22}{77}=\frac{2\cdot\overset{1}{\cancel{11}}}{7\cdot\underset{1}{\cancel{11}}}=\frac{2}{7}$

41. $\frac{72}{64}=\frac{9\cdot\cancel{8}}{8\cdot\cancel{8}}=\frac{9}{8}$

42. $\frac{26}{21}$ is in lowest terms

43. $\frac{36}{225}=\frac{4\cdot\cancel{9}}{25\cdot\cancel{9}}=\frac{4}{25}$

44. $\frac{175}{490}=\frac{5\cdot\cancel{5}\cdot\cancel{7}}{2\cdot\cancel{5}\cdot7\cdot\cancel{7}}=\frac{5}{14}$

45. $\frac{1}{2}\cdot\frac{3}{5}=\frac{1\cdot3}{2\cdot5}=\frac{3}{10}$

46. $\frac{3}{4}\cdot\frac{5}{7}=\frac{3\cdot5}{4\cdot7}=\frac{15}{28}$

47. $\frac{4}{3}\left(\frac{6}{5}\right)=\frac{4\cdot2\cdot3}{3\cdot5}=\frac{4\cdot2\cdot\cancel{3}}{\cancel{3}\cdot5}=\frac{8}{5}$

48. $\frac{7}{8}\left(\frac{6}{15}\right)=\frac{7\cdot2\cdot3}{2\cdot4\cdot3\cdot5}=\frac{7\cdot\cancel{2}\cdot\cancel{3}}{\cancel{2}\cdot4\cdot\cancel{3}\cdot5}=\frac{7}{20}$

49. $\frac{5}{12}\cdot\frac{18}{5}=\frac{5\cdot3\cdot6}{2\cdot6\cdot5}=\frac{\cancel{5}\cdot3\cdot\cancel{6}}{2\cdot\cancel{6}\cdot\cancel{5}}=\frac{3}{2}$

50. $\frac{5}{4}\cdot\frac{12}{10}=\frac{5\cdot3\cdot4}{4\cdot2\cdot5}=\frac{\cancel{5}\cdot3\cdot\cancel{4}}{\cancel{4}\cdot2\cdot\cancel{5}}=\frac{3}{2}$

51. $\frac{10}{21}\cdot14=\frac{2\cdot5\cdot2\cdot7}{3\cdot7}=\frac{2\cdot5\cdot2\cdot\cancel{7}}{3\cdot\cancel{7}}=\frac{20}{3}$

52. $\frac{5}{24}\cdot16=\frac{5\cdot2\cdot8}{3\cdot8}=\frac{5\cdot2\cdot\cancel{8}}{3\cdot\cancel{8}}=\frac{10}{3}$

53. $7\frac{1}{2}\cdot1\frac{2}{5}=\frac{15}{2}\cdot\frac{7}{5}=\frac{3\cdot5\cdot7}{2\cdot5}=\frac{3\cdot\cancel{5}\cdot7}{2\cdot\cancel{5}}=\frac{21}{2}=10\frac{1}{2}$

54. $3\frac{1}{4}\left(1\frac{1}{5}\right)=\frac{13}{4}\cdot\frac{6}{5}=\frac{13\cdot2\cdot3}{2\cdot2\cdot5}=\frac{13\cdot\cancel{2}\cdot3}{2\cdot\cancel{2}\cdot5}=\frac{39}{10}=3\frac{9}{10}$

55. $\frac{3}{5}\div\frac{2}{3}=\frac{3}{5}\cdot\frac{3}{2}=\frac{3\cdot3}{5\cdot2}=\frac{9}{10}$

56. $\frac{4}{5}\div\frac{3}{7}=\frac{4}{5}\cdot\frac{7}{3}=\frac{4\cdot7}{5\cdot3}=\frac{28}{15}$

57. $\frac{3}{4}\div\frac{6}{5}=\frac{3}{4}\cdot\frac{5}{6}=\frac{\cancel{3}\cdot5}{4\cdot2\cdot\cancel{3}}=\frac{5}{8}$

58. $\frac{3}{8}\div\frac{15}{28}=\frac{3}{8}\cdot\frac{28}{15}=\frac{3\cdot4\cdot7}{2\cdot4\cdot5\cdot3}=\frac{\cancel{3}\cdot\cancel{4}\cdot7}{2\cdot\cancel{4}\cdot5\cdot\cancel{3}}=\frac{7}{10}$

59. $\frac{21}{35}\div\frac{3}{14}=\frac{21}{35}\cdot\frac{14}{3}=\frac{\cancel{3}\cdot\cancel{7}\cdot2\cdot7}{5\cdot\cancel{7}\cdot\cancel{3}}=\frac{14}{5}$

60. $\frac{23}{25}\div\frac{46}{5}=\frac{23}{25}\cdot\frac{5}{46}=\frac{\cancel{23}\cdot\cancel{5}}{5\cdot\cancel{5}\cdot2\cdot\cancel{23}}=\frac{1}{10}$

61. $6\div\frac{3}{14}=\frac{6}{1}\cdot\frac{14}{3}=\frac{2\cdot\cancel{3}\cdot2\cdot7}{1\cdot\cancel{3}}=\frac{28}{1}=28$

62. $23\div\frac{46}{5}=\frac{23}{1}\cdot\frac{5}{46}=\frac{\cancel{23}\cdot5}{1\cdot2\cdot\cancel{23}}=\frac{5}{2}$

63. $3\frac{1}{3}\div1\frac{5}{6}=\frac{10}{3}\div\frac{11}{6}=\frac{10}{3}\cdot\frac{6}{11}=\frac{2\cdot5\cdot\cancel{3}\cdot2}{\cancel{3}\cdot11}=\frac{20}{11}=1\frac{9}{11}$

64. $2\frac{1}{2}\div1\frac{5}{8}=\frac{5}{2}\div\frac{13}{8}=\frac{5}{2}\cdot\frac{8}{13}=\frac{5\cdot\cancel{2}\cdot4}{\cancel{2}\cdot13}=\frac{20}{13}=1\frac{7}{13}$

65. $\frac{3}{5}+\frac{3}{5}=\frac{3+3}{5}=\frac{6}{5}$

66. $\frac{4}{13}-\frac{3}{13}=\frac{4-3}{13}=\frac{1}{13}$

67. $\frac{1}{6}+\frac{1}{24}=\frac{1\cdot4}{6\cdot4}+\frac{1}{24}=\frac{4}{24}+\frac{1}{24}=\frac{4+1}{24}=\frac{5}{24}$

68. $\frac{17}{25}-\frac{2}{5}=\frac{17}{25}-\frac{2\cdot5}{5\cdot5}=\frac{17}{25}-\frac{10}{25}=\frac{17-10}{25}=\frac{7}{25}$

69. $\frac{3}{5}+\frac{2}{3}=\frac{3\cdot 3}{5\cdot 3}+\frac{2\cdot 5}{3\cdot 5}=\frac{9}{15}+\frac{10}{15}=\frac{9+10}{15}=\frac{19}{15}$

70. $\frac{4}{3}+\frac{7}{2}=\frac{4\cdot 2}{3\cdot 2}+\frac{7\cdot 3}{2\cdot 3}=\frac{8}{6}+\frac{21}{6}=\frac{8+21}{6}=\frac{29}{6}$

71. $\frac{9}{4}-\frac{5}{6}=\frac{9\cdot 3}{4\cdot 3}-\frac{5\cdot 2}{6\cdot 2}=\frac{27}{12}-\frac{10}{12}=\frac{27-10}{12}=\frac{17}{12}$

72. $\frac{2}{15}+\frac{7}{9}=\frac{2\cdot 3}{15\cdot 3}+\frac{7\cdot 5}{9\cdot 5}=\frac{6}{45}+\frac{35}{45}=\frac{6+35}{45}=\frac{41}{45}$

73. $\frac{7}{10}-\frac{1}{14}=\frac{7\cdot 7}{10\cdot 7}-\frac{1\cdot 5}{14\cdot 5}=\frac{49}{70}-\frac{5}{70}=\frac{49-5}{70}=\frac{44}{70}=\frac{\overset{1}{\cancel{2}}\cdot 22}{\underset{1}{\cancel{2}}\cdot 35}=\frac{22}{35}$

74. $\frac{7}{25}+\frac{3}{10}=\frac{7\cdot 2}{25\cdot 2}+\frac{3\cdot 5}{10\cdot 5}=\frac{14}{50}+\frac{15}{50}=\frac{14+15}{50}=\frac{29}{50}$

75. $\frac{5}{14}-\frac{4}{21}=\frac{5\cdot 3}{14\cdot 3}-\frac{4\cdot 2}{21\cdot 2}=\frac{15}{42}-\frac{8}{42}=\frac{15-8}{42}=\frac{7}{42}=\frac{1\cdot \overset{1}{\cancel{7}}}{6\cdot \underset{1}{\cancel{7}}}=\frac{1}{6}$

76. $\frac{2}{33}+\frac{3}{22}=\frac{2\cdot 2}{33\cdot 2}+\frac{3\cdot 3}{22\cdot 3}=\frac{4}{66}+\frac{9}{66}=\frac{4+9}{66}=\frac{13}{66}$

77. $3-\frac{3}{4}=\frac{3\cdot 4}{1\cdot 4}-\frac{3}{4}=\frac{12}{4}-\frac{3}{4}=\frac{12-3}{4}=\frac{9}{4}$

78. $\frac{17}{3}+4=\frac{17}{3}+\frac{4\cdot 3}{1\cdot 3}=\frac{17}{3}+\frac{12}{3}=\frac{17+12}{3}=\frac{29}{3}$

79. $3\frac{3}{4}-2\frac{1}{2}=\frac{15}{4}-\frac{5}{2}=\frac{15}{4}-\frac{5\cdot 2}{2\cdot 2}=\frac{15}{4}-\frac{10}{4}=\frac{15-10}{4}=\frac{5}{4}=1\frac{1}{4}$

80. $15\frac{5}{6}+11\frac{5}{8}=\frac{95}{6}+\frac{93}{8}=\frac{95\cdot 4}{6\cdot 4}+\frac{93\cdot 3}{8\cdot 3}=\frac{380}{24}+\frac{279}{24}=\frac{380+279}{24}=\frac{659}{24}=27\frac{11}{24}$

81. $8\frac{2}{9}-7\frac{2}{3}=\frac{74}{9}-\frac{23}{3}=\frac{74}{9}-\frac{23\cdot 3}{3\cdot 3}=\frac{74}{9}-\frac{69}{9}=\frac{74-69}{9}=\frac{5}{9}$

82. $3\frac{4}{5}-3\frac{1}{10}=\frac{19}{5}-\frac{31}{10}=\frac{19\cdot 2}{5\cdot 2}-\frac{31}{10}=\frac{38}{10}-\frac{31}{10}=\frac{38-31}{10}=\frac{7}{10}$

83. $\begin{array}{r} 23.45 \\ +\underline{135.2} \\ 158.65 \end{array}$

84. $\begin{array}{r} 345.213 \\ -\underline{27.35} \\ 317.863 \end{array}$

85. $\begin{array}{r} 67.235 \\ -\underline{22.45} \\ 44.785 \end{array}$

86. $\begin{array}{r} 12.17 \\ +\underline{3.457} \\ 15.627 \end{array}$

87. $\begin{array}{r} 3.4 \\ \times\underline{13.2} \\ 68 \\ 1020 \\ \underline{3400} \\ 44.88 \end{array}$

88. $\begin{array}{r} 4.21 \\ \times\underline{2.73} \\ 1263 \\ 29470 \\ \underline{84200} \\ 11.4933 \end{array}$

89. Move decimal point 2 places in both the divisor and dividend.

$$\begin{array}{r} 4.55 \\ 23\overline{)104.65} \\ \underline{92} \\ 12\,6 \\ \underline{11\,5} \\ 1\,15 \\ \underline{1\,15} \\ 0 \end{array}$$

90. Move decimal point 1 place in both the divisor and dividend.

$$\begin{array}{r} 2.31 \\ 47\overline{)108.57} \\ \underline{94} \\ 14\,5 \\ \underline{14\,1} \\ 4\,7 \\ \underline{4\,7} \\ 0 \end{array}$$

91. $2.9517(1{,}000) = 2951.7$

Move decimal 3 places to the right.

92. $100(333.614) = 33{,}361.4$

Move decimal 2 places to the right.

93. $100 \bullet 0.05 = 5$

Move decimal 2 places to the left.

94. $1000 \bullet 0.085 = 85$

Move decimal 3 places to the right.

95.
$$\begin{array}{r} 0.625 \\ 8\overline{)5.000} \\ \underline{4\,8} \\ 20 \\ \underline{16} \\ 40 \\ \underline{40} \\ 0 \end{array}$$

96.
$$\begin{array}{r} 0.09375 \\ 32\overline{)3.00000} \\ \underline{288} \\ 120 \\ \underline{96} \\ 240 \\ \underline{224} \\ 160 \\ \underline{160} \\ 0 \end{array}$$

97.
$$\begin{array}{r} 0.033 \\ 30\overline{)1.000} \\ \underline{90} \\ 100 \\ \underline{90} \\ 10 \end{array} = 0.0\overline{3}$$

98.
$$\begin{array}{r} 0.77 \\ 9\overline{)7.00} \\ \underline{6\,3} \\ 70 \\ \underline{63} \\ 7 \end{array} = 0.\overline{7}$$

99.
$$\begin{array}{r} 0.42 \\ 50\overline{)21.00} \\ \underline{20\,0} \\ 1\,00 \\ \underline{1\,00} \\ 0 \end{array}$$

100.
$$\begin{array}{r} 0.016 \\ 125\overline{)2.000} \\ \underline{1\,25} \\ 750 \\ \underline{750} \\ 0 \end{array}$$

101.
$$\begin{array}{r} 0.4545 \\ 11\overline{)5.0000} \\ \underline{4\,4} \\ 60 \\ \underline{55} \\ 50 \\ \underline{44} \\ 60 \\ \underline{55} \\ 5 \end{array} = 0.\overline{45}$$

102.
$$\begin{array}{r} 0.0166 \\ 60\overline{)1.0000} \\ \underline{60} \\ 400 \\ \underline{360} \\ 400 \end{array} = 0.01\overline{6}$$

APPLICATIONS

103. a. $\frac{5}{32}+\frac{1}{16}=\frac{5}{32}+\frac{1\cdot 2}{16\cdot 2}=\frac{5+2}{32}=\frac{7}{32}$ in.   b. $\frac{5}{32}-\frac{1}{16}=\frac{5}{32}-\frac{1\cdot 2}{16\cdot 2}=\frac{5-2}{32}=\frac{3}{32}$ in.

104. Add the length of each section of the bolt.

$$\frac{5}{8}+4\frac{3}{4}+1\frac{7}{8}+\frac{5}{16}=\frac{5}{8}+\frac{19}{4}+\frac{15}{8}+\frac{5}{16}$$
$$=\frac{5\cdot 2}{8\cdot 2}+\frac{19\cdot 4}{4\cdot 4}+\frac{15\cdot 2}{8\cdot 2}+\frac{5}{16}=\frac{10}{16}+\frac{76}{16}+\frac{30}{16}+\frac{5}{16}=\frac{121}{16}=7\frac{9}{16}\text{ in.}$$

105. Find the perimeter which is 4 times the length of one side. $4\left(10\frac{1}{8}\right)=\frac{4}{1}\cdot\frac{81}{8}=\frac{\cancel{4}\cdot 81}{2\cdot\cancel{4}}=\frac{81}{2}=40\frac{1}{2}$ in.

106. Divide each type of material by the amount needed per pillow. Then compare to find the smallest number.

Lace trim: $135\div\frac{9}{10}=135\cdot\frac{10}{9}=\frac{\cancel{9}\cdot 15\cdot 10}{\cancel{9}}=150$

Corduroy fabric: $154\div\frac{7}{8}=154\cdot\frac{8}{7}=\frac{\cancel{7}\cdot 22\cdot 8}{\cancel{7}}=176$

Cotton filling: $98\div\frac{2}{3}=98\cdot\frac{3}{2}=\frac{\cancel{2}\cdot 49\cdot 3}{\cancel{2}}=147$

The materials in stock would allow 147 pillows to be manufactured.

107. Subtract the sum of the ends of the car from the total length.
$187.8-(43.5+40.9) = 187.8-84.4 = 103.4$ in.

108. Multiply the width of each column by 1.4.
$4.5(1.4) = 6.3$ ft
$3.5(1.4) = 4.9$ ft
$2.5(1.4) = 3.5$ ft

WRITING

109. When an employee is "terminated" his employment ends.

110. Equivalent fractions are fractions whose values are equal but have different numerators and denominators.

111. Factors are numbers which divide into a given number. Prime factors are factors which are also prime numbers which when multiplied together result in the number.
$30 = 3 \cdot 10$ 10 is a factor but 10 is not prime.
$30 = 2 \cdot 3 \cdot 5$ 2, 3, and 5 are prime numbers.

112. Common factors can only be removed if they are part of the same factor or if they are in fractions that are multiplied together, not added.

Appendix II Cumulative Review of Graphing

Appendix II **Cumulative Review of Graphing**

1.

x	$y = 2x + 1$	y
-2	$y = 2(-1) + 1$	-3
0	$y = 2(0) + 1$	1
2	$y = 2(2) + 1$	5

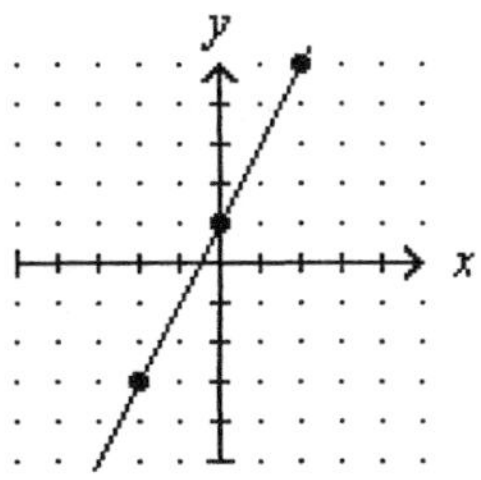

2.

x	$-4y + 5x = -15$	y
-3	$-4(0) + 5x = -15$	0
1	$-4y + 5(1) = -15$	5
-7	$-4y + 5(-7) = -15$	-5

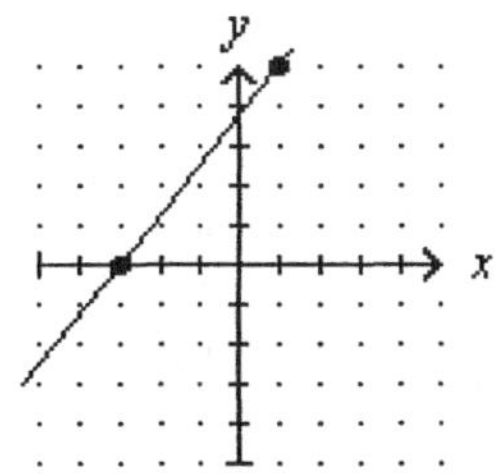

3.

x	$x = 5$	y
5	The value of y can be any number	-1
5		2
5		5

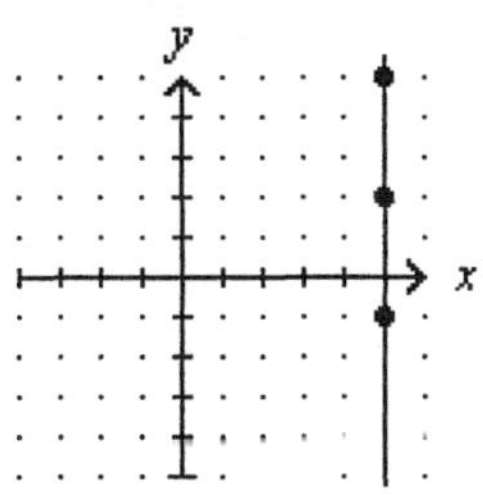

4.

x	$y = -3$	y
-4	The value of x can be any number	-3
-1		-3
0		-3

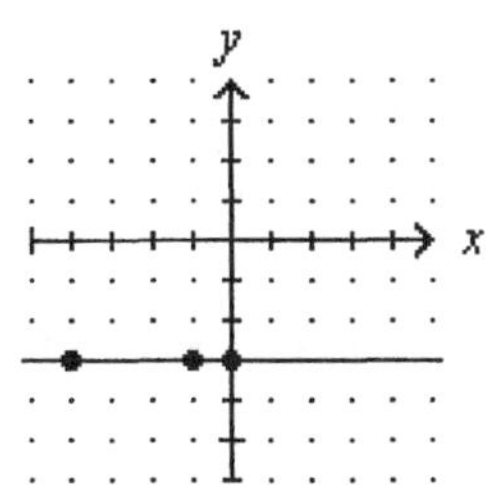

5. Equation #1

$2x + y = 1$

x	y
0	1
-2	5
2	-3

Equation #2

$x - 2y = -7$

x	y
1	4
-5	1
-3	2

The solution is $(-1, 3)$.

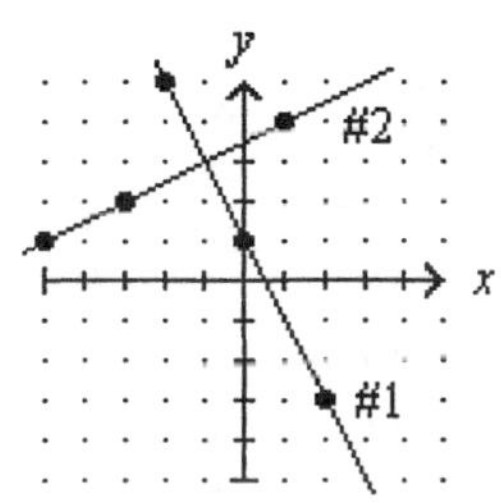

6. Inequality: $y > \frac{x}{3}$

Graph: $3y = x$

x	y
0	0
3	1

Broken line. Test point $(3, 3)$.

$(3) > \frac{3}{3}$

$3 > 1$ True

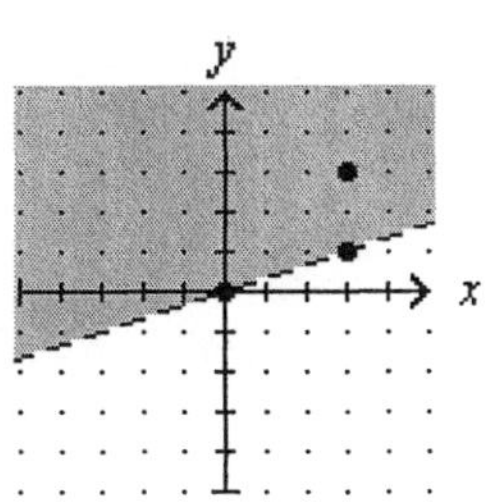

7\. Inequality #1: $3x + 2y > 6$

Graph: $3x + 2y = 6$

x	y
0	3
2	0

Broken line. Test point (0, 0).

$$3(0) + 2(0) > 6$$
$$0 > 6 \quad \text{False}$$

Inequality #2: $x + 3y \le 2$

Graph: $x + 3y = 2$

x	y
−4	2
2	0

Solid line. Test point (0, 0).

$$(0) + 3(0) \le 2$$
$$0 \le 2 \quad \text{True}$$

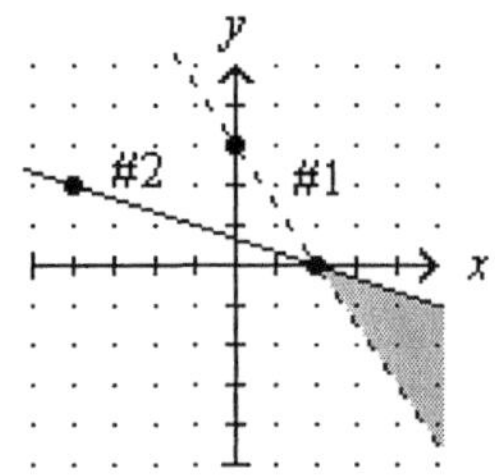

8\. Inequality #1: $y < -2$ or Inequality #2: $y > 3$

Graph: $y = -2$

x	y
0	−2
2	−2

Broken line. Test point (0, 0).

$$(0) < -2 \quad \text{False}$$

Graph: $y = 3$

x	y
−4	3
2	3

Broken line. Test point (0, 0).

$$(0) > 3 \quad \text{False}$$

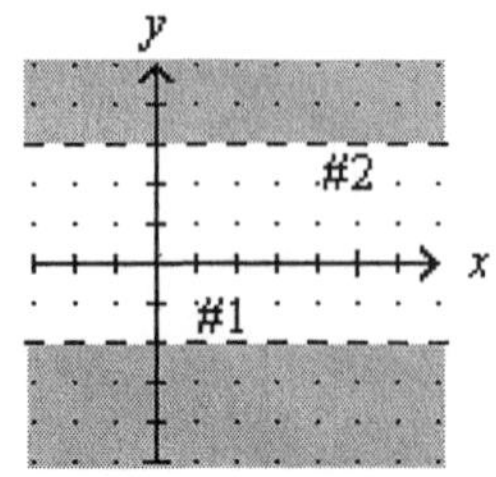

9\.

x	$f(x) = x$	$f(x)$
−2	$f(-2) = -2$	−2
−1	$f(-1) = -1$	−1
0	$f(0) = 0$	0
1	$f(1) = 1$	1
2	$f(2) = 2$	2

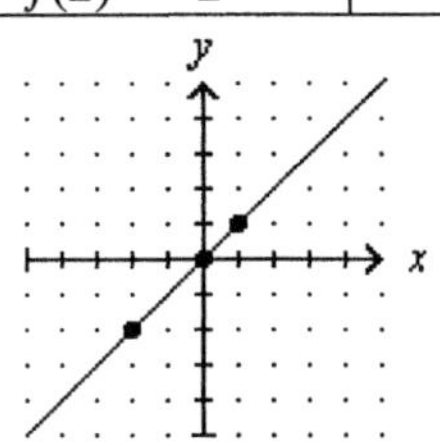

10\.

x	$f(x) = \|x\|$	$f(x)$
−2	$f(-2) = \|-2\|$	2
−1	$f(-1) = \|-1\|$	1
0	$f(0) = \|0\|$	0
1	$f(1) = \|1\|$	1
2	$f(2) = \|2\|$	2

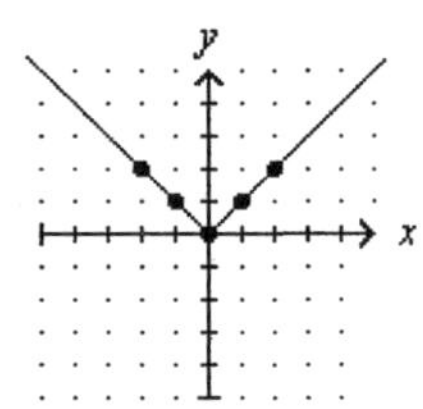

11\.

x	$f(x) = x^2$	$f(x)$
−2	$f(-2) = (-2)^2$	4
−1	$f(-1) = (-1)^2$	1
0	$f(0) = (0)^2$	0
1	$f(1) = (1)^2$	1
2	$f(2) = (2)^2$	4

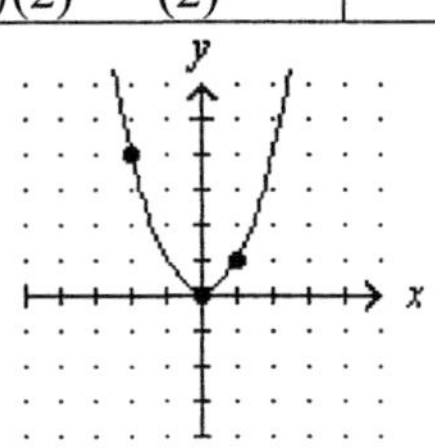

12\.

x	$f(x) = x^3$	$f(x)$
−2	$f(-2) = (-2)^3$	−8
−1	$f(-1) = (-1)^3$	−1
0	$f(0) = (0)^3$	0
1	$f(1) = (1)^3$	1
2	$f(2) = (2)^3$	8

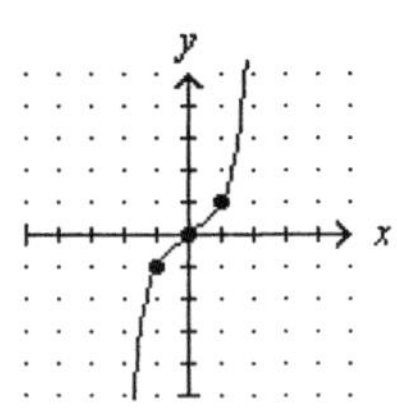

13.

x	$f(x) = x^3 + 4x^2 + 4x$	$f(x)$
−3	$f(x) = (-3)^3 + 4(-3)^2 + 4(-3)$	−3
−2	$f(x) = (-2)^3 + 4(-2)^2 + 4(-2)$	0
−1	$f(x) = (-1)^3 + 4(-1)^2 + 4(-1)$	−1
0	$f(x) = (0)^3 + 4(0)^2 + 4(0)$	0
1	$f(x) = (1)^3 + 4(1)^2 + 4(1)$	9

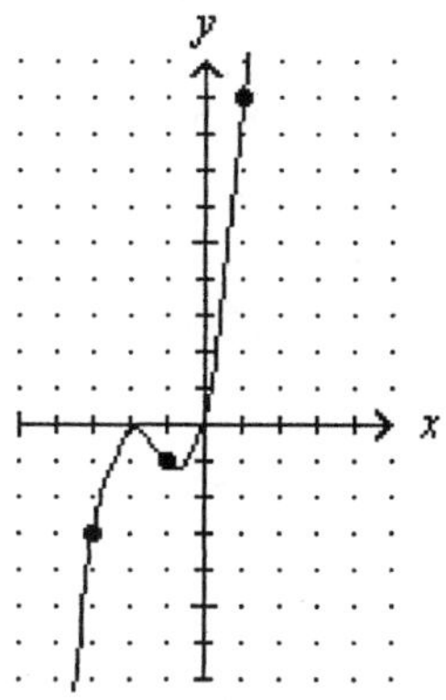

14.

x	$f(x) = \frac{12}{x}$	$f(x)$
0.5	$f(0.5) = \frac{12}{0.5}$	24
1	$f(1) = \frac{12}{1}$	12
4	$f(4) = \frac{12}{4}$	3
12	$f(12) = \frac{12}{12}$	1
24	$f(24) = \frac{12}{24}$	0.5

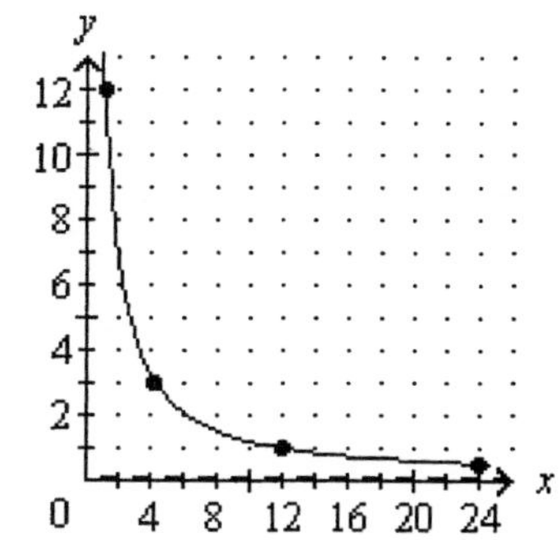

15.

x	$f(x) = \sqrt{x}$	$f(x)$
0	$f(0) = \sqrt{0}$	0
1	$f(1) = \sqrt{1}$	1
4	$f(4) = \sqrt{4}$	2
9	$f(9) = \sqrt{9}$	3

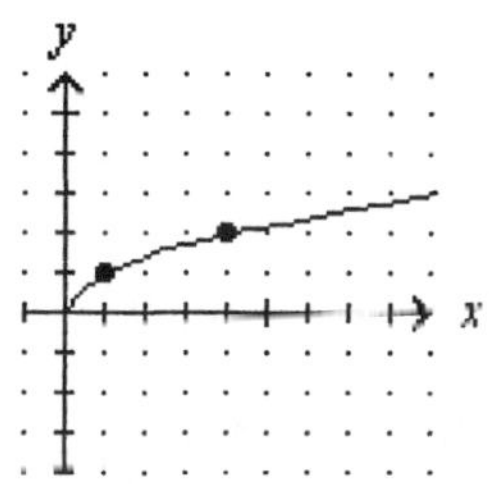

16.

x	$f(x) = \sqrt[3]{x}$	$f(x)$
−8	$f(-8) = \sqrt[3]{-8}$	−2
−1	$f(-8) = \sqrt[3]{-1}$	−1
0	$f(0) = \sqrt[3]{0}$	0
1	$f(0) = \sqrt[3]{1}$	1
8	$f(0) = \sqrt[3]{8}$	**2**

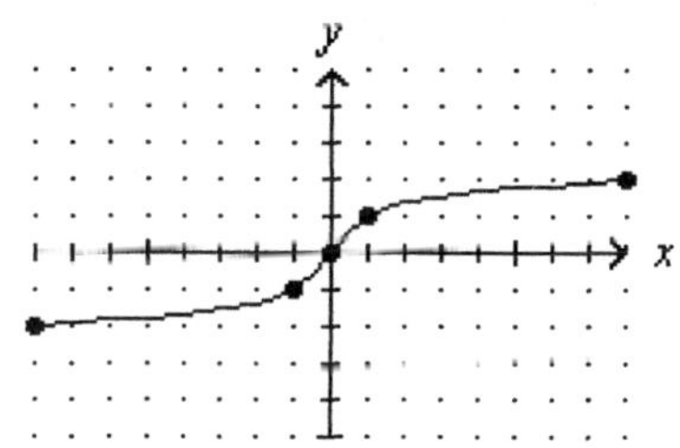

17. $f(x) = 2(x-2)^2 - 4$

Vertex: $(2, -4)$
Axis of symmetry: $x = 2$

x	$f(x)$
0	4
1	−2
2	**−4**
3	−2
4	4

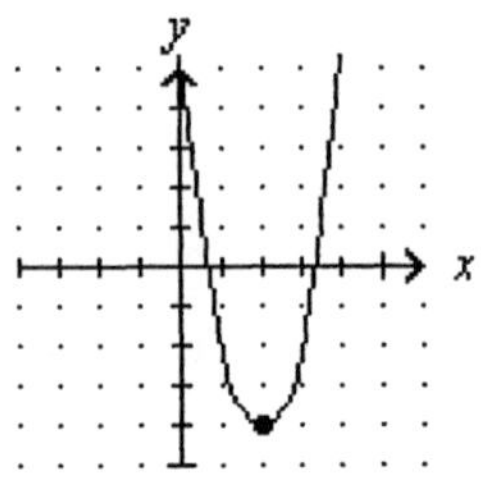

18. $f(x) = -12x^2 - 6x + 6$

$$-\frac{b}{2a} = -\frac{-6}{2(-12)} = -\frac{1}{4}$$

$$f\left(-\frac{1}{4}\right) = -12\left(-\frac{1}{4}\right)^2 - 6\left(-\frac{1}{4}\right) + 6 = 6\frac{3}{4}$$

Vertex $= \left(-\frac{1}{4}, 6\frac{3}{4}\right)$

x	$f(x)$
−2	−30
−1	0
$-\frac{1}{4}$	$6\frac{3}{4}$
0	6
1	−12

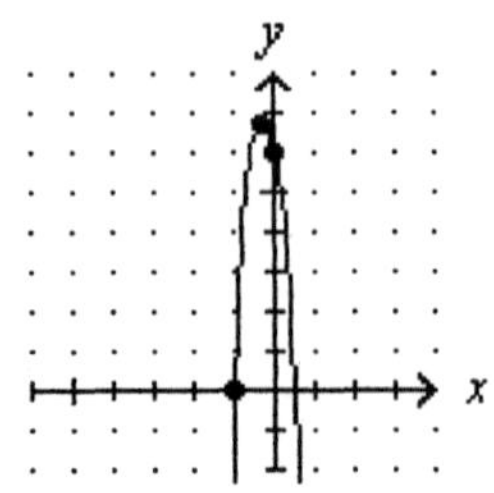

19. $f(x) = 5^x$

x	$f(x)$
−1	$\frac{1}{5}$
0	1
1	5
2	25

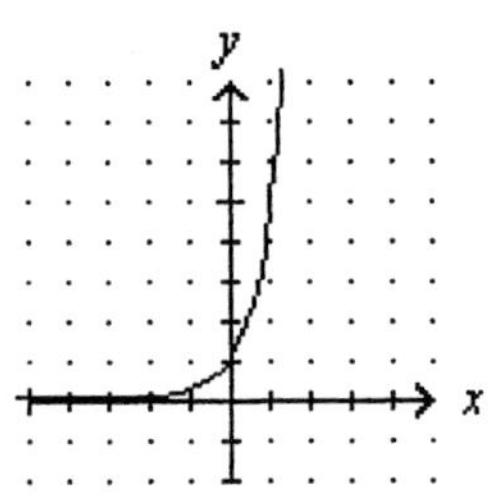

20. $f(x) = \left(\frac{1}{4}\right)^x$

x	$f(x)$
−2	16
−1	4
0	1
1	0.25

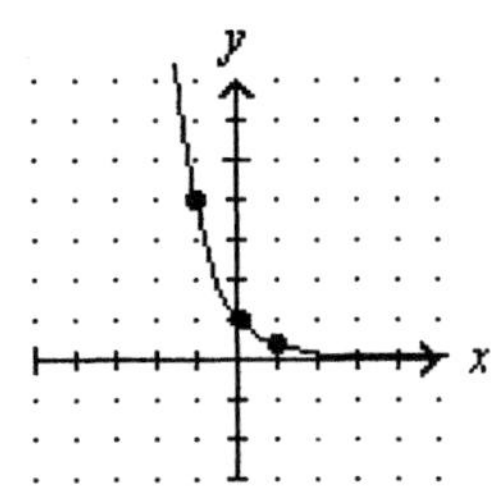

21. $f(x) = e^x$

x	$f(x)$
−1	0.4
0	1
1	2.7
2	7.4

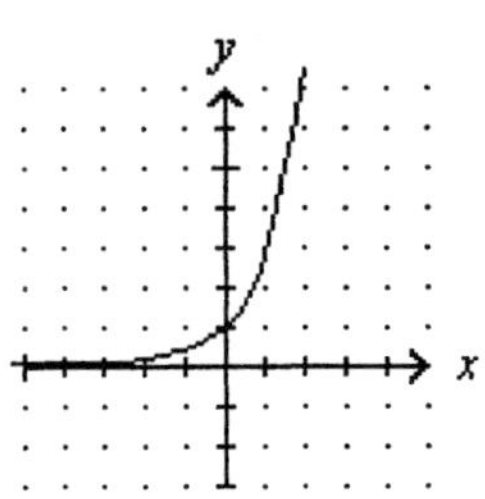

22. $f(x) = \log_3 x$

x	$f(x)$
0.1	−2
0.3	−1
1	0
3	1
9	2

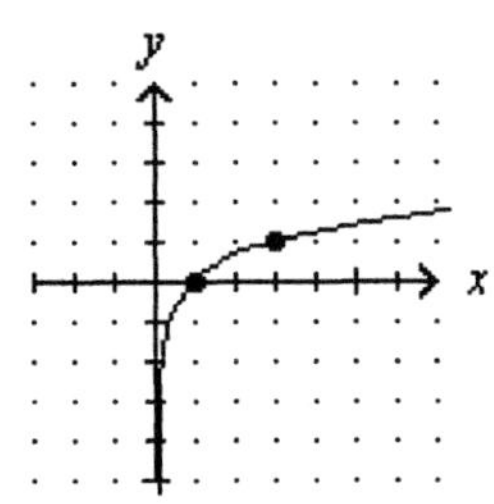

23. $f(x) = \log x$

x	$f(x)$
0.01	−2
0.1	−1
1	0
10	1

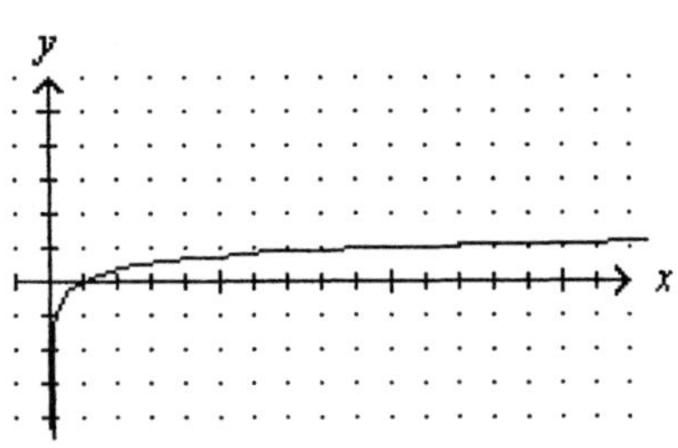

24. $f(x) = \ln x$

x	$f(x)$
0.1	−2
0.4	−1
1	0
2.7	1
7.4	2

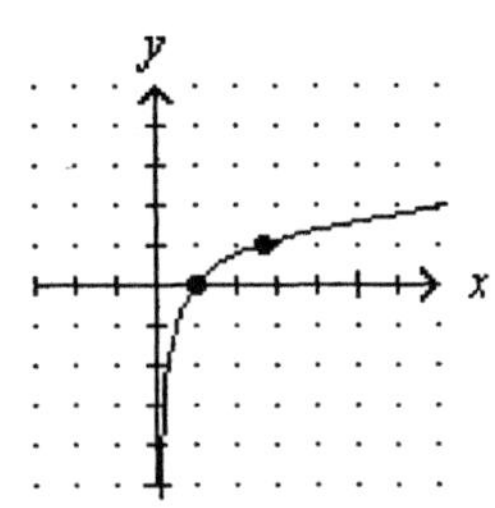

25.
$$\begin{aligned} f(x) &= \tfrac{1}{3}x + \tfrac{1}{3} \\ \text{Function} & \\ x &= \tfrac{1}{3}y + \tfrac{1}{3} \\ x - \tfrac{1}{3} &= \tfrac{1}{3}y \\ 3x - 1 &= y \\ f^{-1}(x) &= 3x - 1 \quad \text{Inverse.} \end{aligned}$$

Function (#1)		Inverse (#2)	
x	y	x	y
−4	−1	−1	−4
−1	0	0	−1
2	1	1	2

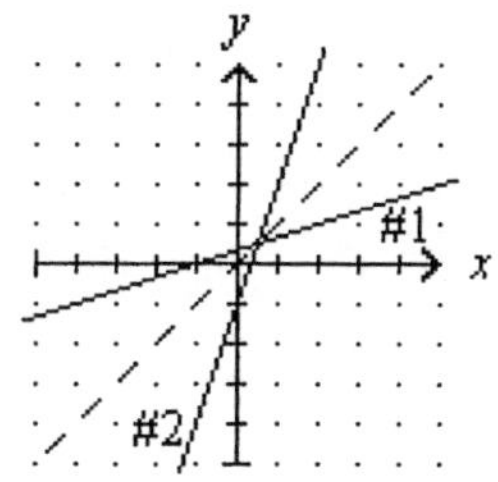

26.
$$\begin{aligned} f(x) &= x^2,\ (x \geq 0) \\ \text{Function} & \\ x &= y^2 \\ \sqrt{x} &= y \\ f^{-1}(x) &= \sqrt{x} \quad \text{Inverse.} \end{aligned}$$

Function (#1)		Inverse (#2)	
x	y	x	y
0	0	0	0
1	1	1	1
2	4	4	2
3	9	9	3

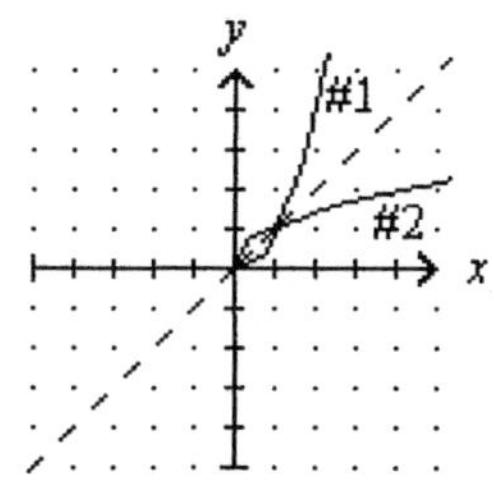

27. $x^2 + y^2 = 16$

Center: (0, 0); Radius = 4

x	y
0	−4
0	4
−4	0
4	0

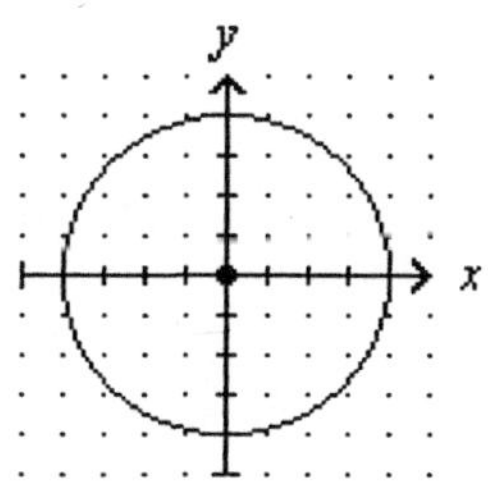

28.
$$\begin{aligned} x^2 + y^2 + 8x + 2y &= -13 \\ x^2 + 8x + y^2 + 2y &= -13 \\ x^2 + 8x + 16 + y^2 + 2y + 1 &= -13 + 16 + 1 \\ (x+4)^2 + (y+1)^2 &= 4 \\ [x - (-4)]^2 + [y - (-1)]^2 &= 2^2 \end{aligned}$$

Center: (−4, −1); Radius = 2

x	y
−4	−3
−4	1
−6	−1
−2	−1

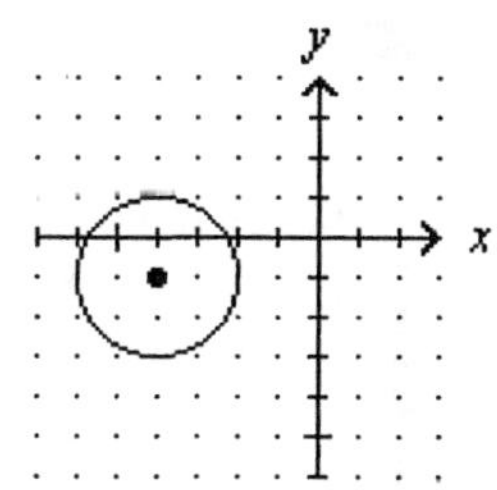

29. $x = -y^2 + 1$

Vertex: (1,0) Opens to the left.

x	y
−3	−2
0	−1
1	**0**
0	1
−3	2

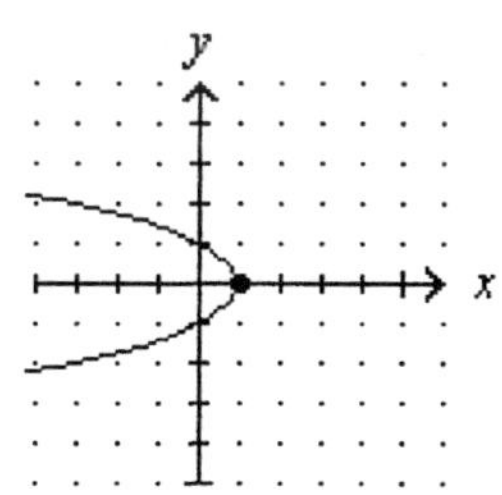

30.
$$\frac{x^2}{4} + \frac{y^2}{9} = 1$$
$$\frac{x^2}{2^2} + \frac{y^2}{3^2} = 1$$

Center: (0, 0)

x	y
0	−3
0	3
−2	0
2	0

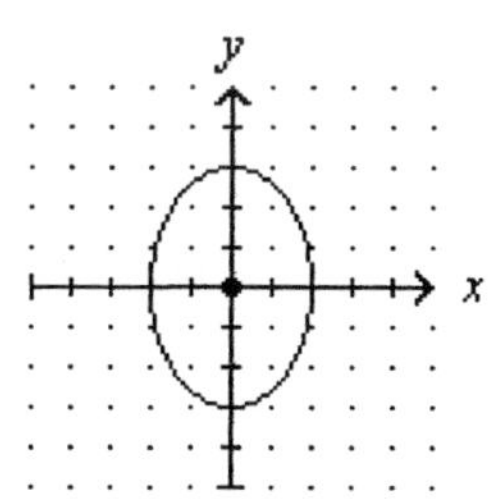

31. $\dfrac{(x+2)^2}{16} - \dfrac{(y-3)^2}{25} = 1$

$\dfrac{(x+2)^2}{4^2} - \dfrac{(y-3)^2}{5^2} = 1$

Center: $(-2, 3)$

Vertices ($y = 3$):

$(x+2)^2 = 16$

$x = -2 \pm 4$

$V_1(-6, 3)$ and $V_2(2, 3)$

Fundamental Rectangle:

Use vertices and $(-2, -2)$ and $(-2, 8)$.

Opens: left and right.

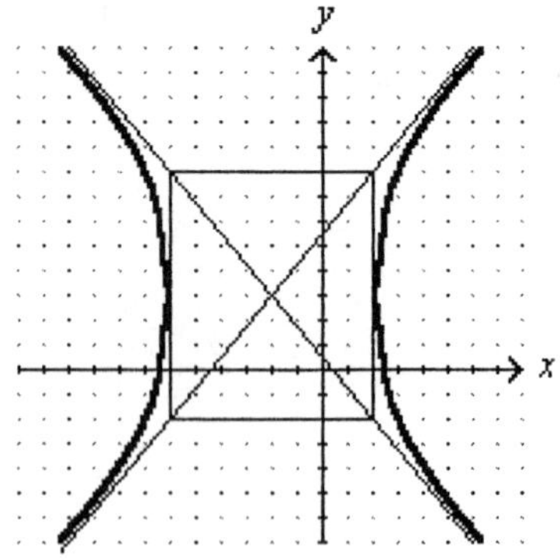

32. $xy = -10$

x	y
-10	1
-5	2
-2	5
-1	10
1	-10
2	-5
5	-2
10	-1

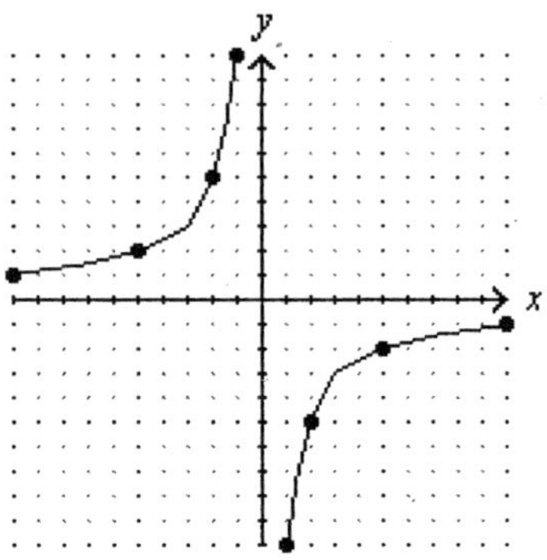

33. $f(x) = |x+2| - 3$

Translate the associated function, $g(x) = |x|$, left 2 units and down 3 units.

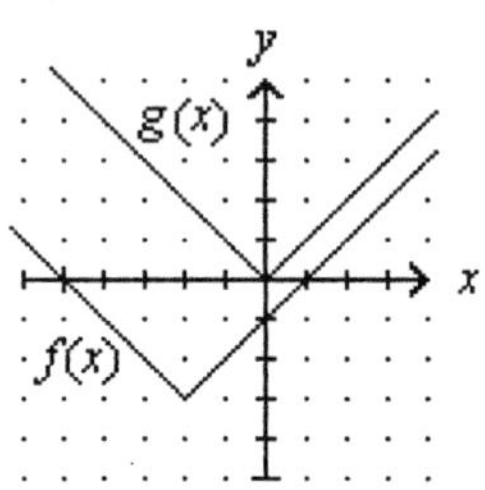

34. $f(x) = -x^3$

Reflect the associated function, $g(x) = x^3$, about the x – axis.

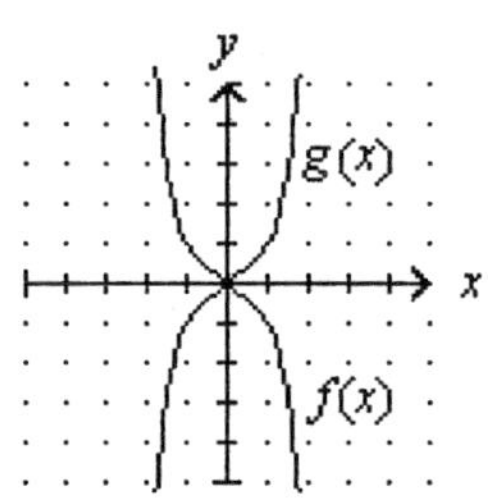